Steel
Structures
Design and Practice

N. SUBRAMANIAN

Consulting Engineer
Maryland
USA

UNIVERSITY PRESS

OXFORD

UNIVERSITY PRESS

YMCA Library Building, Jai Singh Road, New Delhi 110001

Oxford University Press is a department of the University of Oxford.
It furthers the University's objective of excellence in research, scholarship,
and education by publishing worldwide in

Oxford   New York
Auckland   Cape Town   Dar es Salaam   Hong Kong   Karachi
Kuala Lumpur   Madrid   Melbourne   Mexico City   Nairobi
New Delhi   Shanghai   Taipei   Toronto

With offices in
Argentina   Austria   Brazil   Chile   Czech Republic   France   Greece
Guatemala   Hungary   Italy   Japan   Poland   Portugal   Singapore
South Korea   Switzerland   Thailand   Turkey   Ukraine   Vietnam

Oxford is a registered trade mark of Oxford University Press
in the UK and in certain other countries.

Published in India
by Oxford University Press

ISBN-13: 978-0-19-806881-5
ISBN-10: 0-19-806881-6

Typeset in Times
by Pee-Gee Graphics, New Delhi

and published by Oxford University Press
YMCA Library Building, Jai Singh Road, New Delhi 110001

Foreword

Food, water, clothing, and shelter are considered as the basic needs of a human being. Out of these, the last one falls under the ambit of structural engineers. Though primitive civilization used caves, the quest for quality and comfort has resulted in a variety of building types. The desire to bridge lands and reach for the sky has produced several spectacular structures. The invention of the wheel and subsequent developments resulted in several infrastructure constructions, which include bridges, ports, airports, stadiums, and towers as well as design and construction of industrial structural systems matching with the new requirements of different mega- and ultra-mega-capacity projects.

The developments in materials technology have produced a host of building materials, out of which reinforced concrete, steel, and wood are quite popular. Steel with its high strength per unit weight and increased ductility is more efficient than concrete in resisting the loads. Moreover it is environment friendly and 100 per cent recyclable. Additional advantages like speed of erection, easy prefabrication, and dismantling facility make steel an ideal material of choice for several types of structures.

In India, the advantages of steel and its use was not fully realized due to the lack of information and knowledge of its behaviour and design. Hence, as the Joint Director General of the Institute for Steel Development and Growth (INSDAG) [an organization promoted by the Ministry of Steel and major steel producers of India], I was one of the members of the core group involved in developing Teaching Resources for Structural Steel Design, mainly for educating young teachers in various engineering colleges. Many training courses were also conducted for them as well as for professional engineers on 'Limit State Method of Design of Steel Structures'. Realizing that the Indian code, prevailing at that time, was very old (and was in the outdated Working Stress Method format), efforts were made by INSDAG to develop the next edition of the code, using the modern limit states methodology, which is followed in several other countries. To this end, a group was formed under the able guidance of Prof. R. Narayanan, former head of education and publications of the Steel Construction Institute, Ascot, England (which was later headed by Prof. V. Kalyanaraman of the Indian Institute of Technology, Madras). After getting necessary concurrence from the Bureau of Indian Standards, a committee was formed. The committee compared the codes of practices of other national codes, and after fruitful discussions and deliberations with various stakeholders, produced a comprehensive report, which was later published by the Bureau of Indian Standards as the latest version of IS 800 : 2007. The author of this book was a part of the committee that developed the code and actively participated in the deliberations, whenever he was available in India.

This substantial and comprehensive book on the design of steel structures to the Limit States Indian design code IS 800 : 2007 will be a boon to students and design engineers, as it explains the various code clauses with a number of examples. Hence, the book will be quite useful in understanding and applying the code provisions appropriately.

The book in its 750+ pages contains 12 chapters and five useful appendices. The first two chapters of the book cover the basics of materials, the structural design process, loading, and structural analysis, in adequate detail. The next seven chapters on member design discuss in detail tension and compression members, plastic collapse and local buckling, beams, plate girders, gantries, and beam-columns. The chapters on bolted and welded connections follow. The final chapter on industrial buildings includes details about them, and also explains the design of various elements, as learnt in the earlier chapters. The appendices include section properties, soil properties, computer programs for checking beams and beam-columns, and design aids. The computer programs and design aids will help designers in arriving at rapid design of these elements and eliminate the necessity of repetitive calculations.

Most chapters deal with the behaviour of the steel element considered, followed by a treatment of the design rules of IS 800 : 2007, worked examples, a summary, exercises, and review questions. The information on behaviour has been taken or adapted from widespread sources, many of which are cited in the Bibliography. The worked examples are extensive in their scope, while the exercises, review questions, and illustrations will be invaluable to students and teachers alike. The questions at the end of each chapter are very comprehensive and will make the student go over the chapter several times until full understanding is achieved. The case studies included in each chapter add considerable value to this book, making students understand the real structural behaviour, failures, etc.

This book is much more complete than usual, and has a good balance between the processes of design and the practical aspects of steel structures. Students and practising engineers will find it very helpful both in understanding the methods of the modern limit states method of design to the Indian Standard IS 800, and as a design aid and reference book.

The contents reflect the author's long-standing experience in the field, as the head of a well-known consultancy practice, as well as his high-level academic expertise gained in India, Europe, and the USA. As an authentic educator with more than 200 papers and 20 books, Dr Subramanian has produced yet another outstanding book, which will help the students and designers for a long time to come. Further, I would like to take this opportunity to thank him for offering me the privilege and honour of writing this foreword.

Dr T. K. Bandyopadhyay
PhD (Engg.), FICE (UK)
Senior Vice President
ThyssenKrupp Industries India Pvt. Ltd.

Preface

Structural design emphasizes that the elements of a structure are to be proportioned and joined together in such a way that they will be able to withstand all the loads (load effects) that are likely to act on it during its service life, without excessive deformation or collapse. Structural design is often considered as an art as well as a science. It must balance theoretical analysis with practical considerations, such as the degree of certainty of loads and forces, the actual behaviour of the structure as distinguished from the idealized analytical and design model, the actual behaviour of the material compared to the assumed elastic behaviour, and the actual properties of materials used compared to the assumed ones.

Steel is one of the major construction materials used all over the world. It has many advantages over other competing materials, such as high strength to weight ratio, high ductility (hence its suitability for earthquake-resistant structures), and uniformity. It is also a *green material* in the sense that it is fully recyclable. Presently, several grades and shapes of steel products exist.

Structural designers need to have a sound knowledge of structural steel behaviour, including the material behaviour of steel, and the structural behaviour of individual elements and of the complete structure. Unless structural engineers are abreast of the recent developments and understand the relationships between the structural behaviour and the design criteria implied by the rules of the design codes, they will be following the codal rules rigidly and blindly and may even apply them incorrectly in situations beyond their scope.

This text is based on the latest Indian Standard code of practice for general construction using hot-rolled steel sections (IS 800 : 2007) released in February 2008. This third revision of the code is based on the limit state method of design (the earlier versions of the code were based on the working or allowable stress method). The convention for member axis suggested in the code is adopted and SI units have been used throughout the book.

Readers are advised to refer to the latest code (IS 800 : 2007) published by the Bureau of Indian Standards, New Delhi. It is recommended that readers also refer to the latest version of the codes on design loads (IS 875 and IS 1893), dimension of sections (IS 808 or IS Handbook No. 1, IS 1161, IS 12778, IS 4923, and IS 811), specification of steel (IS 2062, IS 8500, IS 6639, and IS 3757), bolts (IS 1364 and IS 4000), and welding (IS 816).

About the Book

The objectives of writing this book are: (a) to explain the provisions of the latest version of IS 800:2007, which has been revised recently based on limit states design, (b) to provide ample examples so that the students understand the concepts clearly, (c) to give information on structural design failures and latest developments in structural steel design, and (d) to provide interested readers with the sources of further reading.

The book completely covers the requirements of undergraduate students of civil and structural engineering for a course on design of steel structures. Each chapter comprises numerous tables, figures, and solved examples to help students understand the concepts clearly. Review questions and exercises given at the end of each chapter will help students assimilate the ideas presented in the chapters and also to apply them to get a feel of the results obtained. Case studies of failures and some important aspects of structural design are sprinkled throughout the text, to enhance the usefulness of the book.

Contents and Coverage

Chapter 1 provides a brief discussion on the historical developments, steel making processes, and the metallurgy of steel.

Chapter 2 introduces the design considerations and the role of structural design in the complete design process as well the loads acting on structures. Many failures are attributed to the lack of determination of the loads acting on different structures. Hence, the various loads that can act on a structure are also briefly discussed, as per the latest Indian codes.

Chapter 3 deals with the design of tension members. Plastic and local buckling behaviour of steel sections are covered in Chapter 4, as they will be useful in understanding the design of axially loaded compression members and flexural members which are covered in Chapters 5 to 8.

The design of beam–columns, which are subjected to both axial loads and bending moments, is discussed briefly in Chapter 9. The two methods used to connect the elements of steel structures, namely, bolted and welded connections are discussed in Chapters 10 and 11.

With the information provided in Chapters 1 to 11, it is possible to design any type of structure consisting of tension members, compression members, flexural members, or beam–columns. To demonstrate this, the design of industrial buildings is dealt with in Chapter 12.

The design aids presented in the appendix (Appendix D) will be quite useful to designers and also to students to check the results.

Though care has been taken to present error-free material, some errors might have crept in inadvertently. I would highly appreciate if these errors and suggestions for improvement are brought to my notice.

Acknowledgements

I am grateful to my teachers, especially Prof. C. Ganapathy of the Indian Institute of Technology, Madras, from whom I learnt the subject. I have also learnt much from Prof. J. Lindner and Prof. Ch. Petersen, while working in Germany as an

Alexander von Humboldt fellow. My understanding of this subject was greatly influenced by the books and publications of several authors. These books and authors are listed in the suggested reading section, at the end of the book. I would like to apologize for any phrase or illustrations used in this book inadvertently without acknowledgement.

While attending the presentations and discussions held in the sub-committee, which prepared the draft of IS 800 : 2007, I received valuable inputs from Dr R. Narayanan (former head of Education and Publication Division of the Steel Construction Institute, UK), Prof. V. Kalyanaraman and Dr S.R. Sathish Kumar (IIT Madras), and other members of the sub-committee. I also learnt a lot from the discussions I had on several occasions with Prof. A.R. Santhakumar (former Dean of Anna University and Professor IIT Madras).

I would like to thank the following organizations/publishers for permitting me to reproduce material from their publications:

- American Institute of Steel Construction Inc., Chicago, for extracts from papers published in Modern Steel Construction and Engineering Journal,
- American Society of Civil Engineers for quotes from their publications,
- Steel Construction Institute, Ascot, UK, for figures from the book *Structural Steel Design* (by P.J. Dowling, P. Knowles, and G.W. Owens (Butterworths, 1988),
- Elsevier, UK, for figures from the book *Structural Steelwork Connections* by G.W. Owens and B.D. Cheal (Butterworths-Heinemann, 1989),
- Canadian Institute of Steel Construction for figures from the book *Limit States Design in Structural Steel* by G.L. Kulak and G.Y. Grondin (2002),
- McGraw-Hill Education for figures from the books *Structural Steel Design* by J.E. Bowles (1980), *Steel, Concrete, and Composite Design of Tall Buildings* by B.S. Taranath and *Handbook of Structural Steel Connection Design and Details* by A.R. Tamboli (1999).

I am privileged and grateful to Dr T.K. Bandyopadhyay (former Joint Director General, Institute for Steel Development and Growth, Kolkata), for writing the Foreword to the book.

I would like to thank all those who assisted me in the preparation of this book. First and foremost, I would like to thank Dr S. Seetharaman, Professor, Sathyabama University, Chennai, and former Deputy Director, Structural Engineering Research Centre (SERC), Madras, for writing the chapters on Beams and Plate Girders; Prof. C. Ganapathy for going through the entire manuscript and offering comments; Ms R. M. Biruntha for her help in working out the examples in the chapters on Plate Girders and Gantry Girders.

I will be failing in my duty if I do not acknowledge the help and assistance I received from Ms S. Chithra at all the stages of this book writing project. Lastly, I acknowledge the excellent support and coordination provided by the editorial team at Oxford University Press, India.

Dr N. Subramanian

Contents

List of Symbols

A area of cross section; surface area of cladding

A_b area of bolt; gross area of horizontal boundary elements in SPSW

A_{br} required bearing area

A_c minor diameter area of the bolt; gross area of vertical boundary elements in SPSW

A_d area of diagonal member

A_e effective cross-sectional area; effective frontal area in wind

A_F moment amplification factor

A_f total flange area of the smaller connected column; floor area; required area of flange plates

A_g gross cross-sectional area

A_{gf} gross cross-sectional area of flange

A_{go} gross cross sectional area of outstanding leg

A_h design horizontal seismic coefficient

A_k design horizontal acceleration spectrum value of mode k

A_n net area of the total cross section

A_{nb} net tensile cross sectional area of bolt

A_{nc} net cross sectional area of the connected leg

A_{ne} effective net area

A_{nf} net cross sectional area of each flange

A_{no} net sectional area of outstanding leg

A_o initial cross-sectional area of tensile test coupon

A_q cross-sectional area of the stiffener in contact with the flange

A_s tensile stress area

A_{sb} shank gross cross sectional area (nominal area) of a bolt

A_{st} area of stiffener

A_t total area of the compartment in fire

A_{tg} gross sectional area in tension from the centre of the hole to the toe of the angle perpendicular to the line of force (block shear failure)

A_{tn} net sectional area in tension from the centre of the hole to the toe of the angle perpendicular to the line of force (block shear failure)

A_v shear area

A_{vg} gross cross sectional area in shear along the line of transmitted force (block shear failure)

A_{vn} net cross sectional area in shear along the line of transmitted force (block shear failure)

A_w effective cross-sectional area of weld; window area; effective area of walls; area of web

a, b larger and smaller projection of the slab base beyond the rectangle considering the column, respectively

a_o peak acceleration

a_1 unsupported length of individual elements being laced between lacing points

B breadth of flange of I-section; length of side of cap or base plate of a column

B_s background factor

b_o outstand/width of the element

b_1 stiff bearing length; stiffener bearing length

b_e effective width of flange between pair of bolts

b_{fc} width of column flange

b_f breadth or width of the flange

b_p panel zone width between column flanges at beam column junction

b_s shear lag distance; stiffener width

b_{sh} average breadth of the structure between heights s and h

b_{0h} average breadth of the structure between heights 0 and h

b_t width of tension field

b_w width of outstanding leg

C centre-to-centre longitudinal distance of battens; coefficient related to thermal properties of wall, floor, etc.; spacing of transverse stiffener; moisture content of insulation

C_1 equivalent uniform moment factor

C_{dyn} dynamic response factor

C_e effective width of interior patch load

C'_f frictional drag coefficient

C_f force coefficient of the structure

C_{fs} cross-wind force spectrum coefficient

C_i specific heat of insulation material

C_{Lh} lateral horizontal load for cranes

C_m coefficient of thermal expansion; equivalent moment factor

C_p cost of purlin

C_{pe} external pressure coefficient

C_{pi} internal pressure coefficient

C_r cost of roof covering

C_s specific heat of steel

C_t cost of truss

c_m moment reduction factor for lateral torsional buckling strength calculation

D overall depth/diameter of the cross section

D_o outer diameter

d depth of web; nominal diameter; grain size of crystals; diagonal length; depth of snow; base dimension of the building

d_2 twice the clear distance from the compression flange angles, plates, or tongue plates to the neutral axis

d_a depth of angle

d_b beam depth; diameter of bolt

d_c column depth

d_g centre-to-centre of the outermost bolt of the end plate connection

d_h diameter of the hole

d_i thickness of insulation

d_o nominal diameter of the pipe column or the dimensions of the column in the depth direction of the base plate

d_p panel zone depth in the beam-column junction

E modulus of elasticity for steel; energy released by earthquake

$E(T)$ modulus of elasticity of steel at T°C

$E(20)$ modulus of elasticity of steel at 20°C

E_e equivalent elastic modulus of rope

EL_x earthquake load in x direction

EL_y earthquake load in y direction

E_p modulus of elasticity of the panel material

E_{sh} strain-hardening modulus

E_t tangent modulus of elasticity

E'_t reduced tangent modulus

e eccentricity, head diagonal of bolt

e_b edge distance of bolt

F net wind force on cladding

F' frictional drag force

F_{br} strength of lateral bracing

F_{cf} flange contribution factor

F_o minimum bolt pretension

F_{bsd} bearing strength of stiffener

F_q stiffener force

F_{qd} design buckling resistance of stiffener

F_w protection material density factor; ultimate web crippling load

F_{xd} design resistance of load carrying web stiffener

F_x external load; force or reaction on stiffener

F_z along-wind equivalent static load at height Z

f actual normal stress range for the detail category, uniaxial stress, frequency of vortex shedding

f_1, f_2, f_3 principal stresses acting in three mutually perpendicular directions

f_a stress amplitude

f_b actual bending stress; bending stress at service load

f_{bc} actual bending stress in compression

f_{bd} design bending compressive stress corresponding to lateral buckling

f_{bt} actual bending stress in tension at service load

f_c average axial compressive stress

f_{cc}, f_{cr} elastic buckling stress of a column or plate; Euler's buckling stress = $\pi^2 E/(KL/r)^2$

f_{ck} characteristic compressive strength of concrete

$f_{cr,b}$ extreme fibre compressive elastic lateral buckling stress

f_{cd} design compressive stress

f_d stress range at constant amplitude

f_e equivalent stress

f_f fatigue strength corresponding to N_{sc} cycle of loading

f_{fd} design normal fatigue strength

f_{feq} equivalent constant amplitude stress range

f_{fmax} highest normal stress range

f_{fn} normal fatigue stress range for 5×10^6 cycles

f_k characteristic strength

f_l fatigue limit

f_L stress range at cut-off limit

f_m mean stress

f_{max} maximum stress

f_{min} minimum stress

f_o proof stress

f_0 first mode natural frequency of vibration of a structure in the along-wind direction

$f_{0.2}$ 0.2% proof stress

f_{pl} steel stress at proportional limit

f_q shear stress

f_R characteristic value of fatigue strength at loading cycle N_R

f_{st} shear stress due to torsion

f_{sw} warping shear stress

f_t tension field strength

f_u characteristic ultimate tensile stress

f_{ub} characteristic ultimate tensile stress of the bolt

f_{up} characteristic ultimate tensile stress of the connected plate

f_{uw} ultimate tensile stress of the weld

f_v yield strength in the panel utilizing tension field action

f_{wd} design strength of weld

f_y characteristic yield stress

$f_y(T)$ yield stress of steel at T°C

$f_y(20)$ yield stress of steel at 20°C

f_{yb} characteristic yield stress of bolt

f_{yf} characteristic yield stress of flange

f_{yn} nominal yield strength

f_{yst} design yield stress of stiffener

f_{yp} characteristic yield stress of connected plate

f_{yq} characteristic yield stress of stiffener material

f_{yw} characteristic yield stress of the web material

f_0 yield strength of very large isolated crystals

G shear modulus of rigidity for steel; thickness of grout

G_v gust factor

G^* design dead load

g gauge length between the centre of the holes perpendicular to the load direction; acceleration due to gravity; gap for clearance and tolerance

g_R peak factor for resonant response

H height of section; transverse load

H_p heated perimeter

H_v calorific value of the vth combustible material

H_w window height

h depth of the section; storey height

h_b total height from the base to the floor level concerned

h_c height of the column

h_e effective thickness

h_{ef} embedment length of anchor bolt

h_o distance between flange centroids of I-section

h_i thickness of protection material; height of floor i

h_L height of the lip

h_s storey height

h_y distance between shear centre of the two flanges of the cross section

I moment of the inertia of the member about an axis perpendicular to the plane of the frame; impact factor fraction; importance factor

I_b moment of inertia of brace

I_f second moment of area of the foundation pad; insulation factor

I_{fc} moment of inertia of the compression flange

I_{ft} moment of inertia of the tension flange

I_h turbulence intensity at height h

I_p polar moment of inertia

I_q moment of inertia of a pair of stiffener about the centre of the web, or a single stiffener about the face of the web

I_s second moment of inertia of stiffener

I_t St. Venant's torsional constant

I_w warping constant

I_y moment of inertia about the minor axis

I_z moment of inertia about the major axis

IF   interference factor

K effective length factor

K_a area averaging factor

K_b effective stiffness of the beam and column; effective length factor for beams against lateral bending

K_c combination factor; stiffness of column

K_d wind directionality factor

K_h reduction factor to account for the bolt holes in HSFG connection

KL   effective length of the member

KL/r appropriate effective slenderness ratio of the section

KL/r_y effective slenderness ratio of the section about the minor axis

KL/r_z effective slenderness ratio of the section about the major axis

(KL/r)$_o$ actual maximum effective slenderness ratio of the laced or battened column

(KL/r)$_e$ effective slenderness ratio of the laced column accounting for shear deformation

K_m mode share correction factor for cross-wind acceleration

K_y, K_z moment amplification factor about respective axes

K_w warping restraint factor

k regression coefficient; constant; mode shape power exponent for the fundamental mode of vibration

k_1 probability factor or risk coefficient

k_2 terrain, height and structure size factor

k_3 topography factor

k_4 importance factor for cyclonic region

k_b, k_{br} stiffness of bracing

k_s modulus of sub-grade reaction

k_{sec} web distortional stiffness

k_{sm} exposed surface area to mass ratio

k_t brace stiffness excluding web distortion, torsion parameter

k_{tb} stiffness of torsional bracing

k_v shear buckling coefficient

L actual length; unsupported length; centre-to-centre distance of the intersecting members; length of the end connection; cantilever length; land in weld

L_b laterally unbraced length or distance between braces

L_c length of end connection measured from the centre of the first bolt hole to the centre of the last bolt hole in the connection; distance between gantry girders

L_{cf} clear distance between flanges of vertical boundary elements

L_e effective horizontal crest length

L_g gauge length of tensile test coupon

L_h measure of the integral turbulence length scale at height h

L_m maximum distance from the restraint at plastic hinge to an adjacent restraint (limiting distance)

L_o length between points of zero moment (inflection) in the span

L_w effective length of weld; length of wall

l_a length of the angle

l_e distance between prying force and bolt centre line

l_g grip length of connection

l_j length of the joint

l_s length between points of lateral support

l_t elongation due to temperature; length of top angle

l_v distance from bolt centre line to the toe of the fillet weld or to half the root radius for a rolled section

M bending moment; magnitude of earthquake

M_{1sway} maximum first order end moment as a result of sway

M_{br} required flexural strength of torsional bracing

M_{cr} elastic critical moment corresponding to lateral torsional buckling

M_d design flexural or bending strength

M_{dv} design bending strength of the section under high shear

M_{dy} design bending strength as governed by overall buckling about minor axis

M_{dz} design bending strength as governed by overall buckling about major axis

M_{eff} reduced effective moment

M_{fr} reduced plastic moment capacity of the flange plate

M_{fd} design plastic resistance of the flange alone

M_{nd} design strength under combined axial force (uni-axial moment acting alone)

M_{ndy}, M_{ndz} design strength under combined axial force and the respective uni-axial moment acting alone

M_o cross-wind base overturning moment; first order elastic moment

M_p plastic moment capacity of the section

M_{pb} moment in the beam at the intersection of the beam and column centre lines

M_{pc} moments in the column above and below the beam surfaces

M_{pd} plastic design strength

M_{pf} plastic design strength of flanges only

M_{pr} reduced plastic moment capacity of the section due to axial force or shear

M_q applied moment on the stiffener due to eccentric load

M_{tf} moment resistance of tension flange

M_u second order elastic moment; factored moment; required ultimate flexural strength of a section

M_y factored applied moment about the minor axis of the cross section; yield moment capacity about minor axis

M_{yq} yield moment capacity of the stiffener about an axis parallel to web

M_z factored applied moment about the major axis of the cross section

m mass; slope of the fatigue strength curve

m_v mass of vth combustible material

m^1 non-dimensional moment parameter = M_u/M_{bp}

N_d design strength in tension or in compression

N_f axial force in the flange

N_{sc} number of stress cycles

n number of parallel planes of battens; mean probable design life of structure in years; reduced frequency; number of cycles to failure; factored applied axial force; number of bolts in the bolt group/critical section; number of stress cycles; number of storeys

n_1, n_2 dispersion length

n_e number of effective interfaces offering frictional resistance to slip

n_n number of shear planes with the threads intercepting the shear plane in a bolted connection

n_s number of shear planes without threads intercepting the shear plane in a bolted connection

n' number of rows of bolts

P factored applied axial force; point load

P_{bf} design strength of column web to resist the force transmitted by beam flange

P_{cc} elastic buckling strength under axial compression

P_{crip} crippling strength of web of I-section

P_d design axial compressive strength

P_{dy}, P_{dz} design compression strength as governed by flexural buckling about the respective axis

P_{dw} design strength of fillet weld

P_e, P_{cr} elastic Euler buckling load; $\pi^2 EI/L^2$

P_f probability of failure

P_k modal participation factor for mode k

P_{min} minimum required strength for each flange splice

P_N probability that an event will be exceeded at least once in N years

P_n nominal axial strength

P_{ny} axial strength of the member bent about its weak axis

P_u maximum load in the column

p pitch length between centres of holes parallel to the direction of the load; pitch of thread in bolt, pressure

p_d design wind pressure

p_s, p_1, p_2 staggered pitch length along the direction of the load between lines of the bolt holes (Fig. 5.21)

p_z wind pressure at height Z

Q prying force; nominal imposed load; static moment of the cross section = $A\bar{y}$

Q^* design imposed load

Q_a accidental load (action)

Q_c characteristic load (action)

Q_d design load (action)

Q_f fire load

Q_i load effect i; design lateral force at floor i

Q_m mean value of load

Q_p permanent loads (action)

Q_v variable loads (action)

q_f fire load/unit floor area

R ratio of the mean compressive stress in the web (equal to stress at mid depth) to yield stress of the web; reaction of the beam at support; stress ratio; response reduction factor; resultant force; root opening of weld; local radius of curvature of beam; return period

R_d design strength of the member at room temperature

R_i net shear in bolt group at bolt i

R_k connection stiffness

R_m mean value of resistance

R_n nominal strength of resistance

R_{nw} design strength of fillet weld per unit length

R_{res} resultant force in the weld

R_r response reduction factor

R_{tf} resultant longitudinal shear in flange

R_u ultimate strength of the member at room temperature; ultimate strength of joint panel

r appropriate radius of gyration

r_a root radius of angle

r_b root radius of beam flange

r_1 minimum radius of gyration of the individual element being laced together

r_f ratio of the design action on the member under fire to the design capacity

r_{vv} radius of gyration about the minor axis (v-v)

r_y radius of gyration about the minor axis

r_z radius of gyration about the major axis

S minimum transverse distance between the centroid of the rivet or bolt or weld group; strouhal number; size reduction factor; spacing of truss

S_a spectral acceleration

S_d spectral displacement

S_p spring stiffness

S_v spectral velocity

S_1, S_2 stability functions

s design snow load; size of weld

s_a actual stiffener spacing

s_c anchorage length of tension field along the compression flange

s_{ii}, s_{jj} stability functions

s_{ii}^*, s_{jj}^* stability function for semi-rigid frames

s_o ground snow load

s_t anchorage length of tension field along the tension flange (distance between adjacent plastic hinges)

T Temperature in °C; factored tension in bolt; natural period of vibration; applied torque

T_a approximate fundamental natural period of vibration

T_b applied tension in bolt

T_d design strength under axial tension

T_{db} design strength of bolt under axial tension; block shear strength of plate/angle

T_{dg} yielding strength of gross section under axial tension

T_{dn} rupture strength of net section under axial tension; design tension capacity

T_{dw} design strength of weld in tension

T_e externally applied tension

T_{eq} equivalent fire rating time

T_f factored tension force of friction type bolt; furnace temperature

$T_{f,max}$ maximum temperature reached in natural fire

T_l limiting temperature of the steel

T_{nb} nominal strength of bolt under axial tension

T_{nd} design tension capacity

T_{ndf} design tensile strength of friction type bolt

T_{nf} nominal tensile strength of friction type bolt

T_o ambient (room) temperature

T_s steel temperature at time t

T_t ambient gas temperature at time t

T_u ultimate net section strength

t thickness of element/angle, time in minutes

t_a thickness of top angle

t_b thickness of base plate

t_{cw} thickness of column web

t_e effective throat thickness of weld

t_f thickness of flange; required fire rating time

t_{fc} thickness of compression flange

t_{fail} time to failure of the element in case of fire

t_{fb} thickness of beam flange

t_p thickness of plate/end plate

t_{pkg} thickness of packing

t_q thickness of stiffeners

t_s thickness of web stiffener; duration of fire

t_v time delay in minutes

t_w thickness of web

U shear lag factor

V factored applied shear force; mean wind speed

V_B total design seismic base shear; basic wind speed

V_b shear in batten plate

V_{bf} factored frictional shear force in HSFG connection

V_{cr} critical shear strength corresponding to web buckling (without tension field action)

V_d design shear strength; design mean wind velocity

V_{db} shear capacity of outstanding leg of cleat

V_{dw} design strength of weld in shear

V_g gradient wind speed; gust speed

V_h design wind speed at height h

V_L longitudinal shear force

V_R vector resultant shear in weld

V_{nb} nominal shear strength of bolt

V_{nbf} bearing capacity of bolt for friction type connection

V_p plastic shear resistance under pure shear or shear strength of web

V_n nominal shear strength or resistance

V_{npb} nominal bearing strength of bolt

V_{nsb} nominal shear capacity of a bolt

V_{nsf} nominal shear capacity of a bolt as governed by slip or friction type connection

V_{sb} factored shear force in the bolt

V_{sd} design shear capacity

V_{sdf} design shear strength of friction type bolt

V_{sf} factored design shear force of friction bolts

V_T applied transverse shear

V_{tf} shear resistance in tension field

$\overline{V}$ average or mean velocity

V_{yw} yield strength of web plate of I-section

V_z mean or design wind speed at height z above the ground

V' Instantaneous velocity fluctuation above the mean velocity

W appropriate load; width; seismic weight; ventilation factor

W_e equivalent cross-wind static force per unit-height

w uniform pressure from below on the slab base due to axial compression under factored load; intensity of uniformly distributed load

w_{tf} width of tension field

X distance from a point to any other point

x_t torsional index

$\overline{x}$ distance from centre of gravity in x direction

Y_s yield stress

$\overline{y}$ distance from centre of gravity in y direction

y_g distance between point of application of the load and shear centre of the cross section

y_s coordinate of the shear centre with respect to centroid

Z section modulus; height above ground; zone factor

Z_e elastic section modulus

Z_g depth of boundary layer

Z_p plastic section modulus

Z_{pr} plastic modulus of the shear area about the major axis; reduced plastic modulus

α coefficient of linear expansion; imperfection factor; power law coefficient; included angle in groove weld

α_{LT} imperfection factor

α_t coefficient of thermal expansion

β reliability index; the ratio of structural damping to critical damping of a structure

β_{lj} reduction factor for overloading of end bolt

β_{lg} reduction factor for the effect of large grip length

β_{pkg} reduction factor for the effect of packing plates

β_M ratio of smaller to the larger bending moment at the ends of a beam column

β_{My}, β_{Mz} equivalent uniform moment factor for flexural buckling for y-y and z-z axes, respectively

β_{MLT} equivalent uniform moment factor for lateral torsional buckling

χ stress reduction factor due to buckling under compression

χ_m stress reduction factor χ at f_{ym}

χ_{LT} strength reduction factor for lateral torsional buckling of a beam

δ, Δ storey deflection or drift; deflection

δ_b moment amplification factor for braced member

δ_L horizontal deflection of the bottom of storey due to combined gravity and notional load

δ_m moment amplification factor

δ_p load amplification factor

δ_s moment amplification factor for sway frame

δ_U horizontal deflection of the top of storey to combined gravity and notional load

ε yield stress ratio; $(250/f_y)^{1/2}$; strain corresponding to stress f, resultant emissivity of surface

ε_p plastic strain

ε_{sh} strain hardening strain

$\varepsilon_u, \varepsilon_{br}$ ultimate strain

ε_y yield strain

υ shape factor

ϕ strength or resistance reduction factor; cumulative distribution function; solidity ratio; inclination of the tension field stress in web; configuration factor; angle of twist

$\phi_{i,k}$ mode shape coefficient at floor i in mode k

ϕ_s sway index

γ unit weight of steel

$\gamma_f, \gamma_{fk}, \gamma_{if}$ partial safety factor for load

γ_{fft} partial safety factor for fatigue load

γ_m partial safety factor for material

γ_{m0} partial safety factor against yield stress and buckling

γ_{ml} partial safety factor against ultimate stress

γ_{mb} partial safety factor for bolted connection with bearing type bolts

γ_{mf} partial safety factor for bolted connection with HSFG bolts

γ_{mft} partial safety factor for fatigue strength

γ_{mi} partial safety factor depending upon the type of failure as prescribed in IS: 800

γ_{mw} partial safety factor for strength of weld

$\bar{\lambda}$ non-dimensional slenderness ratio

$$= \left(KL/r / \sqrt{\pi^2 E/f_y} \right.$$

$$= \sqrt{f_y/f_{cc}} = \sqrt{P_y/P_{cc}} \left. \right)$$

λ_{cr} elastic buckling load factor

λ_e equivalent slenderness ratio

λ_i effective thermal conductivity of insulation

λ_{LT} non-dimensional slenderness ratio

λ_y, λ_z non-dimensional slenderness ratio about respective axis

λ_w non-dimensional web slenderness ratio for shear buckling

μ Poisson's ratio; shape coefficient or factor for snow load

μ_c correction factor; capacity reduction factor for fatigue

μ_f coefficient of friction (slip factor)

μ_r capacity reduction factor for non-redundant load path

θ^l non-dimensional rotation parameter $= \theta_r/\theta_p$

θ ratio of the rotation at the hinge point to the relative elastic rotation of the far end of the beam segment containing plastic hinge, upwind slope of ground

θ_p plastic rotation

θ_r rotation of semi-rigid joint

ρ_s density of steel

ρ_i dry density of insulation

ρ_i' effective density of insulation

s Stefon–Boltzmann constant

τ actual shear stress for the detail category

τ_b shear stress corresponding to buckling

τ_{cr} elastic critical shear buckling stress

τ_e equivalent shear stress

τ_f fatigue shear stress range for N_{sc} cycle

τ_{fd} design fatigue shear strength

τ_{fmax} highest shear stress range

τ_{fn} fatigue shear stress range at 5×10^6 cycles for the detail category

τ_L shear stress range at cut-off limit

τ_o grout-concrete bond strength

τ_{vf} shear stress in the weld due to vertical force

τ_{vfl} shear stress in the weld due to bending moment

τ_w shear stress in weld throat; shear stress due to shear force

τ_y shear yield stress

ξ_i reduction factor for geometric imperfection

ψ ratio of the moment at the ends of the laterally unsupported length of a beam

ω circular natural frequency

Note: The subscripts y and z denotes the y-y and z-z axes of the section, respectively. For symmetrical sections, y-y denotes the minor principle axis whilst z-z denotes the major principal axis.

1

Structural Steel: Types, Properties, and Products

Introduction

In early societies, human beings lived in caves and almost certainly rested in the shade of trees. Gradually, they learnt to use naturally occurring materials such as stone, timber, mud, and biomass (leaves, grass, and natural fibres) to construct houses. Then followed brick making, rope making, glass, and metal work. From these early beginnings, the modern materials manufacturing industries developed.

The principal modern building materials are masonry, concrete (mass, reinforced, and prestressed), glass, plastic, timber, and structural steel (in rolled and fabricated sections). All the mentioned materials have particular advantages in a given situation and hence the construction of a particular building type may involve the use of various materials, e.g., a residential building may be constructed using load-bearing masonry, concrete frame or steel frame. The designer has to think about various possible alternatives and suggest a suitable material which will satisfy economic, aesthetic, and functional requirements.

The main advantages of structural steel as a building material are its strength, speed of erection, prefabrication, and demountability. They are used in load-bearing frames in buildings, and as members in trusses, bridges, and space frames. Steel, however, requires fire and corrosion protection. In steel buildings, claddings and dividing walls are made up of masonry or other materials, and often a concrete foundation is provided. Steel is also used in conjunction with concrete in composite constructions and in combined frame and shear wall constructions. In many cases, the fabrication of steel members is done in the workshop and the members are then transported to the site and assembled. Tolerances specified for steel fabrication and erections are small compared to those for reinforced concrete structures. Moreover, welding, tightening of high-strength friction grip bolts, etc., require proper training. Due to these factors, steel structures are often handled by trained persons and assembled with proper care, resulting in structures with better quality.

Steel offers much better compressive and tensile strength than concrete and enables lighter constructions. Also, unlike masonry or reinforced concrete, steel can be easily recycled.

In this chapter we will discuss the manufacture and those properties of structural steel, which are important in the selection of the material for a particular situation. We will also discuss the various types of steels, the available hot- and cold-rolled sections, and the various types of structures that can be built using these sections.

A brief introduction to the loads to be considered during the analysis of steel structures and a discussion about the methods of analysis are also provided.

1.1 Historical Development

Steel has been known since 3000 BC. Foam steel was used during 500–400 BC in China and then in Europe. The Ashokan pillar made with steel and the iron joints used in Puri temples are more than 1500 years old. They demonstrate that this know-how was available before the modern blast-furnace technology, which was developed in AD 1350 (Gupta 1998).

Structural steel was first introduced in 1740, but was not available in large quantities until Sir Henry Bessemer of England invented and patented the process of making steel in 1855. In 1865, Siemens and Martin invented the open-hearth process and this was used extensively for the production of structural steel till 1980. In steel, the carbon content varies from 0.25% to 1.5%. The first major structure to use the new steel exclusively was Fowler and Baker's Railway Bridge at the Firth of Forth.

Riveting was used as a fastening method until around 1950 when it was superseded by welding. The basic oxygen steel making (BOS) process using the CD converter was invented in Austria in 1953. In the latter part of the nineteenth century and the early twentieth century, newer technologies resulted in better and new grades of steel. Today we have several varieties of steel made with alloying elements such as carbon, manganese, silicon, chromium, nickel, and molybdenum (see Section 1.4). The electric arc furnace is used to make special steels such as stainless steel.

1.2 Processes Used for Steel Making

Currently steel is produced largely by the *basic oxygen steel making* (BOS) process and the *electric arc method*. Steel production is basically a batch process and involves reducing the carbon, sulphur and phosphorous levels and adding, when necessary, manganese, chromium, nickel or vanadium.

Today most structural steel is made in integrated steel plants using the BOS process shown in Fig. 1.1. Iron ore lumps, scrap steel (up to 30%), pellets, coke (made from cooking coal), and fluxes such as limestone and dolomite are used as the major raw materials. The main steps involved in the manufacturing process are as follows.

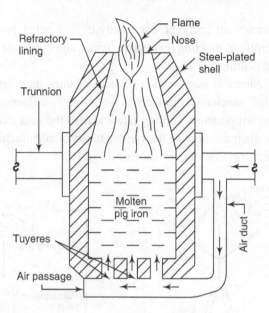

Fig. 1.1 Basic oxygen steel making (BOS) process (Rangwala et al. 1997)

Melting Raw materials are charged in a blast furnace, where hot air is pumped to melt iron and fluxes at 1600°C. The molten metal when cooled and solidified is called *pig iron*. Alternatively, it can be further refined to make steel. The excess carbon and other unwanted impurities are floated off as slag (this slag is blended with clinker to make blast furnace cement, which is used in high-performance concretes).

Refining Molten metal from the blast furnace is taken to the steel melting shop where the impurities are further reduced in a basic-oxygen furnace (LD converter) or an open-hearth furnace. Deoxidizers, such as silicon and/or aluminum are used to control the dissolved oxygen content. Steel which has the highest degree of deoxidation {containing less than 30 parts per million (ppm) of oxygen} is termed *killed steel*. *Semi-killed steel* has an intermediate degree of deoxidation (about 30–150 ppm of oxygen). Steel containing the lowest degree of deoxidation is called *rimmed steel*. During continuous casting, only killed steel is used. Generally structural steel contains carbon (in the range of 0.10–0.25%) manganese (0.4–0.12%), sulphur (0.025–0.05%), and phosphorus (0.025–0.050%) depending on end use and specifications. The crude steel in liquid form is taken in a ladle for further refining/addition of ferro-alloys, etc.

Casting The liquid steel is taken out of the bottom as a continuous ribbon of steel. When sufficiently cooled, it is cut into semi-finished products, such as billets, blooms, and slabs. This process is called *continious casting* or *concast method*.

Hot rolling The semi-finished products, such as billets, blooms, and slabs, are heated at 1200°C to make metal malleable and then rolled into finished products, such as plates, structural sections, bars, and strips. Further processing of steel can include cold rolling, pickling (to remove oxides and mill scale from the surface of the steel), and coating.

Although the chemical composition of steel dictates its potential mechanical properties, its final mechanical properties are strongly influenced by the rolling process, finishing temperature, cooling rate and also the heat treatment (if any).

The schematic diagram showing the various stages of manufacturing of structural steel sections from the iron ore are shown in Fig.1.2.

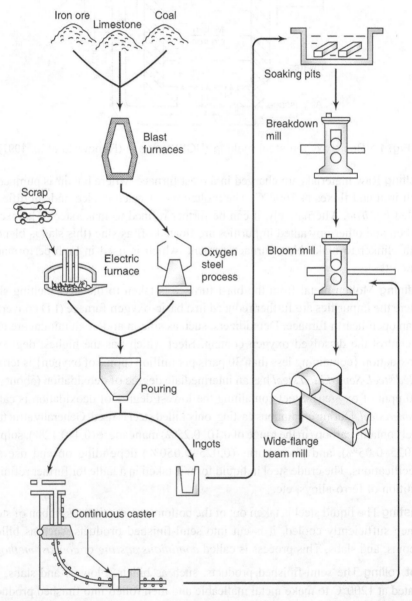

Fig. 1.2 Schematic diagram showing the various stages of manufacturing structural steel sections from iron ore (Kulak & Grondin 2002)

As per the International Iron and Steel Institute, the global production of crude steel during 2005 was estimated as 1129 mt and the consumption was 900 mt.

As early as in 1907, Jamsetji Nusserwanji Tata set up the first integrated steel manufacturing plant at Jamshedpur in Bihar. Then major steel plants were constructed at Bhilai, Rourkela, Durgapur, and Bhadravati steel plant in Karnataka. The steel sector now consists of seven integrated plants and about 180 mini steel plants and rerollers. The Howrah Bridge and second Hooghly cable-stayed bridge in Calcutta are examples of steel-intensive bridge construction. There are numerous bridges (built by the Railways) and industrial buildings exclusively using steel.

In 2004–05, the annual production of steel in India was about 38 million tonnes and is likely to increase in the future. At present, India is the tenth largest producer of steel in the world. However, the per capita consumption of steel in India is low, about 33 kg/person/year as compared to 220 kg in China and 300–600 kg in developed countries like the USA, Germany, the UK, and Japan. In India, a major part of steel is consumed in engineering applications, followed by automobiles and constructions.

1.3 Heat Treatment of Steel

Heat treatment involves the heating and cooling of steel under controlled conditions to change its structural and physical properties. *Annealing* and *normalizing* are the processes used to refine the structure of steel. In the annealing process, the steel is heated to a temperature just greater than 910°C and held at that temperature to achieve uniformity of composition and temperature prior to slow cooling, usually in the furnace. Sufficient cooling time allows the carbon diffusion and transformation process to get completed. This refined pearlite + ferrite microstructure shows both increased strength and ductility.

Normalizing is a process similar to annealing, except that in normalizing the steel is removed from the furnace and allowed to cool in still air. The changes occurring are the same as during annealing but less time at high temperature and the faster cooling rate give a slightly finer grain structure and finer laminations in the pearlite. These finer structures result in slightly improved properties compared to those obtained as a result of annealing. Normalizing is cheaper than annealing since the steel is kept in the furnace for less time. However, it can only be used for fairly uniform sections, where air cooling is unlikely to cause distortion due to differential cooling and contraction.

Mild steel plates, structural sections, etc., show very good properties of strength with ductility in the normalized condition. Heat treatment is costly and in India the heat treated steels amount hardly to about 5% of the steel produced (Rangwala 1997).

1.4 Alloying Elements in Steel

The physical properties of steel such as ductility, elasticity, strength, toughness, etc., are greatly influenced by the following factors.

(a) Carbon content,
(b) Heat-treatment process, and
(c) Alloying elements.

We have already discussed the first two factors in the previous sections. Depending upon the carbon content, the steel is designated as low-carbon steel (carbon content 0.10–0.25%), medium-carbon steel (carbon content 0.25–0.60%) and high-carbon steel (carbon content 0.60–1.10%). Structural steels normally have a carbon content less than 0.25%. As already discussed, increasing the carbon content increases the hardness, yield, and tensile strength of steel. However, it decreases the ductility and toughness. Carbon also has greater influence on weldability. Mild steel is widely used for structural work and will be discussed in detail in the later sections of this chapter.

Manganese, silicon, sulphur, phosphorus, copper, vanadium, nickel, chromium, columbium, molybdenum, and aluminium are some of the other elements that may be restricted in, or added to, structural steel. In recent years, microalloyed steels or high-strength low-alloy (HSLA) steels have been developed. They are basically carbon manganese steels in which small amounts of aluminium, vanadium, mobium, etc., are used to control the grain size. Molybdenum is also added (up to 0.5%) to refine the lamellar spacing in pearlite and to make it evenly distributed. Alloy steels are termed as low-alloy steels (total alloy content < 5%), medium-alloy steels (total alloy content 5–10%) and high-alloy steels (total alloy content > 10.0%). Based on manganese content, steels are also classified as carbon manganese steels (Mn > 1%) and carbon steels (Mn < 1%).

If the silicon content is raised to about 0.30 to 0.40%, the elasticity and strength of steel are considerably increased without serious reduction in ductility. More than 2% of silicon causes brittleness. A sulphur content of more than 0.10% decreases the strength and ductility of steel.

It is desirable to keep the phosphorus content of steel below 0.12%. It reduces the shock resistance, ductility, and strength of steel. If present in quantities between 0.30 and 1.00%, manganese helps to improve the strength and hardness of mild steel in more or less the same way as carbon.

1.5 Weldability of Steel

In most cases, members of steel are welded during fabrication. For good weldability, steel should not show high hardness in welded parts, but should have adequate elongation and notch toughness even in the *heat-affected zone* adjacent to a weld.

A major factor in weldability is the carbon equivalent, C_{eq}, of the chemical components in steel. The smaller this value, the better is the weldability. The carbon equivalent may be calculated by an equation such as that shown below, in which each symbol refers to the proportion of weight of that particular element in percentage (IS 2062 : 1992).

$$C_{eq} = \frac{C + Mn}{6} + \frac{(Cr + Mo + V)}{5} + \frac{(Ni + Cu)}{15} \tag{1.1}$$

where C is carbon, Mn is manganese, Cr is chromium, Mo is molybdenum, V is vanadium, Ni is nickel, and Cu is copper.

High-strength steels tend to have a high carbon equivalent. When the carbon equivalent exceeds a certain limit (C_{eq} = 0.30–0.43), the loss of weldability is compensated by the reheating or post-heating of the weld zone. However, if the carbon content is less than 0.12%, then C_{eq} can be tolerated up to 0.45%.

1.6 Chemical Composition of Steel

Several varieties of steel are produced in India. The Bureau of Indian Standards (BIS) classifies structural steels into different categories based on the ultimate yield strength of the basic material and their use (see IS 7598). They are listed along with the appropriate codes of practice issued by BIS in Table 1.1.

Table 1.1 Types of steel and their relevant IS standards

Type of steel	Relevant IS standards
Structural steel	2062, 1977, 3502, 5517, 8500
Steel for tubes and pipes	1239, 1914, 806, 1161, 10748, 4923
Steel for sheets and strips	277, 1079, 12367, 513, 12313, 14246
Steel for bolts, nuts, and washers	1363, 1364, 1367, 3640, 3757, 6623, 6639, 730, 4000, 5624, 6649, 8412, 10238, 12427
Welding	814, 1395, 816, 819, 1024, 1261, 1323
Steel for filer rods/wires, electrodes	1278, 1387, 7280, 6419, 6560, 2879, 4972, 7280
Steel casting	1030, 2708, 2644, 276

The chemical compositions of some typical steels specified by the Bureau of Indian Standards are listed in Table 1.2. For details of chemical composition of other steels refer IS 1977 [structural ordinary (low tensile) quality], IS 8500 (medium and high strength quality).

Table 1.2 Chemical compositions (in percentage) of some typical structural steels

Type of steel	Designation	IS code	C (max.)	Mn (max.)	S (max.)	P (max.)	Si (max.)		Carbon equivalent
Standard	Fe 410 A[a]	2062	0.23	1.5	0.050	0.050	0.4	SK[b]	0.42
structural	Fe 410 B	2062	0.22	1.5	0.045	0.045	0.4	SK	0.41
steel	Fe 410 C	2062	0.20	1.5	0.040	0.040	0.4	K	0.39
Micro-	Fe 440	8500	0.20	1.3	0.050	0.050	0.45		0.40
alloyed	Fe 540	8500	0.20	1.6	0.045	0.045	0.45		0.44
medium-/									
high-strength									
steel	Fe 590	8500	0.22	1.8	0.045	0.045	0.45		0.48

[a] Fe stands for steel and the number after Fe is the tensile strength in N/mm^2 or MPa
[b] K—killed steel, SK—semi-killed steel (explained in Section 1.2)
C = carbon, Mn = manganese, S = sulphur, P = phosphorus, Si = silicon

1.7 Types of Structural Steel

The structural designer is now in a position to select structural steel for a particular application from the following general categories.

Carbon steel (IS 2062) Carbon and manganese are the main strengthening elements. The specified minimum ultimate tensile strength for these steels varies from about 410 to 440 MPa and their specified minimum yield strength from about 230 to 300 MPa (see Table 1 of IS 800 : 2007).

High-strength carbon steel This steel is specified for structures such as transmission lines and microwave towers, where relatively light members are joined by bolting. Such steels have a specified ultimate tensile strength, ranging from about 480–550 MPa, and a minimum yield strength of about 350–400 MPa.

Medium- and high-strength microalloyed steel (IS 8500) Such steel has a specified ultimate tensile strength ranging from 440 to 590 MPa and a minimum yield strength of about 300–450 MPa.

High-strength quenched and tempered steels These steels are heat treated to develop high strength. Though they are tough and weldable, they require special welding techniques. They have a specified ultimate tensile strength between 700 and 950 MPa and a minimum yield strength between 550 and 700 MPa.

Weathering steels These are low-alloy atmospheric corrosion-resistant steels, which are often left unpainted. They have an ultimate tensile strength of about 480 MPa and a yield strength of about 350 MPa.

Stainless steels These are essentially low-carbon steels to which a minimum of 10.5% (maximum 20%) chromium and 0.50% nickel is added.

Fire-resistant steels Also called thermomechanically treated steels, they perform better than ordinary steel under fire.

1.8 Mechanical Properties of Steel

The mechanical properties of steels depend upon the following factors:
  (a) chemical composition,
  (b) rolling methods,
  (c) rolling thickness,
  (d) heat treatment, and
  (e) stress history.

The important mechanical properties of steel are ultimate strength (also called tensile strength), yield stress (also called proof stress), ductility, weldability, toughness, corrosion resistance, and machinability.

The last four properties are often associated with the fabrication of steel structures and are important for the durability of the material.

1.8.1 Ultimate Strength or Tensile Strength

Ultimate strength, which is the minimum guaranteed *ultimate tensile strength* (UTS) at which the steel would fail, is obtained from a tensile test on a standard specimen, generally called a *coupon*. A typical specimen as per IS 1608 is shown in Fig. 1.3. In this test, the gauge length L_g and the initial cross-sectional area A_0 are important parameters. The dimensions of the specimens are established to ensure that failure occurs within the designated gauge length. The test coupons are actually cut out from a specified portion of the member for which the tensile strength is required. The initial gauge length is taken as $5.65\sqrt{A_0}$ in the case of a specimen with a rectangular cross section and five times the diameter in the case of a circular specimen.

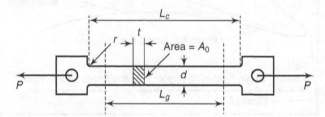

Fig. 1.3 Standard tensile test specimen as per IS 1608

The coupon is fixed in a tensile testing machine, with specified distances between the grips, and tested under uniaxial tension. The loads are applied through the threaded ends. A typical stress–strain curve of ordinary and high-strength steel specimen subjected to a gradually increasing tensile load is shown in Fig. 1.4(a) and the stress–strain curve of mild steel specimen is shown in Fig. 1.4(b).

The ultimate tensile strength is the highest stress at which a tensile specimen fails by fracture and is given by

$$\text{Ultimate tensile strength} = \frac{\text{ultimate tensile load}}{\text{original area of cross section}} \tag{1.2}$$

Note that steel is specified in the code by the characteristic ultimate tensile strength, f_u [which is defined as the minimum value of stress below which not more than a specified percentage (usually 5%) of corresponding stresses of samples tested are expected to occur]. However, in some countries like the USA, steel is specified according to the (characteristic) yield strength.

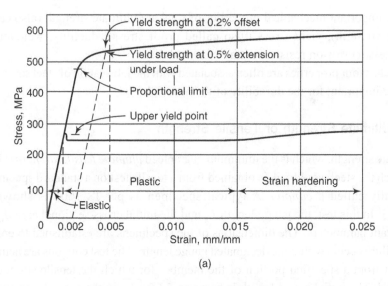

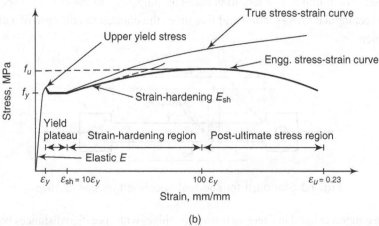

Fig 1.4 Typical stress-strain curves of mild steel-(a) stress-strain curves of ordinary and high-streangth steel, (b) stress-strain curve of a mild steel specimen

Thus, steel is designated in India as Fe 310, Fe 410 WA, Fe 540 B, Fe 590, etc., where Fe stands for the steel and the number after Fe is the characteristic ultimate tensile stress in megapascals. The letter A, B, or C indicates the grade of steel. The letter W denotes that the steel is weldable. (Copper-bearing quality is designated with a suffix Cu, e.g., Fe 410 Cu-WA.) Table 1.3 indicates the minimum ultimate tensile stress and other important mechanical properties of steel produced in India. Grade A steel specified by IS 2062 is intended for use in structures subject to normal conditions and for non-critical applications (for parts not prone to brittle fracture). Grade B is intended for use in structures subject to critical loading applications, where service temperature does not fall below 0°C. Grade B steel is generally specified for those structural parts which are prone to brittle fracture or are subjected to severe fluctuations of stress (for example members in bridges). Naturally, such steel is also specified for structural parts prone to both conditions. Grade C steel has guaranteed low temperature (up to $-40°C$) and impact properties. Grade C steel is used in members or structures where the risk of brittle fracture requires consideration due to their design, size and/or service conditions.

Table 1.3 Mechanical properties of some typical structural steels

(a) Ultimate tensile strength, yield strength, and percentage elongation

Type of steel	Designation	UTS (MPa)	Yield strength (MPa) Thickness (mm)			Min. percentage elongation (gauge length = $5.65\sqrt{A_0}$)	Charpy V-notch impact energy (min.)
			<20	20–40	>40		
Standard	Fe 410 A	410	250	240	230	23	—
structural	Fe 410 B	410	250	240	230	23	27
steel	Fe 410 C	410	250	240	230	23	27
(IS 2062)							
			<16	16–40	41–63		
Micro-alloyed	Fe 440 B	440	300	290	280	22	30
medium-/high-	Fe 540 B	540	410	390	380	20	25
strength steel	Fe 490 B	490	350	330	320	22	25
(IS 8500)	Fe 590 B /570 B	590/570	450	430	420	20	20

(b) Other mechanical properties as per IS 800 : 2007

Property	Value
Modulus of elasticity (E)	2×10^5 MPa
Shear modulus (G)	$E/[2(1 + \mu)] = 0.769 \times 10^5$ MPa for $\mu = 0.3$
Poisson's ratio (μ)	
(i) Elastic range	0.3
(ii) Plastic range	0.5
Unit mass of steel, ρ	7850 kg/m^3
Coefficient of thermal expansion, α_t	12×10^{-6}/°C
Brinell hardness number	150–190
Vickers hardness number	157–190
Approximate melting point	1530°C
Thermal conductivity	0.14 cal/cm^2 s/1°C/cm

After reaching the ultimate tensile stress, a localized reduction in area, called *necking*, begins, and elongation continues with diminishing load until the specimen breaks. After failure, the fractured surface of the two pieces is found to form a cup-and-cone arrangement. Cup-and-cone fracture is considered as an indication of *ductile fracture*.

As shown in Fig. 1.4(a), initially the steel has a linear stress–strain curve whose slope equals Young's modulus of elasticity, *E*. Thus,

$$\text{Modulus of elasticity} = \frac{\text{stress within the proportional limit}}{\text{strain}} \qquad (1.3)$$

This can be expressed as

$$E = \frac{f}{\varepsilon} \qquad (1.4)$$

where f is the uniaxial stress below the proportional limit, and ε is the strain corresponding to the stress f.

The values of E vary in the range 200,000–210,000 MPa and an approximate value of 200,000 MPa is assumed in the code. The steel obeys *Hooke's law* in this linear range. That is, it remains elastic and recovers to the original shape perfectly on unloading. The limit of the elastic behaviour is often closely associated with the yield stress f_y and the corresponding yield strain $\varepsilon_y = f_y/E$. Beyond this limit, the steel flows plastically without any increase in stress until the 'strain hardening' strain ε_{sh} is reached. This *plastic range* is usually considerable, and accounts for the ductility of steel. The stress increases above the yield stress f_y, when the 'strain hardening' strain ε_{sh} is exceeded, until the ultimate tensile stress f_u is reached. As indicated earlier, at this stage, large local reductions in the cross section occur, and the tensile failure takes place.

The yield strain for mild steel is of the order of 0.00125 or 0.125%. Depending on the steel used, ε_{sh} generally varies between 5 ε_y and 15 ε_y. The average value of 10 ε_y is taken as the yield plateau of structural steels. The value of ε_u is taken as 100 ε_y and that of ε_{br} as 0.23 mm/mm. The initial slope of the strain-hardening part of the curve is termed as the *strain-hardening modulus*, E_{sh}. It is much less steep than the elastic part, with E_{sh}/E being typically between 1/30 and 1/100 (Alpsten 1973). The strain-hardening range is not consciously used in design, but some of the buckling limitations are conservatively derived to preclude buckling even at strains well beyond onset of strain hardening.

Yielding is sometimes accompanied by an abrupt decrease in load, as shown in Fig. 1.4(a), which results in upper and lower yield points. The upper yield point (f_{yu}) is influenced by the shape of the test specimen and by the testing machine itself, and is sometimes completely suppressed. The lower yield point (f_{yl}) is much less sensitive and is considered to be more representative. The stress–strain curve shown in Fig. 1.4(b) is typical of low-carbon (mild) steel. Note that the upper as well as lower yield points tend to increase with increase in speed of loading (strain rate). Typical values of the ratio f_{yu}/f_{yl} for normal structural steel range from about 1.05 to 1.10. The term *yield stress* is commonly used to mean either yield point or yield strength when it is not necessary to make the distinction. Steel in compression has the same modulus of elasticity as in tension. The lower yield stress is also the same for tension and compression and there is about the same length of level yielding (contraction).

Parameters that influence yield stress

The strain rates used in tests to determine the yield stress of a particular steel type are significantly higher than the nearly static rates often encountered in actual structures

(McGuire 1968 and Alpsten 1973). However, since the values obtained from the majority of these mill tests are not more than 10% higher or lower than the static rate values, the net effect, when averaged over a complete design, may not be significant (Nethercot 2001). At higher temperatures, the reverse takes place (i.e., at higher strain rates there is reduction in yield strength). This fact needs to be considered only in blast-resistant design. Mild steel and medium-strength steels have clear yield points and should not be stressed beyond the yield point as the deformation will be large and uncontrollable beyond yield. At strain rates characteristic of seismic response (0.01–0.10/s), steel exhibits a significant increase in yield strength (10–20%) above static test values. However, under cyclic straining, i.e., straining under cyclic loads, the effective strain rate decreases, minimizing this effect.

Yield stress may also be influenced by the position from which the test coupons are taken. For example, the webs of the I-section are thinner than those of the flanges and the yield stress will be higher than that at the flange (Alpsten 1973). It has to be noted that in most situations, the flanges of I-sections contribute most to their load-carrying capacity, since most of the area is concentrated in the flanges. Hence, structural designers must be careful in selecting the appropriate value for material strength for use in their calculations. In order to use plastic design or in earthquake-resistant structures, the steel should satisfy the following criteria.

(a) The yield plateau should extend for at least six times the strain at first yield.
(b) The ultimate/yield stress ratio must be greater than 1.25. (To develop an inelastic rotation capacity, a structural member needs adequate length of yield region along the axis of the member. The larger the ultimate to yield ratio, the longer is the yield region.)
(c) The minimum elongation must be 15% on a gauge length of $5.65\sqrt{A_0}$.

It is also preferable that the actual yield strength based on the tensile test of steel does not exceed the specified yield strength by more than 120 MPa. Figure 1.5 shows the stress–strain curves of different types of steel produced in India and the permanent strain line. Fe 410 grade mild steel is the most commonly used in structural applications. Fe 370 grade steel is used in less important works. (Fe 310 mild steel is used primarily for furniture, doors, windows, etc.)

High-carbon steels do not usually have a pronounced yield point. Instead, after a range of linear elastic behaviour, which ends at a point called the *proportional limit*, the rate of increase in stress begins to drop till the tensile strength is reached (the upper curve of Fig. 1.5). In this case, yielding is arbitrarily defined by a yield strength which is usually taken to be that stress which leaves the specimen with a permanent set (plastic elongation) of 0.2% when the specimen is unloaded. It is obtained by drawing a line parallel to the elastic portion at 0.2% strain, which intercepts the stress–strain curve, as shown in Fig. 1.5. However some standards (e.g., ASTM specification, A370) define the yield stress as the stress corresponding to a 0.5% elongation under load. The allowed permanent set in higher tensile bolt is around 0.006.

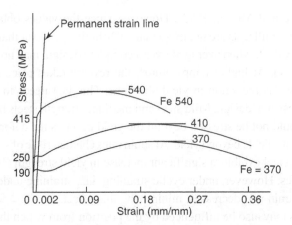

Fig. 1.5 Stress-strain curves of different types of steel produced in India

The yield stress f_y also varies significantly with the chemical constituents of the steel (e.g., the percentage of carbon and manganese), the heat treatment used, and with the amount of working which occurs during the rolling process. Thus, thinner plates which are more worked have higher yield stresses than thicker plates. The yield stress is also increased by cold working.

1.8.2 Inelastic Cyclic Response

The stress–strain response of most materials under cyclic loading is different from that under single (monotonic) loading. For fatigue analysis, it is necessary to consider the cyclic material behaviour for strength and life calculations.

When steel is subjected to cyclic loading in the inelastic range, the yield plateau is suppressed and the stress–strain curve exhibits the *Bauschinger effect*, in which non-linear response develops at a strain much lower than the yield strain, as shown in Fig. 1.6. As seen from this figure, as the amplitude of response increases, the stress level for a given strain also increases and can substantially exceed the stress indicated by the monotonic stress–strain curve.

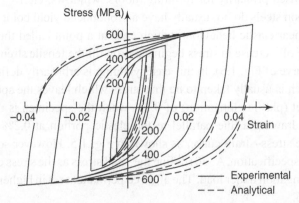

Fig. 1.6 Stress-strain curve of steel subjected to cycle loading

1.8.3 Characteristic Strength

Variations in material properties should be recognized and taken into consideration in the design process. The material properties that are of greatest importance in the design of structures using steel are yield strength, maximum percentage elongation, and Young's modulus. Other properties that are of less importance are hardness, impact resistance, and melting point.

If a number of samples are tested for a particular property (e.g., yield strength) and the number of specimens with the same strength (frequency) are plotted against the strength, then the results approximately fit a *normal distribution curve*, as shown in Fig. 1.7.

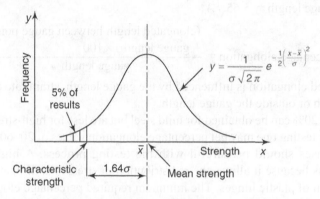

Fig. 1.7 The normal distribution curve

This curve can be mathematically expressed by the equation shown in Fig. 1.7, which can be used to define 'safe' values for design purposes. When defined, this safe value of yield strength, is called *characteristic strength*. If the characteristic strength is defined as the mean strength, then from Fig. 1.7, 50% of the material has a characteristic strength below this value and hence is not acceptable. Hence a characteristic value which has a particular chance (often 95%) of being exceeded in any standard tension test is chosen.

Thus, the characteristic strength is calculated from the equation

$$f_k = f_{mean} - 1.64 \ \sigma \tag{1.5}$$

where σ is the standard derivation for n samples, and is given by

$$\sigma = \left[\frac{\Sigma(f_{mean} - f)^2}{(n-1)} \right] \tag{1.6}$$

The characteristic strength of steel is the value obtained from tests at the rolling mills, but by the time the steel becomes part of the finished structure, its strength might have been reduced (e.g., by corrosion or accidental damage). The strength to be used in design calculations is therefore the characteristic strength divided by a partial safety factor. The value of the partial safety factor adopted for steel is given in Table 5 of IS 800 : 2007 as 1.10 for yielding resistance.

1.8.4 Ductility

Ductility may be described as the ability of a material to change its shape without fracture. In other words, the ductility of a structure or its members is the capacity to undergo large inelastic deformations without significant loss of strength or stiffness. The stress–strain curve of a material also indicates the ductility. It is the amount of permanent strain, i.e. strain exceeding proportional limit, up to the point of fracture. The ductility of the tension test specimen is measured by determining the percentage elongation (comparing the final and the original lengths over a specified gauge distance). The specified gauge length according to the code is as follows.

$$\text{Gauge length} = 5.65\sqrt{A_0} \tag{1.7}$$

$$\text{Percentage elongation} = \frac{\begin{array}{c}\text{(elongated length between gauge point}\\ -\text{ gauge length)} \times 100\end{array}}{\text{gauge length}} \tag{1.8}$$

The measured elongation is influenced by the gauge length, strain rate of test, and failure within or outside the gauge length.

Values of 20% can be obtained for mild steel but are less for high-strength steel. By improper testing one may get percentage elongation values of 20–60 and hence the test houses should be careful with the testing process. A high value is advantageous because it allows the redistribution of stresses at ultimate load and the formation of plastic hinges. The minimum required percentage elongations of steel produced in India as per the code are given in Table 1.3(a) (also see Table 1 of IS 800 : 2007). For most standard mild steels, the values are greater than the minimum required. However, rerolled steel or improperly controlled steel may give higher strength but less percentage of elongation.

1.8.5 Low Temperature and Toughness (Brittle Fracture)

In structural steel design, toughness is a measure of the ability of steel to resist fracture under impact loading, i.e., the capacity to absorb large amounts of energy. Toughness can be an important design criterion, particularly for structures subject to impact loads (e.g., bridges) and for those subject to earthquake loads. Hence both strength and ductility contribute to toughness.

At room temperature, common structural steel is very tough and fails in a ductile manner. At temperatures below 0°C, steel structures sometimes fail suddenly and without warning. A right combination of low temperature, an abrupt change in section size (notch effect) or an imperfection, and the presence of tensile stress can initiate a failure called *brittle fracture*. This may begin as a crack, which may propagate and cause the member to fail. Most brittle fractures occur under static load at stress levels which are not excessive, but they may also be due to the dynamic application of a load or some overload.

One of the best known brittle fractures occurred in Boston in a riveted steel tank of diameter 27.5 m and height 15 m, which contained 7.57 million litres of molasses. The tank failed in an explosive manner on January 15, 1919. Similar sudden failures of steel water tanks, oil tanks, transmission lines, ships, plate girders, and bridges have occurred in the past (McGuire 1968). Most of these failures have occurred under normal service conditions, in welded structures, and at low temperatures. During the cold winter of 1977, several spectacular failures occurred in bridge structures in Illinois, Minnesota and Pennsylvania in the USA. On January 22, 1988, several brittle failures occurred in a bridge in Providence, Rhode Island, the USA (Gaylord et al. 1992).

The *Charpy V-notch test* (CVN test) is commonly used to evaluate the behaviour of metals as they are affected by an abrupt change in cross section (IS 1757). In this test, the temperature of the specimen is varied, the energy absorbed by each specimen is recorded, and the energy–temperature curve is plotted. From this curve, a transition temperature corresponding to some level of energy absorption (usually 20 or 27 J) is selected.

Structural steels vary greatly in toughness. Highly killed, fine grain steel with a suitable chemical composition or specially heat-treated steel exhibit considerable toughness. IS 2026 and IS 1757 codes allow the use of only those steels that exhibit a minimum energy absorption capacity at a predetermined temperature (e.g., 20 J at $23 \pm 5°$C). In addition to the chemistry of steel, size of plates, residual stress, and cold work also affect toughness. (Thick plates, large residual stress, and cold work are detrimental.)

1.8.6 Lamellar Tearing

Lamellar tearing is a form of brittle fracture that may occur in certain welded joints. For example, a tear can occur if a large weld (or welds from both sides) is placed on a thick plate, since the shrinkage strains from the welding operation will be large and restrained. The restraint may be developed due to the weld on the far side or due to the member thickness or due to a combination of both the factors.

Generally I-sections are adequately ductile when loaded either parallel or transverse to the rolling direction. A thin, stiffened column is susceptible to lamellar tearing, since the flange stiffeners that are welded to the column flange produce a restraint. A large overmatch of electrode and base metal in a full penetration butt weld also tends to increase the possibility of tearing.

The use of fillet welds, a joint design that allows weld shrinkage to occur in the rolling direction, and the sequence of welding to minimize shrinkage strains are practical methods used to avoid lamellar tearing.

1.8.7 High-temperature Effects

Steel is not a flammable material. However, its strength reduces with rise in temperature. The yield as well as tensile strength at 500°C are about 60–70% of that at room (about 21°C) temperature. The drop in strength is much higher at still higher temperatures (for example at 800°C it is only 15% of that at room temperature).

Hence, steel frames enclosing materials that are flammable require fire protection, to control the temperature of steel members for a sufficient time for the occupants to seek safety or for the fire be extinguished before the building collapses. In many cases, the building does not collapse even at high temperatures. But the members are deformed beyond acceptable limits, and hence have to be replaced.

The fire-resistance design of steel members, the methods to model real fire, and its effect on steel members are discussed in detail in Subramanian (2008).

1.8.8 Hardness

Hardness is a measure of the resistance of the material to indentations and scratching. Several methods are available to determine the hardness of steel and other metals. In all these methods, an 'indentor' is forced on to the surface of the specimen. On removal, the size of the indentation is measured using a microscope. Based on the size of the indentation, the hardness of the specimen is determined. Brinell hardness (typical value: 150–190) and Vickers hardness tests are used to determine hardness.

1.9 Resistance to Corrosion

Steel readily corrodes in moist air. Exposure to sea water, acid, or alkaline vapours hasten the process. It has been estimated that more than 0.075 mm of the thickness of steel members will be lost every year in an industrial environment in which sulphur dioxide is present. Hence, structural steel members should be protected effectively against corrosion.

The most common method of protecting a steel member involves the use of paint or metallic coating, or a plastic coat in the case of metallic sheeting. Steels with a copper content of 0.2–0.5% have improved resistance to atmospheric corrosion but still need to be protected. Paint systems consist of a zinc- or aluminium-based primary coat on which two or three layers of finishing coats are applied. Metallic coatings involve galvanizing and sheradizing (both of which use zinc), electroplating (usually applied to fasteners), and metal spraying using either zinc or aluminium. For the coating or painting to be effective, the surfaces of the steel members have to be cleaned effectively before treatment. Several methods are available and the usual one consists of blast cleaning the surface using small abrasive particles such as those of iron, which are directed to the surfaces of the members by using compressed air or an impeller. Instead of protective treatment, one can go in for special corrosion-resistant steel, which on exposure to weather forms a protective surface layer of oxide film. Such weathering steels contain a greater amount of phosphorus and some chromium and copper than normal steel. They cost 20% more than normal steel but this initial cost may be offset by savings in weight, protective treatment, and maintenance.

1.10 Fatigue Resistance

Fatigue is the term used in connection with the initiation and propagation of microscopic cracks into macrocracks by the repeated application of alternating stresses. The damage and failure of materials under cyclic loads is called *fatigue damage*. Fatigue need not be considered unless numerous significant fluctuations (usually taken as 2×10^6 to 5×10^6 cycles) of stress are anticipated. As per the code, stress reversals due to wind need not be checked for fatigue. Wind-induced oscillations have to be taken into account in special cases. For instances, oscillations of 2×10^6 cycles can be easily reached in a lighting mast. The code states that the designer should check the following members for fatigue assessment.
(a) Members supporting lifting or moving loads,
(b) Members subjected to wind-induced oscillations of a large number of cycles,
(c) Members subjected to repeated stress cycles from vibrating machinery, and
(d) Members subjected to crowd-induced oscillations.

Thus, fatigue effects are more likely to occur in bridges and gantry girders due to the cyclic nature of loading, which causes reversal of stresses. Welds are susceptible to a reduction in strength due to fatigue because of the presence of small cracks, local stress concentrations, and abrupt changes of geometry. It has to be noted that the incidence of fatigue and fatigue crack growth is independent of steel grade. Fatigue cracks are far more common than brittle fracture. Guidelines for the designer to take into account fatigue loads are discussed in detail in Subramanian (2008).

1.11 Residual Stresses

Higher temperatures in the range of 600–700°C are involved during the rolling of steel sections. Steel members are also subjected to high temperatures, to selected parts of cross-section, while members are fabricated by welding and also when material or members are cut by flame-cut. Cooling of these members or materials always takes place unevenly, for example, the flange tips of an I-section, cool faster than the flange-to-web junctions. Similarly the central portion of the web tends to cool faster than the junctions. Due to this uneven heating and cooling, structural members normally contain *residual stresses*. Residual stresses may also result from the cold straightening of bent members (Gaylord et al. 1992). Although it is possible to remove these stresses by subsequent reheating and slow cooling, this process is not attempted in normal structural engineering applications.

The typical distribution of residual stresses in a standard I-section is shown in Fig. 1.8(a). Residual stresses tend to increase in magnitude with increase in size of the element. The magnitude of tensile residual stresses may reach up to $0.3f_y$ and the compression residual stress up to $0.5f_y$ in rolled I-sections (Trahair et al. 2001).

A welded I-section fabricated from rolled plates has a different residual stress distribution from that of an I-section welded from plates flame-cut to width. The typical distribution of residual stresses in welded sections made of plates with rolled edges is shown in Fig. 1.8(b).

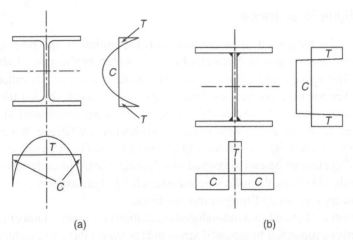

Fig. 1.8 Residual stress distribution in standard and welded I-beams

Fabricating operations such as cambering and straightening by cold bending also induce residual stresses. These stresses are superimposed on the thermal residual stresses. Because residual stresses must themselves be in equilibrium, their effect on structural behaviour is limited. However, residual stresses play an important role in the design of compression members (columns), since under compressive stress, the regions that contain residual compressive stress yield earlier (at loads which produce an applied stress less than f_y). Similarly, members in bending (beams) also yield early and hence tend to deflect more (Tall 1974). Residual stresses are also to be considered in the design of members subject to fluctuating loads (fatigue).

1.12 Stress Concentration

Steel structures often have connected elements which may have abrupt change in geometry and may contain holes for bolts. These features produce *stress concentrations*, which are localized stresses, greater than the average stress in the member (tensile stresses adjacent to a hole are often three times the average tensile stress).

The stress concentration at the holes is usually neglected in structural design and it is assumed that the stress is uniformly distributed over the net area of cross section. Since structural steel is sufficiently ductile to equalize the stress over the area, this assumption is justified in most cases. However, if the average stress in a member is high, the stress concentration effect should not be ignored. Stress concentration effects have been found to be critical in the webs of plate girders. Stress concentrations are also associated with fatigue and can also affect brittle fracture.

1.13 Structural Steel Products

Structural steel products of interest to designers can be divided into the following categories.
(a) Flat hot-rolled products—plates, flat bars, sheets, and strips,
(b) Hot-rolled sections—rolled shapes and hollow structural sections,

(c) Bolts,
(d) Welding electrodes, and
(e) Cold-rolled shapes.

1.13.1 Hot-rolled Sections

The hot-rolled sections and products consist of the following (see Fig. 1.9).
- Rolled beams
  - Junior beams (ISJB, meaning Indian Standard Junior Beams)
  - Lightweight beams (ISLB)
  - Medium-weight beams (ISMB)
  - Wide-flange beams (ISWB)
  - Heavyweight beams/columns (ISHB)
  - Column sections (ISSC)
- Channels: Junior, light, and medium and parallel flange (ISJC, ISLC, ISMC, ISMCP)
- Equal angles (ISEA or ISA)
- Unequal angles (ISA)
- T sections (ISJT, ISLT, ISST, ISNT and ISHT)
- Rolled bars
  - Round (ISRO)
  - Square (ISSQ)
- Tubular sections (ISLT, ISMT, ISHT)
- Plates (ISPL)
- Strips (ISST)
- Flats (ISFl)

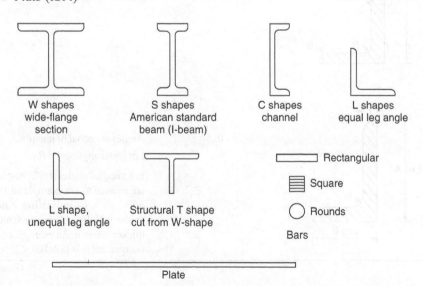

| W shapes
wide-flange
section | S shapes
American standard
beam (I-beam) | C shapes
channel | L shapes
equal leg angle |

| L shape,
unequal leg angle | Structural T shape
cut from W-shape | Rectangular
Square
Rounds
Bars |

Plate

Fig. 1.9 Types of hot-rolled sections produced by steel producers

The standard rolled steel shapes and the nomenclature used in relation to them are mentioned in Table 1.4. Such nomenclature should be properly understood since it will be followed in the subsequent chapters.

Table 1.4 Rolled-steel structural shapes

Rolled section	Indian standard designation	Remarks
	ISJB ISLB ISMB ISWB ISHB	A beam section referred to as ISMB 400 at 0.616 kN/m is an ISMB with a depth of 400 mm and a weight of 0.616 kN per metre length.
	ISJC ISLC ISMC	Channel sections are referred to, for example, as ISMC 200 at 0.221 kN/m.
	ISA	Angles are equal or unequal. For equal angles, $A = B$. For unequal angles $A > B$. Angles are referred to, for example, as ISA $60 \times 60 \times 6$, indicating equal angles with legs 60 mm each and thickness 6 mm. An example of an unequal angle is ISA $100 \times 75 \times 6$.

(contd)

(contd)

Rolled section	Indian standard designation	Remarks

	ISNT	Tee sections are referred to, for ex-
	ISHT	ample, as ISNT 100 at 0.147 kN/m,
	ISST	indicating that the depth of the
	ISLT	section is 100 mm.
	ISJT	

Plates	$t \geq 5$ mm	Plates are referred to in terms of width × thickness, e.g., 900 × 10 indicates a plate 900 mm wide and 10 mm thick.
Strips	$t < 5$ mm	Strips are referred to in terms of width × thickness.
Flats	$t \geq 5$ mm $b \leq 250$ mm	Flats are referred to in terms of $b \times t$.

A square bar is referred to in terms of its sides, e.g., a 20-mm square bar.

A round bar is referred to in terms of its diameter, e.g., a 20-mm diameter bar.

Steel tubes are designated in terms of their nominal bore in millimetres and self-weight. Rolled-steel circular or square rods are designated, respectively, in terms of diameter or side (e.g., ISRO 10 mm or ISSQ 10 mm).

IS Handbook No. 1 and IS 808, published by the Bureau of Indian Standards, provide the dimensions, weights, and geometrical properties of steel beam, column, channel and angle sections. Since these publications do not contain the plastic modulus and shape factor for steel I-beams, they are provided in Annex H of IS 800 : 2007.

Choice of section

The design of steel sections is governed by the cross-sectional area, *section modulus*, and *radius of gyration*. Though IS 808 and IS Handbook No.1 list the properties of various sections, due to the limitations of rolling mills only a few sections are available in the market. Therefore, design is governed by not only sectional properties but also the availability of the section. Another factor governing choice is the ease with which sections can be connected. In India ISMB beams are the most commonly produced. So are limited number of ISHB sections. Also, only medium channels are available. Only a limited number of unequal angles are available in the market. Also, not all the equal-angle sections are available readily in the market. Hence it will be a good idea to get a list of the available sections from steel producers like SAIL and plan the design accordingly. Appendix A gives the sectional properties of some of the sections manufactured in India.

Channels are used mainly as purlins in industrial buildings and angles in trusses and towers. The structural tee is commonly used for chord members in trusses, though in India double angles are also used for the purpose.

IS 1852 specifies allowable rolling tolerances, including amount of flange and web warping, and the deviation of web depth permitted for the section to be satisfactory. Designers should be careful about these tolerances, especially while using smaller sections which may be produced by small rerollers. It is because the designed cross-sectional properties may not match with the actual cross-section properties. Also, as mentioned previously, the rerolled sections may tend to have higher strength at the risk of reduced ductility.

Though the latest IS 800 : 2007 code has removed the minimum thickness requirements, it is advisable to use a minimum thickness of 6 mm for the main members and 5 mm for secondary members exposed to the atmosphere, especially in coastal areas.

1.13.2 Wide-flange Sections

As mentioned in Section 1.13.1, ISMB sections are the only I-sections that are normally produced in India on account of the calibre rolling method. These sections are used for beams as well as columns. Such sections have relatively narrow and sloping flanges and a thick web compared to wide-flange sections (see Fig. 1.10). ISMB beams are not economical, especially for compression members, because of excessive material in the web and the lack of lateral stiffness due to the narrow flanges. Also, since the available sections are limited, when a section is slightly inadequate, the choice is limited to either the next available section (which may be 25–45% heavier in weight) or built-up sections through welding, using which involves extra time and cost.

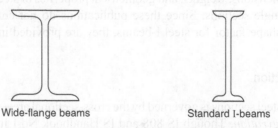

Wide-flange beams   Standard I-beams

Fig. 1.10 Wide-flange sections

The main features of wide-flange beams which make them more popular than Indian standard I-beams are the followings.

(a) Wide-flange beams provide excellent sectional performance, with high bending and buckling resistance due to the H-shaped arrangement of flanges and the web.

(b) The use of such beams reduces fabrication difficulties—since there is no taper in the H-beam flange, no tapered washer is necessary for bolting, and the gussets can be welded to the inner surface of the beam flange. Unlike tapered-flange beams, H-beams can be readily butt welded, and a sound welding is assured.

(c) Since H-beams have a higher section modulus for the same weight, using them is economical. (A saving of the order of 10–24% can be achieved.)

Using a new manufacturing technology, it is now possible to have beams with the same depth but with different flange and web thickness, and also flange width. This facilitates simple design and improves fabrication efficiency. These wide parallel-flange beams and columns are manufactured in India by M/s Jindal Steel and Power Limited (JSPL) at Raigarh, Chhattisgarh. The sectional properties of some of these beams are discussed in detail in Subramanian (2008).

1.13.3 Welded and Hybrid Sections

Hot-rolled plates or flame-cut plates can be welded together to form I-sections or box girders (see Fig. 1.8). Such built-up sections can also be made by using riveting or bolting. Welded I-beams with top and bottom plates and welded stiffeners are often used as plate girders. Welding makes it possible to combine any structural shape to get the desired properties. *Tapered girders* are fabricated either by welding two flange plates to a tapered web plate or by cutting a rolled I-beam lengthwise along its web at an angle, turning one half end for end, and then welding the two halves back together again along the web as shown in Fig. 1.11(a). Tapered girders are widely used in the framing of roofs over large areas, where it is desirable to minimize the number of interior columns or to eliminate them altogether.

Similarly, *castellated beams* (also called *cellular beams*) can be made economically by flame-cutting a rolled I-beam web in a zigzag pattern along its centreline [see Fig. 1.11(b) and (c)]. One of the two equal halves is turned end for end and is welded to the other half. The result is a deeper beam, stronger and stiffer than the original. Castellated beams have more section modulus and moment of inertia and result in greater economy. Figure 1.12 shows a parking structure with castellated beams, that led to significant savings in construction costs. Properties of castellated beams made of ISMB beams and channels are discussed in detail in Subramanian (2008).

Since the web of a beam or a plate girder contributes only a little to the bending resistance and because its strength in shear depends on its slenderness ratio h/t, it is economical to have the web of a lower-strength steel than the flange. Beams with stronger steel in the flanges than in the web are called *hybrid beams*. Such beams are often fabricated by welding plates of different steel strengths.

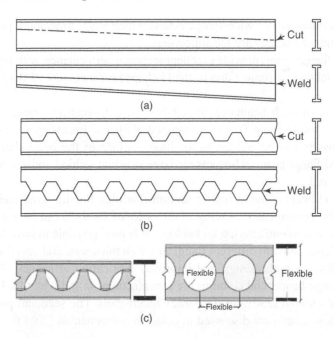

Fig. 1.11 Tapered girders and castellated beams

Fig. 1.12 Cellular beams over Rangers FC, an indoor training facility in Scotland (Courtesy: ASD Westok Ltd, UK)

With the development of high-performance steels (see Section 1.14), it is now possible to fabricate built-up I-beams with corrugated webs. The corrugations can have a trapezoidal shape or a sine-wave cross section. Corrugated webs allow deeper girders that are not susceptible to web stability problems, resulting in thinner webs and smaller flanges.

1.13.4 Hollow Steel Sections

Tubular members are being used extensively in plane and space trusses as tubes are more efficient in compression. The advent of welding has made the connection between tubular members using gusset plates possible and resulted in the widespread use of tubular sections. Welded connections without gusset plates require proper planning for profiling and welding the ends, and have to be executed carefully. Recently, square and rectangular tubes have been introduced in India and these, of course, are much easier to connect because of their flat surfaces. With square or rectangular hollow sections, the smaller tube can be simply sawed with a single cut at the required angle and welded to the bigger tube.

Flowdrill and Hollo-bolt Systems

The use of standard bolt is often impossible in steel hollow sections, as there is no access to the inside of the tube to allow for tightening. The use of gusset plates and brackets overcomes this problem but is not aesthetically pleasing. Several techniques have been developed which permit bolt installation and tightening from one side of the connection only; they are called *blind bolts*. Commercially available examples include Flowdrill and Hollo-bolt.

Flowdrilling is basically a thermal drilling process which makes a hole through the wall of an SHS without removing the metal normally associated with a drilling process. The formed hole is then threaded by the use of a special tool, leaving a threaded hole which will accept a standard fully threaded bolt. In the flowdrill system, the tungsten carbide bit of the Flowdrill is brought into contact with the RHS wall, where it generates sufficient heat to soften the steel. The bit is then advanced through the wall and when it is done, the metal 'flows' to form an internal bush. The tool also removes any surplus material which may arise on the outside of RHS section. In the next stage, the flowdrill bush is threaded with a coldform flowtap so that the connection can be made with bolts. The complete cycle is shown in Fig. CS1. Note that the Flowdrill method cannot be used with pre-galvanized sections.

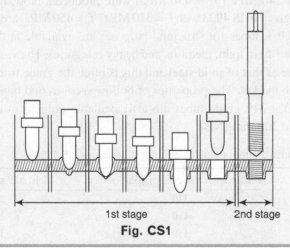

1st stage   2nd stage

Fig. CS1

In the hollo-bolt system, a pre-assembled unit is inserted through normal tolerance holes in both the RHS and attachment plate. As the bolt is tightened, the cone is drawn into the body, spreading the legs, and forming a secure fixing. After installation, the hollo-bolt head and collar only are visible as shown below.

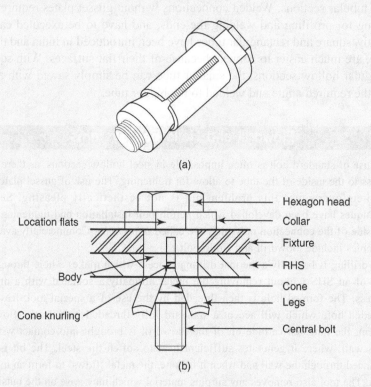

Fig. CS2 (a) Hollo-bolt and (b) hollo-bolt assembly

In India, till recently only two grades of RHS, i.e., $f_y = 210$ MPa ($f_u = 330$ MPa) and $f_y = 240$ MPa ($f_u = 450$ MPa) were produced. Now, HSS as per the specifications given in IS 4923 with $f_y = 310$ MPa ($f_u = 450$ MPa) are also available.

As per IS 1161, tubes for structural purposes are available in three grades of steel (see Table 1.5) in light, medium, and heavy categories. The yield stress of Yst 240 is the same as that of mild steel and this is often the grade (medium-section) available in the market. The properties of hollow sections and tubes are given in Appendix A. Yst 310 grade tubes are also available ranging in diameter from 15 mm (NB) to 300 mm (NB).

Table 1.5 Strength of steel used for circular tubes as per IS 1161

Grade	Ultimate tensile stress, f_u (MPa) (min.)	Yield stress, f_y (MPa) (min.)
Yst 210	330	210
Yst 240	410	240
Yst 310	450	310

It has to be noted that tubes having less than 3.25 mm have to be carefully welded. If the tubes and hollow sections are not plugged at the ends properly, they retain the moisture inside and corrosion starts from the inside, which will not be noticed at the early stages.

1.14 High-performance Steel

A new grade of steel, called high-performance steel, HPS-485W was developed by a cooperative program between the Federal Highway Administration, the American Iron and Steel Institute, and the Department of the Navy in August 1994. Three grades of HPS are available now: HPS-345W, HPS-485W, and HPS-690W, out of which HPS-485W (yield strength = 485 MPa), is widely used. The benefits related to HPS include enhancements in weldability, toughness, corrosion resistance, ductility, fatigue and fire resistance, formability, and strength. These factors combined led to construction elements of higher economic efficiency, ease of maintenance, and longer service life.

The Nebraska Department of Transportation was the first to use HPS 485W in the design and construction of the 45-m simple span Snyder bridge, a welded plate girder (1.37 m) bridge, which opened to traffic in October 1997. In April 2002, there were 21 HPS 70W steel bridges in service and 13 more bridges are under various stages of construction in the USA alone. More details about HPS may be found in Jamshidi et al. (1997), and Gross and Lwin (2002).

High-performance Steel Bridge in Tennessee, USA

The bridge in Tennessee is a two-span continuous structure located on State Route 53 over the Martin Creek in Jackson County. The structure provides an 8.5-m wide roadway over two spans, both 72 meters long. This bridge consists of three continuous welded plate girders, fabricated from HPS-485W, 2-m deep, and spaced at 3.2 m centers. These plate girders act compositely with a cast-in-place concrete deck slab of thickness 212.5 mm.

The bridge is jointless, having integral, pile-supported abutments. The design was fully optimized for the 485-MPa steel, using the AASHTO Load and Resistance Factor (LRFD) Bridge Design Specifications. In the positive moment region, the top flanges are 450 mm wide and vary from 25 to 50 mm thickness in four increments. In the negative moment region, the top flanges vary in four increments of 28 mm by 450 mm to 62.5 mm by 750 mm. The webs vary from 11 mm by 1800 mm in the positive moment region to 12.5 mm by 1800 mm over the negative moment region.

Cost estimates prepared by the Tennessee Department of Transportation indicate that the steel weight was reduced by almost 25% compared to the original grade 345W design. Because HPS-485W costs slightly more than grade 345W steel, this resulted in a 16% reduction in the total cost to fabricate and erect this bridge.

1.15 Stainless Steel

Stainless steel is essentially a low-carbon steel to which chromium has been added. It is this addition of chromium, in amounts greater than 10.5% by weight, that gives the steel its unique 'stainless' corrosion-resistant properties.

Of the various available grades, SS 304, SS 304L, SS 306, SS 409, and SS 430 are suitable for structural applications. The stainless steel production in India was only 20,000 tonnes in 1978; it increased to around 1.7 million tonnes in 2006.

The advantages of stainless steel include its aesthetic appearance, corrosion resistance, high tensile strength, high toughness, impact, and heat resistance. Hence, several countries have developed separate guidelines for the design of stainless steel structural members (Euro Inox 1994, ASCE/ANSI-8-90, ENV 1993-1-4). But Indian Code provisions for the design of stainless steel members are not available. This is one of the reasons, in India, for the non-application of stainless steel members in engineering structures.

The first application of stainless steel in a space frame roof, designed by the author, is at the entrance of M/s. Jindal Strips Limited at Hissar, Haryana (Subramanian, 2002), and is shown in Fig.1.13

Fig. 1.13 Stainless steel double-layer grid roof at Hissar, Haryana

In addition to the above different products, *cold formed steel sections*, made from light-gauge steel strips of thickness 2 to 4 mm thick (20-8 B.G.) are also available. Cold-rolled C, Z, and Sigma sections are often used as purlins and are economical than hot-rolled channels. They are designed using IS 801 code (which is under revision), and is outside the scope of this book. Details of design may be found in Yu (2000) and the IS handbook on light gauged sections (SP 6.5).

1.16 Composite Construction

The properties of reinforced concrete (strong in compression, greater rigidity) can be advantageously combined with the properties of structural steel to produce *composite constructions*. These constructions include (a) concrete–encased steel columns, (b) concrete-filled steel columns, (c) concrete-encased steel beams, and (d) steel beams supporting concrete slabs. Composite construction results in savings in time and material.

The design of composite construction is done by using IS 11384-1985, and is outside the scope of this book. The details of design may be found in Nethercot (2001), Martin and Purkiss (1992), and Kulak and Grondin (2002).

1.17 Advantages of Steel as a Structural Material

Structural steel offers several advantages over other competing materials. These advantages are as follows.

High strength The high strength of steel per unit weight means that structures made of steel sections weigh less than those made of other materials.

High ductility In structures built with structural steel, occasional human errors such as accidental overloading do not cause problems, due to the ductility of steel. Steel building are preferred in earthquake zones due to their greater ductility.

Uniformity The quality of steel-intensive construction is invariably superior, when compared with that of construction involving other materials. This is especially important in India, where quality control in construction sites is poor (resulting in poor performance, especially in concrete structures where water–cement ratio and curing are not controlled properly at site). Moreover, the properties of steel do not change appreciably with time as do those of reinforced concrete.

Environment-friendly Structural steel is recyclable and environment-friendly. Over 400 million tonnes of steel are recycled annually worldwide, which represents 50% of all steel produced. Steel is the world's most versatile material to be recycled. Another characteristic of a steel structure is that it can readily be disassembled at the end of its useful life. It means that the steel components can be reused in future structures without the need for recycling, resulting in the saving of energy and avoidance of CO_2 emitted from the steel production processes.

Versatility Using structural steel, it is possible to fasten different members together by simple connection techniques such as welding, bolting, and riveting. Steel members can also be rolled into a wide variety of sizes and shapes.

Prefabrication Often, steel components are manufactured at the factory (which means that they are produced using strict supervision and quality control), transported to the site, and erected using bolting and a minimum amount of welding. The prefabrication of steel structures results in the proper planning of construction, saving in time and money, speedy erection, and better quality of finished structures. Lighter steel members facilitate easy handling and erection.

Permanence Steel frames that are properly maintained last indefinitely. Several structures are available to testify the durability of steel structures (e.g., the Eiffel Tower and the Railway Bridge across the Firth of Forth, both built in 1890). Under certain conditions, weathering steels do not require any painting or maintenance. In Belgium and Japan, it has been found that steel bridges outlast prestressed concrete bridges by 15–26 years.

Additions to existing structures The repair and retrofit of steel members and their strengthening at a future date (for example, to take into account enhanced loading) is simpler than in concrete members. Thus, new bays or even entire new wings, can be added to existing steel-frame buildings, and steel bridges may often be widened. Of course, special precautions have to be taken while welding on a member already carrying loads.

Least disturbance to the community Steel-intensive construction causes the least disturbance to the community in which the structure is located. Fast-track construction techniques developed in recent years have demonstrated that steel structures cause the least disruption to traffic and minimize financial losses to the community and business. Such construction also results in far less environmental pollution.

Fracture toughness Due to its toughness and ductility, steel members can be subjected to large deformations during fabrication and erection without fracture, thus allowing them to be bent, hammered, sheared, and have holes punched in them without visible damage.

Elasticity Steel behaves closer to design assumptions than most materials because it follows Hooke's law up to fairly high stresses.

Though steel has several advantages, it also has the following disadvantages.

Maintenance costs Most steels are susceptible to corrosion when freely exposed to air and water, and must therefore be periodically painted. However, the use of weathering steels and modern coatings tend to eliminate/reduce this cost. Steel members in the interior of buildings (not exposed to rain) do not corrode quickly.

Fireproofing costs Although structural steel members are incombustible, their strength is tremendously reduced at temperatures commonly reached in fires when other materials in a building burn. Furthermore, since steel is an excellent heat conductor, non-fireproofed steel members may transmit enough heat from a burning section or compartment of a building to ignite materials which come into contact with them in adjoining sections of the building. Due to these factors, the steel frame of a building may have to be protected by materials with certain insulating characteristics. In addition, the building may have to include a sprinkler system, in order to meet the building code requirements.

Susceptibility to buckling The longer and more slender the compression members, the greater is the danger of buckling. Though steel has a high strength per unit of weight, steel columns have to be stiffened against buckling.

Fatigue Another undesirable property of steel is that its strength may be reduced if it is subjected to a large number of stress reversals or several variations of tensile stress. (There are fatigue problems only when tension is involved.) Hence we often reduce the estimated strengths of such members if more than a prescribed number of cycles of stress variation are anticipated.

1.18 Types of Steel Structures

The structural engineer will be concerned with the design of a variety of structures, which may include the following.

Buildings These may include rigid, semirigid, or simple connected frames, load-bearing walls, cable-stayed and cantilevered structures. Buildings may be simple or multi-storeyed, with single or many spans. For multi-storeyed buildings, several lateral bracing systems have been developed, such as trussed, staggered truss, rigid central core, etc. (see Chapter 2 for more details on these structural systems). Buildings are also classified according to use, such as residential, commercial, office, industrial, etc. These buildings may include a steel frame as shown in Fig. 1.14 or have a steel roof supported by load-bearing walls. The steel skeleton of the buildings may be rigid or pinned, a two- or three-hinged arch or a truss-on-column system. The truss also may be rigid or pin-connected, and may assume various shapes or have several bracing systems (see Chapter 12).

The building frame is actually a three-dimensional skeletal system, but in practice it is usually taken as rigid in only one plane. Some buildings are rigid in the *XY* as well as *YZ* planes, but this type of frame will not be considered in this text [see Subramanian (1999) for the details of three-dimensional frameworks]. The planar frame resulting from considering only the principal frame elements and/or the rigidity is termed as a *bent* and may have one or several stories. The spacing between bents in the third dimension is called the bay or *bay width*. *Spandrels* or floor beams are used to span the bays in multistoryed buildings with girders (usually heavier members than floor beams) spanning between columns or bents.

Figure 1.15 illustrates additional terms or members used in one-storey industrial buildings, where lateral bracings are often provided only in selected bays.

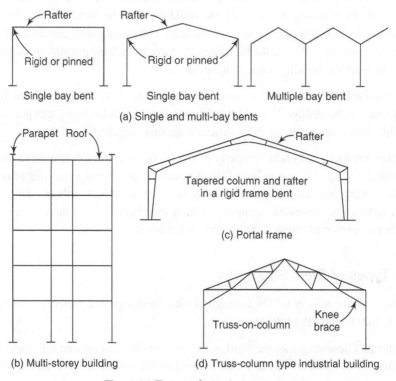

(a) Single and multi-bay bents

(b) Multi-storey building

(c) Portal frame

(d) Truss-column type industrial building

Fig. 1.14 Types of steel structures

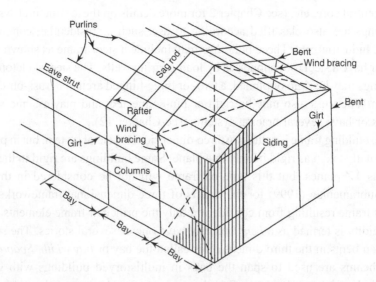

Fig. 1.15 Different structural members in industrial buildings

The roof systems of all buildings consist of a framing, some kind of decking, and a waterproof covering. The main roof framing consists of the *rafters* or the *truss* in any bent. Spanning the bay width are *purlins* spaced at about 0.6–2 m at centres, depending on the type of roof decking or sheeting used. *Sag rods* are provided as additional support for the purlins used on sloping roofs. Purlin design is rather complex for sloping roofs due to the unsymmetrical bending (see Section 6.11). The roof deck rests on purlins and may be a metal deck, precast concrete slabs, wood planking, gypsum sheets, GI sheets, polycarbonate sheets, fibre-glass sheets, or asbestos (which has been banned in several countries due to health hazards).

The siding may be made up of metal sheets, metal sandwich sheets (sometimes called curtain walls) consisting of two metal sheets with some insulating filler, brick, and precast or poured concrete, or asbestos sheets. A lightweight siding is carried by girts and spandrel beams in high-rise buildings and by *eaves struts* and *girts* in industrial buildings (see Fig. 1.14). Note that a spandrel beam is similar to a girt and is located at floor level (as the most exterior floor beam), and carries a proportion of live load. It also carries a part of the load acting on the siding. If the siding is heavy (e.g., brick or concrete blocks), built-up sections (using channels or angles depending on the load) can be used for the spandrel beam. It is necessary to establish the floor framing system very early in the design process, so that the load flow is identified and the sizes of different members are approximately fixed for computer analysis.

Bridges Bridges may be classified as truss, plate-girder, arch, cantilever, cable-stayed, or suspension (using cables as principal load-carrying members). The truss and plate-girder bridges are commonly adopted for small to moderate spans and cable-stayed and suspension bridges for long spans. Many types of trusses like Pratt, Warren, Parker, Baltimore, K-type, Whipple truss, etc., are used in bridges. The truss types include both deck (traffic on top of truss) and through types (traffic passes between the trusses of the bridge). The deck type is preferred if the clearance below the truss is not a critical factor, because the piers could be made shorter. Many truss bridges combine both types. A combination of trusses (for long spans) and girders (for shorter or approach spans) is also adopted in practice. Bridges may also be classified as railroad, highway, or road and pedestrian bridges, depending upon use.

Towers Towers may be of different types, such as lighting towers, power transmission towers, observation towers, towers for radar and TV installation, telephone relay towers, and windmill towers. Towers may be self-supporting or cable-stayed. Most towers are made of steel angles or tubes, which are bolted at site.

Water tanks They may be rectangular, circular, or spherical. They can be used to store oil or water. They may rest on the ground or be elevated on a staging.

Other structures Silos, bunkers, domes, folded plates, offshore platforms, chimneys, cooling towers.

In addition to these structures, the structural engineer may be called upon to design ships, parts of various machines and other mechanical equipment, and automobiles (bus and car bodies, chassis), etc. Figure 1.16 shows the photographs of some different types of steel structures.

(a) The 2,065-m long Pamban railway bridge, on the Palk Strait, connects Rameswaram to mainland India. Constructed in 1914, it is still in good condition and considered as one of the marvels of engineering. The double-leaf bascule bridge section shown in this picture can be raised to let ships pass under the bridge. The bridge is located at the world's second highly corrosive environment (next to Miami, US), which is also a cyclone-prone, high wind velocity zone.

(b) The 110-stories high, 442-m tall Willis Tower (formerly known as Sears Tower), in Chicago, Illinois, USA. Designed by architects, Skidmore, Owings, and Merrill and structural engineer Fazlur Khan, and completed in 1973, it is the tallest building in the western hemisphere. The superstructure consisting of nine interlocking rigid steel framed tubes that terminate at different heights, efficiently counteract all lateral and gravity loads.

(c) Beijing National Stadium, also known as Bird's Nest, built for 2008 Summer Olympics in Beijing, China, and is the world's largest steel structure. This 91,000-seat stadium was designed by Swiss architects Herzog and de Meuron, ArupSport, and China Architectural Design & Research Group. It is 330-m long, 220-m wide, 69.2-m tall and contains 110,000 tonnes of steel.

(d) A typical pre-engineered building for the American Royal Exhibition Hall and Arena, Kansas City, MO. They are adopted for low-rise buildings with an eave height of up to 30 m and can be very economical and speedy. (Photo: www.butlermfg.com)

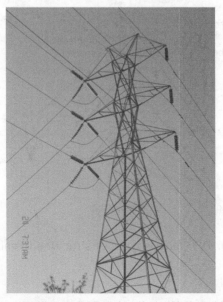

(e) Transmission line towers are often made of steel angles and bolted at site (Photo: Binod Therat)

(f) Steel-elevated water tanks are used to store water and provide pressure in the water distribution system. A tank elevated to 21 m creates about 0.207 MPa of discharge pressure, which is sufficient for most applications. The shapes of the tanks and their surface appearance can be varied according to the needs.

Fig. 1.16 Photographs of some of the different types of steel structures

In this book, we will confine our attention to the design of elements which are encountered in buildings. The various structural members used in buildings can be classified into the following according to the method by which they transmit forces in the structures (see Fig. 1.17).

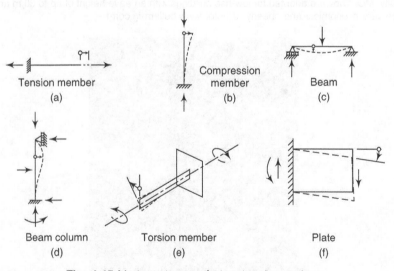

Fig. 1.17 Various types of structural members

Beams *Beams* (also called *girders*) primarily carry bending stresses which are the largest at the extreme fibre. Thus, most of the material of the beam remains understressed, and is not fully utilized, especially if it is of rectangular section. Also the bending moment reduces towards the support and if the beam section is uniform, even the extreme fibres carry smaller stresses. Thus, the overall efficiency of the use of material in beams is low. Yet beams are used so commonly because they give flat floors and roofs, are readily available, and require less fabrication at site. The design of beams is covered in Chapters 6 to 8.

Tension members or ties *Tension members* are structural elements that are subjected to direct axial forces, which try to elongate the members. They occur as components of trusses, hangers and cables for floors or roofs, in bracing systems, as tie rods, and similar members. An axially loaded tension member carries uniform stress and hence would give 100% efficiency in the use of materials. But in practice the efficiency is below 100% because the effective area of a tension member is less than the gross area due to the presence of rivet/bolt holes, shear lag effect, etc. A suspension cable, however, has a structural efficiency of 100% since it is flexible and can carry only tension. However, the use of a cable may require strong anchorage, and a stiffening arrangement to check large deformation under load, vibration due to wind, and other effects. The design of tension members is covered in Chapter 3.

Compression members (columns and struts) *Compression members* are those members in a structure that are subjected to loads that tend to decrease their length. Such members are used as the vertical load-resisting elements of a building structure, called columns, as the posts that resist compressive components of a load in a truss, and as bridge piers. In a building structure, forces and moments are transmitted to the columns through beams at each floor or the roof level of the structure. If the column is required to resist a load acting concentric to the original longitudinal axis of the member, it is termed an *axially loaded column* or simply a *column*. Even axially loaded columns will not be able to use 100% of the material, due to the phenomenon of *buckling*. The design of compression members is discussed in Chapter 5 and that of beam columns (which carry bending moments in addition to compressive forces) in Chapter 9.

Torsional members Such members are often encountered in shafts of machines. Of course, torsional effects should be considered in spandrel beams also. For design of members subjected to torsion, refer Subramanian (2008).

Plates Some built-up members and several light-gauge cold-formed members are made up of slender plate elements, which are liable to buckle locally. Local buckling is prevented by limiting the b/t ratios of plates.

Often, a combination of actions, such as bending and compression, bending and tension, compression bending and torsion, etc., occurs in frameworks where several members are joined together. In many cases, one action may be predominant and the other may be secondary. In the design, such secondary action is usually not taken into account. However, if the secondary action is considerable, the member has to be designed for the combined actions.

As already indicated, these members (beams, columns, tension members, etc.) are joined together in a framework by means of welding, bolting, or riveting. These fastening techniques and designs of connections are covered in Chapters 10 and 11.

1.19 Fabrication and Erection

According to the design of various members of a structure, the various required sections are procured and are fabricated at site or factory. Tolerances for the fabrication of steel structures should conform to IS 7215. The various activities in the fabrication shop include the following.
(a) Exact cutting of length by sawing, shearing, cropping, thermal cutting, or machining, based on the fabrication drawing of the structure,
(b) Straightening of members,
(c) Cambering of beams,
(d) Drilling or punching of holes,
(e) Welding of gusset plates,
(f) Machining of butt joints, caps, and bases,
(g) Surface preparation, such as shot blasting,
(h) Painting or galvanizing after pickling in acids,
(i) Marking,
(j) Shop assembly and erection,
(k) Inspection and testing, and
(l) Packing.

The numbered parts are then transported to the site and the structure is erected following the erection tolerances specified in IS 12843. The normal tolerances after erection are given in Table 33 of IS 800.

The straightness tolerances incorporated in the design rules are given in Table 34 of IS 800. When the actual curvature exceeds these values, the effect of additional curvature on the design calculations should be reviewed. A tension member should not deviate from its correct position by more than 3 mm.

After the structure is erected, the specified protective treatment should be applied on the surfaces of the steel members and joints. No painting should normally be used on the contact surfaces in the friction connection. More guidelines for the fabrication and erection of steel structures are given in Section 17 of IS 800 : 2007. Some recommendations for steel work tenders and contracts are also given in Annex G of the code.

1.19.1 Errors that Lead to Failures

To err is human but the consequences of an error in structural design can lead to loss of life and damage to property. Hence it is necessary to appreciate where errors can occur. Small errors can occur due to rounding of figures but these generally do not lead to failures. The majority of structural failures (whether it is the collapse of the structure or functional failure) are due to errors in design, construction, or operation. It has been reported that 85% of building failures occur due to human errors (Brown and Yin, 1988). Hence it is imperative for engineers and contractors to consciously avoid these errors.

The common errors that occur in the planning and design phase are due to the following (Martin and Purkiss 1992, and Subramanian 1984, 1989).

(a) Ignorance of the physical behaviour of the structure under load, which leads to errors in basic assumptions used in theoretical analysis
(b) Errors in selecting and estimating the loads, especially the erection forces
(c) Numerical errors in calculations—these could be eliminated by proof-checking; however when speed is of paramount importance, checking of calculations is often neglected
(d) Lack of consideration for certain effects such as fatigue, brittle fracture, residual stresses, etc.
(e) Insufficient allowances for temperature strains, tolerances, etc.
(f) Insufficient information about new materials, methods of analysis and design, detailing, erection procedures, etc.

Nowadays computer programs are being used as black boxes; that is, without knowing the limitations of these programs. Such usage leads to erroneous results. Errors that may occur during fabrication and erection are as follows.

(a) Using the wrong grade of steel or wrong types of electrodes for welding,
(b) Using the wrong weight of section,
(c) Errors in fabrication (holes not matching, oversized holes, lack of fit, improper welding, welding distortions, etc.), and
(d) Errors due to improper quality control.

Errors can also occur during the life of the structure, which affect the safety of its occupants. Such errors include

(a) overloading due to change of occupancy,
(b) loading which is not expected during the design stage (an earthquake of greater magnitude, flood, tsunami, etc.),
(c) alteration of the structural system (removal of the web of the flange to provide service ducts, addition of heavy partitions, balconies, etc.), and
(d) poor maintenance.

A study of the various types of failure has been provided by Levy and Salvadori (1992), Kaminetzky (1991), and Feld and Carper (1997). Case studies of building failures provide opportunities to learn from previous mistakes (Subramanian 1999, 2000; Subramanian & Mangalam 1997, 1998; Kevin et al. 2000). A few case studies of some important failures are also included in this book, which will help designers to avoid failures in their designs.

1.20 Aesthetics of Steel Structures

Although architects pay attention to the aesthetic qualities of structures like theatres, stadia, office buildings, residential units, etc., industrial buildings and non-habitat structures, which are made of steel members, are not aesthetically designed. It is very difficult to assess the aesthetic qualities of any structure, because taste differs from person to person. Aesthetics may be viewed as a composite impression of all visual aspects of design. In a normal building, aesthetic qualities may be judged with reference to its site, the physical layout and environment, the view from within and from without, and the appearance of the building with reference to its surroundings. Of course, these aesthetic factors should be considered in addition to functionalism, perfect manufacture, construction safety, stability, durability, disassembly, and reusability.

However, whether a person finds a structure beautiful depends upon whether it is able to evoke an emotion in him or her, which in itself is extremely complex in nature. The magnitude and combinations of the qualities listed previously obviously cannot be specified, since they evoke different responses in different individuals and are beyond the scope of our discussion. Nowadays, designers have moved away from notions of symmetry and order towards more expressive and dynamic structures, as seen in Fig. 1.18, which shows the Walt Disney Concert Hall in Downtown Los Angeles, USA. More details about aesthetic design may be found in Kavanagh (1975), Subramanian (1987, 2003), and Billington (1983).

Fig. 1.18 The Walt Disney Concert Hall in Downtown Los Angeles, California, designed by Architect Frank O. Gehry

Summary

This chapter begins with a brief historical review of steel. The steel making process along with the heat treatments given to steel is briefly outlined.

The chemical composition and mechanical properties of different kinds of steel are included. (It is quite important to specify the steel for the project at hand.) The important properties of structural steel are its ultimate and yield stress, ductility, toughness and weldability. We should remember that the yield stress, as measured by the tension coupons, is affected by several factors, such as the rate of loading and position from where the test coupons are taken. Variation in material properties can be incorporated in the design by the concept of characteristic strength. Ductility and toughness are very important when a steel structure is subjected to earthquake loads or impact loads. Special steels such as stainless steel and high-performance steel are also included.

In any design, the engineer's first task is to think about material selection. First, he or she should consider the working temperature range and form of the structure, and decide whether the components require thick material. A simple yardstick is that satisfactory performance against brittle fracture can be expected if the Charpy impact toughness exceeds 27 J at the lowest working temperature. Of course the quality of design, detailing, and fabrication are also equally important in the prevention of brittle fracture. One should avoid severe stress raisers to achieve smooth stress flow. In critical cases, this is as important as material selection.

Secondly, one must consider fatigue and ascertain whether the incidence of high levels of fluctuating stresses warrant further consideration. In general, this has to be checked in crane supporting structures, bridges, and structures supporting rotating machinery. Residual stresses and stress concentration effects should be properly accounted for in the design.

Though, mainly Fe 410 grade mild steel is used in India, it is prudent to check whether it is possible to use high-strength steel. Though the cost of high-strength steel may be 10–20% higher, its use may result in reduction in steel weight and subsequent foundation cost. Of course, while using high-strength steels, the deflection criteria may become critical due to the reduced section sizes. Other special types of steels are also available and may be specified, depending upon site conditions and cost considerations.

A brief introduction to corrosion, fire protection, and fatigue resistance of steel is also given. An overview of the different types of structures, types of sections (circular, I, angle, channel, hollow sections, etc.) and types of members (compression, tension, bending), is presented, which will be quite useful in understanding the concepts discussed in later chapters.

Review Questions

1. What are the properties of steel that make it better suited for structural applications?
2. Write a short note on the historic development of steel as a structural material.
3. Describe the production of structural steel in integrated steel plants.
4. What is the difference between killed and semi-killed steel?
5. What are the heat treatments employed to improve the properties of steel?
6. Differentiate between normalizing and annealing.
7. How does carbon content affect the properties of steel?
8. What is the relation between weldability and carbon equivalent?
9. List the IS codes that are used for the following applications:
   (a) material for structural steel,
   (b) welding,
   (c) ordinary bolts, and
   (d) high-strength bolts.

10. List the important factors that influence the mechanical properties of steel.
11. Describe the test to predict the tensile strength of steel.
12. Give the values of yield strength, Young's modulus, coefficient of thermal expansion, and ultimate tensile strength for mild steel as per IS 800.
13. Sketch the typical stress-strain curve of steel, indicating the three important regions.
14. What are the parameters that influence the yield stress of steel?
15. How the yield stress of high-carbon steels is determined for design purposes?
16. How the characteristic strength of steel is defined? What is the partial safety factor used with characteristic strength in the IS 800 code?
17. What is meant by ductility? Why and where is it important?
18. How the toughness of steel is measured?
19. What is brittle failure? How can it be controlled?
20. What is Charpy V-notch test? What is the CVN value that provides safety against brittle facture?
21. Write short notes on
    (a) lameller tearing,
    (b) effect of high temperature on steel,
    (c) corrosion resistance of steel,
    (d) hardness, and
    (e) fatigue resistance.
22. How are residual stresses induced in steel sections? Sketch the typical residual stress distribution in a rolled I beam and a welded I beam.
23. What is the effect of residual stress in compression members and beams?
24. What is meant by stress concentration? Where do we have to consider stress concentration effects?
25. Name and sketch some of the hot-rolled steel sections used in practice.
26. What are the problems associated with rerolled sections?
27. Are wide flange beams better than ordinary-rolled ISMB beams? Why?
28. What are castellated/cellular beams? How are they produced? What are the advantages of using them?
29. Write short notes on hollow steel sections and cold-formed steel sections.
30. Explain the flow-drill and hollo-bolt connection systems.
31. Write a short note on high-performance steel.
32. What are the advantages of using stainless steel?
33. State the main advantages of steel as a structural material.
34. List the types of structures that can be built using steel.
35. Discuss the following:
    (a) Fabrication and erection of steel structures,
    (b) Aesthetics of steel structures, and
    (c) Advantages of composite construction.

2

The Basis of Structural Design

2.1 Design Considerations

Structural design, though reasonably scientific, is also a creative process. The aim of a structural designer is to design a structure in such a way that it fulfils its intended purpose during its intended lifetime and be adequately safe (in terms of strength, stability, and structural integrity), and have adequate serviceability (in terms of stiffness, durability, etc.). In addition, the structure should be economically viable (in terms of cost of construction and maintenance), aesthetically pleasing, and environment friendly.

Safety is of paramount importance in any structure, and requires that the possibility of collapse of the structure (partial or total) is acceptably low not only under normal expected loads (service loads), but also under less frequent loads (such as due to earthquakes or extreme winds) and accidental loads (blasts, impacts, etc.). Collapse due to various possibilities such as exposure to a load exceeding the load-bearing capacity, overturning, sliding, buckling, fatigue fracture, etc. should be prevented.

Another aspect related to safety is structural integrity and stability—the structure as a whole should be stable under all conditions. (Even if a portion of it is affected or collapses, the remaining parts should be able to redistribute the loads.) In other words, *progressive failure* should be minimized.

Serviceability is related to the utility of the structure—the structure should perform satisfactorily under service loads, without discomfort to the user due to excessive deflection, cracking, vibration, etc. Other considerations of serviceability are durability, impermeability, acoustic and thermal insulation, etc. It may be noted that a design that adequately satisfies the safety requirement need not satisfy the serviceability requirement. For example, an I-beam at the roof level may have sufficient stiffness for applied loads but may result in excessive deflections, leading to cracking of the slab it is supporting, which will result in loss of permeability (leaking). Similarly, exposed steel is vulnerable to corrosion (thereby affecting durability).

Increasing the design margins of safety may enhance safety and serviceability, but increase the cost of the structure. For overall economy one should look into not only the initial cost but also the life-cycle cost and the long-term environmental effects. For example, using a very-high-strength steel to reduce weight often will not reduce cost because the increased unit price of high-strength steel will make the lighter design more costly. In bridges and buildings the type of corrosion and fire protection selected by the designer will greatly influence the economy of the structure.

While selecting the material and system for the structure the designer has to consider the long-term environmental effects. Such effects considered include maintenance, repair and retrofit, recycleability, environmental effects of the demolished structure, adoptability of fast track construction, demountability, and dismantling of the structure at a future date.

2.2 Steps Involved in the Design Process

The construction of any structure involves many steps. Although the structural designer is not responsible for each of these steps, he should be involved in most of them so that the resulting structure is safe, stable, serviceable, durable, and is economically viable and aesthetically pleasing, and does not have an adverse impact on the environment. The necessary steps may be listed as follows.

1. After receiving the plan and elevation of the building from the architect and the soil report from the geotechnical engineer, the structural engineer estimates the probable loads (dead, live, wind, snow, earthquake, etc) that are acting on the structure. Normally, the material of construction is chosen by the owner in consultation with the architect.
2. The structural engineer arrives at the structural system after comparing various possible systems. In a building, heating and air-conditioning requirements or other functional requirements may dictate the use of a structural system that is not the most efficient from a purely structural viewpoint, but which is the best bearing the total building in mind.
3. A suitable structural analysis, mostly with the aid of computers, is done to determine the internal forces acting on various elements of the structural system, based on the various loads and their combinations.
4. Considering the critical loading conditions, the sizes of various elements are determined following the codal provision.
5. The detailed structural drawings are then prepared once again following codal provisions and approved by the structural engineer.
6. The estimator arrives at the quantities involved and the initial cost of construction.
7. The contractor, based on the structural drawings, prepares the fabrication and erection drawings and a bill of quantity of materials (BOQ). The structural engineer again approves these drawings.

8. The contractor constructs the building based on the specifications given by the architect/project manager.
9. The structural engineer, with the help of quality control inspectors, inspects the work of the fabricator and erector to ensure that the structure has been fabricated/erected in accordance with his or her designs and specifications.
10. After the structure is constructed and handed over to the owner, the owner, by appointing suitable consultants and contractors, maintains the building till its intended age.

From these steps, it may be clear that accurate calculations alone may not produce safe, serviceable, and durable structures. Suitable materials, quality control, adequate detailing, good supervision, and maintenance are also equally important.

While executing the various steps, the structural engineer has to interact with the architect/project manager and also with others (electrical engineers, mechanical engineers, civil engineers, geotechnical engineers, surveyors, urban planners, estimators, etc.) and incorporate their requirements into the design (e.g., load due to mechanical and electrical systems). It has to be noted that steps 1 to 6, which are followed mainly in the design office, are not straightforward operations but are iterative (see Fig. 2.1). This book mainly covers only step 4—the design of structural elements to safely carry the expected loads and to ensure that the elements and the structure perform satisfactorily.

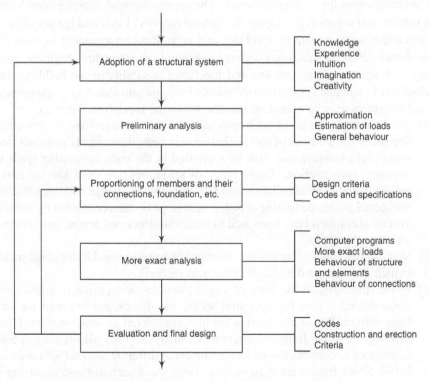

Fig. 2.1 Iterative structural design process

Compared to analysis (where all the parameters are known), design is a creative process. It involves the selection of the span, assessment of loads, and the choice of material, cross section, jointing method and systems, etc. Hence, there is no unique solution to a design problem. The designer has to make several decisions, which will affect the final construction and its cost. Therefore, the designer has to use his engineering judgment in order to reduce the cost and arrive at an efficient solution to the problem.

Today's structural engineer has several aids such as computer programs, handbooks, and charts, and hence should spend more time on thinking about design concepts and select the best structural system for the project at hand.

For most structures, the designer should specify a grade of structural steel that is readily available, keep the structural layout and structural details (e.g., connections) as simple as possible, use sections that are readily available, and use the maximum possible repetition of member sections and connection details. It is preferable for the designer to have a knowledge of fabricating shop capabilities (e.g., size of available zinc baths for galvanization) and erection techniques.

2.3 Structural Systems

The art of structural design is manifested in the selection of the most suitable structural system for a given structure. The arrangement of beams/girders/joints or trusses, and columns to support the vertical (gravity) loads and the selection of a suitable bracing system or a column and beam/truss arrangement to resist the horizontal (lateral) loads poses a great challenge to the structural engineer, since they will determine the economy and functional suitability of the building. The selection of a suitable system is made mainly based on previous data or experience. Steel structures may be classified into the following types (see Fig. 2.2).

(a) Single-storey, single, or multi-bay structures may have truss or stanchion frames, or rigid frames of solid or lattice members. Beams and open-web steel joists (light trusses) may also be supported at the ends by bearing walls of masonry construction. These types of structures are used for industrial buildings, commercial buildings, schools, and some residential buildings. Pitched roof portal frames consisting of rolled-steel sloped beams connected by welding to vertical columns have been used as industrial structures, arenas, auditoriums, and churches.

(b) Multi-storey, single, or multi-bay structures of braced or rigid frame construction (which are discussed in detail in the next section).

(c) Space structures, in the form of single-, double- or multi-layer grids, steel-frame folded plates, braced barrel walls, and domes, are required for very large column-free areas. Towers are also considered as space trusses. Often they require three-dimensional computer analysis. They also require special connectors to connect the various members at different angles (Subramanian 2006). Space frames are used to cover large spans such as those occurring in large arenas, auditoriums, swimming pools, theatres, airport hangars, tennis or baseball grounds, ballrooms, etc.

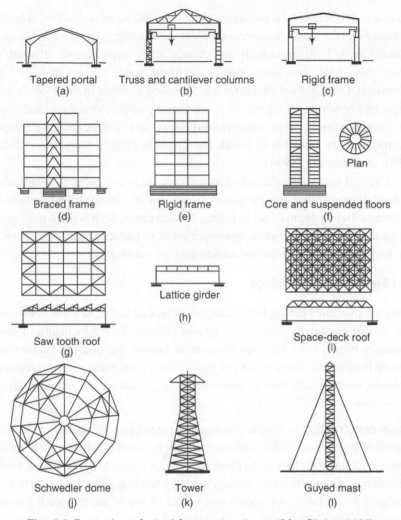

Tapered portal   Truss and cantilever columns   Rigid frame
(a)   (b)   (c)

Braced frame   Rigid frame   Core and suspended floors
(d)   (e)   (f)

Plan

Lattice girder
(h)

Saw tooth roof   Space-deck roof
(g)   (i)

Schwedler dome   Tower   Guyed mast
(j)   (k)   (l)

Fig. 2.2 Examples of steel-frame structures (MacGinley 1997)

(d) Tension structures, tensegritic and cable-supported (cable-suspended or cable-stayed) roof structures (Subramanian 2006).

(e) Stressed skin structure, where the cladding is also designed as a load-bearing member, thus stabilizing the structure; in such structures special shear connections are necessary, in order that the sheeting acts integrally with the main frames of the structure (Subramanian 2006).

(f) High-rise constructions: Tall buildings with more than 20 storeys are often considered in large cities where land costs are very high. In the design of such structures, the designer should pay attention to the system resisting the lateral loads (MacGinley 1999). Several interesting systems have been developed, and a few are discussed in the next section.

It has to be noted that combinations with concrete (in the form of shear walls or floor slabs) are structurally important in many buildings. If adequate interconnection between the concrete slab and the steel beam is provided in buildings and bridges (in the form of *shear connectors*), the resulting system, called *composite construction*, is both structurally and economically advantageous. Braced, rigid frame, truss roof, and space-deck construction are shown in Fig. 2.2 for comparison. Only framed structures are discussed in detail in the book. Analysis, design, and construction aspects of space frames, tension structures, and stressed skin systems are available in Subramanian (2006). Details of the design of composite constructions are available in Kulak and Grondin (2002), Salmon and Johnson (1996), and Johnson (1994).

For framed structures, the main elements are the beam, column, beam-column, tie, and lattice member. For long-span constructions, normal rolled sections may not have sufficient depth to act as beams. In such cases, deep welded plate girders, box girders, castellated girders, open-web joists or trusses may replace them. For very long spans, deep trusses or arches may be necessary.

2.3.1 Steel-Framed Buildings

Most steel structures belong to the category of braced and rigid frame construction. They consist essentially of regularly spaced columns joined by beams or girders. Secondary beams span between these main beams and provide support to the concrete floor or roof sheeting. Depending on the type of beam-column connections employed, such systems may be classified as *simple construction* or as *continuous construction*.

Simple construction In simple construction (see clause 4.2 and F.4 of the code), the ends of beams and girders are connected to transmit transverse shear only and are free to rotate under load in the plane of bending. Hence hinged ends are assumed for the beams. Connections are usually made by welding plates or angles to a beam or column in the fabricator's shop and bolted at site to the connecting beam or column (see Fig. 2.3).

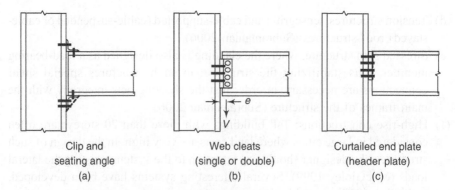

Clip and seating angle	Web cleats (single or double)	Curtailed end plate (header plate)
(a)	(b)	(c)

Fig. 2.3 Simple beam to column connections

These simple constructions are statically determinate and hence the beams are designed as simply supported and the columns are designed for the axial loads (due to the reaction from the beams) and the moments produced by the eccentricity of the beam reactions as shown in Fig. 2.3(b). (A minimum distance of 100 mm from the face of the column is specified in the code clause 7.3.3.1.) In such frames lateral forces due to the wind or earthquake are generally resisted by bracings (usually made of angles), forming vertical or horizontal trusses as required.

The braced bays can be grouped around a central core, distributed around the perimeter of the building, or staggered through various elevations as shown in Fig. 2.2(d) and Fig. 2.4. The floors act as horizontal diaphragms to transmit load to the braced bays. Bracing must be provided in two directions and all connections are taken as pinned. The bracing should be arranged to be symmetrical with respect to the building plan, to avoid twisting. The unbraced portion of the building frame in effect 'leans' on the braced portion to keep from falling over. In multi-storey buildings, reinforced concrete shear walls may replace the vertical steel bracing trusses. This type of construction is used in frames up to about five storeys in height, where strength rather than stiffness governs the design. Manual analysis can be used for the whole structure.

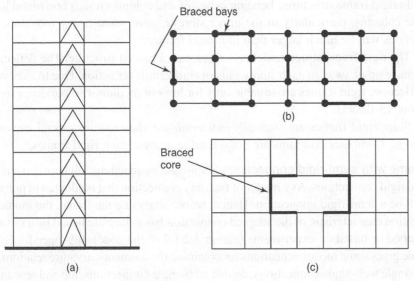

Fig. 2.4 All-steel braced structures: (a) vertical bracing, (b) bracing on perimeter/ interior walls, and (c) bracing around core

Continuous construction or rigid frame structures Continuous construction (also called rigid frame structures) assume sufficient rigidity in the beam-column connections, such that under the action of loads the original angles between intersecting members are unchanged (see clause 4.2 and F4.2 of the code). Connections are usually made in the fabricator's shop as well as the site, by welding and bolting. The connections shown in Fig. 2.5 can be adopted in rigid frame construction, which transfer both shear and moment from beam to column. Fully welded connections can also be considered as rigid beam-to-column connections. Such connections naturally involve additional fabrication and higher erection costs. However, the greater rigidity produced in the structure will result in reduced member sizes and the elimination of bracings. This form of construction is used for low-rise industrial buildings [Figs 2.2(a) and (c)] and for multi-storey buildings [Fig. 2.2(e)].

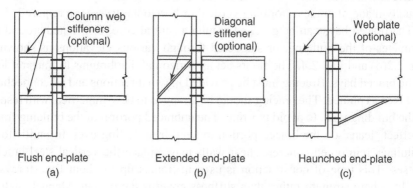

Flush end-plate   Extended end-plate   Haunched end-plate
(a)   (b)   (c)

Fig. 2.5 Rigid beam-to-column connections

In rigid frame structures, bending in beams and columns resists horizontal load. The columns, particularly in the upper storeys, must resist heavy moments. So sections will be much larger than in braced frames.

The rigid frame structure deflects more than a braced structure. The deflection is made up of sway in each storey plus overall cantilever action. Due to excessive deflection, rigid frames are suitable only for low- or medium-rise buildings (up to about 15 floors).

Since rigid frames are *statically indeterminate*, they require several cycles of design. Computer programs are often used to analyse such rigid frames.

Frame with semi-rigid connections Semi-rigid connections fall between simple and rigid connections. As a matter of fact, any connection that is adopted in practice will be a semi-rigid connection. Hence, before analysing the frame, the moment-rotation characteristic of the adopted connection has to be established by a rational method or based on experiments (clause 4.2.1.2 of the code). In Appendix F, the code gives some recommendations for obtaining the moment-curvature relationship of single web-angle connections, double web-angle connections, top and seat angle connections (without double web-angle connections), and header plate connections. Computer programs are available for the analysis of frames with semi-rigid connections (Chen et al. 1995). Research is still being done on semi-rigid connections. In practice, most connections are either assumed as simple connections or rigid connections only.

Composite structures The composite steel-shear wall structure consists of a steel-framed building braced with vertical reinforced concrete shear walls, as shown in Fig. 2.6. The shear walls placed in two directions at right angles carry vertical and horizontal loads. The shear walls replace the braced bays in the all-steel building.

Lifts

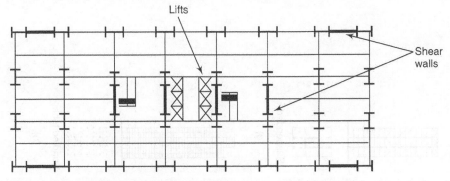

Shear walls

Fig. 2.6 Steel building with concrete shear walls

The shear walls can be located at the ends or sides or in appropriate locations within the building. They should be arranged so as to be symmetrical with respect to the plan, otherwise twisting will occur. They provide fire-proof walls at the lifts and staircase.

2.3.2 High-Rise Structural Systems

As pointed out previously, to build tall buildings economically, the designer must pay attention to the resistance of lateral forces. Several excellent systems have been invented in the past and are shown in Figs 2.7 and 2.8. They include outrigger and belt lattice girders systems, framed tube, braced tube, tube in tube, and bundled tube systems (Subramanian 1995). These systems are discussed briefly in this section.

Outrigger and belt truss system In tall buildings, the lateral deflection can be excessive if the bracing is provided around the core only. This deflection can be reduced by bringing the outside columns also into action to resist the lateral loads by the provision of outrigger and belt lattice girders, as shown in Fig. 2.9. The tension and compression forces in the outer columns apply a couple to the core, which acts against the cantilever bending under wind loads. The belt truss surrounding the building brings all external columns into action (MacGinley 1997). A single outrigger and belt lattice girder system at the top or additional systems at different heights of a very tall building can be provided.

Tube structures The tube type of structure was developed by Dr Fazlur Khan of the USA for very tall buildings, say over 80 storeys in height. If the core type of structure were used, the deflection at the top would be excessive. The tube system is very efficient with respect to structured material used and results in a considerable saving in material when compared with conventional designs.

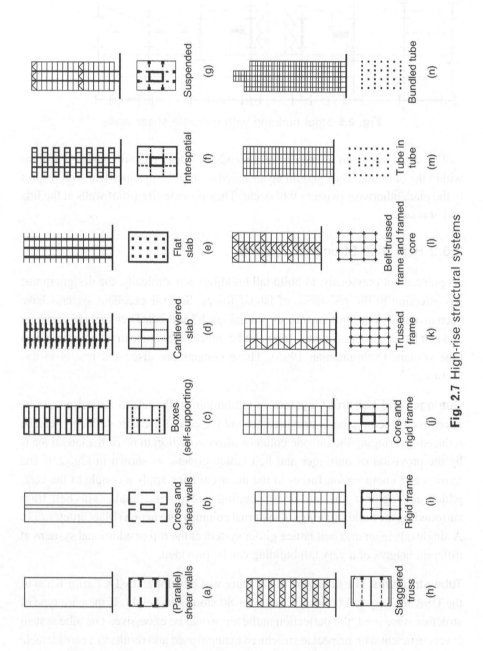

Fig. 2.7 High-rise structural systems

(a) (Parallel) shear walls
(b) Cross and shear walls
(c) Boxes (self-supporting)
(d) Cantilevered slab
(e) Flat slab
(f) Interspatial
(g) Suspended
(h) Staggered truss
(i) Rigid frame
(j) Core and rigid frame
(k) Trussed frame
(l) Belt-trussed frame and framed core
(m) Tube in tube
(n) Bundled tube

Twin Towers of World Trade Centre – Framed Tube Structure

The twin 110-story towers of the World Trade Center in New York City, USA, were designed by Architect Minoru Yamasaki and Structural Engineers Leslie E. Robertson Associates in the early 1960s using a *framed tube* structural system. The North Tower was completed in December 1970 and the South Tower was finished in July 1971.

Both had a 63 m² plan, 350,000 m² of office space, and the facades were sheathed in aluminum-alloy. The buildings were designed with narrow office windows 45-cm wide which also reflected Yamasaki's fear of heights as well as his desire to make building occupants feel secure.

The World Trade Center towers had high-strength, closely spaced load-bearing perimeter steel columns (acting like *Vierendeel trusses*) that supported all lateral loads such as wind loads, and shared the gravity load with the core columns [see Fig. 2.10(a)]. The 59 perimeter columns (per side) were constructed using prefabricated modular pieces, each consisting of three columns, three-storeys tall, connected by spandrel plates. Adjacent modules were bolted together with the splices occurring at mid-span of the columns and spandrels. The spandrel plates were located at each floor, transmitting shear stress between columns, allowing them to work together in resisting lateral loads. The joints between modules were staggered vertically, so the column splices between adjacent modules were not at the same floor.

The core of the towers housed the elevator and utility shafts, restrooms, three stairwells, and other support spaces. The core (a combined steel and concrete structure) of each tower was a rectangular area of 27 by 41 m and contained 47 steel columns running from the bedrock to the top of the tower. The large, column-free space between the perimeter and the core was bridged by prefabricated floor trusses. The floor trusses provided lateral stability to the exterior walls and distributed wind loads among the exterior walls. The trusses supported 100-mm thick lightweight concrete slabs that were laid on a fluted steel deck. The floors were connected to the perimeter spandrel plates with *viscoelastic dampers*, in order to reduce the sway of the buildings.

Outrigger truss systems were provided from the 107th floor to the top of the building. There were six such trusses along the long axis of the core and four along the short axis. This truss system allowed some load redistribution between the perimeter and the core columns and also supported the tall communication antenna on the top of each building. When completed in 1972, the North Tower became the tallest building in the world for two years, surpassing the Empire State Building after a 40-year reign.

On September 11, 2001, terrorists crashed a hijacked plane into the northern facade of the North Tower, impacting between the 93rd and 99th floors. Seventeen minutes later, a second team of terrorists crashed another hijacked plane into the South Tower, impacting between the 77th and 85th floors. After burning for 56 minutes, the South Tower collapsed due to the plane impact and as a result of buckling of steel columns due to the ensuing fire. The North Tower collapsed after burning for approximately 102 minutes. The attacks on the World Trade Center resulted in 2,750 deaths and a huge financial loss.

On the northwest corner of the WTC site, a new 551-m tall Freedom Tower, designed by David M. Childs of Skidmore, Owings & Merrill, is being built from April 2006.

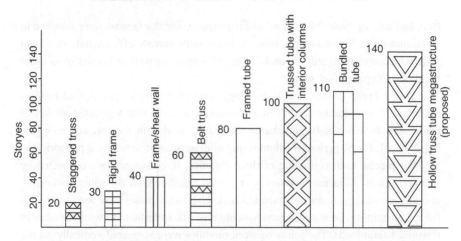

Fig. 2.8 Tall building system for steel structures

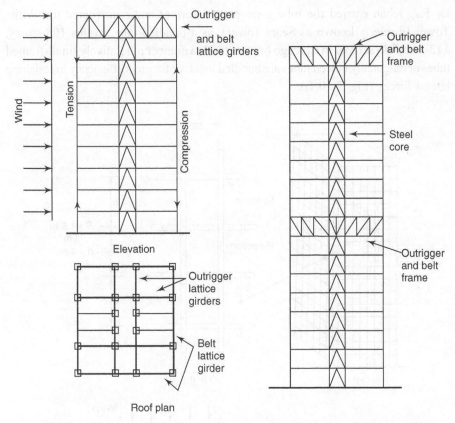

Fig. 2.9 Outrigger and belt lattice girder system (MacGinley 1997)

In this system the perimeter walls are so constructed that they form one large rigid or braced tube, which acts as a unit to resist horizontal load, as shown in Fig. 2.10(a) and (b). The small perforations form spaces for windows, and the normal curtain walling is eliminated.

In the single-tube structure, the perimeter walls carry the entire horizontal load and their share of the vertical load. Internal columns and/or an internal core, if provide, cary vertical loads only. In the tube-within-a tube system shown in Fig. 2.10(e), the internal tube can be designed to carry part of the horizontal load. Very tall stiff structures have been designed on the bundled tube system shown in Fig. 2.10(f) which consists of a number of tubes constructed together. This reduces the *shear lag* problem that is more serious if a single tube is used (MacGinley 1997). Shear lag is a phenomenon in which the stiffer (or more rigid) regions of the structure or structural component attract more stresses than the more flexural regions. Shear lag causes stresses to be unevenly distributed over the cross section of the structure or structural component, as shown in Fig. 2.10(c).

The framed tube can be relatively flexible and bracing the tube, as shown in Fig. 2.10(d), provides additional stiffness. This helps reduce shear lag in the flange tube faces as the diagonal members make all exterior columns act together as a rigid tube.

Dr F.Z. Khan carried the tube concept still further and constructed the Willis Tower (formerly known as Sears Tower), as a bundled tube. In this 108-storey, 442-m tall skyscraper in Chicago built in 1973, a number of relatively small-framed tubes or diagonally braced tubes are bundled together for great effeciency in resisiting lateral forces (Fig. 2.10(f)).

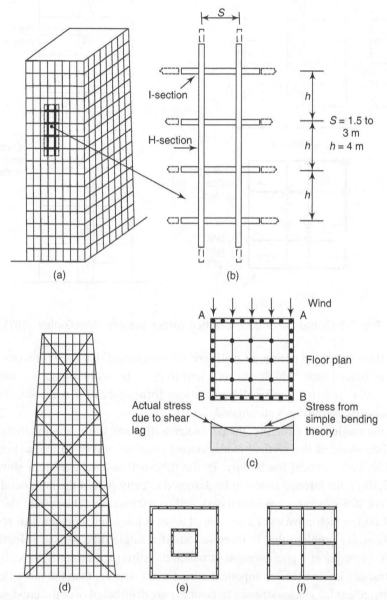

Fig 2.10 Tube structures-(a) framed tube, (b) prefabricated 'tree' unbit (c) stress distribution in walls AA, BB (d) braced tube, (e) tube in tube, and (f) bundled tube (Building may have more storeys than shown.)

John Hancock Center–Braced Tube Structure

The 100-storey, 344-m tall John Hancock Center of Chicago, Illinois, was a collaborative effort of Skidmore, Owings, and Merrill, and structural engineers Fazlur Khan and Bruce Graham. When completed in 1969, it was the tallest building in the world outside New York City.

The distinctive X-bracing at the exterior of the building, is a part of its 'trussed or braced tubular system'. The braced tube system, developed by engineer Fazlur Khan, helps the building stand upright during wind and earthquake loads as the X-bracing provides additional stiffness to the tube. The external bracings of the tube reduce the shear lag in the flange tube faces, as shown in Fig. 2.10(c), as the bracings make all the exterior columns to act together as a rigid tube. In addition to higher performance against lateral loads, it also increases the usable floor space. This original concept made the John Hancock Center an architectural icon.

2.4 Seismic Force Resisting Systems

Moment resisting frames, moment resistant frames with shear walls, braced frames with horizontal diaphragms or a combination of the above systems, may be provided to resist seismic forces (see Fig. 2.11). Out of these, moment resisting frames may be economical for buildings with only up to 5 to 10 storeys (the infill walls of non-reinforced masonry also provides some stiffness). Shear wall and braced systems (which are more rigid than moment resisting frames) are economical up to 15 storeys. When frames and shear walls are combined, the system is called a *dual system*. A moment resisting frame, when provided with specified details for increasing the ductility and energy absorbing capacity of its components, is called a *special moment resisting frame (SMRF)*, otherwise it is called an *ordinary moment resisting frame (OMRF)*.

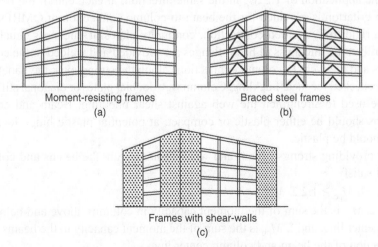

Moment-resisting frames
(a)

Braced steel frames
(b)

Frames with shear-walls
(c)

Fig. 2.11 Lateral-force-resisting system

The design engineer should not consider the structure as composed of a summation of different parts (such as beams, columns, trusses, walls, etc.) but as a completely integrated system, which has its own properties with respect to lateral force response. Thus, he or she should follow the flow of forces through the structure into the foundation and make sure that every connection along the path of stress is adequate to maintain the integrity of the system. It is also necessary to provide adequate redundancy in the structure. When a primary system yields or fails, the redundancy will allow the lateral forces to be redistributed to a secondary system to prevent progressive collapse (see Section 2.5). It is also important to note that the forces due to earthquakes are not static but dynamic, (cyclic and repetitive) and hence the deformations will be well beyond those determined from the elastic design.

2.4.1 Moment Resisting Frames

According to clause 12.10 of the code, ordinary moment resisting frames (OMRF) should be able to withstand inelastic deformation corresponding to a joint rotation of 0.02 radians (for special moment resisting frames it is 0.04 radians) without degradation in strength and stiffness below the full yield value (M_p). OMRFs should not be used in Seismic zones IV and V, and for buildings with an importance factor greater than unity in Seismic zone III. OMRFs and SMRFs with rigid moment connections should be designed to withstand a moment of at least 1.2 times the full plastic moment of the connected beam. For OMRFs, a semirigid moment connection is permitted. In such a case, the connection should be designed to withstand the lesser of the following: a moment of at least 0.5 times either the full plastic moment of the connected beam or the maximum moment that can be delivered by the system. In semi-rigid joints, the design moment should be achieved with a rotation of 0.01 radians.

The beam-to-column connection of SMRFs should be designed to withstand a shear resulting from the load combination of 1.2DL + 0.5LL plus the shear resulting from the application of 1.2 M_p in the same direction, at each end of the beam. A similar criterion is provided for the beam-to-column connection of OMRFs.

In a rigid, fully welded connection, continuity plates of thickness equal to or greater than the thickness of beam flanges are provided and welded to the column flanges and the web. In column connections along the strong axis, the panel zone is to be checked for shear buckling. Column web doubler plates or diagonal stiffeners may be used to strengthen the web against shear buckling. Beam and column sections should be either plastic or compact; at potential plastic hinge locations they should be plastic.

For providing strong-column and weak-beam design, the beams and columns should satisfy

$$\Sigma \, M_{pc} \geq 1.2 \, \Sigma \, M_{pb} \tag{2.1}$$

where $\Sigma \, M_{pc}$ is the sum of the moment capacity in columns above and below the beam center line, and $\Sigma \, M_{pb}$ is the sum of the moment capacity in the beams at the intersection of the beam and column center lines.

Note that engineers, during 1980s, tried to economize their designs by providing only a single bay of moment resistant framing on either side of buildings. The 1994 Northridge earthquake and the 1995 Kobe earthquake showed that such buildings are prone to brittle fracture at their welded-beam to column connections. Research conducted after this earthquake, resulted in several special provisions and some approved connections, which are discussed in the chapter on welded connections. Other guidelines may be found in Section 12 of IS 800.

2.4.2 Braced Frames

The members of braced frame act as a truss system and are subjected primarily to axial stress. Current research shows that significant inelastic deformation occurs in the beams and columns of braced frames in addition to the buckling of the brace. Depending on the diagonal force, length, required stiffness, and clearances, the diagonal members can be made of double angles, channels, tees, tubes or even wide flange shapes. Besides performance, the shape of the diagonal is often based on connection considerations. The braces are often placed around service cores and elevators, were frame diagonals may be enclosed within permanent walls. The braces can also be joined to form a closed or partially closed three dimensional cell so that torsional loads can be resisted effectively. A height to width ratio of 8 to 10 is considered to form a reasonably effective bracing system.

Braced frames may be grouped into *concentrically braced frames* (CBFs), and *eccentrically braced frames* (EBFs), depending on their ductility characteristics. In addition, concentrically braced frames are subdivided into two categories, namely, *ordinary concentrically braced frames* (OCBFs) and *special concentrically braced frames* (SCBFs).

Concentrically braced frames In CBFs, the axes of all members, i.e., columns, beams and braces intersect at a common point such that the member forces are axial. The Chevron bracing, cross bracing (X bracing), and diagonal bracing (single diagonal or K bracing) are classified as concentrically braced and are shown in Fig. 2.12(a)–(d).

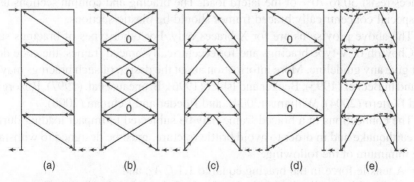

(a)   (b)   (c)   (d)   (e)

Fig. 2.12 Types of bracings and the load path (a) single diagonal bracing, (b) X-bracing, (c) chevron bracing, (d) single-diagonal, alternate direction bracing, and (e) knee bracing.

On the other hand EBFs utilize axial offsets to deliberately introduce flexure and shear into framing beams to increase ductility. For example, in the knee bracing shown in Fig. 2.12(e), the end parts of the beam are in compression and tension with the entire beam subject to double curvature bending. [Note that in all the frames shown in Fig. 2.12(a), a reversal in the direction of horizontal load will reverse all actions and deformations in each of the members]. EBFs are discussed in detail in the next section.

The inability to provide reversible inelastic deformation is the principle disadvantage of CBFs. After buckling, an axially loaded member loses strength and does not return to its original straight configuration. To reduce the possibility of this occurring during moderate earthquakes, more stringent design requirements are imposed on bracing members. Thus, ordinary concentrically braced frames are not allowed in Seismic zones IV and V and for buildings with an importance factor greater than unity (I > 1.0) in zone III; a K bracing is not permitted in earthquake zones by the code (the inelastic deformation and buckling of K bracing members may produce lateral deflection of the connected columns, causing collapse).

Ordinary concentrically braced frames should be designed to withstand inelastic deformation corresponding to a joint rotation of 0.02 radians without degradation in strength and stiffness, below the yield value. The slenderness of bracing members should not exceed 120 and the required compressive strength should not exceed $0.8 \, P_d$, where P_d is the design strength in axial compression. Along any line of bracing, braces shall be provided such that for lateral loading in either direction, the tension braces will resist 30 to 70% of the lateral load. This is to prevent an accumulation of inelastic deformation in one direction and to preclude the use of tension only diagonal bracing.

Special concentrically braced frames should be designed to withstand inelastic deformation corresponding to a joint rotation of 0.04 radians without degradation in strength and stiffness below the full yield value. They are allowed to be used in any zone and for any building. The slenderness ratio of the bracing members should not exceed 160 and the required compressive strength should not exceed the design strength in axial compression, P_d. Along the line of bracing, braces should be provided such that the lateral loading in either direction, the tension braces resist 30 to 70% of the laterd load. The bracing and column sections used in special concentrically braced frames should be plastic sections.

The above provisions are for X braces only. For other types of bracings such as Chevron or V-type bracings and for eccentrically braced frames, the code does not give any guideline. More information about the design of such bracings may be found in Becker (1995), Becker and Ishler (1996), Bruneau, et al. (1997), Bozorgnia and Bertero (2004), Williams (2004), and Roeder and Lehman (2008).

The connections in a braced frame may be subjected to impact loading during an earthquake and in order to avoid brittle fracture, must be designed to withstand the minimum of the following:

(a) A tensile force in the bracing equal to $1.1 \, f_y \, A_g$, and
(b) The force in the brace due to the following load combinations,

$$1.2 \, \text{DL} + 0.5 \text{LL} \pm 2.5 \, \text{EL},$$
$$0.9 \, \text{DL} \pm 2.5 \text{EL}$$

(c) The maximum forces that can be transferred to the brace by the system.

The connection should be checked to withstand a moment of 1.2 times the full plastic moment of the braced section about the buckling axis and for tension rupture, and block shear under the above loading. The gusset plates should be checked for buckling out of their plane, and sufficient length should be provided for plastic hinge formation. Recent research has shown that the current practice of providing a linear clearance of twice the thickness of gusset plates [see Fig. 2.13(a)], leads to thicker and larger size of gusset plates. This creates a rotationally stiff joint, which limits the rotation of the connection and leads to extensive frame yielding. Based on the recent research, Roeder and Lehman (2008) suggest to provide an elliptical clearance of eight times the thickness of the gusset plate [see Fig. 2.13(b)]. This will not only result in smaller, thinner, and compact gusset plates, but also greater ductility and inelastic deformation of the system. Welds joining the gusset plate to the beam and column should be sized using the plastic capacity of the gusset plate rather than the expected resistance of the brace.

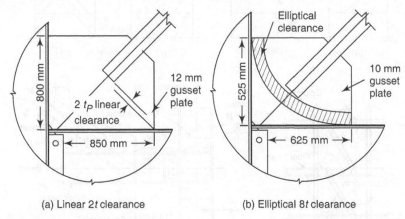

(a) Linear 2*t* clearance   (b) Elliptical 8*t* clearance

Fig. 2.13 Improved connection detail for CBFs (Roeder and Lehman 2008)

Eccentrically braced frames The bracing member in an EBF is connected to the beam so as to form a short link beam between the braces and the column or between two opposing braces (see Fig. 2.14). Thus, the eccentric bracing is a unique structural system that attempts to combine the strength and stiffness of a braced frame with the inelastic behaviour and energy dissipation characteristics of a moment frame. The link beam acts as a fuse to prevent buckling of the brace from large overloads that may occur during major earthquakes. After the elastic capacity of the system is exceeded, shear or flexural yielding of the link provides a ductile response in contrast to that obtained in an SMRF. In addition, EBFs may be designed to control frame deformations and minimize damage to architectural finishes during seismic loading (Williams 2004). Note that the connection between the column and beam are moment resistant [see Fig. 2.14(f)] to achieve brace action. The web buckling is prevented by providing adequate stiffeners in the link. Links longer than twice the depth of the beam tend to develop plastic hinges, while shorter links tend to yield in shear. Buildings using eccentric bracings are lighter than MRFs and, while retaining the elastic stiffness of CBFs, are more ductile. Thus, they provide an economical system in seismic zones. A premature failure of the link does not cause the structure to collapse, since the structure continues to retain its vertical load carrying capacity and stiffness. This facilitates easy repair of the system after a severe earthquake. The design and other details of eccentrically braced systems are provided by Williams (2004) and Bruneau, et al. (1997).

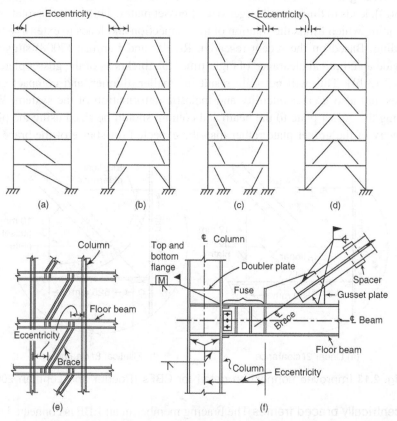

Fig. 2.14 Eccentric bracing system: (a–d) common types of bracing, (e) elevation, and (f) detail (Taranath 1998)

2.5 Structural Integrity

To reduce the risk of localized damage spreading to all parts, buildings should be effectively tied together at each principal floor level. It is important to effectively hold each column in position by means of horizontal ties in two directions (preferably at right angles), at each principal floor level supported by the column. Horizontal ties are also required at roof level, except where the steel work supports only cladding weighing 0.7 kN/m^2 or less and carrying only imposed roof loads and wind loads. At re-entrant corners the tie member nearest to the edge should be anchored into the steel framework, as shown in Fig. 2.15. All these horizontal tie members should be capable of resisting a minimum factored tensile load (should not be considered as additive to other loads) of 75 kN at floor level and 40 kN at roof level. A minimum tie strength of 0.5 $W_fS_tL_a$ for internal ties and one of 0.25 $W_fS_tL_a$ for external ties is also suggested in BS 5950, 2000 (where W_f is the total factored load/unit area, S_t is the tie spacing and L_a is the distance between columns in the direction of the ties). Note that these integrity considerations have a direct influence on connection design, since the tying action of beams requires the connection to possess adequate direct tension capacity. Experimental work done in the UK has established that end plates and web cleats of 8 mm thickness fastened to column flanges by two M20 grade 8.8 bolts will meet the requirement of ties resisting 75 kN factored tensile load (Nethercot 2001).

Each portion of a building between expansion joints should be treated as a separate building. By tying the structure together as shown in Fig. 2.15, alternative load paths, which enhance the safety, may be made available. To ensure sway resistance, no portion of the structure should be dependent on only one bracing system. All columns should be continuous vertically through the floors.

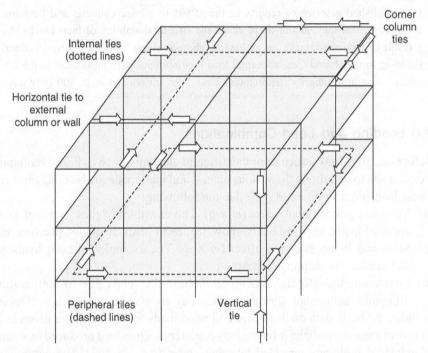

Fig. 2.15 Tying columns of building to achieve structural integrity

Precast concrete or other heavy floor or roof units must be properly anchored at both ends. At the edge of the structure, horizontal ties capable of resisting 1% of the maximum factored column loads should restrain columns. Key elements that would risk the collapse of greater area (greater than 15% of the floor area or 70 m², whichever is less) should be identified and designed for accidental loading (see clause 5.1.2 of IS 800 : 2007 for more on this).

Ronan Point Collapse

Progressive collapse provisions were introduced in the British code as early as 1970. This was a direct result of the Ronan Point collapse in 1968. This involved a 23-storey tower block in Newham, East London, which suffered a partial collapse when a gas explosion demolished a load-bearing wall, causing the collapse of one entire corner of the building. Four people were killed in the incident, and 17 were injured. (Ronan Point was repaired after the explosion; it was demolished in 1986 for a new low-rise housing development project.)

Due to the failure of Ronan Point apartment, many other similar large panel system buildings were demolished. The Building Research Establishment, UK, published a series of reports in the 1980s to advise councils and building owners on what they should do to check the structural stability of their blocks. As a result of terrorist attacks on embassies abroad, along with the Murrah Federal Building in Oklahoma City, abnormal load requirements were introduced in the US codes. Structural integrity requirements have been introduced in IS 800 only now.

2.6 Loading and Load Combinations

Before designing any structure or the different elements, such as beams, columns, etc., one has to determine the various natural and man-made loads acting on them. These loads on a structure may be due the following:

(a) Mass and gravitational effect ($m \times g$). The examples of these types of loads are dead loads, imposed loads, snow, ice, earth loads, hydraulic pressure, etc.
(b) Mass and its acceleration effect ($m \times a$). The examples of such loads are earthquake, wind, impact, blast, etc.
(c) Environmental effects. Examples include the loads due to temperature difference, settlement, shrinkage, etc. They are also termed as *indirect loads*.

In India, the basic data on dead, live and wind loads for buildings are given in IS 875, with more specialized information on matters such as load produced by cranes in industrial buildings provided by other codes (e.g., IS 807). The earthquake loads are specified in IS: 1893. For towers and other forms of structures, the necessary loading data are provided in the code of practice appropriate to that type of structure (e.g. IS 802, IS 9178, IS 6533). We will briefly discuss about a few important loads in this section.

2.6.1 Dead Loads

The load fixed in magnitude and in position is called a *dead load*. Determination of the dead load of the structure requires the estimation of the weight of the structure together with its associated *non-structural* components. Thus, we have to calculate and include the weight of bare steelwork (including items, such as bolts, nuts, and weld material) slabs, beams, walls, columns, partition walls, false ceiling, facades, cladding, water tanks, stairs, plaster finishes, and other services (cable ducts, water pipes, etc). After the design process, the initially assumed dead weight of the structure (based on experience), has to be compared with the actual dead load. If the difference between the two loads is significant, the assumed dead load should be revised and the structure redesigned. Dead weights of different materials are provided in the code IS 875 (part 1 – dead loads). The weights of some important building materials are given in Table 2.1.

Table 2.1 Weights of some building materials

Material	Unit Weight
Brick masonry in CM 1:4	20 kN/m^3
Plain concrete	24 kN/m^3
Reinforced cement concrete	25 kN/m^3
Stone masonry	$20.4 \text{ to } 26.5 \text{ kN/m}^3$
Cement Mortar	20.4 kN/m^3
Steel	78.5 kN/m^3
20-mm cement plaster	450 N/m^2
Roofing sheets	
(a) GI sheet 1.6mm thick	156 N/m^2
(b) Steel sheet 1mm thick	77.5 N/m^2
(c) AC sheet 6mm thick	$160–170 \text{ N/m}^2$
5-mm glass	125 N/m^2
Floor finishes	$600–1200 \text{ N/m}^2$

2.6.2 Imposed Loads

Imposed loads (previously referred to as *live loads*) are gravity loads other than dead loads and cover items, such as occupancy by people, movable equipment and furniture within the buildings, stored materials such as books, machinery, and snow. Hence, they are different for different types of buildings: domestic, office, warehouse, etc. Thus, they vary often in space and time.

Imposed loads are generally expressed as static loads for convenience, although there may be minor dynamic forces involved. The code gives uniformly distributed loads as well as concentrated loads for various occupational categories. The magnitude of a few imposed loads are as given below:

(a) Residential buildings : 2 kN/m^2
(b) Office buildings   : $3–4 \text{ kN/m}^2$
(c) Storage facilities   : $5–7.5 \text{ kN/m}^2$

Note that live load may change from room to room. For considering the load due to partition, increase the floor load by 33.3% per meter run of partition wall subject to a minimum of 1 kN/m^2; the total weight per meter run must be less than 4 kN/m. For complete guidance, the engineer should refer to IS 875 (Part 2).

When large areas are considered, the code allows for a reduction in live load; for single beam or girders, a reduction of 5% for each 50 m^2 floor area, subjected to a maximum of 25% is allowed. In multi-storey buildings, the probability that all the floors will be simultaneously loaded with the maximum live loads is remote, and hence reduction to column loads is therefore allowed. Thus, the live loads may be reduced in the design of columns, walls and foundations of multi-storey buildings, as given in Table 2.2. Note that such reduction is not permissible, if we consider earthquake loads.

Table 2.2 Reduction in live load applicable to columns

Floor measured from top	Percentage
1 (top or roof)	0
2	10
3	20
4	30
5–10	40
11 to ground floor	50

The imposed loads on roofs as per IS 875 (Part 2) are given in Table 2.3.

Table 2.3 Imposed loads on various types of roofs

Type of roof	Uniformly distributed imposed load measured on the plan area	Minimum imposed load measured on plan
Flat, sloping, or curved roof with slopes up to and including 10 degrees		3.75 kN uniformly distributed over any span of 1 m width of the roof slab and 9 kN uniformly distributed over the span of any beam or truss or wall.
(a) Access provided	1.5 kN/m^2	
(b) Access not provided (except ladder for maintenance)	0.75 kN/m^2	Half of case (a) above
Slooping roof with a slope greater than 10 degrees	For roof membrane sheets or purlins, 0.75 kN/m^2 less 0.02 kN/m^2 for every degree increase in the slope over 10 degrees	0.4 kN/m^2

Note: 1. The loads given above do not include the loads due to snow, rain, dust collection, etc. The roof should be designed for imposed loads given above or for snow/rain load, whichever is greater.

2. All roof covering (other than glass) should be capable of carrying an incidental load of 900 N concentrated over an area of 125 mm^2.

3. Trusses, beams, columns, and girders excluding purlins can be designed for 2/3 of the live load on the roof.

IS 875 (Part 2) also gives *horizontal loads* acting on parapets, parapet walls, and balustrades. These loads should be assumed to act at handrail or coping level.

2.6.3 Crane and Impact Loads

In the design of crane runway girders (see Fig. 2.16) and their connections, the horizontal forces caused by moving crane trolleys must be considered, in addition to the vertical and impact loads. The intensity of the horizontal load (also called *lateral load*) is a function of the weight of the trolley, lifting load, and the acceleration of trolley. As per IS 875 – Part 2, the lateral load may be taken as,

(a) $C_{Lh} = 10\%$ of weight of trolley and lifted load in the case of electrically operated cranes (EOT) with a trolley having a rigid mast for the suspension of lifted weight, and

(b) $C_{Lh} = 5\%$ of weight of trolley and lifted load for all other EOT and hand-operated cranes.

The above force should be applied at the top of the rail acting in either direction normal to the runway rails, and should be distributed amongst all the wheels on one side of the rail track.

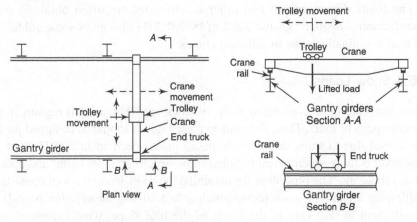

Fig. 2.16 Loads due to crane movement

In addition, due to acceleration and deceleration of the entire crane, a longitudinal tractive force is transmitted to the gantry girder through the friction of the end track wheels with the crane rail. IS 875 (Part 2) specifies that 5% of the maximum static wheel load of the crane is to be applied as longitudinal force, at the top of the rail.

The impact due to vertical crane loads is converted empirically into equivalent static loads through an impact factor, which is normally a percentage of the crane load. Table 2.4 shows the impact factors as suggested by IS 875 code for cranes and lifts. Thus, if the impact is 25%, the live load is multiplied in the calculation of the forces by 1.25.

Table 2.4 Additional impact loads on buildings

Structure	Impact allowance in percentage
Lifts and hoists	
(a) Frames	100
(b) Foundations	40
Reciprocating machinery—frames and foundations	50
Light machinery—structure and foundation	20
Electric overhead cranes	
(a) Girders	25
(b) Columns (class III and IV cranes)	25
(c) Columns (class I and II cranes)	10
(d) Foundations	0
Hand-operated cranes	
(a) Girders	10
(b) Columns and foundations	0

Note: In addition to the impact allowance in the vertical direction, additional loads in lateral and longitudinal directions must be applied on beams and columns as a percentage of the static load.

The loads due to cranes and other machineries are often obtained from manufacturers/suppliers. Clause 3.5.4 of IS 800:2007 also gives some guidelines for load combinations due to different cranes.

2.6.4 Snow Loads

Snow loads are to be considered in the mountainous (Himalayan) regions in the northern parts of India. Thus, the roofs in these regions should be designed for the actual load due to snow or for the imposed loads specified in IS 875 (Part 2), whichever is more severe. Freshly fallen snow weighs up to 96 kg/m^3 and packed snow 160 kg/m^3. The procedure for obtaining the snow load on a roof consists of multiplying the ground snow (corresponding to a 50-year mean return period) by a coefficient to take care of the effect of the roof slope, wind exposure, non-uniform accumulation of snow on pitched or curved roofs, multiple series roofs or multi-level roofs and roof areas adjacent to projections on a roof level. Although maximum snow and maximum wind load are not considered to act simultaneously, it is important to consider drift formation due to wind, since a majority of snow-related roof damage is due to drifted snow. The reader is advised to consult IS 875 (Part 4) for more information on snow loading.

2.6.5 Wind Loads

A difference in atmospheric pressures generates wind flow, which is primarily due to differences in temperature. The wind flow manifests itself into gales, cyclones, hurricanes, typhoons, tornadoes, thunderstorms, and localized storms. Tropical cyclone is the generic term used to denote hurricanes and typhoons. The wind speeds of cyclones can reach up to 30–36 m/s and in severe cases, it may reach up to 90 m/s. Cyclones in India exceed the wind speed for the design given in the code. The horizontal wind flow exerts lateral pressure on the building envelope and hence has to be considered in the design.

Hurricane Katrina

Hurricane Katrina struck the New Orleans area of USA on the early morning of August 29, 2005. It developed into a powerful Category–5 hurricane, on the Saffir–Simpson scale of hurricane intensity, with a highest wind speed of 280 km/h. It caused severe destruction along the Gulf coast from central Florida to Texas, much of it due to the storm surge. The storm surge breached the Louisiana's levees at 53 different points, leaving 80% of the city submerged, leaving several victims clinging to rooftops, and sending several others to shelters around the country. At least 1,836 people lost their lives in the actual hurricane and in the subsequent floods, making it the deadliest hurricane in the United States since the 1928 Okeechobee hurricane. NASA satellite images showed that the floods that buried up to 80% of New Orleans subsided only by September 15, 2005. Katrina redistributed over one million people from the central Gulf coast to elsewhere across the United States, which became the largest diaspora in the history of the United States.

(a)

(b)

(a) Satellite image of the 25 mile-wide eye of Hurricane Katrina (b) Flooded I-10/I-610/West End Blvd interchange and surrounding area of northwest New Orleans and Metairie, Louisiana

The total damage from Katrina is estimated at $81.2 billion, nearly double the cost of the previously most expensive storm, Hurricane Andrew, when adjusted for inflation. Federal disaster declarations covered 233,000 km² of the United States, an area almost as large as the United Kingdom. The hurricane left an estimated three million people without electricity. The Superdome, which was sheltering many people who were not evacuated, sustained significant damage.

It is to be noted that many of the broken levees have been reconstructed after the Katrina damage. While reconstructing them, measures were taken to bring them compatible to modern building code standards in order to ensure their safety. For example, in many locations, I-walls were replaced with T-walls by the Corps of Engineers. Since T-walls have a horizontal concrete base, the soil erosion underneath the floodwalls will be contained.

Tornadoes consist of a rotating column of air, accompanied by funnel-shaped downward extension of a dense cloud having a vortex of about 60–240 m diameter, whirling destructively at speeds of 75–135 m/s. Tornadoes are the most destructive of all wind forces and in the USA alone, the damage is in excess of $100 million per year. More details of tornado-resistant design may be found in Whalen et al. (2004) Minor (1982), and Coulbourne et al. (2002).

IS 875:1987 (Part 3) gives the basic wind speeds, averaged over a short interval of 3 seconds and having a 50-year return period at 10 m height above the ground level in different parts of the country. The entire country is divided into six wind zones (see Figure 1 of this code). The wind pressure/load acting on the structural system and the structural or non-structural component being considered depends on the following:

(a) velocity and density of air,
(b) height above the ground level,
(c) shape and aspect ratio of the building,
(d) topography of the surrounding ground surface,
(e) angle of wind attack,
(f) solidity ratio or openings in the structure, and
(g) the susceptibility of the structural system under consideration to steady and time-dependent (dynamic) effects induced by the wind load.

Depending on the above factors, the wind can create either a positive or a negative pressure (suction) on the sides of the building.

The design wind speed is obtained from the basic wind speed, as per IS 875 (Part 3), after modifying it to include the risk level, terrain roughness, the height and size of structure, and the local topography, as

$$V_z = V_b k_1 k_2 k_3 \qquad (2.2)$$

where

V_z = design wind speed at any height z in m/s
V_b = basic wind speed (given in Fig. 1 of the code)
k_1 = probability factor or risk coefficient (given in Table 1 of the code)
k_2 = terrain, height, and structure size factor (given in Table 2 of the code), and
k_3 = topography (ground contours) factor (given in section 5.3.3 of the code)

Note that the design wind speed up to 10 m height from the mean ground level is considered constant.

The design wind pressure p_d is obtained from the design wind velocity, as

$$p_d = 0.6 \, V_z^2 \tag{2.2a}$$

Wind load on a building can be calculated for
(a) the building as a whole,
(b) individual structural elements, such as roofs and walls, and
(c) individual cladding units including glazing and their fixings.

The code gives pressure coefficients (derived on the basis of models tested on wind tunnels) for a variety of buildings. Force coefficients are also given for (a) clad buildings, (b) unclad structures, and (c) structural elements.

Wind causes pressure or suction normal to the surface of a structure. Pressures are caused both on the exterior as well as interior surfaces, the latter being dependent on the openings (or permeability) in the structure, mostly in the walls. Wind pressure acting normal to the individual element or cladding unit is given by

$$F = (C_{pe} - C_{pi}) \, A p_d \tag{2.3}$$

where

F = net wind force on the element
A = Surface area of element or cladding
C_{pe} = external pressure coefficient
C_{pi} = internal pressure coefficient, and
p_d = design wind pressure

The wind pressure coefficients depend on the following:
(a) shape of the building or roof,
(b) slope of the roof,
(c) direction of wind with respect to the building, and
(d) zone of the building.

A typical industrial bulding elevation is shown in Fig. 2.17 along with its wind pressure coefficients, C_{pe} and C_{pi}. The building is divided into four zones and four local zones. External pressure coefficients (C_{pe}), for walls and pitched roofs of rectangular clad building, are given in Tables 4 and 5 respectively of this code. The internal pressure is considered positive if acting from inside to outside, while the external pressure coefficient is considered positive when acting outside to inside, as shown in Fig. 2.17. All buildings are classified into four types depending on permeability and the corresponding internal pressure coefficients are listed in Table 2.5. Figure 3 of this code gives internal wind pressure coefficients in buildings with large openings, exceeding 20% permeability.

Table 2.5 Internal pressure coefficients C_{pi}

Type of building	C_{pi}
Buildings with low permeability (less than 5% openings in wall area)	± 0.2
Buildings with medium permeability (5 to 20% openings in wall area)	± 0.5
Buildings with large permeability (openings in wall area > 20%)	± 0.7
Buildings with one side opening	See Fig. 3 of code

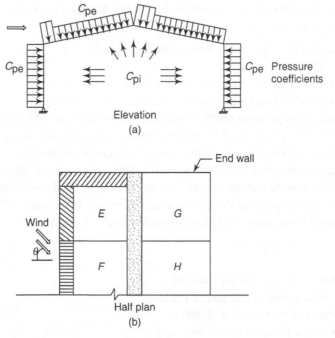

Fig. 2.17 Typical elevation with wind pressure coefficients C_{pe} and C_{pi}

Note that in addition to C_{pe}, local pressure coefficients are also given in Tables 4 and 5 of the code. These local pressure coefficients should not be used for calculating force on structural elements, such as roofs, walls, or the structure as a whole. They should be used only for the calculation of forces on these local areas affecting roof sheeting, glass panels, and individual cladding units including their fixtures.

The code [IS 875 (Part 3)] gives external coefficients for mono-slope and hipped roofs, canopy roofs (e.g. open air-parking garages, railway platforms, stadiums, theatres, etc.), curved roofs, pitched and saw-tooth roofs of multi-span buildings, overhangs from roofs, cylindrical structures, roofs and bottom of cylindrical structures, combined roofs, roofs with skylight, grand stands, and spheres.

The total wind load for a building as a whole is given by the code, as

$$F = C_f A_e p_d \tag{2.4}$$

where

F = force acting in the specified direction,

C_f = force coefficient of the structure,

A_e = effective frontal area, and

p_d = design wind pressure.

The code gives force coefficients for rectangular clad buildings in uniform flow and for other clad buildings of uniform sections. Force coefficients for free standing walls and hoardings, solid circular shapes mounted on a surface, unclad buildings and frame works (individual members of infinite length, flat sided members, circular sections, wires and cables, single frames, multiple frame buildings and lattice towers) are also given.

Force coefficients for latticed towers of square or equivalent triangle section with flat sided members for wind blowing at any face is as given in Table 30 of the code. The solidity ratio, φ, is equal to the effective area (projected area of all the individual members) of a tower normal to the wind direction divided by the area enclosed by the boundary of the tower (or section of tower) normal to the wind direction. Such towers often taper from the base towards the top. The frontal area exposed to the wind, and hence the solidity ratio changes from bottom to top of these towers. Thus, it may be necessary to divide the tower into several sections along the height and compute forces on each part separately.

In certain buildings, a force due to frictional drag is taken into account in addition to those loads specified for rectangular clad buildings. This addition is necessary only where the ratio d/h or d/b is greater than 4. The frictional drag force, F', in the direction of the wind is given by the following formulae.

If $h \leq b$,

$$F' = C_f'(d - 4h)\, bp_d + C_f'\, (d - 4h)\, 2\, hp_d \qquad (2.5\text{a})$$

If $h > b$,

$$F' = C_f'(d - 4b)\, bp_d + C_f'\, (d - 4b)\, 2\, hp_d \qquad (2.5\text{b})$$

The first term in each case gives the drag on the roof and the second that on the walls. The value of C_f' has the following values:

(a) $C_f' = 0.01$ for smooth surfaces without corrugations or ribs across the wind direction,
(b) $C_f' = 0.02$ for surfaces with corrugations across the wind direction, and
(c) $C_f' = 0.04$ for surfaces with ribs across the wind direction.

For other buildings, the frictional drag has been indicated, where necessary, in the tables of pressure coefficients and force coefficients.

For the following cases of buildings (flexible slender buildings), the dynamic effects of winds (excitations along and across the direction of wind) should be studied.

(a) Buildings and closed structures with a height to minimum lateral dimension ratio of more than 5.0 ($h/b > 5.0$)
(b) Buildings and close structures whose fundamental natural frequency (first mode) is less than 1.0 Hz

For these buildings, the calculated wind pressure at height z should be multiplied by the gust factor G.

Note that IS 875 (Part 3) is under revision, and the draft code contains several major changes, based on recent research, especially for calculating the dynamic effects of wind. The draft code stipulates that the wind pressure for flexible buildings should be multiplied by the dynamic response factor, C_{dyn}, instead of the gust factor, G (see http://www.iitk.ac.in/nicee/IITK-GSDMA/W02.pdf).

2.6.6 Earthquake Loads

The crest of the Earth is composed of about 13 large plates and several small ones ranging in thickness from 32 to 240 km. The plates are in constant motion. When they collide at their boundaries, earthquakes occur. Some consider that earthquakes may also be caused by actions, such as underground explosions due to the testing of nuclear bombs, construction of dams, etc. Though most of the earthquakes have occurred in well-defined 'earthquake-belts', a few earthquakes have hit seismically inactive parts of the world. Hence it is important to incorporate some measure of earthquake resistance into the design of all structures, since failures of structures due to earthquakes are catastrophic. Note that tall buildings may be at greater risk than single-storey buildings.

Earthquakes cause the ground to shake violently in all directions, lasting for a few seconds in a moderate earthquake or for a few minutes in very large earthquakes. Earthquakes are recorded using *accelerographs* or *seismographs*. The intensity of an earthquake reduces gradually as we move away from its epicenter. (Epicenter is the location on the surface of the Earth that is above the focal point of an earthquake.) The magnitude and intensity of an earthquake are of interest to the structural engineer. The magnitude is a measure of the amount of energy released by the earthquake, while intensity is the apparent effect of the earthquake. Unlike the intensity, which can vary with the location, the magnitude is constant for a particular earthquake. The magnitude is measured by the Richter scale which is a logarithmic scale. Thus, an earthquake of magnitude 6 is 31.6 times more powerful than the one measuring 5. Earthquakes of Richter magnitude 6, 7, and 8 are categorized as moderate, major, and great earthquakes, respectively.

2008 Sichuan Earthquake

The Sichuan earthquake was a 7.9 magnitude earthquake that hit the Sichuan Province in Western China on May 12, 2008, and had a duration of about 2 minutes. Its epicenter was in Wenchuan County, 80 km west/ northwest of the provincial capital city of Chengdu. It killed more than 87,400, injured 374,176, and left 4.8 million people homeless (estimates range from 4.8 million to 11 million). Approximately 15 million people lived in the affected area.
It was the deadliest earthquake to hit

Damage to steel buildings during the 2008 Sichuan earthquake
(Photo: Archey Firefly)

China since the 1976 Tangshan earthquake, which killed at least 240,000 people, and the strongest since the 1950 Chayu earthquake in the country, which was registered at 8.5 on the Richter magnitude scale. Strong aftershocks, some exceeding the magnitude of 6, continued to hit the area, months after the main quake, causing more casualties and damage.

Using the Modified Mercalli intensity scale, this earthquake was classified as XI– very disastrous. Over 7,000 inadequately engineered schoolrooms and numerous buildings collapsed in the earthquake (see photo).

The intensity at a place is evaluated considering the three features of shaking– perception by people, performance of buildings, and changes to natural surroundings. Two commonly used intensity scales are the Modified Mercalli Intensity (MMI) scale and the MSK scale. Both the MMI and MSK scales are quite similar and range from I (least perceptive) to XII (most severe).

In addition to the peak ground acceleration of an earthquake, the following factors also influence the seismic damage: (a) amplitude, (b) duration and frequency of ground vibration, (c) magnitude, (d) distance from epicenter, (e) geographical conditions between the epicenter and the site, (f) soil properties at the site and foundation type, and (g) the building type and characteristics.

Soil liquefaction is another effect caused by earthquakes, which produces quick-sand type condition, resulting in a loss of the bearing capacity of soil. Soil liquefaction may result in settlement and total collapse of structures.

Earthquake loads are dynamic and produce different degrees of response in different structures. When the ground, under a structure having a mass, suddenly moves, the inertia of the mass tends to resist the movement, as shown in Fig. 2.18, and creates forces, called *inertia forces*, which are equal to the product of mass of the structure times acceleration ($F = ma$). The mass is equal to the weight (W) divided by the acceleration due to gravity, i.e. $m = W/g$.

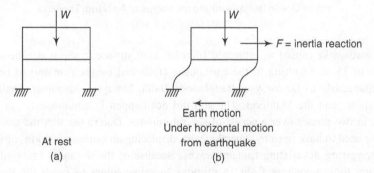

Fig. 2.18 Force developed by earthquake

In the IS 1893 (Part 1) code, the following seismic design philosophy is adopted.
• Minor and frequent earthquakes should not cause any damage to the structure.
• Moderate earthquakes should not cause significant structural damage but could have some non-structural damage (the structure should become operational once the damaged main members are repaired and strengthened).

• Major and infrequent earthquakes should not cause collapse (the structure will become dysfunctional for further use, but will stand so that people can be evacuated and property recovered).

Hence the structures are designed for much smaller forces than actual seismic loads during strong ground shaking. Note that this approach is different than that adopted in the case of wind, dead, live, and other loads, where the structure is designed for the anticipated loads.

The Indian Ocean Tsunami

The Indian Ocean Tsunami was due to an under-sea earthquake (Sumatra–Andaman earthquake) that occurred on December 26, 2004, with an epicenter off the west coast of Sumatra, Indonesia. The quake, with a Richter magnitude of 9.1 to 9.3, is the second largest earthquake ever recorded on a seismograph. This earthquake had the longest duration of faulting ever observed, between 8.3 to 10 minutes.

Indian Ocean tsunami hitting the shores of Ao Nang, Thailand
(Photo: David Rydevik)

This earthquake caused an estimated 1600 km fault surface to slip under the ocean to about 15 m, resulting in the earthquake (followed by the tsunami) to be felt simultaneously as far away as Bangladesh, India, Malaysia, Myanmar, Thailand, Singapore, and the Maldives. The slip did not happen instantaneously but took place in two phases over a period of several minutes. Due to the slip, the sea floor is estimated to have risen by several meters, displacing an estimated 30 km^3 of water and triggering devastating tsunami waves. Because of the distances involved, the tsunami took anywhere from 15 minutes to seven hours to reach the various coastlines. In many places, the waves reached as far as 2 km inland. The wave reached a height of 24 m when coming ashore along large stretches of the coastline, rising to 30 m in some areas when traveling inland.

An analysis by the United Nations found that a total of 229,866 people were lost, including 186,983 dead and 42,883 missing, due to the tsunami in towns and villages along the coast of the Indian Ocean. The livelihoods of over 3 million survivors were destroyed. Beyond the heavy toll on human lives, it had caused an enormous environmental impact that will affect the region for many years to come.

Guidelines for the design of structures against tsunami are scarce (It is generally not feasible or practical to design normal structures to withstand tsunami loads), but warning and evacuation systems have been developed (see FEMA 55 and FEMA P646 at www.fema.gov).

For the purpose of determining seismic forces, the country is classified into four seismic zones (zones II to V) by IS 1893 (Part 1) code (see Fig. 1 of the code). The code requires that the designer either (a) use a dynamic analysis of the structure, (b) for usual generally rectangular medium-height buildings (regular buildings) use an empirical lateral base shear force (see Table 2.6). The dynamics of earthquake action on structures is outside the scope of this book, and the reader may refer to Chopra (2000), Clough and Penizien (1993) and Mazzolani and Piluso (1996) for the details of dynamic analysis methods.

Table 2.6 Requirement of dynamic analysis as per IS:1893 (Part 1)

Seismic zone	Regular buildings	Irregular buildings
II and III	Height > 90 m	Height > 40 m
IV and V	Height > 40 m	Height > 12 m

Note: 1. Large-span industrial buildings may also require dynamic analysis.

2. Buildings with high level of torsion irregularity are prohibited in zones IV and V.

For regular buildings, the IS 1893 (Part 1) suggests that the design horizontal seismic coefficient A_h for a structure is determined by the following expression:

$$A_h = ZI (S_a/g)/2R \tag{2.6}$$

where Z = zone factor, given in Table 2 of IS 1893 (Part 1), for the maximum considered earthquake (MCE). (The factor of 2 in the denominator of Eqn (2.6) is used to reduce the MCE to the design basis earthquake.)

I = importance factor, depending on the functional use of the structure. It is given in Table 2 of IS 1893 (Part 1).

R = response reduction factor, depending on the perceived seismic damage performance of the structure, characterized by ductile or brittle deformations (The values of R for steel buildings are given in Table 23 of IS 800:2007.)

S_a/g = response acceleration coefficients as given by Fig. 2 of IS 1893 (Part 1) or Eqs (2.7) to (2.9), based on appropriate natural periods and the damping of the structure.

A plot of the maximum response (for example, acceleration, velocity, or displacement) against the period of vibration or the natural frequency of vibration is called a *response spectrum*. Using several earthquake spectra, a smooth spectrum representing an upper bound response to ground motion is normally used in the codes. Figure 2 of the code shows such a spectrum adopted by the code. The values given in Fig. 2 of the code can be represented mathematically by the following equations.

(a) For rocky or hard soil sites,

$$\frac{S_a}{g} \begin{cases} = 2.50, \quad 0.0 \le T \le 0.10 \text{ (for fundamental mode)} \\ = 1 + 15T, \quad 0.0 \le T \le 0.10 \text{ (for higher modes)} \\ = 2.50, \quad 0.10 \le T \le 0.40 \\ = \dfrac{1.0}{T}, \quad 0.40 \le T \le 4.00 \\ = 0.25, \quad T > 4.00 \end{cases} \tag{2.7}$$

(b) For stiff soil sites,

$$\frac{S_a}{g} \begin{cases} = 2.50, \quad 0.0 \le T \le 0.10 \text{ (for fundamental mode)} \\ = 1 + 15T, \quad 0.0 \le T \le 0.10 \text{ (for higher modes)} \\ = 2.50, \quad 0.0 \le T \le 0.55 \\ = \dfrac{1.36}{T}, \quad 0.55 \le T \le 4.0 \\ = 0.34, \quad T > 4.0 \end{cases} \tag{2.8}$$

(c) For soft soil sites,

$$\frac{S_a}{g} \begin{cases} = 2.50, \quad 0.0 \le T \le 0.10 \text{ (for fundamental modes)} \\ = 1 + 15T, \quad 0.0 \le T \le 0.10 \text{ (for higher modes)} \\ = 2.5, \quad 0.0 \le T \le 0.67 \\ = \dfrac{1.67}{T} \quad 0.67 \le T \le 4.00 \\ = 0.42, \quad T > 4.00 \end{cases} \tag{2.9}$$

The multiplying factors for obtaining S_a/g values for other damping (these should not be applied to the point at zero period) are given in Table 3 of IS 1893 (Part 1).

Natural frequencies

A structure with N degrees of freedom has N natural frequencies and N mode shapes. The lowest of the natural frequencies of the structure is called its *fundamental natural frequency*, expressed in hertz. The associated natural period is called the *fundamental natural period*, which is the reciprocal of natural frequency and is expressed in seconds. Where a number of modes are to be considered for dynamic analysis, the value of A_h [see Eqn (2.6)] for each mode should be determined using the natural period of vibration of that mode. For underground structures and foundations at depths of 30 m or below, the design horizontal acceleration spectrum value should be taken as half the value obtained from Eqn 2.6. For structures and foundations in between 30 m below ground level and the ground level, the value should be linearly interpolated between A_h and $0.5A_h$. For vertical motions, the value should be taken as two-thirds of the design horizontal acceleration spectrum (IS 1893:2002).

The approximate fundamental natural period of vibration, T_a, in seconds, for a moment-resisting frame without brick infill panels is given by the code (FEMA 450 2001) as

$$T_a = 0.085 \ h^{0.75} \tag{2.10a}$$

where h is the height of the building in metres.

For all other buildings, including moment-resisting frame buildings with brick infills,

$$T_a = \frac{0.09h}{\sqrt{d}} \tag{2.10b}$$

where d is the base dimension of the building at the plinth level, along the considered direction of the lateral force, in metres.

It is suggested by the code to adopt the approximate natural period [given by Eqn (2.10)] only in the calculations, even though one may obtain an exact value, especially for irregular structures (which may be more than this value) by using a dynamic analysis computer program (Jain 1995). It is to safeguard the application of lower design seismic forces calculated using the large natural period obtained by the programs.

In the *equivalent static method* (also referred to as *seismic coefficient method*), which accounts for dynamics of the building in an approximate manner, the total design seismic base shear is determined by

$$V_B = A_h \ W \tag{2.11}$$

where A_h is the design horizontal acceleration spectrum value as per Eqn (2.6) using the approximate fundamental natural period T_a, as given in Eqn (2.10) in the considered direction of vibration, and W is the seismic weight of the building.

Buildings provide a certain amount of damping due to internal friction, slipping, etc. It is usually expressed as a percentage of critical damping. A damping of 2% is considered for steel structures.

The seismic weight of each floor is calculated as its full dead load plus an appropriate amount of imposed load (Table 2.7). While computing the seismic weight of each floor, the weight of columns and walls in any storey should be appropriately apportioned to the floors above and below the storey. It has to be noted that buildings designed for storage purposes are likely to have large percentages of the service load present at the time of earthquake. Other appropriate loads such as snow or permanent equipment should also be considered.

Table 2.7 Percentage of imposed load to be considered while calculating seismic weight

Imposed uniformly distributed floor load (kN/m²)	Percentage of imposed load
Up to and including 3.0	25
Above 3.0	50

Note: The imposed load on the roof need not be considered. No further reduction for large areas or for the number of storeys above the one under consideration [as envisaged in IS 875 (Part 2)] for static load cases is allowed.

After the base shear force V_B is determined, it should be distributed along the height of the building (to the various floor levels) using the following expression.

$$Q_i = V_B \left(\frac{W_i h_i^k}{\sum\limits_{j=1}^{n} W_j h_j^k} \right)$$

(2.12)

where Q_i is the design lateral force at floor i, W_i is the seismic weight of floor i, h_i is the height of floor i measured from the base, and n is the number of storeys in the building. The value of k equal to 2 is adopted in the Indian Code. The use of equivalent static method is explained in Example 2.5.

After obtaining the seismic forces acting at different levels, the forces and moments in different members can be obtained by using any standard computer program for the various load combinations specified in the code. The structure must also be designed to resist overturning effects caused by seismic forces. Also, storey drifts, member forces, and moments due to P-delta effects must be determined. Note that all cantilever vertical projections are to be designed for five times the design horizontal seismic coefficient A_h and horizontal projections should be checked for stability for five times the design vertical component (i.e. $10/3 \, A_h$).

In tall buildings, the contribution of higher modes may be important. In irregular buildings, the mode shape may not be regular, and in industrial buildings with large spans and heights, the assumptions of the static procedure (the fundamental mode of vibration is the most dominant and mass and stiffness are evenly distributed) may not be valid. Hence, for these buildings, dynamic analysis methods are suggested by the code. Such methods are grouped into the *response spectrum method* (multistory buildings, irregular buildings, overhead water ranks, and bridge piers are often designed using this method) and time-history response analysis (most important structures such as nuclear reactors, large span structures, or very tall buildings are designed using this method). The understanding of these methods requires some knowledge of structural dynamics and it is beyond the scope of this book.

Rules to be followed for buildings in seismic areas

For better seismic response, proper precautions have to be taken at the planning stage itself. It is preferable to select a site where bedrock is available close to the surface, so that foundations can be laid directly on the rock. The differential movement of foundation due to seismic motions is an important cause of structural damage, especially in heavy, rigid structures that cannot accommodate these movements. Hence if the foundation is on soft soil with spread footings, adequate plinth or tie beams should be provided to counter differential settlements. If the loads are heavy, pile foundations with strong pile caps may be provided. Raft foundation is also good to resist differential settlements, but may prove to be expensive. In sandy or silty soils, if the water table is near the foundation level, then appropriate methods must be adopted to prevent liquefaction.

To perform well in an earthquake, a building should posses the following four main attributes: (i) simple and regular configuration, (ii) adequate lateral strength, (iii) adequate stiffness, and (iv) adequate ductility. The openings in walls should be located centrally and should be of small size so that the wall is not unduly weakened. (Ventilators provided near the edges of walls, adjacent to columns, will create a short column effect and result in the failure of the column. There will be a similar effect if openings are provided from column to column.) Long cantilevers and floating columns should be avoided. Appendages like sunshades (*chajjas*) and water tanks should be designed for higher safety levels.

Concrete stairways often suffer seismic damage due to their inhibition of drift between connected floors. This can be avoided by providing a slip joint at the lower end of each stairway to eliminate the bracing effect of the stairway or by tying stairways to stairway shear walls.

Masonry and infill (non-structural) walls should be reinforced by vertical and horizontal reinforcing bands to avoid their failure under a severe earthquake. Other non-structural elements should be carefully detailed or tied so that they may not fall under severe shaking.

It has to be noted that the failure of a beam causes localized effect, whereas that of a column may affect the stability of the whole building. Hence, it makes good sense to make columns stronger than beams. This can be achieved by appropriate sizing of the member and detailing. This concept is called *strong-column-weak-beam design*.

When buildings are too close, they may pound on each other. Connections and bridges between buildings should be avoided and buildings with different sizes and shapes should have adequate gap between them to avoid pounding. When building heights do not match, the roof of the shorter building may pound the mid-height of the columns of the taller one, resulting in dangerous consequences. The buildings or two adjacent units of the same building should be separated by a distance equal to R times the sum of the calculated storey displacements to avoid pounding [the value of R is given in Table 7 of IS 1893 (Part 1)]. This value may be multiplied by a factor of 0.5 if the two units have same floor elevation.

Buildings with a simple regular geometry and uniformly distributed mass and stiffness in plan and elevation (regular structures) have been found to suffer less damage in earthquakes than those with irregular structures. Hence, columns and walls should be arranged in grid fashion and should not be staggered in plan. The effect of asymmetry will induce torsional oscillations in structures and stress concentrations at re-entrant corners. These irregularities may be grouped as *plan irregularities* and *vertical irregularities*.

Guidelines for special loads due to temperature effects, differential settlement, soil and hydrostatic pressure, erection loads, accidental loads, etc., are provided by IS 875 (Part 5). It is important for the engineer to accurately calculate the different loads, as per the codal provisions, acting on a structure, as an overestimation of loads will result in uneconomical structures and an underestimation will result in sudden failures.

2.6.7 Load Combinations

The IS 800:2007 gives values for partial safety factors for various load combinations as given below (also see Table 4 of the code).
(a) 1.5 (DL + IL) + 1.05(CL or SL),
(b) 1.2 (DL + IL) + 1.05(CL or SL) ± 0.6(WL or EL),
(c) 1.2 (DL + IL ± WL or EL) + 0.53 (CL or SL),
(d) 1.5(DL ± WL or EL),
(e) 0.9 DL ± 1.5 (WL or EL),
(f) 1.2 (DL + ER),
(g) 0.9DL + 1.2 ER, and
(h) DL + 0.35(IL + CL or SL) + AL.
where DL = dead load, IL = imposed load (live load), WL = wind load, SL = snow load, CL = crane load (vertical/horizontal), AL = accidental load, ER = erection load, and EL = earthquake load.

If we do not consider crane load, snow load, and accidental load, we should consider 13 loading cases for a building in which horizontal (lateral) load is resisted by frames or walls oriented in two orthogonal directions, say X and Y. In structures with a non-orthogonal lateral load resisting system, the lateral load resisting elements may be oriented in a number of directions. In such buildings, eight additional load combinations must be considered as per IS 1893 (Part 1).

2.7 Structural Analysis

In the structural design process, the term *analysis* refers to the determination of the axial forces, bending moments, shears, torsional moments, etc., acting on different members of a structure, due to the applied loads and their combinations (static or dynamic). For the design engineer, *design* involves the selection of sizes of members to resist the forces and moments determined in the analysis phase, safely and economically. In the design phase, we will normally design not only the members but also their connections and the foundations, so that the loads are transmitted to the soil.

For statically determinate structures (simply supported beams, cantilevers, trusses, etc.), the analysis is relatively simple and the laws of statics can be used to determine the forces and moments on each member. The relative stiffness of intersecting members does not affect analysis. After the analysis is completed and the critical moments and forces in the different members are tabulated, the design of the member is a straightforward process using an appropriate method (e.g., the working stress method, the limit states method, etc.). For statically determinate structures, there is no need for re-analysis or redesign of the members.

However the analysis of indeterminate structures, (such as portal frames, multistory bends, etc.) requires the cross-sectional properties of various members, which are not known initially. Hence, a preliminary analysis is done using assumed member sizes and then, the members and the details of structural joints are designed as per the codal rules. Reanalysis is often required, if the assumed member sizes do not satisfy the strength, stability, deflection, or other criteria as specified in the codes of practice. Hence, in these types of structures, analysis and design are interactive.

An overview of the analysis methods to calculate the frame response is shown in Fig. 2.19 using the load-displacement curves of a statically loaded frame. It can be seen from this figure that it is very difficult to model all sources of non-linearity and portray the actual behaviour of a practical structure. Though the degree to which the methods can model true behaviour differs, each method can yield valuable information to the design engineer.

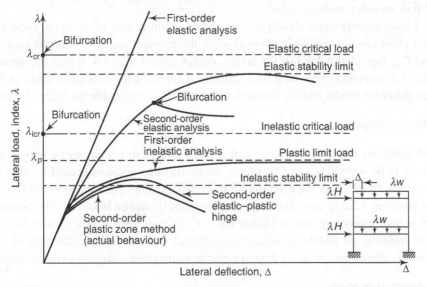

Fig. 2.19 Load–displacement characteristics of different analysis methods

The code suggests the use of any one of the following methods of analysis (see section 4.4 of IS 800:2007).

First-order elastic analysis

This is the most basic method of analysis, in which the material is modelled as linear-elastic and satisfies the requirements of compatibility and internal force equilibrium on the geometry of the un-deformed structure. This assumption is valid when the elastic displacements are small compared to the dimensions of the structure. Thus by definition, this method excludes non-linearity, but it generally represents conditions at service loads very well. Because of the assumption of linear elastic behaviour, the principle of superposition may be used to combine load cases. A first-order elastic analysis is sufficient for normal framed structures, which are braced against sway. Though the first-order-elastic analysis provides an 'exact solution', the solution does not provide any information about the influence of plasticity and stability on the behaviour of the structure. Hence, these influences are normally provided indirectly in member capacity checks through the use of column effective-length and moment-amplification factors.

Second-order elastic analysis

This analysis considers linear-elastic material behaviour and satisfies the requirements of compatibility and internal force equilibrium on the geometry of the deformed structure. Thus, P-Δ and P-δ effects are considered in the analysis. Horizontal frame deflections and gravity loads acting on the displaced structure, result in additional moments in the frame, and is called the P-Δ or *frame stability effect*. Axial compression in the columns, reducing their flexural stiffness is called the P-δ or *member stability effect*.

Linear superposition should not be used to combine load cases. In the code, the first-order elastic analysis is allowed when the moment amplification factors, C_{my} and C_{mz} (see clause 4.4.3 of the code), are not greater than 1.4. If they are greater than 1.4, a second-order analysis is required. Generally, the difference between the second-order results and the first-order results will be negligible for braced frames.

Plastic analysis

For applying this analysis, the members in a frame, in which plastic hinges are likely to be formed, should be plastic sections; other sections should be at least compact. In addition, the ratio of ultimate tensile strength to yield strength should be less than 1.2 and the percentage of elongation should be less than 15%. Several other conditions as given in clause 4.5.2 of the code should also be satisfied. In this analysis, all instability effects are ignored and the collapse strength of the frame is determined by using the rigid-plastic assumption and collapse mechanisms.

Advanced analysis

This analysis takes into account the relevant material properties, residual stresses, geometric imperfections, second-order effects, three-dimensional effects, erection procedures, and interaction with the foundations. Thus, advance analysis methods incorporate both strength and stability behaviour in such a way that separate member designs are not required. The advantage of advanced analysis is that it predicts the exact behaviour of structures and eliminates tedious and often confusing member capacity checks including the calculation of effective length factors. Moreover, the advanced methods of analysis integrate the analysis and design process into a single interactive activity. Advanced analysis methods are described in Chen and Toma (1994) and Chen and Kim (1997).

Dynamic analysis

Dynamic analysis may be performed either by the time-history method or by the response-spectrum method. Refer Table 2.6 for the requirements of dynamic analysis. More details on dynamic analysis procedures may be obtained from Clough and Penizien (1993) and Chopra (2000).

Note that the results of any analysis are dependent on the boundary conditions. Often problem arises while choosing the rigidity of foundations (clause 4.3.4 of code offers some suggestions), and the rigidity of connections (Annex F of code provides some guidelines for rigid and semi-rigid connections).

2.7.1 Software

Various computer programs are available for the analysis and design of different types of structures. They include SAP 2000 and ETABS (www.csiberkeley.com), STAAD III and STAAD PRO (www.reiworld.com), and STRUDS (www.softtech-engr.com). The above list is not exhaustive. These programs are quite general in terms of loading, geometric configurations, and support conditions.

With these programs, it is now possible to analyse any structure with any complicated geometry subjected to any pattern of loading (static or dynamic), and having any boundary condition or discontinuity.

Analytical methods and modelling techniques used by these computer programs offer various levels of sophistication and refinement. However, while using these programs, the designer should be aware of any assumptions used and limitations of these programs. It is because any amount of mathematical precision cannot make up for the use of an analytical method that is not applicable to the structure being designed.

It may be of interest to note that many of the steel buildings in India are analysed using first-order elastic analysis programs only (e.g., STAAD III), though non-linear analysis methods have been employed in a few important high-rise structures.

2.8 Codes and Specifications

A structural engineer is often guided in his efforts by the *code of practice*. Codes are basically written for the purpose of protecting the public. They are revised at regular intervals to reflect new developments. Codes contain recommended loads for a given locality, and recommended fire and corrosion protection. They also contain rules governing the ways in which loads are to be applied, and design rules for steel, concrete, and other material. These rules may be in the form of detailed recommendations or by reference to other standards that provide specific design rules. Often the designer must exercise judgement in interpreting and applying the requirements of a code.

The codes serve at least four distinct functions.

(a) They ensure adequate structural safety, by specifying certain essential minimum requirements for the design.
(b) They aid the designer in the design process. Often the results of sophisticated analysis are made available in the form of simple formulae or charts.
(c) They ensure consistency among different engineers.
(d) They protect the structural engineer from disputes, though codes in many cases do not provide legal protection.

Project specifications, along with design drawings, are given to the builder by the architect/project manager. These specifications and the way in which the drawings are prepared and presented vary from organization to organization. They include the following items.

(a) Materials that must be used in the structure,

(b) Sizes of structural members,

(c) Joint details,

(d) Expected quality, tolerance, and

(e) Instructions on how the construction work is to be done.

Whoever writes the specification, the structural engineer should be involved in preparing or approving its technical contents.

In India, the Bureau of Indian Standards issues the codes and standard handbooks. It is strongly advised that the reader should possess a copy of the following codes, which will be referred frequently in the book.

(a) IS 800 : 2007—Code of practice for general construction in steel

(b) IS 875 : 1987—Code of practice for design loads for buildings and structures (Part 1—Dead loads, Part 2—Imposed loads, Part 3—Wind loads, Part 4—Snow loads and Part 5—Special loads and load combinations)

(c) IS 1893 (Part 1) : 2002—Criteria for earthquake-resistant design of structures

(d) IS 808 : 1989—Dimensions for hot-rolled steel beams, columns, channels and angle sections.

In addition, the designer may need to refer to a number of other codes covering topics such as steel properties, welding of structural steel works, properties of fasteners (bolts), and also codes for the design of bridges, towers, silos, and off-shore structures.

The IS 800 : 2007 code also lists other codes published by the Bureau of Indian Standards that may be useful to the steel designer. It is to be noted that the codes IS 801, IS 806, and IS 4000 are still in the working stress method format and will be revised shortly in the limit states design format.

2.9 Design Philosophies

Over the years, various design philosophies have evolved in different parts of the world, with regard to structural steel design.

The earliest codified design philosophy is the *working stress method of design* (WSM). In IS 800 : 2007, the provisions relating to the WSM design procedure are given in section 11 (with only a few pages devoted to it) so as to give greater emphasis to limit states design.

The WSM was followed by the *ultimate strength design (USD) or plastic design,* which was developed in the 1950s. It was based on the ultimate strength of steel at ultimate loads. This method was introduced as an alternative to WSM in the IS 800 code in 1962.

For codification, the probabilistic 'reliability method' approach was simplified and reduced to a deterministic format involving multiple (partial) *safety factors* (rather than probability of failure). This *limit states design* was first adopted for steel structures in the Canadian Code in 1974, and only in 2007 by the Indian Code.

2.9.1 Working Stress Method (WSM)

This was the traditional method of design not only for structural steel, but also for reinforced concrete and timber design. The conceptual basis of WSM is simple. The method basically assumes that the structural material behaves in a linear elastic manner, and that adequate safety can be ensured by suitably restricting the stresses in the material due to the expected *working loads* (service loads) on the structure.

The stresses under the working loads are obtained by applying the methods of 'strength of materials', such as the simple bending theory. The first attainment of yield stress of steel is taken to be the onset of failure. The limitations due to non-linearity (geometric as well as material) and buckling are neglected.

The stresses caused by the 'characteristic' loads are checked against the *permissible (allowable) stress*, which is a fraction of yield stress. Thus the permissible stress may be defined in terms of a *factor of safety*, which takes care of the overload or other unknown factors. Thus,

$$\text{Permissible (allowable) stress} = \frac{\text{yield stress}}{\text{factor of safety}} \qquad (2.13a)$$

Thus, in the working stress method

$$\text{Working Stress} \leq \text{permissible stress} \qquad (2.13b)$$

Each member of the structure is checked for a number of different combinations of loadings. Usually, a factor of safety of 1.67 is adopted for tension members and beams. A value of 1.92 is used for long columns and 1.67 for short columns. A value of 2.5–3 is used for connections. (However using the WSM, the 'real' safety against 'failure' is unknown.) Since dead load, live load, and wind load are all unlikely to act on the structure simultaneously the stresses are checked as follows.

Stress due to dead load + live load < permissible stress

Stress due dead load + wind load < permissible stress

Stress due to dead load + live load + wind load < 1.33 (permissible stress)

There are many limitations and shortcomings of the working stress method. First, the main assumption of linear elastic behaviour and the implied assumption that the stresses under working loads can be kept within the 'permissible stresses' are not found to be realistic. Many factors are responsible for this, such as the effects of stress concentrations, the long-term effects of creep and shrinkage, residual stresses, and other secondary effects. Moreover, in actual structures, after the first yield at the extreme fibres in a section, the other fibres yield to form plastic hinges. Intermediate structures will not fail after a plastic hinge is formed at a location. They will fail only after a sufficient number of plastic hings are formed to create a collapse mechanism.

All such effects result in significant local increases in and redistribution of the calculated stresses. The WSM does not consider the consequences of material non-linearity and the non-linear behaviour of members in the post-buckled state. Moreover, steel components have the ability to tolerate high elastic stress by yielding locally and redistributing the loads. WSM does not provide a realistic measure of the actual factor of safety underlying a design. It also fails to discriminate between different types of load that act simultaneously, but have different degrees of uncertainty. This can, at times, result in very unconservative designs, particularly when two different loads (say, dead loads and wind loads) have counteracting effects.

In spite of these shortcomings, it may be stated that most structures designed in accordance with WSM have generally performed satisfactorily for many years. The size of tension members is about the same in both LRFD and WSM when the live load to dead load ratio (LDR) is about 3. When the dead load becomes more predominant, there will be economy in using LRFD. With LDR greater than 3, WSM will be slightly (about 3%) more economical (Salmon and Johnson 1996). However, for other members, WSM results in relatively larger member sizes and hence in less deflections. There are instances where WSM results in considerable overdesign and where it is not safe (Allen 1972 and Gordon 1978). The WSM method is notable for its essential simplicity, in concept as well as application.

2.9.2 Limit States Method

In limit states design, we prefer to use the term 'limit states', rather than 'failure'. Thus, a *limit state* is a state of impeding failure, beyond which a structure ceases to perform its intended function satisfactorily.

The limit states usually considered relevant for structural steel work are listed in Table 2.3. They are normally grouped into the following two types.
1. Ultimate (safety) limit states (ULS) which include the following:
    (a) Loss of strength (including yielding, buckling, and transformation into a mechanism),
    (b) Loss of stability against overturning and sway,
    (c) Failure by excessive deformation or rupture,
    (d) Fracture due to fatigue, and
    (e) Brittle fracture.
2. Serviceability limit states (SLS) which deal with discomfort to occupancy and/or malfunction, caused by the following:
    (a) Excessive deformation and deflection,
    (b) Vibration (e.g. wind-induced oscillations, floor vibration, etc.),
    (c) Repairable damage due to fatigue (cracking),
    (d) Corrosion (and subsequent loss of durability), and
    (e) Fire.

The attainment of one or more ultimate limit states (ULS) may be regarded as the inability to sustain any increase in load, whereas the serviceability limits states (SLS) denote a need for remedial action or some loss of utility. Hence ULS are conditions to be avoided and SLS are conditions that are undesirable.

Design for the ultimate limit state may be conveniently explained with reference to the type of diagram shown in Fig. 2.20. This figure shows the hypothetical frequency distribution curves for the effect of loads (Q) on the structural element and the resistance (strength) of the structural element (R). When the two curves overlap, shown by the shaded area, the effect of the loads is greater than the resistance of the element and the element fails. Thus, the structure and its elements should be proportioned in such a way that the overlap of the two curves is small, which means that the probability of failure is within the acceptable range.

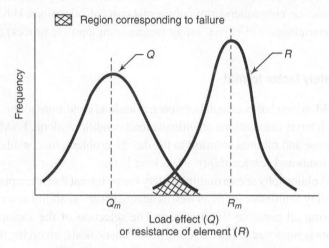

Fig. 2.20 Frequency distribution curves

The values of the reliability index β corresponding to various failure probabilities P_f can be obtained from the standardized normal distribution function of the cumulative densities, and are given in Table 2.8.

Table 2.8 Reliabilty index for various failure probabilities

β	2.32	3.09	3.72	4.27	4.75	5.2	5.61
$P_f = \phi(-\beta)$	10^{-2}	10^{-3}	10^{-4}	10^{-5}	10^{-6}	10^{-7}	10^{-8}

It has to be noted that the values given in Table 2.8 are valid only if the safety margin is normally distributed.

For code use, the method must be as simple as possible, using deterministic rather than probabilistic data. Such a method, called *Level I reliability method* or first-order second moment reliability method, is used in our code to obtain a probability-based assessment of structural safety.

Characteristic load and characteristic strength

The characteristic strength of steel, R_c, is defined as that value of strength below which more than a prescribed percentage of test results will fall. This prescribed percentage is normally taken as 95. Thus the characteristic yield strength of steel is the value of yield strength below which not more than 5% of the test values may fall.

Similarly, the characteristic load, Q_c, is defined as that load which is not expected to be exceeded with more than 5% probability, during the life span of a structure. Thus, the characteristic load will not be exceeded 95% of the time.

The design values are derived from the characteristic values through the use of partial safety factors, both for material and for loads. The acceptable failure probability, P_f, for particular classes of structures is generally derived bearing in mind past practice, consequences of failure, and cost considerations. Having chosen P_f and β, determination of partial safety factors is an iterative process (Galambos 1981).

Multiple safety factor format

Unlike WSM, which bases calculations on service load conditions alone, and unlike ULM, which bases calculations on ultimate load conditions alone, LSM aims for a comprehensive and rational solution to the design problem, by considering *safety* at ultimate loads and *serviceability* at working loads.

The LSM philosophy uses a multiple safety factor format that attempts to provide adequate safety at ultimate loads as well as adequate serviceability at service loads, by considering all possible 'limit states'. The selection of the various multiple safety factors is supposed to have a sound probabilistic basis, involving the separate consideration of different kinds of failure, types of materials, and types of loads.

With every code revision, conscious attempts are made to specify more rational reliability-based safety factors, in order to achieve practical designs that are satisfactory and consistent in terms of the degree of safety, reliability, and economy. The reliability index chosen for different loads in most of the cases is shown in Table 2.9.

Table 2.9 Reliability index associated with different loads

Load combination	Reliability index, β
Dead load + live load (or snow load)	3.0 for members, 4.5 for connection
Dead load + live load + wind load	2.5 for members
Dead load + live load + earthquake load	1.75 for members

Because of the lower probability of wind or earthquake loads occurring simultaneously with the live load, a lower value of β was chosen. Similarly, in keeping with the tradition of making connections stronger than members, a higher value of β was chosen for the connection.

The multiple safety factor adopted by the code is in the so-called *partial safety factor format*, which is expressed as

$$R_d \geq \Sigma \gamma_{if} Q_{id} \tag{2.14}$$

where R_d is the design strength (or resistance) computed using the reduced material strength R_u/γ_m, where R_u is the characteristic material strength and γ_m is the partial safety factor for the material, and allows for uncertainties of element behaviour and possible strength reduction due to manufacturing tolerances and imperfections in the material. The values of γ_m adopted by the code are given below (see also Table 5 of the code):

(a) For resistance, governed by yielding, $\gamma_{m0} = 1.10$
(b) For resistance of member to buckling, $\gamma_{m0} = 1.10$
(c) For resistance, governed by ultimate stress, $\gamma_{m1} = 1.25$
(d) For resistance of connections–bolts and welds, except for site welds (γ_{mf}, γ_{mb}, and γ_{mw}) = 1.25; site welds = 1.50

The partial safety factors for loads, γ_f, makes allowance for possible deviation of loads and the reduced possibility of all loads acting together. The values of γ_f for different load combinations are given in Table 4 of the code. A few important γ_f values are given below:

(a) For dead load, $\gamma_{Df} = 1.50$
(b) For live load, $\gamma_{Lf} = 1.50$
(c) For dead load + live load + wind/earthquake load = 1.2
(d) For dead load + wind/earthquake load = 1.5 or 0.9
(e) For limit state of serviceability = 1.0

All the load factors are generally greater than unity, because overestimation usually results in improved safety. However, one notable exception to this rule is the dead load factor γ_{Df}, which is taken as 0.9 whenever dead load contributes to stability against overturning or sliding, or while considering reversal stresses when dead loads are combined with wind/earthquake loads. In such cases, underestimating the counter-acting effects of dead load results in greater safety.

Note that the load factors are reduced when different types of loads (DL, LL, and EL) are acting simultaneously at their peak values. (This is referred to sometimes as the *load combination effect*.) This is because of the reduced probability of all the loads acting concurrently (see also Section 2.6.7).

2.9.3 Factors Governing Ultimate Strength

According to clause 5.5 of the code, the factors that govern ultimate strength are: stabilty, fatigue, and plastic collapse.

It should be ensured that the structure as a whole and each of its elements remains stable from the commencement of erection until demolition. Sufficient external bracings should be provided for stability.

Stability against overturning

The structure as a whole or any part of it should be designed to prevent instability due to overturning or sliding, while designing tall or cantilever structures. The code suggests the following while checking for stability against overturning:

(a) The loads and forces should be divided into components aiding instability and those resisting instability.

(b) The forces and loads causing instability should be combined using the appropriate load factors given in the code.

(c) The permanent loads and effects causing resistance should be multiplied by a partial safety factor of 0.9 and added together with design resistance (after being multiplied by the appropriate partial safety factor). Note that a low factor is applied to those forces which provide restraint, to guard against them being overestimated.

(d) The resistance effect should be greater than or equal to the destabilizing effect.

Sway stability

This condition imposes that there must not be excessive lateral deformation under applied loads and that the structure is adequately stiff to resist the lateral loads.

All structures should be checked for a minimum notional horizontal load. The notional loads may also arise from practical imperfections.

These notional loads, equal to 0.5% of the factored vertical dead and imposed load at any level, should be applied at each floor level of the building, as shown in Fig. 2.21.

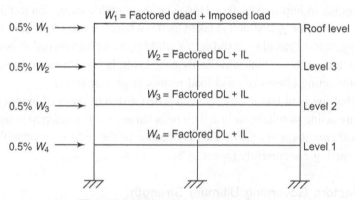

Fig. 2.21 Notional horizontal forces

As per clasuse 4.3.6.2 of the code, this notional horizontal force should not be

(a) applied while considering overturning,

(b) combined with other horizontal loading such as wind or earthquake,

(c) combined with temperature effects, and

(d) taken to contribute to net shear on the foundation.

The notional horizontal load should be applied on the whole structure, in both orthogonal directions, in one direction at a time, and should be taken as acting simultaneously with a factored gravity load. Moreover when the ratio of the height to the lateral width of a building is less than unity, such notional loads need not be considered. More details about these notional loads are provided by Structural Engineering Institute/ASCE 1997.

2.9.4 Serviceability Limit States

In such limit states, the variable to be considered is a serviceability parameter Δ (representing deflection, vibration, etc.). A limit state or failure is considered to occur when a specified maximum limit of serviceability, Δ_{all} is exceeded (Fig. 2.22). In Fig. 2.22, P_f is the probability of failure and $f_\Delta(\Delta)$ is the frequency distribution curve for Δ. It may be noted that unlike the strength limit state shown in Fig. 2.20, the limit defining failure is deterministic and not probabilistic. Serviceability limit states relate to satisfactory performance and correspond to excessive deflection, vibration, local deformation, durability, and fire resistance.

The load factor, γ_f, should be taken as unity for all serviceability limit state calculations, since they relate to the criteria governing normal use.

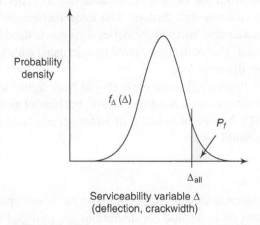

Probability
density

$f_\Delta(\Delta)$

P_f

Δ_{all}

Serviceability variable Δ
(deflection, crackwidth)

Fig. 2.22 Reliability model for serviceability design

Deflections

The maximum deflection affecting the strength and stability of the structure is controlled by the strength limit state. However, excessive deflection should not cause sagging, plaster cracking, or failure to align plant and machinery (e.g., in lifts). Excessive deflection of beams causes damage to supported non-structural elements such as partitions, excessive vibrations of floors, or impairs the usefulness of the structure (e.g., it may distort door frames so that doors do not open or close). On roofs, a major deflection-related concern is ponding of water. Excessive deflections are often indicative of excessive vibration and noise transmission, both serviceability problems. Deflections are to be calculated for all combinations of loads specified in the code, by using an elastic analysis and checked for the maximum values specified in Table 6 of the code. Some of the vertical deflection limits specified by IS 800 : 2007 are shown in Table 2.10.

Table 2.10 Vertical deflection limits

Structure	Member	Loading	Maximum deflection	
			Elastic cladding	Brittle cladding
Industrial	Purlins	Imposed/wind load	$L/150$	$L/180$
buildings	Simply supported beam	Imposed load	$L/240$	$L/300$
	Cantilevers	Imposed load	$L/120$	$L/150$
Other	Floor and roof beams	Imposed load	$L/300$	$L/360$
buildings	cantilevers	Imposed load	$L/150$	$L/180$

Note: L is the span in mm. Brittle cladding are those which are susceptible to cracking

Serviceability, instead of strength, may and often does control the design of beams. In order to avoid the sagging appearance due to deflection, we may pre-camber the beam, trusses, and girders. The code recommends that for spans greater than 25 m, a camber equal to the deflections due to dead load plus half the live load may be used. This deflection should be calculated without considering the effect of impact or dynamic loads.

The code states that very flexible roofs should be designed to withstand loads due to ponding of water, or accumulation of snow, but has not given any guidelines. According to AISC, to prevent ponding of water accumulated on flat roofs, the support member should satisfy

$$\frac{L}{d} \le \frac{4137}{f_b} \tag{2.15}$$

where f_b is the computed service load bending stress in megapascals.

More discussions on deflection considerations are provided in the chapter on beams.

Vibration

With the development of lighter construction (using high-strength steels, cold-rolled sections, etc.) and use of longer spans, there is a higher risk of vibrations becoming critical in a number of situations. Vibration has to be checked when vibrating loads such as due to machinery, cranes, etc., are applied to plant platforms. Activities such as dancing, marching, and drilling and impact loads also produce vibration. In these cases, care must be taken to ensure that the structural response will not amplify the disturbing motion. The code also recommends that flexible structures (with height to effective width ratio exceeding 5:1) should be investigated for lateral vibration under dynamic loads. The code gives guidelines to estimate floor frequency, damping, and acceleration due to vibration, in Appendix D. When vibration becomes a problem, we may have to change the natural frequency of the structure by some means (e.g. by using dampers). More information regarding building vibration may be found in Hatfield (1992), and Allen and Murray (1993).

Software packages (e.g., Floorvibe—www.floorvibe.com) are also available, using which the frequency and amplitude resulting from transient vibration caused by human activity can be quickly estimated.

Vibration of the London Millennium Footbridge

The London Millennium Footbridge is a pedestrian-only steel suspension bridge crossing the river Thames in London, England, linking the Bankside with the City. The bridge was designed by Arup, Foster and Partners, and Sir Anthony Caro. Due to height restrictions, and to improve the view, the bridge's suspension design had the supporting cables below the deck level, giving it a very shallow profile. Its construction began in 1998, and it was opened for public use on 10 June 2000.

The London Millennium Bridge

Londoners nicknamed the bridge 'The Wobbly Bridge' after they felt an unexpected and uncomfortable swaying motion on the first two days after the bridge opened. The bridge's movements were caused by a 'positive feedback' phenomenon, known as *synchronous lateral excitation*. The natural sway motion of people walking caused small sideways oscillations in the bridge, which in turn caused people on the bridge to sway in step, increasing the amplitude of oscillations further.

Any bridge with lateral frequency modes of less than 1.3 Hz, and sufficiently low mass, could witness this phenomenon with sufficient pedestrian loading. The greater the number of people, the greater will be the amplitude of the vibrations.

After extensive analysis by the engineers, the problem was fixed by the retrofitting of 37 fluid-viscous dampers (energy dissipating) to control horizontal movement and 52 tuned mass dampers (inertial) to control vertical movement. After a period of testing, the bridge was successfully re-opened on 22 February 2002. The bridge has not been subject to significant vibration since then.

Durability and corrosion

A durable steel structure is one that performs satisfactorily the desired functions during its service life under the anticipated environment and exposure conditions. This implies that the member's designed cross-sectional area should not deteriorate and its strength should not be reduced mainly due to *corrosion*.

Several factors that may affect the durability of the steel construction may include

(a) the environment,
(b) the degree of exposure,
(c) the shape of the member,
(d) the structural detailing,
(e) the protective coating employed, and
(f) the ease of maintenance.

The methods of corrosion protection are governed by actual environmental conditions as per IS 9172. Section 15 of IS 800 gives the requirements for durability and also specifications for different coating systems under different atmospheric conditions.

Painting is the normal method of protection used to prevent corrosion. The durability of the painting system is enhanced when the steel surface is cleaned before painting by methods such as blast cleaning. Good designers will also try to 'design for prevention' by avoiding traps for dirt and moisture [Subramanian (2000), British Steel Corporation (1996)].

It has to be noted that the code does not specify any minimum thickness for members (though the earlier revision of the code, IS 800 : 1984, specified a minimum thickness of 6 mm for members directly exposed to weather and fully accessible for cleaning and repainting and 8 mm for members directly exposed to weather but not accessible for cleaning and repainting). It is assumed that if the durability requirements as given in Section 15 are followed, it will ensure adequate protection for all thickness.

Painting of the Golden Gate Bridge

Prior to 1965, coatings were generally oil- or alkyd-based and contained pigments using lead and/or chromium compounds as corrosion inhibitor. In addition, they were often applied on steel surfaces that were prepared by power tool cleaning (often using rotary impact tools). These old type of coatings were expected to last about 8 to 10 years. As a result, many bridges have several layers of coating on them, which may fall off due to their own weight. Coatings with an overall thickness of 6 mm or more have been encountered in the past.

The current standard for bridge coating uses a three-layer coating system consisting of an inorganic zinc-rich primer, an epoxy mid-coat, and a urethane topcoat. The primer is applied on the blast-cleaned surface.

The Golden Gate Bridge

The 2737 m long Golden Gate Bridge in San Francisco, USA, when built during 1933 to 1937, was coated with lead-based paint. During 1965, the lead paint was removed and an inorganic zinc-rich paint system was applied. When the bridge was examined recently, it revealed that that the coating system is in excellent overall condition, even after 44 years.

It has to be noted that the metallic zinc pigment in the zinc-rich paint will be able to provide galvanic protection to the steel surface, until the zinc itself is consumed. When the zinc is depleted, the steel will eventually rust. The time it takes for the zinc to be consumed is dependent on the weather, the duration of wetness, and the number of wet/dry cycles encountered. The additional two layers of coating protect the zinc by limiting the entry of moisture and oxygen. It is estimated that after about 30 to 55 years, another overcoat is possibly required. Thus, the new coating system may increase the life of steel structures to more than 100 years!

Richard J. Daley Center, Chicago

Weathering steel is a steel alloy containing copper, chromium, nickel, and other alloying elements to enhance corrosion resistance. When this steel rusts under normal atmospheric exposure, it forms a rusty-orange-brown protective impervious layer that prevents further corrosion. Any minor damage to the coating is self-healing. Over time, as the thickness of coating increases, the texture roughens and the oxide coating changes from rusty-orange-brown to dark purple-brown patina. Thus, weathering steel does not require any painting to protect it and provides significant life-cycle cost savings.

The Richard J. Daley Center, Chicago, built in the year 1965, was the first building to be constructed entirely with a type of weathering steel, called Cor-Ten. The Pablo Picasso sculpture outside the center is also made of weathering steel. As the picture shows, even after more than 40 years of existence, they look great due to proper maintenance of the building and the sculpture.

The Richard J. Daley Center and the Pablo Picasso Sculpture
© Jeremy Atherton (2006)

Of course weathering steel is not a material of choice in environments that are constantly wet or humid or for coastal structures. It has been used in several short-span bridges such as the award-winning Box Elder Creek Bridge in Colorado. More details about weathering steel are available online at: www.aisc.org

Fatigue

The fatigue limit state is important in structures where distress to the structure due to repeated loading is to be considered. Thus, fatigue is an important criterion for bridges, crane girders, platforms carrying vibrating machinery, etc. In such structures, failure may occur at stress levels well below normal yield. The code states that stress changes due to fluctuations in wind loading need not be considered as fatigue. Though fatigue failure is usually considered as an ultimate limit state, fatigue checks are carried out at working loads (with load factor γ_f equal to unity), considering the stress variation, which occur in the normal working cycle. This apparent anomaly (i.e., checking the fatigue limit state at working load) exists because fatigue failure occurs due to the very large number of application of loads normally expected to act on the structure (i.e., the specified load). Other ultimate limit state failures, such as the failure of a column, could occur due to a single application of a greater load than that normally expected to act on the structure; thus a factored load is used in such design calculations. Section 13 of the code gives guidelines for fatigue.

For the purpose of design against fatigue, the code classifies different details (of members and connections) under different fatigue classes. The design stress range corresponding to various numbers of cycles are given in the code for each fatigue class. The requirements of the code are to be satisfied at each critical location of the structure subjected to cyclic loading, considering the relevant number of cycles and magnitudes of stress range expected during the life of the structure.

Brittle fracture

As with fatigue, brittle fracture will rarely occur in building construction. Such fracture is the sudden failure of the material under service condition, caused by low temperature or sudden change in stress. Since thick material is more prone to brittle fracture than thin material, limiting thicknesses are often prescribed by the codes for the various members. In addition to the thickness, the stress in the material, and the type of details employed in the construction also have significant effect on the risk of brittle fracture. Brittle fracture can be avoided by using a steel quality and welding electrodes that have a specified Charpy impact value. (IS 1757 : 1988 allows the use of only those steels that exhibit a minimum energy absorption capacity at a predetermined temperature, say 20 J at 23 ±5°C.)

Fire resistance

Fire resistance of a steel member is a function of its mass, geometry, support conditions, type of fire and the fire protection methods adopted. Section 16 of the code gives guidelines on fire protection. More information on fire protection are provided by Buchanan (2001) and Subramanian (2008).

2.10 Failure Criteria for Steel

The minimum yield strength is the design strength in IS 800, and is an important failure criterion for steel. However, only when the load-carrying member is subject to uniaxial tensile stress, the properties obtained from the tension test will be identical with those of the structural member. In a real structure, the yielding will not be well defined, unlike as observed in the tension test. Hence in these cases, yielding is assumed to be achieved when any one component of stress reaches the uniaxial value f_y.

For all states of stress other than uniaxial, a definition of yielding is needed. These definitions are called *yield criteria* and are expressed as equations of interaction between the stresses acting.

The most commonly accepted theory, developed by Huber, von Mises, and Hencky, gives the uniaxial yield stress in terms of the principle stresses f_1, f_2, and f_3 as follows.

$$2Y_s^2 = (f_1 - f_2)^2 + (f_2 - f_3)^2 + (f_3 - f_1)^2 \qquad (2.16)$$

In the above equation, f_1, f_2, and f_3 are the stresses that act in three mutually perpendicular planes of zero shear, and Y_s is the 'yield stress' that may be compared with the uniaxial value, f_y.

For most structural design situations, one of the principal stresses is either zero or small enough to be neglected. Thus, Eqn (2.16) can be reduced to the following form,

$$f_y^2 = f_{bc}^2 + f_b^2 - f_{bc} f_b + 3f_q^2 \qquad (2.17)$$

The above yield criterion will be used in the book, wherever needed.

2.10.1 Shear Yield Stress

The yield point for pure shear can be determined from a stress–strain curve with shear loading, or if the multiaxial yield criterion is known, it can be used. It can be shown that by using Eqn (2.17), the yield condition for shear stress acting alone is (Salmon and Johnson 1996)

$$\tau_y = \frac{f_y}{\sqrt{3}} = 0.58 f_y \qquad (2.18)$$

Examples

Example 2.1 *A roof truss has a span of 16 m and a pitch of 4 m which is placed at 5 m c/c. Calculate the live load on the roof truss.*

Solution
Span of the roof truss = 16 m
Pitch of the truss = 4 m
Spacing of truss = 5m c/c
Pitch of truss, $\theta = \tan^{-1} (4/8) = 26.57°$
Live load for inaccessible roof = 0.75 kN/m^2

Live load deduction for slopes more than $10° = (0.75 - 0.02 \times 16.57) \times \dfrac{2}{3}$

$$= 0.28 \text{ kN/m}^2$$

However as per IS 875 (Part 2),

Minimum LL to be considered for design = 0.40 kN/m^2

Therefore, Design LL = 1.5 × 0.40 kN/m^2

= 0.6 kN/m^2

Example 2.2 *A commercial building, shown in Fig. 2.23, has seven storeys. The roof is accessible and all the floors are used as offices. Calculate the load on the interior column AB on the first floor, assuming the spacing of columns in the perpendicular direction as 4 m.*

Live load on each floor = 4000 N/m^2

Live load on roof, with access = 1500 N/m^2

Assuming 150 mm thick slabs, dead load = 3750 N/m^2

Add dead load of floor finish, etc. (say) = 1000 N/m^2

Total dead load = 4750 N/m^2

Height of each storey = 3 m

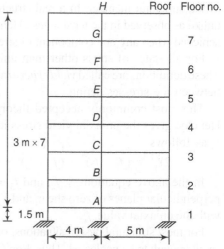

Fig. 2.23 Live load reduction on a seven-storey building

Solution

Loads from the various floor levels are computed as below. The live load has been reduced as per IS 875 (Part 2).

$$\text{Area contributing load to each column} = 4 \times \left(\frac{4+5}{2}\right) = 18\,\text{m}^2$$

Live load on roof = 1500 N/m^2 = 1.5 kN/m^2

Live load on floors = 4000 N/m^2 = 4.0 kN/m^2

Table 2.11

Column	Floor	Live load kN	Dead load kN	Load on column, kN
GH	Roof	1.5 × 18 = 27	4.75 × 18 = 85.5	27 + 85.5 = 112.5
FG	7th floor	4 × 18 = 72	4.75 × 18 = 85.5	(27 + 72) × 0.9 + 85.5 × 2 = 260.10
EF	6th floor	4 × 18 = 72	4.75 × 18 = 85.5	(27 + 2 × 72) × 0.8 + 85.5 × 3 = 393.3
DE	5th floor	4 × 18 = 72	4.75 × 18 = 85.5	(27 + 3 × 72) × 0.7 + 85.5 × 4 = 512.1
CD	4th floor	4 × 18 = 72	4.75 × 18 = 85.5	(27 + 4 × 72) × 0.6 + 85.5 × 5 = 616.5
BC	3rd floor	4 × 18 = 72	4.75 × 18 = 85.5	(27+ 5 × 72) × 0.6 + 85.5 × 6 = 745.2
AB	2nd floor	4 × 18 = 72	4.75 × 18 = 85.5	(27 + 6 × 72) × 0.6 + 85.5 × 7 = 873.9

Design load on column AB = 1.5 × 873.9 = 1310.85 kN

Note that if the live load reduction is not considered, the load on column AB will be 1.5(112.5 + (72 + 85.5) × .6) = 1586.25 kN.

Thus, there is an increase of 21% in the load. Also, note that the dead load on the roof in a real structure may be more due to the type of weathering course adopted.

Example 2.3 *A rectangular industrial building situated in an industrial area is to be designed in Chennai. The height of the building is 4.5 m, and a size of the building is 10 m × 40 m. The walls of the building have 20 openings of size 1.2 m × 1.5 m. The building has a flat roof supported on load-bearing walls. [See Fig. 2.24(a).] Compute the design wind pressure and design forces on the walls and roofs of the building.*

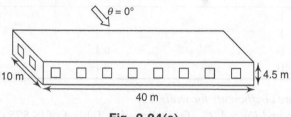

Fig. 2.24(a)

Solution
Basic wind speed in Chennai (from wind zone map or Appendix A of IS 875 (Part 3)],

$$V_b = 50 \text{ m/s}$$

An industrial building is to be designed for a 50-year life, so the risk coefficient from Table 1 of IS 875 (Part 3) is

$$k_1 = 1$$

Since the building is proposed to be erected in an industrial area, it is considered to belong to category 3. The terrain factor, from Table 2 of IS 875 (Part 3), for a height of 4.5m, is

$$k_2 = 0.91$$

The ground is assumed to be plain, and hence the topography factor is

$$k_3 = 1$$

Design wind speed, $V_z = V_b k_1 k_2 k_3 k_4$
$$= 50 \times 1 \times 0.91 \times 1 = 45.5 \text{ m/s}$$

Wind pressure, $p_z = 0.6 \ V_z^2 = 0.6(45.5)^2$
$$= 1242 \text{ N/m}^2 = 1.242 \text{ kN/m}^2$$

Permeability of the building:
Area of the walls $= 4.5(2 \times 10 + 2 \times 40) = 450 \text{ m}^2$
Area of all the openings $= 20 \times 1.5 \times 1.2 = 36 \text{ m}^2$
% opening area $= 8\%$, between 5% and 20%. Hence the building is of medium permeability.

Wind load calculations
$$F = (C_{pe} - C_{pi}) A p_d$$
Internal pressure coefficient (Table 2.5)
$$C_{pi} = \pm 0.5$$

External pressure coefficient
On Roof: Using Table 5 of IS 875 (Part 3), with roof angle 0° without local
coefficients, for $h/w = 0.45$, the following coefficients are obtained.

Table 2.12(a) External pressure coefficients

Portion of roof	Wind incidence angle	
	0°	90°
E	−0.8	−0.8
F	−0.8	−0.4
G	−0.4	−0.8
H	−0.4	−0.4

Design pressure coefficients for walls
For $h/w = 0.45$ and $l/w = 4$, C_{pe} for walls, using Table 4 of IS 875 (Part 3), we get
the values for external pressure for coefficients for walls as shown in Table 2.12(b).

Table 2.12(b) External pressure coefficients for walls

Wall	Wind incidence angle	
	0°	90°
Wall A	+0.7	−0.5
Wall B	−0.25	−0.5
Wall C	−0.6	+0.7
Wall D	−0.6	−0.1

Note that the pressure coefficients are given only for buildings with l/w ratio up
to 4. For longer buildings, i.e. $l/w > 4.0$, the values given in the table up to $l/w = 4.0$
should be used.

The above values have to be combined with the internal pressure coefficients
$C_{pi} = \pm 0.5$.
Thus,
$C_{p\,net}$ for walls A or B
$$= 0.7 - (-0.5) = +1.2 \quad \text{(pressure)}$$
$$= -0.5 - (+0.5) = -1.0 \quad \text{(suction)}$$
$C_{p\,net}$ for walls C or D [See Fig. 2.24(b)],
$$= 0.7 - (-0.5) = +1.2 \quad \text{(pressure)}$$
$$= -0.6 - (+0.5) = -1.1 \quad \text{(suction)}$$
Design pressure for walls
For long walls,
$$F = C_{p\,net} \times p_d$$
$$= 1.2 \times 1.242 = 1.4904 \text{ kN/m}^2 \quad \text{(pressure)}$$
$$= -1 \times 1.242 = -1.242 \text{ kN/m}^2 \quad \text{(suction)}$$
For short walls,
$$F = 1.2 \times 1.242 = 1.4904 \text{ kN/m}^2 \quad \text{(pressure)}$$
$$= -1.1 \times 1.242 = -1.3662 \text{ kN/m}^2 \quad \text{(suction)}$$

For 0° wind incidence, for *E/G* (end zone)

For 90° wind incidence, for *E/G* (end zone)

For 0° wind incidence, for *F/H* (mid-zone)

For 90° wind incidence, for *F/H* (mid-zone)

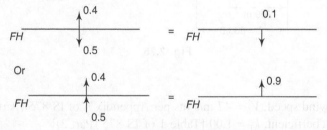

Fig. 2.24(b) Net roof pressure coefficients for different zones and combinations

For the roof,

$$F = 1.3 \times 1.242 = 1.6146 \text{ kN/m}^2 \quad \text{(pressure)}$$
$$= -0.1 \times 1.242 = -0.1242 \text{ kN/m}^2 \quad \text{(suction)}$$

Calculation of force due to frictional drag

Since $40/4.5 = 8.8 > 4.0$ (even though $40/10 = 4.0$), the frictional drag due to wind has to be considered. This will act in the longitudinal direction of the building along the wind. Here $h < b$, and hence Eqn 2.5(a) is used.

$$F' = 0.01 (40 - 4 \times 4.5) 10 \times 1.242 + 0.01 (40 - 4 \times 4.5)2 \times 4.5 \times 1.242$$
$$= 2.7324 + 2.4592 = 5.192 \text{ kN/m}^2$$

This frictional drag will act on the roof of the building.
Alternate calculation using force coefficients given in the code
Size of the building = 40 m × 10 m × 4.5 m
Therefore, $h/b = 4.5/10 = 0.45$

$$\frac{a}{b} = \frac{10}{40} = 0.25$$

and

$$\frac{b}{a} = \frac{40}{10} = 4$$

As per Fig. 4 of the code [IS 875 (Part 3)],

$$C_{f1} = 1.2, \quad C_{f2} = 1.0$$

The force acting on the building = $C_f A_e p_a$
  For 0° wind,
    Force = $1.2 \times (40 \times 4.5) \times 1.242 = 268.27$ kN
  For 90° wind,
    Force = $1 \times (10 \times 4.5) \times 1.242 = 55.89$ kN

Example 2.4 *A building is proposed to be built at Amritsar on a hillock. The height of the hill is 250 m and the slope is 1 in 3. The building is proposed at a distance of 200 m from the base of the hill (Fig. 2.25). Find the design wind pressure. The height of the building is 20 m.*

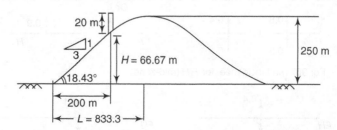

Fig. 2.25

Solution
  Basic wind speed, $V_b = 47$ m/s [as per Appendix A of IS 875 (Part 3)]
  Risk coefficient, $k_1 = 1.00$ [Table 1 of IS 875 (Part 3)]
Terrain category 2 is assumed and the terrain coefficient, from Table 2 of IS 875 (Part 3), is
  $$k_2 = 1.00$$
Topography factor, k_3:
  Z = effective height of the hill = 250 m
  $\theta = 1$ in $3 = \tan^{-1}(1/3) = 18.43°$
  L = Length of upwind slope = 250(3) = 750m
  L_e = effective horizontal crest length of the hill
    = $Z/0.3$ for $\theta > 17°$
    = 250/0.3 = 833.3 m
For $\theta = 18.43°$, $C = 0.36$

Height of the building, $H = 20$ m, $H/L_e = 66.67/833.3 = 0.08$
X is the horizontal distance of the building from the crest measured considered positive towards the leeward side and negative towards the windward side.

$$X = - (L_e - 200) = - (833.3 - 200) = -633.3 \text{ m}$$
$$X/L_e = -633.3/833.3 = - 0.76$$

From Fig. 15 of IS 875 (Part 3),

$$X/L_e = -0.76 \text{ and } H/L_e = 0.08$$
$$s = 0.12$$
$$k_3 = 1+ Cs$$
$$= 1 + 0.36 (0.12) = 1.04$$

Design wind speed,

$$V_z = k_1 k_2 k_3 V_b$$
$$= 1 \times 1.0 \times 1.04 (47) = 48.88 \text{ m/s}$$

Wind pressure, $p_z = 0.6 \ V_z^2 = 0.6(48.88)^2 = 1433 \text{ N/m}^2$

Example 2.5 *Consider a three-storey steel building shown in Fig. 2.26. The building is located in Roorkee (Seismic zone IV). Soil conditions are medium stiff and the entire building is supported on a raft foundation. The steel frames are infilled with unreinforced brick masonry. Determine the seismic load on the structure as per IS 1893 (Part 1). The seismic weights as shown in the figure have been calculated considering 50% of the live load lumped at the floors and no live load at roof.*

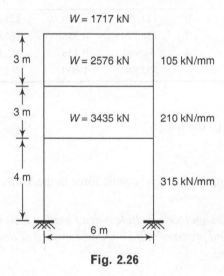

Fig. 2.26

Solution
For Seismic zone IV, the zone factor is 0.24 [Table 2 of IS 1893(Part 1)]. Being an office building, the importance factor is 1.0 (Table 6 of IS 1893). The building has a special moment resisting frame and hence $R = 5$.
Total seismic weight of the structure $= \Sigma W_i = 1717 + 2576 + 3435 = 7728$ kN
$h = 4 + 3 +3 = 10$ m

Assume the depth of building to be 15 m.

Fundamental period

The lateral load resistance is provided by moment resisting frames infilled with brick masonry panels. Hence for EL in the X-direction,

$$T = 0.09h/\sqrt{d} \text{ (Clause 7.6.2 of IS 1893)}$$

$$= 0.09 \times 10/\sqrt{15} = 0.23 \text{ s}$$

From Fig. 2 of IS 1893, for $T = 0.23$ s

$$S_a/g = 2.5$$

$$A_h = ZI(S_a/g)/(2R) \text{ (Clause 6.4.2 of IS 1893)}$$

$$= 0.24 \times 1 \times 2.5/(2 \times 5) = 0.06$$

Design base shear, $V_B = A_h W$

$$= 0.06 \times 7728 = 463.68 \text{ kN}$$

Force distribution with building height

The design base shear is distributed with height as per clause 7.7.1 and the relevant calculations are shown in Table 2.13.

Table 2.13 Lateral load distribution as per static method

Storey level	W_i (kN)	h_i (m)	$W_i h_i^2$	$(W_i h_i^2)/(\Sigma W_i h_i^2)$	Lateral force at i-th level for EL in direction (kN)	
					X	Y
3	1717	10	171,700	0.486	225.35	225.35
2	2576	7	126,224	0.358	166.00	166.00
1	3435	4	54,960	0.156	72.33	72.33
Σ	7728	—	352,884	1.000	463.68	463.68

EL in Y-direction

$$T = 0.09 \times 10/\sqrt{6} = 0.367 \text{ s}$$

Therefore, $S_a/g = 2.5$

and $A_h = 0.06$

Hence for this building, the design seismic force in the Y-direction is same as that in the X-direction.

Example 2.6 *Roof design loads include a dead load of 1.60 kN/m², a live load of 1.15 kN/m², and a wind pressure of 0.70 kN/m² (upward or downward). Determine the governing loading.*

Solution

The load combinations are

(a) 1.5 (DL + LL) = 1.5 (1.6 + 1.15) = 4.125 kN/m²
(b) 1.2 (DL + LL + WL) = 1.2 (1.6 + 1.15 + 0.70) = 4.14 kN/m²
(c) 1.2 (DL + LL − WL) = 1.2 (1.6 + 1.15 − 0.70) = 2.46 kN/m²
(d) 0.9 DL + 1.5 WL = 0.9 × 1.6 + 1.5 × 0.70 = 2.49 kN/m²
(e) 0.9 DL − 1.5 WL = 0.9 × 1.6 − 1.5 × 0.70 = 0.39 kN/m²
(f) 1.5 (DL + WL) = 1.5 (1.6 + 0.70) = 3.45 kN/m²
(g) 1.5 (DL − WL) = 1.5 (1.6 − 0.70) = 1.35 kN/m²

The second load combination governs. Hence the roof has to be designed for a total factored load of 4.14 kN/m². It may be noted that the fifth load case produces the minimum load. When the dead load is comparatively small, it will result in a negative value for the combination, which will be critical for the overturning or stability checks. Also, since it is a simple calculation, we are in a position to find the governing load combination. In a complex structural system, it may not be simple to evaluate the governing loading condition. Moreover, one loading combination may be critical for one set of members (say columns), and another combination may be critical for another set of members (say bracings). Hence, in these cases, a computer program will be quite useful to calculate the critical forces in any member due to any combination of loads.

Summary

Structural design is considered a science as well as an art. The aim of any structural designer should be to design a structure in such a way that it will fulfill its intended purpose during its intended lifetime and be adequately safe (in terms of strength, stability, and structural integrity), and have adequate serviceability (in terms of stiffness, durability, etc). In addition, the structure should be economically viable, (in terms of cost of construction and maintenance), aesthetically pleasing, and environment friendly.

Brief descriptions of the various structural systems that have evolved over the past have been given. Depending on the particular situation, and the requirement of the client, a structural system may be chosen. The systems to be adopted in earthquake zones assume more importance, as more than 60% of our country fall under moderate-to-strong earthquake zones. The concept of structural integrity and the new provisions in the code for the same are explained.

The four main phases in a structural design process are (a) determination of the structural system, (b) calculation of the various loads acting on the system, (c) analysis of the structural system for these loads, and (d) design of the various members as per the codal provisions. Out of these phases, the determination of various loads is the most difficult and important phase, since the final design is based on these loads. Several failures have been reported in the past which clearly show that one of the main reasons for these failures is the lack of consideration of the loads acting on the structures. Hence, a brief review of all the loads that may act on any structure is given. Out of the several natural and man made loads, the following loads are considered important: (a) dead loads, (b) imposed loads (live and snow loads), (c) wind loads, and (d) earthquake loads.

Some loads such as impact loads due to traffic on a bridge, crane loads, wind loads and earthquake loads are dynamic in nature. However, most often they are converted to equivalent static loads. Dynamic analysis is resorted to only in the case of flexible structures, whose natural frequency in the first mode is less than 1.0 Hz or whose height to least lateral dimension ratio is more than about 5. Complicated structures should be avoided especially in earthquake zones, since their analysis and modeling is difficult. It is very important to realize that the earthquake codes require the designer to design the structure only to a fraction of the load that may act on the building. Hence the designer has to detail the structure in such a way that during a major earthquake, the structure may damage but the occupants are able to escape on account of the ductility of the material and overstrength factors.

Once the loads acting in the structure have been calculated, it may be analysed for the given loads.

Analysis refers to the determination of the internal forces acting on different members of a structure, due to the application of external actions (forces). A brief description of the methods of analysis and the available computer programs are given.

A discussion on codes and specifications which will guide the designer in the design process has been provided. The codes published by the Bureau of Indian Standards for the design of steel structures are listed and the reader is advised to obtain a copy of these codes, since they may be required to understand the material provided in the chapters to follow.

The two main design philosophies that have been evolved in the past, namely, the working stress method, and the limit states design, are briefly discussed. The various limit states that have to be considered in design are explained.

The terms characteristic load and characteristic strength are explained. The basis of the limit states method is discussed along with the various partial safety factors adopted by the code for materials and loads. A brief introduction to stability is provided. The various serviceability limit states, related to the satisfactory performance of the structure (as opposed to ultimate/safety limit states, which are concerned with strength, stability, fatigue fracture, etc) are also briefly explained along with the deflection limits specified by the code. The failure criteria, which are required to define the yielding of the material, when subjected to multiaxial stresses, are also discussed.

From the next chapter, we will discuss the design of various elements, starting with the design of tension members.

Exercises

1. A roof having a span of 18 m and rise of 3.5 m is spaced at 4 m apart. Assuming ACC sheeting, estimate the dead and live loads on the purlins, assuming purlin spacing of 1.40 m.
2. An industrial building is supported by columns spaced at 4 m in both X and Y direction and is covered with 150 mm thick RC slab. Calculate the factored dead and imposed load on the slab. Assume suitable tile load.
3. A six-storey building is to be used for office purposes. Calculate the load on an interior column in the ground floor, assuming that the columns are placed in a grid of 5 m × 5 m. Consider live load reduction as per IS 875 (Part 2) and assume a live load of 3 kN/m^2 and 150 mm thick concrete slab with tiles. Also assume some partition load.

4. A tall building is proposed in Mumbai where there are some existing tall buildings. Determine the design wind pressure. Use the following data:
   (a) Level ground,
   (b) Design for a return period of 50 years,
   (c) Basic wind speed = 44 m/s,
   (d) Size of the building = 20 m × 60 m and height = 20 m.
   Estimate the risk, topography, terrain coefficients, and compute design wind speed and pressure as per IS 875 (Part 3).
5. A hospital is proposed to be built on a hill top near Bhuj. The height of the hillock above the surroundings is 500 m with a slope of 1 in 4. The basic wind speed in Bhuj is 50 m/s. The size of the building is 30 m × 60 m with a height of 15 m. Compute the design wind pressure on the building.
6. Compute the design wind pressure and design forces on walls and roofs of a two-storey building which is 6.5 m high and has a size of 20 m × 40 m. Assume there are five openings on each floor of size 1.2 m × 1.2 m in the wall of length 40 m and two similar openings in each floor in the wall of length 10 m. The building has a flat roof and is supported on load-bearing walls.
7. Consider the four-storey office building shown below, located in Shillong (Seismic zone V). The soil condition is medium stiff and the entire building is supported on a raft foundation. The RC frames are infilled with brick masonry. The lumped weight due to dead loads is 12 kN/m^2 on the floors, and 10 kN/m^2 on the roof. The floors carry a live load of 4 kN/m^2 and the roof 1.5 kN/m^2. Determine the design seismic load on the structure by the equivalent static method. Assume that the frames are steel moment resisting frames with $R = 5$.

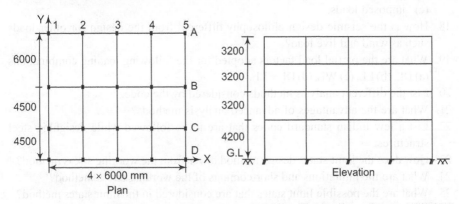

[*Ans*: Design base shear = 1560 kN]

8. Roof design loads include a dead load of 1.75 kN/m^2, a live (or snow) load of 1.25 kN/m^2, and a wind pressure of 0.75 kN/m^2 (upward or downward). Determine the governing loading.

Review Questions

1. What are the main objectives of a designer while designing structures?
2. Is design an iterative process? Why is it so?
3. List the steps involved in the design process.

4. What kind of knowledge a designer should process and what are the aids that are available to a designer?
5. List the different types of steel structures.
6. Distinguish between simple and continuous construction.
7. Write short notes on the following:
   (a) Outriggers and belt truss systems,
   (b) Framed tube, trussed tube, and bundled tube systems,
   (c) Concentrically braced frames and eccentrically braced frames,
   (d) Moment-resisting frames.
8. List a few seismic force resisting systems.
9. Under what circumstances, moment-resisting frames should not be used?
10. What is the condition specified in the code to be satisfied for achieving strong-column and weak-beam systems?
11. Why are concentrically braced frames inferior to eccentrically braced frames?
12. Discuss the importance of structural integrity. How can it be achieved?
13. List the different loads that may be acting on a steel structure.
14. What is analysis phase and how is it different from the design phase?
15. List a few computer programs that are available for the analysis and design of steel structures.
16. Why are codes of practices necessary and what is the function of these codal specifications?
17. Write short notes on
   (a) wind load,
   (b) earthquake loads, and
   (c) imposed loads.
18. How is the seismic design philosophy different from the design for other loads such as wind and live load?
19. What are the partial load factors adopted for the following loading combinations (a) DL, (b) LL, (c) WL, (d) DL + LL.
20. List the different analysis methods considered by the code.
21. What are the advantages of advanced analysis methods?
22. List a few Indian standard codes that are to be followed while designing steel structures.
23. How does the limit states design method differ from the working stress method?
24. What are the limitations and shortcomings of the working stress method?
25. What are the possible limit states that are considered in the limit states method?
26. What are the partial safety factors for the materials adopted by the IS 800:2007 code?
27. What are the three forms of structural stability considered by the code?
28. State the various serviceability limit states considered by the code.
29. Why is serviceability limit state considered as important as failure limit states?
30. State the Huber, von Mises, and Hencky failure theory.

Design of Tension Members

Introduction

Steel tension members are probably the most common and efficient members in the structural applications. They are those structural elements that are subjected to direct axial tensile loads, which tend to elongate the members. A member in pure tension can be stressed up to and beyond the yield limit and does not buckle locally or overall. Hence, their design is not affected by the classification of sections, for example, compact, semi-compact, etc. as described in Chapter 4. The design stress f_y as determined from Table 1.3, is therefore not reduced.

Tension members occur as components of trusses (bottom chord of roof trusses), bridges, transmission line and communication towers, and wind bracing systems in multi-storey buildings (see Fig. 3.1). Some truss web members and members in towers may carry tension under certain loading cases and may be subjected to compression for other loading cases. Steel cables used in suspension bridges and in cable supported roofs are also examples of tension members. Such cables are also used in guyed towers as well as power line poles where alignment changes occur.

Tension members carry loads most efficiently, since the entire cross section is subjected to uniform stress. The strength of these members is influenced by several factors such as the length of connection, size and spacing of fasteners, net area of cross section, type of fabrication, connection eccentricity, and shear lag at the end connection. To simplify the design procedure of tension members, considerable amount of research has been carried out (Salmon & Johnson 1996; Kulak & Wu 1997). This chapter discusses the effects of these parameters and the design of tension members as per IS 800.

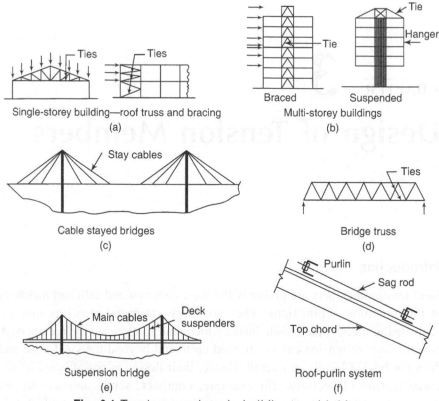

Fig. 3.1 Tension members in buildings and bridges

3.1 Types of Tension Members

Tension members may consist of single structural shape or they may be built using a number of structural shapes. The cross section of some typical tension members are shown in Fig. 3.2. When two elements such as two angles are used as a single member, they should be interconnected at reasonable intervals to enable them to act together as one member. These two separate elements are often placed parallel to each other with a gap of about 6–10 mm between them. They should also have *spacer plates* placed at regular intervals between them, which are connected to these individual elements by tack welds or bolts as shown in Fig. 3.3. Though in Fig. 3.3 only welding is shown, similar rules apply for a bolted spacer (also called as *stitch plates*). Single angle or double angles are either bolted to a single gusset plate at each end or may be welded directly to the webs or flanges of T- or I-chord members.

Structural T-sections may be used as chord members of lightly loaded trusses, instead of the back-to-back two angle sections. The stem of the T-sections may be used to connect the single or double angle web members, thus eliminating the use of any gusset plate, especially in welded connections. Tubular members are also used in roof trusses as tension members.

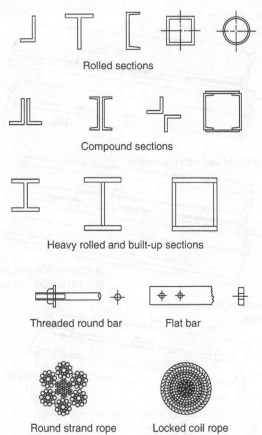

Rolled sections

Compound sections

Heavy rolled and built-up sections

Threaded round bar   Flat bar

Round strand rope   Locked coil rope

Fig. 3.2 Cross section of typical tension members

I-sections, channel sections, and built-up sections using angles, channels, etc. are used when greater rigidity is required and hence are often used in bridge structures.

Rods and bars are used as tension members in the bracing systems. As mentioned earlier, sag rods are used to support purlins, to support girt in industrial buildings, or as longitudinal ties. They are either welded to the gusset plates or threaded and bolted to the main members directly using nuts. When rods are used as wind bracings, they are pre-tensioned to reduce the effect of sway.

3.2 Slenderness Ratio

Although stiffness is not required for the strength of a tension member, a minimum stiffness is stipulated by limiting the maximum slenderness ratio of the tension member. The *slenderness ratio* of a tension member is defined as the ratio of its unsupported length (L) to its least radius of gyration. This limiting slenderness ratio is required in order to prevent undesirable lateral movement or excessive vibration. (As stated already, stability is of little concern in tension members.) The slenderness limits specified in the code for tension members are given in Table 3.1.

Plate to a rolled shape, or
two plates in contact with each other

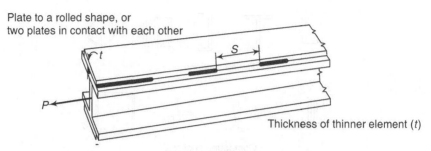

Thickness of thinner element (t)

Two or more shapes in contact with each other

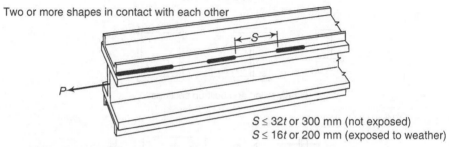

$S \leq 32t$ or 300 mm (not exposed)
$S \leq 16t$ or 200 mm (exposed to weather)

Two or more shapes or plates, separated
by intermittent fillers

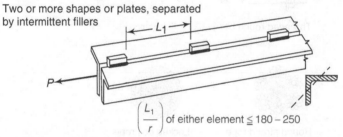

$\left(\dfrac{L_1}{r}\right)$ of either element $\leq 180 - 250$

$L_1 \not> 1000$ mm (tension)
$L_1 \not> 600$ mm (compression)

Tie plates used on open sides of
built-up tension members

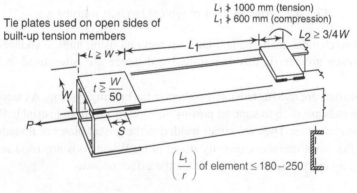

$L_2 \geq 3/4\,W$

$t \geq \dfrac{W}{50}$

$\left(\dfrac{L_1}{r}\right)$ of element $\leq 180 - 250$

Plates with access holes may be used in
built-up tension members

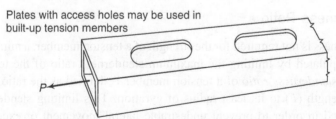

Fig. 3.3 Tack welding of built-up tension members

Table 3.1 Maximum values of effective slenderness ratios as per IS 800

Member	Maximum effective slenderness ratio (L/r)
A tension member in which a reversal of direct stress occurs due to loads other than wind or seismic forces	180
A member subjected to compressive forces resulting only from a combination of wind/earthquake actions, provided the deformation of such a member does not adversely affect the stresses in any part of the structure	250
A member normally acting as a tie in a roof truss or a bracing member, which is not considered effective when subject to reversal of stress resulting from the action of wind or earthquake forces	350
Members always in tension (other than pre-tensioned members)	400

3.3 Displacement of Tension Members

The increase in the length of a member due to axial tension under service loads is

$$\Delta = PL /(EA_g) \tag{3.1}$$

where Δ is the axial elongation of the member (mm), P is the axial tensile force (un-factored) in the member (N), L is the length of the member (mm), and E is the modulus of elasticity of steel = 2.0×10^5 MPa. Note that displacement is a serviceability limit state criterion and hence is checked under service loads and not under factored loads.

3.4 Behaviour of Tension Members

The load–deformation behaviour of an axially loaded tension member is similar to the basic material stress–strain behaviour (see Fig. 1.4). When a member is subjected to tension, the area of cross section and the gauge length continuously change due to the Poisson effect and longitudinal strain, respectively (see Section 1.8.1 also). Stresses and strains may be calculated using the initial area of cross section and the initial gauge length, which is referred to as the engineering stress and engineering strain or using the current area of cross section and the current gauge length, which is referred to as the true stress and true strain.

The engineering stress–strain curve does not give a true indication of the deformation characteristics of a metal because it is based entirely on the original dimensions of the specimen, and these dimensions change continuously as the load increases. In fact, post-ultimate strain softening in engineering stress–strain curve caused by the necking of the cross section is completely absent in the true stress–strain curve. When the true stress based on the actual cross-sectional area of the specimen is used, it is found that the stress–strain curve increases continuously until fracture occurs. The true stress–strain curve is also known as *flow curve* since it represents the basic plastic flow characteristics of the material. Any point on the flow curve can be considered as the local stress for a metal strained in tension by the magnitude shown on the curve. However, since it is difficult to obtain the ordinates of true stress–strain curve, the engineering stress–strain curve is often utilized. As discussed in Section 1.8.1 and shown in Fig. 1.4(a), high-strength steel tension members do not exhibit a well-defined yield point and yield plateau. Hence the 0.2% offset load is usually taken as the yield point for such high-strength steel.

3.5 Modes of Failure

In the following sections, the different modes of failure of tension members are discussed.

3.5.1 Gross Section Yielding

Generally a tension member without bolt holes can resist loads up to the ultimate load without failure. But such a member will deform in the longitudinal direction considerably (nearly 10%–15% of its original length) before fracture. At such a large deformation a structure becomes unserviceable. Hence, code limits design strength in clause 6.2; substituing for γ_{m0}, which is the partial safety factor for failure in tension by yielding ($\gamma_{m0} = 1.10$), we get

$$T_{dg} = 0.909 \, f_y A_g \tag{3.2}$$

where A_g is the gross area of cross section in mm^2, and f_y is the yield strength of the material (in MPa).

3.5.2 Net Section Rupture

A tension member is often connected to the main or other members by bolts or welds. When connected using bolts, tension members have holes and hence reduced cross section, being referred to as the *net area*. Holes in the members cause stress concentration at service loads, as shown in Fig. 3.4(a). From the theory of elasticity, we know that the tensile stress adjacent to a hole will be about two to three times the average stress on the net area, depending upon the ratio of the diameter of the hole to the width of the plate normal to the direction of stress. Stress concentration becomes very significant when repeated applications of load may lead to fatigue failure or when there is a possibility of a brittle fracture of a tension member under dynamic loads. Stress concentration may be minimized by providing suitable joint and member details.

When a tension member with a hole is loaded statically, the point adjacent to the hole reaches the yield stress f_y first. On further loading, the stress at that point remains constant at yield stress and each fibre away from the hole progressively reaches the yield stress f_y [see Fig. 3.4(b)]. Deformations continue with increasing load until finally rupture (tension failure) of the member occurs when the entire net cross section of the member reaches the ultimate stress f_u. The design strength due to net section rupture for plates is given in Section 6.3.1 of the code. Substituting the value for γ_{ml} which is the partial safety factor for failure due to rupture of cross section (= 1.25), we get,

$$T_{dn} = 0.72 f_u A_n \tag{3.3}$$

where A_n is the net effective area of the cross section in mm^2, and f_u is the ultimate strength of the material in MPa. Because of strain hardening, the actual strength of a ductile member may exceed that indicated by Eqn (3.2). However, since there is no reserve of any kind beyond the ultimate resistance an additional multiplier of 0.90 has been introduced in Eqn (3.3). Such a high margin of safety has been traditionally used in design when considering the fracture limit state than for the yielding limit state (Salmon & Johnson 1996). The 0.9 factor was included in the strength equation of Eqn (3.3), based on a statistical evaluation of a large number of test results for net section failure of plates.

Similarly, threaded rods subjected to tension could fail by rupture at the root of the threaded region. Thus, the design strength of the threaded rods in tension is given by Eqn (3.3) where A_n is the net root area at the threaded sections.

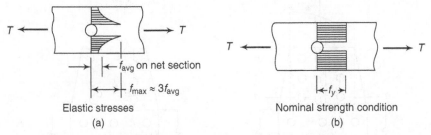

Elastic stresses
(a)

Nominal strength condition
(b)

Fig. 3.4 Stress concentration due to holes

3.5.3 Block Shear Failure

Originally observed in bolted shear connections at coped beam ends, block shear is now recognized as a potential failure mode at the ends of axially loaded tension members also. In this failure mode, the failure of the member occurs along a path involving tension on one plane and shear on a perpendicular plane along the fasteners as shown in Fig. 3.5. Other examples of *block shear failures* including failures in welded connections are given in Fig. 3.6.

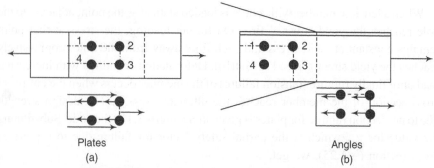

Plates
(a)

Angles
(b)

Fig. 3.5 Block shear failure in plates and angles

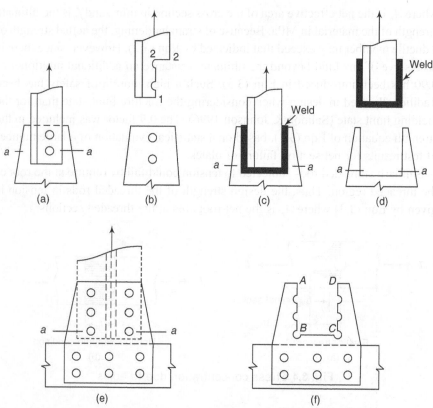

(a)   (b)   (c)   (d)

(e)   (f)

Fig. 3.6 Examples of block shear failure

It can be observed as shown in Fig. 3.6(a) that the gusset plate may fail in tension on the net area of section *a-a*, and in Fig. 3.6(c) it may fail on the gross area of section *a-a*. The angle member in Fig. 3.6(a) may also separate from the gusset plate by shear on net area 1-2 combined with tension on net area 2-2 as shown in Fig. 3.6(b). A similar fracture of the welded connection of Fig. 3.6(c) is shown in Fig. 3.6(d). The fracture of a gusset plate for a double angle member or of one of the gusset plates for an I-Section [Fig. 3.6(e)] is shown in Fig. 3.6(f). The gusset plate in Fig. 3.6(d) may also fail on the net section a-a. All these failures [Figs 3.6(b), (d), and (f)] are called block shear failures.

The block shear phenomenon becomes a possible mode of failure when the material bearing strength and bolt shear strength are higher. As indicated earlier, the appropriate model of the block shear failure is the rupturing of the net tension plane (BC) and yielding on the gross shear plane (AB and CD), as shown in Fig. 3.6(f), which results in rupturing of the shear plane as the connection lengths become shorter.

The block shear strength is given in section 6.4.1 of the code. Substituting the value of γ_{m0} (= 1.1) and γ_{m1} (1.25), we get the following:

(a) Plates: The block shear strength T_{db} of the connection is taken as the smaller of

$$T_{db1} = 0.525 \, A_{vg} f_y + 0.72 \, f_u A_{tn} \tag{3.4a}$$

$$T_{db2} = 0.416 \, f_u A_{vn} + 0.909 \, f_y A_{tg} \tag{3.4b}$$

where A_{vg} and A_{vn} are the minimum gross and net area in shear along a line of transmitted force, respectively [1-2 and 4-3 as shown in Fig. 3.5(a) and 1-2 as shown in Fig. 3.5(b); A_{tg} and A_{tn} are the minimum gross and net area in tension from the hole to the toe of the angle or next last row of bolt in plates, perpendicular to the line of force respectively [2-3 as shown in Figs 3.5(a) and (b)]; and f_u and f_y are the ultimate and yield stress of the material, respectively.

(b) Angles: Strength as governed by block shear failure in angle end connection is calculated using Eqn (3.4) and appropriate areas in shear and tension as shown in Fig. 3.5(b).

The lower values of the design tension capacities, as given by Eqns (3.2) to (3.4), govern the design strength of plates or members with hole and should be greater than the factored design tension. Note that no net areas are involved in the failures of welded connections [see Fig. 3.6(c)]. Therefore, in applying Eqn (3.4) to this case in the second term of Eqn (3.4a), use A_{tg} (instead of A_{tn}) and in the first term of Eqn (3.4b), use A_{vg} (instead of A_{vn}).

Recently, Driver et al. (2006) proposed a unified equation for block shear failure to predict the capacities of angles, tees, gusset plates, and coped beams.

Based on this work, the 2009 version of the Canadian code has adopted the following equation

$$T_{db} = 0.75[U_t A_{tn} f_u + 0.6 \, A_{vg} (f_y + f_u)/2] \tag{3.4c}$$

where U_t is the efficiency factor and equals 1.0 for flange connected tees and for symmetric failure patterns and concentric loading; 0.6 for angles connected by one leg and stem connected tees; 0.9 for coped beams with one bolt line; and 0.3 for coped beams with two bolt lines. For $f_y > 485$ MPa, $(f_y + f_u)/2$ should be replaced by f_y. Other terms are defined already.

3.6 Factors Affecting the Strength of Tension Members

As discussed already, the yielding of the gross section of tension member causes excessive elongation and hence the load corresponding to the yielding of gross section is taken as one limit state. However, the net section through the bolt holes at the ends of the member may be subjected to tensile stresses well in excess of the yield stress to as high as ultimate stress without the member suffering excessive elongation. Hence, the rupture strength of the net section through the bolt holes at the ends is considered another limit state. Several factors affect the rupture strength of the net section of tension members. They are briefly described below.

3.6.1 Effect of Bolt Holes

In order to make connections, tension members are often bolted to adjacent members directly or by using gusset plates. These bolt holes reduce the area of cross section available to carry tension and hence affect the strength as discussed in the following section.

3.6.1.1 Methods of fabrication

There are generally two methods of making holes to receive bolts, namely punching and drilling. Due to punching, the material around the holes is deformed in shear beyond ultimate strength to punch out the hole.

Under cyclic loading the material around the punched holes present the greatest scope for crack initiation due to stress concentration, and hence punched hole is not allowed under fatigue environment.

Presently in many specifications, the punching effect upon the net section strength is accounted for by taking the hole diameter as 2 mm larger than the actual hole size when computing the net area (see clause 3.6.1 of IS 800).

3.6.1.2 Net area of cross section

The presence of a hole tends to reduce the strength of a tension member. When more than one bolt hole is present, the failure paths may occur along sections normal to the axis of the member, or they may include zigzag sections, if the fasteners are staggered (Fig. 3.7). Staggering holes improves the load carrying capacity of the member for a given row of bolts. When the bolts are arranged in a zigzag fashion with a pitch p and gauge g, the net effective area of the plate with a width B and thickness t is given by

$$A_n = [B - nd_h + \Sigma(p^2/4g)]t \tag{3.5}$$

where n is the number of bolt holes in the critical section considered, the summation is over all the paths of the critical sections normal to the direction of the tensile force, and d_h is the diameter of the bolt hole. The above empirical relation was proposed by Cochrane in 1922 based on experimental evidence. All possible failure paths (straight as well as zigzag) are to be considered and the corresponding net areas are to be computed as per Eqn (3.5) to find the minimum net area of the plate.

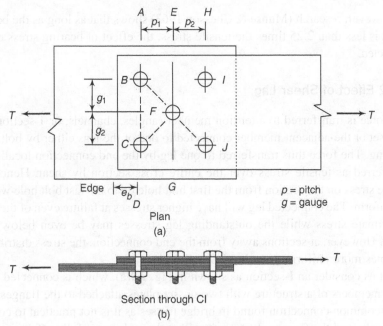

Section through Cl
(b)

Fig. 3.7 Staggered bolt holes

3.6.1.3 Effect of bearing stress

When slip takes place between plates being joined by bolts, one or more fasteners come into bearing against the side of the hole. Consequently bearing stress is developed in the material adjacent to the hole and in the fastener. Initially this stress is concentrated at the point of contact. An increase in load causes local yielding and a larger area of contact resulting in a more uniform bearing stress distribution. The actual failure mode in bearing depends on the end distance, the bolt diameter, and the thickness of the connected material. Either the fastener splits out through the end of the plate because of the insufficient end distance or excessive deformations are developed in the material adjacent to the hole and the elongation of the hole takes place as shown in Fig. 3.8. Often a combination of these failure modes will occur.

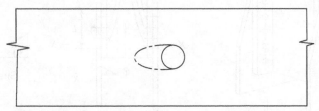

Fig. 3.8 Elongation of bolt hole due to local yielding under bearing stress

However, research (Munse & Chesson 1963) shows that as long as the bearing stress is less than 2.25 times the tensile stress, the effect of bearing stress can be neglected.

3.6.2 Effect of Shear Lag

The force is transferred to a tension member (angles, channels, or T-sections) by a gusset or the adjacent member connected to one of the legs either by bolting or welding. The force thus transferred to one leg by the end connection locally gets transferred as tensile stress over the entire cross section by shear. Hence, the tensile stress on the section from the first bolt hole up to the last bolt hole will not be uniform. The connected leg will have higher stresses at failure even of the order of ultimate stress while the outstanding leg stresses may be even below yield stress. However, at sections away from the end connection, the stress distribution becomes more uniform. (See Fig 3.9 and 3.11).

Let us consider an I-section as shown in Fig. 3.9(a), which is connected to the other members of a structure with two gusset plates attached to the flanges. This is the common connection found in bridge trusses as it is not practical to connect both webs and the flanges. It is obvious that the web is not fully effective in the region of the connection. As shown in Fig. 3.9(b), only a distance away from the connection, there will be uniform stress throughout the section. Because the internal transfer of forces from the flange region into the web region will be by shear (in the case of angles, transfer of force from one leg to the other will be by shear as indicated later in Fig. 3.11) and because one part 'lags' behind the other, the phenomenon is referred to as *shear lag*. The shear lag reduces the effectiveness of the component plates of a tension member that are not connected directly to a gusset plate.

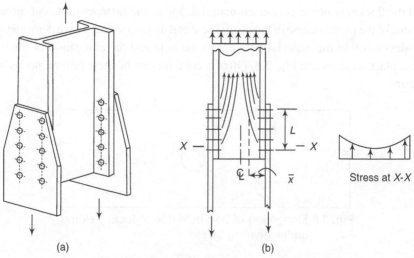

(a)   (b)

Fig. 3.9 Shear lag in tension member.

The shear lag effect reduces with increase in the connection length. In addition to its effect on shear lag, an increase in the connection length of a specimen also allows for a larger restoring moment at the eccentric connection. A longer connection length thus increases the net section efficiency.

The study conducted by Kulak & Wu (1997) revealed that the net section efficiency increases when the number of bolts in a line is increased up to four and after that there is no appreciable increase in the efficiency.

3.6.3 Geometry Factor

Tests on bolted joints show that the net section is more efficient if the ratio of the gauge length g to the diameter d is small (Kulak et al. 1987). The increase in the efficiency due to a smaller g/d ratio is due to the suppression of contraction at the net section.

To account for the effect of gauge or g/d ratio, Munse and Chesson (1963) proposed a geometry factor, K_3, given by Eqn (3.6), which is multiplied with the net section to account for this effect.

$$K_3 = 1.60 - 0.70(A_{ne}/A_g) \qquad (3.6)$$

The value of K_3 generally varies in the range of 0.9 to 1.14.

3.6.4 Ductility Factor

Tension members with bolt holes made from ductile steels have proved to be as much as one-fifth to one-sixth times stronger than similar members made from less ductile steels having the same strengths (Kulak et al. 1987). To account for this effect, Munse and Chesson (1963) proposed a reduction factor

$$K_1 = 0.82 + 0.0032R_a \leq 1.0 \qquad (3.7)$$

where R_a is the area reduction ratio before rupture. In case of commonly used structural steels exhibiting minimum prescribed ductility, the ductility factor K_1 equals 1.0.

3.6.5 Spacing of Fasteners

The closer spacing of fasteners relative to their diameter may sometimes lead to block shear failure at the ends as discussed in Section 3.5.3, which has to be accounted for as a limit state.

3.7 Angles Under Tension

As mentioned earlier angles are used extensively as tension members in towers, trusses, and bracings. Angles, if axially loaded through centroid (as in the case of tower legs), could be designed as in the case of plates. However, in many cases, angles are connected to gusset plates (which in turn are connected to the other members of a structure) by welding or bolting only through one of the two legs (see Fig. 3.10). This kind of connection results in eccentric loading, causing non-uniform distribution of stress over the cross section (see Fig. 3.11). Further, since the load is applied by connecting only one leg of the member, there is a shear lag at the end connections.

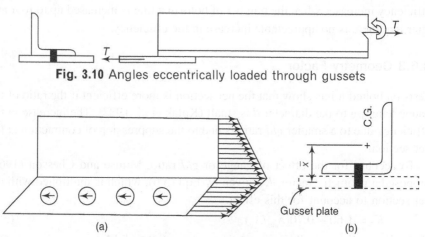

Fig. 3.10 Angles eccentrically loaded through gussets

Fig. 3.11 Distribution of stresses in an angle

When the angles are connected to other angles through the centroid and when the holes are staggered on two legs of an angle (see Fig. 3.12), the gauge length g for use in Eqn (3.5) is obtained by developing the cross section into an equivalent flat plate [see Fig. 3.12(b)] by revolving about the centre lines of the component parts. The critical net section can then be established by the procedure described for plates. Thus referring to Fig. 3.12(b), the gauge distance g^* is obtained as

$$g^* = g_a - (t/2) + (g_b - t/2) = g_a + g_b - t \qquad (3.8)$$

An illustration of the calculations involved is given in Example 3.3. Examples of net section calculations are also provided in Chapter 10.

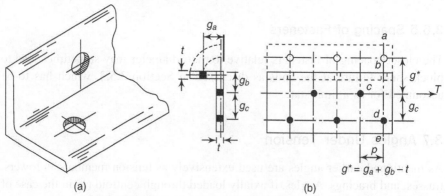

Fig. 3.12 Fasteners in more than one plane

3.7.1 Net Section Design

It was shown in Section 3.6 that the strength of a tension member with bolted or riveted connections can be predicted with good accuracy by taking into account the various factors affecting the strength of the net section. Hence, the following procedure has been suggested for the design of such members (Gaylord et al. 1992; Munse & Chesson 1963). To provide for necessary margin of safety against fracture, the capacity of the member should be determined by multiplying the effective net cross-sectional area with the specified minimum tensile strength f_u divided by the partial safety factor γ_{m1}. Thus, the effective net area is defined by (Gaylord et al. 1992).

$$A_{ne} = K_1 K_2 K_3 K_4 A_n \tag{3.8}$$

where A_n is the net area of the cross-section obtained by $p^2/4g$ rule [see Eqn (3.5)] and A_{ne} is the effective area of the cross-section modified by the other terms in the equation. These terms represent a ductility factor (k_1), a factor for the method of hole forming (k_2), a geometry factor reflecting hole spacing (k_3), and a shear lag factor (k_4).

Kulak & Wu (1997) conducted several tests on bolted angle tension members and studied the effects of various factors such as out-of-plane stiffness of gusset plates, angle thickness, connection by long leg and short leg, and connection length. They proposed the following net section strength formula based on these tests and the finite element analysis done by them (without the partial safety factor for material)

$$T_u = f_u A_{nc} + \beta f_y A_{go} \tag{3.9}$$

where A_{nc} is the net area of the connected leg at the critical cross section, A_{go} is the gross area of the outstanding leg and $\beta = 1.0$ for members with four or more fasteners per line in the connection, or $\beta = 0.5$ for members with two or three fasteners per line in the connection. Kulak & Wu compared the ultimate load predicted by Eqns (3.9) and found that it provides conservative results and falls in a narrower scatter band of results. They also proposed the following equation for single and double angles connected by only one of the legs and made of Fe 410 steel

$$T_u = U A_n f_u \tag{3.10}$$

where A_n is the net area of the critical cross section calculated using a hole diameter 2-mm greater than the nominal diameter of the bolt and using the $p^2/4g$ rule if staggered holes are present and $U = 0.80$ if the connection has four or more member of fasteners per line and 0.60 if there are two or three fasteners per line. They also concluded that Eqn (3.10) gives conservative results if it is applied to steel grades for which f_y/f_u is greater than 0.62. The above recommendations for angles have been adopted in ANSI/AISC code.

The research done by Kulak & Wu (1997) has also shown the following.

(a) The effect of the gusset plate thickness, and hence the out-of-plane stiffness of the end connection, on the ultimate tensile strength is not significant.

(b) The thickness of the angle has no significant influence on the member strength.

(c) When the length of the connection increases, the tensile strength increases up to four bolts and the effect of any further increase in the number of bolts, on the tensile strength of the member, is not significant.

(d) The net section efficiency is higher (7%–10%) when the long leg of the angle is connected, rather than the short leg.

(e) Because of local bending, each angle of a double angle member bends about the bolt line on each side of the gusset plate; thus the double angles seem to act individually rather than as a rigid unit.

3.7.2 Indian Code (IS 800 : 2007) Provisions for Angle Tension Members

The net section strength of single and double angle tension members (either bolted or welded) and connected through one leg (including the shear lag effect) given in the code is based on the research done at IIT, Madras (Usha 2003; Usha & Kalyanaraman 2002). The design strength as governed by the tearing of the net section is given in clause 6.3.3 of the code. Substituting the values of $\gamma_{m1} = 1.25$ and $\gamma_{m0} = 1.1$, we get

$$T_{dn} = 0.72 f_u A_{nc} + 0.909 \ \beta \ A_{go} f_y \tag{3.11}$$

and

$$\beta = 1.4 - 0.076[(b_s/L_c)(w/t)(f_y/f_u)] \leq 0.79 \ f_u/f_y \geq 0.7 \tag{3.12}$$

where f_u and f_y are the ultimate and yield stress of the material, w and t are the size and thickness of the outstanding leg, respectively, b_s is the shear distance from the edge of the outstanding leg to the nearest line of fasteners, measured along the centre line of the legs in the cross section (see Fig. 3.13), L_c is the length of the end connection measured from the center of the first bolt hole to the centre of the last bolt hole in the end connection or length of the weld along load direction and, A_{nc} is the net area of the connected leg at critical cross section, computed after deducting the diameter of hole (the diameter of the holes should be taken as 2-mm larger than the nominal size in the case of punched holes), and A_{go} is the gross area of the outstanding leg.

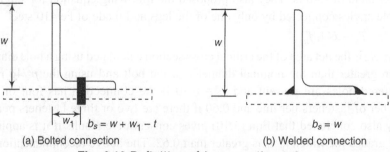

w_1 $b_s = w + w_1 - t$	$b_s = w$
(a) Bolted connection	(b) Welded connection

Fig. 3.13 Definition of b_s as per the code.

Alternatively, the IS code suggests the use of an equation, for preliminary design, similar to Eqn (3.10), with the partial factor of safety for material $\gamma_{m1} = 1.25$, we get

$$T_{dn} = 0.8 \alpha A_n f_u \tag{3.13}$$

with $\alpha = 0.6$ for one or two bolts, 0.7 for three bolts, and 0.8 for four or more bolts in the end connections or equivalent weld length.

It is important to observe that in the case of welds, the determination of the value of α is difficult since the welds may be transverse, longitudinal, or combined. Designers have to use their judgement to arrive at an equivalent number of bolts.

3.8 Other Sections

The tearing strength T_{dn} of double angles, channels, I-sections, and other rolled steel sections, connected by one or more elements to an end gusset is also governed by shear lag effects (see Section 3.6.2). The code suggests that the design tensile strength of such sections, as governed by the tearing of the net section, may also be calculated by using Eqn (3.11) to (3.13), where β is calculated based on the shear lag distance b_s, taken from the farthest edge of the outstanding leg to the nearest bolt/weld line in the connected leg of the cross section. The net effective area of a single channel section connected through the web may be treated as for double angles connected by one leg each to the gusset. When a rolled or built-up channel or I-sections are connected through flanges, the web is found to be partially ineffective in resisting the tensile load. In such cases, the net area may be taken as the total area minus half the web area (Duggal 2000).

3.9 Tension Rods

A common and simple tension member is the threaded rod. Such rods are usually found as secondary members, where the required strength is small. Some examples of tension rods are as follows.

(a) Sag rods are used to help support purlins in industrial buildings. Various arrangements of sag rods are shown in Fig. 3.14. Sag rods reduce the bending moment about the minor axis (channels which are used as purlins are weak about the minor axis), resulting in economy. These rods are threaded at the ends and bolted to purlins. Since individual sag rods are placed between successive pairs of purlins, they may be designed individually, each to carry a tangential component from all the purlins below it. Thus, sag rods just below and suspended from the ridge purlin will be subjected to maximum force.

(b) Vertical ties to help support girts in the walls of industrial buildings [Fig. 3.15(a)].

(c) Hangers, such as tie rods supporting a balcony [Fig. 3.15(b)]. When providing such hangers, proper detailing should be adopted and communicated properly to the fabricator.

(d) Tie rods to resist the thrust of an arch.

As mentioned previously, tie rods are also often used with an initial tension as diagonal wind bracings in walls, roofs, and towers. The initial tension reduces the deflection and vibration and also increases the stiffness of these rods.

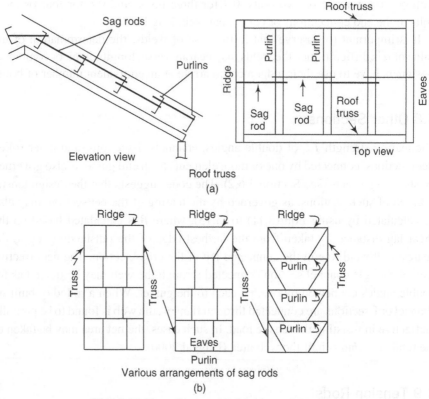

Fig. 3.14 Use of sag rods to support purlins

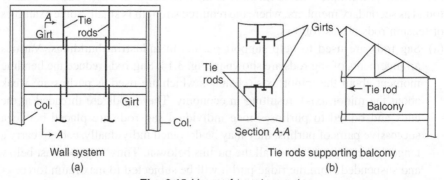

Fig. 3.15 Uses of tension rods

3.10 Design of a Tension Member

In the design of a tension member, based on the tensile force acting on the member, the designer has to arrive at the type and size of the member. The type of member is chosen based on the type of the structure and location of the member (e.g., double angles at the bottom chord or a rafter of roof trusses, angles or pipes for web members of roof trusses, etc.).

The design is iterative, involving a choice of a trial section and an analysis of its capacity. The various steps are as follows:

1. The net area required A_n to carry the design load T is obtained by the equation

$$A_n = T_u/(f_u/\gamma_{m1}) \tag{3.14}$$

2. From the required net area, the gross area may be computed by increasing the net area by about 25% to 40%. The required gross area may also be checked against that required from the yield strength of the gross section as follows

$$A_g = T_u/(f_y/\gamma_{m0}) \tag{3.15}$$

A suitable trial section may be chosen from the steel section tables (IS 808: 1989) to meet the required gross area.

3. The number of bolts or welding required for the connections is calculated. They are arranged in a suitable pattern and the net area of the chosen section is calculated. The design strength of the trial section is evaluated using Eqns (3.2) to (3.4) in the case of plates and threaded bars and additionally using Eqns (3.11) to (3.13) in the case of angles.

4. If the design strength is either small or too large compared to the design force, a new trial section is chosen and Step 3 is repeated until a satisfactory design is obtained.

5. The slenderness ratio of the member is checked as per Table 3.1.

3.11 Lug Angles

When a tension member is subjected to heavy load, the number of bolts or the length of weld required for making a connection with other members becomes large; resulting in uneconomical size of the gusset plates. In such situations, an additional short angle may be used to reduce the joint length and shear lag as shown in Fig. 3.16. Such an angle is called the lug angle. The location of the lug angle is of some importance; it is more effective at the beginning of the connection, as in Fig. 3.16, rather than at the end. The use of lug angles with angles or channels reduces the net area of the main members due to the additional bolt holes in projected members. This reduction in the net area of the member should not be excessive. In the connections of the lug angles to the member or the gusset plate more than two bolts are used. Since both legs of the angles or channels are connected to the lug angles, the net area of the members should be calculated simply as gross area minus the area of the holes.

Lug angles may be eliminated by providing unequal angle sections with the wider leg as the connected leg and using two rows of staggered bolts. In many cases, the cost of providing the lug angles (including their connection and the extra fabrication required to make the holes) may be found to be expensive than providing extra length and thickness of gusset plate. Hence, they are not used in practice.

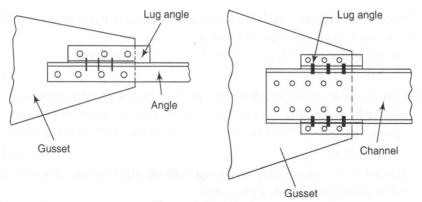

Fig. 3.16 Lug angles

In the case of angle members, the lug angles and their connection to the gusset or other supporting member should be capable of developing a strength of not less than 20% in excess of the force in the outstanding leg of the angle, and the connection of the lug angle to the angle member should be capable of developing 40% in excess of the force (clause 10.12.2).

In channel members, however, the lug angles and their connections to the gusset or other supporting member should be capable of developing a strength of not less than 10% in excess of the force in the flange of the channel and the attachment of the lug angle to the member should have a strength not less than 20% in excess of that force (clause 10.12.3).

3.12 Splices

When the available length is less than the required length of a tension member, splices are provided. The various types of splices that can be provided are shown in Figs 3.17(a) to (c). If the sections are not of the same thickness, packings are introduced, as shown in Fig. 3.17(d). Moreover, reduction in capacity of bolt has to be considered for long joints or if the packing thickness is greater than 6 mm (see Chapter 10 for these calculations).

In the design of a tension splice, the effect of eccentricity is neglected; as far as possible it should be avoided. Thus Fig. 3.17(e) shows an angle section spliced on one leg of the angle only by a plate. Such an arrangement causes eccentricity and introduces bending moments. To overcome this, both the legs of the angle should be spliced, as shown in Fig. 3.17(a). The splice as shown in Fig. 3.17(b) is used in the legs of transmission line or communication towers and aids transfer of tensile loads, without any eccentricity.

The splice cover plates or angles and its connections should be designed to develop the net tensile strength of the main member. The forces in the main member are transferred to the cover plate angle sections through the bolts/welding and carried through these covers across the joint and is transferred to the other portion of the section through the fasteners. For examples of tension splices, see Chapter 10.

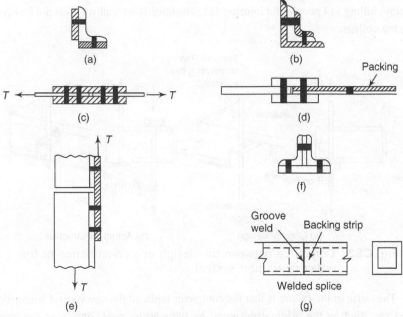

Fig. 3.17 Splices in tension members

Hyatt Regency Walkway Collapse

The 40-story Hyatt Regency Hotel in Kansa City, Missouri, USA, was opened on 1 July 1980. The lobby of the hotel featured a multistory atrium, which had suspended concrete walkways on the second, third, and fourth levels. These three separate pedestrian walkways connected the north and south buildings. The fourth- and second-floor walkways hung one above the other and the third-floor walkway hung offset to one side (see Fig. CS1). These walkways all connected to steel trusses that hung from the atrium ceiling.

The two walkways were suspended from a set of steel tension rods of 32 mm, with the second-floor walkway hanging directly underneath the fourth-floor walkway. The walkway platform was supported on three cross-beams suspended by steel rods retained by nuts. The cross-beams were box beams made from C-channels welded toe-to-toe. The original design called for three pairs of rods running from the second floor all the way to the ceiling.

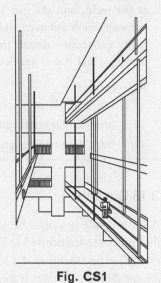

Fig. CS1

On 17 July 1981, when a party was going on, the fourth-floor walkway failed and fell on the lower walkway, both walkways crashing onto the floor three stories below, killing 114 people and injuring 185. The third-floor walkway was not involved in the collapse.

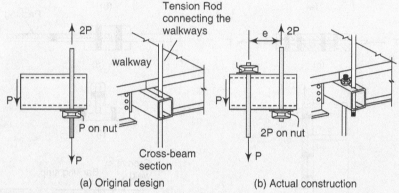

(a) Original design   (b) Actual construction

Fig. CS2 Difference between the design and construction of the walkway support system

The cause of the failure is that the contractor replaced the one vertical suspension rod specified by the original designer, by two shorter rods; one from the upper support to the first walkway, and another from the bottom of the first walkway down to the second walkway (see Figure CS2(b)). Now the nut and washer under the upper rod is subjected to double the design load (in addition the eccentricity created a local bending moment), which led to the failure. Photographs of the wreckage showed excessive deformations of the cross-section; the box beams split at the weld, and the nut supporting them slipped through. Lack of proper communication and overlooking the details were cited as the main problems for the faulty connection detail; the connection that failed was never shown on any drawings, and it was not designed.

References:
1. http://en.wikipedia.org/wiki/Hyatt_Regency_walkway_collapse
2. http://ethics.tamu.edu/ethics/hyatt/hyatt2.htm
3. http://failurebydesign.info/

3.13 Gussets

A gusset plate is a plate provided at the ends of tension members through which the forces are transferred to the main member. Gusset plates may be used to join more than one member at a joint. The lines of action of truss members meeting at a joint should coincide. If they do not coincide, secondary bending moments and stresses are created, which should be considered in the design.

The size and shape of the gusset plates are decided based on the direction of various members meeting at a joint. The plate outlines are fixed so as to meet the minimum edge distances specified for the bolts that are used to connect the various members at a particular joint. The shape of the gusset plate should be such that it should give an aesthetic appearance, in addition to meeting the edge distances of bolts, as mentioned earlier.

It is tedious to analyse the gusset plate for shear stresses, direct stresses, and bending stresses, and hence empirical methods have been used in the past to arrive at the thickness of the gusset plate (e.g., Whitmore method). More details of these methods are discussed in Chapter 10. The block shear model (see Section 3.5.3) could also be used to find the thickness of the gusset plate. It is a usual practice to provide the thickness of the gusset plate equal to or slightly higher than the thickness of members that are connected by the gusset plate. It is interesting to note that the failure of the I-35W bridge at Minneapolis, USA in August 2007, was due to the inadequate thickness of the gusset plate.

Examples

Example 3.1 *What is the net area A_n for the tension member shown in Fig. 3.18, in case of (a) drilled holes, (b) punched holes?*

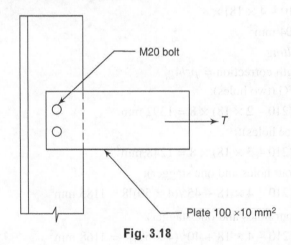

Fig. 3.18

Solution

$$A_g = 100 \times 10 = 1000 \text{ mm}^2$$

(a) Hole made by drilling

Hole for M20 bolt = 22 mm

$A_n = A_g - n$ (hole × thickness of plate)

$$= 1000 - 2\,(22 \times 10) = 560 \text{ mm}^2$$

(b) Holes made by punching

Hole = 22 + 2 = 24 mm

$$A_n = 1000 - 2(24 \times 10) = 520 \text{ mm}^2$$

Example 3.2 *Determine the minimum net area of the plates as shown in Figs. 3.19(a) and (b) with a plate of size of 210 × 8 mm and 16-mm bolts.*

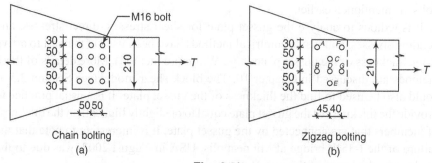

Chain bolting
(a)

Zigzag bolting
(b)

Fig. 3.19

Solution
(a) Chain bolting
For a 16-mm bolt, hole diameter = 18 mm

Net area = $(b - nd)t$

$\qquad = (210 - 4 \times 18) \times 8$

$\qquad = 1104$ mm^2

(b) Zigzag bolting
Staggered length correction = $p_i^2/4g_i$

Path AB and FG (two holes):

$\quad$ Net area = $(210 - 2 \times 18) \times 8 = 1392$ mm^2

Path CDE (three holes):

$\quad$ Net area = $(210 - 3 \times 18) \times 8 = 1248$ mm^2

Path ACDE (four holes and one stagger):

$\quad$ Net area = $[210 - 4 \times 18 + 45^2/(4 \times 50)]8 = 1185$ mm^2

Path FCDE (four holes and one stagger):

$\quad$ Net area = $[210 - 4 \times 18 + 40^2/(4 \times 50)]8 = 1168$ mm^2

Path ACG or FCB (three holes and two staggers):

$\quad$ Net area = $[210 - 3 \times 18 + 45^2/(4 \times 50) + 40^2/(4 \times 50)]8 = 1393$ mm^2

Path FCG (three holes and two staggers):

$\quad$ Net area = $210 - 3 \times 18 + 2 \times 40^2/(4 \times 50)]8 = 1376$ mm^2

The minimum net area is for path FCDE = 1168 mm^2. Note that the minimum net area occurs at a path which has the maximum number of holes and minimum number of staggers.

Example 3.3 *Determine the net area A_n for the 200 × 150 × 10 angle with M20 bolt holes as shown in Fig. 3.20*

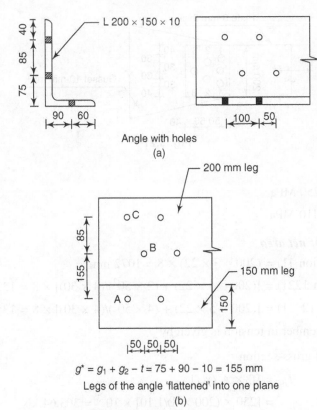

Angle with holes
(a)

$g^* = g_1 + g_2 - t = 75 + 90 - 10 = 155$ mm

Legs of the angle 'flattened' into one plane
(b)

Fig. 3.20 Effective net area of angle

Solution

For an M20 bolt, $d_h = 22$ mm

For net area calculation, the angle may be visualized as being flattened into a plate as shown in Fig. 3.20(b).

$$g^* = g_1 + g_2 - t = 75 + 90 - 10 = 155 \text{ mm}$$

$$A_n = A_g - \Sigma d_h t + \Sigma (p^2/4g)t$$

Gross area of angle = 3430 mm^2

Path AC:

Net area = $3430 - 2 \times 22 \times 10 = 2990$ mm^2

Path ABC:

Net area = $3430 - 3 \times 22 \times 10 + [50^2/ (4 \times 85) + 50^2 / (4 \times 155)] \times 10$

= 2883.85 mm^2

Since the smallest net area is 2883.85 mm^2 for path ABC, therefore, that value governs.

Example 3.4 *Determine the design tensile strength of plate (200 ×8 mm) connected to 10-mm thick gusset using 20 mm bolts as shown in Fig. 3.21, if the yield and the ultimate stress of the steel used are 250 MPa and 410 MPa, respectively.*

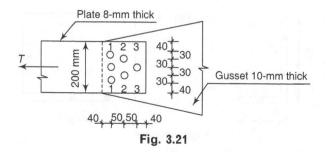

Fig. 3.21

Solution

$$f_y = 250 \text{ MPa}$$
$$f_u = 410 \text{ MPa}$$

Calculation of net area

$$A_n \text{ (Section 11)} = (200 - 3 \times 22) \times 8 = 1072 \text{ mm}^2$$
$$A_n \text{ (Section 1221)} = [(200 - 4 \times 22) + (2 \times 50^2)/(4 \times 30)] \times 8 = 1229.3 \text{ mm}^2$$
$$A_n \text{ (Section 12321)} = [(200 - 5 \times 22) + (4 \times 50^2)/(4 \times 30)] \times 8 = 1386.6 \text{ mm}^2$$

Strength of member in tension is given by

(i) Yielding of gross-section

$$T_{dg} = [f_y \times A_g/\gamma_{m0}]$$
$$= [250 \times (200 \times 8)/1.10] \times 10^{-3} = 363.64 \text{ kN}$$

(ii) Rupture of net section

$$T_{dn} = (0.9 \times f_u \times A_n/\gamma_{ml})$$
$$= (0.9 \times 410 \times 1072/1.25) \times 10^{-3} = 316.45 \text{ kN}$$

Therefore, the design tensile strength of the plate = 316.45 kN

Check for minimum edge distance
Provided edge and end distance = 40 mm > 1.5 × 20 = 30 mm
Hence, the edge distance is as required.

Example 3.5 *A single unequal angle 100 × 75 × 6 is connected to a 10-mm thick gusset plate at the ends with six 16-mm-diameter bolts to transfer tension as shown in Fig. 3.22. Determine the design tensile strength of the angle assuming that the yield and the ultimate stress of steel used are 250 MPa and 410 MPa:*

(i) if the gusset is connected to the 100-mm leg

(ii) if the gusset is connected to the 75-mm leg

Solution
(i) *Gusset is connected to the 100-mm leg of the angle*

$$A_{nc} = (100 - 6/2 - 18) \times 6 = 474 \text{ mm}^2$$
$$A_{go} = (75 - 6/2) \times 6 = 432 \text{ mm}^2$$
$$A_g = 1010 \text{ mm}^2$$

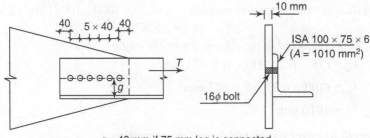

g = 40 mm if 75 mm leg is connected
= 60 mm if 100 mm leg is connected

Fig. 3.22

(a) Strength governed by yielding of gross section

$$T_{dg} = A_g f_y / \gamma_{mo} = (1010 \times 250/1.10) \times 10^{-3} = 229.55 \text{ kN}$$

(b) Strength governed by rupture of critical section

$$T_{dn} = 0.9 f_u A_{nc} / \gamma_{m1} + \beta A_{go} f_y / \gamma_{m0}$$

$$\beta = 1.4 - 0.076(w/t)(f_y/f_u)(b_s/L_c)$$

$$= 1.4 - 0.076[(75 - 3)/6](250/410)[(72 + 60)/(5 \times 40)]$$

$$= 1.4 - 0.367 = 1.033 > 0.7 \text{ and } < 1.30[(0.9 \times 410/250)(1.1/1.25)]$$

$$T_{dn} = [0.9 \times 410 \times 474/1.25 + 1.033 \times 432 \times 250/1.10] \times 10^{-3}$$

$$= 139.92 + 101.42 = 241.34 \text{ kN}$$

Alternatively,

$$T_{dn} = \alpha A_n f_u / \gamma_{m1} = 0.8 \times [(474 + 432) \times 410/1.25] \times 10^{-3} = 237.73 \text{ kN}$$

Hence, take

$$T_{dn} = 241.34 \text{ kN}$$

(c) Strength governed by block shear

$$A_{vg} = 6 \times (5 \times 40 + 40) = 1440 \text{ mm}^2$$

$$A_{vn} = 6 \times [(5 \times 40 + 40) - 5.5 \times 18] = 846 \text{ mm}^2$$

$$A_{tg} = 6 \times 40 = 240 \text{ mm}^2$$

$$A_{tn} = 6 \times (40 - 0.5 \times 18) = 186 \text{ mm}^2$$

$$T_{db1} = A_{vg} f_y / (\sqrt{3} \gamma_{m0}) + 0.9 f_u A_{tn} / \gamma_{m1}$$

$$= [1440 \times 250/(\sqrt{3} \times 1.1) + 0.9 \times 410 \times 186/1.25] \times 10^{-3} = 243.85 \text{ kN}$$

$$T_{db2} = 0.9 f_u A_{vn} / (\sqrt{3} \gamma_{m1}) + f_y A_{tg} / \gamma_{m0}$$

$$= [0.9 \times 410 \times 846/(\sqrt{3} \times 1.25) + 250 \times 240/1.10] \times 10^{-3} = 198.73 \text{ kN}$$

Hence,

$$T_{db} = 198.73 \text{ kN}$$

Thus, the design tensile strength of the angle = 198.73 kN (least of 198.73, 229.55, and 241.34).

The efficiency of the tension member $= 198.73 \times 1000 \times 100/(1010 \times 250/1.10)$
$$= 86.57\%$$

(ii) *Gusset is connected to the 75-mm leg of the angle*

$A_{nc} = (75 - 6/2 - 18) \times 6 = 324$ mm^2

$A_{go} = (100 - 6/2) \times 6 = 582$ mm^2

$A_g = 1010$ mm^2

(a) Strength as governed by yielding of gross-section

$T_{dg} = A_g f_y/\gamma_{m0} = 229.55$ kN

(b) Strength governed by tearing of net section

$T_{dn} = 0.9 f_u A_{nc}/\gamma_{m1} + \beta A_{go} f_y/\gamma_{m0}$

$\beta = 1.4 - 0.076(w/t)(f_y/f_u)(b_s/L_c)$

$\quad = 1.4 - 0.076[(100 - 3)/6](250/410)[(97 + 40)/(5 \times 40)]$

$\quad = 0.8868 > 0.7$

$T_{dn} = [0.9 \times 410 \times 324/1.25 + 0.8868 \times 582 \times 250/1.1] \times 10^{-3}$

$\quad = 212.94$ kN

Alternatively,

$T_{dn} = \alpha A_n f_u/\gamma_{m1}$

$\quad = [0.8 \times (324 + 582) \times 410/1.25] \times 10^{-3}$

$\quad = 237.73$ kN

Hence, take $T_{dn} = 212.94$ kN

(c) Strength governed by block shear

$A_{vg} = 6 \times (5 \times 40 + 40) = 1440$ mm^2

$A_{vn} = 6 \times (5 \times 40 + 40 - 5.5 \times 18) = 846$ mm^2

$A_{tg} = 6 \times 35 = 210$ mm^2

$A_{tn} = 6 \times (35 - 0.5 \times 18) = 156$ mm^2

$T_{db1} = A_{vg} f_y/(\sqrt{3}\gamma_{m0}) + 0.9 f_u A_{tn}/\gamma_{m1}$

$\quad = [1440 \times 250/(\sqrt{3} \times 1.1) + 0.9 \times 410 \times 156/1.25] \times 10^{-3}$

$\quad = 235$ kN

$T_{db2} = 0.9 f_u A_{vn}/(\sqrt{3}\gamma_{m1}) + f_y A_{tg}/\gamma_{m0}$

$\quad = [0.9 \times 410 \times 846/(\sqrt{3} \times 1.25) + 250 \times 210/1.10] \times 10^{-3}$

$\quad = 191.91$ kN

Hence,

$T_{db} = 191.91$ kN

Thus, the design tensile strength of the angle = 191.91 kN (least of 229.55, 212.94 and, 191.91)

The efficiency of the tension member = 191.91 × 1000 × 100/(1010 × 250/1.10)
$$= 83.6\%$$

Hence, in this case, by connecting the short leg, the efficiency is reduced by about 3%. Note that as the outstanding leg increases, gross net area increases and hence block shear may govern.

Example 3.6 *Determine the tensile strength of a roof truss diagonal* 100 × 75 × 6 *mm* (f_y = 250 MPa) *connected to the gusset plate by 4-mm welds as shown in Fig. 3.23.*

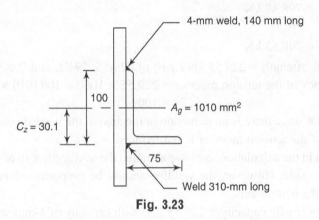

Fig. 3.23

Solution

Area of the connected leg = (100 − 6/2) × 6 = 582 mm²
Area of the outstanding leg = (75 − 6/2) × 6 = 432 mm²
$$A_g = 1010 \text{ mm}^2$$

(a) Strength governed by yielding of cross section

$$T_{dg} = A_g f_y/\gamma_{mo} = (1010 \times 250/1.10) \times 10^{-3} = 229.55 \text{ kN}$$

(b) Strength governed by rupture of critical section

$$T_{dn} = 0.9 f_u A_{nc}/\gamma_{m1} + \beta A_{go} f_y/\gamma_{m0}$$

Assuming average length of weld L_w = 225 mm

$$\beta = 1.4 - 0.076(w/t)(f_y/f_u)(b_s/L_w)$$
$$= 1.4 - 0.076[(75 - 3)/6](250/410)(75/225)$$
$$= 1.215$$

Hence,

$$T_{dn} = [0.9 \times 410 \times 582/1.25 + 1.215 \times 432 \times 250/1.10] \times 10^{-3}$$
$$= 291.1 \text{ kN}$$

Hence, T_{dn} = 291.1 kN

(c) Strength governed by block shear

Since the member is welded to the gusset plate, no net areas are involved and hence A_{vn} and A_{tn} in the equation for T_{db} (Section 6.3.1 of the code) should be taken as the corresponding gross areas (Gaylord et al. 1992). Assuming average length of the weld on each side as 225 mm and the gusset plate thickness as 8 mm,

$$T_{db1} = [8 \times (225 \times 2) \times 250/(\sqrt{3} \times 1.1) + 0.9 \times 410 \times 8 \times 100/1.25] \times 10^{-3}$$
$$= 708.53 \text{ kN}$$

$$T_{db2} = [0.9 \times 410 \times 8 \times 225 \times 2/(\sqrt{3} \times 1.25) + 250 \times 8 \times 100/1.1] \times 10^{-3}$$
$$= 798.38 \text{ kN}$$

Hence,

$$T_{db} = 708.53 \text{ kN}$$

Thus, tensile strength = 229.55 kN (least of 229.55, 291.1, and 708.53)

The efficiency of the tension member = $229.55 \times 1000 \times 10/(1010 \times 250/1.10)$
$$= 100\%$$

It is clear that since there is no reduction in the area in the welded connection, the efficiency of the tension member is not reduced.

Note that in the calculation, we have assumed the average length of weld as 225 mm on each side. However, the welding should be proportioned based on the position of the neutral axis.

Thus, for the tensile capacity = 229.55 kN, with capacity of 4-mm weld = 0.530 kN/mm

Length of the weld at the upper side of the angle

$$= (229.55 \times 30.1/100)/0.530 = 130 \text{ mm, say } 140 \text{ mm}$$

Length of the weld at the bottom side of the angle

$$= [229.55 \times (100 - 30.1)/100]/0.530 = 302 \text{ mm, say } 310 \text{ mm}$$

Example 3.7 *Select a suitable angle section to carry a factored tensile force of 290 kN assuming a single row of M24 bolts and assuming design strength as $f_y = 250$ N/mm²*

Solution

Approximate required area = $1.1 \times 290 \times 10^3/250 = 1276 \text{ mm}^2$

Choose $90 \times 90 \times 8$ angle with $A = 1380 \text{ mm}^2$

Strength governed by yielding = $[1380 \times 250/1.1] \times 10^{-3} = 313.64 \text{ kN}$

A_{nc} = area of connected leg = $(90 - 4 - 22) \times 8 = 512 \text{ mm}^2$

$A_{go} = (90 - 4) \times 8 = 688 \text{ mm}^2$

Required number of M24 bolts (Appendix D) = $313.64/65.3 = 4.8$

Provide five bolts at a pitch of 60 mm

Strength governed by rupture of critical section

$$T_{dn} = 0.9 f_u A_{nc}/\gamma_{m1} + \beta A_{go} f_y/\gamma_{m0}$$

$\beta = 1.4 - 0.076(w/t)(f_y/f_u)\,(b_s/L_c)$

$\quad = 1.4 - 0.076(90/8)\,(250/410)\,(82 + 50)/(4 \times 60)$

$\quad = 1.113 < 1.44$ and > 0.7

$T_{dn} = [0.9 \times 410 \times 512/1.25 + 1.113 \times 688 \times 250/1.10] \times 10^{-3}$

$\quad = 325.18$ kN

Strength governed by block shear

Assuming an edge distance of 40 mm,

$A_{vg} = 8 \times (4 \times 60 + 40) = 2240$ mm^2

$A_{vn} = 8 \times (4 \times 60 + 40 - 4.5 \times 26) = 1304$ mm^2

$A_{tg} = 8 \times 40 = 320$ mm^2

$A_{tn} = 8 \times (40 - 0.5 \times 26) = 216$ mm^2

$T_{db1} = A_{vg}\,f_y\,/(\sqrt{3}\gamma_{m0}) + 0.9f_u\,A_{tn}/\gamma_{m1}$

$\quad = [2240 \times 250/(\sqrt{3} \times 1.1) + 0.9 \times 410 \times 216/1.25] \times 10^{-3} = 357.68$ kN

$T_{db2} = 0.9f_u\,A_{vn}/(\sqrt{3}\,\gamma_{m1}) + f_y A_{tg}/\gamma_{m0}$

$\quad = [0.9 \times 410 \times 1304/(\sqrt{3} \times 1.25) + 250 \times 320/1.10] \times 10^{-3} = 294.97$ kN

Tension capacity of the angle = 294.97 kN > 290 kN

Hence the angle is safe.

Exampe 3.8 *A tie member in a bracing system consists of two angles 75 × 75 × 6 bolted to a 10-mm gusset, one on each side using a single row of bolts [See Fig. 3.24(a)] and tack bolted. Determine the tensile capacity of the member and the number of bolts required to develop full capacity of the member. What will be the capacity if the angles are connected on the same side of the gusset plate and tack bolted [Fig. 3.24(b)]? What is the effect on tensile strength if the members are not tack bolted?*

Solution
(a) Two angles connected to opposite side of the gusset as in Fig. 3.24(a)
(i) Design strength due to yielding of gross section $T_{dg} = f_y(A_g/\gamma_{m0})$

$\quad A_g = 866$ mm^2 (for a single angle)

$\quad T_{dg} = 250 \times 2 \times 866/1.10 \times 10^{-3}$

$\quad T_{dg} = 393.64$ kN

(ii) The design strength governed by tearing at net section

$\quad T_{dn} = \alpha A_n(f_u/\gamma_{m1})$

Assume a single line of four numbers of 20-mm-diameter bolts ($\alpha = 0.8$)

$\quad A_n = [(75 - 6/2 - 22)6 + (75 - 6/2)6]2$

$\quad A_n = (300 + 432)2 = 1464$ mm^2

$\quad T_{dn} = (0.8 \times 1454 \times 410/1.25) = 384.15$ kN

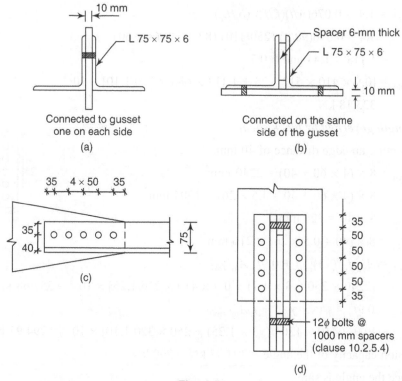

Fig. 3.24

Therefore,

Tensile capacity = 384.15 kN

Design of bolts

Choose edge distance = 35 mm

Capacity of bolt in double shear (Appendix D)

$$= 2 \times 45.3 = 90.6 \text{ kN}$$

Bearing capacity of the bolt does not govern.

Hence,

Strength of a single bolt = 90.6 kN

Provide five bolts. Then,

Total strength of the bolts = $5 \times 90.6 = 453$ kN > 384.15 kN

Hence the connection is safe.

Minimum spacing = $2.5t = 2.5 \times 20 = 50$ mm

Hence, provide a spacing of 50 mm.

The arrangements of bolts are shown in Fig. 3.24(c).

Check for block shear strength: (clause 6.4)

Block shear strength T_{db} of connection will be taken as

$$T_{db1} = [A_{vg}f_y/\sqrt{3}\gamma_{m0}) + (0.9A_{tn}f_u/\gamma_{m1})]$$

or

$$T_{db2} = 0.9f_u A_{vn}/\sqrt{3}\gamma_{m1} + (f_y A_{tg}/\gamma_{m0})$$

whichever is smaller.

$$A_{vg} = (4 \times 50 + 35)6 = 1410 \text{ mm}^2$$
$$A_{vn} = (4 \times 50 + 35 - 4.5 \times 22)6 = 816 \text{ mm}^2$$
$$A_{tn} = (35.0 - 22/2)6 = 144 \text{ mm}^2$$
$$A_{tg} = (35 \times 6) = 210 \text{ mm}^2$$

$$T_{db1} = \{[(1410 \times 250)/(\sqrt{3} \times 1.10)] + [(0.9 \times 144 \times 410)/1.25]\} \times 10^{-3}$$
$$= 227.5 \text{ kN}$$

$$T_{db2} = \{[(0.9 \times 410 \times 816)/(\sqrt{3} \times 1.25)] + [(250 \times 210)/1.10]\} \times 10^{-3}$$
$$= 186.8 \text{ kN}$$

For double angle,
block shear strength = $2 \times 186.8 = 373.6$ kN
Therefore,
   Tensile capacity = 373.6 kN (least of 393.64 kN, 384.15 kN, and 373.6 kN)

(b) Two angles connected to the same side of the gusset plate [Fig. 3.24(b)]
   (i) Design strength due to yielding of the gross section = 393.64 kN
   (ii) Design strength governed by tearing at the net section = 384.15 kN
      Assuming ten bolts of 20 mm diameter, five bolts in each connected leg
      Capacity of an M20 bolt in single shear = 45.3 kN
      Total strength of bolts = $10 \times 45.3 = 453$ kN > 393.64 kN
      Hence the connection is safe.
         The arrangement of bolts is shown in Fig. 3.24(d). Since it is similar to
      the arrangement in Fig. 3.24(c), the block shear strength will be the same,
      i.e., 373.6 kN.
         Hence, the tensile capacity = 373.6 kN
         The tensile capacity of both the arrangements (angles connected on the
      same side and connected to the opposite side of gusset) are same, as per
      the code though the load application is eccentric in this case. Moreover,
      the number of bolts are ten whereas in case (a) we used only five bolts
      since the bolts were in double shear.

(c) If the angles are not tack bolted, they behave as single angles connected to
   gusset plate.
In this case also the tensile capacity will be the same and we have to use ten M20
bolts. This fact is confirmed by the test and FEM results of Usha (2003).

Example 3.9 *Design a single angle to carry 350 kN. Assume that the length of the*
member is 3 m and $f_y = 250$ KPa.

Solution
Required area = $1.1 \times 350 \times 1000/250 = 1540 \text{ mm}^2$
Let us choose an unequal angle of size $150 \times 75 \times 8$ mm with a weight of 13.7 kg/m
and area = 1750 mm^2, $r_{vv} = 16.2$ mm.

(i) Design strength due to yielding of cross-section
$$T_{dg} = f_y\, A_g/\gamma_{m0} = 250 \times 1750/1.10 = 397.7 \text{ kN} > 350 \text{ kN}$$

(ii) Design strength governed by tearing of net section
$$T_{dn} = \alpha A_n\, f_u/\gamma_{m1}$$
Assuming nine M20 bolts, with strength $= 9 \times 45.3 = 407.7$kN
$$A_n = (150 - 4 - 2 \times 22)\,8 + (75 - 4)8 = 1384 \text{ mm}^2$$
$$T_{dn} = (0.8 \times 1384 \times 410/1.25) \times 10^{-3} = 363.1 \text{ kN} > 350 \text{ kN}$$

(iii) Assuming a staggered bolting and block shear failure as shown in Fig. 3.25
$$A_{vg} = (3 \times 50 + 25 + 35) \times 8 = 1680 \text{ mm}^2$$
$$A_{vn} = (3 \times 50 + 25 + 35 - 3.5 \times 22)\,8 = 1064 \text{ mm}^2$$
$$A_{tn} = (65 + 30 + 25^2/(4 \times 65) - 1.5 \times 22)\,8 = 515 \text{ mm}^2$$
$$A_{tg} = (65 + 30 + 25^2/(4 \times 65)) \times 8 = 779 \text{ mm}^2$$

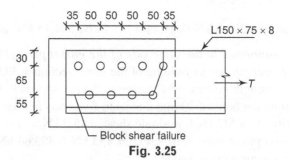

Fig. 3.25

Block shear capacity
$$T_{db1} = A_{vg}\, f_y/(\sqrt{3}\gamma_{m0}) + 0.9 A_{tn}\, f_u/\gamma_{m1}$$
$$= [1680 \times 250/(\sqrt{3} \times 1.1) + 0.9 \times 515 \times 410/1.25] \times 10^{-3} = 372.47 \text{ kN}$$
$$T_{db2} = 0.9 A_{vn}\, f_u/\sqrt{3}\gamma_{m1} + A_{tg}\, f_y/\gamma_{m0}$$
$$= [0.9 \times 1064 \times 410/(\sqrt{3} \times 1.25) + (779 \times 250/1.1) \times 10^{-3} = 358.38 \text{ kN}$$
Hence $T_{db} = 358.38 \text{ kN} > 350 \text{ kN}$

Check for stiffness (Table 3 of code)
$L/r = 3000/16.2 = 185 < 250$
Hence the section is safe.
It is seen from Examples 3.8 and 3.9 that unequal angle section with its long leg connected has a higher load carrying capacity than two equal angles of the same weight connected on the same side or opposite side of gusset plate. Hence wherever possible, unequal angles (with its long leg connected) should be used. But unfortunately unequal angles are not freely available in the market.

Example 3.10 *Design sag rods for consecutive purlins near the supported end of a roof truss system as shown in Fig. 3.26. The purlins are supported at one-third points by sag rods. Also design the ridge rod between ridge purlins. Assume c/c spacing of truss = 6 m, spacing of purlin = 1.4 m, self weight of roofing = 200 N/*

m^2, *intensity of wind pressure* = *1500 N/m², slope of the roof truss* = *25°, and no access is provided to the roof.*

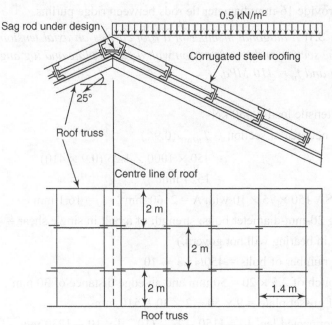

Fig. 3.26 Roof and sag rods for Example 3.10

Solution

Dead load from roofing = 200 × 1.4 = 280 N/m

Self weight of purlin = 100 N/m (assumed)

Live load = (0.75 − 0.02 × 15) × 1000 = 450 N/m² > 0.40 kN/m²

Live load on purlin = 450 × 1.4 = 630 N/m

Total gravity load = 630 + 280 + 100 = 1010 N/m

Wind load = 1500 × 1.4 = 2100 N/m (Normal to roof)

Component of gravity load parallel to roof

$$= 1010 × \sin 25° = 426.8 \text{ N/m}$$

As the sag rods are placed at third points on the purlin

Pull on sag rod = 426.8 × 6/3 = 853.6 N

Factored load = 1.5 × 853.6 = 1280.4 N

Required net area = $T_{dn} × \gamma_{m1}/(0.9\, f_u)$

$$= 1280.4 × 1.25/(0.9 × 410) = 4.34 \text{ mm}^2$$

Provide a 16-mm sag rod with a threaded area of 157 mm² between purlins (Note that the provided rod should not be less than 16-mm diameter).

Tie rod between ridge purlins

Pull in the tie rod = 4 × 853.6 × sec 25° = 3767 N

Factored load = 3767 × 1.5 = 5650 N

Required net area $= T_{dn}\gamma_{m1}/(0.9\ f_u)$

$$= 5650 \times 1.25/\ (0.9 \times 410) = 19.14\ mm^2$$

Hence, provide 16-mm diameter tie rods between ridge purlins.

Example 3.11 *A diagonal member of a roof carries an axial tension of 450 kN. Design the section and its connection with a gusset plate and lug angle. Use* $f_y = 250\ MPa$ *and* $f_u = 410\ MPa$.

Solution

Factored tensile load $= 450$ kN

Required net area of section $= T_u\gamma_{m1}/(0.9f_u)$

$$= 450 \times 1000 \times 1.25/(0.9 \times 410)$$

$$= 1524\ mm^2$$

Choose ISA $150 \times 75 \times 10$ with A $= 2160\ mm^2$, $r_{vv} = 16.1$ mm

Providing 20-mm-diameter bolts; strength of a bolt in single shear $= 45.3$ kN

(Strength in bearing will not govern.)

Required number of bolts $= 450/45.3 \approx 10$

Using a pitch of $2.5 \times 20 = 50$ mm and an edge distance of 30 mm

Length of gusset plate $= 9 \times 50 + 2 \times 30 = 510$ mm

Area of connected leg $A_{nc} = [150 - 22 - (10/2)] \times 10 = 1230\ mm^2$

Area of outstanding leg $A_{go} = [75 - (10/2)] \times 10 = 700\ mm^2$

$\quad A_n = 1230 + 700 = 1930\ mm^2 > 1524\ mm^2$

Tearing strength of the net section

$$T_{dn} = \alpha A_n f_u/\gamma_{m1} = 0.8 \times 1930 \times 410/1.25$$

$$= 506.4\ kN > 450\ kN$$

Hence safe.

Without lug angle, the length of the gusset plate is 510 mm. If the bolts are staggered and arranged in two rows, the length of the gusset plate may be reduced. We will now provide a lug angle (see Fig. 3.27).

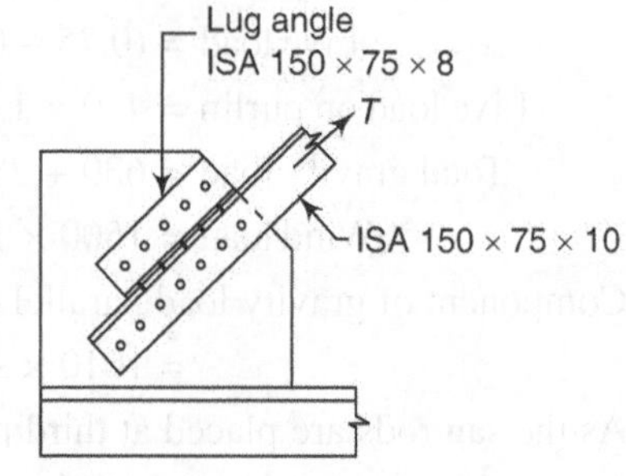

Fig. 3.27

Design of lug angle

Total factored tensile load $= 450$ kN

Gross area of the connected leg $= [150 - (10/2)] \times 10 = 1450\ mm^2$

Gross area of outstanding leg $= [75 - (10/2)] \times 10 = 700\ mm^2$

In an unequal angle, the load gets distributed in the ratio of the gross area of connected and outstanding legs.

Load shared by outstanding leg of main angle

$$= 450 \times 700/(1450 + 700) = 146.5\ kN$$

Load on lug angle $= 1.2 \times 146.5 = 175.8$ (clause 10.12.2)

Required net area for lug angle $= 175.8 \times 10^3 \times 1.25/(0.9 \times 410) = 596$ mm^2

Use ISA $150 \times 75 \times 8$ angle with $A = 1750$ mm^2

Assuming that the section is weakened by one row of 20-mm-diameter bolt

Net area $= 1750 - 22 \times 8 = 1574$ mm$^2 > 596$ mm^2 (for simplicity the effect of $\Sigma p^2/4g$ is not considered here but has to be considered as explained in previous examples).

The lug angle is also kept with its 75-mm long leg as outstanding leg

Number of bolts to connect 150-mm leg of lug angle with gusset plate

$$= 175.8/45.3 \approx 4$$

Provide five bolts of 20 mm diameter to connect lug angle with gusset plate.

Check

Load on connected leg $= 450 \times 1450/(1450 + 700) = 303.5$ kN

Required number of bolts $= 303.5/45.3 \approx 7$

Hence provide seven 20-mm-diameter bolts to connect the diagonal tension member with the gusset.

Required number of bolts to connect outstanding legs of the two angles (clause 10.12.2)

$$= 1.4 \times 146.5/45.3 \approx 5$$

Hence, provide five bolts of 20 mm diameter.

Required length of gusset plate $= 6 \times 50 + 2 \times 30 = 360$ mm (compared with 510 mm without lug angle).

Summary

Steel tension members are the most common and efficient members in structural applications. These members are connected to other members of a structure using gusset plates and by bolts or welds. A variety of cross sections can be used for tension members. The behaviour of tension members has been discussed. A brief discussion about the various factors such as the net area of cross section, method of fabrication, effect of bearing stress, and effect of shear lag have been provided. The various modes of failures of tension members have been identified. It has been found that the rupture of the net section at the end connections where the tensile stresses are the largest or the block shear failure at the end connections or the yield strength of gross section may govern the failure of tension members. The yielding of the gross section may be the governing failure mode of tension members connected by welding at the ends, whereas the other two failure modes will govern when the members are connected at the ends by bolts. Shorter connections are governed by block shear and longer connections by net sections failure. The effect of connecting the end gusset plate to only one element of the cross section has been empirically accounted for by reducing the effectiveness of the outstanding legs, while calculating the net effective area. The methods for accounting for these factors in the design of tension members are discussed with emphasis to the formulae given in the Indian code. The iterative method of design of tension members is presented. The concepts presented are explained with the use of several examples.

Exercises

1. What is the net area A_n for the tension member shown in Fig. 3.28, when (a) the holes are made by drilling, (b) holes are made by punching. Assume M20 bolts.

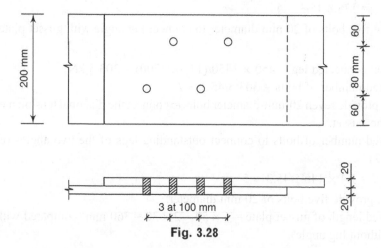

3 at 100 mm

Fig. 3.28

2. Determine the net area A_n for the $150 \times 75 \times 8$ angle with M20 bolt holes as shown in Fig. 3.29.

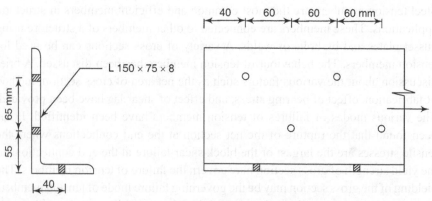

L 150 × 75 × 8

Fig. 3.29

3. Determine the design tensile strength of a plate (160×8 mm) connected to a 10-mm thick gusset using 16 mm-diameter bolts as shown in Fig. 3.30, if the yield and the ultimate stress of the steel used are 250 MPa and 410 MPa, respectively.

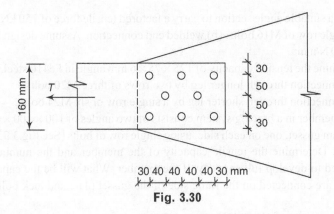

Fig. 3.30

4. What tensile load can an ISA 75 × 75 × 6 carry with the connections shown in Fig. 3.31? Assume that the connection is stronger than the members connected. Assume M30 bolts with an edge distance of 50 mm and a pitch of 75 mm, for bolted connections. The size of the fillet weld is 4 mm and the length on each side is 225 mm. Assume that the yield and the ultimate stress of steel used are 250 MPa and 410 MPa, respectively.

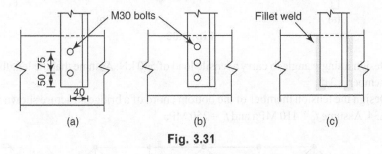

Fig. 3.31

5. Determine the tensile strength of a roof truss diagonal of 150 × 75 × 10 mm (f_y = 250 MPa) connected to the gusset plate by 6 mm welds as shown in Fig. 3.32.

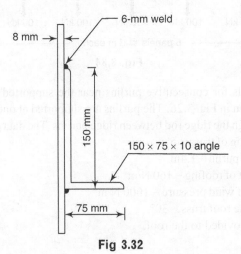

Fig 3.32

6. Select a suitable angle section to carry a factored tensile force of 150 kN, assuming (a) single row of M16 bolts, (b) welded end connection. Assume design strength as $f_y = 250 \text{ N/mm}^2$.

7. Determine the tension capacity of $125 \times 75 \times 6$ mm angle in Fe410 steel, assuming
   (a) Connection through longer leg by two rows of three M20 bolts,
   (b) Connection through shorter leg by a single row of six M24 bolts.

8. A tie member in a bracing system consists of two angles of $100 \times 100 \times 6$ bolted to a 12-mm gusset, one on each side, using single row of bolts (See Fig.3.33) and tack bolted. Determine the tensile capacity of the member and the number of bolts required to develop full capacity of the member. What will be the capacity if the angles are connected on the same side of the gusset plate and tack bolted?

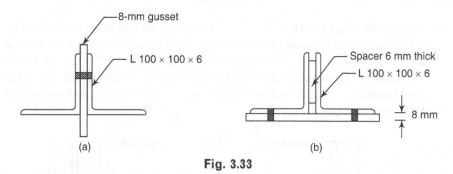

Fig. 3.33

9. Design a single angle to carry a tensile load of 500 kN. Assume that the length of the member is 3 m.

10. Design the tension member of the bottom chord of a bridge structure shown in Fig. 3.34. Assume $f_u = 410$ MPa and $f_y = 250$ MPa.

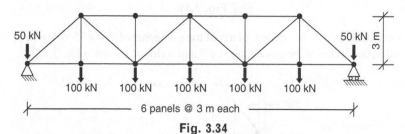

Fig. 3.34

11. Design sag rods for consecutive purlins near the supported end of a roof truss system as shown in Fig. 3.26. The purlins are supported at one-third points by sag rods. Also design the ridge rod between ridge purlins. The data given are as follows:
   C/C Spacing of truss = 5 m
   Spacing of purlin = 1.4m
   Self weight of roofing = 160 N/m^2
   Intensity of wind pressure = 1000 N/m^2
   Slope of the roof truss = 30°
No access is provided to the roof.

12. A diagonal member of a roof carries an axial tension of 300 kN. Design the section and its connection with a gusset plate and lug angle. Use $f_y = 250$ MPa and $f_u = 410$ MPa.

Review Questions

1. List the type of cross section that can be used as tension members and their use in typical structures.
2. Why are rods, which are used as tension members, required to be pre-tensioned?
3. What is the use of spacer plates or stitch plates? At what spacing are they connected to the members?
4. What is meant by slenderness ratio?
5. The maximum slenderness ratio permissible in steel ties is
   (a) 250   (b) 350   (c) 450   (d) 400
   (e) indirectly controlled by deflection
6. The maximum slenderness ratio permissible in steel ties which may be subjected to compression under wind load condition is
   (a) 250   (b) 350   (c) 400   (d) 180   (e) no limit
7. Write down the expression for the axial elongation of the member subjected to a tensile force.
8. List the different modes of failures of a tension member.
9. What is the design stress of a tension member based on
   (a) gross-section yielding   (b) net section rupture
10. Write short notes on block shear failure in plates and angles.
11. Why are drilled holes preferred over punched holes?
12. What are the methods by which the effect of punched holes can be considered in the calculation?
13. Write down the expression to calculate the net area of cross-section of a plate of width B, thickness t, and having staggered holes of pitch p and gauge g.
14. What is meant by shear lag?
15. How can the effects of shear lag be considered in the design calculation?
16. Do the geometry and ductility factors affect the design strength of tension member considerably?
17. How is the net area calculated when the angles are connected through both the legs with staggered bolts?
18. Write down the expression given in IS 800 for the net section design of angle tension members.
19. List the use of tension rods in building structures.
20. List the various steps in the design of a tension member.
21. What is a lug angle? Why is it not used in practice?
22. Write short notes on splices to tension members.
23. How are the sizes of gussets determined?

4

Plastic and Local Buckling Behaviour

Introduction

The two aspects of structural behaviour under stress which are of particular importance and have considerable influence on the design of steel members are the following.

1. Behaviour of steel in the plastic region of the stress–strain curve [see Fig. 1.4(a)].
2. The tendency of unsupported compression members to 'buckle' and become unstable. Buckling may be defined as a structural behaviour in which a deformation develops in a direction or plane perpendicular to that of the load which produced it; this deformation changes rapidly with variations in the magnitude of applied load. Buckling occurs mainly in members that are subjected to compression (Dowling et al. 1988). Its effect is to decrease the load carrying capacity of a structure (i.e., reduce the strength) and also to increase the deformation (i.e., reduce the stiffness).

These two behavioural aspects will be briefly considered in this chapter to provide the necessary theoretical background material for the later design-oriented chapters on columns and beams.

4.1 Plastic Theory

4.1.1 Basis of Plastic Theory

To explain the concepts of plastic analysis, consider an I-beam subjected to a steadily increasing bending moment M as shown in Fig. 4.1. In the service load range the section is elastic as shown in Fig. 4.1(a) and the elastic condition exists until the stress at the extreme fibre reaches the yield stress f_y [Fig. 4.1(b)]. Once the strain ε reaches ε_y (see Fig. 4.2), increasing the strain does not induce any increase in stress. This elastic–plastic stress–strain behaviour is the accepted idealization for structural steel having yield stresses up to about $f_y = 450$ MPa.

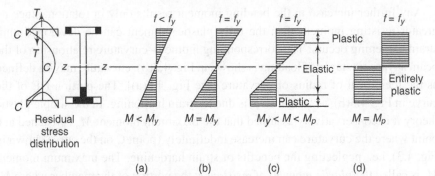

Note: Shaded portion indicates portion that have reached f_y due to residual stress of stage *b*.

Fig. 4.1 Stress distribution at different stages of loading

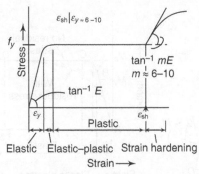

Fig. 4.2 Idealized tensile stress–strain diagram for steel

When the yield stress reaches the extreme fibre as shown in Fig. 4.1(b), the nominal moment strength M_n of the beam is referred to as the *yield moment M_y* and is given by

$$M_n = M_y = Z_e f_y \qquad (4.1)$$

where Z_e is the elastic section modulus.

A further increase in the bending moment causes the yield to spread inwards from the lower and upper surfaces of the beam as shown in Fig. 4.1(c). The spread of yielding in an I-section with residual stresses is also shown in the left hand side of Fig. 4.1. This stage of partial plasticity occurs because of the yielding of the outer fibres without increase of stresses, as shown by the horizontal line of the idealized stress–strain diagram shown in Fig. 4.2. Upon increasing the bending moment further, the whole section yields as shown in Fig. 4.1(d). When this condition is reached, every fibre has a strain equal to or greater than $\varepsilon_y = f_y/E_s$. The nominal moment strength M_n at this stage is referred to as the *plastic moment M_p* and is given by

$$M_p = f_y \int_A y \, dA = f_y Z_p \qquad (4.2)$$

where $Z_p = \int y \, dA$ is the *plastic section modulus*.

Any further increase in the bending moment results only in rotation, since no greater resisting moment than the fully plastic moment can be developed until strain hardening occurs. The corresponding moment–curvature relationship of the beam at various stages of loading is shown in Fig. 4.3(b); curvature may be defined as the reciprocal of radius of curvature [see Fig. 4.3(a)]. The portion DE of the curve in Fig. 4.3(b), a rising curve, is due to strain hardening. In the simple plastic theory it is conservatively assumed that the maximum moment M_p is reached at a point where the curvature can increase indefinitely (point C on the curve shown in Fig. 4.3), i.e., neglecting the benefits of strain hardening. The maximum moment M_p is called the *plastic moment of resistance*; the portion of the member where M_p occurs, termed a *plastic hinge*, can sustain large local increases of curvature [a plastic hinge is shown in Fig. 4.3(c)].

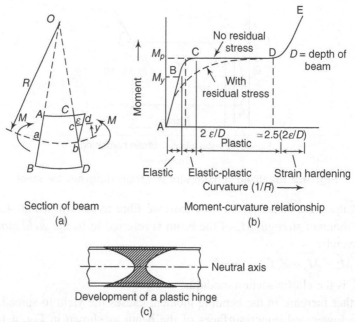

Section of beam
(a)

Moment-curvature relationship
(b)

Development of a plastic hinge
(c)

Fig. 4.3 Moment-curvature relationship at various stages of loading

Thus the plastic hinge may be defined as a *yielded zone*, which can cause an infinite rotation to take place at a constant plastic moment M_p of the section. As shown above, plastic hinges form in a member at the maximum bending moment locations. However, at the intersections of two members, where the bending moment is the same, a hinge forms in the weaker member. Generally, hinges are located at restrained ends, intersections of members, and point loads.

Generally the number of plastic hinges are $n = r + 1$, where r is the number of redundancies or the indeterminacy. (However, there are exceptions, e.g., the partial collapse of a beam in a structure.)

In the fully plastic stage, because the stress is uniformly equal to the yield stress, equilibrium is achieved when the neutral axis divides the section into two equal areas. Thus, considering the general cross section shown in Fig. 4.4 and equating the compressive and tensile forces, we get

$$f_y A_1 = f_y A_2$$

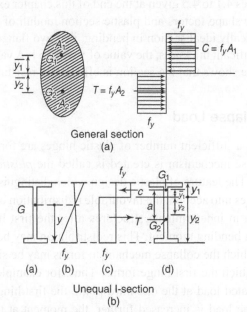

General section
(a)

Unequal I-section
(b)

Fig. 4.4 Force equilibrium of a section

Since $A_1 = A_2$ and $A = A_1 + A_2$,

$$A_1 = A_2 = A/2$$

Plastic moment of resistance

$$M_p = f_y A_1 \bar{y}_1 + f_y A_2 \bar{y}_2$$

$$= f_y A/2 \ (\bar{y}_1 + \bar{y}_2)$$

Thus, $M_p = f_y Z_p$

where $Z_p = (A/2) \ (\bar{y}_1 + \bar{y}_2)$ is the plastic modulus of the section. (4.3)

Thus, the *plastic modulus* of the section may be defined as the combined statical moment of the cross-sectional area above and below the equal-area axis. It is also referred to as the *resisting modulus of the completely plasticized section*.

4.1.2 Shape Factor

The ratio M_p/M_y is a property of the cross-sectional shape and is independent of the material properties. This ratio is known as the shape factor v and is given by

$$v = M_p/M_y = Z_p/Z_e \tag{4.4}$$

For wide flange I-sections in flexure about the strong axis (z-z), the shape factor ranges from 1.09 to about 1.18, with the average value being about 1.14. One may conservatively take the plastic moment strength M_p of I-sections bent about their strong axis to be at least 15% greater than the strength M_y when the extreme fibre alone reaches the yield stress f_y. On the other hand, the shape factor for I-sections bent about their minor axis is about the same as for a rectangle, i.e., about 1.5. Note that when the material at the centre of the section is increased, the value of v increases. Examples 4.1 to 4.5 given at the end of this chapter explain the methods of determining the shape factors and plastic section moduli of different sections.

For the theoretically ideal section in bending, i.e., two flange plates connected by a web of insignificant thickness, the value of v will be 1. A value of shape factor nearly equal to one shows that the section is efficient in resisting bending.

4.2 Plastic-collapse Load

The load at which a sufficient number of plastic hinges are formed in a structure such that a collapse mechanism is created is called the *plastic-collapse load* or *plastic limit load*. The feature of plastic design which distinguishes it from elastic design is that it takes into account the favourable redistribution of bending moment which takes place in indeterminate structures after the first hinge forms at the point of maximum bending moment. This redistribution may be considerable and the final load at which the collapse mechanism forms may be significantly higher than the load at which the first hinge forms. Thus, for example, in a fixed beam having a concentrated load at the one-third point, the first hinge forms at one of the supports; as the load is increased further, the moment at this hinge remains constant at M_p, while the moments at the other support and the load point increase until a second hinge is formed. When the load is increased further, the moments at these two hinges remain constant at M_p, until a third and final hinge is formed to make the beam a mechanism. In this beam, the final ultimate load will be 33% higher than the first hinge load.

A number of examples of plastic-collapse mechanisms in cantilevers and single- and multiple-span beams are shown in Fig. 4.5.

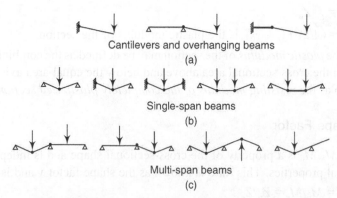

Cantilevers and overhanging beams
(a)

Single-span beams
(b)

Multi-span beams
(c)

Fig. 4.5 Plastic-collapse mechanisms in beams

As noted already, plastic hinges normally occur at supports, points of concentrated load, and points where cross sections change. The location of a plastic hinge in a beam with a uniformly distributed load is, however, not well defined.

4.3 Conditions of Plastic Analysis

The basic conditions that are to be satisfied for any structure in elastic and plastic analysis are shown in Table 4.1.

Table 4.1 Conditions to be satisfied in elastic and plastic analysis

Elastic analysis	Plastic analysis
Equilibrium The summation of the forces and moments acting on any free body must be equal to zero, i.e., $\Sigma F_x = 0$, $\Sigma F_y = 0$, $\Sigma M_{xy} = 0$	*Equilibrium* The bending moment distribution defined by the assumed plastic hinges must be in static equilibrium with the applied loads and reactions.
Compatibility Strain in one section is assumed to be equal to that in the adjoining section.	*Mechanism* There must be sufficient number of plastic and frictionless hinges for the beam/structure to form a mechanism. The ultimate or collapse load is reached when a mechanism is formed.
Moment–curvature relations $M/I = f/y = E/R$ (Navier's theorem) where f = stress at any layer, E = Young's modulus of elasticity, R = local radius of curvature of beam, M = maximum bending moment, I = moment of inertia of the cross section of beam, and y = the distance of the layer under stress from the neutral axis.	*Plasticity* The bending moments in any section of a structure must be less than the plastic moment of the section $-M_p \le M \le M_p$

If all the three conditions are satisfied, the lowest plastic limit load (a unique value) is obtained. If only the equilibrium and mechanism conditions are satisfied (this forms the basis for the mechanism method of plastic analysis), an upper bound solution for the true ultimate load is obtained. A lower bound solution for the true ultimate load is obtained when the equilibrium and plasticity conditions only are satisfied (statical method of plastic analysis). See Section 4.5 for the mechanism and statical methods of analysis.

4.3.1 Principle of Virtual Work

The principle of virtual work may be stated as follows: If a system of forces in equilibrium is subjected to a virtual displacement, then the work done by the external forces equals the work done by the internal forces, i.e.,

$$W_E = W_I \tag{4.5}$$

where W_E and W_I are the work done by the external and internal forces, respectively. This is simply a method to express the equilibrium condition. While applying this method to determine the moment at collapse, an arbitrary displacement is assumed at a plastic hinge location (the arbitrary displacement must be one for which only the internal moments at the plastic hinges contribute to the internal work) and the work done by the external and internal forces is equated. This is accomplished by allowing rotations of the structure only at points of simple support and at points where plastic moments are expected to occur in producing the mechanism.

4.4 Theorems of Plastic Collapse

The plastic analysis of structures is governed by three theorems, which are as follows. The *static* or *lower bound theorem* states that a load computed on the basis of an assumed equilibrium moment diagram, in which the moments are nowhere greater than the plastic moment, is less than, or at the best equal to, the correct collapse load. Hence the static method represents the lower limit to the true ultimate load and has a maximum factor of safety. The static theorem was first suggested by Kist and its proof was given by Gvozder, Greenberg, and Horne (Horne 1979).

The *kinematic* or *upper bound theorem* states that a load computed on the basis of an assumed mechanism will always be greater than, or at the best equal to, the correct collapse load. Hence the kinematic method represents an upper limit to the true ultimate load and has a smaller factor of safety compared to the static method. A proof of this theorem was provided by Gvozder, Greenberg, and Prager (Horne 1979).

The upper and lower bound theorems can be combined to produce the *uniqueness theorem*, which states that the load that satisfies both the theorems at the same time is the correct collapse load.

4.5 Methods of Plastic Analysis

Using the principle of virtual work and the upper and lower bound theorems, a structure can be analysed for its ultimate load by any of the following methods: (1) static method, (2) kinematic method.

In the mechanism or kinematic method, a mechanism is assumed and virtual work equations are used to determine the collapse load. The number of independent mechanisms (n) is related to the number of possible plastic hinge locations (h) and the number of degrees of redundancy (r) of the frame by the equation

$$n = h - r \tag{4.6}$$

In the statical or equilibrium method, an equilibrium moment diagram is obtained such that the moment at any section is less than or equal to the plastic moment capacity. Even though the equilibrium method gives a lower bound solution, the virtual work method is often used due to its simplicity of application in comparison with the equilibrium method.

If the upper and lower bound solutions obtained by the mechanism and statical methods coincide or are sufficiently close, then the assumed plastic hinge locations are correct. If, however, these bounds are not precise enough, then the location of the assumed hinge should be modified (an indication of this will be provided by the bending moments determined in the statical analysis) and the analysis repeated. The use of the mechanism and equilibrium methods of plastic analysis in single-span beams is explained in Examples 4.6–4.10.

A load factor of 1.7 is normally used in plastic design.

4.6 Plastic Design of Portal Frames

Plastic design is used extensively for the design of single-storey portal frame structures. The design of portals is a little complicated because there can be more than one mechanism of failure, and the mechanism giving the least collapse load has to be chosen. Single-bay portal frames with fixed bases have three redundancies and require four hinges to produce a mechanism.

Furthermore, as the number of redundancies increases, so does the number of possible modes of collapse. The possible mechanisms are classified into two groups: elementary and combined mechanisms. Whereas the former are independent of each other, the latter are linear combinations of elementary mechanisms. The elementary and combined mechanisms of a single-storey portal frame are shown in Fig. 4.6. The *beam mechanism* results when three plastic hinges form within any one of the elements of the frame [see Fig. 4.6(b)]. A set of columns in a storey may develop plastic hinges at the top and bottom of each column so as to generate

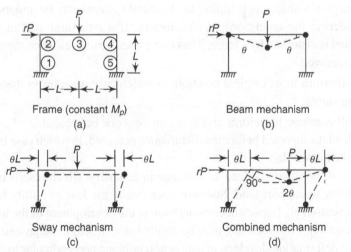

Frame (constant M_p)
(a)

Beam mechanism
(b)

Sway mechanism
(c)

Combined mechanism
(d)

Fig. 4.6 Collapse mechanisms of a single-bay, fixed-base portal frame

a simple sway mechanism of the structure [see Fig. 4.6(c)]. The sway and beam mechanisms may be combined to form a combined mechanism as shown in Fig. 4.6(d). Thus, a structure may collapse partially or as a whole. As mentioned earlier, if the order of indeterminacy is r, then the minimum number of plastic hinges required for total collapse is $r + 1$, whereas the number of hinges required for partial collapse may be less.

Examples 4.11 and 4.12 illustrate the methods used to determine the collapse load of simple portal frames.

4.7 Special Considerations

There are a number of special considerations while attempting a plastic design. These considerations are as follows:

- Local buckling (concerned with the width-to-thickness ratios for the flange and web of a section)
- Flange stability
- Column action (effect of axial force and shear on plastic moment capacity)
- Deflection
- Connection.

Local buckling is discussed in Section 4.8. For the effect of axial force on plastic moment capacity and other considerations/limitations, refer to Davies and Brown (1996), and Horne and Morris (1981).

4.8 Local Buckling of Plates

The concept of stability as it applies to structural systems may be understood best by considering the conditions of equilibrium. If a structural system that is in equilibrium is disturbed by a force, it has two basic alternatives when the disturbing force is removed.

1. It will return to its original position, in which case we refer to the system as being *stable*.
2. It will continue to deform and as a consequence be incapable of supporting the load it supported before the disturbance occurred, in which case the system is called *unstable*.

Instability is thus characterized as a change in geometry, which results in the loss of the ability to support load. Stability, specifically the loss of ability to support load, is an extremely important consideration in the development of the limit states design criterion. Thus *buckling* may be defined as a structural behaviour in which a mode of deformation develops in a direction or plane perpendicular to that of the loading which produces it; such a deformation changes rapidly with increase in the magnitude of the applied loading. It occurs mainly in members or elements that are subjected to compressive forces.

While using plastic design, it is assumed that plastic deformation can take place without the geometry of the structure changing to such an extent that the conditions of equilibrium are significantly modified. Such changes in geometry can arise at three levels, namely (Horne & Morris 1981),

1. deformation within the cross section of a member (resulting from *local buckling* in the plate elements constituting the web or the flange),

2. displacements within the length of the member relative to straight lines drawn between the corresponding points of the end sections (due to the bending and/ or twisting of the member), and

3. overall change of geometry of the structure, causing the joints to displace relative to each other (e.g., the sway deformation in multi-storey frames).

These three levels of deformation are thus associated, respectively, with local, member, and frame instability. The problem of member stability is discussed in Chapter 5. We will discuss local buckling and its consequences in this section.

4.8.1 Elastic Buckling of Plates

Local stability of the compressed elements of a section without transverse stiffeners can be studied by reference to the elastic stability of an infinite plate having width b and thickness t, as shown in Fig. 4.7. Assume that the plate is loaded by compressive forces acting on the simply supported sides having width b. The critical stress of this plate is given by (Timoshenko & Gere, 1961)

$$f_{cr} = (k\,\pi^2\,E)/[12\,(1-\mu^2)(b/t)^2] \qquad (4.7)$$

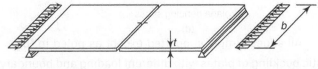

Fig. 4.7 Infinite plate subjected to compressive forces

where μ is Poisson's ratio of the material, b/t is the width-to-thickness ratio of the plate, k is the buckling coefficient, and E is Young's modulus of rigidity of the material.

The value of the coefficient k depends on the constraints along the non-loaded edges of the plate. The way in which the plates buckle and also the value of their critical buckling stress depend on the edge conditions, dimensions, and loading. The buckled plate patterns of four different plates with different loading and boundary conditions are shown in Fig. 4.8. Some values of k for the four plates are given in Table 4.2. It has to be noted that the edge conditions affect not only the critical buckling stress but also the post buckling behaviour.

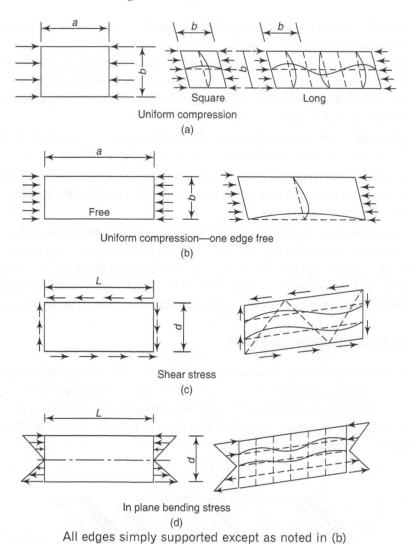

Uniform compression
(a)

Uniform compression—one edge free
(b)

Shear stress
(c)

In plane bending stress
(d)

All edges simply supported except as noted in (b)

Fig. 4.8 Elastic buckling of plates with different loading and boundary conditions

On the basis of the boundary conditions, plate elements in structural members can be divided into two categories: unstiffened elements and stiffened elements. *Unstiffened elements* are supported along one edge parallel to the axial stress (e.g., legs of single angles, flanges of beams, and stems of T-sections). *Stiffened elements* are supported along both the edges parallel to the axial stress (e.g., flanges of square and rectangular hollow sections, perforated cover plates, and webs of I-sections and channel sections (see Fig. 4.9).

4.8.2 Plates with Different Boundary Conditions

The edges of flat plates may be fixed or elastically restrained, instead of being simply supported or free. The elastic buckling loads of flat plates with various support conditions have been determined, and many values of buckling coefficient

Table 4.2 Buckling coefficients and limiting values of width/thickness ratios

Plate and load	$\dfrac{\text{Length}}{\text{Width}} = \dfrac{a}{b}$	Buckling coefficient, k	Limiting value of b/t for $f_{cr} =$ yield stress
(plate with edge loads, width b, length a)	1.0 5.0	4.0 4.0	53.8 53.8
(plate, one edge Free)	1.0 ∞	1.425 0.425	32.1 17.5
(plate with loads)	1.0 ∞	9.35 5.35	108.2 81.9
(plate with loads)	1.0 ∞	25.6 minimum 23.9	136.1 131.4

a = plate length/stiffener spacing on a plate girder
All edges simply supported except as noted

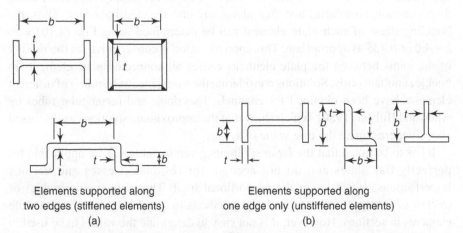

Elements supported along two edges (stiffened elements)
(a)

Elements supported along one edge only (unstiffened elements)
(b)

Fig. 4.9 Stiffened and unstiffened compression elements

k to be used with Eqn (4.7) are given in Timoshenko and Gere (1961), Bleich (1952), and Allen and Bulson (1980). Some of these coefficients are shown in Fig. 4.10.

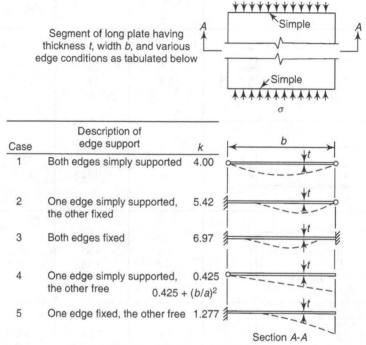

Case	Description of edge support	*k*
1	Both edges simply supported	4.00
2	One edge simply supported, the other fixed	5.42
3	Both edges fixed	6.97
4	One edge simply supported, the other free	0.425 $0.425 + (b/a)^2$
5	One edge fixed, the other free	1.277

Segment of long plate having thickness *t*, width *b*, and various edge conditions as tabulated below

Section *A-A*

Fig. 4.10 Coefficients *k* for Eqn (4.7)

Many structural steel compression members are assemblies of flat plate elements which are rigidly connected together along their common boundaries. The local buckling of such an assembly (e.g., welded box girder, plate girder) can be analysed approximately by assuming that the plate elements are simply supported along their common boundaries and free along any unconnected boundary. Thus, the buckling stress of each plate element can be determined using Eqn (4.10) with $k = 4.0$ or 0.425 as appropriate. This approximation is conservative, as the rigidity of the joints between the plate elements causes all connected plate elements to buckle simultaneously. Solutions considering the simultaneous buckling of different elements have been obtained for channels, I-sections, and rectangular tubes by Allen and Bulson (1980), which show that the approximate solutions as discussed above underestimate the true value of *k*.

It has to be noted that the *b/t* or *d/t* limits given in Table 4.2 are applicable for perfectly flat plates and do not account for residual stresses and various imperfections, which lower the proportional limit. Therefore smaller values of (*b/t*) or (*d/t*) have to be used in design methods to avoid local buckling of plate elements in sections. However, it is not easy to determine the values to be used in design, since there can be considerable variations in out-of-flatness, residual stress, etc. Hence design codes suggest the limiting values based on judgement and experience. Table 4.3 gives the limiting values of b/t as adopted in IS 800 (See also Table 2 of code).

Table 4.3 Cross-sectional limits necessary to prevent local buckling in members

Type of element	Method of manufacture	Ratio	Limiting proportions for sections		
			Plastic	Compact	Semi-compact
Outstand element of compression flange	Welded	b/t_f	8.4	9.4	13.6
	Rolled	b/t_f	9.4	10.5	15.7
Internal element of compression flange	Welded or rolled	b/t_f	29.3	33.5	42
Web of an I, H or box section	Welded or rolled				
	Neutral axis at mid depth	d/t_w	84	105	126
	Generally	d/t_w	$84/(1 + r_1)$ but ≥ 42	$105/$ $(1 + r_1)$	126
	if r_1 is negative				$(1 + 2r_2)$ but ≥ 42
	If r_1 is positive	d/t_w		$105/$ $(1 + 1.5\,r_1)$ but ≥ 42	
Web of channel	—	d/t_w	42	42	42
Angle (both criteria should be satisfied)	—	b/t and d/t	9.4	10.5	15.7
Stem of a T-section (rolled or cut from a rolled I- or H-section)	—	d/t	8.4	9.4	18.9

Notes

1. The above values should be multiplied by $\varepsilon = (250/f_y)^{0.5}$ if f_y is not equal to 250 MPa.
2. The stress ratios r_1 and r_2 are defined as average axial compressive stress/design compressive stress of web alone and average axial compressive stress/design compressive stress of overall section, respectively. r_1 and r_2 are negative if axial stress is tensile.

4.9 Cross Section Classification

Determining the resistance (strength) of structural steel components requires the designer to consider first the cross-sectional behaviour and second the overall member behaviour—whether in the elastic or inelastic material range; cross-sectional resistance and rotation capacity are limited by the effects of local buckling.

In the code (IS 800), cross sections are placed into four behavioural classes depending upon the material yield strength, the width-to-thickness ratios of the individual components (e.g., webs and flanges) within the cross section, and the loading arrangement. The four classes of sections are defined as follows.

(a) *Plastic or class 1* Cross sections which can develop plastic hinges and have the rotation capacity required for the failure of the structure by the formation of a plastic mechanism (only these sections are used in plastic analysis and design).

(b) *Compact or class 2* Cross sections which can develop their plastic moment resistance, but have inadequate plastic hinge rotation capacity because of local buckling.

(c) *Semi-compact or class 3* Cross sections in which the elastically calculated stress in the extreme compression fibre of the steel member, assuming an elastic distribution of stresses, can reach the yield strength, but local buckling is liable to prevent the development of the plastic moment resistance.

(d) *Slender or class 4* Cross sections in which local buckling will occur even before the attainment of yield stress in one or more parts of the cross section. In such cases, the effective sections for design are calculated by deducting the width of the compression plate element in excess of the semi-compact section limit.

The moment–rotation characteristics of these four classes of cross sections are shown in Fig. 4.11. As seen from this figure, class 1 (plastic) cross sections are fully effective under pure compression, and capable of reaching and maintaining their full plastic moment in bending and hence used in plastic design. These sections will exhibit sufficient ductility ($\theta_2 > 6\ \theta_1$), where θ_1 is the rotation at the onset of plasticity and θ_2 is the lower limit of rotation for treatment as a plastic section).

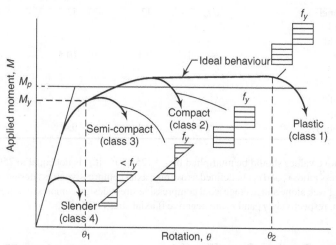

Fig. 4.11 Moment–rotation behaviour of the four classes of cross sections as defined by IS 800

Class 2 (compact) cross sections have lower deformation capacity, but are also fully effective in pure compression and are capable of reaching their full plastic moment in bending (they have ductility in the range $\theta_1 < \theta_2 < 6\theta_1$). Class 3 (semi-compact) cross sections are fully effective in pure compression but local buckling prevents the attainment of the full plastic moment in bending; bending moment resistance in these cross sections is limited to the (elastic) yield moment only. For class 4 (slender) cross sections, the local or lateral buckling of the member occurs in the elastic range. An effective cross section is therefore defined based on the width-to-thickness ratios of the individual plate elements and this is used to determine the resistance of the cross section. The majority of the hot-rolled cross sections belong to class 1, 2, or 3, and hence their resistances may be based on the gross cross-section properties obtained from section tables (IS 808).

For cold-formed cross sections, which predominantly have open cross sections (e.g., channel) and light gauge materials (with thickness less than 6 mm), the design will seldom be based on the gross cross-section properties. The class 4 cross sections of Table 4.3 contain slender elements that are susceptible to local buckling in the elastic range of material behaviour. For the design of slender compression elements considering the strength beyond the elastic local buckling of elements (such as those of cold-formed sections), the reader should refer to IS 801. The effective cross sections are to be calculated in the case of slender welded plate girders (see Example 4.13).

The design moment capacity M_d of each of the four classes of sections defined above may be calculated as follows (of course a partial factor of safety for materials as per IS 800 should also be applied to the design moment capacity):

Plastic:	$M_d = Z_p f_y$	(4.8a)
Compact:	$M_d = Z_p f_y$	(4.8b)
Semi-compact:	$M_d = Z_e f_y$	(4.8c)
Slender:	$M_d < Z_e f_y$	(4.8d)

where Z_p and Z_e are the plastic and elastic section moduli, respectively.

Each compressed (or partially compressed) element is assessed individually against the limiting width-to-thickness ratios for the class 1, 2, and 3 elements defined in Table 4.3. If an element fails to meet the class 3 limits, it should be considered as class 4. The different elements of a cross section can belong to different classes. In such cases, the section is classified based on the least favourable classification. A member with one or more slender elements has its effective area reduced to

$$A_{\text{eff}} = \sum (b_{\text{eff}} t) \tag{4.9}$$

in which the effective width of any slender plate element is determined from

$$B_{\text{eff}} = \lambda_{L3} t \varepsilon \leq b \tag{4.10}$$

where λ_{L3} is the limiting width-to-thickness ratio of the semi-compact section. The limiting width-to-thickness ratios are modified by a factor ε, which takes into account the yield strength of the material (for a circular hollow section, the width-to-thickness ratios are modified by ε^2). The factor ε is defined by

$$\varepsilon = (250/f_y)^{0.5} \tag{4.11}$$

where f_y is the nominal yield strength of the steel. Note that increasing the nominal yield strength results in stricter classification limits.

The terms internal element, outstanding elements, etc., as used in Table 4.3, are described as follows.

(a) *Internal elements* These are elements attached along both longitudinal edges to other elements or to longitudinal stiffeners connected at suitable intervals to transverse stiffeners (e.g., a web of I-sections and a flange and a web of box sections).

(b) *Outstanding elements or outstands* These are elements attached along only one of the longitudinal edges to an adjacent element, the other edge being free to displace out of the plane (e.g., the flange overhang of an I-section, stem of a T-section, and leg of an angle section).

(c) *Tapered elements* These elements may be treated as flat elements having the average thickness defined in IS 808.

Compound elements in built-up section

In the case of a compound element consisting of two or more elements bolted or welded together as shown in Fig. 4.12(i), the following width-to-thickness ratios should be considered:

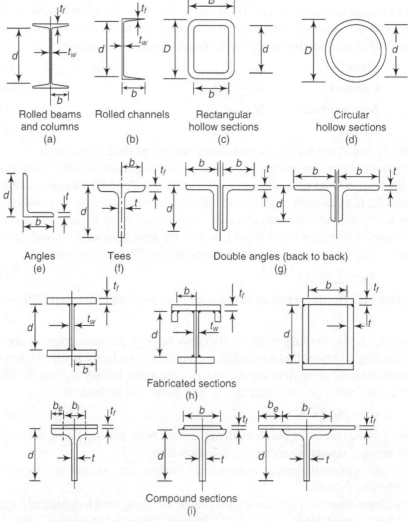

Rolled beams and columns (a) Rolled channels (b) Rectangular hollow sections (c) Circular hollow sections (d)

Angles (e) Tees (f) Double angles (back to back) (g)

Fabricated sections (h)

Compound sections (i)

Fig. 4.12 Definition of b, d, t_f, and t_w as used in Table 4.3

(a) Outstand width of compound element b_e to its own thickness
(b) The internal width of each added plate between the lines of welds or fasteners connecting it to the original section compared to its own thickness
(c) Any outstand of the added plates beyond the line of welds or fasteners connecting it to original section compared to its own thickness.

Note that stricter limits are imposed for welded elements, in recognition of the weakening effect of the more severe residual stresses present.

Hollow section members

A circular hollow section member has no buckling effects when its diameter-to-thickness ratio D/t satisfies

$$D/(t\varepsilon^2) \leq 88 \tag{4.12}$$

A circular hollow section member which does not satisfy Eqn (4.12) is considered a slender member. Its effective area is reduced to

$$A_{eff} = A_g \{88/[D/(t\varepsilon^2)]\}^{0.5} \tag{4.13}$$

The limiting width-to-thickness ratios of hollow rectangular sections are not specified in IS 800. However, for these members, the values shown in Table 4.4, may be used which is based on BS: 5950.

Table 4.4 Limiting width-to-thickness ratios for a circular hollow section (CHS) and a rectangular hollow section (RHS) for $f_y = 250$ MPa

Compression element		Ratio	Limiting value		
			Plastic	Compact	Semi-compact
Circular hollow section	Compression due to bending	D/t	$42\varepsilon^2$	$52\varepsilon^2$	$146\varepsilon^2$
	axial compression	D/t	–	–	$88\varepsilon^2$
Hot-rolled (HR) RHS	Flange: compression due to bending	b/t	29.3ε	33.5ε	42ε
	Web: neutral axis at mid-depth	d/t	67ε	84ε	126ε
	Generally	d/t	$67\varepsilon/(1 + 0.6\,r_1)$ but $\geq 42\varepsilon$	$84\varepsilon/(1 + r_1)$ but $\geq 42\varepsilon$	$126\varepsilon/(1 + 2r_2)$ but $\geq 42\varepsilon$
	Axial compression	d/t	–	–	126ε
Cold formed (CF) RHS	Flange: compression due to bending	b/t	27.3ε but $\leq 75.62 - d/t$	29.3ε but $\leq 56.7\varepsilon - 0.5\,d/t$	36.7ε
	Web: neutral axis at mid-depth	d/t	58.7ε	73.4ε	110.1ε
	Generally	d/t	$58.7\varepsilon/(1 + 0.6r_1)$ but $\geq 35\varepsilon$	$73.4\varepsilon/(1 + r_1)$ but $\geq 35\varepsilon$	$110.1\varepsilon/(1 + 2r_2)$ but $\geq 35\varepsilon$

Notes
1. For a RHS, the dimensions b and d should be as follows:
   For HR RHS: $b = B - 3t$; $d = D - 3t$
   For CF RHS: $b = B - 5t$; $d = D - 5t$
   where B, D, and t are the breadth, depth, and thickness. For a RHS subject to bending, D must always be equal to the web dimension.
2. $\varepsilon = (250/f_y)^{0.5}$

The relationship between the design moment capacity M_d and the compression flange slenderness b/t indicating the λ limits is shown in Fig. 4.13. In this figure, the value of M_d for the semi-compact section is conservatively taken as M_y. Note that in the above classifications, it is assumed that the web slenderness d/t is such that its buckling before yielding is prevented. When a beam is subjected to bending, the entire web may not be in uniform compression, and if the neutral axis lies at mid-depth, half of the web will actually experience tension. In such cases, the slenderness limits for webs are somewhat relaxed.

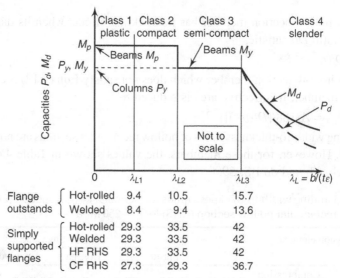

Flange outstands		λ_{L1}	λ_{L2}	λ_{L3}
	Hot-rolled	9.4	10.5	15.7
	Welded	8.4	9.4	13.6
Simply supported flanges	Hot-rolled	29.3	33.5	42
	Welded	29.3	33.5	42
	HF RHS	29.3	33.5	42
	CF RHS	27.3	29.3	36.7

Fig. 4.13 Moment capacities of fully braced beams and columns

As mentioned earlier, only plastic sections should be used in indeterminate frames forming plastic-collapse mechanisms. In elastic design, semi-compact sections can be used with the understanding that the maximum stress reached will be M_y. Slender sections also have stiffness problems and are not preferable for hot-rolled structural steelwork. However, they are used extensively in cold-formed members. Plate girders are usually designed based on the tension field approach (which takes into account the post-buckling behaviour of plates) to achieve economy.

Since the classification into plastic, compact, etc. is based on bending, it cannot be used for a compression member. In compression members, a criterion is used to determine whether the member is slender or not (see Chapter 5). However, in practice, compact or plastic sections are used for compression members, since they have more stiffness than semi-compact or slender members.

4.10 Behaviour and Ultimate Strength of Plates

Unlike columns, plates continue to support loads even after buckling. The post buckling strength has been considered in most of the codes, by using an effective width concept. This difference in behaviour of plates is explained in this section by comparing the bahaviour of plates with that of columns.

4.10.1 Behaviour of Plates

Though the behaviour of columns is discussed in the next chapter, it may be interesting to compare the behaviour of a column and a plate. In the case of an ideal column, as the axial load is increased, the lateral displacement remains zero until the attainment of the critical buckling load called the *Euler load* (see Chapter 5 for more information on column buckling). If the axial load versus lateral displacement is plotted, we will get a line along the load axis up to $P = P_{cr}$ (Fig. 4.14). This is called the 'fundamental path'. When the axial load reaches the Euler buckling load, the lateral displacement increases indefinitely at constant load. This is the 'secondary path', which bifurcates from the fundamental path at the buckling load. The secondary path for columns represents *neutral equilibrium*. A smooth transition from the stable path to the neutral equilibrium path will occur for practical columns, having initial imperfections [as shown by the dashed curves in Fig. 4.14(a)].

The fundamental path for a perfectly flat plate is similar to that for an ideal column. At the critical buckling load, this path bifurcates into a secondary path as shown in Fig. 4.14(b). However in this case, the secondary path shows that the plate can carry loads higher than the elastic critical load. Unlike columns, the secondary path for a plate is stable. Therefore, the elastic buckling of a plate need not be considered a collapse. However, plates having one edge free and other edges (outstands) simply supported have very little post-buckling strength.

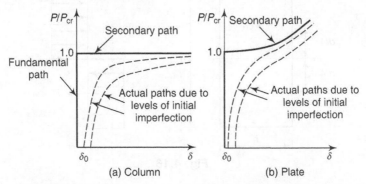

Fig. 4.14 Load versus out-of-plane displacement curves

The actual failure load of columns and plates is reached when the yielding spreads from the supported edges, triggering collapse; thereafter, the unloading occurs.

Local buckling causes loss of stiffness and redistribution of the stresses. Uniform edge compression in the longitudinal direction results in non-uniform stress distribution after buckling as shown in Fig. 4.15, and the buckled plate derives most of its stiffness from the longitudinal edge supports.

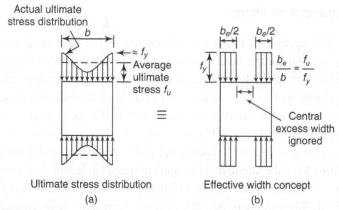

Fig. **4.15** Effective width concept for plates with simply supported edges

Examples

Example 4.1 *Determine the shape factor for a rectangular beam of width b and depth d.*

Solution

For a rectangular section (Fig. 4.16), the elastic modulus

$$Z_e = (bh^3/12)/(d/2) = bd^2/6$$

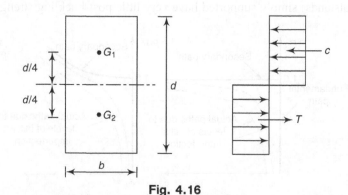

Fig. **4.16**

the plastic modulus

$$Z_p = A/2\,(\bar{y}_1 + \bar{y}_2)$$
$$= bd/2(d/4 + d/4) = bd^2/4$$

Hence the shape factor

$$v = Z_p/Z_e = (bd^2/4)/(bd^2/6) = 1.5$$

Example 4.2 *Determine the plastic moment capacity and shape factor of the I-section shown in Fig. 4.17. This section is ISMB 400 with the root radius omitted. Assume* $f_y = 250$ *MPa.*

Solution

Plastic section modulus about major axes

To determine the plastic section modulus about the z-z axis, divide the section into areas A_1 and A_2 as shown in Fig. 4.17, where

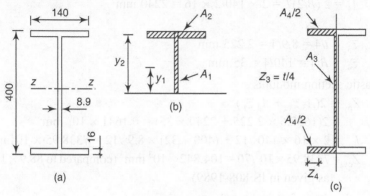

Fig. 4.17

$$A_1 = (D/2)t = (400/2) \times 8.9 = 1780 \text{ mm}^2$$
$$A_2 = (B - t)T = (140 - 8.9) \times 16 = 2097.6 \text{ mm}^2$$

and

$$\bar{y}_1 = D/4 = 400/4 = 100 \text{ mm}$$
$$\bar{y}_2 = (D/2 - T/2) = (400/2 - 16/2) = 192 \text{ mm}$$

The plastic section modulus

$$Z_p = 2(A_1 \bar{y}_1 + A_2 \bar{y}_2)$$
$$= 2(1780 \times 100 + 2097.6 \times 192)$$
$$= 1.1615 \times 10^6 \text{ mm}^3$$

The value given in Annex H of the code is $1176.18 \times 10^3 \text{ mm}^3$, which is slightly greater (1.27%) because of the additional material at the root radius.

$$I_{zz} = 2(140 \times 16^3/12) + 2 \times 140 \times 16 \times (200 - 8)^2 + (400 - 16 \times 2)^3 \times 8.9/12$$
$$= 202.20 \times 10^6 \text{mm}^4$$
$$Z_e = 202.20 \times 10^6/200 = 1011.04 \times 10^3 \text{ mm}^3 \text{(compared to}$$
$$1020.00 \times 10^3 \text{mm}^3 \text{ in Annex H of the code)}$$

Hence the shape factor

$$Z_p/Z_e = 1.1615 \times 10^6/(1011.04 \times 10^3)$$
$$= 1.1488 \text{ (compared to 1.1498 given in Annex H)}$$

Note that this value together with the stress diagrams shown in Fig. 4.1 explain the use of the limit for moment capacity of $M_d \leq 1.2 Z_e f_y/\gamma_{m0}$, which prevents plasticity at service load (clause 8.2.1.2)

Plastic section modulus about minor axis

Similarly, to determine the plastic section modulus about the y-y axis, divide the section into areas A_3 and A_4 as shown in Fig. 4.17(c), where

$$A_3 = (D - 2T)t/2 = (400 - 2 \times 16)8.9/2 = 1637.6 \text{ mm}^2$$
$$A_4 = 2 (B/2)T = 2 \times 140/2 \times 16 = 2240 \text{ mm}^2$$

and

$$\bar{z}_3 = t/4 = 8.9/4 = 2.225 \text{ mm}$$
$$\bar{z}_4 = B/4 = 140/4 = 35 \text{ mm}$$

The plastic section modulus

$$
\begin{aligned}
Z_{py} &= 2(A_3 \bar{z}_3 + A_4 \bar{z}_4) \\
&= 2(1637.6 \times 2.225 + 2240 \times 35) = 0.1641 \times 10^6 \text{ mm}^3 \\
I_{yy} &= 2 \times 16 \times 140^3/12 + (400 - 32) \times 8.9^3/12 = 7338.95 \times 10^3 \text{ mm}^4 \\
Z_{ey} &= 7338.95 \times 10^3/70 = 104.842 \times 10^3 \text{ mm}^3 \text{ (compared to } 88.9 \times 10^3 \text{mm}^3 \\
&\quad \text{as given in IS 808-1989)}
\end{aligned}
$$

Hence the shape factor about the y-y axis

$$
\begin{aligned}
v &= Z_{py}/Z_{ey} = 164.1 \times 10^3/104.842 \times 10^3 \\
&= 1.565
\end{aligned}
$$

and the plastic moment capacity

$$= 164.1 \times 10^3 \times 250 \times 10^{-6} = 41.025 \text{ kN m}$$

Example 4.3 *Determine the shape factor for a triangular section of base b and height h as shown in Fig. 4.18.*

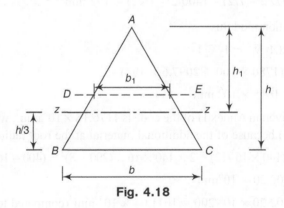

Fig. 4.18

Solution

Moment of inertia

$$I_{zz} = bh^3/36$$

Elastic section modulus

$$
\begin{aligned}
Z_e &= (bh^3/36)/(2h/3) \\
&= bh^2/24
\end{aligned}
$$

Let *DE* be the equal-area axis (see Fig. 4.18). Then,

$$0.5 \, b_1 \, h_1 = 0.5(bh/2) \text{ or } b_1 \, h_1 = bh/2 \qquad (1)$$

From similar triangles *ADE* and *ABC*

$$h_1/b_1 = h/b \qquad (2)$$

From Eqns (1) and (2), we get

$$h_1 = h/\sqrt{2} \text{ and } b_1 = b/\sqrt{2}$$

$$A = bh/2$$

$$\bar{y}_1 = h_1/3 = h/(3\sqrt{2})$$

$$\bar{y}_2 = [(h - h_1)/3] \, [(b_1 + 2b)/(b_1 + b)]$$

$$= [(h - h/\sqrt{2})/3] \, [(2b + b/\sqrt{2})/(b + b/\sqrt{2})]$$

$$= [h \, (\sqrt{2} - 1)/(3\sqrt{2})] \, (2\sqrt{2} + 1)/(\sqrt{2} + 1)$$

$$= 0.1548 \, h$$

$$Z_p = A/2(\bar{y}_1 + \bar{y}_1) = bh/4 \, [h/(3\sqrt{2}) + 0.1548h]$$

$$= 0.0976 \, bh^2$$

The shape factor

$$v = Z_p/Z_e = 0.0976 \, bh^2/(bh^2/24) = 2.343$$

The shape factor about the *y-y* axis will be different and the calculation of the same is left as an exercise to the reader.

Example 4.4 *Find out the collapse load of a simply supported beam subjected to a concentrated load at mid-span as shown in Fig. 4.19.*

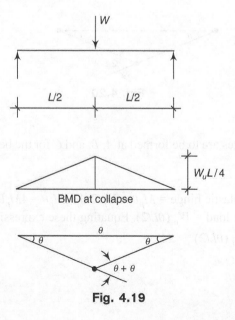

Fig. 4.19

Solution

Kinematic method

Since it is a determinate structure, one hinge is sufficient to make the beam a mechanism. Let Δ be the vertical virtual deflection at the load point. The slope of the beam at the supports

$$\theta = \Delta/(L/2) \quad \text{or} \quad \Delta = \theta L/2$$

Work done by the plastic hinge = $2M_p\theta$; work done by the load = $W_u\theta L/2$. At collapse, work by hinge = work by load

$$\therefore \qquad 2M_p\theta = W_u\theta L/2$$

$$\therefore \qquad W_u = 4M_p/L$$

Static method

$$M_p = W_uL/4 \quad \text{or} \quad W_u = 4M_p/L$$

Example 4.5 *Determine the collapse load of a fixed beam with a concentrated load at mid-span as shown in Fig. 4.20.*

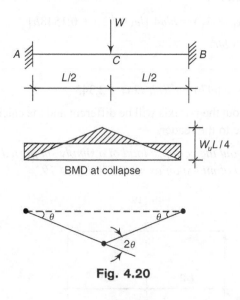

Fig. 4.20

Solution

Note that three hinges are to be formed at A, B, and C for the beam to be converted into a mechanism.

Kinematic method

Work done by the plastic hinge = $M_p\theta + M_p(2\theta) + M_p\theta = 4M_p\theta$; work done by the displacement of the load = $W_u (\theta L/2)$. Equating these expressions, we get

$$4M_p\theta = W_u (\theta L/2)$$

or

$$W_u = 8 M_p/L$$

Static method

$$W_u L/4 = 2M_p \text{ or } W_u = 8 M_p/L$$

Example 4.6 *Determine the collapse load of the fixed beam as shown in Fig. 4.21.*

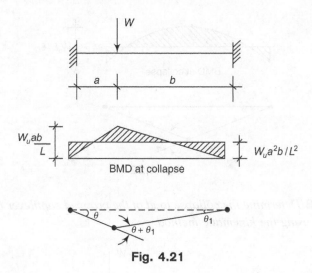

Fig. 4.21

Solution

Kinematic method

Work done by the plastic hinge

$$= M_p\theta + M_p\theta_1 + M_p(\theta + \theta_1)$$
$$= M_p\theta + M_p(a/b)\theta + M_p[\theta + (a/b)\theta]$$
$$= 2M_p\theta (1 + a/b)$$
$$= 2M_p\theta (a + b)/b = 2 M_pL\ \theta/b$$

Work done by the displacement of the load = $W_u a\theta$. Equating the two expressions, we get

$$W_u a\theta = 2M_pL\theta/b \text{ or } W_u = 2M_pL/ab$$

Static method

$$W_u ab/L = 2M_p \text{ or } W_u = 2 M_pL/ab$$

Example 4.7 *Determine the collapse load of the fixed beam shown in Fig. 4.22 using the kinematic method.*

Solution

Kinematic method

Work done by the plastic hinges = $M_p\theta + M_p\theta + M_p(2\theta) = 4M_p\theta$; work done by the displacement of the load = $W_u(1/2)(L/2\theta) = W_uL\theta/4$. Equating the two, we get

$$4M_p\theta = W_uL\theta/4 \text{ or } W_u = 16 M_p/L$$

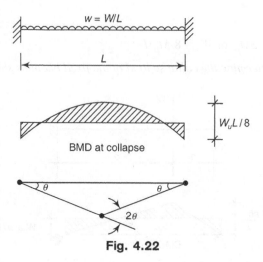

Fig. 4.22

Example 4.8 *Determine the collapse load of the propped cantilever beam shown in Fig. 4.23 using the kinematic method.*

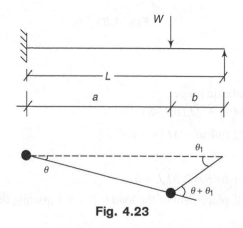

Fig. 4.23

Solution

Since the order of determinacy of this beam is 1, two hinges are required to generate a mechanism. Work done by the plastic hinges

$$= M_p\theta + M_p (\theta + \theta_1)$$
$$= M_p\theta + M_p [\theta + (a/b)\theta]$$
$$= M_p\theta(1 + 1 + a/b)$$
$$= M_p\theta[b + (a + b)]/b$$
$$= M_p\theta(L + b)/b$$

Work done by the displacement of two loads $= W_u a\theta$. Equating the two, we get

$$W_u a\theta = (L + b)M_p\theta/b$$
$$W_u = M_p(L + b)/ab$$

Example 4.9 *Determine the collapse load of a fixed-end beam subjected to a load of W at one-third span as shown in Fig. 4.24.*

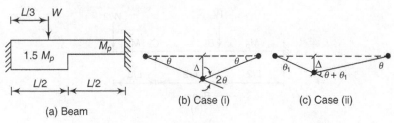

(b) Case (i)   (c) Case (ii)

(a) Beam

Fig. 4.24

Solution

To convert the beam into a mechanism, three plastic hinges are required. A plastic hinge can form at the point where the cross section changes. Hence we have two possible mechanisms.

Case 1 Two plastic hinges at the supports and one plastic hinge at the point where the cross section changes [see Fig. 4.24(b)]:

$$\text{External work done} = \text{load} \times \text{deflection}$$
$$= W_u(L/3)\theta$$
$$\text{Internal work done} = \text{moment} \times \text{rotation}$$
$$= 1.5\ M_p\theta + M_p(\theta + \theta) + M_p\theta$$
$$= 4.5 M_p\theta$$

By the principle of virtual work,

$$4.5 M_p\theta = W_u(L/3)\theta$$

or

$$W_u = 13.5 M_p/L$$

Case 2 Two plastic hinges at the supports and one just below the concentrated load:

$$\Delta = (L/3)\ \theta_1 = (2/3)L\theta \ \text{ or }\ \theta_1 = 2\theta$$
$$\text{External work done} = W_u(L/3)\theta_1$$
$$= W_u(L/3)2\theta = (2/3)W_uL\theta$$
$$\text{Internal work done} = 1.5 M_p\theta_1 + 1.5\ M_p(\theta + \theta_1) + M_p\theta$$
$$= 1.5 M_p(2\theta) + 1.5\ M_p(\theta + 2\theta) + M_p\theta$$
$$= 8.5 M_p\theta$$

By the principle of virtual work,

$$(2/3)W_uL\theta = 8.5 M_p\theta$$

Hence

$$W_u = 12.75 M_p/L$$

The collapse load will be the smaller of the above two values. Hence the collapse load of the beam is $12.75 M_p/L$.

Example 4.10 *Find the collapse load of the beam of uniform cross section shown in Fig. 4.25.*

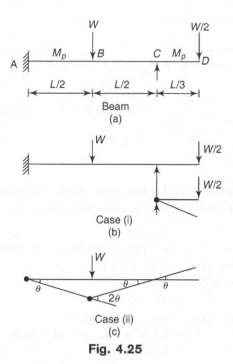

Beam
(a)

Case (i)
(b)

Case (ii)
(c)

Fig. 4.25

Solution

Span CD

Since the end D is free, a plastic hinge will form at C.

External work done = load × deflection

$$= (W/2) \times (L/3)\theta$$
$$= (WL/6)\theta$$

Internal work done = $M_p\theta$

By the principle of virtual work,

$$(WL/6)\theta = M_p\theta$$

or $W_u = 6\,M_p/L$

Span AC

Since the end C is propped, hinges will form at A and B only. The end D will be lifted up when a mechanism is formed and negative work is done by the load acting at D. Thus,

External work done = $W(L/2)\theta + (-W/2)(L/3)\theta$

$$= WL\theta[(1/2) - (1/6)] = (WL/3)\,\theta$$

Internal work done = $M_p\theta + M_p(\theta + \theta) = 3M_p\theta$

By the principle of virtual work,

$$(WL/3)\,\theta = 3M_p\theta$$

or
$$W_u = 9M_p/L$$
Hence the collapse load for the beam is $6M_p/L$.

Example 4.11 *Determine the collapse load for the portal frame shown in Fig. 4.26, assuming that the beams and columns have the same cross section.*

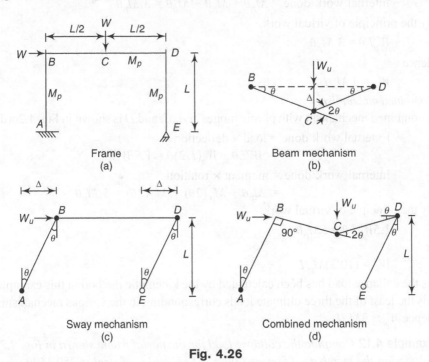

Frame
(a)

Beam mechanism
(b)

Sway mechanism
(c)

Combined mechanism
(d)

Fig. 4.26

Solution

In the case of portal frames, the first step is to find the number of independent mechanisms and then all the possible combined mechanisms. The possible locations for the plastic hinges are A, B, C, and D. Hence the number of possible plastic hinges, $N = 4$; the degree of redundancy, $r = 5 - 3 = 2$; and the number of possible independent mechanisms, $n = N - r = 4 - 2 = 2$. The two independent mechanisms are the (a) beam mechanism and (b) sway mechanism. These two independent mechanisms may be combined to form other combined mechanisms.

Beam mechanism

Assuming that plastic hinges form at B, C, and D to give a beam mechanism,

External work done = $W_u(L/2)\theta$
Internal work done = $M_p\theta + M_p(\theta + \theta) + M_p\theta = 4M_p\theta$

By the principle of virtual work,
$$W_u(L/2)\theta = 4M_p\theta$$
Hence
$$W_u = 8M_p/L$$

Sway mechanism

With a plastic hinge forming at A, B, and D (E is a mechanical hinge with zero moment and hence no plastic hinge will develop at E), the sway mechanism shown in Fig. 4.26(c) will result.

External work done $= W_u L\theta$

Internal work done $= M_p\theta + M_p\theta + M_p\theta = 3\ M_p\theta$

By the principle of virtual work,

$W_u L\theta = 3\ M_p\theta$

Hence

$W_u = 3\ M_p/L$

Combined mechanism

A combined mechanism with plastic hinges at A, C, and D is shown in Fig. 4.26(d).

External work done $=$ load $\times$ deflection

$= W_u L\theta + W_u(L/2)\theta = 1.5\ W_u L\theta$

Internal work done $=$ moment $\times$ rotation

$= M_p\theta + M_p\ (2\theta) + M_p\ (2\theta) = 5\ M_p\theta$

By the principle of virtual work,

$1.5 W_u L\theta = 5 M_p\theta$

or

$W_u = (10/3) M_p/L$

As the collapse load has been calculated by the kinematic method in this example, it is the least of the three ultimate loads corresponding to the various mechanisms. Hence $W_u = 3 M_p/L$.

Example 4.12 *Compute the collapse load for the portal frame shown in Fig. 4.27 and design the members if factored W_u = 72 kN and f_y of steel is 250 MPa.*

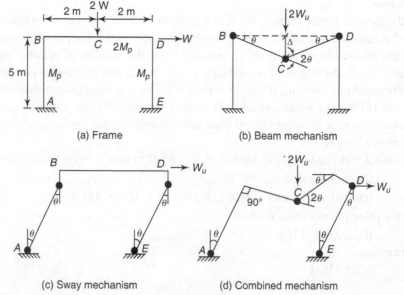

(a) Frame

(b) Beam mechanism

(c) Sway mechanism

(d) Combined mechanism

Fig. 4.27

Solution

The possible locations of the plastic hinges are A, B, C, D, and E.

The number of possible plastic hinges, $N = 5$; degree of redundancy, $r = 6 - 3 = 3$; number of possible independent mechanisms, $n = N - r = 5 - 3 = 2$. As stated in the previous example, the beam and sway mechanisms can be combined to form combined mechanisms.

Beam mechanism This mechanism is generated with three plastic hinges at B, C, and D and is associated with partial collapse [see Fig. 4.27(b)].

External work done = load × deflection

$$= 2W_u \times 2\theta = 4W_u\theta$$

Internal work done = moment × rotation

$$= M_p\theta + 2M_p(\theta + \theta) + M_p\theta = 6M_p\theta$$

By the principle of virtual work, $4W_u\theta = 6M_p\theta$. Hence $W_u = 1.5M_p$.

Sway mechanism The sway mechanism is shown in Fig. 4.27(c), with plastic hinges in the columns at A, B, D, and E.

External work done $= W_uL\theta = W_u5\theta$

Internal work done $= M_p\theta + M_p\theta + M_p\theta + M_p\theta = 4M_p\theta$

By the principle of virtual work, $5W_u\theta = 4M_p\theta$. Hence $W_u = (4/5)M_p$.

Combined mechanism

The combined mechanism is shown in Fig. 4.27(d) with hinges forming at A, C, D, and E.

External work done $= 2W_u \times 2\theta + W_u5\theta = 9W_u\theta$

Internal work done $= M_p\theta + 2M_p(2\theta) + M_p\theta + M_p\theta + M_p\theta = 8M_p\theta$

By the principle of virtual work, $9W_u\theta = 8M_p\theta$. Hence $W_u = (8/9)M_p$.

Therefore, the collapse load for the frame is $(4/5)M_p$ [least of $1.5M_p$, $(4/5)M_p$, and $(8/9)M_p$].

Design

The section must be so designed that it resists the maximum value of M_p. From the preceding calculations,

$$M_p = (9/8)W_u = (9/8) \times 72 = 80 \text{ kN m}$$

Required Z_p for column $= 80 \times 10^6/250 = 320 \times 10^3$ mm^3

Required Z_p for beam $= 2 \times 320 \times 10^3 = 640 \times 10^3$ mm^3

Hence, ISMB 225 (with $Z_p = 348.27$ mm^3) is required for the columns and ISMB 300 (with $Z_p = 651.74$ mm^3) is required for the beam of the portal frame.

Example 4.13 *Check whether the moment capacity of a welded plate girder comprising two 650- × 25-mm flange plates and one 1500- × 15-mm web plate will be affected by flange local buckling assuming (a) Fe 410 steel with a design strength of $f_y = 250$ MPa (b) Fe 540 steel with a design strength of $f_y = 410$ MPa (see Fig. 4.28).*

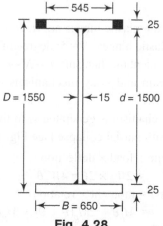

Fig. 4.28

Solution

(a) For $f_y = 250$ MPa, from Table 4.3, (Table 2 of IS 800), maximum outstand b/t for the flange to be compact = 8.4.

Actual b/t, using Fig. 4.28 = (325 – 15/2)/25 = 12.7 > 8.4

Maximum b/t for the flange to be semi-compact = 13.6

Hence the section is semi-compact and $M_d = Z_e f_y$

$$I_z = BD^3/12 - (B - t_w)d^3/12$$

$$= 650 \times 1550^3/12 - (650 - 15) \times 1500^3/12 = 23116.15 \times 10^6 \text{ mm}^4$$

$$Z_{ez} = 23116.15 \times 10^6/(1500/2 + 25) = 29827.3 \times 10^3 \text{ mm}^3$$

$$M_d = 250 \times 29827.3 \times 10^3/10^6 = 7456.8 \text{ kN m}$$

$$Z_p = 2Bt_f(D - t_f)/2 + t_w d^2/4$$

$$= 2 \times 650 \times 25 \ (1550 - 25)/2 + 15 \times 1500^2/4$$

$$= 33218.75 \times 10^3 \text{ mm}^3$$

$$M_p = 33218.75 \times 10^3 \times 250/10^6 = 8304.7 \text{ kN m}$$

Hence reduction in capacity from that corresponding to compact behaviour

$$= (8304.7 - 7456.8)/8304.7 \times 100 = 10.21\%$$

(b) For $f_y = 410$ MPa, maximum b/t for the flange to be semi-compact

$$= 13.6 \times (250/410)^{0.5} = 10.6$$

Therefore the section is slender. Limit for the effective flange width for semi-compact behaviour

$$= 10.6 \times 25 = 265 \text{ mm}$$

Hence effective top width of the flange = 265 × 2 + 15 = 545 mm (see Fig. 4.28)

Location of neutral axis

Taking moments about the top edge, the distance of the neutral axis from the top edge,

$$\bar{y} = [545 \times 25 \times 25/2 + 1500 \times 15 \times (1500/2 + 25) + 650 \times 25 \times$$

$$(1525 + 25/2)]/(545 \times 25 + 1500 \times 15 + 650 \times 25)$$

$$= 813.2 \text{ mm}$$

$$I_z = (545 \times 25) (813.2 - 12.5)^2 + (15 \times 1500^3/12) + (15 \times 1500) \times$$

$$(813.2 - 775)^2 + 650 \times 25 (736.8 - 12.5)^2 + 545 \times 25^3/12$$

$$+ 650 \times 25^3/12$$

$$= 2.1513 \times 10^{10} \text{ mm}^4$$

$$Z_z \text{ (top flange)} = 2.1513 \times 10^{10}/813.2 = 26454 \times 10^3 \text{ mm}^3$$

$$M_d = 410 \times 26454 \times 10^3/10^6 = 10,846.2 \text{ kN m}$$

Capacity, if the whole section were as effective as the semi-compact section,

$$M_{d1} = 410 \times 29827.3 \times 10^3/10^6 = 12,229.2 \text{ kN m}$$

Hence reduction in capacity from that corresponding to semi-compact section behaviour

$$= (12,229.2 - 10,846.2) \, 12,229.2 \times 100 = 11.3\%$$

Capacity, if the whole section is as effective as the plastic section,

$$M_p = 33218.75 \times 10^3 \times 410/10^6 = 13619.7 \text{ kN m}$$

Hence reduction in capacity due to slenderness

$$= (13619.7 - 10,846.2/13619.7) \times 100 = 18.4\%$$

Thus, 18.4% of the capacity of the cross section could not be utilized, due to the slenderness of the cross section.

Summary

The basic concepts of plastic analysis have been introduced in this chapter. The section modulus, either elastic or plastic depending on the design philosophy adopted, is the most appropriate property to be considered while selecting a beam section. Methods for calculating the plastic modulus and shape factor (ratio of M_P/M_y) of any cross section have been explained with a number of examples. The plastic hinge concept, conditions to be satisfied for elastic/plastic analysis, and the methods of plastic analysis (mechanism and statical method) have been discussed. The theorems of plastic collapse and alternative patterns of hinge formation triggering plastic collapse have been discussed. The plastic designs of simply supported beams, fixed beams propped cantilevers, continuous beams, and portal frames have been explained with the use of worked examples. A number of special considerations, such as local buckling, flange stability, deflection, connection, etc., should be satisfied while attempting a plastic design.

Most structural sections are assemblages of plates of slender proportions, which are prone to local buckling. Local buckling has the effect of reducing the overall load carrying capacity of columns and beams due to the reduction in the stiffness and strength of the locally buckled elements. Therefore it is desirable to avoid local buckling before the yielding of the member. The local buckling action can be analysed approximately by considering each plate of a cross section in isolation. The critical stress formulae of plates subjected to different loading and boundary conditions have been given is this chapter. From these formulae it is found that the buckling capacity of a plate is inversely proportional to the square of its width-to-thickness ratio (b/t or d/t). One way of avoiding local buckling is to ensure that the width-to-thickness ratio of each component plate does not exceed a certain value. In IS 800, a comprehensive system has been introduced for classifying the cross section of members that are subjected to compression due to moment and/or axial loading. The classification depends primarily on the geometry of the cross section and has the following four classes: plastic (class 1), compact (class 2), semi-compact (class 3), and slender (class 4). These four classes have been described for various cross sections in this chapter.

Plastic and compact sections are preferable if limit state design is used, and only plastic sections can be used in mechanism-forming indeterminate frames (plastic design). Slender sections are to be avoided even in elastic design. Most hot-rolled steel sections have individual elements of sufficient thickness (may be classified as plastic, compact, or semi-compact) and hence local buckling is avoided before the yielding of the member. However, fabricated (welded) sections and thin-walled cold-formed steel members usually experience local buckling of plate elements before the yield stress is reached. For such sections, only the effective area and moment of inertia should be used for calculating the capacity.

Substantial reserve strength exists in plates beyond the point of elastic buckling. The utilization of this reserve capacity may also be a design objective. Post-buckling reserve strength is normally taken into account in the design of cold-formed sections, approximately by using the empirical effective width concept.

Further details on plastic analysis, plastic design of frames (especially pitched roof portals and multi-storey frames), and buckling of plate elements may be found in the references given at the end of the book.

Exercises

1. Find the shape factor for the following sections:
    (a) square of side a with its diagonal parallel to z-z axis as shown in Fig. 4.29(a) [Ans: $v = 2$]
    (b) Hollow square section of external side D, internal side d and wall thickness t as shown in Fig. 4.29(b) [Ans: $v = 1.12$]
    (c) Circular section of radius r as shown in Fig. 4.29(c) [Ans: $v = 1.7$]
    (d) Hollow tube section of external diameter D and internal diameter d

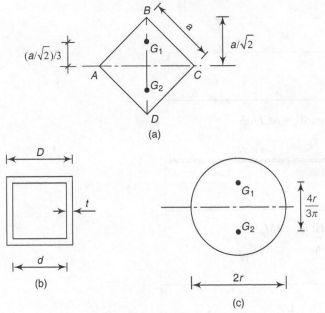

Fig. 4.29

2. Determine the plastic moment capacity of the sections shown in Figs 4.30(a) to (c), assuming $f_y = 250$ MPa

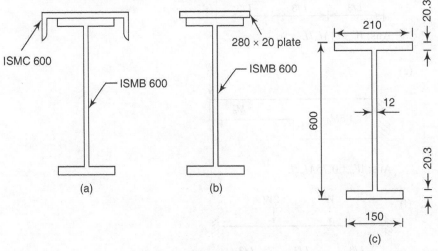

Fig. 4.30

3. Find out the collapse load of the following beams (assume that the beam is of uniform cross section, unless otherwise shown).

(a)

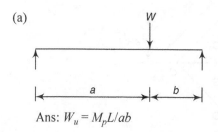

Ans: $W_u = M_p L / ab$

(b)

$w = W/L$

L

Ans: $W_u = 8 M_p / L$

(c)

W

$L/2$ $L/2$

Ans: $W_u = 6 M_p / L$

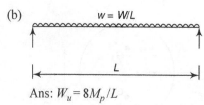

(d)

W W

$L/3$ $L/3$ $L/3$

Ans: $W_u = 4 M_p / L$

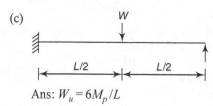

(e)

W

$1.5 M_p$ M_p

Ans: $W_u = 0.75 M_p / L$

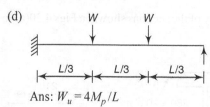

(f)

W $2W$

$L/3$ $L/3$ $L/3$

Ans: $W_u = 3.6 M_p / L$

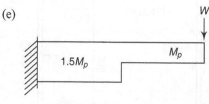

(g)

L L L L

Ans: $W_u = 12 M_p / L^2$

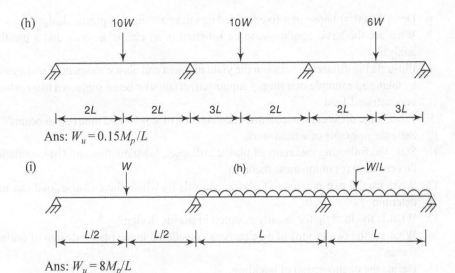

(h)

Ans: $W_u = 0.15 M_p/L$

Ans: $W_u = 8 M_p/L$

4. Find out the collapse load for a portal frame of uniform cross section as shown in Figs 4.31(a) and (b).

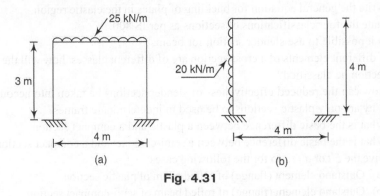

Fig. 4.31

5. Design a suitable section for a two-span continuous beam, each having a span of 10 m and supporting a dead load of 15 kN/m and a live load of 25 kN/m by (i) elastic design procedure and (ii) plastic design procedure.

6. Check whether the moment capacity of a welded plate girder consisting of two 600-× 20-mm flange plates and one 1000-× 15-mm web plate will be affected by flange buckling, assuming (a) Fe 410 grade steel ($f_y = 250$ MPa), (b) Fe 540 grade steel ($f_y = 410$ MPa).

Review Questions

1. Distinguish between the elastic modulus and plastic modulus of a section.
2. Define shape factor.
3. Will the shape factor of rectangular and circular cross sections be higher than that of I-sections? If yes, state the reason.
4. What is a plastic hinge? In what way is it different from an ordinary hinge?
5. What are the points at which a plastic hinge is likely to form?

6. Describe the collapse of a fixed-ended beam, as assumed in plastic design.
7. What are the basic conditions to be satisfied in an elastic analysis and a plastic analysis?
8. Illustrate the difference between the yield moment and plastic moment of resistance by taking an example of a simply supported rectangular beam subjected to a central concentrated load.
9. What is the difference in collapse mechanisms of single- and multi-span beams?
10. State the principle of virtual work.
11. State the following theorems of plastic collapse: (a) static theorem (b) kinematic theorem and (c) uniqueness theorem.
12. What are the two methods of plastic analysis by which the collapse load can be determined?
13. What is the load factor usually adopted in plastic design?
14. What are the two groups of mechanisms considered in the plastic design of portal frames?
15. Define the phenomenon of buckling.
16. What are the three levels of changes in geometry that are associated with local, member, and frame instability?
17. Write the general equation for buckling of plates in the elastic region.
18. State the four classifications of sections as per IS 800.
19. Is it possible to use slender section for beams?
20. If different elements of a cross section are of different classes, how will the whole section be classified?
21. How can the reduced effectiveness of slender sections be taken into account?
22. Why are only plastic sections to be used in indeterminate frames?
23. What is the basic difference between a plastic and a compact section?
24. What is the basic difference between a semi-compact and a compact section?
25. Give the b/t or d/t ratio for the following cases:
    (a) Outstand element (flange) of rolled beam of plastic section
    (b) Outstand element (flange) of rolled beam of semi-compact section
    (c) Web of an I-beam of plastic cross section
    (d) Web of an I-beam of semi-compact section
26. Describe the difference in behaviour of columns and plates.

CHAPTER **5**

Design of Compression Members

Introduction

A structural member which is subjected to compressive forces along its axis is called a *compression member*. Thus, compression members are subjected to loads that tend to decrease their lengths. Except in pin-jointed trusses, such members (in any plane or space structure), under external loads, experience bending moments and shear forces. If the net end moments are zero, the compression member is required to resist load acting concentric to the original longitudinal axis of the member and is termed *axially loaded column,* or simply *column.* If the net end moments are not zero, the member will be subjected to an axial load and bending moments along its length. Such members are called *beam-columns* and are treated in Chapter 9.

Let us consider an example of an axially loaded column shown in Fig. 5.1. For this column, the axial load P, to be resisted by it, is the sum of the beam shears $V_1 + V_2$. For the column shown in Fig. 5.1, the net end moment is assumed to be zero; this is true if the end moments and shears developed by the two beams are equal. Such situations arise in many interior columns of buildings having equal column spacing. Where the beam is not connected rigidly to the column, the beams will not develop significant end moments and in such situations also the column has to resist only the difference in end shears. In several interior columns, the net moment will be small and the member is designed as an axially loaded column.

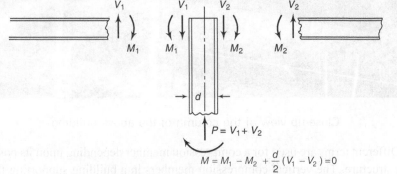

$$P = V_1 + V_2$$

$$M = M_1 - M_2 + \frac{d}{2}(V_1 - V_2) = 0$$

Fig. 5.1 Axially loaded column

Multi-storey steel building under construction in Maryland, USA

Close-up view of the column of the above building

Different terms are used for a compression member depending upon its position in a structure. The vertical compression members in a building supporting floors or girders are normally called as columns (referred sometimes as *stanchions* in UK). They are subjected to heavy loads. Sometimes vertical compression members are called *posts*. The compression members used in roof trusses and bracings are called *struts*. They may be vertical or inclined and normally have small lengths. The top chord members of a roof truss are called the *principal rafter*. The principal compression member in a crane is called the *boom*. Short compression members at the junction of columns and roof trusses or beams are called *knee braces*. Some of these compression members are shown in Fig. 5.2.

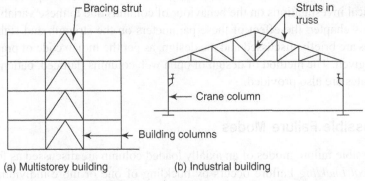

(a) Multistorey building   (b) Industrial building

Fig. 5.2 Types of compression members

It is well known from basic mechanics of materials that only very short columns can be loaded up to their yield stress. For long columns (see Section 5.3.1 also), *buckling* (deformation in the direction normal to the load axis) occurs prior to developing the full material strength of the member. Hence, a sound knowledge of stability theory is necessary for designing compression members in structural steel.

Since compression members have to resist buckling, they tend to be stocky and 'square' and circular tubes are found to be the ideal sections, since their radius of gyration is same in the two axes. This situation is in contrast to the slender and more compact tension members and deep beam sections. Unlike the member subjected to tension, a compression member is designed on the assumption that its gross cross-sectional area will be effective in resisting the applied loads. Bolts may be used to connect columns to adjacent members. As the load is applied, the member will contract. It is assumed that the action of bolts is such that they will replace the material removed for holes. Thus, the bolt holes are often ignored in the design. Since compression members comprise of thin plates, they also experience local buckling, as discussed in Chapter 4.

The strength of a column depends on the following parameters:

- Material of the column
- Cross-sectional configuration
- Length of the column
- Support conditions at the ends (called restraint conditions)
- Residual stresses
- Imperfections

The imperfections include the following:
- The material not being isotropic and homogenous
- Geometric variations of columns
- Eccentricity of load

It is difficult to assess the residual stress acting on each column cross section and also to assess the degree of support condition offered by a variety of connection details adopted in practice. Due to the large number of variables, which influence the strength of columns and beam-columns in the elastic and inelastic ranges, several researchers throughout the world have done extensive experimental and theoretical investigations on the behaviour of columns due to these variables.

In this chapter, the effect of these parameters on the strength and stability of columns are briefly discussed and the design, as per the Indian code of practice IS 800, is given. The method of design of open web columns (latticed, battened) and base plates are also provided.

5.1 Possible Failure Modes

The possible failure modes of an axially loaded column are discussed as follows:

1. *Local buckling* Failure occurs by buckling of one or more individual plate elements. This failure mode may be prevented by selecting suitable width-to-thickness ratios of component plates. Alternatively, when slender plates are used, the design strength may be reduced (see Section 5.2).

2. *Squashing* When the length is relatively small (stocky column) and its component plate elements are prevented from local buckling, then the column will be able to attain its full strength or 'squash load' (yield stress × area of cross section).

3. *Overall flexural buckling* This mode of failure normally controls the design of most compression members. In this mode, failure of the member occurs by excessive deflection in the plane of the weaker principal axis. An increase in the length of the column, results in the column resisting progressively less loads.

4. *Torsional and flexural–torsional buckling* Torsional buckling failure occurs by twisting about the shear centre in the longitudinal axis. A combination of flexure and twisting, called *flexural–torsional buckling* is also possible. Torsional buckling is a possible mode of failure for point symmetric sections. Flexural torsional buckling must be checked for open sections that are singly symmetric and for sections that have no symmetry. Note that open sections that are doubly symmetric or point symmetric are not subjected to flexural-torsional buckling, since their shear centre and centroid coincide. Closed sections are also immune to flexural–torsional buckling.

In addition to the above failure modes, in compound members (two or more shapes joined together to form a lattice cross section), failure of a component member may occur, if the joints between members are sparsely placed. Codes and specifications usually have rules to prevent such failures (see Section 5.11).

5.2 Classification of Cross Section

If individual plate elements which make up the cross section of a compression member, (for example, the web and two flanges in the case of an I-section), are thin, local buckling may occur. For columns, it is frequently possible to eliminate this problem by limiting the proportions of component plates as given in Table 5.1, such that local buckling will not influence the strength of the cross sections.

These limits are applicable to semi-compact sections. When more slender plates, having width-to-thickness ratios higher than the limits given in Table 5.1 are to be used, the strength of the section should be suitably reduced. However, it is a good practice to use plastic or compact sections for compression members, since they provide more stiffness than semi-compact or slender sections.

Table 5.1 Limiting width-to-thickness ratios for axial compression elements

Element	Ratio	Upper limit for semi-compact section
Outstand element of compression flange (I, H, or C)		
Rolled section	b/t_f	15.7ε
Welded section	b/t_f	13.6ε
Internal element of compression flange (box)	b/t_f	42ε
Web of an I-H or box-section	d/t_w	126ε
Single angle or double angle with the components separated (all three criteria should be satisfied)	b/t d/t $(b+d)/t$	15.7ε 15.7ε 25ε
Circular hollow section	D/t	$88\varepsilon^2$
Hot rolled RHS—Flange	b/t	42ε
Web	d/t	126ε
Cold formed RHS—Flange	b/t	36.7ε
Web	d/t	110ε

$\varepsilon = (250/f_y)^{0.5}$
For HR RHS: $b = B - 3t$; $d = D - 3t$
For CF RHS: $b = B - 5t$; $d = D - 5t$

5.3 Behaviour of Compression Members

Before discussing the behaviour of compression members, it may be useful to know about the classification of compression members based on their length.

5.3.1 Long, Short, and Intermediate Compression Members

Compression members are sometimes classified as being long, short, or intermediate. A brief discussion about this classification is as follows:

Short compression members For very short compression members the failure stress will equal the yield stress and no buckling will occur. Note that for a compression member to fall into this classification, it has to be so short (for an initially straight column $L \leq 88.85r$, for $f_y = 250$ MPa) that it will not have any practical application.

Long compression members For these compression members, the Euler formula [see Section 5.4 and Eqn (5.3)] predicts the strength of long compression members very well, where the axial buckling stress remains below proportional limit. Such compression members will buckle elastically.

Intermediate length compression members For intermediate length compression members, some fibres would have yielded and some fibres will still be elastic. These compression members will fail both by yielding and buckling and their behaviour is said to be 'inelastic'. For the Euler formula Eqn (5.3) to be applicable to these compression members, it should be modified according to the reduced modulus concept or the tangent modulus concept (as is done in AISC code formula) to account for the presence of residual stresses. Now let us discuss the behaviour of short and slender columns.

5.3.2 Short Compression Members

Consider an axially compressed member of short length which is initially straight and made of material having the ideal rigid–plastic stress–strain relationship as shown in Fig. 5.3. At low values of external load P, there will be no visible deformation—neither lateral nor axial. Since P is applied at the centroid of the section, apart from possible localized effects at the ends of the member, all parts of the member will experience the same value of compressive stress $f_c = P/A$. Large deformation is possible only when f_c reaches the yield stress f_y. At this stage the member deforms axially. The value of the axial force at which this happens is termed as the 'squash load' and is given by

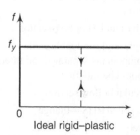

Fig. 5.3 Idealized material behaviour

$$P_y = f_y A \tag{5.1}$$

Typical laboratory compression test on the short length of the rolled section is often referred to as the *stub column test*. Since rolled steel sections have residual stresses, there will be non-uniform yielding of the cross section of the stub column. Thus, those fibres which contain residual compression have their effective yield point reduced; while those containing residual tension have theirs increased. However, in both cases, the squash load of the stub column can be achieved provided there is no local buckling. (Nethercot 2001).

5.3.3 Slender Compression Members

As mentioned earlier, the strength of the compression member decreases as its length increases, in contrast to the axially loaded tension member whose strength is independent of its length. Thus, the compressive strength of a very slender member may be much less than its tensile strength, as shown in Fig. 5.4 (also see Fig. 5.6). This decrease in strength is due to the following parameters, which are often grouped under the heading of imperfections: the initial lack of straightness, accidental

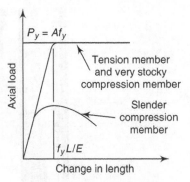

Fig. 5.4 Strengths of axially loaded members

eccentricities of loading, residual stresses, and variation of material properties over the cross section. In order to understand the effect of these parameters, we should study the behaviour of ideal, straight pin-ended columns, which is discussed in the next section.

5.4 Elastic Buckling of Slender Compression Members

Though the first qualitative remarks on column strength and stability were given by Erone of Alexandria (75 B.C.) and similar descriptions of buckled columns by Leonardo Da Vinci (1452–1519), the theory of column buckling was first originated by Euler during 1744–1759 (Timoshenko 1953; Euler 1759). Euler considered an ideal column with the following attributes.

- Material is isotropic and homogenous and is assumed to be perfectly elastic.
- The column is initially straight and the load acts along the centroidal axis (i.e., no eccentricity of loads).
- Column has no imperfections.
- Column ends are hinged.

Such a column is also known as an *Euler column* and is shown in Fig. 5.5. A concentric load P is applied to the upper end of the member, which remains straight until buckling occurs, when it is slightly bent as shown in Fig. 5.5.

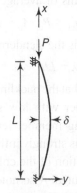

Fig. 5.5 Behaviour of a perfectly straight elastic pin-ended column

The Euler critical load for a column with both ends hinged may be derived as

$$P_{cr} = \pi^2 EI/L^2 \tag{5.2}$$

Buckling of a pin-ended column in a laboratory test
(Courtesy: Late Dr. W. Kurth of Germany)

Or in terms of average critical stress, using $I = A_g r^2$

$$f_{cr} = P_{cr}/A_g = \pi^2 E/\lambda^2 \tag{5.3}$$

where λ is the slenderness ratio defined by

$$\lambda = L/r \tag{5.4}$$

Note that the buckling phenomenon is associated with the stiffness of the member. A member with low stiffness will buckle early than one with high stiffness. Increasing member lengths causes reduction in stiffness. The stiffness of the member is strongly influenced by the amount and distribution of the material in the cross section of the column; the value of r reflects the way in which the material is distributed. Also note that any member will tend to buckle about the weak axis (associated with lesser ability to resist buckling).

Critical loads and critical stresses can be similarly found for struts having other end-restraint conditions. For example, if both ends are prevented from rotation as well as from lateral movement (fixed ends), the critical load may be $4\pi^2 EI/L^2$. In such cases, the slenderness ratio is defined as KL/r, where KL is the *effective length* of the member. This concept is discussed in Section 5.8.

Inspection of Eqn (5.3) indicates that the critical stress is inversely proportional to the slenderness ratio of the column and very large values of f_{cr} can be obtained by using $L/r \to 0$. However, as per the differential equation for bending, stress should be proportional to strain. Thus, the upper limit of validity is the proportional limit, which is often taken as $f_{cr} \rightarrow f_y$, though short columns may be loaded even into the strain-hardening range. Thus, one may write,

$$f_y = \pi^2 E / \lambda_p^2 \tag{5.5}$$

Hence, $\lambda_p = \pi \sqrt{(E/f_y)}$; for $f_y = 250$ MPa, $\lambda_p = 88.85$ $\qquad$ (5.6)

Thus, the changeover from yielding to buckling failure occurs at point C (see Fig. 5.6), defined by the slenderness ratio λ_p. Figure 5.6(a) shows the strength curve for an axially loaded initially straight pin-ended column indicating the plastic yield (stocky columns) and elastic buckling (long column) behaviour. Figure 5.6(a) is often presented in a non-dimensional form as shown in Fig. 5.6(b).

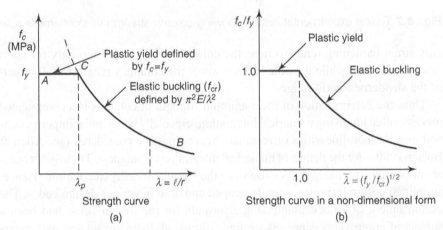

Fig. 5.6 Strength curve for an axially loaded initially straight pin-ended column

5.5 Behaviour of Real Compression Members

Euler's approach was generally ignored for design because test results did not agree with it (see Fig. 5.7). Test results included effects of initial crookedness of the member, accidental eccentricity of load, end restraint, local or lateral buckling, and residual stress. The effects of these parameters are discussed in Galambos (1938), Salmon and Johnson (1996), and Subramanian (2008). Note that columns are usually an integral part of a structure and as such cannot behave independently. Due to the imperfections, the load deflection curve of a real column will be much different than that of an ideal column.

5.6 Development of Multiple Column Curves

In real columns, all the effects mentioned in Section 5.5 occur simultaneously, i.e., out-of-straightness, eccentricity of loading, residual stresses, also depending on the material, lack of clearly defined yield point, and strain hardening. Note that

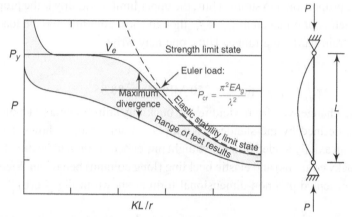

Fig. 5.7 Typical experimental results showing column strength vs slenderness ratio

only strain hardening tends to raise the column strength (that too only for short stocky columns), while the other effects lower the column strength for all or part of the slenderness ratio range.

Thus, the determination of the maximum strength of columns is a complicated process, often involving numerical integration, especially when initial imperfections and material non-linearities or residual stresses must be considered (see Allen & Bulson 1980—for the details of numerical integration techniques). These procedures are not suitable for design office use since they require lengthy calculations. Hence, simplified design formulae were developed and used in various design codes. The establishment of an acceptable single formula for the design stress had been a subject of controversy, since, as we have discussed, there are several parameters that influence the strength.

In the past, four basic methods were used to establish column design formulae, curves or charts (Galambos 1998).

1. Empirical formulae such as Merchant-Rankine formula
2. Formulae based on the yield limit state, e.g. Perry-Robertson formula
3. Formula based on tangent modulus theory, and
4. Formula based on maximum strength

In order to develop a rational basis for predicting the strength of real columns (which have initial imperfections, end restraints, eccentricity of loading, and residual stresses), the committee 8 on instability (headed by Prof. Beer from Austria) of the European Convention for Constructional Steel work (founded in the year 1955 and shortly known as ECCS) launched an extremely large series of experiments on axially loaded columns. Seven European countries were involved and about thousand buckling tests were performed (Sfintesco 1970).

Based on these results, the ECCS committee concluded that in order to represent the real strength of columns, five different curves have to be used, depending on different cross sections of columns. These curves are called the European multiple column curves.

Based on an extensive numerical study, Bjorhovde (1972) generated a set of 112 column curves for members for which measured residual stress distribution were available, assuming that the initial crookedness was of a sinusoidal shape having maximum amplitude of $1/1000^{th}$ of the column length and that the end restraint is zero. These curves were reduced to a set of three column curves and adopted by SSRC. These curves are known as SSRC column strength curves.

Several European countries have adopted the ECCS curves. Eurocode 3 has adopted a modified form of ECCS curves. The Canadian code has adopted the SSRC column curves 1 and 2 and uses the following equation (Galambos 1998).

$$(f_u/f_y) = 0.9A(1 + \bar{\lambda}^{2n})^{-(1/n)} \tag{5.7}$$

with $n = 2.24$ (welded members) and 1.34 (hot rolled members) for curves 1 and 2 respectively.

5.6.1 Multiple Column Curves in the IS Code

The Indian Code (IS 800) has adopted the multiple column curves, as shown in Fig. 5.8, which are similar to the curves given in the British Code BS 5950 (Part 1)-2000, which is based on the Perry–Robertson approach. The design compressive strength is given by

$$P_d = A_e f_{cd} \tag{5.8}$$

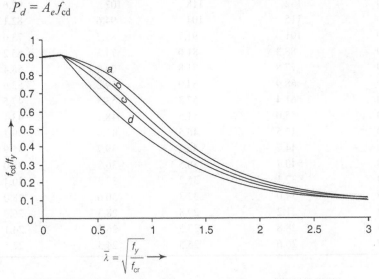

Fig. 5.8 Column buckling curves as per IS 800-2007

where A_e is the effective sectional area.

The design stress in compression f_{cd} is given by (substituting for γ_{m0})

$$f_{cd} = [(0.909 f_y)/\{\phi + (\phi^2 - \bar{\lambda}^2)^{0.5}\}] = 0.909 \, \chi f_y \leq 0.909 \, f_y \tag{5.9a}$$

where $\phi = 0.5[1 + \alpha(\bar{\lambda} - 0.2) + \bar{\lambda}^2]$ (5.9b)

and $\bar{\lambda}$ is the non-dimensional effective slenderness ratio $= \sqrt{(f_y/f_{cr})}$, f_{cr} is the Euler buckling stress $= \pi^2 E/(KL/r)^2$, KL/r is the effective slenderness ratio, α is the imperfection factor (see Table 5.2), and χ is the stress reduction factor.

Table 5.2 Imperfection factor α

Buckling curve	a	b	c	d
α	0.21	0.34	0.49	0.76

The design compressive stress f_{cd} for different buckling curves and effective slenderness ratio are given in Table 5.3 for $f_y = 250$ MPa. For other values of the yield stress, refer IS 800-2007. The classification of different sections under different buckling curves a, b, c, or d is given in Table 5.4.

Table 5.3 Design compressive stress f_{cd} (MPa) for $f_y = 250$ MPa

$KL/r\downarrow$	Curve a	Curve b	Curve c	Curve d
10	227	227	227	227
20	226	225	224	223
30	220	216	211	204
40	213	206	198	185
50	205	194	183	167
60	195	181	168	150
70	182	166	152	133
80	167	150	136	118
90	149	134	121	105
100	132	118	107	92.6
110	115	104	94.6	82.1
120	101	91.7	83.7	73.0
130	88.3	81.0	74.3	65.2
140	77.8	71.8	66.2	58.4
150	68.9	64.0	59.2	52.6
160	61.4	57.3	53.3	47.5
170	55.0	51.5	48.1	43.1
180	49.5	46.5	43.6	39.3
190	44.7	42.2	39.7	35.9
200	40.7	38.5	36.3	33.0
210	37.1	35.2	33.3	30.4
220	34.0	32.3	30.6	28.0
230	31.2	29.8	28.3	26.0
240	28.8	27.5	26.2	24.1
250	26.6	25.5	24.3	22.5

5.7 Sections used for Compression Members

Though numerous sections may be selected to safely resist a compressive load in a given structure, from a practical viewpoint, the possible solutions are limited mainly due to the considerations of availability of sections, connection problems, and the type of structure in which they will be used. Figure 5.9 shows the possible configurations that may be used as compression members. From this figure it can be observed that the sections used as compression members are similar to tension

Table 5.4 Selection of buckling curve depending on cross section

Cross section	Limits		Buckling about axis	Buckling curve
Rolled I-sections	$h/b > 1.2$:	$t_f \leq 40$ mm	z-z	a
			y-y	b
		40 mm $< t_f \leq 100$ mm	z-z	b
			y-y	c
	$h/b \leq 1.2$:	$t_f \leq 100$ mm	z-z	b
			y-y	c
		$t_f > 100$ mm	z-z	d
			y-y	d
Welded I-section		$t_f \leq 40$ mm	z-z	b
			y-y	c
		$t_f > 40$ mm	z-z	c
			y-y	d
Hollow section	Hot rolled	Any		a
	Cold formed	Any		b
Welded box section	Generally (except as below)		Any	b
	Thick welds and	$b/t_f < 30$	z-z	c
		$h/t_w < 30$	y-y	c
Channel, angle, T- and solid sections		Any		c

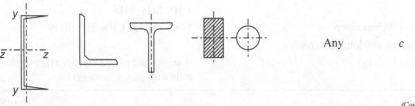

(Contd)

(Contd)

Built-up member		
	Any	c

members with some exceptions. These exceptions are caused by the fact that the strength of compression members vary inversely to the slenderness ratio [see Section 5.4 and Eqn (5.3)]. Thus, rods, bars, and plates are too slender to be used as compression members.

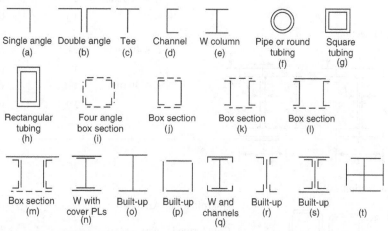

Fig. 5.9 Shapes of compression members

The choice of the sections in a particular situation depends on the following:

- Functional aspects of the structure
- Functional aspects of the member
- Easy connectivity to other members
- High radius of gyration

Table 5.5 gives some recommendations for the above aspects.

Table 5.5 Recommended structural shapes

Structure/member	Type of cross section
Small span roof trusses (up to 20 m span)	
(a) Rafters and bottom ties	Double angles (connected back-to-back), Ts, CHS, SHS, RHS
(b) Web members	Single angles, CHS, SHS, RHS
Medium and long span roofs	
(a) Rafters	4 angles and plate {Fig. 5.9(r)}, two channels with plates, box-section

(Contd)

(Contd)

(b) Web members	4 angles, 2 channels
Small towers (up to 20 m high)	
(a) Leg members	CHS, 2 angles (star formation)
(b) Bracings	CHS or angles
Towers about 150 m high	
(a) Bottom 40 m legs	4 angles (star formation)
(b) Bracings and horizontal members	2 angles (connected back-to-back)
Members of space frames	CHS, SHS
Columns of multi-storey buildings	I-sections, built-up I-sections, box sections (for heavy loads)
Columns of industrial buildings	I-sections, built-up channels or I-sections

5.8 Effective Length of Compression Members

The effect of end restraints on column strength is usually incorporated in the design by the concept of effective length. (It must be emphasized that the concept of effective lenght is only a simple and convenient short cut and the designer is not obliged to use it, if he/she knows a better, yet sufficiently simple solution.) Hence, in this section the effective length concept is first explained for idealized boundary conditions and then extended to cases where intermediate restraints are present. The Julian and Lawrence alignment charts are used in American and Canadian codes. Note that Wood's Curves (used in British and Indian codes), which are used to determine the effective lenght of columns of multi-storey buildings are also explained. The Indian code recommendations for the effective length factors of truss members are also discussed.

5.8.1 Effective Length for Idealized Boundary Conditions

Till now, the discussion in this chapter has been centered around pin-ended columns. Under ideal conditions, the boundary conditions of a column may be idealized in one of the following ways:
- Both ends pinned (as already discussed in Section 5.4)
- Both ends fixed
- One end fixed and the other end pinned
- One end fixed and the other end free

For all these conditions, the differential equations can be set up and the appropriate boundary conditions applied to get the following critical loads (Timoshenko & Gere 1961; Allen & Bulson 1980).

1. For column with both ends fixed:
$$P_{cr} = 4\pi^2 EI/L^2 = 4\pi^2 E/(L/r)^2 \qquad (5.10)$$

2. For column with one end fixed and the other end pinned:
$$P_{cr} = 2\pi^2 E/(L/r)^2 \qquad (5.11)$$

3. For columns with one end fixed and the other end free:
$$P_{cr} = \pi^2 E/[4(L/r)^2] \qquad (5.12)$$

Using the length of the pin-ended column, as the basis for comparison, the critical load in the three cases mentioned earlier can be obtained by employing the concept of *effective length KL* where *K* is called the *effective length ratio* or effective length coefficient. (It is the ratio of the effective length to the unsupported length of the columns.) The value of *K* depends on the degree of rotational and translational restraints at the column ends. The *unsupported length L* is taken as the distance between lateral connections, or actual length in the case of a cantilever. In a conventional framed construction, *L* is taken as the clear distance between the floor and the shallower beam framing into the columns in each direction at the next higher floor level.

In other words, the *effective length* of a column in a given plane may be defined as the distance between the points of inflection (zero moment) in the buckled configuration of the column in that plane (see Fig. 5.10). Note that when there is relative translation at the ends of the column, the points of inflection may not lie within the member. In such a case, they may be located by extending the deflection curve beyond the column end(s) and by applying conditions of symmetry, as shown in Fig. 5.11. Thus, the effective length can also be defined as the length of an equivalent pin-ended column, having the same load-carrying capacity as the member under consideration. The smaller the effective length of a particular column, the smaller is its danger of lateral buckling and the greater its load carrying capacity. It must be recognized that it is very difficult to get perfectly fixed or perfectly hinged end conditions in practice.

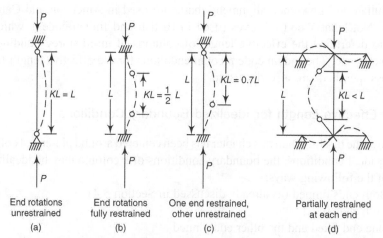

| End rotations unrestrained | End rotations fully restrained | One end restrained, other unrestrained | Partially restrained at each end |
| (a) | (b) | (c) | (d) |

Fig. 5.10 Effective length *KL* when there is no joint translation at the ends

Using *KL*, Eqns (5.10), (5.11), and (5.12) can be written as

$$P_{cr} = \pi^2 E/(KL/r)^2 \tag{5.13}$$

where $K = 1.0$ for columns with both ends pinned,

$K = 0.5$ for columns with both ends fixed,

$K = 0.707$ for columns with one end fixed and the other end pinned,

$K = 2.0$ for columns with one end fixed and the other end free,

$K \leq 1.0$ for columns partially restrained at each end, and

$K \geq 2.0$ for columns with one end unstrained and the other end rotation
   partially restrained.

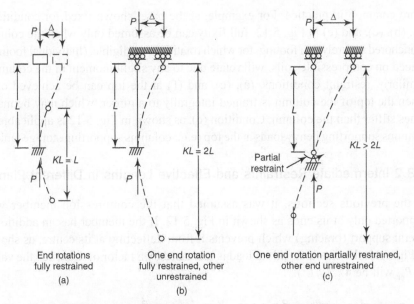

Fig. 5.11 Effective length *KL* when there is joint translation at the ends

Thus, the effective length is important in design calculations because the buckling load is inversely proportional to the square of the effective length.

Approximate values for effective lengths, which can be used in design, are given in IS 800 and are shown in Fig. 5.12 along with the theoretical K values {as given by Eqn (5.13)}. Note that the values given in the code are slightly more than the theoretical values given in (a), (b), and (c) of Fig. 5.12. It is because, fully rigid end restraints are difficult to achieve in practice; partial end restraints are much

	(a)	(b)	(c)	(d)	(e)	(f)
Buckled shape of column is shown by dashed line	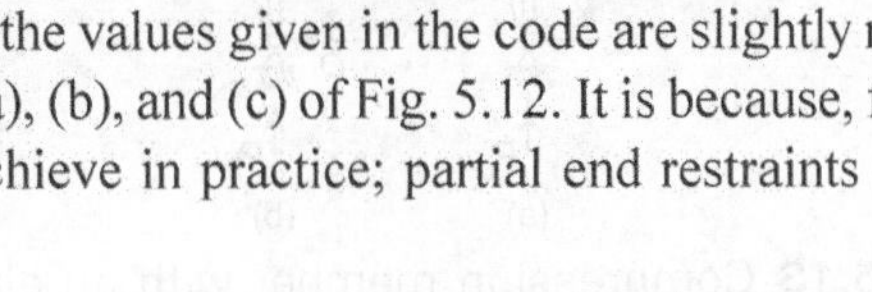					
Theoretical *K* value	0.5	0.7	1.0	1.0	2.0	2.0
Recommended *K* value when ideal conditions are approximated	0.65	0.80	1.2	1.0	2.0	2.0
End condition code		Rotation fixed, translation fixed				
		Rotation free, translation fixed				
		Rotation fixed, translation free				
		Rotation free, translation free				

Fig. 5.12 Effective-length factors *K* for centrally loaded columns with various end conditions

more common in practice. For example, at the base, shown fixed for conditions (a), (b), (c), and (e) in Fig. 5.12, full fixity can be assumed only when the column is anchored securely to a footing, for which rotation is negligible. (Individual footings placed on compressible soils, will rotate due to any slight moment in the column.) Similarly, restraint conditions, (a), (c), and (f) at the top can be achieved only when the top of the column is framed integrally to a girder, which may be many times stiffer than the column. Condition (c), as shown in Fig. 5.12 is applicable to columns supporting heavy loads at the top (e.g., columns supporting storage tanks).

5.8.2 Intermediate Restraints and Effective Lengths in Different Planes

In the previous sections, it was assumed that the compression member was supported only at its ends as shown in Fig. 5.12. If the member has an additional lateral support (bracing) which prevents it from deflecting at its centre, as shown in Fig. 5.13, the elastic buckling load is increased by a factor of four (i.e., the value of P_{cr} will be $4\pi^2 EI/L^2$).

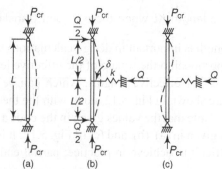

Fig. 5.13 Compression member with an elastic intermediate restraint

This restraint need not be completely rigid, but may be elastic, provided its stiffness exceeds a certain minimum value.

Many specifications and codes suggest a rule-of-thumb of using a bracing having a strength of 2 to 2.5% of the compressive strength of the compression elements being braced. This seems to be a conservative alternative to a rigorous analytical study (Lay & Galambos 1966). Such a clause is also available in the code (IS 800 clause 7.6.6.1).

When such restraints are provided, the buckling behaviour will be different about the two column axes. For example, consider a pin-ended column which is braced about the minor axis against lateral movement (but not rotationally restrained) at a spacing of $L/4$. Now the minor axis buckling mode will be such that $K = 1/4$. If there is no bracing in the major axis, the effective length for buckling about major axis will be L. Hence, the slenderness ratio about the major and minor axis will be L/r_z and $L/4r_y$ respectively. For ISHB sections $r_z < 4r_y$ and hence the major axis slenderness ratio will be greater, giving lower value of critical load and failure will occur by major axis buckling. If this is not the case, checks have to be carried out about both the axes.

In many situations, it may not be possible to use bracing in more than one plane. For example, columns are often used with walls, where these walls can serve as lateral bracing in one plane; in other planes for functional reasons it may not be possible to provide bracings. In such situations, non-symmetric members could be employed advantageously with their strong axis oriented in the out-of-braced-plane buckling mode and their weak axis in the in-plane mode. A great level of efficiency may be obtained by keeping the buckling load about one axis exactly equal to the buckling load about the other axis. The number of braces required to achieve this may be found out by the relationship $r_z/r_y = L_z/L_y$. (This relationship may also be useful in the design of built-up columns—see Example 5.14.)

5.8.3 Columns in Multi-storey or Framed Buildings

As mentioned already, isolated columns are rare and they normally form a part of any framework. Moreover, their end conditions are influenced by the members to which they are connected. The more accurate determination of K for such a compression member as part of any framework requires the application of methods of indeterminate structural analysis, modified to take into account the effects of axial load and inelastic behaviour on the rigidity of the members. These procedures are not directly applicable to routine design and hence simple models are often used to determine the effective length factor for framed members. In such simplified approaches, a distinction is always made about sway (unbraced) and non-sway (braced) frames because of their distinct buckling modes (see Fig. 5.14). In non-sway (braced) frames, the columns buckle in single curvature and hence their effective length factor will always be less than unity; whereas the columns in sway frames buckle in double curvature and hence their effective length factor will always be greater than unity.

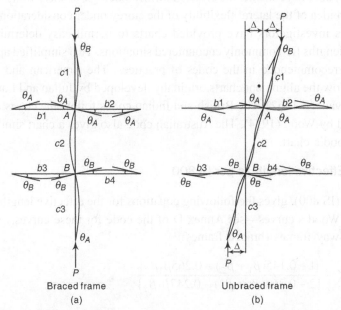

Braced frame   Unbraced frame
(a)   (b)

Fig. 5.14 Sub-assemblage model for braced and unbraced frames

5.8.3.1 Braced Frames

A braced frame is considered to be a frame in which lateral stability is provided by diagonal bracing, shear walls, in-fill walls, or by any other equivalent means. It is not necessary that every bay of a multi-storey building has to be braced, since the unbraced bays are restrained by those that are braced. The code (IS 800) gives criteria for considering the frame as a non-sway frame in clause 4.1.2. The ACI code (ACI 318-2008) suggests the calculation of the stability index Q, which is given by

$$Q = [(\Sigma P_u)/h_s](\Delta_u/H_u) \tag{5.14}$$

where ΣP_u is the sum of axial loads on all columns in the storey, h_s is the height of the storey, Δ_u is the elastic first-order lateral deflection of the storey, and H_u is the total lateral force acting on the storey. If the value of Q is less than 0.05, the column may be assumed as braced.

In the absence of bracing elements, the *lateral flexibility* measure of the storey Δ_u/H_u (storey drift per unit storey shear) may be taken for a typical intermediate storey as (Taranath 1988)

$$(\Delta_u/H_u) = h_s^2/[(12E\ \Sigma(I_c/h_s)] + h_s^2/(12E\Sigma I_b/L_b) \tag{5.15}$$

where $\Sigma I_c/h_s$ is the sum of ratios of the second moment of area to the height of all columns in the storey in the plane under consideration, $\Sigma I_b/L_b$ is the sum of ratios of the second moment of the area to the span of all floor beams in the storey in the plane under consideration, and E is Young's modulus of elasticity of steel.

The application of this concept is demonstrated in Example 5.6. Note that Eqn (5.15) does not consider the effect of infills, bracings, or shear walls. However, it gives an idea of the lateral flexibility of the storey under consideration.

Various investigators have provided charts to permit easy determination of effective lengths for commonly encountered situations. Two simplified approaches are often recommended in the codes of practices. The American and Canadian codes follow the alignment charts originally developed by Julian and Lawrence in 1959 (Kavanagh 1962). The British and Indian codes follow the charts originally developed by Wood (1974). The Australian code also gives a chart similar to that of the Wood's chart.

5.8.3.2 Effective length as per IS 800

The code (IS 800), gives the following equations for the effective length factor K, based on Wood's curves—see Annex D of the code for these curves.
For non-sway frames (braced frames):

$$K = \frac{[1 + 0.145(\beta_1 + \beta_2) - 0.265\beta_1\beta_2]}{[2 - 0.364(\beta_1 + \beta_2) - 0.247\beta_1\beta_2]} \tag{5.16a}$$

For sway frames (moment-resisting frames):

$$K = \left\{ \frac{[1 - 0.2(\beta_1 + \beta_2) - 0.12\beta_1\beta_2]}{[1 - 0.8(\beta_1 + \beta_2) + 0.6\beta_1\beta_2]} \right\}^{0.5} \tag{5.16b}$$

with $\beta_i = \Sigma K_c / (\Sigma K_c + \Sigma K_b)$ (5.16c)

where K_c and K_b are the effective flexural stiffness of the columns or beams meeting at the joint at the ends of the columns and rigidly connected at the joints.

$$K_c \text{ or } K_b = C(I/L)$$

where I is the moment of inertia about an axis perpendicular to the plan of the frame, L is the length of the member, taken as centre-to-centre distance of the intersecting member, and C is the connection factor as shown in Table 5.6.

Table 5.6 Connection factor C

Fixity conditions at far end	Connection factor C	
	Braced frame	Unbraced frame
Pinned	$1.5(1 - \bar{n})$	$0.5(1 - \bar{n})$
Rigidly connected column	$1.0(1 - \bar{n})$	$1.0(1 - 0.2\bar{n})$
Fixed	$2.0(1 - 0.4\bar{n})$	$0.67(1 - 0.4\bar{n})$

Note $\bar{n} = P/P_{cr}$, where P is the applied load and P_{cr} is the elastic buckling load $= \pi^2 EI/(KL)^2$

Note that for calculating C we need the effective length and hence the determination of effective length is an interactive process. Initially, we can assume $K = 1$ for calculating the value of C.

A review of the IS code provisions for effective length of columns in frames is provided by Dafedar et al. (2001).

5.8.4 Compression Members in Trusses

The general recommendations for effective length of compression members in trusses are given in Table 5.7 (see also clauses 7.2.4 and 7.5 of the code).

Table 5.7 Effective length for angle struts as per IS 800

Type	End connection	Effective length
Discontinuous single angle	(a) One bolt or rivet	Distance between the centre of end fastening (as per IS 800-1984)
	(b) Two or more bolts or Rivets or equivalent Welding	0.85 of distance between node points (as per IS 800-1984)
Discontinuous double angle, stitched together by bolts or welding at regular intervals	(a) Same side of gusset	
	(i) One bolt or rivet	Distance between centre of end fastenings
	(ii) Two bolts or rivets or equivalent welding	0.7–0.85 of distance between nodes

(Contd)

(Contd)

	(b) Both sides of gusset	
	(i) One bolt or rivet 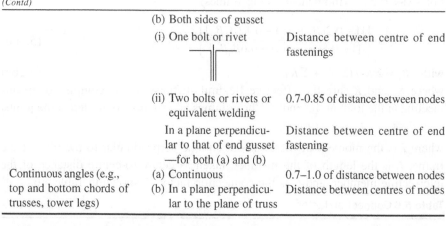	Distance between centre of end fastenings
	(ii) Two bolts or rivets or equivalent welding	0.7-0.85 of distance between nodes
	In a plane perpendicular to that of end gusset —for both (a) and (b)	Distance between centre of end fastening
Continuous angles (e.g., top and bottom chords of trusses, tower legs)	(a) Continuous	0.7–1.0 of distance between nodes
	(b) In a plane perpendicular to the plane of truss	Distance between centres of nodes

The effective length factor of single angle discontinuous struts connected by single or more bolts/rivets or equivalent welding may be determined as discussed in Section 5.9.1.

For the design of tapered columns, column cross sections with one axis of symmetry (for example, channel columns), and columns with no axis of symmetry, specialized literature should be consulted.

Hartford Civil Centre Roof Collapse

The Hartford Civil Centre Coliseum, Connecticut, USA, was completed in 1973. The space frame roof structure was 7.6 m high and covered 110 m by 91 m, with clear spans of 64 to 82 m.

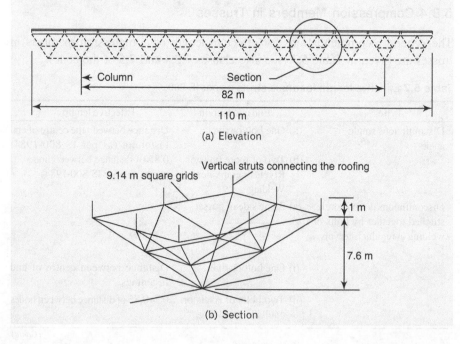

(a) Elevation

(b) Section

(c) View of the collapsed roof
Space frame roof of Hartford Civil Centre Coliseum roof
Fig. CS 1

The space frame consisted of warren trusses with triangular bracing between top and bottom chords (see Fig. CS 1). The two main layers were arranged in 9.14 m by 9.14 m grids composed of horizontal steel bars. The 9.14 m diagonal bars connected the nodes of the upper and lower layers, and in turn, were braced by horizontal bars at a middle layer (see Fig. CS 1). Struts 1 m long attached to the space truss were used to support the 75 mm wood fibre composition roofing.

On January 17, 1978, at 4:15 a.m., the roof crashed down 25.2 m to the floor, due to a large snow storm. Luckily it was empty by the time of the collapse, and no one was hurt. Though there were several causes for the collapse, the main cause was the relatively minor changes in the connections between steel components, i.e., the fabrication deviating from design. A few centimeters shift of the fabricated connection, cut down the axial force capacity to less than tenth of the design value! Some angle sections found at the wreckage were found to have failed in block shear. Epstein and Thacker, in 1991, used finite element analysis and found that block shear was the mode of failure for these angles. This study also established the difference in behaviour of coped beams (where the load is applied to the connection in the plane of the web, which also is the block shear plane) and angles (where the load is applied eccentric to the failure plane).

In addition, the Hartford Civil Centre coliseum roof design was extremely susceptible to the torsional buckling of compression members which, as a mode of failure, was not considered by the computer analysis used by the designers. Had the designers chosen tubular or even I sections, instead of the cruciform section adopted in the roof members, the failure might have been averted (the four steel angles forming the cruciform cross-section has much smaller radius of gyration than tubes or I-sections, and hence not efficient in resisting compressive loads). The failure also showed that computer software should be used only as a software tool, and not as a substitute for sound engineering experience and judgement.

References:
1. Smith, E. and Epstein. H. (1980). Hartford Coliseum Roof Collapse: Structural Collapse and Lessons Learned. Civil Engineering, ASCE, April, pp. 59–62.
2. Epstein, H.I. and Thacker, B. (1991) Effect of bolt stagger for block shear tension failures in angles, Computers and Structures, V.39,N.5, pp. 571–76.
3. Epstein, H.I. and Aleksiewicz , L.J. (2008) Block shear equations revisted.... Again, Engineering Journal, AISC, V.45, No.1, First Quarter, pp.5–12.

5.9 Single Angle Struts

Angles are perhaps the most basic and widely used of all rolled structural steel shapes. There is a wide range of sizes available and end conditions are relatively simple. Single angles are commonly used in many applications such as web members in steel joists and trusses, members of latticed transmission towers and communication structures, elements of built-up columns, and bracing members. In roof trusses, the single angle web members are often connected by one leg (thus introducing eccentricity with respect to the centroid of the cross sections) on one side of the chords and sometimes alternatively on opposite sides of T-sections as shown in Fig. 5.15. In towers, the bracing members meeting at a joint are connected to the opposite sides of the leg member, in order to reduce eccentricity as well as to reduce (or eliminate) gusset plates. The ease with which connections can be made contributes to the popularity of their use.

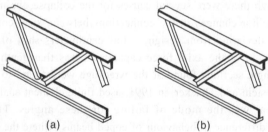

Fig. 5.15 Web members in trusses: (a) same sides and (b) opposite sides

Due to the asymmetry of the angle cross section, the determination of the compression capacity under eccentric loading along with end restraints is complex. Many researchers have studied the load carrying capacity of single angle members (e.g., Madugula & Kennedy 1985; Woolcock & Kitipornchai 1986; Kitipornchai & Lee 1986; Bathon et al. 1993; Adluri & Madugula, 1996). A review of experimental and analytical research on angle members is provided by Galambos (1998).

Angle members loaded through the centroid by a compressive axial force will buckle in flexural buckling about the minor principal axis of the cross section or in a torsional-flexural mode. When the width-to-thickness ratio of the legs of the angle are larger, the greater will be the possibility of torsional or flexural-torsional buckling being the controlling limit states. Note that the end restraint can significantly increase the ultimate strength of single angles of higher slenderness ratios, whereas it may weaken the ultimate strength in the lower slenderness ratio ranges. One of the most difficult tasks for designers is to judge or determine the end restraint and eccentricity condition for their specific application. Failure to consider the end restraints may lead to uneconomical designs whereas ignoring the end eccentricity may result in an unsafe design.

An empirical approach is adopted in ASCE Standard 10-1992, ASCE Manual 52-1988, ECCS recommendations-1985, and BS 5950-2000, to include the effects of end restraints and joint eccentricities by modifying the member's effective slenderness ratio. For example, in BS-5950, for single-angle struts connected to a gusset or directly to another member at each end by two or more fasteners in line along the angle or by equivalent welded connection, the slenderness ratio L/r should not be less than

$$0.85L/r_v \quad \text{or} \quad 0.7L/r_a + 30 \tag{5.17}$$

where r_v is the radius of gyration about the minor principal axis and r_a is the radius of gyration about the axis parallel to the plane of the gusset or the supporting member. Consideration of the rotational restraint permits the use of an effective length factor 0.85 while the end eccentricity effect is taken into account in the out-of-plane buckling. If a single fastener is used at each ends, the L/r should be taken as not less than

$$L/r_v \quad \text{or} \quad 0.7L/r_a + 30 \tag{5.18}$$

Thus, the rotational restraint effect is ignored for this connection.

The complete treatment of single angles is outside the scope of this book and the interested reader may refer to Galambos (1998). A separate publication is available from AISC for the design of single angle members subjected to compression bending and beam-column action (AISC-1993), though these provisions have been integrated in the current version of the code (AISC-360, 2005). Based on an inelastic analysis of steel angle columns with residual stresses Adluri and Madugula (1996) developed an average column curve for steel angles (similar to column c of Fig. 5.8).

5.9.1 Indian Code Provisions

The Indian code (IS 800) suggests that when the single angle is loaded concentrically in compression, the design strength may be evaluated using Fig. 5.8, choosing class c curve. When the single angle is loaded eccentrically through one of its legs, the flexural torsional buckling strength may be evaluated using an equivalent slenderness ratio λ_e given by

$$\lambda_e = \sqrt{[k_1 + k_2\lambda_{vv}^2 + k_3\lambda_\phi^2]} \tag{5.19a}$$

where k_1, k_2, k_3 are constants depending on end conditions as shown in Table 5.8.

$$\lambda_{vv} = (L/r_{vv})/\{88.86\varepsilon\} \tag{5.19b}$$

and

$$\lambda_{\phi} = (b_1 + b_2)/\{88.86\varepsilon \times 2t\} \tag{5.19c}$$

where L is the unsupported length of the member, r_{vv} is the radius of gyration about the minor axis, b_1 and b_2 are the widths of the two legs of the angle, t is the thickness of the leg, and ε is the yield stress ratio $(250/f_y)^{0.5}$.

Table 5.8 Constants k_1, k_2, and k_3

No. of bolts at the end of member	Gusset/connecting member fixity*	k_1	k_2	k_3
≥ 2	fixed	0.20	0.35	20
1	hinged	1.25	0.50	60

*Stiffness of in-plane rotational restraint provided to the gusset/connecting member. For partial restraint, the λ_e value can be interpolated between the λ_e values for fixed and hinged cases.

Equations (5.19a) to (5.19c) are based on the research conducted at the Indian Institute of Technology, Madras.

5.10 Design of Compresssion Members

The strength of a compression member is based on its gross area A_g (for slender cross section, A_{eff} should be used). The strength is always a function of the effective slenderness ratio KL/r, and for short columns the yield stress f_y of the steel. Since the radius of gyration r depends on the section selected, the design of compression members is an iterative process, unless column load tables are available (see Appendix D). The usual design procedure involves the following steps.

1. The axial force in the member is determined by a rational frame analysis, or by statics for statically determinate structures. The factored load P_u is determined by summing up the specified loads multiplied by the appropriate partial load factors γ_f.

2. Select a trial section. Note that the width/thickness limitations as given in Table 4.3 to prevent local buckling must be satisfied (most of the rolled sections satisfy the width-to-thickness ratios specified in Table 4.3). If it is not satisfied and a slender section is chosen, the reduced effective area A_{eff} should be used in the calculation. The trial section may be chosen by making initial guesses for A_{eff}/A, and f_{cd} (say 0.4–$0.6f_y$) and calculating the target area A.

   The following member sizes may be used as a trial section:

   (a) *Single angle size* 1/30 of the length of compression member

   (b) *Double angle size* 1/35 of the length of compression member

   (c) *Circular hollow section* diameter = 1/40 of length

   The slenderness ratios as given in Table 5.9 will help the designer to choose the trial sections.

Table 5.9 Slenderness ratios to be assumed while selecting the trial sections

Type of member	Slenderness ratio (L/r)
Single angles	100–150
CHS, SHS, RHS	90–110
Single channels	90–150
Double angles	80–120
Double channels	40–80
Single I-section	80–150
Double I-sections	30–60

3. Compute KL/r for the section selected. The computed value of KL/r should be within the maximum limiting value given in Table 5.10. Using Fig. 5.8 and Tables 5.3 and 5.4 compute f_{cd} and the design strength $P_d = Af_{cd}$.

Table 5.10 Maximum slenderness ratio of compression members

Type of Member	KL/r
Carrying loads resulting from dead loads and superimposed loads	180
Carrying loads resulting from wind and seismic loads only, provided the deformation of such a member does not adversely affect the stress in any part of the structure	250
Normally acting as a tie in a roof truss or a bracing system but subject to possible reversal of stress resulting from the action of wind or seismic forces	350
Lacing bars in columns	145
Elements (components) in built-up sections	50

4. Compare P_d with P_u. When the strength provided does not exceed the strength required by more than a few per cent, the design would be acceptable; otherwise repeat steps 2 through 4.

5.10.1 Limiting Slenderness Ratio

The maximum slenderness ratio of compression members under axial load is limited by the code as shown in Table 5.10.

The limit of 180 may be taken as normally applicable to primary members in compression, such as columns, compound column sections, etc. The limit of 250 may be applied to secondary members, which themselves do not support any structure, but form a part of one. The limit of 350 may be applied to bracing systems, which lend rigidity to a structure, but by themselves carry only nominal loads.

It may be of interest to note that the latest version of AISC code (AISC: 360-2005) does not specify any upper limit on the slenderness ratio, though the commentary to this code recommends an upper limit of 200. The upper limits provided are based on professional judgement and practical considerations such as economics, ease of handling, and care required to minimize any inadvertent damage during fabrication, transport, and erection.

5.11 Built-up Compression Members

For large loads and for efficient use of material, *built-up columns* (also called as *combined columns* or *open-web columns*) are often used. They are generally made up of two or more individual sections such as angles, channels, or I-sections and properly connected along their length by lacing or battening so that they act together as a single unit. Such laced combined compression members are often used in bridge trusses. According to the type of connection between the chords, built-up members may be classified as follows:

- Laced members {Fig. 5.16(a)}
- Struts with batten plates {Fig. 5.16(b)}
- Battened struts {Fig. 5.16(c)}
- Members with perforated cover plates {Fig. 5.16(f)}

In general, such struts can be considered either as simple or built-up struts depending on the plane of bending. A strut having a cross section as shown in Fig. 5.16(d), for example, must be considered as a simple strut if it bends about the z-axis and considered as a built-up strut if it bends about the y-axis (Ballio & Mazzolani 1983). Struts having cross sections of the type shown in Fig. 5.16(e) behave as built-up struts both in z and y directions.

The effects of shear in built-up columns sets apart the design of these members from that of other columns. The importance of designing the elements connecting the main longitudinal members for shear was tragically demonstrated by the failure of the first Quebec Bridge in Canada during construction in 1907 (Galambos 1998). It has been found that about three fourths of the early failure of laced columns resulted from local rather than general column failure. Moreover, the critical load for a built-up column is less than that of a comparable solid column because the effect of shear on deflections is much greater for the former (Galambos 1998). The shear in column may be due to the following:

1. Lateral loads from wind, earthquake, gravity, or other causes
2. The slope of column with respect to the line of thrust due both to un-intentional initial curvature and the increased curvature during buckling
3. The end eccentricity of the load due to either end connections or fabrication imperfections

The slope effect is most important for slender columns and the eccentricity effect for short columns. Lin et al. (1970) suggested a shear flexibility factor μ, using which the equivalent slenderness λ_e can be computed and using curve c of Fig. 5.8 the design stress can be computed (for more details see Lin et al. 1970; Galambos 1998).

For columns with batten plates, the strut may be designed as a single integral member with a slenderness given by (Bleich 1952)

$$\lambda_e = \sqrt{[\lambda_m^2 + \lambda_c^2(\pi^2/12)]} \tag{5.20}$$

With the limitations $\lambda_e \geqslant 50$ and $\lambda_e \not< 1.4\,\lambda_c$ where $\lambda_m = L/r_{\min}$ = strut slenderness, $\lambda_c = L_o/r_o$ = local chord slenderness between one batten plate and the next, and L_o is the centre-to-centre distance of batten plate.

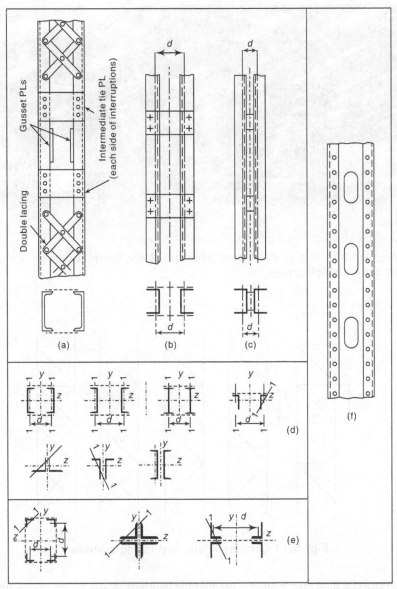

Fig. 5.16 Built-up columns

The most commonly adopted lacing systems are shown in Fig. 5.17. The simplest form of lacing consists of single bars connecting the two components [Fig. 5.17(b)]. The double lacing [Fig. 5.17(d)] is sometimes considered preferable, although a well-designed single lacing is equally effective. In single or double lacing systems, cross members perpendicular to the longitudinal axis of the strut should not be used [see Figs 5.17(c) and (e)]. The 'accordion' like action of the lacing system without cross members permits the lateral-expansion of the column. The introduction of cross members prevents the lateral expansion and thus forces the lacing bar to

Built-up column in a portal frame, Mumbai (Note the stubs in the column, which carry a small crane).

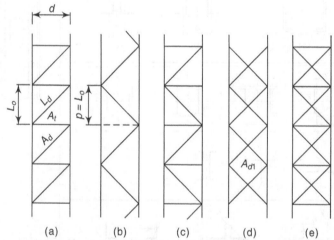

| (a) | (b) | (c) | (d) | (e) |

Fig. 5.17 Commonly adopted lacing systems

share the axial load on the strut. Note that the lacing bars and batten plates are not designed as load carrying elements. Their function is primarily to hold the main component members of the built-up column in their relative position and equalize the stress distribution in them. At the ends and at intermediate points where it is necessary to interrupt the lacing (for example, to admit gusset plates), the open sides are connected with *tie plates* (also called batten plates or stay plates). Tie plates are also provided at the top and bottom of the column (see Fig. 5.18).

It should to be noted that the battened columns have the least resistance to shear compared to columns with lacings and perforated plates, and may experience an appreciable reduction in strength. Hence, they are not generally used in the United States. Columns with perforated plates require no special considerations

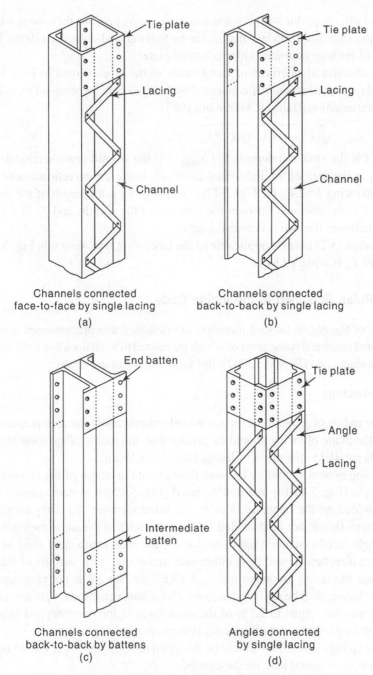

Channels connected
face-to-face by single lacing
(a)

Channels connected
back-to-back by single lacing
(b)

Channels connected
back-to-back by battens
(c)

Angles connected
by single lacing
(d)

Fig. 5.18 Isometric view of built-up columns

for shear effects. These cover plates contain perforations spaced axially, which afford access for welding or painting. After the advent of automatic cutting machines, the production of such perforated plates have become simpler. Hence, they are used extensively in USA. They result in reduction of fabrication and maintenance cost and offer superior stiffness and straightness. At present, they are not used in India and interested readers may refer to Salmon and Johnson (1996) for the design of built-up columns with perforated plates.

The effective slenderness of laced struts of the types shown in Figs 5.17(c), (d), and (e) (with two chords connected by lacings) may be obtained by using the following equation (Ballio & Mazzolani 1983)

$$\lambda_{eq} = \sqrt{[\lambda^2 + \pi^2(A/A_d)L_d^3/(L_o d^2)]} \tag{5.21}$$

where λ is the strut slenderness = L/r_{min}, A is the overall cross-sectional area = $2A_1$, A_1 is the area of the individual chord, A_d is the cross-sectional area of the diagonal lacing {= $2A_{d1}$ for Figs 5.17(a) and (d)}, L_d is the length of the diagonal lacing, d is the distance between the centroid of the chords, and L_o is the chord length between the two successive joints.

Equation (5.21) is also applicable to the laced strut type shown in Fig. 5.17(b), provided L_o is replaced by p.

5.11.1 Rules Specified in the Indian Code

In most of the codes, latticed members are designed and proportioned according to detailed empirical rules, most of which are related to local buckling requirements. In the Indian code (IS 800 : 2007), the following rules are given.

5.11.1.1 Lacings

(a) The radius of gyration of the combined column about the axis perpendicular to the plane of lacing should be greater than the radius of gyration about the axis parallel to the plane of lacing [see Fig. 5.19(a)].
(b) Lacing system should be uniform throughout the length of the column.
(c) Single [Fig. 5.19(b)] and double laced [Fig. 5.19(c)] systems should not be provided on the opposite sides of the same member. Similarly lacings and battens should not be provided on opposite sides of the same member.
(d) Single laced system on opposite sides of the main component shall be in the same direction viewed from either side so that one is the shadow of the other. Thus, the lacing as shown in Fig. 5.19(b), for face cd, is not recommended.
(e) The lacing shall be designed to resist a total transverse shear V_t at any point in the member, equal to 2.5% of the axial force in the member; and this shear shall be divided among the lacing systems in parallel planes.
(f) The lacings in addition should be designed to resist any shear due to bending moment or lateral load on the member.
(g) The slenderness ratio of lacing shall not exceed 145.
(h) The effective length shall be taken as the length between inner end bolts/rivets of the bar for single lacings and 0.7 times the length for double lacings effectively connected at intersections. For welded bars, the effective length is taken as 0.7 times the distance between the inner ends of the welds connecting the single bars to the members.

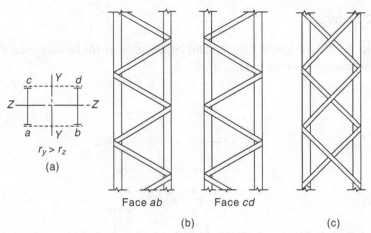

$$r_y > r_z$$

(a)

Face *ab*   Face *cd*

(b)   (c)

Fig. 5.19 Lacing systems—(a) In plan (b) Single lacing (c) Double lacing

(i) The minimum width of the lacing bar shall not be less than approximately three times the diameter of the connecting bolt/rivet; the thickness shall not be less than $1/40^{th}$ of the effective length for single lacing and $1/60^{th}$ for double lacing.

(j) The spacing of lacing bars shall be such that the maximum slenderness ratio of the components of the main member between two consecutive lacing connections is not greater than 50 or 0.7 times the most unfavourable slenderness ratio of the combined column.

(k) When welded lacing bars overlap the main members, the amount of lap should be not less than four times the thickness of the bar and the welding is to be provided along each side of the bar for the full length of lap. Where lacing bars are fitted between main members, they should be connected by fillet welds on each side or by full penetration butt weld.

(l) Where lacing bars are not lapped to form the connection to the components of members, they should be so connected that there is no appreciable interruption in the triangulated system.

(m) Plates shall be provided at the ends of laced compression members and shall be designed as battens.

(n) Flats, angles (normally adopted in practice), channels, or tubes may be used as lacings.

(o) Lacing bars, whether in double or single shear shall be inclined at an angle of 40° to 70° to the axis of the built-up member.

(p) The effective slenderness ratio $(KL/r)_e$ of the laced column shall be taken as 1.05 times $(KL/r)_o$, where $(KL/r)_o$ is the maximum actual slenderness ratio of the column, to account for shear deformation effects.

(q) The required sections of lacing bars as compression/tension members may be determined using the appropriate design stresses f_{cd} as given in Section 5.5.1 and T_d in Sections 3.5 and 3.7.2.

5.11.1.2 Battens

The rules for the design of battens shall be the same as for lacings except for the following conditions (see Fig. 5.20).

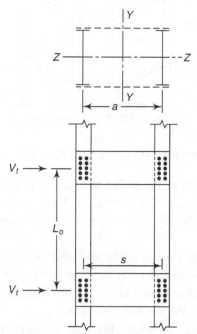

Fig. 5.20 Battening system

(a) The number of battens shall be such that the member is divided into not less than three bays.

(b) Battens shall be designed to resist simultaneously

$$\text{Longitudinal shear } V_b = V_t L_o/ns \tag{5.22}$$

and

$$\text{Moment } M = V_t L_o/2n \tag{5.23}$$

where V_t is the transverse shear force, L_o is the distance between centre-to-centre of battens, longitudinally, n is the number of parallel planes of battens, and s is the minimum transverse distance between the centroids of the bolt/rivet group/welding connecting the batten to the main member.

(c) When plates are used for battens, the effective depth between the end bolts/rivets or welds shall not be less than twice the width of one member in the plane of battens; nor less than three quarters of the perpendicular distance between centroids of the main members for intermediate battens; and not less than the perpendicular distance between the centroids of main members for end battens.

(d) The thickness of batten plates shall not be less than $1/50^{th}$ of the distance between the innermost connecting transverse bolts/rivets or welds.

(e) The requirement of size and thickness does not apply when other rolled sections are used for battens with their legs or flanges perpendicular to the main member.
(f) When connected to main members by welds, the length of the weld connecting each end of the batten shall not be less than half the depth of the plate; at least one third of its length should be placed at each end of the edge; in addition the weld shall be returned along the other two edges for a length not less than the minimum lap (i.e., not less than four times the thickness of the plate). The length of the weld and depth of batten should be measured along the longitudinal axis of the main member.
(g) The effective slenderness ratio of battened column shall be taken as 1.10 times $(KL/r)_o$, where $(KL/r)_o$ is the maximum actual slenderness ratio of the column, to account for shear deformation effects.
(h) Battened compression members, not complying with the preceding rules or those subjected to eccentricity of loading, applied moments, and lateral forces in the plane of the battens, shall be designed according to exact theory of elastic stability (see Bleich 1952; Timoshenko & Gere 1961) or empirically but verified by test results.

It should be noted that in Western countries such as USA and UK, due to the prohibitive labour and fabrication costs and the availability of larger rolled steel sections, built-up columns are seldom used nowadays.

5.12 Compression Members Composed of Two Components Back-to-back

Compression members may also be composed of two angles, channels, or Ts back-to-back in contact or separated by a small distance and connected together by bolting, rivetting, or welding [Fig. 5.16(d)]. In such a case, the code (IS 800 : 2007) gives the following specifications (see clauses 7.8 and 10.2.5 of the code).

(a) The slenderness ratio of each member between the connections should not be greater than 40 or 0.6 times the minimum slenderness ratio of the strut as a whole.
(b) The ends of the strut should be connected with a minimum of two bolts/rivets or equivalent weld length (weld length must not be less than the maximum width of the member) and there should be a minimum of two additional connections in between, spaced equidistant along the length of the member. Where there is small spacing between the two sections, washers (in case of bolts) and packing (in case of welding) should be provided to make the connections. Where the legs of angles or Ts are more than 125-mm wide, or where the web of channels is 150-mm wide, a minimum of two bolts should be used in each connection. Spacing of tack bolts or welds should be less than 600 mm. If bolts are used, they should be spaced longitudinally at less than 4.5 times the bolt diameter and the connection should extend to at least 1.5 times the width of the member.

(c) The bolts/rivets should be 16 mm or more in diameter for a member ≤ 10 mm thick and 20 mm in diameter for a member ≤ 16 mm thick, and 22 mm in diameter for members greater than 16 mm thick.

(d) Such members connected by bolts/welding should not be subjected to transverse loading in a plane perpendicular to the rivetted, bolted, or welded surfaces.

(e) When placed back-to-back, the spacing of bolts/rivets should not exceed 12t or 200 mm and the longitudinal spacing between intermittent welds should not be more than 16t, where t is the thickness of the thinner section.

5.13 Column Bases and Caps

For transmitting the load from columns to its foundations, *base plates* are used. Base plates assist in reducing the intensity of loading and distributing it over the foundations. The area of base plate is so chosen that the intensity of load distributed is less than the bearing capacity of concrete on which it rests.

The safety of a column and thus of a structure depends mainly upon the stability of the foundations and consequently on the bases, in the case of steel columns. Hence, column base plates should be designed with great care. The design of a base plate is generally assumed to be on the condition that the distribution of load under the base is uniform and the outstanding portions of the base plate are treated as cantilevers.

The main types of bases used are shown in Fig. 5.21. These are as follows:
(a) Slab base
(b) gussetted base; and
(c) pocket base

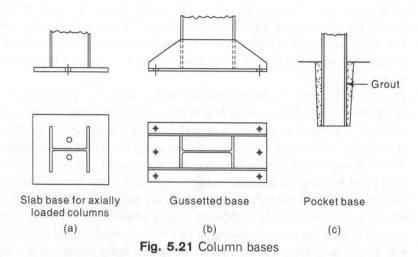

Slab base for axially loaded columns (a)   Gussetted base (b)   Pocket base (c)

Fig. 5.21 Column bases

With respect to slab and gussetted bases, depending on the values of axial load and moment, there may be compression over the whole base or compression over part of the base and tension in the holding-down bolts. Horizontal loads are restricted by shear in the weld between column and base plates, friction, and bond between the base and the concrete. Though the AISC code does not allow the anchor rods to transfer substantial shear, ACI-318, Appendix D gives the limit states to be checked in the anchorage, including the steel strength of the anchor in shear, as well as the various concrete limit states. If the base plate has a grout pad of any substantial thickness and the anchor rod does not bear against the base plate (the base plate holes will be larger than the anchor rod and hence in many cases the base connection will not bear against the side of the hole), then bending will be introduced in the rod in addition to shear. The bending capacity of the anchor rods is limited and hence the AISC code does not allow shear transfer through anchor rods (AISC design guide 1-2005). Hence AISC suggests the use of a shear key or lug or embedded plate with welded side plates to transfer a large horizontal shear force from the column base to the foundation (see Fig. 5.22). Note that the horizontal loads will be substantial for earthquake loading or wind loading. We will consider only the design of base plates with concentric loading here. Base plates subjected to bending moments are covered in Chapter 9.

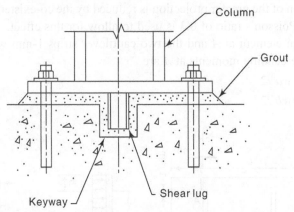

Fig. 5.22 Use of shear lug to transfer heavy shear force

Lightly loaded columns are provided with thick slab bases. The slab base is free from pockets where corrosion may start. Base plates with especially large loads require more than a simple plate. This may result in a double layer of plates, a grillage system, or the use of stiffeners to reduce the plate thickness. The *column caps* serve similar purpose except that they act as a link between load coming on the columns and the column itself (see Fig. 5.23).

The design of slab bases with concentric load is covered in Section 7.4.3 of IS 800 : 2007. This states that where the rectangular plate is loaded by *I*-, *H*-, channel, box, or rectangular hollow sections, the minimum thickness of base plate t_s should be

$$t_s = [2.75w(a^2 - 0.3b^2)/f_y]^{0.5} > t_f \tag{5.24}$$

where w is the pressure on the underside of the slab base due to the factored compressive load on the column (assumed as uniformly distributed over the area

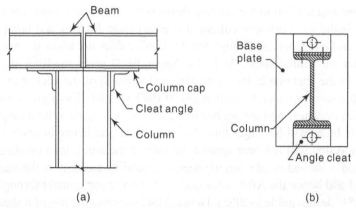

Fig. 5.23 (a) Column cap (elevation), (b) column base (plan)

of the slab base), a and b are the larger and smaller projections of the slab base beyond the rectangle, circumscribing the column, respectively, f_y is the yield strength of the base plate, and t_f is the flange thickness of the compression member.

Equation (5.24) takes into account plate bending in two directions. The moment in the direction of the greater projection is reduced by the co-existence moment at right angles. Poisson's ratio of 0.3 is used to allow for this effect.

Consider an element at A and the two cantilever strips 1-mm wide shown in Fig. 5.24. The bending moments at A are

$$M_x = wa^2/2$$
$$M_y = wb^2/2$$

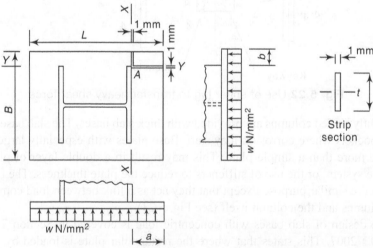

Fig. 5.24 Moment in column base plate

The projection a is greater than b and hence the net moment, with $\mu = 0.3$, is

$$M_x = wa^2/2 - 0.3wb^2/2$$
$$= w/2(a^2 - 0.3b^2) \tag{5.25a}$$

The moment capacity of the plate is given by

$$M_p = 1.2 f_y Z_e$$

The elastic modulus for the cantilever strip is $t^2/6$. Thus,

$$M_p = 1.2 f_y t^2/6 \qquad (5.25b)$$

Equating Eqns (5.25a) and (5.25b), solving for t, and applying the partial factor of safety for material we get Eqn (5.24), given in the code.

5.13.1 Weld: Column to Slab Base

The code states in clause 7.4.3.4 that where the cap or slab base is fillet welded directly to the column, the contact surfaces should be machined to give a perfect bearing and the welding shall be sufficient to transmit the forces. When full strength butt welds are provided, machining of contact surfaces is not required. Also, where the end of the column is connected directly to the base plate, by means of full penetration butt welds, the connection is deemed to transmit to the base all the forces and moments to which the column is subjected.

5.13.2 Design of Base Plate

The design of base plate consists of finding out its size and thickness. The following are the design steps to be followed:

1. Assuming the grade of concrete, calculate the bearing strength of concrete, which is given by $0.45 f_{ck}$ as per clause 34.4 of IS 456-2000
2. Required area of slab base may be computed by

   Required $A = P_u/$bearing strength of concrete $\qquad (5.26)$

   where P_u is the factored concentric load on the column
3. From the above, choose a size for the base plate $L \times B$, so that $L \times B$ is greater than the required area. Though a few designers prefer to have a square base plate, it is advisable to keep the projections of the base plate beyond column edges a and b as equal (see Fig. 5.24). Hence, the size can be worked out as

   $$(D + 2b) (b_f + 2a) = A \qquad (5.27)$$

   where D is the depth of the column section (in mm), b_f is the breadth of flange of column (in mm), and a and b are the larger and smaller projections of the base plate beyond the column (in mm).
4. Calculate the intensity of pressure w acting below the base plate using

   $$w = P_u/A_1$$

   where A_1 is the provided area of the base plate ($L \times B$)
5. Calculate the minimum thickness of slab base as per Eqn (5.24). If it is less than t_f, thickness of flange of column, provide the thickness of slab base = thickness of flange of column
6. Provide nominal four 20-mm holding-down bolts
7. Check the weld length connecting the base plate with the column (this check is required only for fillet welds)

5.13.3 Some Practical Aspects

Anchor bolts vary in size from approximately 20-mm diameter to 60-mm diameter. Anchor rod layouts for base plates should provide ample clearance distance for the washer from the column shaft and its weld, as well as reasonable edge distance. When the hole edge is not subjected to lateral force, even an edge distance that provides a clear dimension as small as 12 mm of material from the edge of the hole to the edge of the plate will normally suffice, though field issues with anchor rod placement may necessitate a larger dimension to allow some slotting of the base plate holes. When the hole's edge is subjected to lateral force, the edge distance must be large enough for the necessary force transfer. Anchor bolts, subject to corrosive conditions may be galvanized.

When ordering galvanized bolts, the threads should be *chased* so the nuts will work freely. Most anchor bolts are made from grade 410 steel though high-strength bolts may also be used. Most anchor bolts come with hexagonal nuts. Because the base plates will be provided with oversize holes, it is necessary to provide thick erection washers under the nuts.

The following three methods can be used effectively to prepare a landing site for the erection of a column:

(a) Leveling plates (for small to medium sized base plates, say up to 560 mm)
(b) Leveling Nuts
(c) Pre-set base plates

The details of these methods may be found in AISC design guide 1-2005.

The grout thickness will be approximately between 20 mm to 40 mm, the lesser figure being common for smaller plates. For the grout to flow laterally there must be ample space between the bottom of the steel base plate and the top of the concrete. About 40 mm would be a minimum space. For large base plates (say over 900 mm it may be necessary to drill a hole in the base plate near the centre but not so as to foul the column section. The grout is fed through this hole and this hole should be approximately 75 mm in diameter. For very large plates or long rectangular base plates, two grout holes may be required (Ricker 1989). Many good quality grouts are available in the market with varying strengths. They should be of the non-shrink variety (Warner 2004). More grout space should be allowed for a inexperienced foundation contractor.

Often anchor bolts are dislocated at sites. In addition to being out of place, the bolts may be tilted, have too much or too little projection above the top of the concrete, or be at the wrong elevation. Since most of the column bases will be having oversized holes, small dislocation of anchor bolts can usually be tolerated. Bolts which are not vertical (tilted) may be straightened with a rod bending device. For anchor bolts dislocated up to about 20 mm, the concrete may be chipped away to a depth of few centimeters and the bending device may be used to bend the bolt into proper position (Ricker 1989). Anchor bolts dislocated over 20 mm usually require that the base plate be slotted. More remedies for large errors are discussed by Ricker (1989).

Buckling-Restrained Braces (BRB)

When a concentrically braced frame (CBF) system, as discussed in Section 2.4.2, is subjected to earthquake loads, the braces will be subjected to alternate compression and tension. Hence traditionally, CBFs have been treated as high-strength, low-ductility systems, because steel braces show significant strength and ductility in tension, whereas deliver only a fraction of this strength and ductility in compression, due to buckling.

Theoretically, buckling in compression members can be eliminated by providing lateral bracings at close intervals, such that the unbraced length of the member approaches a small value. In the 1980s, Prof. Akira Wada of the Tokyo Institute of Technology developed a system called Un-bonded Brace™ in collaboration with Nippon Steel Corporation; his inspiration came from the collarbone of the human body.

In this system, which resembled a typical bone (bigger at the ends and a reduced section in the middle), Euler buckling of the central steel core is prevented by encasing it over its entire length in a steel tube. A typical BRB consists of a steel core surrounded by a hollow steel section, coated with a low-friction material, and then grouted with a specialized mortar. The encasing and mortar prohibit the brace from buckling when in compression, while the coating prevents the axial load from being transferred to the encasement, thus preventing strength loss and allowing for better and more symmetric cyclic performance (see Fig. CS 2 below which shows the WILDCAT™ BRB manufactured by Star Seismic). It also shows the comparison of the behaviour of a BRB and a conventional brace. Most of the BRBs developed to date are proprietary, but all of them are based on the similar concept. Some of them are Unbonded Brace™ manufactured by Nippon Steel Corporation, CoreBrace™ by CoreBrace, WILDCAT™, POWERCAT™ and Star Seismic™ Modular Systems by Star Seismic, and POWERCAT™ by PKM Steel.

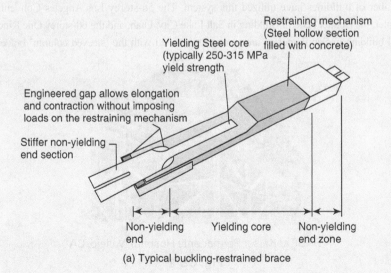

(a) Typical buckling-restrained brace

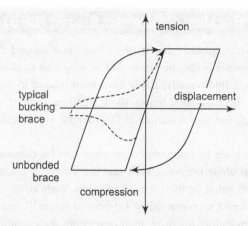

(b) Axial force-displacement behaviour

Fig. CS 2 Typical buckling-restrained brace and its behaviour as opposed to conventional brace

Interestingly, a similar concept of 'core loaded sleeved strut' was originally proposed by B.N. Sridhara of Bangalore which was also experimentally studied for compression load carrying capacity in IIT Madras by Kalyanaraman et al. (1994), Prasad (1992) and Sridhara (1990). Sridhara also obtained an US patent for his invention and Star Seismic[TM] and CoreBrace[TM] are now manufactured in USA using his patent.

A BRBF is a structural steel frame which resists a building's lateral forces by using the BRB. A BRBF is typically a special case of a concentrically braced frame. Advantages of a BRBF over other concentrically braced frames are that they exhibit higher ductility and energy dissipation. BRBs most commonly brace a bay diagonally or in a chevron pattern. As the maximum tension and compression forces in a BRB are much closer than in a standard brace, there is much less imbalance of force in the chevron configuration.

The first BRBF system was installed in the United States at UC-Davis in 2000. Now a number of buildings have utilized this system. The 56-storey Los Angeles Convention Center, the Bennet Federal Building in Salt Lake City, Utah, and the 60-storey One Rincon Hill building in San Francisco are among those fitted with the 'sleeved column' braces.

BRBF at Kaiser Permanente Hospital, Vallejo, CA

Fig. CS 3

Unlike the massive gusset plates required on the special concentric braced frame, the connections for the BRBF system are of smaller size and require less welding/ bolting. The ANSI/AISC 341-05 code contains provisions for BRBF design. Recent testing has demonstrated that gusset plate connections may be a critical aspect of the design of BRBF. The BRBF design procedure requires the columns to have the strength to resist the vertical component of the expected yield strength of each brace in a frame. This design philosophy allows the column to remain elastic during a seismic event while the BRBs are yielding and absorbing seismic energy. The braces are also designed for deformations corresponding to two times the design storey drift. Hence BRBF performs with a higher degree of ductility than conventional braced frames. Thornton and Muir (2009) provide detailing of the gusseted joint with beam hinges, to take into account the potential distortional forces induced by large seismic drifts.

References:

1. Thornton, W.A., Muir, L.S. Design of vertical bracing connections for high-seismic drift, Modern Steel Construction, Vol. 49, No.3, Mar 2009, pp. 61–65
2. http://www.starseismic.net
3. Sabelli, R., 2004, "Recommended Provisions for Buckling-Restrained Braced Frames," *AISC Engineering Journal*, 4th Quarter, pp. 155–175.
4. Kalayanaraman, V., Sridhara, B.N., and Mahadevan, K., Sleeved Column System, Proc. of the SSRC Task Group Meetings and Task Force Sessions, Lehigh Univ., Bethlehem. Pa., 1994

5.14 Displacement

The decrease in length of a member due to axial compression under service loads is given by

$$\Delta = PL/(A_g E) \tag{5.28}$$

where Δ is the axial shortening of the member (in mm), P is the service (unfactored) axial compressive force in the member (in N); L is the length of the member (in mm), E is the Young's modulus of rigidity (in N/mm^2), and A_g is the gross area of cross section (in mm^2).

It should be noted that columns in tall buildings are subjected to large axial displacements because they accumulate loads from a large number of floors and are also relatively long. For example, a 60-storey interior column in a steel building may shorten as much as 50 to 75 mm because of dead and live loads. Due to this shortening, problems may develop in the performance of curtain walls and levelness of floor system, certain wall panels may bow or even pop-off the building. Frame shortening may also force the mechanical and plumbing lines that are attached rigidly to the structure, to act as structural column, resulting in their distress. Also since the axial loads in all columns in a tall building are seldom the same, there will be *differential-shortening* of columns. Hence, it may be necessary to appropriately correct column lengths (say, at every sixth floor level). Certain columns may be made slightly longer than their nominal length to account for axial shortening. More details about procedures to calculate the corrections and making the corrections at construction site may be found in Taranath (1998).

Examples

Example 5.1 *A wide flanged section W 360 × 370 × 147 is to be used as a short column carrying axial load. Will its compressive strength likely to be affected by local buckling, assuming Fe 410 steel with f_y = 250 MPa?*

Solution
From Section tables, B = 370 mm, t_f = 19.8 mm, H = 360 mm, R = 20.0 mm, t_w = 12.3 mm, and A = 18790 mm².
For f_y = 250 MPa, from Table 2 of the code, IS 800, (b/t) allowable = 15.7
Actual (b/t) for flange = (370/2)/19.8 = 9.34 < 15.7
d/t for web = [360 − 2 (19.8 + 20)]/12.3 = 22.79
From Table 2 of code, (d/t) max = 42
Thus full cross-section is effective and design strength = 250 × 18790/(1.1 × 1000) = 4270 kN

Example 5.2 *Determine the design axial load on the column section ISMB 400, given that the height of column is 5.0 m and that it is pin-ended. Also assume the following: f_y = 250 N/mm²; f_u = 410 N/ mm²; E = 2 × 10⁵ N/mm².*

Solution
Unless the axis about which buckling will occur is obvious, all possibilities must be checked. For ISMB section, r_y is normally between one-fourth to one-fifth of r_z and hence the likely mode of failure is by buckling about the minor axis. However, in case where different effective lengths apply for the two planes, both possibilities should normally be checked. Here both the possibilities are shown just for illustration.
Cross-section properties
Flange thickness t_f = 16 mm
Thickness of web, t_w = 8.9 mm
Flange width, b_f = 140 mm
Self weight, w = 615 N/m
Cross-sectional area, A = 7840 mm²
r_z = 162 mm
r_y = 28.2 mm
(i) Bucking curve classification
$$h/b_f = 400/140 = 2.86 > 1.2$$
$$t_f = 16 \text{ mm} < 40 \text{ mm}$$
Hence, we should use buckling curve 'a' about z-z axis and 'b' about y-y axis.
(ii) Effective length
Since both ends are pinned, effective length, KL_y = KL_z = 5.0 m

(iii) Non–dimensional slenderness ratio

About z-z axis: $\alpha = 0.21$

$$\lambda_z = Kl/r \sqrt{[f_y / (\pi^2 E)]}$$

$$= (5000/162) \sqrt{[250/(\pi^2 \times 2 \times 10^5)]} = 0.3473$$

$$\phi = 0.5[1 + \alpha(\lambda - 0.2) + \lambda^2]$$
$$= 0.5[1 + 0.21(0.3473 - 0.2) + 0.3473^2]$$
$$= 0.5758$$

$$f_{cd} = (f_y/\gamma_{m0})/\{\phi + [\phi^2 - \lambda^2]^{0.5}\} = (250/1.10)/\{0.5758 + [0.5758^2 - 0.3473^2]^{0.5}\}$$
$$= 219.6 \text{ N/mm}^2$$

About y-y axis: $\alpha = 0.34$

$$\lambda_y = (KL/r) \sqrt{[f_y / (\pi^2 E)]}$$

$$= (5000/28.2) \sqrt{[250/(\pi^2 \times 2 \times 10^5)]}$$

$$= 1.9954$$

$$\phi = 0.5[1 + \alpha(\lambda - 0.2) + \lambda^2]$$
$$= 0.5[1 + 0.34(1.9954 - 0.2) + 1.9954^2]$$
$$= 2.796$$

$$f_{cd} = (f_y/\gamma_{m0})/\{\phi + [\phi^2 - \lambda^2]^{0.5}\}$$
$$= (250/1.10)/\{2.796 + [2.796^2 - 1.9954^2]^{0.5}\}$$
$$= 47.8 \text{ N/mm}^2$$

The same result may be obtained by using Table 9b of IS 800. Thus, for $KL/r = 177.3$ and $f_y = 250$ MPa, we get

$$f_{cd} = 47.8 \text{ N/mm}^2$$

(iv) Design stresses

In z direction, $f_{cd} = 219.6$ MPa

In y direction, $f_{cd} = 47.8$ MPa

Hence design axial compressive stress, $f_{cd} = 47.8$ MPa

The design strength, $P_d = 7840 \times 47.8/1000$
$$= 374.7 \text{ KN}$$

Example 5.3 *Determine the design axial load on the column section ISMB 400. The height of the column is 6.0 m as shown in Fig. 5.25. It is effectively restrained at mid-height by a bracing member in the z-z direction, but is free to move in the y-y direction and both the ends of the column are pinned. Also assume: $f_y = 250$ N/mm^2; $f_u = 410$ N/mm^2; E and $= 2 \times 10^5$ N/mm^2.*

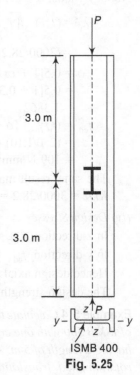

Fig. 5.25

Solution

Properties of column section are same as in Example 5.2

(i) *Bucking curve classification*

$h/b_f = 400/140 = 2.86 > 1.2$

$t_f = 16$ mm < 40 mm

Hence we should use Buckling Curve '*a*' about *z-z* axis '*b*' about *y-y* axis

(ii) *Effective Length*

Effective length in *z* direction, $l_z = 6.0$ mm

Since effectively restrained in *z* direction at mid-height, effective length $l_y = 3.0$ mm

(iii) *Non-dimensional slenderness ratio*

About *z-z* axis: $\alpha = 0.21$

$$\lambda_z = (KL/r) \sqrt{[f_y/(\pi^2 E)]}$$

$$= (6000/162) \sqrt{[250^2/(\pi^2 \times 2 \times 10^5)]} = 0.4168$$

$$\phi = 0.5[1 + \alpha(\lambda - 0.2) + \lambda^2]$$

$$= 0.5[1 + 0.21(0.4168 - 0.2) + 0.4168^2]$$

$$= 0.6096$$

$$f_{cd} = (f_y/\gamma_{m0})/\{\phi + [\phi^2 - \lambda^2]^{0.5}\}$$

$$= (250/1.10)/\{0.6096 + [0.6096^2 - 0.4168^2]^{0.5}\}$$

$$= 215.5 \text{ N/mm}^2$$

About *y-y* axis: $\alpha = 0.34$

$$\lambda_y = (L/r) \sqrt{[f_y/(\pi^2 E)]}$$

$$= (3000/28.2) \sqrt{[250/(\pi^2 \times 2 \times 10^5)]} = 1.1972$$

$$\phi = 0.5[1 + \alpha(\lambda - 0.2) + \lambda^2]$$

$$= 0.5[1 + 0.34(1.1972 - 0.2) + 1.1972^2]$$

$$= 1.3862$$

$$f_{cd} = (f_y/\gamma_{m0})/\{\phi + [\phi^2 - \lambda^2]^{0.5}\}$$

$$= (250/1.10)/\{1.3862 + [1.3862^2 - 1.1972^2]^{0.5}\}$$

$$= 109 \text{ N/mm}^2$$

The same result may be got from the Table 9b of the code for $f_y = 250$ and $KL/r = 3000/28.2 = 106.38$, $f_{cd} = 109$ N/mm²

(iv) *Design Stresses*

In z-direction, $f_{cd,z} = 215.5$ MPa

In y-direction, $f_{cd,y} = 109$ MPa

Hence design axial compressive stress, $f_{cd} = 109$ MPa

The design strength $= 7840 \times 109/1000 = 854.6$ KN

Example 5.4 *Calculate the compressive resistance of a compound column consisting of ISMB 500 with one cover plate 350 × 20 mm on each flange (see Fig.5.26) and having a length of 5 m. Assume that the bottom of the column is hinged and top is rotation fixed, translation free, and $f_y = 250$ MPa.*

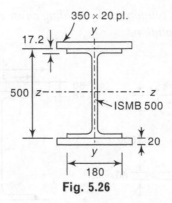

350 × 20 pl.

17.2

500 |z

ISMB 500

z

20

y

180

Fig. 5.26

Solution

From section tables for ISMB 500

Area = 11100 mm^2; I_{zz} = 45200 × 10^4 mm^4; I_{yy} = 1370 × 10^4 mm^4

(i) Determining the of radii of gyration for the compound section

Area = 11100 + 350 × 20 × 2 = 25,100 mm^2

I_{zz} for plates = 2[$I + Ay_1^2$]

= 2 × [350 × 20^3/12 + 350 × 20 × 260^2]

= 94686.7 × 10^4 mm^3

Total I_{zz} = 45200 × 10^4 + 94686.7 × 10^4

= 139886.7 × 10^4 mm^4

Hence $r_z = \sqrt{(139886.7 \times 10^4 / 25100)]}$ = 236 mm

I_{yy} of ISMB 500 = 1370 × 10^4 mm^4

I_{yy} of plates = 2 × 20 × 350^3/12

= 14291.7 × 10^4 mm^4

Total I_{yy} = 1370 × 10^4 + 14291.7 × 10^4

= 15661.7 × 10^4 mm^4

$r_y = \sqrt{(15661.7 \times 10^4 / 25100)]}$ = 79 mm

(ii) Buckling curve classification

From Table 10 of the code for buckling about any axis use curve c.

(iii) Since r_y is small, the buckling will be about the y-y axis

Effective length = 2 × 5000 = 10,000 mm (see Fig. 5.12)

$\lambda = KL/r_y$ = 10,000/79 = 126.6 < 180

From Table 9c of the code, for

f_y = 250 MPa and λ = 126.6,

f_{cd} = 77.50 N/mm^2

Hence design strength = 25100 × 77.50/1000 = 1945 kN

Example 5.5 *A heavy column is required to support a gantry girder and a special H-section is to be fabricated. The trial section is shown in Fig. 5.27(a). Check its suitability to support a factored load of 11,000 kN, assuming both ends are pinned and a length of 8 m. Steel of design strength 250 N/mm^2 is to be used. Could a* *rolled section be suitably reinforced (by welding cover plates to its flanges) so as to provide an alternate solution?*

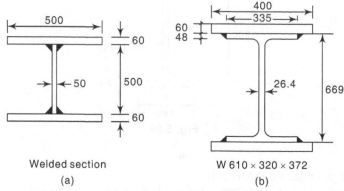

Welded section
(a)

W 610 × 320 × 372
(b)

Fig. 5.27

Solution

(a) For the welded section shown in Fig. 5.27(a)

$A = (500 \times 60) \times 2 + 500 \times 50 = 85{,}000$ mm^2

$I_y = 2(60 \times 500^3)/12 + 500 \times 50^3/12 = 125520.8 \times 10^4$ mm^4

$r_y = \sqrt{(I_y/A)} = \sqrt{(125520.8 \times 10^4/85{,}000)} = 121.5$ mm

$\lambda = KL/r_y = 1 \times 8000/121.5 = 65.83$

(i) *Section classification* (Table 2 of IS 800)

Flange $b/t_f = (500/2)/60 = 4.1 < 8.4$

Web $d/t_w = 500/50 = 10 < 42$

Hence, the section is not slender.

(ii) *Buckling curve classification*

From Table 10 of the code, $t_f > 40$ mm

For z-z axis use curve '*c*'

For y-y axis use curve '*d*'

(iii) *Design strength*

From Table 9d, for $KL/r = 65.83$, and $f_y = 230$ N/mm^2, (As per Table 1 of code)

$f_{cd} = 133.25$ N/mm^2

Hence design strength $= f_{cd} \times A = 133.25 \times 85{,}000/1000 = 11{,}326$ kN $> 11{,}000$ kN

Hence, the section is suitable.

(b) The heaviest rolled section is wide flange W 610×320 with the following properties:

$A = 47630$ mm^2, $I_y = 30200 \times 10^4$ mm^4, $r_y = 79.6$ mm

Hence we have to provide substantial cover plates to make the area about 85,000 mm^2. Let us use 400×60 mm plates on both flanges as shown in Fig. 5.27(b).

$A = 47630 + 2 (400 \times 60) = 95630$ mm^2

$I_y = 30200 \times 10^4 + 2 (60 \times 400^3)/12 = 94200 \times 10^4$ mm^4

$r_y = \sqrt{(94200 \times 10^4/95{,}630)} = 99.25$ mm

From Table 10 of the code, since $t_f > 40$ mm use curve 'd'.

$$= KL/r_y = 1 \times 8000/99.25 = 80.60$$

From Table 9d, for

$\qquad f_y = 230$ N/mm^2 and $\lambda = 80.6$,

$\qquad f_{cd} = 112.28$ N/mm^2

Capacity of the section $= 112.28 \times 95630/1000$

$$= 10737 \text{ kN} \approx 11,000 \text{ kN}$$

Hence the section is suitable.

Note that table 1 of IS 800:2007 indicates that for thicknesses greater than 20 mm, reduced f_y should be used. Hence using Table 9d of the code, for $f_y = 230$ N/mm^2 and $KL/r = 80.6$, we get $f_{cd} = 112.28$ N/mm^2.

Thus, design strength $= 112.28 \times 95,630/1000 = 10,737$ kN $\approx 11,000$ kN

Example 5.6 *The framing plan of a multi-storey building is given in Fig. 5.28. Assume that all the columns have a size of ISHB 400; the longitudinal beams (z-direction) have a size of ISMB 600 and the transverse beams (y-direction) have a size of ISMB 400. The storey height is 3.5 m, and the columns are assumed to be fixed at the base; For a column in a typical lower floor of the building, determine the effective length KL_z and KL_y. For the purpose of estimating the total axial loads on the columns in the storey, assume a total distributed load of 35 kN/m^2 from all the floors above (combined). Assume Fe 410 grade steel.*

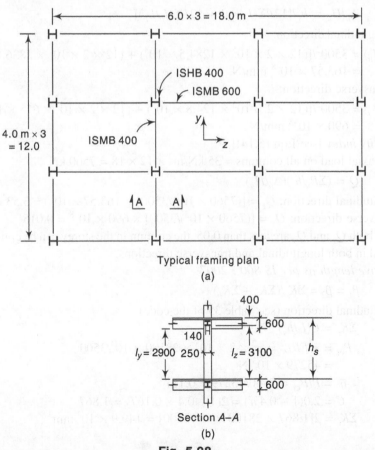

Typical framing plan
(a)

Section *A–A*
(b)

Fig. 5.28

Solution

Unsupported length of the column, in z-direction,

$$L = 3500 - 400 = 3100 \text{ mm}$$

In y-direction, $L = 3500 - 600 = 2900$ mm

(i) *Relative stiffness of columns and beams*

ISHB 400: I_z of column $= 28100 \times 10^4$ mm^4; $I_y = 2730 \times 10^4$ mm^3;

$$r_z = 169 \text{ mm}; \ r_y = 52.6 \text{ mm}$$

ISMB 400: $I_z = 20500 \times 10^4$ mm^4; $I_y = 622 \times 10^4$ mm^4

ISMB 600: $I_z = 91800 \times 10^4$ mm^4; $I_y = 2650 \times 10^4$ mm^4

- Columns, 16 in number, ISHB 400, $h_s = 3500$ mm

  In z-direction

  $$\Sigma I_c/h_s = 16 \times 28100 \times 10^4/3500 = 1284.5 \times 10^3 \text{ mm}^3$$

  In y-direction $= 16 \times 2730 \times 10^4/3500 = 124.8 \times 10^3$ mm^3

- Longitudinal beams, 12 Nos, ISMB 600, $L_b = 6000$ mm

  $$\Sigma(I_b/L_b)_z = 12 \times 91800 \times 10^4/6000 = 1836 \times 10^3 \text{ mm}^3$$

- Transverse beams

  $$\Sigma(I_b/L_b)_y = 12 \times 20500 \times 10^4/4000 = 615 \times 10^3 \text{ mm}^3$$

(ii) *Determination of whether column is braced or unbraced*

Ignoring the contribution of in-fill walls, as per Eqn (5.15)

$$\Delta_u/H_u = h_s^2/[12E\Sigma(I_c/h_s)] + 12E\Sigma(I_b/L_b)]$$

In longitudinal direction:

$(\Delta_u/H_u) = 3500^2/[(12 \times 2 \times 10^5 \times 1284.5 \times 10^3) + (12 \times 2 \times 10^5 \times 1836 \times 10^3)]$

$$= 163.57 \times 10^{-8} \text{ mm/N}$$

In transverse direction:

$(\Delta_u/H_u) = 3500^2/[(12 \times 2 \times 10^5 \times 124.8 \times 10^3) + (12 \times 2 \times 10^5 \times 615 \times 10^3)]$

$$= 690 \times 10^{-8} \text{ mm/N}$$

Stability index [see Eqn (5.14)]

Total axial load on all columns ≈ 35 kN/m$^2 \times 12 \times 18 = 7560$ kN

$$Q = (\Sigma P_u/h_s)(\Delta_u/H_u)$$

Longitudinal direction: $Q_z = [(7560 \times 10^3)/3500] \times 163.57 \times 10^{-8} = 3.53 \times 10^{-3}$

Transverse direction: $Q_y = [(7560 \times 10^3)/3500] \times 690 \times 10^{-8} = 0.015$

Since both Q_z and Q_y are less than 0.05, the column in the storey can be considered braced in both longitudinal and transverse directions.

Effective length as per IS 800 : 2007

$$\beta_1 = \beta_2 = \Sigma K_c/(\Sigma K_c + \Sigma K_b)$$

Longitudinal direction (see Table 35 of the code)

$$\Sigma K_c = C(I_c/h_s)$$

$$P_{cr} = \pi^2EI/L^2 = \pi^2 \times 2 \times 10^5 \times 28100 \times 10^4/3500^2$$

$$= 45279 \times 10^3 \text{ N}$$

$$\bar{n} = P/P_{cr} = 7560/45279 = 0.167$$

$$C = 2.0(1 - 0.4\bar{n}) = 2(1 - 0.4 \times 0.167) = 1.867$$

$$\Sigma K_c = 2[1.867 \times 28100 \times 10^4/3500] = 149.9 \times 10^3 \text{ mm}^3$$

$\Sigma K_b = 2 \times [91800 \times 10^4/6000] = 306 \times 10^3 \text{ mm}^3$

$\beta_2 = 0.0$ (assumed fixed at base)

$\beta_1 = 149.9 \times 10^3/(149.9 \times 10^3 + 306 \times 10^3) = 0.328$

Hence,

$$K = \{[1 + 0.145(\beta_1 + \beta_2) - 0.265\beta_1\beta_2]\}/[2 - 0.364(\beta_1 + \beta_2) - 0.247\beta_1\beta_2)]$$

(5.16a)

Substituting the values of β_1 and β_2, we get the effective length factor

$K = 0.557$

Thus,

$KL_z = 0.557 \times 2900 = 1615.3 \text{ mm}$

Transverse direction

$\Sigma K_c = 2(0.625 \times 2730 \times 10^4/3500) = 4875 \text{ mm}^3$

$\Sigma K_b = 2 \times (20500 \times 10^4/4000) = 102.5 \times 10^3 \text{ mm}^3$

$\beta_1 = 4875/(4875 + 102.5 \times 10^3) = 0.045$

$\beta_2 = 0.0$ (fixed at base)

Hence $K = \{[1 + 0.145 \times 0.045 - 0.265 \times 0]/[2 - 0.364 \times 0.045 - 0.247 \times 0]\} = 0.51$

Hence $KL_y = 0.51 \times 3100 = 1581 \text{ mm}$

Example 5.7 *Calculate the compressive resistance of the leg of a microwave tower consisting of a $150 \times 150 \times 16$ angle section of height 3.5 m. Assume that the conditions at both ends of the z-z and y-y planes are such as to provide simple support. Take the design strength of steel as 250 N/mm² and that the load is concentrically applied to the angle.*

Solution

Since the principal axes for an angle section do not coincide with the rectangular y-y and z-z axes, the buckling strength about the minor principal axis v-v should normally be checked. From Table 5.4, the curve c is appropriate for buckling about any axis. Therefore, only the axis about which the slenderness is greatest needs to be considered.

Check for limiting thickness

$d/t = b/t = 150/16 = 9.4 < 15.7$

$(b + d)/t = 300/16 = 18.75 < 25$

Hence the section is not slender.

From section tables,

Area = 4560 mm²

$r_{xx} = r_{yy} = 45.8 \text{ mm}, r_{vv} = 29.4 \text{ mm}$

Maximum $\lambda = 3500/29.4 = 119 < 180$

From Table 9c of the code for

$\lambda = 119$ and $f_y = 250$ MPa,
$f_{cd} = 84.79$ N/mm^2

Hence design strength $= 84.79 \times 4560/1000 = 386.64$ kN

Example 5.8 *Calculate the compressive resistance of a 150 × 150 × 16 angle assuming that the angle is loaded through only one leg, when*
(a) it is connected by two bolts at the ends
(b) it is connected by one bolt at each end and
(c) it is welded at each end.
Assume that the member has a length of 3 m and $f_y = 250$ MPa.

Solution

(a) Connected by two bolts at the ends
   (i) For two bolts at each end, for fixed condition,
       $k_1 = 0.20$, $k_2 = 0.35$ and $k_3 = 20$ (from Table 12 of code)
       $\varepsilon = (250/f_y)^{0.5} = 1.0$

$$\lambda_{vv} = (3000/29.4)/\sqrt{(\pi^2 E / 250)} = 1.1484$$

$$\lambda_\varphi = (150 + 150)/[(2 \times 16)\sqrt{(\pi^2 \times 2 \times 10^5 / 250)}\,] = 0.1055$$

$$\lambda_e = \sqrt{[k_1 + k_2\lambda_{vv}^{\,2} + k_3\lambda^2_{\,\varphi}]}$$

$$= \sqrt{[0.20 + 0.35 \times 1.1484^2 + 20 \times 0.1055^2]} = 0.94$$

From Table 10 of code, select buckling curve 'c'.
From Table 7 of code, $\alpha = 0.49$

$f_{cd} = (f_y/\gamma_{mo})/[\varphi + (\varphi^2 - \lambda_e^2)^{0.5}]$
$\varphi = 0.5[1 + 0.49\,(0.94 - 0.2) + 0.94^2] = 1.1231$
$f_{cd} = (250/1.1)/[1.1231 + (1.1231^2 - 0.94^2)^{0.5}]$
$= 130.79$ N/mm^2
$P_d = 130.79 \times 4560/1000 = 596.4$ kN

   (ii) If we consider the fixity condition as hinged (Table 12 of code)
       $k_1 = 0.70$, $k_2 = 0.60$, and $k_3 = 5$

Hence $\lambda_e = \sqrt{(0.70 + 0.60 \times 1.1484^2 + 5 \times 0.1055^2)} = 1.2438$
$\varphi = 0.5[1 + 0.49\,(1.2438 - 0.2) + 1.2438^2] = 1.5293$
$f_{cd} = (250/1.1)/[1.5293 + (1.5293^2 - 1.2438^{\,2})^{0.5}] = 93.95$ N/mm^2
$P_d = 93.95 \times 4560/1000 = 428.41$ kN

(b) Connected by only one bolt at the ends and the fixity condition taken as hinged,
   $k_1 = 1.25$, $k_2 = 0.50$, $k_3 = 60$

Hence $\lambda_e = \sqrt{(1.25 + 0.50 \times 1.1484^2 + 60 \times 0.1055^2)} = 1.605$
$\varphi = 0.5[1 + 0.49\,(1.605 - 0.2) + 1.605^2] = 2.1322$
$f_{cd} = (250/1.1)/[2.1322 + (2.1322^2 - 1.605^2)^{0.5}] = 64.28$ N/mm^2

$P_d = 64.28 \times 4560/1000 = 293.12$ kN

(c) When the strut is welded at each end, it is similar to case (a) and the strength will be equal to 596.4 kN.

Note that when two bolts are provided at the ends, depending on the assumed fixity condition, we get the capacity as 596.4 kN or 428.41 kN. Note that for this angle, with concentric loading, the capacity works out to 574 kN, using Table 9c for an effective length factor of 0.85 ($KL/r = 86.73$) and 476 kN for an effective length of 1.0 ($KL/r = 102$).

Example 5.9 *Design a column to support a factored axial load of 1850 kN. The column has an effective length of 6 m with respect to the z-axis and 4 m with respect to the y- axis. Use Fe 410 grade steel.*

Solution
Assume the design compressive stress as $f_{cd} = 0.5 f_y = 125$ MPa.
Required area = $1900 \times 10^3/125 = 15200$ mm^2
Try ISMB 600 at 1230 N/m. The relevant properties of the section are as follows (IS 808-1989):
$A = 15600$ mm^2; $r_z = 242$ mm; $r_y = 41.2$ mm; $h/b = 600/210 = 2.85$, $t_f = 20.3$ mm
$L/r_z = 6000/242 = 24.8 < 180$. Hence ok.
$L/r_y = 4000 / 41.2 = 97.09 < 180$. Hence ok.
From Table 1 of code, $f_y = 240$ MPa as $t_f > 20$ mm
From Tables 10 and 9b of the code, for
$\quad L/r_y = 97.09$ and $f_y = 240$ MPa
$\quad f_{cd} = 120.37$ N/mm^2
Therefore strength of the column = $120.37 \times 15600/1000 = 1877$ kN < 1850 kN
Hence, the chosen section is safe.
Note: If the beams in this example are rigidly connected to the columns, then the Wood's curves should be used to determine the effective length factor K.

Example 5.10 *The strut of a space frame member having a length of 3.5 m has to carry a factored load of 280 kN. Assuming $f_y = 220$ N/mm^2, design a circular tube to carry the load. Assume that the ends are simply supported.*

Solution
Since tubes are efficient in resisting compression, let us assume the design compressive stress as $f_{cd} = 0.7 f_y = 0.7 \times 220 = 154$ MPa
Required area = $P_u/(0.7 f_y) = 280 \times 10^3/154 = 1818$ mm^2
Assume medium class pipe nominal bore 110 mm with an outside diameter 127 mm with thickness 4.85 mm, weight = 146 N/m, $r = 43.2$ mm, and area = 1860 mm^2

(i) *Check for limiting width-to-thickness ratio* (Table 2 of code)
$\quad D/t = 127/4.85 = 26.2 < 88\varepsilon^2$ ($88 \times 1.066^2 = 100$)
$\quad\quad\quad\quad\quad\quad\quad\quad$ Note: $\varepsilon = (250/220)^{0.5} = 1.066$
Hence, the tube is not slender.

(ii) *Buckling curve classification*
$\quad$ From Table 10 of the code, for hot-rolled tubes, use buckling curve 'a'.

(iii) *Resistance to flexural buckling*

$KL/r = 1 \times 3500/43.2 = 81$

From Table 9a of the code, for

$f_y = 220$ N/mm^2 and $KL/r = 81$,

$f_{cd} = 152.6$ N/mm^2

Design strength $= A \times f_{cd} = 1860 \times 152.6/1000$

$= 283.8$ kN > 280 kN

Hence, the assumed section is safe.

Example 5.11 *Repeat the problem given in Example 5.10 with a square tube, with* $f_y = 240$ N/mm^2

Solution

Required area $= P_u/(0.6 f_y) = 280 \times 10^3/(0.6 \times 240) = 1944$ mm^2

Assume a SHS of size 113.5×113.5 mm with a thickness of 4.8 mm, area $= 2028$ mm^2, weight $= 159.2$ N/m, $r = 44$ mm, and $\varepsilon = (250/f_y)^{0.5} = (250/240)^{0.5} = 1.042$

(i) *Check for limiting width-to-thickness ratio*

$D/t = 113.5/4.8 = 23.6 < 42\varepsilon$ ($42 \times 1.042 = 43.7$)

Hence, the section is not slender.

(ii) *Buckling curve classification*

From Table 10 of the code, for hot-rolled tubes, use buckling curve 'a'.

(iii) *Resistance to flexural buckling*

$KL/r = 1 \times 3500/44 = 79.54$

From Table 9a of the code, for

$f_y = 240$ N/mm^2 and $KL/r = 79.54$

$f_{cd} = 163.6$ N/mm^2

Hence design strength $= 163.6 \times 2028/1000 = 331$ kN > 280 kN

Hence, the assumed section is safe.

Example 5.12 *Design a double angle discontinuous strut to carry a factored load of 175 kN. The length of the strut is 3.0 m between intersections. The two angles are placed back-to-back and are tack bolted. Consider the following cases.*

(a) Angles are placed on opposite sides of the gusset plate.

(b) Angles are placed on the same side of the gusset plate.

(c) Two angles in star formation.

Assume grade Fe 410 steel with $f_y = 250$ MPa.

Solution

Let us assume the design compressive stress is $0.4 f_y = 100$ MPa.

Required area $= 175 \times 1000/100 = 1750$ mm^2

(a) *Angles placed on opposite sides of the gusset plate*

Let us try 2 ISA $70 \times 70 \times 8$ mm at 83 N/m with $A = 2 \times 1060$ mm^2,

$I_y = 47.4 \times 10^4$ mm^4, $r_z = 21.2$ mm, $c_y = 20.2$ mm

$A^* = 2 \times 1060 = 2120$ mm^2

Assuming a 10-mm thick gusset plate

$I_y = 2[I_y + A(c_y + t_g/2)^2]$

$= 2[47.4 \times 10^4 + 1060(20.2 + 5)^2]$

$= 229.42 \times 10^4$ mm^2

$$r_y^* = \sqrt{(229.42 \times 10^4/2120)} = 32.90 \text{ mm}$$

$$r_z^* = r_z = 21.2 \text{ mm}$$

Note The asterisk sign * is used for the properties of double angle strut.
Minimum radius of gyration

$$r^* = r_z = 21.2 \text{ mm}$$

Effective length factor could be between 0.7 and 0.85 (clause 7.5.2.1)
Assume $K = 0.85$.

Hence, effective length = $0.85 \times 3000 = 2550$ mm

$L/r = 2550/21.2 = 120.28 < 180$. Hence, the selected section is ok.

For $L/r = 120.28$ and $f_y = 250$ MPa, using Tables 10 and 9c of the code,

$$f_{cd} = 83.44 \text{ N/mm}^2$$

Hence strength of the member = $83.44 \times 2120/1000 = 176.89$ kN > 175 kN

Note ISA 70 × 70 × 8 is not commonly rolled and hence the designer may choose 75 × 75 × 8 (89 N/m) or 80 × 80 × 6 (73 N/m).

For 80 × 80 × 6, $L/r = 103.65$, $f_{cd} = 102.47$ MPa

Hence capacity = $102.47 \times 2 \times 929/1000 = 190.4$ kN.

Hence, provide 2 ISA 80 × 80 × 6 as shown in Fig. 5.29(a).

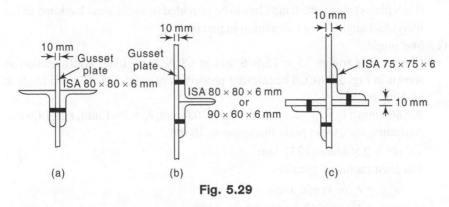

Fig. 5.29

(b) *Angles placed on the same side of the gusset plate [Fig. 5.29(b)]*

Try 80 × 80 × 6 at 73 N/m, with $A = 929$ mm^2 and $r_z = 24.6$ mm.

Now, $L/r = 2550/24.6 = 103.66$.

From Table 10 and Table 9c of the code, for $L/r = 103.66$ and $f_y = 250$ MPa,

$$f_{cd} = 102.47 \text{ MPa}$$

Capacity = $102.47 \times 929 \times 2 = 190.4$ kN > 175 kN. Hence the assumed angle is safe.

Let us try an unequal angle of size 90 × 60 × 6 at 68 N/m, with 10 mm gap, with $A = 865$ mm^2, $r_z = 28.6$ mm, $r_y = 17.1$ mm, $I_y = 25.2 \times 10^4$ mm^4,

$c_y = 13.9$ mm.

Radius of gyration about z-z axis for the combination is same as for one angle

$$r_z^* = 28.6 \text{ mm}$$

Radius of gyration r_y is obtained by finding the moment of inertia about y-y axis

$$I^*_{yy} = 2 \times 25.2 \times 10^4 + 2 \times 865 \, (13.9 + 10/2)^2$$
$$= 50.4 \times 10^4 + 61.79 \times 10^4$$
$$= 112.19 \times 10^4 \text{ mm}^4$$

$$r^*_y = \sqrt{[112.19 \times 10^4 / (2 \times 865)]} = 25.46 \text{ mm}$$

Hence $r^*_{min} = 25.46$ mm

$L/r = 2550/25.46 = 100.13$

From Tables 10 and 9c of the code for $L/r = 100.13$ and $f_y = 250$ MPa,

$f_{cd} = 106.84$ MPa

Capacity $= 106.84 \times 2 \times 865/1000 = 184.83$ kN > 175 kN

Hence, the section is safe.

Note From these calculations, it is seen that the designers' task is to design an angle that will be available in the market and also to choose a section that will provide an optimum section. It is clear that two unequal angles with the longer leg connected to the gusset plate provides the economic solution. Two angles on the opposite side of the gusset plate will be economical than two angles on the same side of the gusset plate. Tack welding at a spacing of 550 mm (using stitch plates) along the length has to be provided to avoid local buckling of the individual angles {see calculation in part (c)}:

(c) *Star angle*

Choose two angles 75 × 75 × 6 mm at 68 N/m in a star configuration as shown in Fig. 5.29(c). The relevant properties of single 75 × 75 × 6 angle at 68 N/m are

$A = 866$ mm^2; $r_z = r_y = 23$ mm; $c_z = c_y = 20.6$ mm; $r_u = 29.1$ mm; $r_v = 14.6$ mm

Assuming the gusset plate thickness as 10 mm,

$$A^* = 2 \times 866 = 1732 \text{ mm}^2$$

The least radius of gyration

$r^*_{min} = r_u$ of single angle $= 29.1$ mm

Hence $\lambda = L/r = 2550/29.1 = 87.62 < 180$

For $\lambda = 87.62$ and $f_y = 250$ MPa from Tables 10 and 9c of the code,

$f_{cd} = 124.57$ N/mm^2

Capacity of the section $= 124.57 \times 1732/1000$
$$= 215.75 \text{ kN} > 175 \text{ kN}.$$

The chosen section has a capacity much higher than the required strength and hence a lower available angle section may be chosen. The reader may do it as an exercise and select a suitable angle.

Note that the star type configuration results in a smaller section compared to the other two angle configurations in (a) and (b). Such star angles are often used for the legs of transmission or communication towers (see Subramanian 1992). They are also used as bracings in industrial buildings.

Tack Welding

Tack welding along the length should be provided to avoid local buckling of each of the elements. The spacing of the welds to be adjusted based on (clause 7.8.1 of the code)

$\lambda_e \le 0.6\lambda = 0.6 \times 87.62 = 52.05$ or ≤ 40

Selecting the lowest value,

$\lambda_e = 40$

The spacing of the tack weld is given by

$\lambda_e = S/r_v \le 40$

Hence $S = 40r_v = 40 \times 14.6 = 584$ mm

The welding should be designed to resist a transverse load of 2.5% of the axial load, i.e., $175 \times 1000 \times 2.5/100 = 4375$ N.

Assuming a 3 mm fillet weld (as per Table 11.2),

Design strength of the weld (Appendix D) = 158 N/mm^2

Hence length of weld = $4375/(0.7 \times 3 \times 158) = 13.2$ mm

Hence provide a 3 mm tack welding of 15-mm length at 550 mm spacing (see Fig. 3.3 given in Chapter 3 for the details of stitch plates).

Example 5.13 *Design a single angle discontinuous strut to carry a factored load of 65 kN. Assume that the distance between its joints is 2.5 m and is loaded through one leg. Use f_y = 250 MPa.*

Solution

Take effective length = L = 2500 mm

Assume a slenderness ratio of 120 and the corresponding f_{cd} = 83.7 MPa (from Table 9c of code)

Area required = $65 \times 1000/83.7 = 777$ mm^2

Choose ISA $75 \times 75 \times 6$ at 68 N/m with

Check for section classification

$d/t = b/t = 75/6 = 12.5 < 15.7$

$(b + d)/t = 150/6 = 25 < 25$

Hence section is semi-compact

$A = 866$ mm^2, $r_{vv} = 14.6$ mm

Assuming two bolts at each end and fixed condition (from Table 12 of the code)

$k_1 = 0.2$, $k_2 = 0.35$, and $k_3 = 20$

$\varepsilon = (250/f_y)^{0.5} = 1.0$

$\lambda_{vv} = (L/r_{vv})/[\varepsilon\sqrt{(\pi^2 E/250)}] = (2500/14.6)/[1.0 \times \sqrt{(\pi^2 \times 2 \times 10^5/250)}\,]$
$= 171.23 \times 0.01125 = 1.926$

$\lambda_\phi = (b_1 + b_2)/[\varepsilon\sqrt{(\pi^2 E/250)} \times 2t] = 150 \times 0.01125/(2 \times 6) = 0.1406$

$\lambda_e = \sqrt{[k_1 + k_2\lambda_{vv}^2 + k_3\lambda_\phi^2]}$

$= \sqrt{[0.2 + 0.35 \times 1.926^2 + 20 \times 0.1406^2]} = 1.376$

$f_{cd} = (f_y/\gamma_{mo})/[\phi + (\phi^2 - \lambda_e^2)^{0.5}]$

Using Tables 10 and 7 of the code

$\phi = 0.5[1 + 0.49(1.376 - 0.2) + 1.376^2] = 1.735$

$$f_{cd} = (250/1.1)/[1.735 + (1.735^2 - 1.376^2)^{0.5}]$$
$$= 81.4 \text{ N/mm}^2$$
$$P_d = 81.4 \times 866/1000 = 70.497 \text{ kN} > 65 \text{ kN}$$

Hence, the assumed section is safe.

Example 5.14 *Design a laced column 10-m long to carry a factored axial load of 1100 kN. The column is restrained in position but not in direction at both ends. Provide single lacing system with bolted connection.*
(a) Design the column with two channels back-to-back
(b) Design the column with two channels placed toe-to-toe
(c) Design the lacing system with welded connections for channels back-to-back.

Solution
Design of column
$$P = 1100 \times 10^3 \text{ N}$$
$$L = 1.0 \times 10 = 10 \text{ m}$$

Assume a design stress of 125 MPa,
Required area $= 1100 \times 10^3/125 = 8800 \text{ mm}^2$
Select two ISMC 300 at 363 N/m. The relevant properties of ISMC 300 are (IS 808-1989)

$$A = 4630 \text{ mm}^2, r_{zz} = 118.0 \text{ mm}, r_{yy} = 26.0 \text{ mm}, t_f = 13.6 \text{ mm}$$
$$c_{yy} = 23.5 \text{ mm}, I_{zz} = 6420 \times 10^4 \text{ mm}^4, I_{yy} = 313 \times 10^4 \text{ mm}^4$$

Area available $= 2 \times 4630 = 9260 \text{ mm}^2$

Built-up sections will be economical, when the radius of gyration of the *y-y* axis is increased in such a way that it is more or less equal to the radius of gyration about the *z-z* axis. This is achieved by spacing the sections in such a way that r_{zz} becomes r_{min}. Let us first check the safety of the section and then workout the required spacing between the two channels.

$$L/r_{zz} = 10 \times 10^3/118.0 = 84.74$$

The *L/r* of the built-up column should be taken as $1.05 \times (L/r_{zz}) = 1.05 \times 84.74 = 88.98$
For $L/r_{zz} = 88.98$ and $f_y = 250$ MPa, using Table 9c of the code, $f_{cd} = 122.53$ MPa.
Load carrying capacity $= A_e f_{cd} = 9260 \times 122.53/1000$
$$= 1135 \text{ kN} > 1100 \text{ kN}$$

Hence the column is safe.
(a) Let us provide two channels back-to-back and connect them by lacing and denote *S* as the spacing between two channels [see Fig. 5.30(a)].
   Spacing of channels
$$2I_{zz} = 2[I_{yy} + A(S/2 + c_{yy})^2]$$
   Thus, $2 \times 6420 \times 10^4 = 2[313 \times 10^4 + 4630(S/2 + 23.5)^2]$
   or $(S/2 + 23.5)^2 = 13190$
$$S = 182.70 \text{ mm}$$

   Let us keep the channels at a spacing of 183 mm.
   Lacing system Using a single lacing system with the inclination of lacing bar = 45° (gauge length for a 90 mm flange = 50 mm)

Spacing of lacing bars, $L_o = 2(183 + 50 + 50)$ cot $45°$
$$= 2 \times 283 \times 1 = 566 \text{ mm}$$
L_o/r_{yy} should be $< 0.7 \times L/r$ of whole column
$$21.77 < 0.7 \times 88.98 = 62.3$$
Hence safe.

Maximum shear $= (2.5/100) \times 1100 \times 10^3 = 27,500$ N

Transverse shear in each panel $= V/N = 27,500/2 = 13,750$ N

Compressive force in the lacing bar $= (V/N)$ cosec $45°$
$$= 13750 \times 1.414 = 19445 \text{ N}$$

Assuming 16-mm diameter bolts,

Minimum width of lacing flat (clause 7.6.2 of the code) $= 3 \times 16$, say 50 mm

Minimum thickness $= (1/40)(183 + 50 + 50)$ cosec $45° = 10.01$ mm

Provide 12-mm thick plate with a width of 50 mm

Minimum $r = t/\sqrt{12} = 12/\sqrt{12} = 3.464$ mm

L/r of the lacing bar $= 283 \times$ cosec $45°/3.464 = 115.5 < 145$

Hence safe.

For $L/r = 115.5$ and $f_y = 250$ MPa, using Table 9c of the code
$$f_{cd} = 88.6 \text{ MPa}$$

Load carrying capacity $= 88.6 \times 50 \times 12 = 53,163$ N $> 19,445$ N

Hence the lacing bar is safe.

Tensile strength of the lacing flat $= 0.9(B - d)t f_u/\gamma_{m1}$ or $f_y A_g/\gamma_{m0}$

Thus $0.9 (50 - 18) \times 12 \times 410/1.25$ or $250 \times 50 \times 12/1.1$

113,356 N or 136,363

Thus, the tensile strength of the lacing flat $= 113,356$ N $> 19,445$ N

Hence the lacing flat is safe.

Check

r_{min} of the built-up column $= 118$ mm,
$$L/r = 10,000/118 = 84.74$$

r_{min} of the individual chords $= 26.0$ mm,
$$L_0/r = 566/26 = 21.77$$

λ of the built-up column [Eqn (5.21)],
$$\lambda_e = \sqrt{\{84.74^2 + 3.14^2(9260/600) \times 400.22^3/(566 \times 230^2)\}}$$
$$= 86.64 < 88.98$$

Hence, the column is safe.

Connection Assuming that the 16 mm bolts of grade 4.6 are connecting both the lacing flats with the channel at one point and that the shear plane will not pass through the threaded portion of bolt.

Strength of bolt in double shear $= 2 \times A_{sb} (f_u/\sqrt{3})/\gamma_{mb}$
$$= 2 \times \pi \times 16^2/4 \times (400/\sqrt{3})/1.25 = 74,293 \text{ N}$$

Strength in bearing $= 2.5 k_b d t f_u/\gamma_{mb}$ (with $k_b = 0.49$)
$$= 2.5 \times 0.49 \times 16 \times 12 \times 410/1.25$$
$$= 77,145 \text{ N}$$

Hence, strength of bolt = 74,293 N > 19,445 N

Hence one 16-mm diameter bolt of grade 4.6 is required.

Tie plates Tie plates must be provided at the ends of the laced column.

Effective depth = $183 + 2 \times c_{yy} > 2 \times b_f$

$$= 183 + 2 \times 23.5 = 230 \text{ mm} > 2 \times 90 = 180 \text{ mm}$$

Hence

Required overall depth of tie plate = $230 + 2 \times 25 = 280$ mm (edge distance of 16-mm diameter bolts = 25 mm)

Provide a tie plate of 300-mm depth.

Length of tie plate = $183 + 2 \times 90 = 363$ mm

Required thickness of tie plate = $1/50(183 + 2g) = 1/50(183 + 2 \times 50)$ = 5.66 mm

(where g = gauge distance—see Appendix D)

Hence, provide a tie plate of 6-mm thickness.

Provide a tie plate of size $363 \times 300 \times 6$ mm at both ends with six 16-mm diameter bolts.

(b) Consider the case of laced column with the two channels provided toe-to-toe.

Spacing [Fig. 5.30(b)]

$$2I_{zz} = 2[I_{yy} + A(S/2 - c_{yy})^2]$$

$$2 \times 6420 \times 10^4 = 2[313 \times 10^4 + 4630(S/2 - 23.5)^2]$$

or $(S/2 - 23.5)^2 = 13190$

$$S = 276.7 \text{ mm}$$

Let us place the channel at a spacing of 280 mm.

Connecting system Assuming single lacing system is provided with an inclination of 45°; gauge length for 90 m flange = 50 mm

$$L_o = 2(280 - 50 - 50) \cot 45° = 360 \text{ mm}$$

$$L_o/r_{yy} = 360/26 = 13.8 < 50$$

Hence L_o/r_{yy} ratio is fine.

$0.7(L/r)$ of combined channel = $0.7 \times 88.98 = 62.3 > 13.8$

Hence the L/r ratio is ok.

Compressive force in lacing bar = 19,445 N

Minimum width of lacing flat for 16 mm bolt (clause 7.6.2 of the code) = 50 mm

Minimum thickness = $1/40(280 - 50 - 50) \times$ cosec 45°

$$= 6.36 \text{ mm}$$

Hence, provide a 50×8 mm flat.

Check $r_{min} = t/\sqrt{12} = 8/\sqrt{12} = 2.309$ mm

$$L/r = 180 \times \text{cosec } 45°/2.309 = 110.2 < 145$$

Hence the chosen flat is safe.

For $L/r = 110.2$ and $f_y = 250$ MPa, from Table 9c of the code,

$$f_{cd} = 94.4 \text{ N/mm}^2$$

Capacity of the lacing flat = $94.4 \times 50 \times 8$

$$= 37,760 \text{ N} > 19,445 \text{ N}$$

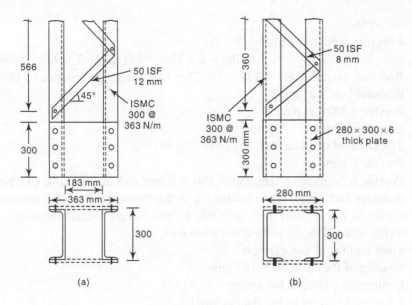

(a)

(b)

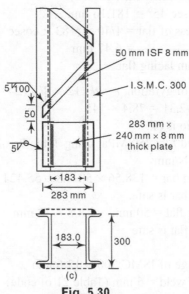

(c)

Fig. 5.30

Tensile strength of the lacing flat = $0.9(B - d)tf_u/\gamma_{m1}$ or f_yA_g/γ_{m0}
   = $0.9(50 - 18) \times 8 \times 410/1.25$ or $250 \times 50 \times 8/1.1$
   = 75,571 N or 90,909 N both > 19,445 N
Hence the lacing flat is safe.
Connection
Strength of bolt in double shear {from (a)} = 74,293 N
Strength in bearing = $2.5k_bdtf_u/\gamma_{mb}$ = $2.5 \times 0.49 \times 16 \times 8 \times 410/1.25$ = 51,430 N
Hence, strength of bolt = 51,430 N > 19,445 N
Therefore, provide one 16-mm diameter bolt of grade 4.6.

Tie plates

Effective depth of tie plate = $S - 2c_{yy}$

$$= 280 - 2 \times 23.5 = 233 \text{ mm} > 2 \times 90 = 180 \text{ mm}$$

Required overall depth = $230 + 2 \times 25 = 280$ mm (edge distance of 16-mm diameter bolt = 25 mm)

Provide a 300 mm plate

Length of tie plate = 280 mm

Thickness of tie plate = $(1/50)(280 - 2 \times 50) = 3.6$ mm

Provide 6 mm.

Provide a tie plate of size $280 \times 300 \times 6$ mm and use six bolts of 16-mm diameter and grade 4.6 to connect it to the channels. The arrangement is shown in Fig. 5.30(b). It is seen that by providing channels toe-to-toe, the lacing size and the tie plate size are reduced.

(c) From part (a) of this example,

Spacing of the channels = 183 mm

Compressive force in the lacing = 19,445 N

Effective length of lacing flat (welded)

= $0.7 \times 183 \times$ cosec $45° = 181.16$ mm

Minimum thickness of flat = $1/40 \times (183 \times$ cosec $45°)$

$$= 6.47 \text{ mm}$$

Provide 50×8 mm lacing flat.

Minimum radius of gyration, $r = t/\sqrt{12} = 8/\sqrt{12} = 2.31$ mm

$L/r = 181.16/2.31 = 78.4 < 145$

Hence the L/r ratio is ok.

For $L/r = 78.4$ and $f_y = 250$ MPa, using Table 9c of the code

$f_{cd} = 138.56$ N/mm^2

Capacity of lacing bar = $138.56 \times 50 \times 8 = 55,424$ N $> 19,445$ N

Hence the lacing bar is safe.

Overlap of lacing flat = 50 mm $> 4 \times 8 = 32$ mm

Hence the lacing flat is safe.

Connection

Thickness of flange of ISMC 300 = 13.6 mm

Minimum size of weld = 5 mm (Table 21 of code)

Strength of weld/unit length = $0.7 \times 5 \times 410/(\sqrt{3} \times 1.5)$

$$= 552 \text{ N/mm}$$

Required length of weld = $19445/552 = 35.2$ mm

Adding extra length for ends, the weld length to be provided

= $36 + 2(2 \times 5) = 56$ mm

Provide 100 mm weld length at both ends.

Tie plate

Overall depth of plate = $183 \times 2 \times c_{yy}$

$$= 183 + 2 \times 23.5$$

$$= 230 \text{ mm} > 2 \times 90 \text{ mm}$$

Let length of tie plate = 183 + 2 × 50 = 283 mm
Thickness of tie plate = 1/50(183 + 2 × 50) = 5.66 mm
Provide a 8 mm plate to accommodate a 5 mm weld.
Provide a tie plate of 283 × 240 × 8 mm size and connect it with 5 mm welds
as shown in Fig. 5.30(c).

Example 5.15 *Design a batten system for the column in Example 5.14. Assume
that the two channels are kept back-to-back. (See Fig. 5.31)*

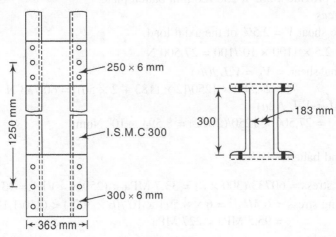

Fig. 5.31

Solution
From the previous example: with two ISMC 300 channels,
Spacing of channel = 183 mm, L/r = 84.74
L/r of the battened column = 1.1 × 84.74 = 93.21
For L/r = 93.21 and f_y = 250 MPa, from Table 9c of the code,

f_{cd} = 116.5 MPa

Capacity of the built-up column = 2 × 4630 × 116.5/1000 = 1078.8 kN ≈ 1100 kN
Hence the column is safe.
Spacing of battens
(L_o/r_{yy}) should be less than 0.7 times the slenderness of the built-up column.
i.e., L_o/r_{yy} < 0.7 (L/r)

L_o < $0.7r_{yy}$ (L/r) = 0.7 × 26 × 93.21 = 1696.4 mm

Also, L_o/r_{yy} < 50
or L_o < 50 × 26 = 1300 mm
Hence provide battens at a spacing of 1250 mm.
Size of end battens
Provide 20 mm bolts.
Edge distance (Table 10.6) = 33 mm
Effective depth = 183 + 2 × 23.5 = 230 mm > 2 × 90 mm
Hence, chosen effective depth is safe.
Overall depth = 230 + 2 × 33 = 296 mm

Required thickness of batten = $1/50(183 + 2 \times 50) = 5.66$ mm
Length of the batten = $183 + 2 \times 90 = 363$ mm
Provide $363 \times 300 \times 6$ mm end batten plates.
Size of intermediate batten
Effective depth = $3/4 \times 230 = 172.5$ mm $> 2 \times 90$ mm
Hence adopt an effective depth of 180 mm
Overall depth = $180 + 2 \times 33 = 246$ mm
Therefore, provide a $363 \times 250 \times 6$ mm batten plate.
Design forces
Transverse shear $V = 2.5\%$ of the axial load
$$= 2.5 \times 1100 \times 10^3/100 = 27,500 \text{ N}$$
Longitudinal shear $= V_b = V_t L_o/(ns)$
$$= 27,500 \times 1250/[2 \times (183 + 2 \times 50)] = 60,733 \text{ N}$$
Moment $M = V_t L_o/(2n)$
$$= 27,500 \times 1250/(2 \times 2) = 8.594 \times 10^6 \text{ Nmm}$$
Check
(a) For end battens

Shear stress = $60733/(300 \times 6) = 33.7$ MPa $< (250/\sqrt{3})/1.1 = 131.2$ MPa
Bending stress = $6 \, M/td^2 = 6 \times 8.594 \times 10^6/(6 \times 300^2) < (250/1.1)$
$$= 95.5 \text{ MPa} < 227 \text{ MPa}$$

Hence safe.
(b) For intermediate battens
Shear stress = $60,733/(250 \times 6) = 40.44$ MPa < 131.2 MPa
Bending stress = $6 \times 8.594 \times 10^6/(6 \times 250^2)$
$$= 137.52 < 227 \text{ MPa}$$

Hence safe.
Check
L/r of the built-up column = 84.74
r_{min} of the individual chords = 26.0 mm
$$L/r = 1250/26 = 48.08$$
λ of the built-up column [Eqn (5.20)],

$$\lambda_e = \sqrt{\{84.74^2 + (3.14^2/12) \times 48.08^2\}} = 95.3 > 93.21 \ (1.1 \times 84.74)$$

However, the difference is only 2.2% and hence the built-up column is safe.
Connection
The connections should be designed to transmit both shear and bending moment.
Assuming 20-mm diameter bolts.
Strength of bolt in bearing will not govern. Hence, strength of bolt (Table D.3)
= 45.3 kN
Required number of bolts = $60.733/45.3 = 1.34$
Let us provide three bolts to take into account the stresses due to bending moments.
Check
Force in each bolt due to shear = $60733/3 = 20244$ N
Adopt a pitch of 120 mm.

Force due to moment $= Mr/\Sigma r^2$

$$= 8.594 \times 10^6 \times 120/(120^2 + 120^2) = 35,808 \text{ N}$$

Resultant force $= \sqrt{(20244^2 + 35,808^2)}$

$$= 41,134 \text{ N} < 45,300 \text{ N}$$

Hence safe.

Number of batten plates on one face = (length of column/spacing) + 1

$$= (10 \times 10^3)/1250 + 1 = 9 > 3$$

Hence the number chosen is ok.

Example 5.16 *Design a column having an effective length of 6 m and subjected to a factored axial load of 2400 kN. Provide the channels back-to-back connected by welded battens. Assume Fe 410 grade steel.*

Solution

Required area of cross section $= 2400 \times 10^3/(0.6 \times 250) = 16,000 \text{ mm}^2$

Let us assume two channels ISMC 400 at 501 N/m. Relevant properties of ISMC 400 as per IS 808 – 1989 are

$A = 6380 \text{ mm}^2$; $b_f = 100 \text{ mm}$; $t_f = 15.3 \text{ mm}$; $r_z = 154 \text{ mm}$, $r_y = 28.2 \text{ mm}$;
$I_{zz} = 15,200 \times 10^4 \text{ mm}^4$; $I_{yy} = 508 \times 10^4 \text{ mm}^4$; $c_{yy} = 24.2 \text{ mm}$
Area provided $= 2 \times 6380 = 12,760 \text{ mm}^2$

$$\lambda = L/r = 1.1 \times 6.0 \times 10^3/154 = 42.86$$

For $L/r = 42.86$ and $f_y = 250$ MPa, using Table 9c of the code,

$$f_{cd} = 193.71 \text{ MPa}$$

Capacity of the built-up column $= 193.71 \times 2 \times 6380/1000 = 2471 \text{ kN} > 2400 \text{ kN}$
Hence the column is safe.

Spacing of channels

$$2I_{zz} = 2[I_{yy} + A(S/2 + c_{yy})^2]$$
$$2 \times 15200 \times 10^4 = 2[508 \times 10^4 + 6380 (S/2 + 24.2)^2]$$
$$(S/2 + 24.2)^2 = 23028$$

Hence $S = 255.1$ mm. Therefore, provide two ISMC 400 at a spacing of 256 mm back-to-back as shown in Fig. 5.32.

Spacing of battens

L_o/r_{yy} should be less than $0.7 \times$ slenderness ratio of the built-up column. Hence,

$$L_o/r_{yy} < 0.7(L/r)$$
$$L_o < 0.7(L/r)r_{yy}$$
$$< 0.7 \times 42.86 \times 28.2$$
$$< 846 \text{ mm}$$

Also, L_o/r_{yy} should be less than 50. Therefore,

$$L_o/r_{yy} < 50$$

i.e., $L_o < 50 \times 28.2 = 1210$ mm
Provide the batten at a spacing of 840 mm.

Size of end battens

Overall depth of batten $= 256 + 2 \times c_{yy}$

$$= 256 + 2 \times 24.2 = 304.4 \text{ mm}$$

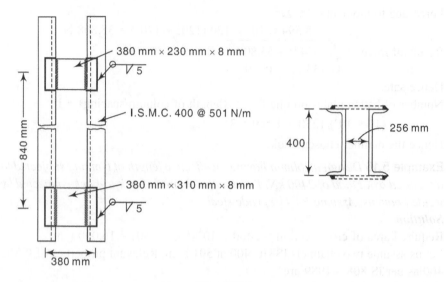

Fig. 5.32

Provide a 62 mm overlap of batten on channel flange for welding.
Length of batten = 256 + 2 × 62 = 380 mm
Thickness of batten = 1/50 × 380 = 7.60 mm
Provide 380 × 310 × 8 mm end batten plate.
Size of intermediate battens
Overall depth = 3/4 × 304.4 = 228.3 mm
Provide 380 × 230 × 8 mm intermediate battens.
Design forces
Transverse shear = 2.5% of axial load
$$= 2.5 \times 2400 \times 10^3/100 = 60,000 \text{ N}$$
Longitudinal shear $V_b = V_t L_o/ns$
$$= 60,000 \times 840/[2(256 + 2 \times 62/2)]$$
$$= 79,245 \text{ N}$$
Moment $M = V_t L_o/2n$
$$= 60,000 \times 840/(2 \times 2) = 12.6 \times 10^6 \text{ N mm}$$
Check
(a) For end battens,
Shear stress = 79,245/(310 × 8)
$$= 31.9 \text{ MPa} < 250/(\sqrt{3} \times 1.1) = 131.2 \text{ MPa}$$
Bending stress = $12.6 \times 10^6 \times 6/(8 \times 310^2)$
$$= 98.3 \text{ MPa} < 250/1.1 = 227 \text{ MPa}$$
(b) For intermediate battens,
Shear stress = 79,245/(230 × 8)
$$= 43.1 \text{ MPa} < 131.2 \text{ MPa}$$
Bending stress = $12.6 \times 10^6 \times 6/(8 \times 230^2)$
$$= 178.6 \text{ MPa} < 227 \text{ MPa}$$

Hence the battens are safe.

Design of weld

Let the welding be done on all the four sides at each end of batten plate as shown in Fig. 5.32.

Let t be the throat thickness of weld.

$I_{zz} = 2[(62 \times t^3/12) + (62 \times t)(230/2)^2] + 2 \times t \times 230^3/12$

Neglecting $62 \times t^3/12$, which will be insignificant, we get

$I_{zz} = 366.77 \times 10^4 t \text{ mm}^4$

$I_{yy} = 2[t \times 62^3/12] + 2 \times 230 \times t^3/12 + 2 \times 230 \times t \times 31^2$

Again neglecting $2 \times 230 \times t^3/12$, which will be very small, we get

$I_{yy} = 48.18 \times 10^4 \, t \text{ mm}^4$

$I_p = I_{zz} + I_{yy} = t(366.77 \times 10^4 + 48.18 \times 10^4) = 414.95 \times 10^4 t \text{ mm}^4$

The same result may be obtained by using Table 11.4

$$I_p = t\frac{(b+d)^3}{6} = \frac{(62+230)^3}{6} t = 414.95 \times 10^4 t \text{ mm}^4$$

$$r = \sqrt{[(230/2)^2 + (62/2)^2]} = 119.1 \text{ mm}$$

$\cos = 31/119.1 = 0.260$

Direct shear stress $= 79245/(2 \times 62 + 2 \times 230)t$

$= 135.7/t \text{ N/mm}^2$

Shear stress due to bending moment

$= 12.6 \times 10^6 \times 119.1/(414.95 \times 10^4 \times t)$

$= 361.65/t \text{ N/mm}^2$

Combined stress (clause 10.5.10 of code)

$[3 \times (135.7/t)^2 + (361.65/t)^2]^{0.5}$

$= 431.2/t < 410/(1.25 \times \sqrt{3}) = 189.4$

or $t = 431.2/189.4 = 2.28$ mm

Size of weld $= 2.28/0.7 = 3.26$ mm

The size of weld should not be less than 5 mm for 15.3 mm flange. Hence, provide a 5 mm weld to make the connections.

Example 5.17 *Design the base plate for an ISMB 500 column to carry a factored load of 1500 kN. Assume Fe 410 grade steel and M25 concrete.*

Solution

For ISMB 500 section,

$h = 500$ mm; $b_f = 180$ mm; $t_f = 17.2$ mm; $t_w = 10.2$ mm

Bearing strength of concrete $= 0.45 f_{ck} = 0.45 \times 25 = 11.25 \text{ N/mm}^2$

Required area of base plate $= 1500 \times 10^3/11.25 = 1,33,333 \text{ mm}^2$

Use a base plate of size 640×320 mm with area $= 2,04,800 \text{ mm}^2$

If the ISMB 500 is cept at centre, the projection will be 70 mm on each side as shown in Fig. 5.33.

$w = (1500 \times 10^3)/(640 \times 320) = 7.32$ MPa

$$t_{bp} = \sqrt{[2.5 \, w(a^2 - 0.3b^2)\gamma_{mo} / f_y]}$$

$$= \sqrt{[2.5 \times 7.32 \times (70^2 - 0.3 \times 70^2) \times 1.10 / 250]}$$

Hence provide $640 \times 320 \times 18$ mm plate. Also provide four nos. 20-mm diameter and 300-mm long anchor bolts to connect the base plate to the foundation concrete.

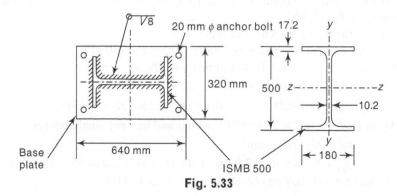

Fig. 5.33

Weld connecting base plate to column

Use an 8-mm fillet weld all round the column section to hold the base plate in place. Note that the surfaces are to be machined for direct bearing.

Total length available for welding along the periphery of ISMB 500
$$= 2[180 + (180 - 10.2) + (500 - 17.2)] = 1665.2 \text{ mm}$$
After deducting end returns of the weld, at the rate of two times the size of the weld at each end, we get,
$$L_{\text{eff}} = 1665.2 - 2(4 + 2)2a = 1665.2 - 24 \times 8 = 1473.2 \text{ mm}$$
Capacity of the weld $= 0.7 \times 8 \times 189/1000 = 1.0584$ kN/mm
Required length of weld $= 1500/1.0584 = 1417.2$ mm < 1473.2 mm
Hence provide an 8-mm weld.
Note: This is a fillet weld on the edge and not on the rounded ends of the member. Hence the limitation 0.75 times thickness (clause 10.5.8.2 of the code) will not apply here.

Summary

A structural member subjected to compressive forces along its axis is termed as a compression member. Compression members in buildings are called as columns and those found in trusses are called struts or rafters. Although the structural column is a simple structural member, the following parameters have an effect on the column strength:

1. Material properties (stress–strain relationship, yield stress)
2. Manufacturing method (hot-rolled/welded and cold-straightened shape resulting in different residual stress patterns)
3. Shape of the cross section (area, cross section geometry and bending axis)
4. Length
5. Initial out-of-straightness (maximum value and shape of the distribution) and other imperfections

6. End support conditions (pinned-with or without sway and restrained ends, with or without sway)

All these parameters that affect the strength of columns are discussed. The possible failure modes of columns are identified. The behaviour of compression members differ based on their length. Short or stocky columns can be loaded up to their yield stress and can attain their squash loads, provided the elements that make up the cross section are prevented from local buckling. Long compression members buckle elastically and hence their strength may be predicted by the Euler's formula. Intermediate length compression members fail both by yielding and buckling and hence their behaviour is inelastic. The behaviour of these slender columns is affected by the various parameters as listed above. Since there is a wide scatter in the strength of columns with different cross sections, many international codes have gone in for multiple column curves. Multiple column curves were first suggested by ECCS, which were followed by the American SSRC column strength curves. The Indian code has adopted the column curves similar to those given in the British code. The various individual and built-up or combined cross sections which can be used as compression members are discussed.

The effect of end supports (end restraints) on column strength is usually incorporated in the design by the concept of effective length. The effective length of columns with idealized boundary conditions are discussed and also the effect of intermediate restraints and the calculation of effective lengths of compressive members in different planes. The effective length of columns in multi-storey buildings is difficult to calculate, since the end restraints depend on the stiffness of the beams meeting at column ends and also on whether the frames are braced or unbraced. Columns in sway frames buckle in double curvature (hence the effective length factor will always be greater than unity) and the columns in braced (non-sway) frames, buckle in single curvature (hence the effective length factor will be less than unity). The IS code provisions on effective length are based on the Wood's curves, with slight modifications.

The compression members in trusses have been considered separately since the loads are usually applied at the joints and the joints are often considered as pinned. Similarly single-angle struts should also be treated separately, since they will be connected in most of the cases through one of their legs to the other members. The IS code provisions for such continuous and discontinuous angle members are summarized.

The various steps involved in the design of compression members are also given. Some specifications for the compression members composed of two components back-to-back are also included. The design of base plates, used as a link between the compression members and the concrete foundation, is discussed with some practical aspects. Most of the concepts are explained with illustrative examples.

Exercises

Note Use grade Fe 410 steel ($f_y = 250$ MPa) and $E = 2.0 \times 10^5$ MPa, if not specified in the problem.

1. Check whether an ISHB 400 in Fe 540 steel would be affected by local buckling effects when used as a column.
2. A column consisting of ISHB 400 at 774 N/m has an unsupported length of 3.8 m. It is effectively held in position at both ends, restrained against rotation at one end. Calculate the axial load this column can carry.
3. Determine the design axial load of a column section ISHB 300. The column is having a height of 9 m and is effectively restrained by two bracings in the *y-y* direction at 3 m and 6 m and by one bracing member in the *z-z* direction at mid-height. Assume pinned condition at both ends of the column.
4. Determine the capacity of a column of size (wide flange section) W $250 \times 250 \times 101$ (with weight = 1010 N/m, $r_z = 112.7$ mm, $r_y = 65.6$ mm, area = 12900 mm^2) with bracing and end condition as shown in Fig. 5.34.

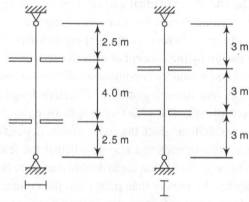

Fig. 5.34

5. For the same bracing and end condition as shown in Fig. 5.34 determine the capacity of ISMB 550 (weight = 1040 N/m) and compare the capacity with the W250 × 250 × 101 section of the previous example. (*Note:* Both the sections are having more or less equal weight.)
6. Determine the axial load capacity of the following section, assuming that the column has pinned ends and has a length of 3 m.
   (a) ISMB 400 with one cover plate of size 250 × 12 mm on each flange [see Fig. 5.9(n)]
   (b) Two ISMC 300 back-to-back with a distance of 200 mm between them [see Fig. 5.9(k)]
7. A single angle discontinuous strut ISA 90 × 90 × 6 is 2.5-m long. It is connected by one bolt at each end. Calculate the safe load this strut can carry.
8. A double angle discontinuous strut consists of two ISA 75 × 75 × 6 connected back-to-back to both sides of the 10-mm thick gusset plate with two bolts. The length of the strut is 4.5 m. Calculate the safe load carrying capacity of the section.

9. A double angle discontinuous strut 2.25 m long consists of two ISA $120 \times 120 \times 8$ with longer leg connected to the same side of the 8-mm thick gusset plate by two bolts. Calculate the load that this strut can carry.

10. An ISHB 400 column is an interior column with strong axis buckling in the plane of the frame as shown in Fig. 5.35. The columns above and below are also ISHB 400. The beams framing at the top are ISMB 400 and those at the bottom are ISMB 450. The columns are 4-m high and the beam span is 7 m. Determine the effective length of this column (*AB* in Fig. 5.35) as per Wood's curves of IS 800 : 2007. Assuming that the column carries a factored load of 1200 kN, determine whether the column is safe to carry this load. Assume that the frame is braced by shear walls.

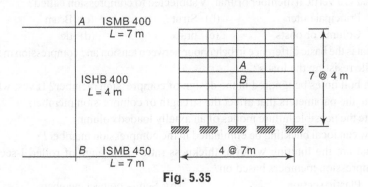

Fig. 5.35

11. For the same building shown in Fig. 5.35, determine the effective length and load carrying capacity of column *AB*, if the frame is a sway frame. Compare the results with those of Exercise 10.

12. Design a column to support an axial load of 2500 kN, assuming that the column has an effective length of 8 m with respect to the *z*-axis and 4 m with respect to the *y*-axis. Since the maximum available rolled section (ISHB 450) may not be sufficient to support this load, design a built-up I-section with two flange plates.

13. Design a hollow circular tubular strut of a truss having an effective length of 2.5 m and to carry a factored load of 250 kN.

14. For the problem in Exercise 13, design a suitable hollow rectangular section and compare the weights.

15. Design a suitable section for a column 5-m long which is effectively held in position and restrained against rotation at both ends, in order to carry a factored load of 600 kN.

16. Design a single angle strut carrying a factored compressive load of 65 kN with length between centre-to-centre of intersection as 3 m. Design also the bolted end connections.

17. Design a double angle section to act as a compression member in a truss having 2.1 m length and to carry a factored load of 150 kN.
    (a) Connected back-to-back to both sides of a 12-mm thick gusset plate.
    (b) Connected back-to-back to one side of the 12-mm thick gusset.
    (c) Angles in star formation

18. Design a built-up column with two channels back-to-back having an effective length of 6.5 m and carrying a factored load of 1000 kN. Also design the lacings.

19. Design a built-up column with two channels toe-to-toe to carry a factored load of 2000 kN. Take the effective length as 5.5 m.
    (a) Design it as a laced column and also design the lacings.
    (b) Design it as a battened column and also design the battens.
20. Design a base plate for a column of ISMB 400 carrying a factored load of 1500 kN. Assume that the column is supported on a concrete of grade M25.

Review Questions

1. Distinguish between column and beam column.
2. What is a vertical member primarily subjected to compression called?
    (a) Principal rafter   (b) Strut   (c) Beam
    (d) Column or posts   (e) brace   (d) tie
3. What is the basic difference in behaviour between tension and compression members, while resisting the loads?
4. Can bolt holes be ignored in the design of compression member? If yes, why?
5. State the parameters that affect the strength of compression members.
6. State the possible failure modes of an axially loaded column.
7. How can local buckling be eliminated in the compression member?
8. What are the limiting width-to-thickness ratios for flange of rolled I-section of compression members based on?
    (a) Plastic section   (b) Semi-compact members
    (c) Slender members   (d) Compact members
9. Why is it better to choose plastic or compact sections for columns?
10. The effective width of an outstand in a rolled I-section under compression is
    (a) Equal to the flange width   (b) Equal to half of flange width
    (c) Equal to half of (flange width—thickness of web)
11. The effective width of an outstand in a rolled angle under compression is
    (a) Equal to the width of the leg   (b) Width of leg—thickness of leg
    (c) Depends on slenderness ratio
12. How does the behaviour of a compression member differ based on its length?
13. Will a short column attain its squash load? Under what conditions?
14. What is the difference in behaviour of a long column and an intermediate length column?
15. What is the critical load of a fixed-fixed column according to Euler's theory?
16. Why is Euler's formula not adopted for practical designs?
17. What are multiple column curves? Why are several curves necessary to determine the strength of compression members?
18. State the formula adopted in the code for predicting the strength of compression members.
19. For rolled I-section, the following curve is used for major axis buckling
    (a) Curve A   (b) Curve B   (c) Curve C
20. Why a curve that will predict less strength is used for the welded I-section?
21. State the curve to be used for the following sections:
    (a) Hot rolled CHS
    (b) Cold rolled RHS

(c) Major axis buckling of welded box section
(d) Built-up section
(e) Channel, angle, T-, and solid sections
22. What is the necessity of using the effective length factor?
23. As per the code, the effective length of a column fixed at both ends is (L = Length of the column)
    (a) 0.50L (b) 0.65L (c) 0.85L (d) L
24. As per the code, the effective length of a cantilever column is
    (a) 1.2L (b) 0.65L (c) 2L (d) 3L
25. The effective buckling length of a strut made of double angles back-to-back connected by bolting is
    (a) 1.2L (b) 0.7–0.85L (c) 0.85L
    (d) 0.7L (e) L
26. State the equation for finding the stability index.
27. What is the method used in the Indian code for calculating the effective length of multi-storey frames?
28. Why is a different formula suggested for single angle struts in the Indian code?
29. What are the steps involved in the design of a compression member?
30. The lacing bars in built-up columns should be designed to resist a total transverse shear equal to
    (a) 2.5% of the column load
    (b) 1.5% of the column load
    (c) 3.5% of the column load
    (d) 2.5% of the column load + additional shear due to B.M. or lateral load
31. What is the main purpose of lacings and battens?
32. The slenderness of lacing should not exceed
    (a) 180   (b) 145   (c) 200   (d) 140
33. State some of the important empirical rules that are to be followed while designing
    (a) Lacing bars   (b) End battens   (c) Intermediate battens
34. Which built-up column is stronger (a) one with lacings (b) one with battens?
35. Why are double lacings with horizontal members not advisable?
36. What are the types of base plates used in practice?
37. State the different steps to be followed while designing a slab base.

CHAPTER 6

Design of Beams

Introduction

Beams are structural members that support loads which are applied transverse to their longitudinal axes. They are assumed to be placed horizontally and subjected to vertical loads. Beams have a far more complex load-carrying action than other structural elements such as trusses and cables.

The load transfer by a beam is primarily by bending and shear. Any structural member could be considered as a beam if the loads cause bending of the member. If a substantial amount of axial load is also present, the member is referred to as a beam-column (beam-columns are covered in Chapter 9). Though some amount of axial effects will be present in any structural member, in several practical situations, the axial effect can be neglected and the member can be treated as a beam.

If we consider a normal building frame, the beams that span between adjacent columns are called *main* or *primary beams/girders*. The beams that transfer the floor loading to the main beams are called *secondary beams/joists*. Beams may also be classified as follows:

1. *Floor beams* A major beam of a floor system usually supporting joists in buildings; a transverse beam in bridge floors
2. *Girder* In buildings, girders are the same as floor beams; also a major beam in any structure. Floor beams are often referred to as girders
3. *Girt* A horizontal member fastened to and spanning between peripheral columns of an industrial building; used to support wall cladding such as corrugated metal sheeting
4. *Joist* A beam supporting floor construction but not a major beam
5. *Lintels* Beam members used to carry wall loads over wall openings for doors, windows, etc.
6. *Purlin* A roof beam, usually supported by roof trusses
7. *Rafter* A roof beam, usually supporting purlins
8. *Spandrels* Exterior beams at the floor level of buildings, which carry part of the floor load and the exterior wall

9. *Stringers* Members used in bridges parallel to the traffic to carry the deck slab; they will be connected by transverse floor beams

Beams may be termed as *simple beams* when the end conditions do not carry any end moments from any continuity developed by the connection. A beam is called a *continuous beam*, when it extends continuously across more than two supports. A *fixed beam* has its ends rigidly connected to other members, so that the moments can be carried across the connection. In steel frames, the term 'fixed end' is somewhat a misnomer, since the ends of rigid connections are not fixed but allow some joint rotation.

Beam loadings will consist of both dead and live loads. A part of the beam dead load is its own weight. When long, heavy members are used, the self weight may be a significant part of the total load. Hence, the beam section should be checked for adequacy for the applied loads, including the beam weight. For the usual structural frame of a steel building, bending effects are predominant in beams with negligible torsional effects.

In order to resist the bending effects in the most optimum way, much of the material in a beam has to be placed as far away as possible from the neutral axis. The web area of the beam has to be adequate for resisting the shear. Usually, maximum bending moment and maximum shear occur at different cross sections. However, in continuous beams, they may occur at the same cross section near the interior supports. Commonly used cross-sectional shapes are the rolled I-sections and the built-up I-sections or box sections, though channel sections are also often used as purlins (Fig. 6.1). Doubly symmetric shapes such as the standard rolled I-shapes are the most efficient.

Generally a beam may be subjected to simple, unsymmetrical, or biaxial bending. Simple bending takes place if the loading plane coincides with one of the principal planes of a doubly symmetric section such as an I-section or, in the case of a singly symmetric open section (C-section), the loading passes through the *shear centré* and is parallel to the principal plane. Unsymmetrical bending occurs if the plane of loading does not pass through the shear centre. In this case, bending is coupled with torsion. If simple bending takes place about both the principal planes without torsion, the beam is said to undergo biaxial bending. A typical example for biaxial bending in a beam is the roof purlin.

Several complications may arise in a beam design due to a variety of reasons, such as complex stress conditions, tendency of the member to buckle, and conflicting design requirements. Complex stress states may arise when loads are inclined to the principal axes, when unsymmetrical sections are used or where large values of shear and bending moment occur at a section. Buckling may take place in many ways: (a) lateral buckling of the whole beam between supports, (b) local buckling of flanges, and (c) longitudinal buckling of the web and buckling in the depth direction under concentrated loads or near the supports. As far as design is concerned, the depth of the beam is a very important parameter. While a large depth is desirable for moment resistance, the resulting thin web may not be able to resist lateral or web buckling.

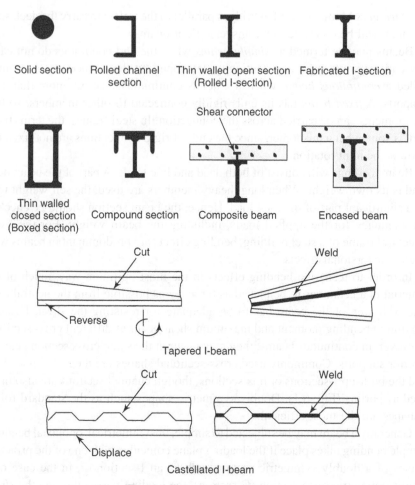

Solid section   Rolled channel section   Thin walled open section (Rolled I-section)   Fabricated I-section

Shear connector

Thin walled closed section (Boxed section)   Compound section   Composite beam   Encased beam

Cut   Weld

Reverse

Tapered I-beam

Cut   Weld

Displace

Castellated I-beam

Fig. 6.1 Types of steel beams

In the design of a beam, two aspects are of primary consideration: (a) strength requirement, that is, the beam has adequate strength to resist the applied bending moments and accompanying shear forces and (b) stability consideration, that is, the member is safe against buckling. Another requirement is adequate bending stiffness, which is often satisfied by members meeting the requirement of strength. Beams develop higher stresses compared to axial members for the same loads and bending deflections are considerably high. Bending effects rather than shear govern the proportions of a rolled beam section, except in cases where shear-to-moment ratio is very large. Occasionally, bending deflections may form the primary design consideration. Shear deflections are usually very small and hence neglected except in the case of deep beams with very short spans.

6.1 Beam Types

There are various forms of beam cross sections used in practice, as shown in Fig. 6.1. The selection of a section would depend upon the use for which it is intended and on the overall economy. The beam chosen should possess required strength and it should not deflect beyond a limit. Members fabricated out of rolled plates are called *plate girders*, which are dealt in detail in the next chapter. Compound sections may also be fabricated out of rolled sections for special purposes. Similarly, in *hybrid sections*, the flanges of higher yield stress compared to the web can also be formed. Certain specific features, such as the passage of services below the floor of a building, may necessitate the use of sections with openings in the web. Other economical types of beams that can be fabricated out of rolled sections are the *castellated* and *tapered beams*. Composite steel beams and concrete-encased steel beams, derive added strength, economy, and in the latter case fire protection also. Table 6.1 gives the important types of steel beams used in practice with their optimum span range.

Table 6.1 Beam types

Type of beam	Optimum span range (m)	Application
Angles	3–6	For lightly loaded beams such as roof purlin and sheeting rail
Rolled I-sections	1–30	Most frequently used as a beam
Castellated beams	6–60	Long spans and light loads
Plate girders	10–100	Long spans with heavy loads such as bridge girders
Box girders	15–200	Long spans and heavy loads such as bridge girders

6.2 Section Classification

The bending strength of a section depends on how well the section performs in bending. When a stocky beam (thick-walled beam) (see Fig. 4.1) is subjected to bending, extreme fibres in the maximum moment region reach the yield stress [see Fig. 4.1(b)]. Upon further loading, more and more fibres reach the yield stress and the stress distribution is as shown in Fig. 4.1(c). The bending moment that causes the whole cross section to reach yield stress [Fig. 4.1(d)], is termed as the plastic moment of the cross section M_p. The cross section is incapable of resisting any additional moment, but may maintain the plastic moment, acting as a plastic hinge, for some more amount of rotation. If the section is thin-walled (slender), it may fail by local buckling, even before reaching the yield stress. The ratio of ultimate rotation to yield rotation is called the rotation capacity. Four classes of section have been identified: (a) plastic (class 1), (b) compact (class 2), (c) semi compact (class 3), and (d) slender (class 4) based on their yield and plastic moments along with their rotation capacities. Refer to Section 4.9 and Fig. 4.11 for more details of this classification.

Local buckling can be prevented by limiting the width–thickness ratio. In the case of rolled sections, higher thickness of the plate is adopted to prevent local buckling. For built-up and cold-formed sections, longitudinal stiffeners are provided to reduce the width into smaller sizes. It is evident that only plastic sections can be used in indeterminate frames which form collapse mechanisms. Compact sections can be used in simply supported beams which fail after reaching M_p at one section. Semi-compact sections can be used for elastic design, where the section fails after reaching M_y at the extreme fibres. Slender sections are not preferred in hot-rolled structural steelwork; but they are extensively used in cold-formed members.

6.3 Lateral Stability of Beams

A beam loaded predominantly in flexure would attain its full moment capacity if the local and lateral instabilities of the beam are prevented. To ensure the first condition, the cross sections of flanges and web of the beam chosen must be *plastic* or *compact*. If significant ductility is required, sections must invariably be plastic. If the laterally unrestrained length of the compression flange of the beam is relatively long, then a phenomenon, known as *lateral buckling* or *lateral torsional buckling* of the beam may take place. The beam would fail well before it attains its full moment capacity. This phenomenon (see Section 4.1.1 for the behaviour of beams, which are laterally restrained) has a close similarity to the Euler buckling of columns, where collapse is triggered before the column attains its squash load (full compressive yield load).

In steel structures, especially in buildings, beams are restrained laterally by the floor decks, which are placed on top of them. During the construction stage, before the floor decks are in place, if the beams are not adequately supported laterally, they may be susceptible to lateral buckling. Therefore, during construction stage, they may need special attention with regard to their lateral stability. If adequate lateral restraints are not provided to beams in the plane of their compression flanges, the beams would buckle laterally resulting in a reduction of their maximum moment capacity. Lateral buckling can be prevented, if adequate restraints are provided to the beam in the plane of the compression flange; such beams are called *laterally restrained beams*.

Beams may also fail by local buckling or local failure (shear yielding of web, local crushing of web or buckling of thin flanges). These may be prevented by providing stiffeners/additional flange plates. It is interesting to explain the phenomenon of lateral buckling. Consider a simply supported, but laterally unsupported (except at the ends of the beam) beam subjected to incremental transverse loads at its mid-section as shown in Fig. 6.2(a).

In the case of axially loaded columns, the deflection takes place sideways and the column is said to buckle in a pure flexural mode. A beam, under transverse loads, has a part of its cross section in compression and the other in tension. The part under compression becomes unstable while the portion under tension tend to stabilize the beam and keep it straight. Thus, beams when loaded exactly in the plane of the web, at a particular load, will fail suddenly by deflecting sideways and then twisting about its longitudinal axis. This form of instability is more complex (compared to column instability which is two-dimensional) since the lateral buckling

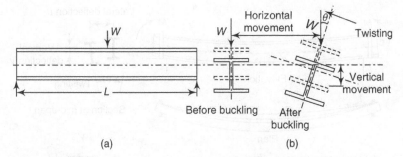

Fig. 6.2 (a) Long span beam (b) Laterally deflected shape of the beam

problem is three-dimensional in nature. It involves coupled lateral deflection and twist. Thus, when the beam deflects laterally, the applied moment exerts a torque about the deflected longitudinal axis, which causes the beam to twist. The bending moment at which a beam fails by lateral buckling when subjected to a uniform end moment is called its *elastic critical moment* (M_{cr}). In the case of lateral buckling of beams, the elastic buckling load provides a close upper limit to the load carrying capacity of the beam. Lateral instability occurs only if the following two conditions are satisfied.

- The section possesses different stiffness in the two principal planes.
- The applied loading induces bending in the stiffer plane (about the major axis).

Similar to columns, the lateral buckling of unrestrained beams is also a function of its slenderness. Several factors influence the buckling capacity of beams. Some of them are cross-sectional shape of the beam, type of support, type of loading and position of the applied load.

6.3.1 Lateral Torsional Buckling of Symmetric Sections

As explained earlier, when a beam fails by lateral torsional buckling, it buckles about its weak axis, even though it is loaded in the strong plane as shown in Figs. 6.3(a) and (b)].

Let us consider the lateral torsional buckling of an ideal I-section with the following assumptions.

1. The beam has no initial imperfections.
2. Its behaviour is elastic.
3. It is loaded by equal and opposite end moments in the plane of the web.
4. The beam does not have residual stresses.
5. Its ends are simply supported vertically and laterally.

Note, in practice, the above ideal conditions are seldom met.

The differential equation for the angle of twist for the beam shown in Fig. 6.3 is

$$EI_w \frac{d^4\phi}{dx^4} - GI_t \frac{d^2\phi}{dx^2} - \frac{M^2}{EI_y}\phi = 0$$

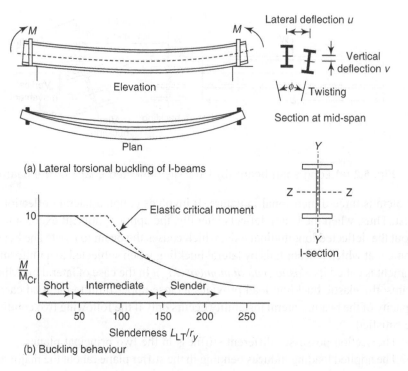

Fig. 6.3 Lateral torsional buckling of I-section beams

The solution to this differential equation is given by (Bleich 1952; Timoshenko & Gere 1961; Salmon & Johnson 1996)

$$M_{cr} = (\pi/L)\sqrt{EI_yGI_t + \left(\frac{\pi E}{L}\right)^2 I_wI_y} \qquad (6.1a)$$

where EI_y is the minor axis flexural rigidity, GI_t is the torsional rigidity, and EI_w is the warping rigidity.

Equation (6.1a) may be rewritten as

$$M_{cr} = \frac{\pi}{L}(EI_yGI_t)^{0.5}\left[1 + \frac{\pi^2 EI_w}{L^2 GI_t}\right]^{0.5} \qquad (6.1b)$$

The cross-sectional shape is a particularly important parameter in assessing the lateral buckling capacity of a beam. Conversely, the problem of lateral instability can be minimized or even eliminated by a judicious choice of the section as shown in Table 6.2. Although the five cross sections shown are having the same cross-sectional area, the values of their flexural and torsional properties relative to those of the square cross section show considerable variation. It is observed from Table 6.2 that the flat and deep I-sections, which have largest in-plane bending stiffness, have less buckling strength compared to the box section (Kirby & Nethercot 1979).

Table 6.2 Types of cross-section and their relative values of section properties (Kirby & Nethercot 1979)

Section type	Square	Flat	Shallow beam (typical)	Deep beam (typical)	RHS (typical)
Section properties					
A	1	1	1	1	1
I_z	1	25	12.45	45.59	16.94
I_y	1	0.04	3.20	3.20	8.10
I_t	1	0.04	0.034	0.033	4.731

For a beam with central load acting at the level of centroidal axis, the critical buckling moment is given by

$$M_{cr} = \frac{4.24}{L} \sqrt{(EI_y GI_t)} \sqrt{[1 + \pi^2 EI_w/(L^2 GI_t)]} \tag{6.2}$$

The ratio of the two constants $4.24/\pi = 1.35$ is often termed as 'equivalent uniform moment factor' C_1. Its value is a direct measure of the severity of a particular pattern of moments relative to the basic case given in Eqn (6.1). Taking into account C_1, Eqn (6.1) may be rewritten as

$$M_{cr} = C_1 (EI_y GI_t)^{1/2} \left[\frac{\pi}{L} \left(1 + \frac{\pi^2}{B^2} \right)^{1/2} \right] \tag{6.3}$$

where $B^2 = L^2 G I_t/EI_w$ or

$$M_{cr} = C_1 (EI_y GI_t)^{1/2} \gamma \tag{6.4}$$

where $\gamma = \pi/\ell (1 + \pi^2/B^2)^{1/2}$

Equation (6.4) is a product of three terms: the first term C_1 varies with the loading and support conditions; the second term varies with the material properties and the shape of the beam; and the third term γ varies with the length of the beam. This equation is regarded as the basic equation for lateral torsional buckling of beams.

Lateral Torsional Failure of Marcy Bridge

A pedestrian footbridge near Marcy, New York, collapsed during construction on October 12, 2002, killing one and injuring nine workers. The bridge was designed as a single-span composite box girder bridge consisting of a trapezoidal steel tub girder with stay-in-place forms and a concrete deck. The Marcy Bridge was designed to span approximately 52 m across an extension of the Utica-Rome expressway (NYS Route 49). At the time of collapse, the concrete deck was being placed, starting at the north end and moving south, using an automatic screeding machine to distribute the concrete. Eyewitnesses reported that the bridge had been noticeably 'bouncy' while the concrete was placed, and when the concrete placement had progressed to about mid-span, the bridge suddenly twisted and rolled to the east, tipping off the abutments.

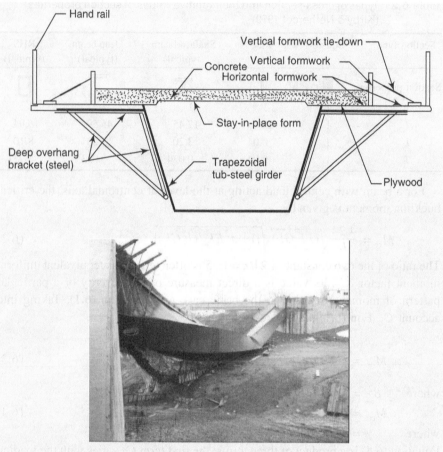

Fig. CS1 Cross-section of Marcy Bridge and the failure due to lateral-torsional buckling (source: Corr, et al, 2009)

Prior to and during the placement of the concrete deck, the tub cross-section was closed only by stay-in-place forms, which were nominally connected to closure angles on each top flange and to each other by using self-tapping screws (see Fig. CS1). The stay-in-place forms do not provide sufficient torsional strength or stiffness, thus the section is open and torsionally flexible. Computer analysis conducted by Exponent Engineering and Scientific Consulting confirmed that the tub girder did not have sufficient strength to carry the wet concrete without collapse. The mode of failure was considered as lateral-torsional buckling of the entire girder cross-section. Such bridges are particularly susceptible to this type of failure, because they are very flexible in a twisting mode before the concrete deck hardens. (After curing, the concrete deck would have provided a closed section with increased torsional stiffness and would have prevented the lateral torsional buckling failure.) The standard practice at that time, as recorded in bridge design specifications and national design standards, did not recognize this failure mode, and there were no requirements in those codes or standards to provide bracing, falsework, or other measures to prevent such a collapse. After the collapse, and a very similar accident in Sweden that occurred just months before this, bridge design standards in New York and the American Association of State Highway and Transportation Officials (AASHTO) were updated with provisions intended to prevent further accidents. Now these codes explicitly require diagonal bracing between the top-flanges in order to suppress this buckling mode.

Reference:

Corr, D.J., McCann, D.M., and McDonald, B.M., Lessons Learned from Marcy Bridge Collapse, Forensic Engineering, 2009, pp. 395–403.

6.4 Effective Length

The concept of effective length of the compression flange incorporates the various types of support conditions. For the beam with simply supported end conditions and no intermediate lateral restraint, the effective length is equal to the actual length between the supports. When a greater amount of lateral and torsional restraints is provided at the supports, the effective length is less than the actual length and alternatively, the length becomes more when there is less restraint. The effective length factor would indirectly account for the increased lateral and torsional rigidities provided by the restraints.

Thus, there are two K factors, K_b (restraint against lateral bending) and K_w (restraint against warping). The values of K_b and K_w are 1.0 for free bending and free warping (simply supported) and 0.5 for bending and warping prevented (fixed) support condition (Kirby & Nethercot 1979). The values of K_b and K_w vary with the proportions of beams and the accurate assessment of the degree of restraint provided by practical forms of connection is difficult. Hence in the code, this problem is treated in an approximate way by considering $K_b = K_w$ and treating the effective length as $L = KL$ (Kirby & Nethercot 1979). The effective lengths KL of the compression flange for different end restraints according to IS 800 : 2007 are given in Table 6.3 (see also Table 15 of the code, which is based on BS 5950–1 : 2000).

For cantilevers, the most severe loading condition is the point load acting at the tip. Nethercot (1983) has shown that for most applications, the simple effective length method (similar to that used for struts) is satisfactory and the critical buckling moment is given by Eqn. (6.16) with L replaced by KL.

Table 6.3 Effective length of compression flanges

Effective length KL for beams, between supports	
Condition at supports (see the following figures)	Effective length, KL
Compression flange at the ends unrestrained against lateral bending (free to rotate in plan)	$L*$
Compression flange partially restrained against lateral bending (Partially free to rotate in plane at the bearings)	$0.85L*$
Compression flange restrained fully against lateral bending (rotation fully restrained in plan)	$0.7\ L*$

*When the ends of the beam are not restrained against torsion, or where the loading condition is destabilizing or when flanges are free to move laterally, these values have to be increased as per Table 15 of the code.

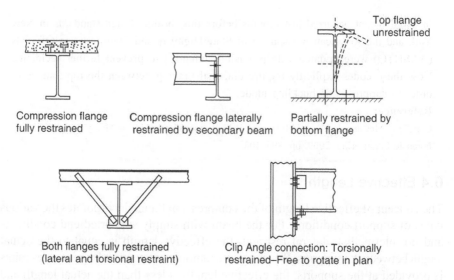

Compression flange
fully restrained

Compression flange laterally
restrained by secondary beam

Partially restrained by
bottom flange

Top flange
unrestrained

Both flanges fully restrained
(lateral and torsional restraint)

Clip Angle connection: Torsionally
restrained–Free to rotate in plan

The effective length factors for various restraint conditions at the tip and at the root of the cantilever are given in Table 16 of the code.

6.4.1 Intermediate Braces

Provision of intermediate lateral supports (bracings) can increase the lateral stability of a beam. For the bracings to be effective, the designer should make certain that the braces themselves are prevented from moving in their axial direction. For example, for the beam shown in Fig. 6.4(a), the lateral buckling of the beams will not be prevented by the braces, and hence the unbraced length should be taken as L. To prevent this type of system-buckling, the bracings should be either anchored into a wall [Fig. 6.4(b)] or provided with diagonal bracings, which effectively transfer the loads to the columns [Fig. 6.4(c)]. It has to be noted that the diagonal bracing need not be provided in all the bays [see Fig. 6.4(d)]. Generally, even a light bracing has the ability to provide substantial increase in stability. Lateral bracing may be either discrete (e.g., cross beams) or continuous [(e.g., beam encased in concrete floors).

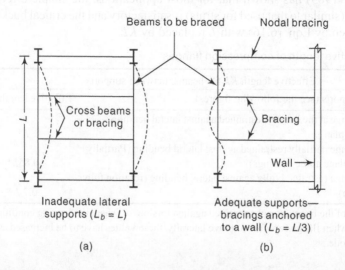

Beams to be braced   Nodal bracing

Cross beams
or bracing

Bracing

Wall →

Inadequate lateral
supports ($L_b = L$)

Adequate supports—
bracings anchored
to a wall ($L_b = L/3$)

(a)   (b)

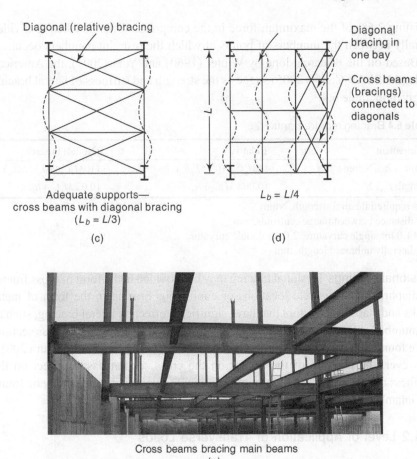

Diagonal (relative) bracing

Adequate supports—
cross beams with diagonal bracing
($L_b = L/3$)

(c)

Diagonal
bracing in
one bay

Cross beams
(bracings)
connected to
diagonals

$L_b = L/4$

(d)

Cross beams bracing main beams
(e)

Fig. 6.4 Lateral bracing systems for roof and floor beams

It is also important to check the adequacy of beams, during the intermediate stages of construction, as the amount of bracing provided during these stages may be low.

For bracing to be effective, it has to be provided in the compression flange. Bracing provided below the point of application of the transverse load would not be able to resist the twisting and hence the full capacity of the beam cannot be achieved (Kirby & Nethercot 1979). Similarly lateral bracing system, attached near the beam centroid are ineffective. The two requirements for effective bracing are as follows.

- It should have sufficient stiffness so that buckling of the beam occurs in between the braces.
- It should have sufficient strength to withstand the force transferred to it by the beam.

The effective length of the compression flange is the maximum distance centre-to-centre of the restrained members. In the code only the strength aspect is covered. Thus it states that the effective lateral restraints for a beam should be capable of resisting 2.5% of the maximum force in the compression flange taken as divided equally between the numbers of points at which the restraint members occur.

Based on the research done by Winter (1960) and Yura (2001), the American code (ANSI/AISC: 360-2005) suggests the strength and stiffness of lateral bracing as given in Table 6.4.

Table 6.4 Bracing requirements

Requirement	Relative	(Nodal) Discrete
Stiffness k_{br}, N/mm	$(4M_uC_d)/(0.75L_bh_o)$	$(10M_uC_d)/(0.75L_bh_o)$
Strength F_{br}, N	$(0.008\ M_uC_d)/h_o$	$(0.02\ M_uC_d)/h_o$

M_u = required flexural strength, Nmm
h_o = distance between flange centroids, mm
C_d = 1.0 for single curvature, 2.0 for double curvature
L_b = laterally unbraced length, mm

Torsional bracings Torsional bracing may be provided in the form of cross frames or diaphragms at discrete locations or continuous bracing in the form of metal decks and slabs. The factors that have significant effect on lateral bracing, such as the number of bracings, top flange loading and brace location on the cross section, were found to be relatively unimportant while sizing a torsional brace (Yura 2001). However, the position of the braces on the cross section has an effect on the stiffness of the brace itself. More information on torsional bracings may be found in Galambos (1998).

6.4.2 Level of Application of Transverse Loads

The lateral stability of a transversely loaded beam is dependent on the arrangement of the loads as well as the level of application of the loads with respect to the centroid of the cross section. IS 800 takes into account the destabilising effect of the top flange loading by using a notional effective length of 1.2 times the actual span to be used in the calculation of effective length (see Table 15 of the code.)

6.4.3 Influence of Type of Loading

To take into account the non-uniform loading, a simple modifier is applied to Eqn. (6.1) in many codes (Salvadori, 1956) as given below

$$M_{cr} = C_1 M_{ocr} \tag{6.5}$$

where M_{ocr} is the critical moment obtained from Eqn. (6.1) and C_1 is the equivalent uniform moment factor. Various lower-bound formulae have been proposed for C_1.

The Indian code gives various values of C_1 in a table in Annex E for different moment diagrams. This table is based on the Eurocode 3. The values of C_1 given in Table Annex E of the code for end moment loading maty be approximated by the following equation (Gardner & and Nethercot 2005) when $k = 1$

$$C_1 = 1.88 - 1.40\psi + 0.52\psi^2 \text{ but } C_1 \le 2.70 \tag{6.6}$$

where ψ is the ratio of the end moments defined in Table 42 of the code, It should be noted that the Table 42 given in the Indian code considers the end rotational restraint against lateral bending K and suggests that Kw values be taken as equal to unity.

The Wind-induced Collapse of Tacoma Narrows Bridge

The original Tacoma Narrows Bridge opened on July 1, 1940, and dramatically collapsed into Puget Sound on November 7 of the same year. This suspension bridge spanned the Tacoma Narrows strait between Tacoma and the Kitsap Peninsula, in the United States. At the time of its construction (and its destruction), Galloping Gertie (nick named due to its oscillations) was the third longest suspension bridge in the world, behind the Golden Gate Bridge and the George Washington Bridge. 'Longest' is a comparison of the main spans in suspension bridges. It had a total length of 1,810.2 m with a central (longest) span of 853.4 m.

Collapse of the Tacoma Narrows Bridge, (Source: Smithsonian Institution)

The failure of the bridge occurred due to the twisting of the bridge deck in mild winds of about 64 km/h. This failure mode is termed as the torsional vibration mode (which is different from the transversal or longitudinal vibration mode). Thus, when the left side of the roadway went down, the right side would rise, and vice versa, with the center line of the road remaining still. Specifically, it was the 'second' torsional mode. In fact, two men proved this point by walking along the center line, unaffected by the flapping of the roadway rising and falling to each side. This vibration was caused by aeroelastic fluttering (a phenomenon in which aerodynamic forces on an object couple with a structure's natural mode of vibration to produce rapid periodic motion). The wind-induced collapse occurred on November 7, 1940. The collapse of the bridge was filmed by Mr Barney Elliott (who was travelling on the bridge), which helped engineers to study the behaviour of the bridge.

Suspension bridges consist of cables anchored in heavy foundations at their ends and supported by towers at intermediate points. From these cables, the deck is suspended. Thus, they are more flexible than other types of bridges, and require bracings to reduce vertical and torsional motions. In the Tacoma Narrows bridge, instead of the usual deep open trusses, narrow and shallow solid I-beams were used in the decks, which resulted in the build-up of wind loads.

Fig. CS2 Solid I-beams in the deck of the original bridge and trussed deck of the reconstructed bridge (www.failurebydesign.info)

The bridge collapse boosted research in the field of bridge aerodynamics which resulted in better designs. After the collapse, two bridges were constructed in the same general location. The first one, now called the Tacoma Westbound Bridge, is 1822 m long –12 m longer than the Galloping Gertie. The second one, the Tacoma Eastbound Bridge, opened in 2007.

Reference: http://en.wikipedia.org/wiki/Tacoma_Narrows_Bridge_%281940%29

6.5 Buckling of Real Beams

The theoretical assumptions made in Section 6.3 are generally not realised in practice. In order to understand the behaviour of real beams, it is necessary to consider the combined effects of instability and plasticity and also the role of factors such as residual stress and geometrical imperfections. When all these effects are considered, it may be found that slender beams fail more or less elastically by excessive deformation at loads that are close to M_{cr}, beams of intermediate slenderness fail inelastically by excessive lateral deformation and stocky beams will attain M_p with negligible lateral deformation. This kind of behaviour is shown in Fig. 6.5. The behaviour of beams in bending has already been explained in Section 4.1.1.

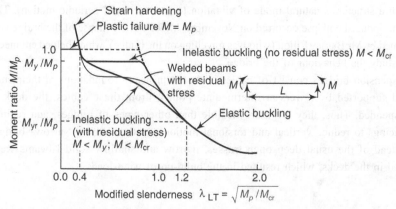

Fig. 6.5 Interaction between instability and plasticity

6.6 Design Strength of Laterally Supported Beams in Bending

For laterally supported beams, the factored design moment M at any section in a beam, due to external actions, satisfy the relationship $M < M_d$, where M_d is the design bending strength of the section. The design bending strength of a laterally supported beam is governed by the yield stress. For a laterally unsupported beam, the design strength is most often controlled by the lateral torsional buckling strength. The above relationship is obtained with the assumption that the beam web is stocky. When the flanges are plastic, compact, or semi-compact but the web is slender (i.e., $d/t_w > 67\varepsilon$), the design bending strength may be calculated using one of the following methods.

- The flanges resist the bending moment and the axial force acting on the section and the web resists only the shear.
- The whole section resists the bending moment and the axial force and therefore the web has to be designed for combined shear and its share of normal stresses. This is done by using simple elastic theory in the case of semi-compact webs and simple plastic theory in the case of compact and plastic webs.

Shear force does not have any influence on the bending moment for values of shear up to $0.6V_d$ (called the low shear load), where V_d is the design shear strength. The provision for determining V_d is explained a little later. When the design shear force V is less than $0.6V_d$, the design bending strength M_d will be taken as

$$M_d = 0.909\beta_b\, Z_p f_y \leq 1.09 Z_e f_y \leq 1.36\, Z_e f_y \text{ (for cantilevers)} \qquad (6.7a)$$

where $\beta_b = 1.0$, for plastic and compact sections and $\beta_b = Z_e/Z_p$, for semi-compact sections.

Thus, for class 3 semi-compact sections,

$$M_d = 0.909 f_y Z_e \qquad (6.7b)$$

For class 4 slender sections, $M_d = f_y' Z_e$, where f_y' is the reduced design strength for slender sections. Z_p and Z_e are the plastic and elastic section moduli of the cross section, respectively, and f_y is the yield stress of the material.

The additional check ($M_d < 1.09 Z_e f_y$) is provided to prevent the onset of plasticity under unfactored dead, imposed, and wind loads. For most of the I-beams and channels given in IS 808, Z_{pz}/Z_{ez} is less than 1.2 and hence the plastic moment capacity governs the design. For sections where $Z_{pz}/Z_{ez} > 1.2$, the constant 1.2 may be replaced by the ratio of factored load/unfactored load (γ_f). Thus the limitation $1.2 Z_e f_y$ is purely notional and becomes in practice $\gamma_f Z_e f_y$ (Morris & Plum 1996).

When the design shear force (factored) V exceeds $0.6V_d$ (called the high shear load), where V_d is the design shear strength of the cross section, the design bending strength M_d will be taken as

$$M_d = M_{dv}$$

where M_{dv} is the design bending strength under high shear and it is calculated as follows:

(a) *Plastic or Compact Section*

As the shear force V is increased from zero, no reduction in plastic moment is assumed below the value of M_p until V reaches $0.5V_p$, where V_p is the shear strength of the web given by $A_v f_y$, where A_v is the shear area taken as Dt_w for rolled sections and dt_w for built up sections ($d = D - 2t_f$). The design shear strength f_v is taken as $0.6f_y$, where f_y is the design strength in tension or compression, i.e., slightly greater than the true von Mises value of $f_y/\sqrt{3}$. When the full capacity in shear V_p is reached, the shear area is assumed to be completely ineffective in resisting the moment, and hence the reduced plastic moment M_{dv} becomes M_{fd}, where $M_{fd} = M_p - (D^2t_w/4)f_y$ for rolled sections and $M_{fd} = M_p - (d^2t_w/4)f_y$ for built-up sections. Between $V = 0.5V_p$ and V_p, M_{dv} is assumed to reduce for plastic and compact sections according to the following equation

$$M_{dv} = M_p - \beta (M_p - M_{fd}) \leq 1.09 \, Z_e f_y \qquad (6.8a)$$

where $\beta = (2V/V_p - 1)^2$ $\qquad\qquad\qquad\qquad\qquad\qquad\qquad\qquad$ (6.8b)

In Eqn (6.8), M_p is the design plastic moment of the whole section disregarding high shear force effect, but considering web buckling effects, V is the factored applied shear force, V_p is the design shear strength as governed by web yielding or web buckling $= 0.6Dt_w f_y$, and M_{fd} is the plastic design strength of the area of the cross section excluding the shear area, considering partial safety factor γ_{m0}.

(b) *Semi-compact section*

$$M_{dv} = 0.909 \, Z_e f_y \qquad\qquad\qquad\qquad\qquad\qquad\qquad (6.8c)$$

where Z_e is the elastic section modulus of the whole section.

6.6.1 Holes in the Tension Zone

The effect of holes in the tension flange, on the design bending strength need not be considered if

$$(A_{nf}/A_{gf}) \geq 1.26(f_y/f_u) \qquad\qquad\qquad\qquad\qquad\qquad (6.9)$$

where A_{nf}/A_{gf} is the ratio of net and gross area of the flange and f_y/f_u is the ratio of yield and ultimate strength of the material.

When A_{nf}/A_{gf} does not satisfy Eqn (6.9), the reduced flange area A_{nf} satisfying Eqn (6.9) may be taken as the effective flange area in tension (Dexter and Altstadt 2003).

The effect of holes in the tension region of the web on the design flexural strength need not be considered, if the limit given in Eqn (6.9) is satisfied for the complete tension zone of the cross section, comprising the tension flange and tension region of the web. Fastener holes in the compression zone of the cross section need not be considered in the design bending strength calculation, except for oversize and slotted holes, or holes without any fastener.

6.6.2 Shear Lag Effects

The simple theory of bending is based on the assumption that plane sections remain plane after bending. In reality, shear strains cause the sections to warp. The effect in the flange is to modify the bending stresses obtained by simple bending theory. Thus higher stresses are produced near the junction of a web and lower stresses at points away from the web as shown in Fig. 6.6. This phenomenon is known as *shear lag*. It results in a non-uniform stress distribution across the width of the flange. The shear lag effects are minimal in rolled sections, which have relatively narrow and thick flanges. For normal dimensions of the flanges, the effects are negligible. But if the flanges are unusually wide, (as in plate girders or box girders), these shear strains influence the normal bending stresses in the flanges.

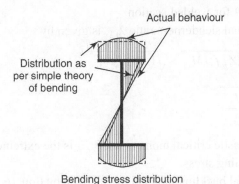

Fig. 6.6 Shear lag effects for an I-section

The shear lag effects in flanges may be disregarded provided, for outstand elements (supported along one edge), $b_o \leq L_0/20$ and for internal elements (supported along two edges), $b_i \leq L_0/10$ where L_0 is the length between points of zero moment (inflection) in the span, b_o is the width of the outstand, and b_i is the width of an internal element.

When these limits are exceeded, the effective width of the flange for design strength may be taken conservatively as the values satisfying the limits given above.

6.7 Design Strength of Laterally Unsupported Beams

The effect of lateral torsional buckling on flexural strength need not be considered when $\lambda_{LT} \leq 0.4$ (see Fig. 6.5) where λ_{LT} is the non-dimensional slenderness ratio for lateral torsional buckling. The design bending strength of a laterally unsupported beam as governed by lateral torsional buckling as per clause 8.2.2 of the Indian code (IS 800 : 2007) is given by

$$M_d = \beta_b \, Z_p \, f_{bd} \qquad (6.10)$$

with

$\beta_b = 1.0$ for plastic and compact sections

$\quad = Z_e/Z_p$ for semi-compact sections

where Z_p and Z_e are the plastic section modulus and elastic section modulus with respect to extreme compression fibre and f_{bd} is the design bending compressive stress. The design bending compressive stress is given by

$$f_{bd} = 0.909 \, \chi_{LT} \, f_y \qquad (6.11a)$$

where χ_{LT} is the reduction factor to account for lateral torsional buckling given by

$$\chi_{LT} = \frac{1}{\phi_{LT} + (\phi_{LT}^2 - \lambda_{LT}^2)^{0.5}} \leq 1.0 \qquad (6.11b)$$

in which $\phi_{LT} = 0.5[1 + \alpha_{LT}(\lambda_{LT} - 0.2) + \lambda_{LT}^2]$.

The values of imperfection factor α_{LT} for lateral torsional buckling of beams is given by

$\alpha_{LT} = 0.21$ for rolled section

$\alpha_{LT} = 0.49$ for welded section

The non-dimensional slenderness ratio λ_{LT}, is given by

$$\lambda_{LT} = \sqrt{\beta_b Z_p f_y / M_{cr}} \leq \sqrt{1.2 Z_e f_y / M_{cr}} \tag{6.12}$$

$$= \sqrt{\frac{f_y}{f_{cr,b}}}$$

where M_{cr} is the elastic critical moment and $f_{cr, b}$ is the extreme fibre compressive elastic lateral buckling stress.

The elastic lateral buckling moment M_{cr} is given by Eqn. (6.1a) with L replaced by KL. Kerensky et al. (1956) simplified Eqn (6.1a) by introducing the following approximations for doubly symmetric sections.

$I_y = b_f^3 t_f / 6$, $I_w = I_y h^2 / 4$, $I_t = 0.9\, b_f t_f^3$, $b_f = 4.2 r_y$ and $E = 2.6G$

Thus, Eqn (6.1a) is reduced to

$$M_{cr} = \frac{\beta_{LT} \pi^2 E I_y h}{2(KL)^2} \left[1 + \frac{1}{20} \left[\frac{KL/r_y}{h/t_f} \right]^2 \right]^{0.5} \tag{6.13}$$

$$= \beta_b\, Z_e f_{cr, b}$$

where I_t is the torsional constant, I_w is the warping constant, I_y is the moment of inertia about the weak axis, r_y is the radius of gyration of the section about the weak axis, KL is the effective laterally unsupported length of the member, h is the overall depth of the section, t_f is the thickness of the flange, and $\beta_{LT} = 1.20$ for plastic and compact sections with $t_f / t_w \leq 2.0$ and 1.00 for semi-compact sections or sections with $t_f / t_w > 2.0$.

Using the same approximations, the extreme fibre compressive elastic buckling stress may be obtained as (with $E = 2.0 \times 10^5$ MPa and $Z_z = 1.1 b_f d_f h$)

$$f_{cr, b} = \left(\frac{1625}{KL/r_y} \right)^2 \left[1 + \frac{1}{20} \left[\frac{KL/r_y}{h/t_f} \right]^2 \right]^{0.5} \tag{6.14}$$

IS 800 : 2007 uses a similar expression but the coefficient of 1625 is replaced by 1473.5. Table 6.5 gives the value of $f_{cr, b}$ for various values of KL/r_y and h/t_f based on IS 800 : 2007. Table 6.6 gives the values of design bending compressive strength corresponding to lateral torsional buckling [based on Eqn (6.11)] for $\alpha_{LT} = 0.21$ and $\alpha_{LT} = 0.49$ for $f_y = 250$ MPa. Intermediate values may be obtained by interpolating the values given in these tables.

Table 6.5 Critical Stress $f_{cr,b}$ (in N/mm²)

KL/r_y															
							h/t_f								
	8	10	12	14	16	18	20	25	30	35	40	50	60	80	100
10	22545.42	22249.41	22086.95	21988.41	21924.22	21880.10	21848.49	21799.88	21773.43	21757.47	21747.10	21734.90	21728.27	21721.68	21718.63
20	6218.90	5946.41	5793.01	5698.53	5636.36	5593.33	5562.35	5514.47	5488.28	5472.43	5462.12	5449.97	5443.36	5436.78	5433.73
30	3148.51	2905.13	2763.96	2675.23	2616.04	2574.67	2544.67	2497.92	2472.16	2456.49	2446.27	2434.19	2427.61	2421.04	2418.00
40	2035.61	1820.71	1692.57	1610.39	1554.72	1515.38	1486.60	1441.31	1416.11	1400.69	1390.59	1378.62	1372.07	1365.53	1362.49
50	1492.54	1302.79	1187.08	1111.50	1059.56	1022.43	995.02	951.42	926.88	911.76	901.82	889.98	883.48	876.97	873.94
60	1177.68	1009.25	904.72	835.39	787.13	752.25	726.28	684.51	660.71	645.94	636.17	624.48	618.04	611.57	608.55
70	973.68	823.07	728.32	664.69	619.91	587.23	562.70	522.81	499.81	485.42	475.85	464.33	457.96	451.53	448.52
80	831.04	695.29	609.01	550.48	508.90	478.32	455.18	417.18	395.00	381.01	371.65	360.33	354.03	347.65	344.65
90	725.66	602.40	523.41	469.41	430.74	402.10	380.28	344.13	322.79	309.22	300.08	288.96	282.74	276.42	273.44
100	644.57	531.86	459.18	409.17	373.13	346.27	325.70	291.31	270.81	257.66	248.76	237.86	231.72	225.45	222.49
110	580.18	476.47	409.26	362.77	329.09	303.86	284.44	251.74	232.05	219.33	210.66	199.99	193.94	187.74	184.80
120	527.75	431.79	369.35	325.97	294.42	270.67	252.31	221.20	202.30	190.00	181.57	171.13	165.18	159.04	156.12
130	484.20	394.96	336.71	296.10	266.45	244.05	226.67	197.04	178.90	167.01	158.82	148.62	142.77	136.70	133.80
140	447.43	364.07	309.51	271.36	243.42	222.25	205.77	177.53	160.11	148.63	140.68	130.70	124.95	118.96	116.08
150	415.95	337.76	286.48	250.53	224.14	204.08	188.43	161.48	144.75	133.66	125.94	116.21	110.56	104.64	101.79
160	388.68	315.08	266.72	232.76	207.76	188.72	173.82	148.08	132.01	121.29	113.79	104.29	98.75	92.91	90.08
170	364.82	295.32	249.58	217.40	193.67	175.56	161.36	136.73	121.28	110.92	103.65	94.38	88.94	83.18	80.38
180	343.76	277.93	234.56	203.99	181.42	164.16	150.60	127.01	112.14	102.13	95.07	86.03	80.70	75.02	72.24
190	325.04	262.52	221.28	192.18	170.66	154.18	141.22	118.60	104.27	94.60	87.74	78.93	73.70	68.10	65.35
200	308.27	248.76	209.46	181.70	161.14	145.38	132.97	111.25	97.44	88.08	81.42	72.83	67.70	62.19	59.46
210	293.17	236.39	198.86	172.33	152.66	137.56	125.65	104.77	91.45	82.39	75.93	67.55	62.52	57.09	54.39
220	279.49	225.21	189.30	163.90	145.04	130.56	119.12	99.02	86.16	77.39	71.11	62.93	58.01	52.67	50.00
230	267.04	215.05	180.64	156.27	138.17	124.25	113.25	93.89	81.46	72.96	66.86	58.88	54.06	48.80	46.16
240	255.67	205.78	172.75	149.34	131.94	118.54	107.95	89.27	77.25	69.01	63.08	55.30	50.58	45.39	42.78
250	245.23	197.29	165.53	143.00	126.25	113.35	103.13	85.10	73.47	65.47	59.70	52.11	47.48	42.38	39.80

Table 6.6 Design Bending Stress f_{bd} (in N/mm²) for $f_y = 250$ MPa

f_{cr} N/mm²	λ_{LT}	ϕ_{LT}	χ_{LT}	f_{bd} N/mm²	ϕ_{LT}	χ_{LT}	f_{bd} N/mm²
		$\alpha = 0.21$			$\alpha = 0.49$		
10000	0.1581	0.508	1.000	227.27	0.502	1.000	227.27
9000	0.1667	0.510	1.000	227.27	0.506	1.000	227.27
8000	0.1768	0.513	1.000	227.27	0.510	1.000	227.27
7000	0.1890	0.517	1.000	227.27	0.515	1.000	227.27
6000	0.2041	0.521	0.999	227.07	0.522	0.998	226.79
5000	0.2236	0.527	0.995	226.09	0.531	0.988	224.54
4000	0.2500	0.537	0.989	224.76	0.544	0.975	221.49
3000	0.2887	0.551	0.980	222.76	0.563	0.955	217.03
2000	0.3536	0.579	0.965	219.23	0.600	0.922	209.46
1000	0.5000	0.657	0.924	210.06	0.699	0.843	191.59
900	0.5270	0.673	0.916	208.10	0.719	0.828	188.12
800	0.5590	0.694	0.905	205.65	0.744	0.809	183.95
700	0.5976	0.720	0.891	202.48	0.776	0.787	178.82
600	0.6455	0.755	0.872	198.16	0.817	0.758	172.30
500	0.7071	0.803	0.844	191.90	0.874	0.720	163.70
450	0.7454	0.835	0.825	187.59	0.911	0.696	158.28
300	0.9129	0.992	0.725	164.87	1.091	0.592	134.53
150	1.2910	1.448	0.475	108.05	1.601	0.393	89.24
90	1.6667	2.043	0.310	70.49	2.248	0.266	60.49
80	1.7678	2.227	0.279	63.45	2.447	0.242	54.92
70	1.8898	2.463	0.247	56.22	2.700	0.216	49.11
60	2.0412	2.777	0.215	48.78	3.034	0.189	43.05
50	2.2361	3.214	0.181	41.16	3.499	0.162	36.72
40	2.5000	3.867	0.147	33.34	4.189	0.132	30.11
30	2.8868	4.949	0.112	25.34	5.325	0.102	23.19
20	3.5355	7.100	0.075	17.14	7.567	0.070	15.94
10	5.0000	13.504	0.038	8.73	14.176	0.036	8.28

6.7.1 Elastic Critical Moment of a Section Symmetrical about Minor Axis

In the case of a beam which is symmetrical only about the minor axis (see Fig. 6.7) with bending about major axis, the elastic critical moment for lateral torsional buckling can be calculated by the equation given in Section E-1.2 (Annexe) of the code, which also gives the values of c_1, c_2, and c_3 which are factors to take into account the loading and end restraint conditions.

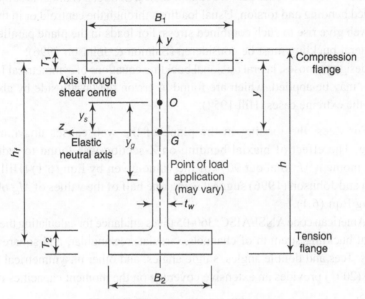

Fig. 6.7 Monosymmetric I-beams

The effective length factors K and K_w vary from 0.5 for full fixity (against warping) to 1.0 for free (to warp) case and 0.7 for the case of one end fixed and other end free. It is analogous to the effective length factors for compression members with end rotational restraint.

The K_w factor refers to the warping restraint. Unless special provisions to restrain warping of the section at the end lateral supports are made, K_w should be taken as 1.0.

The torsion constant I_t is given by

$$I_t = \sum b_i t_i^3 / 3 \quad \text{for open sections} \tag{6.15a}$$

$$= \frac{4 A_e^2}{\sum (b/t)} \quad \text{for hollow sections} \tag{6.15b}$$

where A_e is the area enclosed by the section and b and t are the breadth and thickness of the elements of the section respectively.

The warping constant I_w, for I-sections mono-symmetric about the weak axis, is given by

$$I_w = (1 - \beta_f)\beta_f I_y h_f^2 \tag{6.16}$$

$= 0$ for angle, T-, narrow rectangle sections, and approximately for hollow sections

$$\beta_f = I_{fc} / (I_{fc} + I_{ft}) \tag{6.17}$$

where I_{fc} and I_{ft} are the moment of inertia of the compression and tension flanges, respectively, about the minor axis of the entire section. Note that for equal flange beams $\beta_f = 0.5$.

6.7.2 Beams with Other Cross Sections

Channels Unless the loads pass through the shear centre, a channel is subjected to combined bending and torsion. Usual loadings through the centroid or in the plane of the web give rise to such combined stress. For loads in the plane parallel to the web, lateral buckling must be considered (Salmon & Johnson 1996).

For design purposes, lateral torsional buckling equations of symmetrical I-shaped sections may be applied, which are found to err on the unsafe side by about 6% only in the extreme cases (Hill 1954).

Zees For Zees, the loading in the plane of the web causes unsymmetrical bending. The effect of biaxial bending on Z-section was found to reduce the critical moment M_z to about 90% of the value given by Eqn (6.1) (Hill 1954). Salmon and Johnson (1996) suggest using one half of the values of M_{cr} obtained by using Eqn (6.1).

The American code ANSI/AISC: 360-05 gives guidance for calculating the lateral-torsional buckling strength of channels, circular, rectangular, and square hollow sections, Tees and double angles, single angles, and other unsymmetrical shapes. Trahair (2001) provides an extensive coverage on the moment capacities of angle sections.

6.7.3 Compound Beams

(a) *Section classification* Compound sections are classified into plastic, compact, and semi-compact in the same way as discussed for rolled beams in Section 6.2. However, compound beams are treated as a section built-up by welding in the British code. (The Indian code has not specified this clearly.) The limiting width-to-thickness ratios have to be checked as follows (see Fig. 6.8).
   (i) Whole flange consisting of flange plate and rolled beam flange is checked using b_1/t_f, where b_1 is the total outstand of the compound beam flange and t_f is the thickness of the flange of the rolled section.
   (ii) The outstand b_2 of the flange plate from the rolled beam flange is checked using b_2/t_p, where t_p is the thickness of the flange plate.
   (iii) The width-to-thickness ratio of the flange plate between welds b_3/t_p is checked, where b_3 is the width of the flange of rolled sections.
   (iv) The rolled beam flange itself and the web must be checked.
(b) *Moment capacity* The area of flange plates to be added to a given rolled sections to increase the strength by the required amount may be determined as below, for a laterally restrained beam.
   Total plastic modulus required

$$Z_{p,z} = M/f_y \qquad (6.18)$$

where M is the applied factored moment. If $Z_{p,rb}$ is the plastic modulus of the rolled beam, the additional plastic modulus required is

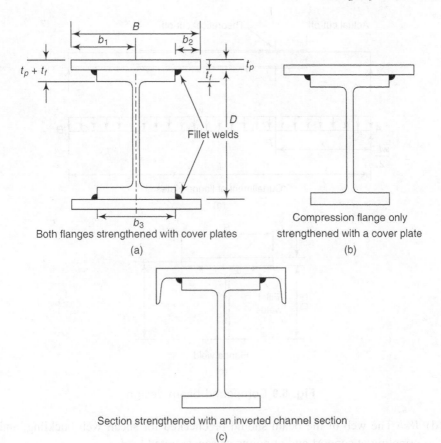

Both flanges strengthened with cover plates
(a)

Compression flange only strengthened with a cover plate
(b)

Section strengthened with an inverted channel section
(c)

Fig. 6.8 Typical compound beam cross sections

$$Z_{p,\, az} = Z_{p,\, z} - Z_{p,\, rb}$$
$$= 2\, Bt_p(D + t_p)/2 \tag{6.19}$$

where Bt_p is the area of the flange plate and D is the depth of the rolled beam. Using Eqn. (6.19), dimensions of the flange plates can be quickly obtained. If the beam is not restrained, successive trials are required.

(c) *Curtailment of flange plates* For a restrained beam with a uniformly distributed load, the theoretical cut-off points for the flange plates can be determined as follows [see Fig. 6.9(a)]. The moment capacity of the rolled beam

$$M_d = 0.909\, f_y Z_p \beta_b \le 1.09 Z_e f_y \tag{6.20}$$

Equate M_d to the moment at P at a distance x from the support

$$wLx/2 - wx^2/2 = M_d$$

where w is the factored uniform load and L is the span of the beam. Solving for x, we will get the theoretical cut-off point. The flange plate should be continued beyond this point so that the weld on the extension can develop the load in the plate at the cut-off point.

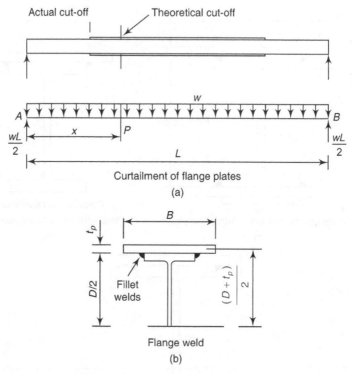

Fig. 6.9 Compound beam design

(d) *Web* The web of the beam should be checked for shear, web buckling, and crippling at support and at points of concentrated loads.

(e) *Welds connecting flange plates and beam flange* The fillet welds between flange plates and rolled beams should be designed to resist horizontal shear using elastic theory [Fig. 6.9(b)].

Horizontal shear in each fillet weld $= V_u B t_p (D - t_p)/4I_Z$ (6.21)

where V_u is the factored shear force, I_Z is the moment of inertia about the Z-Z axis. The other terms have been defined earlier. The leg length can be selected using the minimum recommended size. Intermittent welds may be specified, but continuous automatic welding considerably reduces the likelihood of failure due to fatigue or brittle fracture (MacGinley and Ang 1992).

Slim Floor Construction

In the early 1990s, engineers in Scandinavia developed the 'slim floor system', which is similar to concrete flat slabs, with 5 to 9 m spans. The essential feature of this system is that the steel beam is contained within the depth if the slab. In the earlier systems, precast concrete slabs were used. Later, deep composite slabs have been developed in UK, Sweden, and Germany.

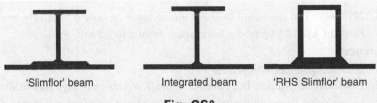

'Slimflor' beam   Integrated beam   'RHS Slimflor' beam

Fig. CS3

The various forms of slim floor beams are shown in Fig. CS3. The 'Slimflor' system developed by British Steel uses an I-section with welded bottom plate. Another system developed by Arbed consists of a section cut from an I-section to which bottom flange plate is welded. Even hollow sections, which have better torsional resistance, with a welded bottom plate have been used. The Corus Corporation developed another system called SLIMDEK®, which uses an asymmetric beam (see Fig. CS4).

Fig. CS4

Almost the whole steel section is protected from the fire by the floor slab. Hence periods of fire resistance up to 60 minutes are achievable without any protection to the exposed bottom plate. Service ducts up to 160 mm deep × 320 mm long can be accommodated within the depth of Slimdek, by penetrating the web with circular or elongated openings. Extensive tests conducted at City University, London, has shown that a 30-mm concrete cover to the top flange provides sufficient shear bond and hence, welded shear studs are not necessary.

The Slimfloor system optimizes the effective volume of the building and offers the following advantages:

1. *The floor thickness is reduced:* This can be advantageous in tall buildings, where extra floors can be added for the same total height.
2. *Easy installation of equipment:* The integrated beam makes it easier to build under-floor equipment (air-conditioning, piping, electrical networks, etc.) and simplify the fitting of false ceilings.
3. *Increased fire resistance:* It eliminates the need for passive fire protection, resulting in savings in cost and time. Built-in fire resistance of up to 60 minutes, for beams without web-opening.

4. *Economy:* The amount of steel per square metre of floor is relatively low (in the range of 15 to 25 kg/m^2 for beam spans from 5 to 7.5 m)

References

1. www.corusconstruction.com
2. Lawson R.M., Mullett, D.L., and Rackham, J.W. Design of asymmetric Slimflor beams using deep composite decking, The Steel Construction Institute, U.K., SCI-P-175-1997

6.8 Shear Strength of Steel Beams

Since shear force generally exists with bending moments, the maximum shear stress in a beam is to be compared with the shear yield stress. Though bending will govern the design in most steel beams, shear forces may control in cases where the beams are short and carry heavy concentrated loads. The pattern of shear stress distribution in I-section is shown in Fig. 6.10. It may be seen that shear stress varies parabolically with depth, with the maximum occurring at the neutral axis.

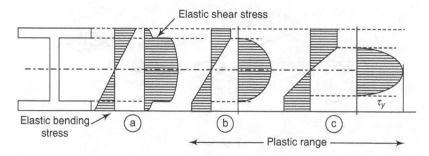

Fig. 6.10 Combined bending and shear in beams

Let us take the case of an I-beam subjected to the maximum shear force (at the support of a simply supported beam). The external shear V varies along the longitudinal axis x of the beam with bending moment as $V = dM/dx$. For beams of open cross section subjected to no twisting, the internal shear stresses   which resist the external shear V can be written as,

$$\tau = \frac{VQ}{I_z t} \tag{6.22}$$

where V is the shear force at the section, I_z is the moment of inertia of the entire cross section about the neutral axis, Q (= $A\bar{y}$) is the static moment of the cross section (above the location at which the stress is being determined) about the neutral axis and t is the thickness of the portion at which   is calculated.

Using the above equation, the maximum shear stress at the centroidal axis can be evaluated. For the purpose of design, we can assume without much error, the average shear stress for most commonly adopted sections (such as I, channel, T, etc.) as

$$\tau_{av} = \frac{V}{t_w d_w} \qquad (6.23)$$

where t_w is the thickness of the web and d_w is the depth of the web. Whenever there are bolt holes in the web, this stress is multiplied by the ratio of gross web area/net web area. The nominal shear yielding strength of webs (based on the von Mises yield criterion) is given by

$$\tau_y = \frac{f_y}{\sqrt{3}} = 0.58 f_y \qquad (6.24)$$

where f_y is the yield stress.

Taking the shear yield stress as 60% of the tensile yields stress, V_p can be written as

$$V_p \approx 0.6 f_y t_w d_w \qquad (6.25)$$

This expression gives the nominal shear strength provided by the web when there is no shear buckling of the web. Whether that occurs will depend on d_w/t_w, the depth-thickness ratio of the web. If this ratio is too large, i.e., the web is too slender, the web may buckle in shear either inelastically or elastically.

When the shear capacity of the beam is exceeded, *shear failure* occurs by excessive shear yielding of the gross area of the webs. Shear yielding is very rare in rolled steel beams. Shear is rarely a problem in rolled steel beams; the usual practice being to design the beam for flexure and then to check it for shear.

6.8.1 Shear Buckling of Beam Webs

Since the web of an I-beam is essentially a plate, it may buckle due to shearing stresses which are less than the shearing yield strength of steel. In a plate subjected to pure shear, the shear stresses are equivalent to principal stresses of the same magnitude, one tension and another compression, acting at 45° to the shear stresses. This is shown in the Fig. 4.8(c) given in Chapter 4. Buckling takes place in the form of waves or wrinkles inclined at around 45°.

As discussed in Chapter 4, the shear stress at which buckling of a perfect plate takes place is given by

$$\tau_{cr,\, v} = \frac{k_v \pi^2 E}{12(1-\mu^2)(d/t_w)^2} \qquad (6.26)$$

where $\tau_{cr,\, v}$ is the elastic critical shear buckling stress of the web and k_v is the buckling coefficient. For a plate with all four edges simply supported, $k_v = 5.35$. When stiffeners are provided only at supports

$$k_v = 4 + \frac{5.34}{(L/d)^2} \qquad L/d \le 1 \qquad (6.27a)$$

$$= 5.35 + \frac{4}{(L/d)^2} \qquad L/d \ge 1 \qquad (6.27b)$$

where (L/d) is the aspect ratio of the plate with L being the length and d the depth of the beam.

There is some approximation in using Eqn (6.27) as some bending stresses are always present. But these stresses will be very small at the ends of simply supported beams. Note that web buckling due to shear is not a design consideration for rolled beams. However, shear strength of thinner webs of beams of high-strength steels may be less than the yield strength.

6.8.2 Bend Buckling of Webs

The web may undergo local buckling due to the compressive part of the bending stresses along the depth of the beam. The buckling may occur overall or in multiple lengthwise waves. Tests have shown that beam webs are partially restrained against rotation by the flanges and this restraint raises the critical stress by at least 30%. There is no likelihood of bend buckling of the webs of rolled steel I shapes as the d/t ratio is less than 67. However, in plate girders having much thinner webs than rolled sections, bend buckling deserves attention. This is discussed in Chapter 7.

6.8.3 Design for Shear

As per IS 800 the factored design shear force V in a beam due to external actions should satisfy (clause 8.4)

$$V \leq V_d \tag{6.28a}$$

where V_d, the design strength, is given by

$$V_d = 0.909 \, V_n \tag{6.28b}$$

The nominal shear strength of a cross section V_n may be governed by plastic shear resistance or the strength of the web governed by shear buckling. The nominal plastic shear resistance under pure shear is given by $V_n = V_p$, where

$$V_p = \frac{A_v f_{yw}}{\sqrt{3}} \tag{6.29}$$

A_v is the shear area and f_{yw} is the yield strength of the web. The shear area may be calculated as given in Table 6.7 for different cross sections (Nethercot 1991).
Note Fastener holes need not be accounted for in the plastic design shear strength calculation provided that

$$A_{vn} \geq 1.26 \, (f_y/f_u) A_v$$

If A_{vn} does not satisfy the above condition, the effective shear area may be taken as that satisfying the above limit. Block shear failure criteria may be verified at the end connections.

6.9 Maximum Deflection

A beam designed to have adequate strength may become unsuitable if it deflects excessively under the loads. Beams that deflect too much may not normally lead to a structural failure, but nevertheless endanger the functioning of the structure. For

Table 6.7 Shear areas

Section	Shear area A_v
Hot-rolled (a)	ht_w
	ht_w
Built-up (b)	dt_w
Built-up sections or rolled (minor axis bending) (c)	$2bt_f$
Rectangular hollow section (d)	$Ah/(b+h)$
Rectangular hollow section (loaded parallel to width) (e)	$Ab/(b+h)$
Circular hollow section (f)	$2A/\pi$
Plates and solid bar (g)	A

example, excessive deflection in a floor not only gives a feeling of insecurity, but also damages the non-structural components (cracking of plaster ceilings) attached to it. Excessive deflections in industrial structures often cause misalignment of the supporting machinery and cause excessive vibration. Similarly high deflections in purlins may cause damage to the roofing material. Excessive deflection in the case of a flat roof results in accumulation of water during rainstorms called *ponding*.

Crane misalignment is another consequence of excessive deflection. Since these affect the performance of the structure in its working conditions, the serviceability check is done at working load levels. The deflection in beams is restricted by codes of practice by specifying deflection limitations which are usually in terms of deflection to span ratio (see Table 6 of the code).

The code also recommends that when the deflection, due to dead load plus the live load combination, is likely to be excessive, it is desirable to pre-camber the beams, trusses, and girders. Often larger beam sections are used instead of cambering.

Deflections for some common load cases for simply supported beams together with the maximum moments are given in Fig. 6.11. Some guidelines about design against floor vibration are given in Annex C of the code.

Beam and Load	Maximum Moment	Deflection at Centre
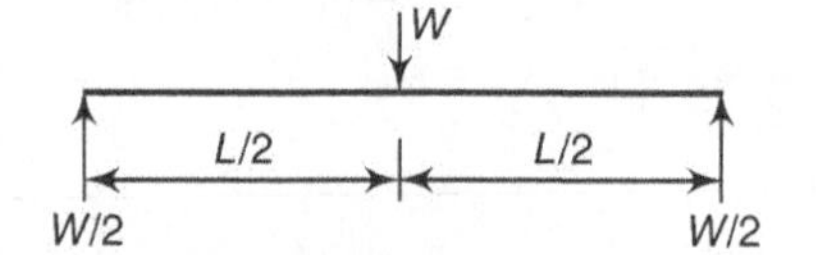	$WL/4$	$\dfrac{WL^3}{48EI}$
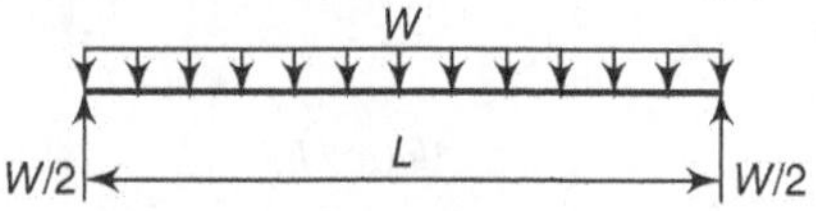	$WL/8$	$\dfrac{5WL^3}{384EI}$
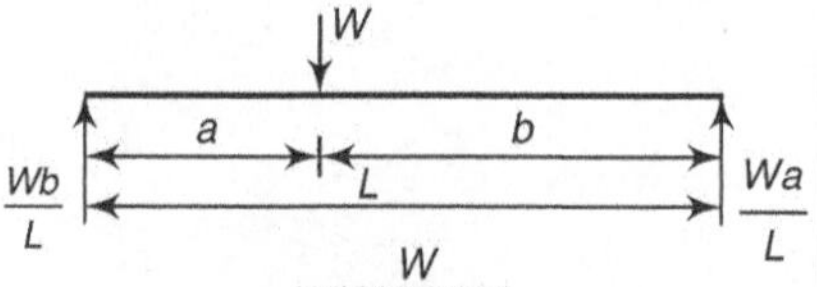	Wab/L	$\dfrac{WL^3}{48EI}\left[\dfrac{3a}{L} - 4\left(\dfrac{a}{L}\right)^3\right]$
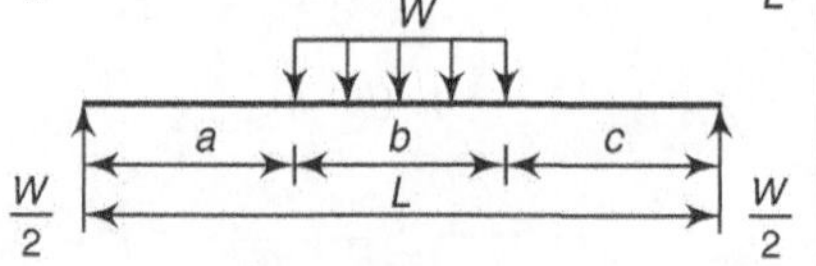	$W\left(\dfrac{a}{2}+\dfrac{b}{8}\right)$	$\dfrac{W}{384EI}[8L^3 - 4Lb^2 + b^3]$
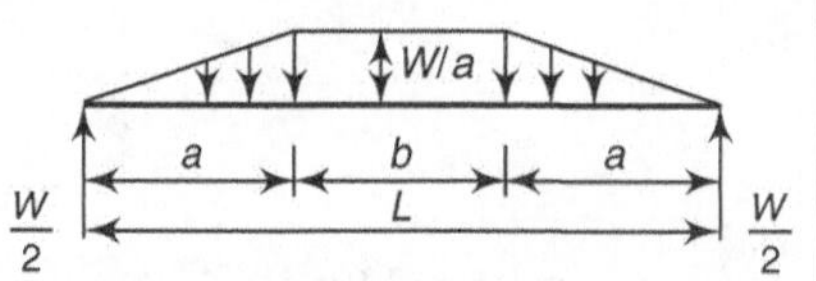	$Wa/3$	$\dfrac{W}{120EI}[16a^2 + 20ab + 5b^2]$
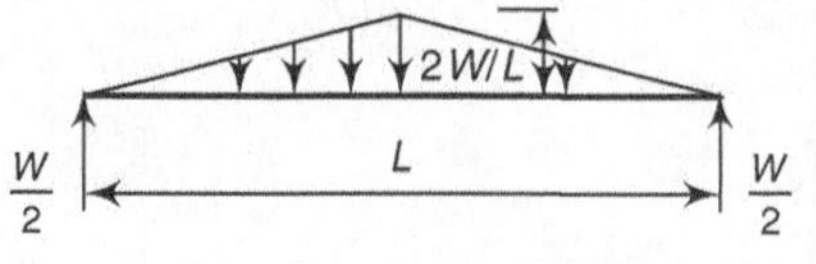	$WL/6$	$\dfrac{WL^3}{60EI}$
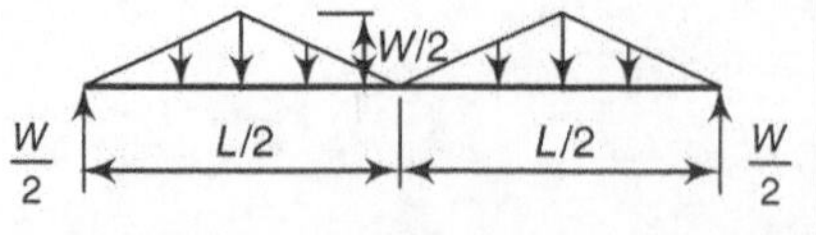	$WL/8$	$\dfrac{WL^3}{73.14EI}$

Fig. 6.11 Simply supported beams: maximum moments and deflections

6.10 Web Buckling and Web Crippling

A heavy load or reaction concentrated on a short length produces a region of high compressive stresses in the web either under the load or at the support. This may cause web failures such as web crippling, web buckling, or web crushing, as shown in Fig. 6.12. Web buckling or vertical buckling occurs when the intensity of vertical compressive stress near the centre of the section becomes greater than the critical buckling stress for the web acting as a column. Tests indicate that for rolled steel beams, the initial failure is by web crippling rather than by buckling. But for built-up beams having greater ratios of depth to thickness of web, failure by vertical buckling may be more probable than failure by web crippling. Provision of web stiffeners at points of load, and reaction or thickening of the web will solve this problem. Since web crippling occurs only at few sections, it is economical to provide stiffeners at these sections.

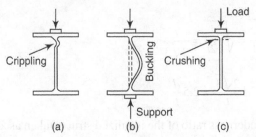

Fig. 6.12 Three possible modes of failures of the web

In the case of web buckling, the web may be considered as a strut restrained by the beam flanges. Such *idealised struts* should be considered at the points of application of concentrated load or reactions at the supports as shown in Fig. 6.13 and Fig. 6.14.

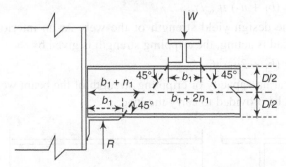

Fig. 6.13 Dispersion of concentrated loads and reactions for evaluating web buckling

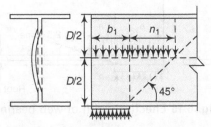

Fig. 6.14 Effective width for web buckling

In both the cases, the load is spread out over a finite length of the web called despersion length, as shown in Fig. 6.13. The dispersion length is taken as $(b_1 + n_1)$ where b_1 is the stiff bearing length and n_1 is the dispersion of 45° line at the mid depth of the section as shown in Fig. 6.14. Hence the *web buckling strength* at the support is given by

$$F_{wb} = (b_1 + n_1)tf_c \tag{6.30}$$

where t is the web thickness and f_c is the allowable compressive stress corresponding to the assumed *web strut*. The effective length of the strut is taken as $L_E = 0.7d$ where d is the depth of the *strut* in between the flanges. Thus, the slenderness ratio is

$$\lambda = \frac{L_E}{r_y} = \frac{0.7d}{r_y}$$

Since,
$$r_y = \sqrt{\frac{I_y}{A}} = \sqrt{\frac{t^3}{12t}} = \frac{t}{2\sqrt{3}} \tag{6.31}$$

$$\frac{L_E}{r_y} = 0.7d\frac{2\sqrt{3}}{t} \approx 2.5\frac{d}{t}$$

Hence, the slenderness ratio of the idealized strut is taken as $\lambda = 2.5d/t$. For web crippling, an empirical dispersion length of $b_1 + n_2$ is assumed, where n_2 is the length obtained by dispersion through the flange, to the flange to web connection (web toes of fillets), at a slope of 1:2.5 to the plane of the flange (i.e. $n_2 = 2.5d_1$) as shown in Fig. 6.15. As before, the *crippling strength* of the web (also called as the *web bearing capacity*) at supports is calculated as

$$F_{crip} = (b_1 + n_2)\, tf_{yw}/\gamma_{mo} \tag{6.32}$$

where f_{yw} is the design yield strength of the web. At an interior point where concentrated load is acting, the crippling strength is given by

$$F_{crip} = (b_1 + 2n_1)\, tf_{yw}/\gamma_{mo} \tag{6.33}$$

If the above bearing capacity or crippling strength of the beam web is exceeded, stiffeners must be provided to carry the load.

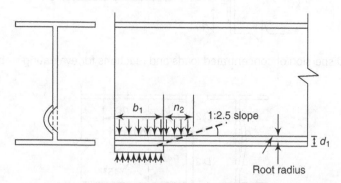

Fig. 6.15 Effective width of web bearing

6.11 Purlins

Purlins are beams used on trusses to support the sloping roof system between the adjacent trusses. Channels, angle sections, and cold formed C- or Z-sections are widely used as purlins. They are placed in an inclined position over the main rafters of the trusses. To avoid bending in the top chords of roof trusses, it is theoretically desirable to place purlins only at panel points. For larger trusses, however, it is more economical to space purlins at closer intervals. In India, where asbestos cement (AC) sheets are used, the maximum spacing of purlins is also restricted by the length of these sheets. AC sheets (though banned in many countries for the risk of lung cancer while working with them) provide better insulation to sun's heat (compared to GI sheets), which can be further improved by painting them white on the top surface. The maximum permissible span for these sheets is 1.68 m. A longitudinal overlap of not less than 150 mm is provided for AC sheets. The purlin spacing is so adjusted with lengths of the sheets that the longitudinal overlaps fall on the purlins to which they are directly bolted. Spacing of purlins should be so fixed that the cutting of sheets is avoided. Hence in practice when AC sheets are used, the purlin spacing is kept between 1.35 to 1.40 m. But in general, purlins are spaced from 0.6 m to about 2 m and their most desirable depth to span ratio is about 1/24. While the dead loads act through the centre of gravity of the purlin section, the wind loads act normal to the roof trusses. Thus, the purlin section is subjected to bending and twisting resulting in unsymmetrical bending.

Purlins may be designed as simple, continuous, and cantilever beams. The simple beam design yields the largest moments and deflections. For simply supported purlins the maximum moment will be $WL^2/8$ and if they are assumed as continuous, the moment will be $WL^2/10$. While erecting angle, channel- or I-section purlins, it is desirable that they are erected over the rafter with their flange facing up slope [see Fig. 6.16(a)]. In this position, the twisting moment does not cause instability. If the purlins are kept in such a way that the flanges face the downward slope, then the twisting moment will cause instability [Fig. 6.16(b)].

As we know channels are very weak about their web axes and have a tendency to sag in the direction of the sloping roof and often sag rods are provided midway or at one-third points between the roof trusses to take up the sag. If sag rods are used they will provide lateral support with respect to y-axis bending. Consequently, the moment M_{yy} is reduced and thereby the required purlin section is smaller. The code also permits to take advantage of the continuity of purlins over supports (clause 8.9.1). Note that if sag rods are not used, the maximum moment about the web axis would be $w_{uy}L^2/8$. Thus, when one sag rod is used the moments are reduced by 75% $w_{uy}L^2/32$ and when two sag rods are used at one-third points, the moments are reduced by 91% $w_{uy}L^2/90$. In addition to providing lateral supports to purlins, sag rods also help to keep the purlins in proper alignment during erection until the roof deck is installed and connected to the purlins. Sag rods are often used with channel and I-section purlins, but are very rarely used with angle purlins.

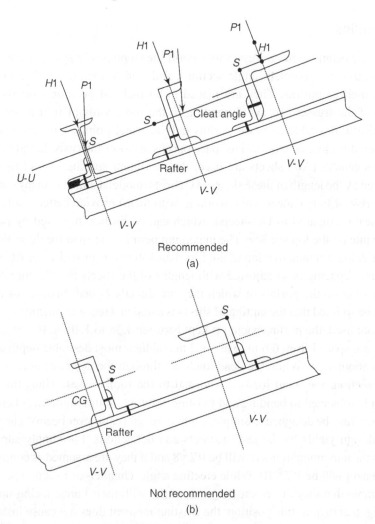

Fig. 6.16 Orientation of purlins

The design of a purlin is a trial and error process (see next section). Taking advantage of the continuity of the purlin over the supports, the maximum bending moments about the two axes M_u and M_v are calculated separately and checked according to the biaxial bending requirements. This is applicable to channel and I-sections. The recommended bending moments are

$$M_u = PL/10 \text{ and } M_v = HL/10$$

where P acts along v-v axis, H acts along u-u axis, and L is the span of the purlin. It has to be noted that the purlins at the edges or end spans be designed considering local wind effects.

Various linear and non-linear formulae have been suggested for different types of cross sections. Horne and Morris (1981), recommend the following formulae which are safe and sufficiently accurate.

(a) for I-sections
$$[M_z/(Z_{pz}f_y)]^2 + M_y/(Z_{py}f_y) \leq 1.0 \tag{6.34a}$$
(b) For all solid sections and closed hollow sections
$$[M_z/(Z_{pz}f_y)]^{5/3} + [M_y/(Z_{py}f_y)]^{5/3} \leq 1.0 \tag{6.34b}$$
(c) For all other sections
$$M_z/(Z_{pz}f_y) + M_y/(Z_{py}f_y) \leq 1.0 \tag{6.34c}$$

However, the following equation may be used conservatively for biaxial moment of channel and I-sections.

$$\frac{M_u}{M_{du}} + \frac{M_v}{M_{dv}} \leq 1.0 \tag{6.34d}$$

6.11.1 Design Procedure for Channel/I-section Purlins

As mentioned earlier, the design of purlins is a trial and error procedure and the various steps involved in the design are as follows:

1. The span of the purlin is taken as the centre-to-centre distance of adjacent trusses.
2. The gravity loads P, due to sheeting and live load, and the load H due to wind are computed. The components of these loads in the direction perpendicular and parallel to the sheeting are determined. These loads are multiplied with partial safety factors γ_f for loads (see Table 4 of the code) for the various load combinations.
3. The maximum bending moments (M_z and M_y) and shear forces (F_z and F_y) using the factored loads are determined.
4. The required value of plastic section modulus of the section may be determined by using the following equations

$$Z_{pz} = M_z\, \gamma_{m0}/f_y + 2.5(d/b)[M_y\, \gamma_{m0}/f_y] \tag{6.35}$$

where γ_{m0} is the partial safety factor for material = 1.1, d is the depth of the trial section, b is the breadth of the trial section, M_z and M_y are the factored bending moments about the Z and Y axes, respectively, and f_y is the yield stress of steel.

Since the above equation involves b and d of a section, we must use a trial section and from the above equation find out whether the chosen section is adequate or not.

5. Check for the section classification (Table 2 of the code).
6. Check for the shear capacity of the section for both the z and y axes (for purlins shear capacity will always be high and may not govern the design. The shear capacity in z and y axes is taken as (Morris & Plum 1996)

$$V_{dz} = f_y/(\sqrt{3}\gamma_{m0})A_{vz}$$

$$V_{dy} = f_y/(\sqrt{3}\gamma_{m0})A_{vy}$$

and

$$A_{vz} = ht_w$$
$$A_{vy} = 2b_f t_f$$

where h is the height, t_w is the thickness of the web, b_f is the breadth of the flange, and t_f is the thickness of the flange of I-channel section, respectively.

7. Compute the design capacity of the section in both the axes.

$$M_{dz} = Z_{pz} f_y / \gamma_{m0} \leq 1.2 Z_{ez} f_y / \gamma_{m0}$$
$$M_{dy} = Z_{py} f_y / \gamma_{m0} \leq \gamma_f Z_{ey} f_y / \gamma_{m0}$$

Note that in the second equation 1.2 is replaced by γ_f. It is because, in the y-direction, the shape factor Z_p/Z_e will be greater than 1.2 and hence if we use the factor as 1.2 we cannot prevent the onset of yielding under unfactored loads.

8. Check for local capacity by using the interaction equation

$$(M_z/M_{dz}) + (M_y/M_{dy}) \leq 1.0$$

9. Check whether the deflection is under permissible limits (Table 6 of the code).

10. Under wind suction (combined with dead load), the bottom flange of the purlin, which is laterally unsupported will be under compression. Hence, under this loading case, the lateral-torsional buckling capacity of the section has to be calculated using Eqns (6.11) to (6.14) and an overall member buckling check is to be made by using the interaction equation

$$(M_z/M_{dz}) + (M_y/M_{dy}) \leq 1.0$$

where M_{dz} is the lateral-torsional buckling strength of the member.

6.11.2 Empirical Design of Angle Purlins

An angle purlin section is unsymmetrical about both axes. Angle purlin may be used wherever the slope of the roof is less than 30°. The vertical and normal loads acting on the purlin are determined and the maximum bending moment is calculated as $PL/10$ and $HL/10$, where P and H are the respective vertical and horizontal loads. A trial angle section may be assumed with depth as 1/45 of the span and width as 1/60 of the span. The trial depth and width are arrived to ensure that the deflection is within limits. Equation (6.34d) is again used to satisfy the adequacy of the section.

BS 5950-1:2000 gives the following empirical design method for purlins with the following general requirements.

(a) Unfactored loads are used in the design.
(b) The span should not exceed 6.5 m.
(c) If the purlin spans one bay only, it should be connected at the ends by at least two bolts.
(d) If the purlins are continuous over two or three spans, with staggered joints in adjacent lines of members, at least one end of any single bay member should be connected by not less than two bolts.

The rules for empirical design of angle purlins are as follows:

(a) The roof slope should not exceed 30°.

(b) The loading should be substantially uniformly distributed. Not more than 10% of the total load should be due to other type of load.

(c) The elastic modulus about the axis parallel to the plane of cladding should not be less than $W_pL/1800$ mm^3 where W_p is the total unfactored load on one span (N) due to dead and imposed load or dead minus wind load and L is the span of purlin in mm.

(d) Dimension D perpendicular to the plane of the cladding should not to be less than $L/45$ and dimension B parallel to the plane of cladding should not be less than $L/60$.

When sag rods are used, the sag rod spacing may be used to determine B.

6.12 Design of Beams

Laterally supported beams

The design of laterally supported beams consists of selecting a section on the basis of the modulus of the section and checking it for shear, deflection, and web crippling. The steps to be followed are as follows.

1. The loads that may be acting on the beam are ascertained. The design loads are obtained by summing up the loads multiplied by the appropriate partial load factors as given in Table 4 of the code.
2. A trial beam section is assumed and the distribution of the bending moment along the length of the beam is determined by an elastic analysis (if the beam is statically indeterminate) or by statics (if the beam is statically determinate). The maximum bending moment and shear force are calculated.
3. The required section modulus may be determined using Eqn (6.7).
4. From section tables (IS 808-1989), a suitable section is selected, which has a section modulus equal to or more than the calculated modulus. ISMB sections may be preferred as they are readily available in the market.
5. The beam is checked for shear as per Eqn (6.28).
6. The beam is checked for deflection, as per Table 6 of the code.
7. The beam is checked for web crippling as per Eqns (6.32) and (6.33).

Laterally unsupported beams

As discussed in this chapter, when the compression flange of the beam is laterally unsupported, lateral torsional buckling may take place, leading to failure at or below the elastic critical moment.

The design of laterally unsupported beams with equal flanges (e.g., I-sections and channel sections) or mono symmetric beams is essentially a trial and error process, since the section dimensions are not initially known. Hence we have to assume a section in order to compute the strength of that section. The design procedure is essentially the same as described for a laterally supported section, except that in step 3, the design strength M_d is computed based on Eqn (6.11) or Eqn (E.1.2 of the code) (for mono symmetric section) and compared with the factored design moment *M*. If it is not equal to or greater than the factored design moment, next higher section is chosen and the process is repeated till a section which satisfies the condition $M \le M_d$, is found.

Examples

Example 6.1 *Design a simply supported beam of span 4 m carrying a reinforced concrete floor capable of providing lateral restraint to the top compression flange. The uniformly distributed load is made up of 20 kN/m imposed load and 20 kN/m dead load (section is stiff against bearing). Assume Fe 410 grade steel.*

Solution
Step 1: Calculation of factored loads

Dead load = 1.5 × 20 = 30 kN/m

Live load = 1.5 × 20 = 30 kN/m

Total factored load on the beam = 60 kN/m (see Fig.6.17)

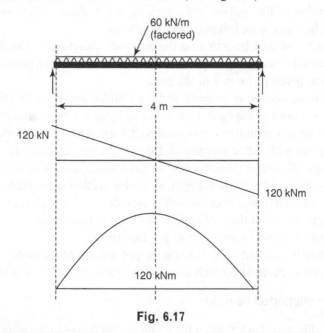

60 kN/m (factored)

120 kN

4 m

120 kNm

120 kNm

Fig. 6.17

Step 2: Calculation of maximum bending moment and shear force
Maximum bending moment = 60 × 4²/8 = 120 kNm

Step 3: Section modulus required (Z_p)

$$Z_p \text{ (required)} = \frac{M \times \gamma_{m0}}{f_y} = \frac{120 \times 10^6 \times 1.1}{250} = 528 \times 10^3 \text{ mm}^3$$

Step 4: Selection of suitable section

Choose a trial section of ISMB 300 @ 0.442 kN/m

The properties of the section are as follows:

Depth of section (D) = 300 mm

Width of flange (b_f) = 140 mm

Thickness of flange (t_f) = 12.4 mm

Depth of web (d) = $D - 2(t_f + R)$ = 300 - 2(12.4 + 14) = 247.2 mm

Thickness of web (t_w) = 7.5 mm

Moment of Inertia about major axis I_{zz} = 8990 × 10^4 mm^4

Elastic section modulus (Z_e) = 573.6 × 10^3 mm^3

Plastic section modulus (Z_p) = 651.74 × 10^3 mm^3

Section classification

$$\varepsilon = \sqrt{\frac{250}{f_y}} = \sqrt{\frac{250}{250}} = 1$$

$$\frac{b}{t_f} = \frac{(140/2)}{12.4} = 5.64 < 9.4$$

$$\frac{d}{t_w} = \frac{247.2}{7.5} = 32.96 < 84$$

Hence the section is classified as plastic section.

Step 5: Adequacy of the section including self weight of the beam

Factored self weight of the beam = 1.5 × 0.442 = 0.663 kN/m

Total load acting on the beam = 60.663 kN/m

Maximum bending moment = 121.3 kN m

$$\text{Plastic section modulus required} = \frac{121.3 \times 10^6 \times 1.1}{250} = 534 \times 10^3 \text{ mm}^3$$

Since it is less than 651.74 × 10^3 mm^3, hence the chosen section is adequate.

Step 6: Calculation of design shear force

$$\text{Design shear force, } V = \frac{wl}{2} = \frac{60.663 \times 4}{2} = 121.33 \text{ kN}$$

Step 7: Design shear strength of the section

$$\text{Design shear strength, } V_d = \frac{f_y}{\gamma_{m0} \times \sqrt{3}} \times h \times t_w = \frac{250}{1.1 \times \sqrt{3}} \times 300 \times 7.5 \times 10^{-3}$$

$$= 295.2 \text{ kN} > 121.33 \text{ kN}$$

Also

$$0.6 V_d = 177$$

Therefore, the design shear force $V < 0.6 V_d$

Step 8: Check for design capacity of the section

$$\frac{d}{t_w} = 32.96 \text{ (which is less than } 67\varepsilon)$$

Hence,

$$M_d = \beta_b Z_p \times \frac{f_y}{\gamma_{m0}}$$

$\beta_b = 1.0$, since the section is plastic section.
Therefore,

$$M_d = \frac{1.0 \times 651.74 \times 10^3 \times 250}{1.1 \times 10^6} = 148.12 \text{ kNm}$$

$$148.12 \text{ kNm} \le \frac{1.2 \times Z_e \times f_y}{\gamma_{m0}} = \frac{1.2 \times 573.6 \times 10^3 \times 250}{1.1 \times 10^6} = 156.4 \text{ kNm}$$

Hence the design capacity of the member is more than maximum bending moment M_d (148.12 kNm > 121.33 kNm)

Step 9: Check for deflection

Deflection (which is a serviceability limit state) must be calculated on the basis of the unfactored imposed loads.

$$\delta = \frac{5wl^4}{384EI} = \frac{5 \times 20 \times 4000^4}{384 \times 2 \times 10^5 \times 8990 \times 10^4} = 3.7 \text{ mm}$$

Allowable maximum deflection, $\delta_{max} = \dfrac{L}{300} = \dfrac{4000}{300} = 13.33 \text{ mm}$

The deflection is less than the allowable maximum deflection.

Example 6.2 *Design a simply supported beam of 6 m span carrying a reinforced concrete floor capable of providing lateral restraint to the top compression flange. The total udl is made up of 80 kN dead load plus 120 kN imposed load. In addition, the beam carries a point load at mid span made up of 50 kN dead load and 50 kN imposed load (assuming a stiff bearing length of 65 mm).*

Solution
Step 1: Calculation of factored load
Udl:

      Dead load = 1.5 × 80 = 120 kN
      Live load = 1.5 × 120 = 180 kN
Total udl acting on the beam = 300 kN
Concentrated load:

      Dead load = 1.5 × 50 = 75 kN
      Live load = 1.5 × 50 = 75 kN
Total concentrated load acting on the beam = 150 kN (see Fig. 6.18)

Total concentrated load acting on the beam = 150 kN (see Fig. 6.18)

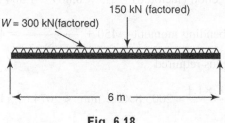

Fig. 6.18

Step 2: Calculation of maximum bending moment

$$\text{Bending moment} = \frac{WL}{8} + \frac{WL}{4} = \frac{300 \times 6}{8} + \frac{150 \times 6}{4}$$

$$= 450 \text{ kNm}$$

Step 3: Calculation of required plastic section modulus

$$Z_p = \frac{M \times \gamma_{m0}}{f_y} = \frac{450 \times 10^6 \times 1.10}{250} = 1980 \times 10^3 \text{ mm}^3$$

Step 4: Selection of suitable section

Choose a trial section of ISMB 500 @ 0.869 kN/m.

The properties of the section are as follows:

Overall depth (D) = 500 mm

Width of flange (b)= 180 mm

Thickness of flange (t_f) = 17.2 mm

Depth of web (d) = $D - 2(t_f + R)$ = 500 − 2(17.2 + 17) = 431.6 mm

Thickness of web (t_w) = 10.2 mm

Moment of inertia about major axis I_{zz} = 45200 × 10⁴ mm⁴

Elastic section modulus (Z_e) = 1808.7 × 10³ mm³

Plastic section modulus(Z_p) = 2074.67 × 10³ mm³

Section classification

$$\varepsilon = \sqrt{\frac{250}{f_y}} = \sqrt{\frac{250}{250}} = 1$$

$$\frac{b}{t_f} = \frac{(180/2)}{17.2} = 5.23 < 9.4\varepsilon$$

$$\frac{d}{t_w} = \frac{431.6}{10.2} = 42.31 < 84\varepsilon$$

Hence the section is classified as a plastic section.

Step 5: Adequacy of the section including self weight of the beam

Self weight of the section (factored) = $1.5 \times 0.869 = 1.304$ kN/m

$$\text{Maximum bending moment} = 450 + \frac{1.304 \times 6^2}{8} = 455.87 \text{ kNm}$$

Plastic section modulus required

$$\frac{455.87 \times 10^6 \times 1.1}{250} = 2005.8 \times 10^3 \text{ mm}^3 < 2074.67 \times 10^3 \text{ mm}^3$$

Hence the chosen section is adequate.

Step 6: Calculation of design shear force

$$\text{Shear force } V = \frac{1.304 \times 6}{2} + \frac{300}{2} + \frac{150}{2} = 228.9 \text{ kN}$$

Step 7: Design shear strength of the section

$$\text{Design shear strength } V_d = \frac{f_y}{\gamma_{m0} \times \sqrt{3}} \times h \times t_w$$

$$= \frac{250}{1.1 \times \sqrt{3}} \times 500 \times 10.2 \times 10^{-3} = 669.2 \text{ kN}$$

Also,

$$0.6V_d = 401.5 \text{ kN}$$

The design shear force $V < 0.6V_d$

Step 8: Check for design capacity of the section

$$\frac{d}{t_w} = 42.31 \text{ which is less than } 67\varepsilon$$

$$M_d = \beta_b Z_p \times \frac{f_y}{\gamma_{m0}} < \frac{1.2 \times Z_e \times f_y}{1.10}$$

$\beta_b = 1.0$, since the section is plastic section

$$M_d = \frac{1.0 \times 2074.67 \times 10^3 \times 250}{1.1 \times 10^6} = 471.5 \text{ kNm}$$

$$\leq \frac{1.2 \times Z_e \times f_y}{\gamma_{m0}} = \frac{1.2 \times 1808.7 \times 10^3 \times 250}{1.1 \times 10^6} = 493.28 \text{ kNm}$$

Hence the design capacity of the member is more than the maximum bending moment M_d (471.5 kNm > 455.87 kNm).

Step 9: Check for deflection

Actual deflection

$$\delta = \delta_{\text{udl}} + \delta_{\text{pl}}$$

$$\delta = \frac{5Wl^4}{384EI} + \frac{Wl^3}{48EI}$$

$$= \frac{5 \times 120 \times 6000^4}{384 \times 2 \times 10^5 \times 45200 \times 10^4} + \frac{50 \times 10^3 \times 6000^3}{48 \times 2 \times 10^5 \times 45200 \times 10^4} = 6.22 \text{ mm}$$

Allowable maximum deflection $\delta_{max} = \dfrac{L}{300} = \dfrac{6000}{300} = 20$ mm

Actual deflection is less than the allowable maximum deflection.

Step 10: Check for web buckling at support

Assume stiff bearing length, $b_1 = 65$ mm

Depth of web, $d = 431.6$ mm

$$I_{eff} \text{web} = \frac{b_1 \times t_w^3}{12} = \frac{65 \times 10.2^3}{12} = 5748.2 \text{ mm}^4$$

$$A_{eff} \text{web} = b_1 \times t_w = 65 \times 10.2 = 663 \text{ mm}^2$$

$$r = \sqrt{\frac{I_{eff} \text{ web}}{A_{eff} \text{ web}}} = \sqrt{\frac{5748.2}{663}} = 2.94 \text{ mm}$$

Effective length of web $d_{eff} = 0.7$ times the depth of web (d)

$$\lambda = \frac{d_{eff}}{r} = \frac{0.7 \times 431.6}{2.94} = 102.8$$

Therefore, $f_{cd} = 103.5$ N/mm^2 (From Table 9c of the code)

$$n_1 = 500/2 = 250 \text{ mm}$$

$$b_1 + n_1 = 65 + 250 = 315 \text{ mm}$$

$$A_b = 315 \times 10.2 = 3213 \text{ mm}^2$$

Buckling resistance $= f_{cd} \times A_b$

$$= 103.5 \times 3213/1000$$

$$= 332.6 > 228.9 \text{kN}$$

Hence it is safe.

Step 11: Check for web bearing

$$F_w = \frac{(b_1 + n_2) \times t_w \times f_y}{\gamma_{m0}}$$

$$b_1 = 65 \text{ mm}$$

$$n_2 = 2.5 \times (\text{Root radius} + \text{flange thickness})$$

$$= 2.5 (17 + 17.2) = 85.5 \text{ mm (since the angle of dispersion is 1:2.5)}$$

$$t_w = 10.2 \text{ mm}$$

$$f_y = 250 \text{ N/mm}^2$$

$$F_w = \frac{(65 + 85.5) \times 10.2 \times 250}{1.10 \times 10^3}$$

$$= 348.8 \text{ kN} > 228.9 \text{ kN}$$

Hence the web is safe.

Example 6.3 *A proposed cantilever beam is built into a concrete wall. It supports a dead load of 15 kN/m and a live load of 12 kN/m. The length of beam is 4 m. Select a suitable section with necessary checks. Assume stiff bearing length of 80 mm.*

Solution

Step 1: Calculation of load

Dead load = $1.5 \times 15 = 22.5$ kN/m

Live load = $1.5 \times 12 = 18$ kN/m

Total load = 40.5 kN/m

Step 2: Calculation of bending moment and shear force

$$BM = \frac{wl^2}{2} = \frac{40.5 \times 4^2}{2} = 324 \text{ kNm}$$

$$SF = wl = 40.5 \times 4 = 162 \text{ kN}$$

Step 3: Choosing a trial section

$$Z_p = \frac{M \times \gamma_{m0}}{f_y} = \frac{324 \times 10^6 \times 1.10}{250} = 1425.6 \times 10^3 \text{ mm}^3$$

Choose a trial section of ISMB 450 @ 0.724 kN/m

Overall depth (D) = 450 mm

Width of flange (b) = 150 mm

Depth of web (d) = 450 − 2(17.4 + 15) = 385.2 mm

Thickness of flange (t_f) = 17.4 mm

Thickness of web (t_w) = 9.4 mm

Moment of inertia about major axis, I_z = 30400 × 10^4 mm^4

Elastic section modulus (Z_e) = 1350.7 × 10^3 mm^3

Plastic section modulus (Z_p) = 1533.36 × 10^3 mm^3

Section classification

$$\frac{b}{t_f} = \frac{(150/2)}{17.4} = 4.31 < 9.4$$

$$\frac{d}{t_w} = \frac{385.2}{9.4} = 40.98 < 84$$

Hence the section is classified as plastic.

Step 4: Calculation of shear capacity of the section

$$V_d = \frac{f_y}{\gamma_{mo} \times \sqrt{3}} \times h \times t_w$$

$$= \frac{250}{1.1 \times \sqrt{3}} \times 450 \times 9.4 \times 10^{-3} = 555 \text{ kN}$$

$0.6 V_d = 333 \text{ kN} > 162 \text{ kN}$

Step 5: Design capacity of the section

$$M_d = \frac{Z_p \times f_y}{\gamma_{m0}} = \frac{1533.36 \times 10^3 \times 250}{1.10 \times 10^6} = 348.5 \text{ kNm} > 324 \text{ kNm}$$

$$\leq \frac{1.2 \times Z_e \times f_y}{\gamma_{m0}} = \frac{1.2 \times 1350.7 \times 10^3 \times 250}{1.1 \times 10^6} = 460.5 \text{ kNm}$$

Hence the section is safe.

Step 6: Check for deflection

$$\delta = \frac{wl^4}{8EI} = \frac{12 \times 4000^4}{8 \times 2 \times 10^5 \times 30400 \times 10^4} = 6.31 \text{ mm}$$

Allowable deflection $= \dfrac{L}{150} = \dfrac{4000}{150} = 26.67 \text{ mm} > 6.31 \text{ mm}$

Step 7: Check for web buckling

Cross-sectional area of web for buckling, $A_b = (b_1 + n_1) t_w$

$b_1 = 80 \text{ mm}$

$n_1 = 450/2 = 225 \text{ mm}$

$A_b = (80 + 225) \times 9.4 = 2867 \text{ mm}^2$

Effective length of the web $= 0.7 \times d = 0.7 \times 385.2 = 269.64 \text{ mm}$

$$I = \frac{b_1 \times t_w^3}{12} = \frac{80 \times 9.4^3}{12} = 5537 \text{ mm}^4$$

$A = 80 \times 9.4 = 752 \text{ mm}^2$

$$r_{min} = \sqrt{\frac{5537}{752}} = 2.71$$

$$\lambda = \frac{l_{eff}}{r_{min}} = \frac{0.7 \times 385.2}{2.71} = 99.5$$

Allowable stress = 107.7 N/mm^2 (Table 9c of the code)

Capacity of the section = $107.7 \times 2867 = 308.8 \text{ kN} > 162 \text{ kN}$

Hence the section is safe against web buckling.

Step 8: Check for web bearing

$$F_w = \frac{(b_1 + n_2) \times t_w \times f_y}{\gamma_{m0}}$$

$n_2 = 2.5(R + t_f) = 2.5(15 + 17.4) = 81 \text{ mm}$

$$F_w = \frac{(80 + 81) \times 9.4 \times 250}{1.10 \times 10^3} = 344 \text{ kN} > 162 \text{ kN}$$

Hence the section is safe against web bearing.

Example 6.4 *Design a continuous beam of spans 4.9 m, 6 m, and 4.9 m carrying a uniformly distributed load of 32.5 kN/m and the beam is laterally supported.*

Solution

Step 1: Factored load calculation

Factored uniformly distributed load = $1.5 \times 32.5 = 48.75 \text{ kN/m}$

Step 2: The bending moment and shear force distribution are shown in Fig. 6.19.

Maximum bending moment = 146.25 kN m

Maximum reaction = 146.25 kN

Step 3: Plastic section modulus required

$$Z_p = \frac{M \times \gamma_{m0}}{f_y} = \frac{146.25 \times 10^6 \times 1.10}{250} = 643.5 \times 10^3 \text{ mm}^3$$

Step 4: Selection of suitable section

Choose a trial section of ISLB 350 @ 0.495 kN/m.

Overall depth (h) = 350 mm

Width of flange (b) = 165 mm

Thickness of flange (t_f) = 11.4 mm

Depth of web $(d) = h - 2(t_f + R) = 350 - 2(11.4 + 16) = 295.2 \text{ mm}$

Thickness of web (t_w) = 7.4 mm

Moment of inertia about major axis $I_z = 13158.3 \times 10^4 \text{ mm}^4$

Elastic section modulus $(Z_e) = 751.9 \times 10^3 \text{ mm}^3$

Plastic section modulus $(Z_p) = 851.11 \times 10^3 \text{ mm}^3$

Section classification

$$\frac{b}{t_f} = \frac{82.5}{11.4} = 7.23 < 9.4$$

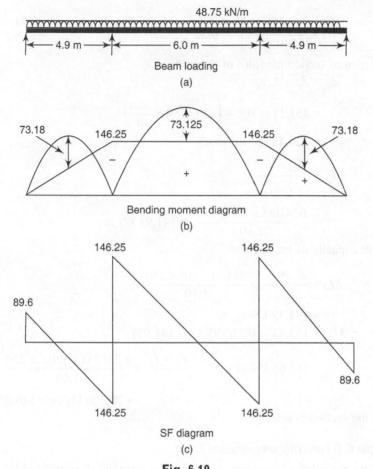

48.75 kN/m

|← 4.9 m →|← 6.0 m →|← 4.9 m →|

Beam loading
(a)

73.18   146.25   73.125   146.25   73.18

Bending moment diagram
(b)

146.25   146.25

89.6

89.6

146.25   146.25

SF diagram
(c)

Fig. 6.19

$$\frac{d}{t_w} = \frac{295.2}{7.4} = 39.9 < 84$$

Hence the section is plastic.
Check for shear capacity of section

$$V_d = \frac{f_y}{\gamma_{m0} \times \sqrt{3}} \times h \times t_w = \frac{250}{1.1 \times \sqrt{3}} \times 350 \times 7.4 = 340 \text{ kN} > 292.5 \text{ kN}$$

But $0.6\ V_d = 204$ kN > 146.25 kN
(For maximum shear, the load has to be very close to the support and not at the support).
Check for moment capacity of the section assuming high shear condition with
$V = 292.5$ [Eqn 6.8(a)]

$$M_{dv} = M_d - \beta(M_d - M_{fd}) \le 1.09 \times Z_e \times f_y$$

where M_{fd} is the plastic design strength of the area of cross section excluding the shear area.

$$\beta = \left[2 \times \left(\frac{V}{V_d}\right) - 1\right]^2$$

$$= \left[2 \times \left(\frac{292.5}{340} \right) - 1 \right]^2 = 0.52$$

Calculation of section modulus of flange

$$Z_{fd} = Z_p - A_w y_w$$

$$= 851.11 \times 10^3 - \left(350 \times 7.4 \times \frac{350}{4} \right)$$

$$= 624.485 \times 10^3 \text{ mm}^3$$

$$\therefore M_{fd} = \frac{Z_{fd} \times f_y}{\gamma_{m0}}$$

$$= \frac{624.485 \times 10^3 \times 250}{1.10} = 141.93 \text{ kN m}$$

Moment capacity of the section

$$M_d = \frac{Z_p \times f_y}{\gamma_{m0}} = \frac{851.11 \times 10^3 \times 250}{1.10}$$

$$= 193.43 \text{ kN m}$$

$$\therefore M_{dv} = 193.43 - 0.52(193.43 - 141.93)$$

$$= 165.65 \text{ kN m} < \frac{1.2 \times Z_e \times f_y}{\gamma_{m0}} = \frac{1.2 \times 751.9 \times 10^3 \times 250}{1.10}$$

$$= 205.06 \text{ kN m} > 146.25 \text{ kN m}$$

Hence the section is safe.

Example 6.5 *Laterally unrestrained beam*

Calculate the moment carrying capacity of a laterally unrestrained ISMB 500 member of length 4 m.

Solution

Section properties:

Overall depth (D) = 500 mm

Width of flange (b) = 180 mm

Thickness of flange (t_f) = 17.2

Height of web$(d) = D - 2(t_f + R) = 500 - 2(17.2 + 15) = 435.6$ mm

Thickness of web (t_w) = 10.2 mm

Moment of Inertia about major axis $I_z = 45200 \times 10^4$ mm^4

Moment of Inertia about minor axis $I_y = 1370 \times 10^4$ mm^4

Elastic section modulus $(Z_e) = 1808 \times 10^3$ mm^3

Plastic section modulus $(Z_p) = 2074.67 \times 10^3$ mm^3

As the beam is laterally unsupported the design bending strength is found by calculating the lateral torsional buckling moment:

$$M_{cr} = \sqrt{\frac{\pi^2 EI_y}{(KL)^2}\left(GI_t + \frac{\pi^2 EI_w}{(KL)^2}\right)}$$

$$G = \frac{E}{2(1+\mu)} = \frac{2 \times 10^5}{2 \times (1+0.3)} = 76.923 \times 10^3 \text{ N/mm}^2$$

$$I_t = \sum \frac{b_i t_i^3}{3} = \left[\frac{2 \times 180 \times 17.2^3}{3} + \frac{(500 - 2 \times 17.2) \times 10.2^3}{3}\right]$$

$$= 7.753 \times 10^5 \text{ mm}^4$$

$KL = 4000$ mm

$h_f = 500 - 17.2 = 482.8$ mm

$I_w = (1 - \beta_f)\beta_f I_y h_f^2$

$$\beta_f = \frac{I_{fc}}{I_{fc} + I_{ft}} = 0.5$$

$I_w = (1 - 0.5) \times 0.5 \times 1370 \times 10^4 \times 482.8^2$

$\quad = 7.984 \times 10^{11} \text{ mm}^6$

$$M_{cr} = \sqrt{\frac{\pi^2 \times 2 \times 10^5 \times 1370 \times 10^4}{4000^2}\left(76.923 \times 10^3 \times 7.753 \times 10^5 + \frac{\pi^2 \times 2 \times 10^5 \times 7.984 \times 10^{11}}{4000^2}\right)}$$

$$= 516.98 \text{ kNm}$$

Section classification

Outstanding element of compression flange $= b/t_f = (180/2)/17.2 = 5.23 < 9.4\varepsilon$

Web with neutral axis at mid section $= d/t_w = 435.6/10.2 = 38.2 < 84\varepsilon$

Hence the section is plastic.

Calculation of moment carrying capacity

$$\lambda_{LT} = \sqrt{\frac{\beta_b Z_p f_y}{M_{cr}}}$$

$$= \sqrt{\frac{1.0 \times 2074.67 \times 10^3 \times 250}{516.98 \times 10^6}} = 1.002$$

$\phi_{LT} = 0.5[1 + \alpha_{LT}(\lambda_{LT} - 0.2) + \lambda_{LT}^2]$

$\alpha_{LT} = 0.21$

$\phi_{LT} = 0.5[1 + 0.21(1.002 - 0.2) + 1.002^2] = 1.0858$

$$\chi_{LT} = \frac{1}{\phi_{LT} + [\phi_{LT}^2 - \lambda_{LT}^2]^{0.5}} \leq 1.0$$

$$= \frac{1}{1.0858 + [1.0858^2 - 1.002^2]^{0.5}} = 0.6645 \leq 1.0$$

$$f_{bd} = \frac{\chi_{LT} f_y}{\gamma_{m0}}$$

$$= \frac{0.6645 \times 250}{1.10} = 151.01 \text{ N/mm}^2$$

$$M_d = \beta_b Z_p f_{bd}$$
$$= 1.0 \times 2074.67 \times 10^3 \times 151.01 \times 10^{-6}$$
$$= 313.31 \text{ kNm}$$

The moment carrying capacity of the section = 313.31 kNm

Example 6.6 *A proposed cantilever beam is built into a concrete wall and free at its end. It supports dead load of 20 kN/m and a live load of 10 kN/m. The length of the beam is 5 m. Select an available section with necessary checks. Assume bearing length of 100 mm.*

Solution
Step 1: Calculation of load

Dead load = $1.5 \times 20 = 30$ kN/m

Live load = $1.5 \times 10 = 15$ kN/m

Total load = 45 kN/m

Step 2: Calculation of bending moment and shear force

$$BM = \frac{wl^2}{2} = \frac{45 \times 5^2}{2} = 562.5 \text{ kN m}$$

$$SF = wl = 45 \times 5 = 225 \text{ kN}$$

Step 3: Selection of initial section

Assume $\lambda = 80$; $h/t_f = 20$

$$f_{cr,\,b} = \left(\frac{1473.5}{\lambda}\right)^2 \left[1 + \frac{1}{20}\left(\frac{\lambda}{h/t_f}\right)^2\right]^{0.5} = \left(\frac{1473.5}{80}\right)^2 \left[1 + \frac{1}{20}\left(\frac{80}{20}\right)^2\right]^{0.5}$$

$$= 455.1 \text{ N/mm}^2$$

$$\lambda_{LT} = \sqrt{\frac{f_y}{f_{crb}}} = \sqrt{\frac{250}{455.1}} = 0.741$$

$$\phi_{LT} = 0.5[1 + \alpha_{LT}(\lambda_{LT} - 0.2) + \lambda_{LT}^2]$$
$$= 0.5[1 + 0.21(0.741 - 0.2) + 0.741^2]$$
$$= 0.831$$

$$\chi_{LT} = \frac{1}{\phi_{LT} + [\phi_{LT}^2 - \lambda_{LT}^2]^{0.5}} \leq 1.0$$

$$= \frac{1}{0.831 + [0.831^2 - 0.741^2]^{0.5}} \leq 1.0$$

$$= 0.828$$

$$f_{bd} = \frac{\chi_{LT} f_y}{\gamma_{m0}} = \frac{0.828 \times 250}{1.10} = 188.27 \text{ N/mm}^2$$

Therefore required section modulus $= \dfrac{562.5 \times 10^6}{188.27} = 2987.6 \times 10^3 \text{ mm}^3$

Choose a section of ISWB 600 @ 1.337 kN/m

Overall depth (h) = 600 mm

Width of flange (b) = 250 mm

Thickness of flange (t_f) = 21.3 mm

Thickness of web (t_w) = 11.2 mm

Depth of web $(d) = h - 2(t_f + R) = 600 - 2(21.3 + 17) = 523.4$ mm

Moment of inertia about major axis $I_z = 106199 \times 10^4 \text{ mm}^4$

Moment of inertia about minor axis $I_y = 4702.5 \times 10^4 \text{ mm}^4$

Elastic section modulus $(Z_e) = 3540 \times 10^3 \text{ mm}^3$

Plastic section modulus $(Z_p) = 3986.6 \times 10^3 \text{ mm}^3$

Section classification

Outstand of compression flange = 125/21.3 = 5.86 < 9.4ε

Web with N.A at mid depth = 523.4/11.2 = 46.74 < 83.9ε

Therefore, the section is plastic.

Step 4: Calculation of lateral torsional buckling moment

$$M_{cr} = \sqrt{\frac{\pi^2 EI_y}{(KL)^2}\left(GI_t + \frac{\pi^2 EI_w}{(KL)^2}\right)}$$

$$KL = 0.85L = 0.85 \times 5000 = 4250 \text{ mm}$$

$$G = \frac{E}{2(1+\mu)} = \frac{2 \times 10^5}{2 \times (1+0.3)} = 76.923 \times 10^3 \text{ N/mm}^2$$

$$I_t = \sum \frac{b_i t_i^3}{3} = \left[\frac{2 \times 250 \times 21.3^3}{3} + \frac{(600-21.3) \times 11.2^3}{3}\right] = 1.88 \times 10^6 \text{ N/mm}^2$$

$$I_w = (1-\beta_f)\,\beta_f I_y h_f^2$$

$$\beta_f = \frac{I_{fc}}{I_{fc} + I_{ft}} = 0.5$$

$$I_w = (1-0.5) \times 0.5 \times 4702.5 \times 10^4 \times 578.7^2 = 3.94 \times 10^{12} \text{ mm}^6$$

$$M_{cr} = \sqrt{\frac{\pi^2 \times 2 \times 10^5 \times 4702.5 \times 10^4}{4250^2}}$$

$$\sqrt{\left(76.923 \times 10^3 \times 1.88 \times 10^6 + \frac{\pi^2 \times 2 \times 10^5 \times 3.94 \times 10^{12}}{4250^2}\right)}$$

$$= 1719 \text{ kN m}$$

Step 5: Calculation of moment carrying capacity of a section

$$\lambda_{LT} = \sqrt{\frac{\beta_b Z_p f_y}{M_{cr}}}$$

$$= \sqrt{\frac{1.0 \times 3986.6 \times 10^3 \times 240}{1719 \times 10^6}} = 0.746$$

Note: As per Table 1 of IS 800, f_y has been reduced to 240 MPa, as the flange thickness is greater than 20 mm.

$$\phi_{LT} = 0.5[1 + \alpha_{LT}(\lambda_{LT} - 0.2) + \lambda_{LT}^2]$$
$$\alpha_{LT} = 0.21$$
$$\phi_{LT} = 0.5[1 + 0.21(0.746 - 0.2) + 0.746^2]$$
$$= 0.8356$$

$$\chi_{LT} = \frac{1}{\phi_{LT} + [\phi_{LT}^2 - \lambda_{LT}^2]^{0.5}} \leq 1.0$$

$$= \frac{1}{0.8356 + [0.8356^2 - 0.746^2]^{0.5}} \leq 1.0$$

$$= 0.825 \leq 1.0$$

$$f_{bd} = \frac{\chi_{LT} f_y}{\gamma_{m0}}$$

$$= \frac{0.825 \times 240}{1.10} = 180 \text{ N/mm}^2$$

$$M_d = \beta_b Z_p f_{bd}$$
$$= 1.0 \times 3986.6 \times 10^3 \times 180$$
$$= 717.6 \text{ kN m} > 562.5 \text{ kN m}$$

Step 6: Calculation of shear capacity of section

$$V_d = \frac{f_y}{\gamma_{m0}\sqrt{3}} \times h \times t_w$$

$$= \frac{250}{1.1 \times \sqrt{3}} \times 600 \times 11.2$$

$$= 881.7 \text{ kN}$$

$$0.6 \, V_d = 529 \text{ kN} > 225 \text{ kN}$$

Step 7: Calculation of deflection

$$\delta_b = \frac{wl^4}{8EI}$$

$$w = 30 \text{ kN/m}$$

$$\delta_b = \frac{30 \times 5000^4}{8 \times 2 \times 10^5 \times 106199 \times 10^4} = 11.03 \text{ mm}$$

$$\text{Allowable deflection} = \frac{l}{150} = \frac{5000}{150} = 33.33 \text{ mm}$$

Hence the section is safe against deflection.

Step 8: Check for web buckling

$$A_b = (b_1 + n_1)t_w$$
$$b_1 = 100 \text{ mm}$$
$$n_1 = 600/2 = 300 \text{ mm}$$
$$\therefore A_b = (100 + 300) \times 11.2 = 4480 \text{ mm}^2$$

Effective length = $0.7d = 0.7 \times 523.4 = 366$ mm

$$I = \frac{bt^3}{12} = \frac{100 \times 11.2^3}{12} = 11.707 \times 10^3 \text{ mm}^4$$

$$A = 100 \times 11.2 = 1120 \text{ mm}^2$$

$$r_{min} = \sqrt{\frac{I}{A}} = \sqrt{\frac{11.707 \times 10^3}{1120}} = 3.23 \text{ mm}$$

$$\lambda = \frac{l_{eff}}{r_{min}} = \frac{366}{3.23} = 113.31$$

$\therefore f_{cd} = 91$ N/mm^2 (From Table 9c of the code with $f_y = 240$ MPa)

Strength of the section against web buckling = $91 \times 4480 = 407.68$ kN > 225 kN

Hence the section is safe against web buckling.

Step 9: Check for web bearing

$$F_w = (b_1 + n_2)t_w f_y / \gamma_{m0}$$
$$b_1 = 100 \text{ mm}$$
$$n_2 = 2.5(R + t_f) = 2.5(17.0 + 21.3) = 95.75 \text{ mm}$$
$$F_w = (100 + 95.75) \times 11.2 \times 250/1.10$$
$$= 498.27 \text{ kN} > 225 \text{ kN}$$

Hence safe.

Example 6.7 *Design a continuous beam of span 4.9 m, 6 m and 4.9 m carrying a total uniformly distributed load of 32.5 kN/m and laterally unrestrained with a bearing length of 100 mm.*

Solution

Step 1: Load calculation

Factored load = 1.5×32.5
$$= 48.75 \text{ kN/m}$$

For the bending moment and shear force diagram see Fig. 6.19.

Maximum bending moment = 146.25 kN m

Maximum reaction = 146.25 + 146.25 = 292.5 kN m

Step 2: Selection of initial section

Assume $\lambda = 100$ and $h/t_f = 25$. Therefore $f_{crb} = 291.4$ N/mm^2 (from Table 14 of the code).

$$\lambda_{LT} = \sqrt{\frac{f_y}{f_{crb}}} = \sqrt{\frac{250}{291.4}} = 0.926$$

$$\phi_{LT} = 0.5[1 + \alpha_{LT}(\lambda_{LT} - 0.2) + \lambda_{LT}^2]$$
$$= 0.5[1 + 0.21(0.926 - 0.2) + 0.926^2]$$
$$= 1.005$$

$$\chi_{LT} = \frac{1}{\phi_{LT} + [\phi_{LT}^2 - \lambda_{LT}^2]^{0.5}} \leq 1.0$$

$$= \frac{1}{1.005 + [1.005^2 - 0.926^2]^{0.5}} \leq 1.0$$

$$= 0.716 \leq 1$$

$$f_{bd} = \frac{\chi_{LT} f_y}{\gamma_{m0}} = \frac{0.716 \times 250}{1.10} = 162.7 \text{ N/mm}^2$$

Therefore required Z of section $= \dfrac{146.25 \times 10^6}{162.7} = 898.7 \times 10^3 \text{ mm}^3$

Choose a section of ISLB 500 @ 0.75 kN/m.

    Overall depth $(D) = 500$ mm

    Width of flange $(B) = 180$ mm

    Thickness of flange $(t_f) = 14.1$ mm

    Depth of the web $(d) = D - 2(t_f + R) = 500 - 2(14.1 + 17) = 437.8$ mm

    Thickness of web $(t_w) = 9.2$ mm

    Moment of inertia about major axis $I_z = 38579 \times 10^4 \text{ mm}^4$

    Moment of inertia about minor axis $I_y = 1060 \times 10^4 \text{ mm}^4$

    Elastic section modulus $(Z_{ez}) = 1543 \times 10^3 \text{ mm}^3$

    Plastic section modulus $(Z_{pz}) = 1770.8 \times 10^3 \text{ mm}^3$

Section classification

    Outstand of compression flange $= b/t_f = 90/14.1 = 6.38 < 9.4$

    Web with N.A at mid depth $= d/t_w = 437.8/9.2 = 47.58 < 83.9$

Therefore the section is plastic.

Step 3: Calculation of moment carrying capacity of the section

$$M_{cr} = \sqrt{\frac{\pi^2 EI_y}{(KL)^2}\left(GI_t + \frac{\pi^2 EI_w}{(KL)^2}\right)}$$

$$G = \frac{E}{2(1 + \mu)} = \frac{2 \times 10^5}{2 \times (1 + 0.3)} = 76.923 \times 10^3 \text{ N/mm}^2$$

$$I_t = \sum \frac{b_i t_i^3}{3} = \left[\frac{2 \times 180 \times 14.1^3}{3} + \frac{485.9 \times 9.2^3}{3}\right] = 4.63 \times 10^5 \text{ mm}^4$$

$$I_w = (1 - \beta_f)\beta_f I_y h_f^2$$

$$h_f = D - t_f = 500 - 14.1 = 485.9$$

$$\beta_f = \frac{I_{fc}}{I_{fc} + I_{ft}} = 0.5$$

$$I_w = (1 - 0.5) \times 0.5 \times 1060 \times 10^4 \times 485.9^2$$

$$= 6.256 \times 10^{11} \text{ mm}^6$$

$$M_{cr} = \sqrt{\frac{\pi^2 \times 2 \times 10^5 \times 1060 \times 10^4}{6000^2} \left(\begin{array}{c} 76.923 \times 10^3 \times 4.63 \times 10^5 \\ + \dfrac{\pi^2 \times 2 \times 10^5 \times 6.256 \times 10^{11}}{6000^2} \end{array} \right)}$$

$$= 201.6 \text{ kN m}$$

$$\lambda_{LT} = \sqrt{\frac{\beta_b Z_p f_y}{M_{cr}}}$$

$$= \sqrt{\frac{1.0 \times 1770.8 \times 10^3 \times 250}{201.6 \times 10^6}} = 1.48$$

$$\phi_{LT} = 0.5[1 + \alpha_{LT}(\lambda_{LT} - 0.2) + \lambda_{LT}^2]$$

$$\alpha_{LT} = 0.21$$

$$\phi_{LT} = 0.5[1 + 0.21(1.48 - 0.2) + 1.48^2] = 1.7296$$

$$\chi_{LT} = \frac{1}{\phi_{LT} + [\phi_{LT}^2 - \lambda_{LT}^2]^{0.5}} \leq 1.0$$

$$= \frac{1}{1.7296 + [1.7296^2 - 1.48^2]^{0.5}} = 0.381 \leq 1.0$$

$$f_{bd} = \frac{\chi_{LT} f_y}{\gamma_{m0}} = \frac{0.381 \times 250}{1.10} = 86.6 \text{ N/mm}^2$$

$$M_d = Z_p f_{bd}$$

$$= 1770.8 \times 10^3 \times 86.6 = 153.34 \text{ kN m} > 146.25 \text{ kN m}$$

Step 4: Calculation of shear capacity of section

$$V_d = \frac{f_y}{\gamma_{m0}\sqrt{3}} \times D \times t_w = \frac{250}{1.1 \times \sqrt{3}} \times 500 \times 9.2 = 603.59 \text{ kN}$$

$0.6 \ V_d = 362.15 \text{ kN} > 292.5 \text{ kN}$

Step 5: Calculation of deflection

$$\delta_b = \frac{5wl^4}{384EI}$$

$$w = 32.5 \text{ kN/m}$$

$$\delta_b = \frac{5 \times 32.5 \times 6000^4}{384 \times 2 \times 10^5 \times 38579 \times 10^4} = 7.10 \text{ mm}$$

Allowable deflection $= \dfrac{l}{300} = \dfrac{6000}{300} = 20 \text{ mm}$

Hence safe against deflection.

Step 6: Check for web buckling

$$A_b = (b_1 + n_1)t_w, \quad b_1 = 100 \text{ mm}, \quad n_1 = D/2 = 500/2 = 250 \text{ mm}$$

$$\therefore A_b = (100 + 250) \times 9.2 = 3220 \text{ mm}^2$$

Effective length $= 0.7d = 0.7 \times 437.8 = 306.46$ mm

$$I = \frac{bt^3}{12} = \frac{100 \times 9.2^3}{12} = 6489 \text{ mm}^4$$

$$A = 100 \times 9.2 = 920 \text{ mm}^2$$

$$r_{\min} = \sqrt{\frac{I}{A}} = \sqrt{(6489/920)} = 2.66 \text{ mm}$$

$$\lambda = \frac{L_{\text{eff}}}{r_{\min}} = \frac{306.46}{2.66} = 115.21$$

$\therefore$ From Table 9c of the code, $f_{cd} = 88.9$ N/mm^2

Strength of the section against web buckling $= 88.9 \times 3220 = 286.26$ kN ≈ 292.5 kN

Hence the section is safe against web buckling.

Step 7: Check for web bearing

$$F_w = (b_1 + n_2)t_w f_y/\gamma_{m0}$$

$$b_1 = 100 \text{ mm}$$

$$n_2 = 2.5(t_f + R) = 2.5 \times (14.1 + 17) = 77.75$$

$$F_w = (100 + 77.75) \times 9.2 \times 250/1.10 = 371.65 \text{ kN} > 292.5 \text{ kN}$$

Hence the section is safe against web bearing.

Example 6.8 *Design a laterally unrestrained beam to carry a uniformly distributed load of 30 kN/m. The beam is unsupported for a length of 3 m and is simply placed on longitudinal beams at its ends.*

Solution

Step 1: Calculation of load

Factored load $= 1.5 \times 30 = 45$ kN/m

Step 2: Calculation of bending moment and shear force

$$BM = \frac{wl^2}{8} = \frac{45 \times 3^2}{8} = 50.625 \text{ kN.m}$$

$$SF = \frac{wl}{2} = \frac{45 \times 3}{2} = 67.5 \text{ kN}$$

Step 3: Initialization of section

Assume $\lambda = 100$; $h/t_f = 25$ and hence from Table 6.5, $f_{cr, b} = 291.31$ N/mm^2

$$\lambda_{\text{LT}} = \sqrt{\frac{f_y}{f_{crb}}} = \sqrt{\frac{250}{291.31}} = 0.926$$

$$\phi_{LT} = 0.5[1 + \alpha_{LT}(\lambda_{LT} - 0.2) + \lambda_{LT}^2]$$
$$= 0.5[1 + 0.21(0.926 - 0.2) + 0.926^2] = 1.005$$

$$\chi_{LT} = \frac{1}{\phi_{LT} + [\phi_{LT}^2 - \lambda_{LT}^2]^{0.5}} \leq 1.0$$

$$= \frac{1}{1.005 + [1.005^2 - 0.926^2]^{0.5}} = 0.716 \leq 1.0$$

$$f_{bd} = \frac{\chi_{LT}f_y}{\gamma_{m0}} = \frac{0.716 \times 250}{1.10} = 162.7 \text{ N/mm}^2$$

Therefore required z of section $= \dfrac{50.625 \times 10^6}{162.7} = 311.1 \times 10^3 \text{ mm}^3$

Choose a section of ISMB 225 @ 0.312 kN/m.

Overall depth $(D) = 225$ mm
Width of flange $(b_f) = 110$ mm
Thickness of flange $(t_f) = 11.8$ mm
Thickness of web $(t_w) = 6.5$ mm
Depth of web $(d) = D - 2(t_f + R) = 225 - 2(11.8 + 12) = 177.4$ mm
Moment of inertia about major axis $I_{zz} = 3440 \times 10^4 \text{ mm}^4$
Moment of Inertia about minor axis $I_{yy} = 218 \times 10^4 \text{ mm}^4$
Elastic section modulus $(Z_{ez}) = 305.9 \times 10^3 \text{ mm}^3$
Plastic section modulus $(Z_{py}) = 348.27 \times 10^3 \text{ mm}^3$
Minimum radius of gyration $(r_y) = 23.4$ mm

Section classification

Outstand of compression flange $= (110/2)/11.8 = 4.66 < 9.4$
Web with neutral axis at mid depth $= 177.4/6.5 = 27.3 < 84$
Therefore the section is plastic.

Step 4: Calculation of lateral-torsional buckling moment

$$M_{cr} = \sqrt{\frac{\pi^2 EI_y}{(KL)^2}\left(GI_t + \frac{\pi^2 EI_w}{(KL)^2}\right)}$$

$$G = \frac{E}{2(1 + \mu)} = \frac{2 \times 10^5}{2 \times (1 + 0.3)} = 76.923 \times 10^3 \text{ N/mm}^2$$

$$I_t = \sum \frac{b_i t_i^3}{3} = \left[\frac{2 \times 110 \times 11.8^3}{3} + \frac{(225 - 2 \times 11.8) \times 6.5^3}{3}\right]$$

$$= 138.926 \times 10^3 \text{ mm}^3$$
$$I_w = (1 - \beta_f)\beta_f I_y h_f^2$$

$$\beta_f = \frac{I_{fc}}{I_{fc} + I_{ft}} = 0.5$$

$h_f = 225 - 11.8 = 213.2$ mm

$I_w = (1 - 0.5) \times 0.5 \times 218 \times 10^4 \times 213.2^2 = 24.77 \times 10^9$ mm^6

$$M_{cr} = \sqrt{\frac{\pi^2 \times 2 \times 10^5 \times \frac{218 \times 10^4}{3000^2}}{} \left(76.923 \times 10^3 \times 138.926 \times 10^3 + \frac{\pi^2 \times 2 \times 10^5 \times 24.77 \times 10^9}{3000^2} \right)}$$

$= 87.79$ kNm

$$\lambda_{LT} = \sqrt{\frac{Z_p f_y}{M_{cr}}} = \sqrt{\frac{348.27 \times 10^3 \times 250}{87.79 \times 10^6}} = 0.9959$$

$\phi_{LT} = 0.5[1 + \alpha_{LT}(\lambda_{LT} - 0.2) + \lambda_{LT}^2]$

$\alpha_{LT} = 0.21$

$\phi_{LT} = 0.5[1 + 0.21(0.9959 - 0.2) + 0.9959^2] = 1.0794$

$\chi_{LT} =$

$$\frac{1}{\phi_{LT} + [\phi_{LT}^2 - \lambda_{LT}^2]^{0.5}} = \frac{1}{1.0794 + [1.0794^2 - 0.9959^2]^{0.5}} = 0.6685 \le 1.0$$

$$f_{bd} = \frac{\chi_{LT} f_y}{\gamma_{m0}} = \frac{0.6685 \times 250}{1.10} = 151.93 \text{ N/mm}^2$$

$M_d = Z_p f_{bd} = 348.27 \times 10^3 \times 151.93 = 52.91$ kNm > 50.625 kNm

Step 5: Calculation of shear capacity of section

$$V_d = \frac{f_y}{\gamma_{m0}\sqrt{3}} \times D \times t_w = \frac{250}{1.1 \times \sqrt{3}} \times 225 \times 6.5 = 191.9 \text{ kN}$$

$0.6 \, V_d = 115$ kN > 67.5 kN

Step 6: Calculation of deflection

$$\delta_b = \frac{5wl^4}{384EI}$$

$w = 30$ kN/m

$$\delta_b = \frac{5 \times 30 \times 3000^4}{384 \times 2 \times 10^5 \times 3440 \times 10^4} = 4.6 \text{ mm}$$

Allowable deflection $= \dfrac{l}{300} = \dfrac{3000}{300} = 10$ mm

Hence the section is safe against deflection.

Step 7: Check for web buckling
Assuming that longitudinal beams are of the same size,

$$A_b = (b_1 + n_1)t_w$$
$$b_1 = (b_f - t_w)/2 = (110 - 6.5)/2 = 51.75 \text{ mm}$$
$$n_1 = D/2 = 225/2 = 112.5 \text{ mm}$$
$$\therefore A_b = (51.75 + 112.5) \times 6.5 = 1067.6 \text{ mm}^2$$

$$I = \frac{bt_w^3}{12} = \frac{51.75 \times 6.5^3}{12} = 1184 \text{ mm}^4$$

$$A = 51.75 \times 6.5 = 336.4 \text{ mm}^2$$

$$r_{min} = \sqrt{\frac{I}{A}} = \sqrt{\frac{1184}{336.4}} = 1.88 \text{ mm}$$

$$\lambda = \frac{l_{eff}}{r_{min}} = \frac{0.7 \times 177.4}{1.88} = 66.18$$

$\therefore f_{cd} = 158.36 \text{ N/mm}^2$ (From Table 9c of the code)

Strength of the section against web buckling $= 158.36 \times 1067.6 = 169.07 \text{ kN} > 67.5 \text{ kN}$

Step 8: Check for web bearing

$$F_w = (b_1 + n_2)t_w f_y/\gamma_{m0}$$
$$b_1 = 51.75 \text{ mm}$$
$$n_2 = 2.5(t_f + R) = 2.5(11.8 + 12) = 59.5 \text{ mm}$$
$$F_w = (51.75 + 59.5) \times 6.5 \times 250/(1.10 \times 10^3) = 164.35 \text{ kN} > 67.5 \text{ kN}$$

Hence the section is safe against web bearing.

Example 6.9 *Design a steel beam having a clear span of 9 m are resting on 150 mm wide end bearings. The beams spacing is 3 m and the beams carries a dead load of 5 kN/m², including the weight of the section. The imposed load on the beam is 15 kN/m². The beam depth is restricted to 575 mm and the yield strength of the steel is 250 N/mm².*

Solution
Factored loads
Dead load $= 1.5 \times 5 = 7.5 \text{ kN/m}^2$
Imposed load $= 1.5 \times 15 = 22.5 \text{ kN/m}^2$
The beams are spaced at 3 m intervals, therefore the load per meter
Dead load $= 7.5 \times 3 = 22.5 \text{ kN/m}$
Imposed load $= 22.5 \times 3 = 67.5 \text{ kN/m}$
Total load $= 90 \text{ kN/m}$
Effective span = clear span + 2 × half the width of the end bearing
$$= 9 + 2 \times 0.075 = 9.15 \text{ m}$$

Reactions at support $= \dfrac{90 \times 9.15}{2} = 411.75$ kN

Mid span moment $= \dfrac{wl^2}{8} = \dfrac{90 \times 9.15^2}{8} = 941.878$ kN m

Selection of the section

Plastic section modulus required

$$Z_p = \frac{M \times \gamma_{m0}}{f_y} = \frac{941.878 \times 10^6 \times 1.10}{250} = 414.43 \times 10^4 \text{ mm}^3$$

The section with the largest plastic modulus under 575 mm depth restriction is ISWB 550 @ 1.125 kN/m. The plastic section modulus of this section is 306.6×10^4 mm^3 which is less than the required value. The section must be strengthened with additional plates to provide the required plastic section modulus. The stiffness required to be provided to restrict the deflection problem can be calculated as follows:

Maximum deflection = Effective span/360 = 9150/360 = 25.42 mm.

The moment of inertia of the required beam section that can satisfy the deflection requirement mentioned earlier due to unfactored imposed load

$$I_z = \frac{5}{384} \times \frac{45 \times 9150^4}{2 \times 10^5 \times 25.42} = 80784 \times 10^4 \text{ mm}^4$$

Hence provide ISWB 550 with 200 mm $\times$ 10 mm cover plates which gives $I_z = 106,266.1 \times 10^4$ mm^4.

Check for shear

Shear capacity of the section

$$V_d = \frac{f_y}{\gamma_{m0} \times \sqrt{3}} \times h \times t_w = \frac{250}{1.1 \times \sqrt{3}} \times 550 \times 10.5 = 757.7 \text{ kN}$$

$0.6V_d = 0.6 \times 757.7 = 454.663$

The maximum shear force is 411.75 kN and it is less than $0.6V_d$.

Check for plastic modulus

The plastic modulus required for the section is 414.43×10^4 mm^3. ISWB 550 provides 306.6×10^4 mm^3. Assumed thickness of the plate is 10 mm and the total depth of the beam is 570 mm. Distance between the c/c of the plates is 560 mm.

Additional plastic section modulus to be provided by the plate $= 107.83 \times 10^4$ mm^3

Required area of the plate $= \dfrac{107.83 \times 10^4}{560} = 1925.535$ mm^2

The provided area of the plate = 2000 mm^2. The plastic modulus Z_p of the compound section $= 306.6 \times 10^4 + 2000 \times (550 + 10) = 418.6 \times 10^4$ mm^3, which is more than the required plastic section modulus.

Check for deflection

Minimum I_z required is 80784×10^4 mm^4

I_z provided by ISWB 550 = 74906.1×10^4 mm^4

I_z to be provided by the plates = 5878×10^4 mm^4

I_z provided by the plates = $2 \times 200 \times 10 \times 280^2 = 31360 \times 10^4$ mm^4.

Total I_z provided = 106266.1×10^4 mm^4; greater than the I_z required.

Curtailment of flange plates

In the case of plated beams, the flange plates may be curtailed near the supports of the beam to save some steel. The theoretical cut off points may be calculated using the algebraic method or by a geometrical construction. For the above problem the theoretical cut off points are computed using the algebraic method. The bending moment capacity for the beam ISWB 550

$$M = 306.6 \times 10^4 \times 250 = 766.5 \text{ kN m}$$

At any point, distance x from the support

$$M_z = Rx - \frac{wx^2}{2}, \text{ where } M_z = 766.5 \text{ kN m}, R = 411.75 \text{ kN, and } w = 90 \text{ kN/m}$$

Substituting the values

$$766.5 \times 10^6 = 411.75 \times 10^3 \times x - \frac{90 \times x^2}{2}$$

Solving this equation $x = 6500$ mm or 2600 mm. Hence the theoretical cut-off point is 2600 from either side of the support.

Example 6.10 *Design a purlin on a sloping roof truss with the dead load of 0.15 kN/m^2 (cladding and insulation), a live load of 2 kN/m^2 and wind load of 0.5 kN/m^2 (suction). The purlins are 2 m centre to centre and of span 4 m, simply supported on a rafter at a slope of 20 degrees (see Fig. 6.20).*

(a) Provide channel section purlin

(b) Provide channel purlin with a sag rod at mid span

(c) Provide angle purlin

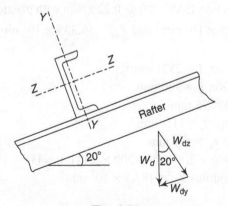

Fig. 6.20

Solution

Load calculation

Dead load = $0.15 \times 2 = 0.3$ kN/m

Live load = $2 \times 2 = 4$ kN/m

Wind load = $0.5 \times 2 = 1$ kN/m (suction)

$W_{dz} = 0.3 \times \cos 20° = 0.282$ kN/m

$W_{iz} = 4 \times \cos 20° = 3.76$ kN/m

$W_{wz} = -1$ kN/m

$W_{dy} = 0.3 \times \sin 20° = 0.103$ kN/m

$W_{iy} = 4 \times \sin 20° = 1.37$ kN/m

Note that W_{wy} is zero as wind pressure is perpendicular to the surface on which it acts, i.e., normal to the rafter.

Factored load combination

Z-direction:

$WL + DL + LL = (1.2 \times 1.0) + (1.2 \times 0.282) + (1.2 \times 3.76) = 6.0552$ kN/m

$DL + LL = (1.5 \times 0.282) + (1.5 \times 3.76) = 6.063$ kN/m

Y-direction:

$DL + LL = (1.5 \times 0.103) + (1.5 \times 1.37) = 2.21$ kN/m

Bending moment and shear force calculation:

$M_z = 6.063 \times 4^2/8 = 12.126$ kN m

$M_y = 2.21 \times 4^2/8 = 4.42$ kN m

$F_z = 6.063 \times 4/2 = 12.126$ kN

$F_y = 2.21 \times 4/2 = 4.42$ kN

(a) Channel section purlin

Assume an ISMC 200 channel.

Plastic section modulus required

$$= \frac{M_z \times \gamma_{m0}}{f_y} + 2.5 \times \frac{d}{b} \times \frac{M_y \times \gamma_{m0}}{f_y}$$

$$= \frac{12.126 \times 10^6 \times 1.1}{250} + 2.5 \times \frac{200}{75} \times \frac{4.42 \times 10^6 \times 1.1}{250} = 183 \times 10^3 \text{ mm}^3$$

Choose a channel section ISMC 200 @ 0.22 kN/m with plastic section modulus of

$Z_{pz} = 211.25 \times 10^3$ mm^3 and $Z_{py} = 56.339 \times 10^3$ mm^3.

Section properties

Cross sectional area $A = 2821$ mm^2

Depth of the section $h = 200$ mm

Width of flange $b = 75$ mm

Thickness of flange $t_f = 11.4$ mm

Thickness of web $t_w = 6.1$ mm

Depth of web $d = h - 2(t_f + R) = 200 \times 2 \ (11.4 + 11) = 155.2$ mm

Elastic section modulus $Z_{ez} = 181.7 \times 10^3$ mm^3

Elastic section modulus $Z_{ey} = 26.3 \times 10^3$ mm^3
Plastic section modulus $Z_{pz} = 211.25 \times 10^3$ mm^3
Plastic section modulus $Z_{py} = 56.339 \times 10^3$ mm^3
Moment of inertia $I_{zz} = 1830 \times 10^4$ mm^4
Moment of inertia $I_{yy} = 141 \times 10^4$ mm^4

Section classification

$$\frac{b}{t_f} = \frac{75}{11.4} = 6.58 < 9.4$$

$$\frac{d}{t_w} = \frac{155.2}{6.1} = 25.44 < 42$$

Hence the section is plastic.
Calculation of shear capacity of the section
Z-direction

$$V_d = \frac{f_y}{\gamma_{m0} \times \sqrt{3}} \times h \times t_w = \frac{250}{1.1 \times \sqrt{3}} \times 200 \times 6.1 = 160.18 \text{ kN}$$

$0.6V_d = 96 \text{ kN} > 12.126 \text{ kN}$
Y-direction

$$\text{Shear capacity} = \frac{250}{1.1 \times \sqrt{3}} \times 2 \times 75 \times \frac{11.4}{10^3} = 224.4 \text{ kN} > 4.42 \text{ kN}$$

Note that in purlin design, the shear capacity is usually high relative to the shear force.
Design capacity of the section

$$M_{dz} = \frac{Z_{pz} \times f_y}{\gamma_{m0}} = \frac{211.25 \times 10^3 \times 250}{1.1 \times 10^6} = 48 \text{ kN m}$$

$$\leq \frac{1.2 \times Z_{ez} \times f_y}{\gamma_{m0}} = \frac{1.2 \times 181.7 \times 10^3 \times 250}{1.1 \times 10^6} = 49.55 \text{ kN m}$$

Hence $M_{dz} = 48$ kN m > 12.126 kN m

$$M_{dy} = \frac{Z_{py} \times f_y}{\gamma_{m0}} = \frac{56.339 \times 10^3 \times 250}{1.10} = 12.8 \text{ kN m}$$

$$\leq \frac{\gamma_f \times Z_{ey} \times f_y}{\gamma_{m0}} = \frac{1.5 \times 26.3 \times 10^3 \times 250}{1.1 \times 10^6} = 8.96 \text{ kN m}$$

Since the ratio Z_p/Z_e is greater than 1.2, the constant in the preceding equation is replaced by the ratio of $\gamma_f = 1.5$. Hence $M_{dy} = 8.96$ kN m > 4.42 kN m.
Overall member strength (local capacity)
To ascertain the overall member strength, the following interaction equation should be satisfied.

$$\frac{M_z}{M_{dz}} + \frac{M_y}{M_{dy}} \leq 1$$

$$\frac{12.126}{48} + \frac{4.42}{8.96} = 0.75 \leq 1$$

Hence the overall member strength is satisfactory.

Check for deflection

$$\delta = \frac{5wl^4}{384EI} = \frac{5 \times 3.76 \times 4000^4}{384 \times 2 \times 10^5 \times 1830 \times 10^4} = 3.42 \text{ mm}$$

Allowable deflection $= \dfrac{l}{180} = \dfrac{4000}{180} = 22.22$ mm (Table 6 of IS 800)

Hence the section is safe.

Check for wind suction

The effect of wind suction has not been considered till now; it can become critical in some situations. It has to be combined with dead load.

Factored wind load $W_z = 0.9 \times 0.282 - 1.5 \times 1 = -1.246$ kN/m

$$W_y = 0.9 \times 0.103 = 0.0927 \text{ kN/m}$$

Buckling resistance of section

Equivalent length $l_e = 4$ m

Moment $= M_z = wl^2/8 = -1.246 \times 4^2/8 = -2.492$ kN m

$$M_y = 0.0927 \times 4^2/8 = 0.1854 \text{ kN m}$$

The value of M_z is much lower than the value 12.126 kN m earlier, but the negative sign indicates that the lower flange of the channel is in compression and this flange is unrestrained. Hence the buckling resistance of the channel must be found.

$$M_{cr} = \sqrt{\frac{\pi^2 EI_y}{(KL)^2}\left(GI_t + \frac{\pi^2 EI_w}{(KL)^2}\right)}$$

$$G = \frac{E}{2(1+\mu)} = \frac{2 \times 10^5}{2 \times (1+0.3)} = 76.923 \times 10^3 \text{ N/mm}^2$$

$$I_t = \sum\frac{b_i t_i^3}{3} = \left[\frac{2 \times 75 \times 11.4^3}{3} + \frac{(200-11.4) \times 6.1^3}{3}\right] = 88346.7 \text{ mm}^4$$

$$I_w = (1 - \beta_f)\beta_f I_y h_f^2$$

$$h_f = 200 - 11.4 = 188.6 \text{ mm}$$

$$\beta_f = \frac{I_{fc}}{I_{fc} + I_{ft}} = 0.5$$

$$I_w = (1 - 0.5) \times 0.5 \times 141 \times 10^4 \times 188.6^2$$
$$= 1.2538 \times 10^{10} \text{ mm}^6$$

$$M_{cr} = \sqrt{\frac{\pi^2 \times 2 \times 10^5 \times}{4000^2} \frac{141 \times 10^4}{} \left(76.923 \times 10^3 \times 88346.7 + \frac{\pi^2 \times 2 \times 10^5 \times}{4000^2} \frac{1.2538 \times 10^{10}}{} \right)}$$

$$= 38.09 \text{ kN m}$$

$$\lambda_{LT} = \sqrt{\frac{\beta_b Z_p f_y}{M_{cr}}}$$

$$= \sqrt{\frac{1.0 \times 211.25 \times 10^3 \times 250}{38.09 \times 10^6}} = 1.1775$$

$$\phi_{LT} = 0.5[1 + \alpha_{LT}(\lambda_{LT} - 0.2) + \lambda_{LT}^2]$$
$$= 0.5[1 + 0.21(1.1775 - 0.2) + 1.1775^2]$$
$$= 1.296$$

$$\chi_{LT} = \frac{1}{\phi_{LT} + [\phi_{LT}^2 - \lambda_{LT}^2]^{0.5}} \leq 1.0$$

$$= \frac{1}{1.296 + [1.296^2 - 1.1775^2]^{0.5}} \leq 1.0$$

$$= 0.544 \leq 1$$

$$f_{bd} = \frac{\chi_{LT} f_y}{\gamma_{m0}} = \frac{0.544 \times 250}{1.10} = 123.71 \text{ N/mm}^2$$

$$M_{dz} = Z_p f_{bd}$$
$$= 211.25 \times 10^3 \times 123.71$$
$$= 26.13 \text{ kN m} > 2.492 \text{ kN m}$$

The buckling resistance M_{dy} of the section need not be found out, because the purlin is restrained by the cladding in the z-plane and hence instability is not considered for a moment about the minor axis.

Overall member strength

To ascertain the overall member buckling strength, the following interaction should be satisfied.

$$\frac{M_z}{M_{dz}} + \frac{M_y}{M_{dy}} \leq 1$$

$$\frac{2.492}{26.13} + \frac{0.1854}{8.96} = 0.097 < 1$$

Hence the overall member strength is satisfactory.

It has to be noted that the maximum bending moment occurs at the centre of the beam and the maximum shear force at the supports. Hence it is not necessary to check the moment capacity in the presence of shear force. Also purlins are not normally checked for web bearing and crippling as the applied concentrated loads are low (note the low value of shear force in this example).

(b) Channel Section purlin with one sag rod at mid span.

Since the channel section purlin is provided with a sag rod at mid-span, the bending moment in the y-direction will be reduced considerably.

$$M_y = 2.21 \times 4^2/32 = 1.105 \text{ kN m}$$
$$M_z = 12.126 \text{ kN m}$$

Required section modulus = $(M_z \times \gamma_{m0}/f_y) + 2.5(d/b)(M_y\gamma_{m0}/f_y)$

Assuming ISMC 100 with $d = 100$ mm and $b = 50$ mm,

Required $Z = (12.126 \times 10^6 \times 1.1/250) + 2.5 \times (100/50) \times (1.105 \times 10^6 \times 1.1/250)$
$$= 77.66 \times 10^3 \text{ mm}^3$$

Provide ISMC 150 with the following section properties.

Depth of section $h = 150$ mm; $r_y = 22$ mm
Width of flange $b = 75$ mm
Thickness of flange $t_f = 9.0$ mm
Thickness of web $t_w = 5.7$ mm
Elastic section modulus $Z_{ez} = 105 \times 10^3 \text{ mm}^3$
Elastic section modulus $Z_{ey} = 19.5 \times 10^3 \text{ mm}^3$
Plastic section modulus $Z_{pz} = 119.82 \times 10^3 \text{ mm}^3 > 77.66 \times 10^3 \text{ mm}^3$
Moment of Inertia $I_z = 788 \times 10^4 \text{ mm}^3$

Section classification
$$b/t_f = 75/9.0 = 8.33 < 9.4$$
$$d/t_w = [150 - 2 (9.0 + 10)]/5.7 = 19.65 < 42$$

Hence the section is plastic. Shear capacity is not being checked since the shear force is small and hence the section will be adequate.

Design capacity of the section
$$M_{dz} = (Z_{pz} \times f_y/\gamma_{m0}) = (119.82 \times 10^3 \times 250/1.1 \times 10^6) = 27.23 \text{ kN m}$$
$$\leq (1.2 \times Z_{ez} f_y/\gamma_{m0}) = [(1.2 \times 105 \times 10^3 \times 250)/(1.1 \times 10^6)] = 28.63$$
$$Z_{py} = 2t_f b_f^2/4 + (h - 2t_f) t_w^2/4 = 2 \times 9.0 \times 75^2/4 + (150 - 2 \times 9.0) \times 5.7^2/4$$
$$= 26384.6 \text{ mm}^3$$
$$M_{dy} = (Z_{py}f_y/\gamma_{m0}) = 26384.6 \times 250/(1.1 \times 10^6) = 6.0 \text{ kN m}$$
$$\leq 1.5 Z_{ey} f_y/\gamma_{m0} = 1.5 \times 19.5 \times 10^3 \times 250/(1.1 \times 10^6) = 6.6 \text{ kN m}$$

Hence the section is safe.

Overall member strength

For overall member strength, the following interaction equation must be satisfied.
$$(M_z/M_{dz}) + (M_y/M_{dy}) \leq 1.0$$
$$(12.126/27.23) + (1.105/6.0) = 0.629 < 1.0$$

Hence the member strength is satisfactory.

Check for deflection
$$\delta = (5wl^4/384EI) = (5 \times 3.76 \times 4000^4)/(384 \times 2 \times 10^5 \times 788 \times 10^4)$$
$$= 7.95 \text{ mm} < 22.22 \text{ mm}$$

Hence the section is safe.

Check for wind suction

From Example 6.10(a)

$$M_z = 2.492 \text{ kN m}$$
$$M_y = 0.0927 \times 4^2/32 = 0.0464 \text{ kN m}$$
$$f_{cr} = [1473.5/(KL/r_y)]^2 \{1 + (1/20) [(KL/r_y)/(h/t_f)]^2\}^{0.5}$$
$$KL/r_y = 4000/22 = 181.8$$
$$h/t_f = 150/9.0 = 16.67$$

Thus,
$$f_{cr} = (1473.5/181.8)^2 \{1 + (1/20) [181.8/16.67]^2\}^{0.5}$$
$$= 173.1 \text{ N/mm}^2$$
$$f_{bd} = 120.0 \text{ N/mm}^2 \text{ (from Table 13a of the code)}$$
$$M_{dz} = Z_{pz} f_{bd} = 119.82 \times 10^3 \times 120.0/10^6 = 14.38 \text{ kN m}$$

Overall buckling strength

For overall buckling strength, the following interaction equation should be satisfied.

$$(M_z/M_{dz}) + (M_y/M_{dy}) = (2.492/14.38) + (0.0464/6.0) = 0.18 < 1.0$$

Hence the overall buckling strength is satisfactory.

Hence by using one sag rod, it was possible to reduce the section from ISMC 200 to ISMC 150 (about 25% reduction in weight).

(c) Angle Section Purlin (as per BS 5950-1:2000)

From Example 6.10(a) $M_z = 12.126$ kN m; $W_p = (1.0 + 0.282 + 3.76) \times 4$
$$= 20.168 \text{ kN}$$

Moment at working load $= 12.126/1.5 = 8.084$ kN m

Let us assume that bending about z-z axis resists the vertical loads and the horizontal component is resisted by the sheeting.

Design strength $f_y = 250$ MPa

Applied moment = moment capacity of single angle

$$8.084 \times 10^6 = 250 \times Z_{ez}$$

Required $Z_{ez} = 8.084 \times 10^6/250 = 32.33 \times 10^3 \text{ mm}^3$

Provide ISA $150 \times 75 \times 10$ angle @ 0.17 kN/m,

with $Z_{ez} = 51.9 \times 10^3 \text{ mm}^3 > 20.168 \times 4 \times 10^6/1800 = 44.817 \times 10^3 \text{ mm}^3$

$$d/t = 150/10 = 15.0 > 10.5 \text{ but} < 15.7$$

The section is *semi-compact*.

Leg length perpendicular to plane of cladding
$$= 4000/45 = 88.88 \text{ mm} < 150 \text{ mm}$$

Leg length parallel to the plane of cladding
$$= 4000/60 = 66.66 \text{ mm} < 75 \text{ mm}$$

Deflection need not be checked in this case.

Summary

Beams are structural members that support loads which are applied transverse to their longitudinal axis. They resist the load primarily by bending and shear. Beams may be classified as lintels, spandrels, stringers, floor or roof beams, and purlins. Several types of beams exist, viz. angles, rolled I-sections, compound beams, castellated beams, plate girders, box girders, and joists. As in the case of columns, the sections are classified as plastic, compact, semi-compact, and slender depending on the width-to-thickness ratios of the individual plate elements. Local buckling can be prevented by limiting the width-to-thickness ratios. In many situations, the compression flanges of the beams are restrained by roofing elements in the form of trapezoidal sheeting or by reinforced concrete decking. In such cases, the beams attain their full plastic moment capacity, provided the beams chosen are plastic or compact.

When the laterally unrestrained length of the compression flange of the beam is long, then lateral buckling (also called lateral-torsional buckling) of the beam takes place. This phenomenon is similar to that of Euler buckling of columns. However, this form of instability is more complex compared to column instability, since the lateral buckling problem is three-dimensional is nature. The elastic lateral torsional buckling moment of symmetrical sections is derived based on several simplifying assumptions. It is observed that box sections provide greater resistance to lateral torsional buckling compared to I-sections, which are normally used as beams.

Several factors affect the lateral buckling behaviour, which include the type of cross section, type of loading, support conditions, restraint from other members, level of application of transverse load, effects of plasticity, residual stresses, and imperfections. The effect of these parameters is briefly discussed. We have to consider two effective length factors in the case of lateral torsional buckling of beams (one concerned with the restraint against lateral buckling and the other with the restraint against warping). Since this is complicated, in the code a single effective length factor only is specified for restraints against lateral bending, which has to be increased by 20% when restraints are not provided for warping. Intermediate restraints (in the form of bracings) and continuous restraints (in the form of decking) to the compression flange of the beams may be advantageously used to alter the lateral torsional buckling behaviour of beams. The stiffness and strength required for the bracings are also provided based on the American code provisions. Discussions about cantilever beams and continuous beams are also included. The effect of type of loading can be included by specifying an equivalent uniform moment factor.

The inelastic buckling is usually taken into account by specifying some type of empirical curve which give essentially the elastic solution for long beams and which terminate at $M_{cr} = M_p$, the plastic moment for short beams. Similar to the empirical multiple column curves, multiple beam curves could be generated by using an equivalent slenderness ratio $\lambda_{LT} = \sqrt{(M_p/M_{cr})}$. The provisions in the Indian code for the design strength of laterally supported and laterally unsupported beams are presented for both bending and shear.

The moment capacity of laterally continuous beams may be obtained by calculating all segmental critical loads individually (the distances between cross-beams may be considered as segments) and choosing the lowest value, assuming each segment is simply supported at ends. Shear forces may control short beams which carry heavy concentrated loads. Hence methods to consider shear strength of beams are given based on the provisions of the Indian code.

The beams should not excessively deflect or vibrate during the service life of the structure (limit state of serviceability). Hence some expressions are provided to check the beams for deflection. Information about and expressions for web buckling, web crippling, purlins, and biaxial bending are also included.

Exercises

1. Determine the buckling moment of an ISMB 250 section with $f_y = 250$ MPa when it is simply supported over a span of 3.5 m.
2. Select a rolled I-section to safely carry end moments of 500 kN m and 75 kN m over a laterally supported span of 5.5 m assuming the moments produce single curvature bending.
3. A simply supported beam spanning 6 m carries a udl of 3 kN/m including its self weight. Floor construction restrains it against lateral buckling. What size of beam is required with $f_y = 250$ MPa?
4. Calculate the maximum value of a mid span concentrated load W that can be safely applied on a laterally unsupported beam of length 4 m. The beam is simply supported at either end and is free to rotate in plane but restrained torsionally at their ends. The beam section is ISMB 400 and the load is applied at the top flange at mid span. f_y of steel is 250 MPa. See Fig. 6.21.

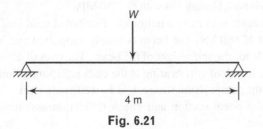

Fig. 6.21

5. Select a standard hot-rolled section of steel with $f_y = 250$ MPa for the beam shown in Fig. 6.22. The beam has continuous lateral support. The maximum permissible live load deflection is $L/360$.

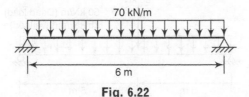

Fig. 6.22

6. The beam shown in Fig. 6.23 must support two concentrated live loads of 90 kN each at the quarter points. The maximum deflection must not exceed $L/240$. Lateral support is provided at the ends of the beam. Use $f_y = 250$ MPa steel and select a rolled section.

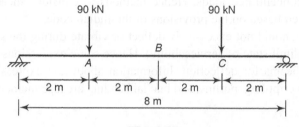

Fig. 6.23

7. Using steel with $f_y = 250$ MPa, select a rolled I-section for the beam shown in Fig. 6.24. The loads shown are service loads. Lateral bracing is provided by a concrete slab. There is no restriction on deflection.

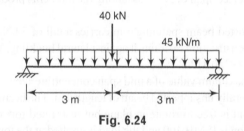

Fig. 6.24

8. Compute the moment carrying capacity and shear strength of a laterally unrestrained ISMB 500 member of length 5 m with $f_y = 250$ MPa.
9. A compound beam is to carry a uniformly distributed dead load of 400 kN and an imposed load of 600 kN. The beam is simply supported and has a span of 8 m. Assume 30 kN for the self weight of the beam. The overall depth must not exceed 500 mm. The length of stiff bearing at the ends is 215 mm. Full lateral support is provided for the compression flange. Use Fe 410 grade steel.
    (a) Design the beam section and check deflection assuming uniform section throughout.
    (b) Determine the theoretical cut-off points for the flange plates.
10. The beam shown in Fig. 6.25 has continuous lateral support. Using steel with $f_y = 250$ MPa, select a rolled section. There is no limit on the deflection.

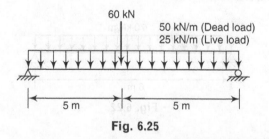

Fig. 6.25

11. The cantilever beam shown in Fig. 6.26 is laterally supported only at the fixed end. The concentrated load is service live load. Use steel with $f_y = 250$ MPa and select a rolled section. There is no limit on deflection.

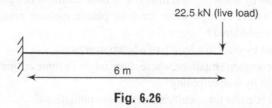

Fig. 6.26

12. Design an I-section purlin for an industrial building situated in Cuttack to support a 18 gauge galvanized corrugated iron sheet roof (with a dead weight of 131 N/m^2) with the following data:
   Spacing of the truss c/c = 5.0 m
   Span of truss = 12 m
   Spacing of purlin c/c = 1.5 m
   Intensity of wind pressure = 2.3 kN/m^2
   Yield stress of steel = 250 MPa

13. Design a purlin for a roof truss having the following data:
   Span of truss = 16 m
   Spacing of truss = 4.0 m centre-to-centre
   Inclination of the roof = 30°
   Spacing of purlin = 1.6 m centre-to-centre
   Wind suction = 1.4 kN/m^2
   Roof covering is AC sheets (assume a weight of 1.68 kN/m^2). Take live load as 0.75 kN/m^2 on plan area less 0.02 kN/m^2 for every degree increase in slope over 10°, subject to a minimum 0.40 kN/m^2.
   (a) Provide channel purlin
   (b) Provide channel purlin with a sag rod at mid-span
   (c) Provide angle purlin and design by the empirical method.

Review Questions

1. Why are rolled I-sections widely used as beam members?
2. What are the checks to be performed for beam member design?
3. What are castellated sections and under what conditions are they used?
4. What is the difference between bending and buckling of a beam member?
5. What is meant by lateral torsional buckling of a beam member?
6. Under what conditions can lateral buckling occur?
7. Under what conditions can a beam member be assumed as laterally restrained?
8. What is the difference between column buckling and beam buckling?
9. An I-beam loaded perpendicular to the minor axis does not buckle. Why?
10. What is local buckling of a beam member?
11. Differentiate between local and lateral buckling of beams.

12. How can the lateral buckling behaviour be improved in a beam member?
13. Under what conditions should a beam be checked for shear?
14. Which section performs best in torsion and why?
15. What is meant by shear lag and how is it accounted for in design?
16. What is the effect of high shear force on plastic moment resistance and when should it be considered?
17. What is meant by effective length of a beam member?
18. Write down common situations where shear might become critical?
19. What is meant by web crippling?
20. What are the reasons for specifying deflection limitations?
21. What is meant by camber and why is it provided?

CHAPTER 7

Design of Plate Girders

Introduction

A plate girder is basically an I-beam built up from plates using riveting or welding. It is a deep flexural member used to carry loads that cannot be economically carried by rolled beams. Standard rolled sections may be adequate for many of the usual structures; but in situations where the load is heavier and the span is also large, the designer has the following choices (Fig. 7.1).

- Use two or more regularly available sections, side-by-side.
- Use a cover-plated beam, i.e., weld a plate of adequate thickness to increase the bending resistance of the flange.
- Use a fabricated plate girder, which provides the freedom (within limits) to choose the size of web and flanges.
- Use a steel truss or a steel-concrete-composite truss.

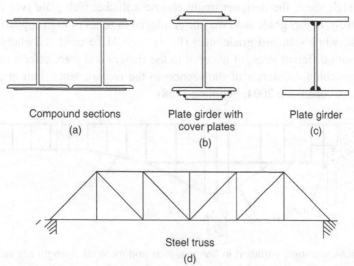

Compound sections
(a)

Plate girder with cover plates
(b)

Plate girder
(c)

Steel truss
(d)

Fig. 7.1 Options for beams of long spans to carry heavy loads

Among the options available, the first is usually uneconomical and does not satisfy deflection limitation. The second option is advantageous where the rolled section is marginally inadequate. Therefore, the real choice lies between the plate girder and the truss girder. The moment resisting capacity of plate girders lies between rolled I-sections and truss girders. Truss girders involve higher costs of fabrication and erection, problems of vibration and impact, and require higher vertical clearance. For short spans (<10 m), plate girders are uneconomical due to higher connection costs and rolled I-sections are the preferred choice.

7.1 Plate Girders

As stated earlier, plate girders are deep flexural members used to carry loads that cannot be carried economically by rolled beams. Plate girders provide maximum flexibility and economy. In the design of a plate girder, the designer has the freedom to choose components of convenient size, but has to provide connections between the web and the flanges. Plate girders offer a unique flexibility in fabrication and the cross section can be uniform or non-uniform along the length. It is possible to provide the exact amount of steel required at each section along the length of the girder by changing the flange areas and keeping the same depth of the girder. In other words, it can be shaped to match the bending moment curve itself. Thus, a plate girder offers limitless possibilities to the creativity of the engineer.

It is a normal practice to fabricate plate girders by welding together three plates, though (see Fig. 7.6) in the past, plate girders were constructed by riveting or bolting. It is possible to have tapered, cranked, and haunched girders (see Fig. 7.2). We can also have holes in the webs to accommodate the services (see Fig. 7.3). The designer may choose to reduce flange thickness (or breadth) in the zone of low applied moment, especially when a field-splice facilitates the change. Similarly in the zone of high shear, the designer might choose a thicker web plate (see Fig. 7.4). Alternatively, higher grade steel may be employed in zones of high applied moment and shear, while standard grade steel (Fe 410) could be used elsewhere. Hybrid girders, with different strength material in the flanges and web, offer a method of closely matching resistance of the section to the requirements (Fan et al. 2006; Veljkovic & Johansson 2004; Schilling 1968).

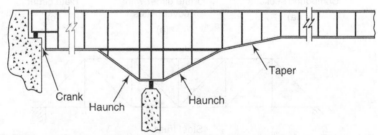

Note Splices, camber, variation in flanges, web and material strength are not shown

Fig. 7.2 Plate girder with haunches, tapers, and cranks

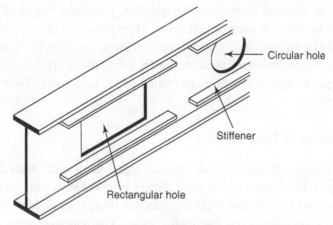

Fig. 7.3 Plate girder with holes for services

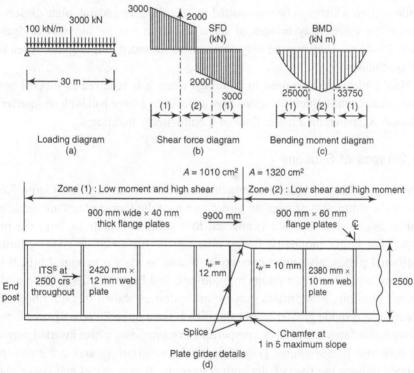

Fig. 7.4 Plate girder with splice and variable cross section (ESDEP 2006)

For making the cross section efficient in resisting the in-plane bending, it is required that maximum material is placed as far away from the neutral axis as possible. From this point of view it is economical to keep the flanges as far apart as possible. The axial forces in the flanges decrease, as the depth of the girder increases. Thus a smaller area of cross section would suffice than would be the case if a smaller depth were chosen. However, this would also mean that the web would be thin and deep. In such a situation, premature failure of the girder due to web buckling in shear might occur. Here there is a choice between thin web provided with vertical and horizontal stiffeners, and a thicker web requiring no stiffening (and therefore avoiding costly fabrication). The choice between the two depends upon a careful examination of the total costs of both forms of construction.

7.1.1 Examples of Plate Girders

Most steel highway bridges in USA for spans less than about 24 m are steel rolled beam bridges. For longer spans, plate girders compete very well economically. Similarly, when loads are extremely large, as in railway bridges, plate girders may prove to be economical even for smaller spans. The upper economical limit of plate girder spans depend on the following factors: (a) whether the bridge is simple or continuous, (b) whether it is a highway or a railway bridge, and (c) the length of the section which can be transported in one piece. In general, plate girders are economical for railway bridges of spans 15–40 m and for highway bridges of spans 24–46 m. They may be very competitive for much longer spans, when they are continuous.

Plate girders are also used in buildings when it is required to support heavy concentrated loads. Such situations may arise when a large hall with no interfering columns is desired on a lower floor of a multi-storey building.

7.1.2 Types of Sections

Several possible plate girder arrangements are shown in Fig. 7.6. Figure 7.6(a) shows the simplest type of plate girder in which the flange plates are made of a pair of angles and they are connected to a solid web plate to form the plate girder. For larger moments, flange area can be increased by riveting/bolting additional plates, also known as cover plates, as shown in Fig. 7.6(b). When there are head room restrictions in buildings, and larger depths of plate girders are not possible, box girders may be an option as shown in Fig. 7.6(c). Box girders also provide greater lateral stability. If the number of cover plates is to be reduced, the flange area can be compensated by providing plates inserted between the web and flange angles [Fig. 7.6(d)]. These arrangements are sometimes adopted to keep the rivets/bolts both connecting flange angles and cover plates shorter. Figures 7.6(e) and 7.6(f) show typical welded plate girders. The form shown in Fig. 7.6(e) is the most commonly used type of welded plate girder. Here, flange angles are not required and instead of using a number of cover plates, a thick plate is used as the flange. As discussed earlier, a thinner cover plate can be used where lesser flange area is desired such as at regions away from the maximum bending moment and the plates of different thicknesses can be joined by butt welding.

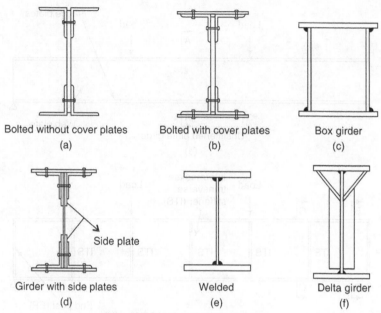

Bolted without cover plates	Bolted with cover plates	Box girder
(a)	(b)	(c)

Girder with side plates	Welded	Delta girder
(d)	(e)	(f)

Side plate

Fig. 7.6 Types of plate girders

7.1.3 Elements of a Plate Girder

The various components of welded stiffened and unstiffened plate girders, shown in Fig. 7.7, are as follows:

(a) Web plate

(b) Flange plate with or without cover plates (with curtailment of cover plate)

(c) Bearing stiffeners or end post (EP)

(d) Intermediate transverse stiffeners (ITS)

(e) Longitudinal stiffeners (LS)

(f) Web splices

(g) Flange splices

(h) Connection between flange and the web

(i) End bearing or end connections

Slender webs (with large d/t values) would buckle at relatively low values of applied shear loading.

A girder of high strength to weight ratio can be designed by incorporating the post buckling strength of the web in the design method employed. This would be particularly advantageous where the reduction of self-weight is of prime importance. Examples of such situations arise in long span bridges, ship girders, transfer girders in buildings, etc.

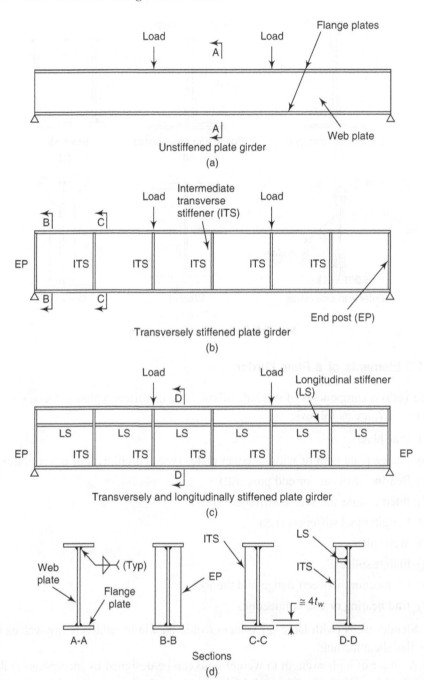

Fig. 7.7 Stiffened and unstiffened plate girders

7.2 General Considerations

Problems arise in plate girders because of the deep thin webs. One way of improving the load carrying capacity of a slender web plate is to provide stiffeners; selection of the right kind of stiffening is an important consideration in a plate girder design.

The modes of failure of a plate girder are by yielding of the tension flange and buckling of the compression flange. The compression flange buckling can take place in various ways, such as vertical buckling into the web, flange local buckling, or lateral torsional buckling.

Plate girders depend on the post-buckling strength of the webs. At high shear locations in the girder web, normally near the supports and the neutral axis, the principal planes would be inclined to the longitudinal axis of the member. Along the principal planes, the principal stresses would be diagonal tension and diagonal compression. Though diagonal tension does not cause any problem, diagonal compression causes the web to buckle in a direction perpendicular to its action. This problem can be solved by any of the following three ways:

- Reduce the depth to thickness ratio of the web such that the problem is eliminated.
- Provide web stiffeners to form panels that would enhance the shear strength of the web.
- Provide web stiffeners to form panels that would develop tension field action to resist diagonal compression.

Figure 7.8(a) shows the tension field action in a panel. As the web begins to buckle, the web loses its ability to resist the diagonal compression. The diagonal compression is then transferred to the transverse stiffeners and the flanges. The vertical component of the diagonal compression is supported by the stiffeners and the flanges resist the horizontal component. The web resists only the diagonal tension and this behaviour of the web is called *tension field action*. The behaviour is very similar to a Pratt truss, in which the vertical members carry compression and the tension is carried by the diagonals, as shown in Fig. 7.8(b). The contribution of the tension field action is realized only after the web starts to buckle. The total strength of the web is made up of the web strength before buckling and the post buckling strength developed due to tension field action. Thus, the provision of suitably spaced stiffeners, called *intermediate stiffeners,* can be used to develop tension field action and improve the shear capacity of the web. The main purpose of these stiffeners is to provide stiffness to the web rather than to resist the applied loads.

Additional stiffeners, called *bearing stiffeners*, are provided at points of concentrated loads, to protect the web from the direct compressive loads. They can simultaneously act as intermediate stiffeners also.

Application of concentrated loads on the top flange may produce web yielding, web crippling, and side-sway web buckling. This form of buckling occurs when the compression in the web causes the tension flange to buckle laterally. In order to prevent this, the relative movements of the flanges should be restrained by lateral bracing. General welding procedures are followed for connecting the components during the design of welds.

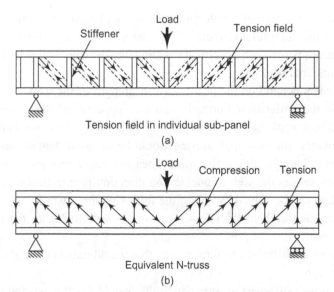

Tension field in individual sub-panel
(a)

Equivalent N-truss
(b)

Fig. 7.8 Tension field action and the equivalent N-truss

7.3 Preliminary Design Procedure

General dimensions of a plate girder are fixed based on its optimum behaviour. Some guidelines for proportioning the flanges and webs are given in Section 7.8. One condition that may limit the proportions of the girder is the largest size that can be fabricated in shop and transported to the site. There may be transportation problems such as clearance requirements (of overhead bridges) that limit maximum depths to about 3–3.6 m. A plate girder design begins with values of live load shears and bending moments and estimated dead loads. After the selection of the girder depth, the web may be designed to resist the shear, taking into account the minimum thickness requirements.

A major portion of the moment of resistance in a I-section plate girder is provided by the flanges; the contribution of the web being very small. On the tension side, provision of bolt holes may reduce the effective flange area, which may be compensated by adding more flange area. In the design of a plate girder, a trial section needs to be assumed initially and the amount of load to be carried includes the self weight of the girder also. A preliminary estimate of the girder weight can be obtained using the simplified flange area method.

The design requirements for plate girders are addressed in Section 8 of IS 800 : 2007 specification. The identification of a laterally supported member as a beam or a plate girder is decided based on the d/t_w ratio, where d is the depth of the beam and t_w is the thickness of the web. If d/t_w is less than 67ε, the member is classified as a beam and the respective clauses applicable to beams are used, irrespective of whether the member is made up of plates or a rolled section. If the ratio is greater than 67ε, the clauses relating to plate girders are invoked.

7.3.1 Minimum Web Thickness

The thickness of the web in a section should satisfy the following requirements with respect to serviceability (clause 8.6.1 of the code):

(a) When transverse stiffeners are not provided:

$$\frac{d}{t_w} \leq 200\varepsilon \text{ (web connection to flanges along both longitudinal edges)}$$

$$\frac{d}{t_w} \leq 90\varepsilon \text{ (web connection to flanges along one longitudinal edge only)}$$

(b) When only transverse stiffeners are provided:

(i) When $3d \geq c \geq d$

$$\frac{d}{t_w} \leq 200\varepsilon_w$$

(ii) When $0.74d \leq c < d$

$$\frac{c}{t_w} \leq 200\varepsilon_w$$

(iii) When $c < 0.74d$

$$\frac{d}{t_w} \leq 270\varepsilon_w$$

(iv) When $c > 3d$, the web is considered as unstiffened.

(c) When transverse stiffeners and longitudinal stiffeners are provided at one level only, at $0.2d$ from the compression flange:

(i) When $2.4d \geq c \geq d$

$$\frac{d}{t_w} \leq 250\varepsilon_w$$

(ii) When $0.74 d \leq c \leq d$

$$\frac{c}{t_w} \leq 250\varepsilon_w$$

(iii) When $c < 0.74 d$

$$\frac{d}{t_w} \leq 340\varepsilon_w$$

(d) When there is a second longitudinal stiffener (provided at neutral axis):

$$\frac{d}{t_w} \leq 400\varepsilon_w$$

where d is the depth of the web, t_w is the thickness of the web, c is the spacing of the transverse stiffener (see Fig. 7.9), $\varepsilon_w = \sqrt{250/f_{yw}}$, and f_{yw} is the yield stress of the web.

These criteria are to ensure that the web will not buckle under normal service conditions. It should also be ensured that the web is strong enough so that the flange will not buckle into the web. It is also evident that a thinner web can be used if the *c/d* ratio is less than one and that such a web panel has a higher strength.

7.3.2 Flexural Strength

The method by which the calculation of the bending capacity of a plate girder is determined depends upon the classification of the flanges and the web as calculated from b/t_f and d/t_w ratios. The limiting ratios are presented in the chapter on beams. As mentioned earlier, if the web slenderness ratio is less than 67ε, then the member should be designed as a rolled beam section. The flexural strength of a plate girder is based on the tension flange yielding or otherwise on the compression flange buckling. The compression flange buckling strength will be decided by either flange local buckling or lateral torsional buckling (see Section 6.7).

For sections with slender webs, when their flanges are plastic, compact, or semi-compact (i.e., $d/t_w > 67\varepsilon$), the design bending strength is calculated using one of the following methods:

(a) The applied moment is resisted by the flanges with a uniform stress distribution and the applied shear is resisted by the web. Using this assumption, it is convenient to make preliminary sizes of the girder but the final design may turn out to be uneconomical. The method is very simple and applicable only to laterally fully restrained beams.

(b) The bending moment and axial force acting on the section may be assumed to be resisted by the whole section, and the web designed for combined shear and normal stresses, by using simple elastic theory in case of semi-compact flanges and simple plastic theory in the case of compact and plastic flanges.

For sections with slender flanges, moment capacity is calculated based on the approach similar to that of a slender rolled section. In the design of plate girders, a minimum weight solution by adopting thin webs may not be economical if we consider the fabrication cost of the stiffeners.

7.3.3 Lateral Torsional Buckling of Plate Girders

The same procedure as adopted for rolled sections (Section 6.7) may be used except that different values of the imperfection coefficient α_{LT} have to be used to account for higher residual stresses in welded plate girders. Approximate formulae may be used to arrive at the sections properties of the members.

7.3.4 Shear Strength

The shear strength of a plate girder depends upon the depth to thickness ratio of the web and the spacing of the intermediate stiffeners provided. Webs of plate girders are usually stiffened transversely as shown in Fig. 7.7. This helps to increase the ultimate shear resistance of the webs. The shear capacity of the web has two components, namely, strength before the onset of buckling and strength after post-buckling. As the shear load is increased on a stiffened web panel, the web panel buckles (see Fig. 7.9). This load does not indicate the maximum shear capacity of the web. The load can be still increased and the web panel continues to carry further load relying on the tension field action. Part of the buckled web takes the load in tension. This tension member action is across the web panel in an inclined direction to the web panel diagonal as shown in Fig. 7.10.

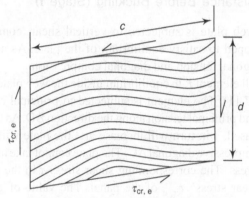

Fig. 7.9 Buckling of the web panel

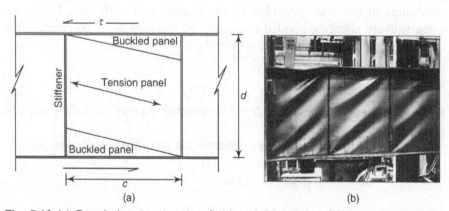

(a)   (b)

Fig. 7.10 (a) Panel showing tension field and (b) tension field in a plate girder

At this stage, the girder acts like an N-truss (Fig. 7.8) with the compression forces being carried by the flanges and the intermediate stiffeners and the buckled web resisting the tension. This additional reserve strength is termed as *tension field action*. It may be noted here that if no intermediate stiffeners are present or their spacing is large, it is not possible for tension field action to take place and the shear capacity is restricted to the strength before buckling. The stiffener spacing c influences both buckling and post-buckling behaviour of the web under shear. The various stages of shear resistance are explained in the following section.

A long span plate girder may have various web panels, each panel having different combinations of bending moments and shear forces. Panels, which are closer to the support will be predominantly under shear, and those close to the centre will be under predominant bending moments. The effect of the shear alone will be considered first, followed by the case where the effect of combined bending moment and shear forces would be treated.

7.4 Web Panel Subjected to Shear

7.4.1 Shear Resistance Before Buckling (Stage 1)

When a square web plate is subjected to vertical shear, complementary shear stresses are developed to satisfy equilibrium of the plate. As a consequence, the plate develops diagonal tension and diagonal compression.

Consider a small element E in equilibrium inside the web plate subject to a shear stress τ (see Fig. 7.11). The element is subjected to principal compression along the direction AC and principal tension along the direction BD. As the applied loading is gradually increased, τ in turn will increase and the plate will buckle along the direction of the compressive diagonal AC. The plate cannot take any further increase in compressive stress. The corresponding shear stress τ in the plate is called the 'elastic critical shear stress' $\tau_{cr,\,e}$ of the panel. The value of $\tau_{cr,\,e}$ can be easily determined from the classical stability theory, if the boundary conditions of the plate can be established. However, it is very difficult to know the true boundary conditions of the plate provided by the flanges and stiffeners. Therefore, conservatively, simply supported conditions are assumed at the boundaries. The critical shear stress in such a case is given by

$$\tau_{cr,\,e} = k_v \frac{\pi^2 E}{12(1-\mu^2)}\left(\frac{t_w}{d}\right)^2 \tag{7.1}$$

where k_v is a coefficient which depends upon the support conditions and μ is Poisson's ratio.

$k_v = 5.35$, when transverse stiffeners are provided only at support
$ = 4.0 + 5.35/(c/d)^2$ for $c/d < 1.0$
$ = 5.35 + 4.0/(c/d)^2$ for $c/d \geq 1.0$

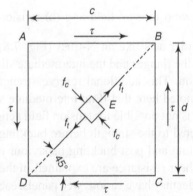

Fig. 7.11 Unbuckled shear panel ($\tau < \tau_{cr}$)

where c and d are the spacing of transverse stiffeners and the depth of the web, respectively.

The limiting value of d/t_w is given by

$$\frac{d}{t_w}\sqrt{\frac{f_y}{250}} = 82.0$$

Thus, it is seen that for $f_y = 250$ MPa, elastic buckling due to shear will occur when the d/t ratio of the web is greater than 82.

The elastic buckling stress may be significantly increased either by using intermediate transverse stiffeners to decrease the aspect ratio c/d (thereby increasing the value of buckling coefficient k_v), or by using longitudinal stiffeners to decrease the depth-to-thickness ratio d/t_w. The aspect ratio of each panel in the range of 0.5 to 2 is found to be more efficient.

IS 800 : 2007 provisions for Resistance to Shear Buckling   Resistance to shear buckling shall be verified when

$$\frac{d}{t_w} > 67\varepsilon \text{ for an unstiffened web}$$

$$> 67\varepsilon\sqrt{\frac{k_v}{5.35}} \text{ for a stiffened web}$$

where k_v is the shear buckling coefficient and $\varepsilon = \sqrt{250/f_y}$.

7.4.2 Shear Buckling Design

The nominal shear strength V_n of webs with or without intermediate stiffeners as governed by buckling may be evaluated using one of the following methods:
* Simple post-critical method
* Tensile field theory

These two methods are discussed in the following sections.

7.4.2.1 Simple Post-critical Method

The simple post-critical method is a general method and can be applied to the design of all girders. The simple post critical method can be used for webs of I-section girders, with or without intermediate transverse stiffener, provided that the web has transverse stiffeners at the supports. The nominal shear strength is given by

$$V_n = V_{cr} \tag{7.2a}$$

where V_{cr} is the shear force corresponding to web buckling.

$$V_{cr} = A_v \tau_b \tag{7.2b}$$

where A_v is the shear area (see Table 6.7) and τ_b is the shear stress corresponding to web buckling, determined as follows:

Table 7.1 Shear Buckling Strength τ_{au} (N/mm^2) of Webs for f_y = 250 MPa

d/t	Stiffener spacing ratio c/d														
	0.4	0.5	0.6	0.7	0.8	0.9	1.0	1.2	1.4	1.6	1.8	2.0	2.5	3.0	Infinity
55	144.3	144.3	144.3	144.3	144.3	144.3	144.3	144.3	144.3	144.3	144.3	144.3	144.3	144.3	144.3
60	144.3	144.3	144.3	144.3	144.3	144.3	144.3	144.3	144.3	144.3	144.3	144.3	144.3	144.3	144.3
65	144.3	144.3	144.3	144.3	144.3	144.3	144.3	144.3	144.3	144.3	144.3	144.3	144.3	144.3	144.3
70	144.3	144.3	144.3	144.3	144.3	144.3	144.3	144.3	144.3	144.3	144.3	144.3	143.4	141.9	138.0
75	144.3	144.3	144.3	144.3	144.3	144.3	144.3	144.3	144.3	143.7	141.4	139.6	136.8	135.1	131.0
80	144.3	144.3	144.3	144.3	144.3	144.3	144.3	144.3	140.7	137.5	135.0	133.2	130.1	128.3	123.9
85	144.3	144.3	144.3	144.3	144.3	144.3	144.3	139.5	134.7	131.3	128.7	126.7	123.4	121.5	116.9
90	144.3	144.3	144.3	144.3	144.3	144.3	140.7	133.7	128.7	125.1	122.3	120.2	116.8	114.8	109.8
95	144.3	144.3	144.3	144.3	144.3	141.6	135.4	128.0	122.7	118.9	116.0	113.8	110.1	108.0	102.8
100	144.3	144.3	144.3	144.3	143.9	136.6	130.0	122.3	116.7	112.7	109.6	107.3	103.4	101.2	95.7
105	144.3	144.3	144.3	144.3	139.3	131.5	124.7	116.6	110.7	106.5	103.3	100.8	96.8	94.4	87.8
110	144.3	144.3	144.3	143.8	134.7	126.5	119.4	110.9	104.7	100.3	98.4	94.9	89.6	86.6	80.0
115	144.3	144.3	144.3	139.6	130.0	121.5	114.0	105.1	98.7	94.6	90.1	86.9	81.9	79.3	73.2
120	144.3	144.3	144.3	135.4	125.4	116.5	108.7	99.4	92.9	86.8	82.7	79.8	75.3	72.8	67.2
125	144.3	144.3	142.8	131.2	120.7	111.5	103.4	94.1	85.6	80.0	76.2	73.5	69.4	67.1	61.9
130	144.3	144.3	139.1	126.9	116.1	106.5	100.1	87.0	79.1	74.0	70.5	68.0	64.1	62.0	57.3
135	144.3	144.3	135.3	122.7	111.5	101.5	92.8	80.7	73.4	68.6	65.4	63.0	59.5	57.5	53.1
140	144.3	144.3	131.6	118.5	106.8	97.9	86.3	75.0	68.2	63.8	60.8	58.6	55.3	53.5	49.4
145	144.3	142.9	127.8	114.3	102.2	91.2	80.5	69.9	63.6	59.5	56.7	54.6	51.5	49.9	46.0
150	144.3	139.6	124.1	110.0	99.4	85.3	75.2	65.4	59.4	55.6	52.9	51.1	48.2	46.6	43.0
155	144.3	136.4	120.3	105.8	93.1	79.9	70.4	61.2	55.7	52.1	49.6	47.8	45.1	43.6	40.3
160	144.3	133.2	116.6	101.6	87.3	74.9	66.1	57.4	52.2	48.8	46.5	44.9	42.3	40.9	37.8

(contd)

165	144.3	129.9	112.8	99.1	82.1	70.5	62.1	54.0	49.1	45.9	43.8	42.2	39.8	38.5	35.6
170	144.3	126.7	109.0	93.4	77.4	66.4	58.5	50.9	46.3	43.3	41.2	39.7	37.5	36.3	33.5
175	143.4	123.5	105.3	88.1	73.0	62.6	55.2	48.0	43.7	40.8	38.9	37.5	35.4	34.2	31.6
180	140.8	120.2	101.5	83.3	69.0	59.2	52.2	45.4	41.3	38.6	36.8	35.5	33.4	32.4	29.9
185	138.1	117.0	99.7	78.9	65.3	56.1	49.4	43.0	39.1	36.5	34.8	33.6	31.7	30.6	28.3
190	135.4	113.8	94.5	74.8	61.9	53.1	46.9	40.7	37.0	34.6	33.0	31.8	30.0	29.0	26.8
195	132.8	110.5	89.7	71.0	58.8	50.5	44.5	38.7	35.2	32.9	31.3	30.2	28.5	27.6	25.5
200	130.1	107.3	85.3	67.5	55.9	48.0	42.3	36.8	33.4	31.3	29.8	28.7	27.1	26.2	24.2
205	127.4	104.0	81.2	64.2	53.2	45.7	40.2								
210	124.8	100.8	77.4	61.2	50.7	43.5	38.4								
215	122.1	99.4	73.8	58.4	48.4	41.5	36.6				Not applicable				
220	119.4	94.9	70.5	55.8	46.2	39.6	34.9								
225	116.8	90.8	67.4	53.3	44.2	37.9	33.4								
230	114.1	86.9	64.5	51.0	42.3	36.3	32.0								
235	111.4	83.2	61.8	48.9	40.5	34.7	30.6								
240	108.8	79.8	59.2	46.9	38.8	33.3	29.4								
245	106.1	76.6	56.8	45.0	37.2	32.0	28.2								
250	103.4	73.5	54.6	43.2	35.8	30.7	27.1								

Note Intermediate values may be obtained by linear interpolation.

(a) When $\lambda_w \leq 0.8$

$$\tau_b = f_{yw}/\sqrt{3} \qquad\qquad (7.3a)$$

(b) When $0.8 < \lambda_w < 1.2$

$$\tau_b = [1 - 0.8(\lambda_w - 0.8)](f_{yw}/\sqrt{3}) \qquad\qquad (7.3b)$$

(c) When $\lambda_w \geq 1.2$

$$\tau_b = f_{yw}/(\sqrt{3}\lambda_w^2) \qquad\qquad (7.3c)$$

where λ_w is the non-dimensional web slenderness ratio for shear buckling stress, given by

$$\lambda_w = \sqrt{f_{yw}/(\sqrt{3}\tau_{cr,e})} \qquad\qquad (7.4)$$

The elastic critical shear stress of the web $\tau_{cr,\,e}$ is given by Eqn (7.1)

Table 7.1 provides the values of shear buckling strength τ_b for various values of stiffener spacing ratio c/d and depth-to-thickness ratio d/t_w.

7.4.2.2 Web Strength using Tension Field Theory

As stated earlier, a plate girder with intermediate stiffeners ($1.0 \leq c/d \leq 3.0$) derives considerable post-buckling shear strength due to the tension field action. Tension field action can be applied to a certain range of girders only and will produce efficient designs of girders due to its taking advantage of the post-buckling reserve of resistance. According to the theory of tension field action, there are three components contributing to the predicted shear strength: the first is the primary buckling strength of the plate, the second is due to the tension field action of the web, and the third arises from the plastic moment capacity of the flanges. Basler (1961) was the first to formulate a successful model for tension-field action for plate girders of the type used in civil engineering structures. Basler's approach (1961, 1963) disregarded the flexural rigidity of the flanges and hence did not reflect the true anchorage of the tension field along the flanges.

The model as suggested by Porter et al. (1975), is often referred to as the *Cardiff model*. It takes into account the effect of the bending stiffness of the flanges on the width of the diagonal tension band. Thus the diagonal tension field is considered to be composed of a central part that anchors on the transverse stiffeners, and two additional parts anchored on the flanges. It has been found that the shear model as proposed by Porter et al. (1975) provides a fairly good correlation between theory and test results (Maquoi 1992; Valitnat 1982).

7.4.3 Provisions of IS 800 : 2007

The tension field method may be used for girders having transverse stiffeners at the supports and at intermediate points. In the case of end panels, the end posts provide anchorage for the tension fields. It is also necessary that $c/d \geq 1.0$, where c and d are the spacing of transverse stiffeners and the depth of the web, respectively.

In the tension field method, the nominal shear resistance V_n is obtained from $V_n = V_{tf}$ with

$$V_{tf} = [A_v \tau_b + 0.9 w_{tf} t_w f_v \sin \phi] \leq V_p \qquad (7.5)$$

where τ_b is the buckling strength, as obtained by Eqn (7.3), A_v is the shear area (see Table 6.7), f_v is the yield strength of the tension field obtained from

$$f_v = [f_{yw}^2 - 3\tau_b^2 + \psi^2]^{0.5} - \psi \qquad (7.6a)$$

In Eqn (7.6a),

$$\psi = 1.5 \tau_b \sin 2\phi \qquad (7.6b)$$

$$\phi = \text{inclination of the tension field} = \tan^{-1}\left(\frac{d}{c}\right) \qquad (7.6c)$$

The above equation given in the code is correct for θ (slope of the panel diagonal). The value of ϕ ranges between $\theta/2$ to θ. The minimum value of $\theta/2$ applies when the flanges are fully utilized in resisting the applied bending moment, and the maximum value of θ applies when there is complete tension field condtion with $s = c$. Any assumed value between $\theta/2$ and θ is conservative. Eurocode suggests an approximate value of $\phi = \theta/1.5$. We will use this value in the calculations. The width of the tension field w_{tf} is given by

$$w_{tf} = d\cos\phi - (c - s_c - s_t) \sin\phi \qquad (7.7)$$

and f_{yw} is the yield stress of the web, d is the depth of the web, c is the spacing of stiffeners in the web, and s_c and s_t are the anchorage lengths of the tension field along the compression and tension flange respectively, obtained from

$$s = \frac{2}{\sin\phi}\left[\frac{M_{fr}}{f_y t_w}\right]^{0.5} \leq c \qquad (7.8)$$

In this equation, M_{fr} is the reduced plastic moment of the respective flange plate (disregarding any edge stiffener), after accounting for the axial force N_f in the flange, due to the overall bending, and any external axial force in the cross section, as follows

$$M_{fr} = 0.25 b_f t_f^2 f_{yf} [1 - \{N_f/(0.909 b_f t_f f_{yf})\}^2] \qquad (7.9)$$

where b_f and t_f are the width and thickness of the relevant flange, respectively.

In the case of a stiffened web of a plate girder, it would be of interest to consider the roles played by the stiffeners in the behaviour of the girders. The code provides various methods of designing the end panel depending upon whether tension field action can be developed in the end panel or not. These methods are explained in the next section.

7.4.4 End Panel Design Without Tension Field Action

A tension field cannot normally be fully developed in an end panel. This is evident by looking at the horizontal components of the tension fields shown in Fig. 7.12.

The vertical components are resisted by the stiffeners. Tension field in panel CD is balanced by the tension field in panel BC. Therefore the interior panels are anchored by the neighboring panels. The end panel AB has no anchorage on the left side. Anchorage can be provided by an end stiffener that is specially designed to resist the bending induced by the tension field.

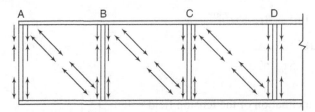

Fig. 7.12 Tension fields in a plate girder

The end panel (panel *B* in Fig. 7.13) may be designed according to the post-critical method, explained earlier. Additionally, the end panel along with the stiffeners should be checked as a beam spanning between the flanges of the girder to resist a shear force R_{tf} and a moment M_{tf} [as given by Eqn. (7.10a)]. The end stiffener should also be capable of resisting the reaction plus a compressive force due to the moment M_{tf}.

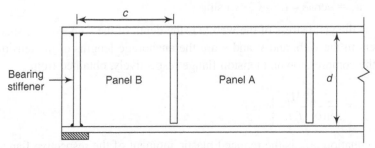

Fig. 7.13 End panel design without tension field action

End panels designed using tension field action (Figs 7.14 and 7.15) In this case, the interior panels are designed as per the provisions explained earlier. The end panel is provided with an end post consisting of a single or double stiffener. If a single stiffener is provided, it should play the roles of a bearing stiffener and end post (Fig. 7.14). These stiffeners need to satisfy the following criteria.

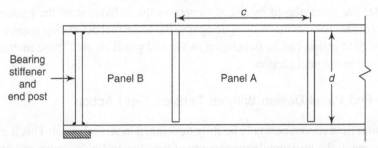

Fig. 7.14 End panel designed with tension field action (single stiffener)

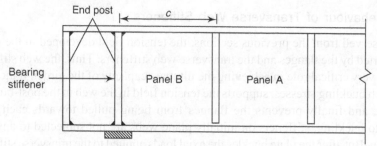

Fig. 7.15 End panel with double stiffener

In the case of a single stiffener, the top of the end post should be rigidly connected to the flange using full strength welds. The end post should be capable of resisting the reaction plus a moment from the anchor forces equal to $2/3M_{tf}$, where M_{tf} is obtained by Eqn (7.10a). The width and thickness of the end post should not exceed the width and thickness of the flange.

The anchor field forces developed in the panel A creates a resultant longitudinal shear, R_{tf}, plus a moment M_{tf} on the panel B. They are evaluated as follows.

$$R_{tf} = \frac{H_q}{2} \text{ and } M_{tf} = \frac{H_q d}{10} \tag{7.10a}$$

where $H_q = 1.25 V_{dp} \left(1 - \frac{V_{cr}}{V_{dp}}\right)^{1/2}$ (7.10b)

$$V_{dp} = \frac{d t_w f_y}{\sqrt{3}} \tag{7.10c}$$

If the actual factored shear force V in the panel designed using tension field approach is less than the nominal shear strength V_{nt}, as determined earlier, then the values of H_q may be reduced by the ratio

$$\frac{V - V_{cr}}{V_{tf} - V_{cr}} \tag{7.10d}$$

where d is the web depth, V_{tf} is the the basic shear strength for the panel utilizing tension field action [see Eqn (7.5)], and V_{cr} is the critical shear strength for the panel designed without utilizing tension field action [see Eqn (7.2)].

Alternatively, a double stiffener can be provided for the end panel as shown in Fig. 7.15. Panels A and B are designed using tension field action. The bearing stiffener is designed for compressive stresses due to the external reaction. In this case, the two stiffeners and the portion of the web projecting beyond the end support form a rigid end post to provide necessary anchorage for the tension field in the end panel. Adequate space must be available to allow the girder to project beyond its support. The end post needs to act like a beam spanning between the flanges of the girder that should resist the shear force R_{tf} and bending moment M_{tf} due to anchor forces (Martin & Purkiss 1992).

7.5 Behaviour of Transverse Web Stiffeners

As observed from the previous sections, the tension field developed in the web is supported by the flanges and the transverse web stiffeners. Thus, the web stiffeners play a very critical role in achieving the ultimate capacity of the girder. It increases the web buckling stresses, supports the tension field in the web in the post-buckling stages, and finally prevents the flanges from being pulled towards each other, Prior to buckling, stiffeners on initially plane webs are not subjected to any axial loading. But after the plate buckles the axial loads applied to the transverse stiffeners steadily increase, as the web plate develops a membrane tension field (Rockey et al. 1981). Thus, the stiffeners must have sufficient rigidity to remain straight and to contain the web buckling to individual web panels.

From experimental studies, the following set of behaviours is deduced:

1. A portion of the web plates on either side of the stiffener acts with the stiffener, forming a T or a cruciform in resisting the vertical load (see Fig. 7.16), though the plate has theoretically buckled due to tension field action. Rockey et al. (1981) suggested that when a width of $40t_w$ of the web is assumed to act with the stiffener, the design procedure proposed by them are in agreement with the test data.

2. In practical cases, eccentricity of loading is unavoidable. This gives rise to additional bending moment. Further, any imperfections in the stiffener also result in bending moments.

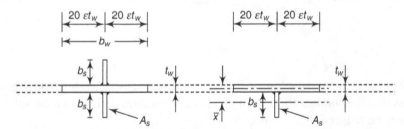

Fig. 7.16 Effective cross section of stiffeners

Design codes simplify this procedure for enabling quick sizing of the stiffeners, by assuming that the compressive force in the stiffener is constant and considering the whole depth of the web as a compression member.

7.6 Design of Plate Girders using IS 800 : 2007 Provisions

In the design of plate girders, required sections are proportioned for various elements. In general, bending moments are assumed to be carried by the flanges and the shear by the web. The depth of the section is chosen in such a way that the flanges are able to carry the design bending moment.

For arriving at a good plate girder design, IS 800 : 2007 has made several provisions with respect to its sectional requirements. The code specifies minimum web thickness requirement from a serviceability point of view for different situations with regard to provision of stiffeners (see Section 7.3.1). Similarly, the code specifies web thickness requirement in order to avoid buckling of the compression flange into the web as follows:

(a) When transverse stiffeners are not provided

$$\frac{d}{t_w} \le 345\varepsilon_f^2 \qquad (7.11a)$$

(b) When transverse stiffeners are provided and
  (i) When $c \ge 1.5d$:

$$\frac{d}{t_w} \le 345\varepsilon_f^2 \qquad (7.11b)$$

  (ii) When $c < 1.5d$:

$$\frac{d}{t_w} \le 345\varepsilon_f \qquad (7.11c)$$

where d is the depth of the web, t_w is the thickness of the web, c is the spacing of the transverse stiffener, $\varepsilon_f = \sqrt{250/f_{yf}}$, and f_{yf} is the yield stress of the compression flange.

It should be noted that when these two sets of criteria are used, the serviceability criteria will be more critical for Fe 410 grade steel and the buckling criteria becomes more important for higher grade steels. This is because buckling strengths are dependent on Young's modulus, which is the same for all grades of steel.

7.6.1 Sectional Properties

In the design of a plate girder, the gross cross-sectional area may not be available for resistance in several cases due to various reasons. In some situations, the plates may be too wide and may be classified as slender. In such cases, the excess width of plates has to be deducted to get the effective area (see Examples 4.13 and 7.1). For some reasons, if open holes are to be provided in the plane perpendicular to the direction of stress, they also have to be deducted for arriving at the effective area. In a similar manner, the effective area of cross section in the tensile flange will have to be determined after deducting the area for holes. This aspect has been elaborately treated in Chapter 6. The calculation of the effective sectional area for shear is also dealt in detail in Chapter 6.

7.6.2 Flanges

When more than one plate is used in the flange, they are to be stitched together by tacking bolts or other means to act as a single piece. At regions where bending moment is less than that at the critical sections, the number of plates can be reduced or curtailed. While curtailing the plates, the curtailed plates have to be extended slightly beyond the cut-off point and the extended plate has to be sufficiently connected to develop the required resistance at the cut-off point. It is also essential that the outstand of the flange plates in a plate girder satisfies the minimum section classification requirement, as described in Chapter 6.

In instances of a plate girder design where the sections are available only in small lengths or where sectional requirements are changing along the span, provision of flange splices are permitted. Splices should connect the two adjoining sections with adequate bolting or welding. It is always prudent to provide splicing only at points away from the locations of maximum stress. The cross section of the splice plate is normally kept 5% more than the area of the flange plates that are spliced. Also the connections need to develop at least 5% more load than the load at the spliced portion or at least 50% of the effective strength of the spliced section. In the case of a welded construction, the flange plate should be welded together using full penetration butt weld that is capable of developing full strength of the plates connected.

In the fabrication of the plate girder, the web and flange plate are joined together with adequate bolts or welds such that they can transmit horizontal shear arising out of the bending moments that develop in the girder and also due to any applied vertical load on the flange. When fillet welding is adopted for the flange-web connection, the web and flange plates are brought as close as possible to each other before welding and the maximum gap between them is limited to 1 mm. The effective sectional area for the web of a plate girder is arrived by considering the full depth of web plate and its uniform thickness. If the thickness of the web is varying in the depth of the section due to provision of tongue plates or in cases where the web contributes 25% or more of its depth to the flange area, the above provision is not applicable. In such cases the exact area has to be calculated on a case-to-case basis.

Similar to splices in flanges, webs also may need splicing or cut outs may become necessary in webs for routing of services. Such web splices and cut outs are best provided at locations away from high shear and large concentrated loads. Splices in the web are designed to withstand the shears and moments at the spliced section. Splice plates are often provided on both sides of the web in equal thickness and they are connected together. If welding is resorted to, complete full penetration butt welding is recommended. In locations where extra plate is added to the web, to increase its strength, they may be provided equally on both sides of the web plate. These plates also extend beyond the length of the requirement as arrived by calculations. The amount of shear force that these reinforcing plates can carry is restricted to the extent of horizontal shear that they transmit to the flanges by way of the fastening provided.

7.6.3 Stiffeners

In a plate girder, the web is often made very thin to derive maximum economy in weight. The webs, when they are inadequate to carry the load, are made strong and stable by the provision of a wide variety of stiffeners. Sometimes double stiffeners are adopted near the bearing [see Fig. 7.15] and in such cases the over hangs beyond the support should be limited to 1/8 of the depth of the girder. The stiffeners are classified based on the role they play in strengthening the web and they are mentioned as follows.

• The function of a *bearing stiffener* is to preclude any crushing of the web at locations of heavy concentrated loads. Thus, bearing stiffeners transfer heavy reactions or concentrated loads to the full depth of the web. They are placed in pairs on the web of plate girders at unframed girder ends and where required for concentrated loads. They should fit tightly against the flanges being loaded and extend out towards the edges of flange plates as far as possible. If the load normal to the girder flange is tensile, the stiffeners must be welded to the loaded flange. If the load is compressive, it is necessary to obtain a snug fit. To accomplish this goal, the stiffeners may be welded to the flange or the outstanding leg of the stiffener may be milled.

- A *load carrying stiffener* prevents local buckling of the web due to any concentrated load.
- An *intermediate transverse web stiffener* mainly improves the buckling strength of the web due to shear. They continue to remain effective after the web buckles to provide anchorage for the tension field and finally they prevent the flanges from moving towards one another. They are also called as non-bearing stiffeners or stability stiffeners.
- *Torsional stiffeners* are provided at supports to restrain the girders against torsional effects.
- Local strengthening of the web under the combination of shear and bending is provided by *diagonal stiffener*.
- The tensile forces from the flange are transmitted to the web through the *tension stiffener.*
- A *longitudinal stiffener* increases the buckling resistance of the web.

The same stiffener may perform more than one function and in such cases, their design should comply with the requirement of those functions.

In the design of various stiffeners, it is important to know the effective dimensions of the stiffener that needs to be considered for the design. The outstand of the stiffener from the face of the web is restricted to a value of $20t_q\varepsilon$. If the outstand value is anywhere between $14t_q\varepsilon$ and $20t_q\varepsilon$, the value of $14t_q\varepsilon$ is chosen as the outstand value. In the above expressions, t_q refers to the thickness of the stiffener plate. Another important parameter of the stiffener is its *stiff bearing length* b_1. It is defined as that length that does not deform appreciably in bending. The stiff bearing length b_1 is determined by considering the dispersion of the load at $45°$ through a solid material such as the bearing or flange plates as shown in Fig. 7.17.

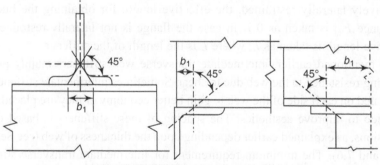

Fig. 7.17 Stiff bearing length (b_1)

The eccentricity of loading with respect to the stiffener centroid is to be considered in the design of a stiffener. The eccentricity arises when the load or reaction is eccentric to the centre line of the web or when the stiffener centroid does not coincide with the centre line of the web. Buckling of the stiffeners is also a matter of serious concern and hence adequate buckling resistance needs to be provided for stiffeners. In order to evaluate the buckling resistance of the stiffener, the stiffener is considered as a strut with its radius of gyration taken about the axis parallel to the web. The effective sectional area of the stiffener is the area of the core section as determined below. The effective width of web acting together with the stiffener is taken as 20 times the web thickness on each side of the centre line of the stiffener, i.e., $40t_w$ for interior stiffeners (see Fig. 7.18), and as $20t_w$ for end stiffeners. The buckling resistance of the stiffener is obtained on the basis of the design compressive stress, f_{cd} of the strut using buckling curve c (see Section 5.6.1). In arriving at the buckling resistance, the lower value between the design strength of the web material and the stiffener material is to be adopted.

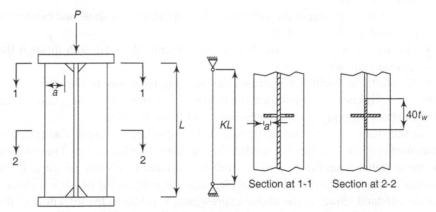

Fig. 7.18 Design of intermediate transverse web stiffener

The effective length of a intermediate transverse web stiffener, which is provided mainly to improve the shear buckling resistance of the web F_{qd} is taken as 0.7 times the length of the stiffener. For other load carrying stiffeners, with the assumptions that the flange, through which the load or reaction is applied, is effectively laterally restrained, the effective length for obtaining the buckling resistance F_{xd} is taken as 0.7. In case the flange is not laterally restrained, the effective length is taken as L, where L is the length of the stiffener.

As mentioned earlier, intermediate transverse web stiffeners mainly provide buckling resistance to the web due to shear. Sometimes intermediate stiffeners are alternated on each side of the web to gain better economy or they are placed all on one side to improve aesthetics. The spacing of these stiffeners is based on the provisions, as explained earlier depending upon the thickness of web (see Sections 7.3.1 and 7.6). The minimum requirement for intermediate transverse stiffener, when they are subjected to external loads or moments is considered in terms of a second moment of area I_s about the centre line of the web (Rockey et al. 1981).

$$\text{if } \frac{c}{d} \geq \sqrt{2}; \ I_s \geq 0.75 dt_w^3 \qquad (7.12a)$$

$$\text{if } \frac{c}{d} < \sqrt{2}; \ I_s \geq \frac{1.5 d^3 t_w^3}{c^2} \qquad (7.12b)$$

where d is the depth of the web, t_w is the minimum thickness required for tension field action, and c is the stiffener spacing.

Transverse web stiffeners are also checked for buckling resistance for a stiffener force

$$F_q = 0.909 \ (V - V_{cr}) \leq F_{qd} \qquad (7.13)$$

where F_{qd} is the design buckling resistance of the intermediate stiffener, V is the factored shear force adjacent to the stiffener, and V_{cr} is the shear buckling resistance of the web panel without considering tension field action [Eqn (7.2b)].

Transverse web stiffeners that are subjected to external load and moments have to satisfy requirements of load carrying stiffeners. They have to also meet the interaction equation

$$\frac{F_q - F_x}{F_{qd}} + \frac{F_x}{F_{xd}} + \frac{M_q}{M_{yq}} \leq 1 \qquad (7.14)$$

If $F_q < F_x$, then

$$F_q - F_x = 0$$

where F_q is the stiffener force [in N, see Eqn (7.13)], F_{qd} is the design buckling resistance of intermediate web stiffener about an axis parallel to the web (in N), F_x is the external load or reaction at the stiffener (in N), F_{xd} is the design resistance of load carrying stiffener, buckling about an axis parallel to the web (in N), M_q is the moment at the stiffener due to eccentric load (in N mm), and M_{yq} is the yield moment capacity of the stiffener about an axis parallel to the web (in N mm).

An important aspect of the intermediate transverse stiffener is with respect to their connection with the web. If the stiffeners are not subjected to external load, they may be connected to the web in such a way that they withstand a shear of not less than $t_w^2 /$ $5b_s$, where t_w is the web thickness and b_s is the stiffener outstand width.

If the stiffeners are subjected to external loading, the resulting shear to be resisted is the total shear due to loads and shear specified earlier. If the stiffeners are not subjected to external load, they may be provided in such a way as to be clear of the tension flange and the gap provided is $4t_w$. With such a gap, the fabrication problems associated with close fit can be avoided.

7.6.4 Load Carrying Stiffeners

As defined previously, these stiffeners are provided at locations where the compressive forces that are applied through the flange by an external load or reactions exceed the buckling resistance F_{cdw} of the web alone. The buckling resistance of the web alone may be calculated by using the effective length and the sectional area of the stiffener. The sectional area of the stiffener is taken as $(b_1 + n_1)t_w$, where b_1 is the stiff bearing length and n_1 is the dispersion length of the load through the web at 45° to the level of half the depth of the cross section. The buckling strength of the web is calculated about an axis parallel to the web thickness using buckling curve c. Although the web can be proportioned to resist any applied concentrated loads, bearing stiffeners are generally provided. If stiffeners are used at each concentrated load, the limit states of web yielding, web crippling, and side sway web buckling need not be checked.

7.6.5 Bearing Stiffeners

Bearing stiffeners are provided where forces, applied through a flange by loads or reaction, are in excess of the local capacity of the web at its connection to the flange, F_w calculated as follows:

$$F_w = 0.909 \, (b_1 + n_2) t_w f_{yw} \qquad (7.15)$$

where b_1 is the stiff bearing length, n_2 is the length obtained by dispersion through the flange to the web junction at a slope of 1:2.5 to the plane of the flange, t_w is the thickness of the web, and f_{yw} is the yield stress of the web.

7.6.6 Design of Load Carrying Stiffeners

Load carrying stiffeners are checked for their buckling and bearing resistances. For buckling check

$$F_x \leq F_{xd} \qquad (7.16)$$

where F_x is the external load or reaction and F_{xd} is the buckling resistance.

In the event that the load carrying stiffeners also act as intermediate transverse web stiffeners, the bearing strength should be checked for the combined loads as given in Eqn (7.14). The bearing strength of the stiffener F_{bsd} is obtained as follows:

$$F_{bsd} = 1.136 \, A_q f_{yq} \qquad (7.17)$$

where A_q is the area of the stiffener in contact with the flange in mm² {the actual area A is generally less than the full cross-sectional area of the stiffener as the corners of the stiffener are often coped to clear the web-to flange fillet weld (see section 1-1 of Fig. 7.18)} and f_{yq} as the yield stress of the stiffener in N/mm².

The bearing strength of the web stiffener F_{bsd} as calculated above should be greater than or equal to F_x, the load transferred.

7.6.7 Design of Bearing Stiffener

Bearing stiffeners are designed for the difference in force between the applied load or reaction and the local capacity of the web calculated in Eqn (7.15). In all cases, if the web and stiffener are of materials with different strength, the lower of the two has to be used for the strength equation. A bearing stiffener should be as wide as the overhang of the flange through which the load is applied. The weld connecting the stiffener to the web should have the capacity to transfer the unbalanced shear force. Conversely, the weld can be designed to carry the entire concentrated load.

7.6.8 Design of Diagonal and Tension Stiffeners

As in the case of bearing stiffeners, the diagonal stiffeners are designed for the portion of the combination of applied shear and bearing that exceeds the capacity of the web. Tension stiffeners are designed to carry the portion of the applied load or reaction that exceeds the capacity of the web.

7.6.9 Design of Torsional Stiffener

Torsional stiffeners provide torsional restraint at the supports and they should satisfy the following:
(a) the local capacity of the web is exceeded as calculated in Eqn (7.15)
(b) second moment of area of the stiffener section about the centre line of the web I_s should satisfy the following relation

$$I_s \geq 0.34 \alpha_s D^3 t_{cf} \qquad (7.18)$$

where $\quad \alpha_s = 0.006$ for $KL/r_y \leq 50$

$\alpha_s = 0.3/(KL/r_y)$ for $50 < KL/r_y \leq 100$

$\alpha_s = 30/(KL/r_y)^2$ for $KL/r_y > 100$

and D is the depth of the beam at support (in mm), t_{cf} is the maximum thickness of the compression flange of the girder (in mm), KL is the effective length of the laterally unsupported compression flange of the girder (in mm), and r_y is the radius of gyration about minor axis (in mm).

7.6.10 Connection of Load Carrying Stiffeners and Bearing Stiffeners to Web

Stiffeners which are subjected to load or reaction applied through a flange have to be connected to the web adequately. This connection is designed to transmit a force equal to the lesser of
(a) the tension capacity of the stiffener or
(b) the sum of the forces applied at the two ends of the stiffener when they act in the same direction or the larger of the forces when they act in different directions.

It should also be ensured that the shear stress in the web due to the design force transferred by the stiffener is less than the shear strength of the web. The stiffeners which resist tension should be connected to the flange with continuous welds or non-slip fasteners. Stiffeners resisting compression can either be fitted against the loaded flange or be connected by continuous welds or non-slip fasteners. The stiffener has to be fitted or connected to both flanges if

- the load is applied directly over the support
- the stiffener is an end stiffener of a stiffened web
- the stiffener also acts as torsional stiffener

7.6.11 Longitudinal Stiffeners

In order to obtain greater economy and efficiency in the design of plate girders, slender webs are often stiffened both longitudinally and transversely. The longitudinal stiffeners are generally located in the compression zone of the girder. The main function of the longitudinal stiffeners is to increase the buckling resistance of the web. The longitudinal stiffener remains straight, thereby, subdividing the web and limiting the web buckling to smaller web panels (see Fig. 7.19). In the past it was usually thought that the resulting increase in the ultimate strength could be significant, but recent studies have shown that this is not always the case, as the additional cost of welding the longitudinal stiffeners invariably offsets any economy resulting in their use. Longitudinal stiffeners, thus, are not as effective as the transverse ones, and are frequently used in bridge girders, because many designers feel that they are more attractive. Clause 8.7.13 of the code contains provisions for horizontal stiffeners.

Fig. 7.19 Tension field in a plate girder with longitudianl stiffener (EDEP 2006)

7.7 Welding of Girder Components

The fillet weld between the flange and the web is generally kept as small as possible depending upon the thickness of the flange and the allowable stresses. Continuous welds are desirable preferably with automatic welding equipment. Bearing stiffeners must fit tightly against the flange that transmits the load and must be attached to the flange by full penetration groove welds. The connection of the stiffener to the web may be by intermittent or continuous welds. Appropriate small openings are provided at the weld junctions to avoid interaction of different fillet welds. Transverse stiffeners are solely provided to increase the shear buckling resistance of the web and hence are fitted tightly with the compression flange but welded only to the web. But one-sided transverse stiffeners must be welded to the compression flanges also. Transverse intermediate stiffeners are stopped short of the tension flange by a distance of about $4t_w$. Welding transverse stiffeners to tension flanges may reduce the fatigue strength and may lead to brittle fracture.

7.8 Proportioning of the Section

The cross section of the girder must be selected such that it adequately performs its functions and requires minimum cost. The functional requirements are as follows:

(a) Strength to carry bending moment (adequate Z_p)
(b) Vertical stiffness to satisfy any deflection limitation (adequate moment of inertia I_z)
(c) Lateral stiffness to prevent lateral-torsional buckling of compression flange (adequate lateral bracing or low values of (L_b/r_t))
(d) Strength to carry shear (adequate web area)
(e) Stiffness to improve buckling or post-buckling strength of the web (related to d/t_w and c/d ratios)

One method of determining the initial dimensions is to use a minimum weight analysis (Schilling 1974). However, it has to be emphasized that a minimum weight design may not result in minimum cost, unless fabrication cost is also included in the optimization process.

7.8.1 Optimum Girder Depth

If the moment M is assumed to be resisted entirely by the flanges, then for a I-section beam,

$$M = f_y b_f t_f d \tag{7.19}$$

where f_y is the design strength of the flanges, b_f and t_f are the width and thickness, respectively of the flanges, and d is the distance between centre of flanges.

The gross-sectional area of the beam is given by

$$A = 2\,b_f t_f + d t_w \tag{7.20}$$

where t_w is the thickness of web. Eliminating $b_f t_f$ using Eqn (7.19), we get

$$A = 2M/(df_y) + d t_w \tag{7.21}$$

Defining the web slenderness as $k = d/t_w$, we get

$$A = 2M/(k t_w f_y) + k t_w^2 \tag{7.22}$$

Differentiating the above equation with respect to t_w and setting the result equal to zero, we get the optimum value of t_w as

$$t_w = [M/(f_y k^2)]^{0.33} \tag{7.23}$$

The optimum value of the depth is

$$d = (Mk/f_y)^{0.33} \tag{7.24}$$

Thus the optimum values produce a beam which has the area of single flange and the web as equal (Martin & Purkiss 1992). Extended treatment of this subject has been provided by Salmon and Johnson (1996), who also propose the following simple expression for the required area of one flange plate

$$A_f = M/(f_y d) - A_w/6 \tag{7.25}$$

where A_w is the area of web plate. The above equation may be used for preliminary design purposes.

7.8.2 Preliminary Sizing

Depth of Plate Girder The depth of plate girder is of paramount importance because it governs the stiffness and bending resistance. The design has to comply with both strength and deflection requirements. The full girder depth D, in terms of the equivalent span L, may be chosen in the following range:

D/L = 1/15 to 1/25 for girders in buildings,

D/L = 1/12 to 1/18 for girders in highway bridges, and

D/L = 1/10 to 1/15 for girders in railway bridges.

The equivalent span L may be taken as the distance between the adjacent points of contraflexure under a uniformly distributed load.

The preceding values may be reduced when the top flange of the steel girder is attached to a reinforced concrete deck (composite girder) or to an orthotropic steel deck plate. A constant depth girder is chosen when a launching procedure is used for erection. However, when unequal spans are present in a continuous plate girder, it may be economical to provide variable depth plate girder.

Web thickness From a corrosion standpoint, the usual practice is to choose a minimum thickness of 6–8 mm. For large depth bridge girders, a minimum web thickness of 8–10 mm may be required because of handling considerations. Though thin stiffened webs have been used in the past, the tendency now is to avoid stiffeners and use thick webs, which reduce the fabrication time and cost. The web thickness may be kept constant over a length of about 4–6 m from the supports. For webs with no longitudinal stiffeners, the maximum d/t_w ratio of the web may be kept below 360 for f_y = 250 MPa and below 240 for f_y = 360 MPa. When fatigue strength governs the design, these values may be reduced (Maquoi & Skaloud 2000). For stiffened webs the recommended web thickness is given in Sections 7.3.1 and 7.6.

Flanges Flanges are designed for both strength and rigidity requirements. For a non-composite plate girder, the width of the flange plate may be chosen as 0.3 times the depth of the section. For a composite plate girder, it may be kept as 0.2 times the depth of the girder. As indicated already the b_f/t_f ratio of the flange should be such that the flange is either plastic or compact or semi-compact. This is to avoid local buckling before reaching the yield stress. For preliminary sizing the flange width to thickness ratio may be limited to 24 for grade 410 steel. The thickness and width of the flange rarely exceeds 80–100 mm and 1 m respectively. The width of tension flange (i.e., bottom flange of a simply supported beam) may be increased by about 30%. Erection requirements may govern the size of flanges. Pre-heating before welding is sometimes adopted to prevent brittle fracture in the heat-affected zone.

Stiffeners Though transverse stiffeners are provided symmetrically on both the faces of the web at the supports, intermediate stiffeners are used only on one side of web plate. Longitudinal stiffeners, when necessary, may be located on the opposite side of the transverse stiffeners. Web stiffeners are welded continuously to the web plate. When stiffeners are provided, the panel aspect ratio c/d may be chosen in the range of 1.2 to 1.6. When end panel is designed near support, without using the tension field action, the aspect ratio should be in the range 0.6–1.0.

7.8.3 Steps Involved in the Design of Plate Girders

The following steps are involved in the design of plate girders.
1. Assume the self weight of the girder. (The self weight w may be assumed as equal to $W/200$ where w is in kN/m and W is the total factored load applied to the girder in kN.) Estimate the live loads. Calculate the maximum bending moment and shear force in the plate girder.
2. Using Eqns (7.23) and (7.24), assume depth and thickness of the web. Check the web thickness as per clause 8.6.1.1 and 8.6.1.2 of the code and adopt a suitable web thickness.
3. Select suitable flange plate thickness and width (approximately 0.3 times the depth of web). The flange plates should be so proportioned that it should satisfy the requirements of plastic/compact/semi-compact section. Note that due to the slender web, the plate girder cross section will always fall in the semi-compact category.
4. Check for moment capacity as per clause 8.2.1 or 8.2.2 depending on whether the plate girder is laterally supported or unsupported. It is generally advantageous to provide lateral support at sufficiently close intervals such that lateral-torsional buckling will not govern the design.
5. Check for shear resistance of the web. One may use either clause 8.4.2.2 (a) simple post–critical method or clause 8.4.2.2 (b) by considering tension field method.
6. Design of the connection between the flange plate and the web plate.
7. Design of bearing stiffeners and their connections (clauses 8.7.4, 8.7.5, and 8.7.9 of the code)
8. Design of load carrying stiffeners; if required (clause 8.7.5) and their connections.
9. Design of intermediate stiffeners, if required (clauses 8.7.2 and 8.7.1.2) and their connections (clause 8.7.2.6)
10. Design of web splice and its connections.
11. Design of flange splice and its connections.
The design of web and flange splices is covered in Sections 10.10.1 and 11.15.1.

Plate Girders with Corrugated Webs

To increase the shear capacity of large steel plate girders, shaped webs may be used; in this situation, the web plate is first cold formed with waves or corrugations usually parallel to the web depth and then welded to the flanges. Although many types of corrugations are possible, trapezoidal and sinusoidal corrugations (see

Fig. CS 1) have received the most attention. Corrugated webs, though not yet commonly used for highway bridges in India and North America, have been incorporated into several highway bridges constructed in Europe and Japan.

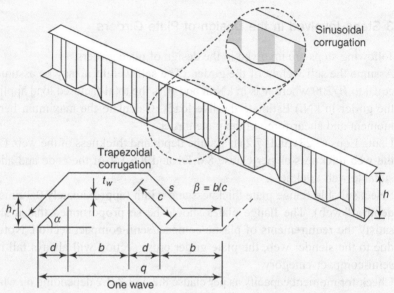

Fig. CS 1 Profile and geometric notation of a girder with corrugated web

The results of studies on girders with corrugated web indicate that the fatigue strength of these girders can be 50% higher compared to girders with flat stiffened webs (Machacek & Tuma 2006). In addition to the improved fatigue life, these girders could be 30 to 60% lighter than the girders with flat webs and have the same capacity (Elgaaly & Dagher 1990). Thus, larger spans can be achieved with less weight. Moreover, corrugated webs improve the aesthetics of the structure. Beams used in Germany for buildings have web thicknesses that vary between 2 and 5 mm and the corresponding web height-to-thickness ratios varying between 150 and 260 mm. The corrugated webs of two bridges built in France were 8-mm thick and the web height-to-thickness ratio was in the range of 220 to 375 mm.

Steel Plate Shear Walls (SPSW)

Although the post-buckling behaviour of plates under monotonic load has been investigated by several researchers for more than half a century, post-buckling strength of plates under cyclic loading has not been investigated till now (Caccese et al. 1993; Driver et al. 1998; Kulak et al. 2001). The results of these investigations revealed that plates can be subjected to a few reversed cycles of loading in the post-buckling domain, without damage. However, steel plate shear walls (SPSW) have been used as the primary lateral load resisting system in buildings for more than three decades in United States, Canada, and Japan. Some of these buildings include the following (Seilie & Hooper 2005):

- United States Federal Courthouse, Seattle, WA: A 23 storey building with a height of 107 m.
- Canam Manac Headquarters Expansion, St George, Quebec, Canada: A six storey building. Figure CS 2 shows the Planar SPSW system adopted for this building.

Fig. CS 2 Planar SPSW system of the Canam HQ building, Canada (Seilie & Hooper 2005)

- The 35-storey (130 m tall) Kobe office building, Kobe, Japan
- Shinjuku Nomura building, Tokyo, Japan: 51-storey building with a height of 211 m

Among the listed buildings, the Kobe office and Sylmar Hospital buildings have withstood fairly significant earthquakes (the Kobe earthquake in 1995 and Northridge earthquake in 1994, respectively) and survived without any structural damage (Seilie & Hooper 2005).

There are three different types of SPSW systems:

(a) Unstiffened, thin SPSW

(b) Stiffened SPSW

(c) Composite concrete SPSW

In North America unstiffened, thin SPSW are common while in Japan, the stiffened SPSW system is often used. In this section, we will confine our discussion to unstiffened thin SPSW system only.

The following are the advantages of the SPSW system.

- SPSW systems allow for less structural wall thickness in comparison to the thickness of concrete shear walls. This results in saving of rentable floor area.
- Steel savings of as much as 50% has been achieved in structures employing a steel plate shear wall system rather than a moment resisting frame.
- The use of SPSW system reduces construction time. It is fast to erect and does not require a curing period.

- As mentioned earlier, the thin plates have excellent post-buckling capacity. It has been found that the system can survive up to 4% drift without experiencing significant damage.

Large scale laboratory tests have been conducted to verify the behaviour of SPSW systems (Driver et al. 1998). These tests confirmed the high initial stiffness, large energy dissipation capacity, and great ductility of these systems, even after a large number of extreme load cycles. The Canadian standard for structural steel design (CAN/CSA-S16 2001) is the first standard to include provisions for unstiffened SPSW design, though it included analysis and design provisions in the 1994 edition itself in a non-mandatory appendix. Similar provisions have been incorporated in the AISC seismic provisions for steel buildings (ANSI/AISC 341-05).

Examples

Example 7.1 *The plate girder shown in Fig 7.20 is required to carry a factored shear of 1600 kN.*
(a) Assuming that tension field action is not utilized in the design, determine whether intermediate stiffeners are necessary. Assume $f_y = 250 N/mm^2$.
(b) How thick must the web be in order that this same load may be carried without the need for intermediate stiffeners?

Solution

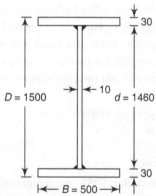

Fig. 7.20 Cross section of the girder

(a) Determination of required stiffeners

Using $d/t = 1460/10 = 146$

From Table 7.1, for no stiffeners ($c/d = \infty$)

$$t_b = 44.4 \ N/mm^2$$

Hence $V_n = V_{cr} = dt_w\tau_b = 1460 \times 10 \times 44.4/10^3 = 648.24$ kN < 1600 kN

Hence stiffening is required.

Required $\tau_b = 1600 \times 10^3/(1460 \times 10) = 109.6 \ N/mm^2$

From Table 7.1, for $d/t = 146$, max c/d corresponding to this strength = 0.7
Hence stiffeners at 1020 mm (= 0.7×1460) intervals should be provided.

(b) Thickness of web without stiffeners
A trial and error approach is necessary to determine the web thickness which will support the load without any intermediate stiffener. Obviously it must be greater than 10 mm.

$$\text{Try } t = 14 \text{ mm}; \quad d/t = 1460/14 = 104.3$$
$$\text{Hence } \tau_b \text{ (for } c/d = \infty) \text{ from Table 7.1} = 88.9 \text{ N/mm}^2$$
$$\text{Hence } V_n = 1460 \times 14 \times 88.9/10^3 = 1817 \text{ kN} > 1600 \text{ kN}$$

Note: On many occasions, it may be advisable to increase the thickness of web instead of providing stiffeners; it is because, stiffeners will increase the cost of fabrication, in addition to extra weight of stiffeners.

Example 7.2 *Design a load carrying stiffener for a load of 700 kN for the section ISMB 500 (see Fig. 7.21).*

Solution
For ISMB 500, $t_w = 10.2$ mm, $t_f = 17.2$ mm, $R = 17$ mm
Factored load = $1.5 \times 700 = 1050$ kN

$$\text{Shear capacity of the web of ISMB 500} = \frac{A_v f_{yw}}{\gamma_{m0}\sqrt{3}} = \frac{500 \times 10.2 \times 250}{1.1 \times 1000 \times \sqrt{3}} = 669.2 \text{ kN}$$

Load to be resisted by the stiffener = $1050 - 669.2 = 380.8$ kN
Assume the thickness of the stiffener plate as 6 mm. The maximum allowable breadth (outstand) of stiffener (clause 8.7.1.2 of code) = $20 \times t_q \varepsilon = 20 \times 6 = 120$ mm (since $\varepsilon = 1$, for $f_y = 250$ MPa)
Take the breadth of the stiffener plate as 12.5 times the thickness of the plate ($12.5 \times 6 = 75$ mm). Provide two plates of size 75 mm $\times$ 6 mm on either side of the web.

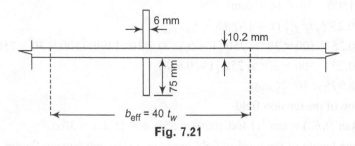

6 mm

10.2 mm

75 mm

$b_{eff} = 40\, t_w$

Fig. 7.21

As per clause 8.7.1.5 of the code,
$b_{eff} = (40 \times 10.2) = 408$ mm
Effective area = $(b_{eff} \times t_w) + (2 \times b_s \times t_s) = (408 \times 10.2) + (2 \times 75 \times 6) = 5061.6$ mm^2

$$\text{Moment of inertia of stiffener} = \frac{b_s \times d_s^{\,3}}{12} = 6 \times (2 \times 75)^3/12 = 1.6875 \times 10^6 \text{ mm}^4$$

Assuming $c/d = 1.5$ (clause 8.7.2.4),

Minimum $I_s = 0.75\ dt_w^3 = 0.75 \times (500 - 2 \times 17.2) \times 10.2^3$
$$= 0.370 \times 10^6\ \text{mm}^4 < 1.6875 \times 10^6\ \text{mm}^4$$

Hence the adopted size of the stiffener is fine.

Check for buckling resistance

$$I = \frac{6 \times 150^3}{12} + \frac{408 \times 10.2 \times 10.2^3}{12} = 2.056 \times 10^6\ \text{mm}^4$$

$$r = \sqrt{\frac{I}{A}} = \sqrt{\frac{2.056 \times 10^6}{5061.6}} = 20.15\ \text{mm}$$

Depth of web $= 500 - 2(t_f + R) = 500 - 2(17.2 + 17) = 431.6$ mm

Effective length of member $= 0.7 \times d = 0.7 \times 431.6 = 302.12$ mm

$$\lambda = \frac{l}{r} = \frac{302.12}{20.15} = 15$$

Therefore, compressive stress, $f_{cd} = 225.5$ N/mm² (From Table 9c of the code).
Buckling resistance of the stiffener $= 225.5 \times 5061.6/10^3 = 1141.4$ kN > 380.8 kN
Hence the stiffener is safe.

Example 7.3 *Assuming stiffener spacing equal to 1020 mm, determine the shear capacity of the girder of Example 7.1, including the tension field action, assuming a factored bending moment of 3500 kNm.*

Solution

As per clause 8.4.2.2d of the code,

$V_n = V_{tf}$

$V_{tf} = [A_v\ \tau_b + 0.9\ w_{tf}\ t_w\ f_v \sin \phi] \le V_{np}$

From Table 7.1, for $d/t = 146$ and $c/d = 0.70$, $\tau_b = 113.44$ N/mm²

Reduced plastic moment of the flange plate

$h_f = 1500 - 30 = 1470$ mm

$M_{fr} = 0.25\ b_f\ t_f^2\ f_{yf}\ [1 - \{N_f /(b_f\ t_f f_{yf} /\gamma_{mo})\}^2]$

$\qquad = 0.25 \times 500 \times 30^2 \times 250\ [1 - \{(3500 \times 10^6/1490)/(500 \times 30 \times 250/1.1)\}^2]$

$\qquad = 0.25 \times 500 \times 30^2 \times 250\ [1 - 0.688]$

$\qquad = 8.775 \times 10^6$ N mm

Inclination of the tension field

$\theta = \tan^{-1}(d/c) = \tan^{-1}(1460/1020) = 55°;\ \phi = 55/1.5 = 36.67°$

Anchorage length of the tension field in compression and tension flange

$s = 2/\sin\phi\ [M_{fr}/(f_y\ t_w)]^{0.5} \le c$

$\qquad = 2/\sin 36.67°\ [(8.775 \times 10^6)/(250 \times 10)]^{0.5} = 198.4$ mm < 1500 mm

The width of the tension filed

$w_{tf} = d \cos \phi - (c - s_c - s_t) \sin \phi$

$\qquad = 1460 \times \cos 36.67° - (1020 - 2 \times 198.4) \sin 36.67°$

$\qquad = 798.87$ mm

Yield strength of tension field

$$f_v = [f_{yw}^2 - 3\ \tau_b^2 + y^2]^{\,0.5} - \psi$$
$$\psi = 1.5\tau_b \sin 2\phi$$
$$= 1.5 \times 113.44 \times \sin 73.34° = 163 \text{ N/mm}^2$$
$$f_v = [250^2 - 3 \times 113.44^2 + 163^2]^{0.5} - 163$$
$$= 61.64 \text{ N/mm}^2$$

The shear resistance

$$V_{tf} = [1460 \times 10 \times 113.44 + 0.9 \times 798.87 \times 10 \times 61.64 \times \sin 36.67°]/1000$$
$$= 1656.2 + 264.67 = 1920.87 \text{ kN}$$

Example 7.4 *Design a welded plate girder for a simply supported bridge deck beam with a clear span of 20 m, subjected to the following:*

Dead load including self weight = 20 kN/m

Imposed load = 10 kN/m

Two moving loads = 150 kN each spaced 2 m apart

Assume that the top compression flange of the plate girder is restrained laterally and prevented from rotating. Use mild steel with $f_y = 250$ *MPa. Design as an unstiffened plate girder with thick webs.*

Solution

(1) Loading

Total factored udl on girder = $1.5 \times (20 + 10) = 45$ kN/m

Factored moving loads = $1.5 \times 150 = 225$ kN

(2) Maximum Bending Moment

$$R_A \text{ and } R_B \text{ due to udl} = \left(\frac{45 \times 20}{2}\right) = 450 \text{ kN}$$

The maximum bending moment due to UDL occurs at mid span; whereas the maximum bending moment due to the two moving loads will occur under one of the loads, when one of the loads is placed such that the middle of the beam divides the distance between this load (under which maximum bending moment occurs) and CG of the moving loads equally [see Fig.7.22(b)].

CG of the loads = $a/2$ (from the loads under consideration)'

where a is the distance between the moving loads. The distance of the loads from middle of the beam = $a/4$. Hence,

Distance of the CG of loads from right support = $L/2 - a/4$

= 20/2 − 2/4 = 9.5m.

Since the maximum BM under UDL is at a different location from that of maximum BM under moving loads, we have to compute the bending moment in both the cases and choose the higher of the two. However, in our case the bending moment occurred under one of the moving loads.

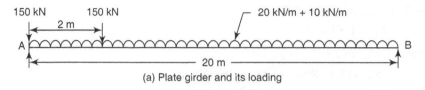

(a) Plate girder and its loading

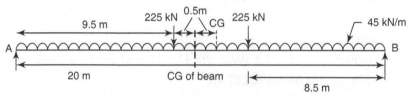

(b) Loading for maximum bending moment

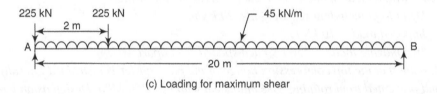

(c) Loading for maximum shear

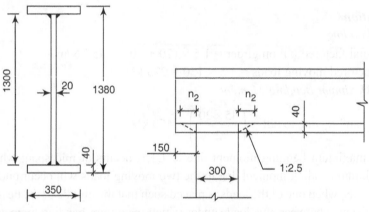

(d) Cross section of the plate girder   (e) Dispersion length

Fig.7.22

$$R_A \text{ due to moving load } = \frac{1}{20}\left[(225\times10.5)+(225\times8.5)\right] = 213.75 \text{ kN}$$

R_B due to moving load $= (2 \times 225) - 213.75 = 236.25$ kN

Total $R_B = 450 + 236.25 = 686.25$ kN

Total $R_A = 450 + 213.75 = 663.75$ kN

$$\text{Maximum } BM = (663.75\times9.5)-\left(\frac{45\times9.5^2}{2}\right) = 4275 \text{ kNm}$$

Maximum BM, $M_z = 4275$ kNm

Note: Impact due to moving load should also be considered in practice.

(3) Maximum Shear Force
For maximum shear force, the wheel loads should be placed such that one of the wheel loads should be very close to the supports [see Fig. 7.22(c)]

$$R_A \text{ due to moving load } = \frac{1}{20}\left[(225\times20)+(225\times18)\right] = 427.5 \text{ kN}$$

Total $R_A = 427.5 + 450 = 877.5$ kN

∴ Maximum shear force, $V_z = 877.5$ kN

If stiffeners are to be avoided, $k = d/t_w$ should be less than 67ε.

Optimum depth of plate girder

$$d = (Mk/f_{yf})^{0.33} = (4275 \times 10^6 \times 67 /250)^{0.33} = 976 \text{ mm}$$

Assume $d = 1300$mm

Optimum value of thickness of web

$$t_w = [M/(f_y k^2)]^{0.33} = [4275 \times 10^6/(250 \times 67^2)]^{0.33} = 16.3 \text{ mm}$$

There are two design requirements regarding the minimum web thickness for the condition of no intermediate stiffeners (clause 8.6.1.1 and 8.6.1.2 of code)

$d/t_w < 200\varepsilon$ (for serviceability)

$d/t_w \le 345\ \varepsilon_f^2$ (to avoid flange buckling).

If the web is deliberately made thick, i.e. $d/t < 67\varepsilon$ (clause 8.4.2.1), these requirements are automatically met. Thus the minimum web thickness should be as given below,

$$t_w > \frac{1300}{67\times1} \text{ or } t_w > 19.403 \text{ mm}$$

Hence, provide $t_w = 20$ mm

In order to maximize the moment capacity, the cross section of the plate girder should be so proportioned that it satisfies the requirements of plastic/compact section. Thus b/t_f should be less than 8.4ε or 9.4ε for plastic and compact sections respectively (see Table 2 of the code). Assuming b_f as 0.3 times the depth of web $b_f = 0.3 \times 1300 = 390$mm. Provide $b_f = 350$mm; $t_f > b_f/(2 \times 8.4) = 20.8$ mm. Provide $t_f = 40$mm

Use 1300 mm × 20 mm web plates with flange plates of 350 mm × 40 mm. As discussed above, it is a *plastic section*.

(4) Shear capacity
As per Clause 8.4 of the code,

$$V \le V_d$$

$$V_d = \frac{V_n}{\gamma_{m0}}$$

V_n = nominal plastic shear resistance = $V_p = \dfrac{A_v f_{yw}}{\sqrt{3}}$

As per Clause 8.4.1.1 of the code, for welded section, $A_v = dt_w$

$$\therefore \quad V_n = \frac{dt_w f_{yw}}{\sqrt{3}} = \frac{1300 \times 20 \times 250}{\sqrt{3} \times 1.1 \times 10^3} = 3411.6 \text{ kN} > V_z = 877.5 \text{ kN}$$

Hence safe in shear.

(5) Moment Capacity

According to clause 8.2.1.2 of the code, design bending strength,

$$M_d = \beta_b Z_p \frac{f_y}{\gamma_{m0}} \leq 1.2 Z_e \frac{f_y}{\gamma_{m0}}$$

Plastic modulus, $Z_p = 2 b_f t_f (D - t_f)/2 + t_w d^2/4$
$$= 2 \times 350 \times 40 \times 1340/2 + 20 \times 1300^2/4 = 27.21 \times 10^6 \text{ mm}^3$$

$$M_d = 1.0 \times 27.21 \times 10^6 \times \frac{250}{1.10 \times 10^6} = 6184.09 \text{ kNm} > M_z = 4275 \text{ kNm}$$

Hence safe to carry the applied moment.

As the compression flange is restrained, there is no need to check for lateral-torsional buckling.

(6) Check for bearing stiffeners

(a) At the supports [see Fig.7.22e]

Assume that the width of support is 300 mm and that the minimum stiff bearing provided by the support $b_1 = 300/2 = 150$mm,

Dispersion length (1:2.5), $n_2 = 2.5 \times 40 = 100$ mm

Local capacity of the web, $F_w = (b_1 + n_2) t_w \dfrac{f_{yw}}{\gamma_{m0}}$

$$F_w = (100 + 150) \times 20 \times \frac{250}{1.10 \times 10^3} = 1136.35 \text{ kN} > \text{support reaction (877.5 kN)}$$

(b) At position of moving wheel loads

$$F_w = (b_1 + n_2) t_w \frac{f_{yw}}{\gamma_{m0}}$$

Assuming, $b_1 = 0$

$$F_w = (0 + 2.5 \times 2 \times 40) \times 20 \times \frac{250}{1.10 \times 10^3} = 909.09 \text{ kN} > 225 \text{ kN}$$

The associated buckling resistance F_{qd} is dependent on the slenderness of the unstiffened web (clause 8.7.1.5).

Slenderness ratio of the web $= L_e/r_y = 0.7 L/r_y$ [with $r_y = t/2\sqrt{3}$ – see Eqn. (6.31)]
$$= 2.5 d/t = 2.5 \times 1300/20 = 162.5$$

From Table 9c of the code, $f_{cd} = 52 \text{ N/mm}^2$

As per clause 8.7.1.3 of the code,

Stiff bearing length $= 0$ (assumed)

45° dispersion length (to the level of half the depth of beam) = 1380/2 = 690 mm

$$F_{qd} = (0 + 690) \times 20 \times 52/10^3 = 717.6 \text{ kN} > 225 \text{ kN}$$

The web is adequate at both supports and positions of concentrated loads. Hence, there is no need to provide bearing stiffeners.

Thus by using thick webs, the use of load bearing stiffeners may be eliminated which will minimize the fabrication.

Example 7.5 *Redesign the plate girder given in Example 7.4 using thin web plates.*

Solution

From Example 7.4,

$$M_z = 4275 \text{ kN m}$$
$$F_z = 877.5 \text{ kN m}$$

As shown in Example 7.4,

optimum depth = 976 mm and t_w = 16.3 mm

Let us assume web plate of 1200 mm × 12 mm size. Assuming that the bending moment is carried by the flanges,

$$A_f = M_z \times \gamma_{m0}/(f_y \times D) = 4275 \times 10^6 \times 1.1/(250 \times 1200) = 15675 \text{ mm}^2$$

Limiting 8.4ε for the plastic section, the approximate flange thickness is

$$t_f = \sqrt{15675/(2 \times 8.4)} = 30.5 \text{ mm}.$$

Provide t_f = 35 mm.

$$b_f = 15675/35 = 447.86 \text{ mm}$$

Provide b_f = 450 mm.

The cross section of the plate girder is shown in Fig. 7.23.

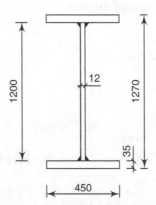

Fig. 7.23 Cross section of plate girder

According to clause 8.2.1 of the code,

$$\frac{d}{t_w} = \frac{1200}{12} = 100 > 67\varepsilon$$

(1) Moment capacity check
As discussed earlier, the flanges are *plastic*

Moment of resistance (clause 8.2.1.2) = $\dfrac{\beta_b Z_p f_y}{\gamma_{m0}}$

Considering that the flanges only resist the bending moment,

$$Z_p = 2b_f t_f (D - t_f)/2 = 2 \times 450 \times 35(1270 - 35)/2 = 19.45 \times 10^6 \text{ mm}^3$$

$$M_d = 1 \times 19.45 \times 10^6 \times \frac{250}{1.10 \times 10^6} = 4420 \text{ kN m} > 4275 \text{ kN m}$$

(2) Shear resistance of the web
As per clause 8.6.1.1a of the code, for serviceability

$$\frac{d}{t_w} = \frac{1200}{12} = 100 < 200\varepsilon$$

As per clause 8.6.1.2a of the code to avoid flange buckling

$$\frac{d}{t_w} \le 345\varepsilon_f^2$$

$$\frac{1200}{12} = 100 < 345\varepsilon_f^2$$

Hence the minimum web thickness requirements are met.
Let us consider the simple post-critical method. According to clause 8.4.2.2(a) of the code,

Nominal shear strength $V_n = V_{cr}$

$k_v = 5.35$, when transverse stiffeners are provided only at supports.
The elastic critical shear stress of the web

$$\tau_{cr} = \frac{k_v \pi^2 E}{12(1 - \mu^2)\left(\dfrac{d}{t_w}\right)^2}$$

With $\mu = 0.3$,

$$\tau_{cr} = \frac{5.35 \times \pi^2 \times 2 \times 10^5}{12(1 - 0.3^2) \times 100^2} = 96.7 \text{ N/mm}^2$$

The non-dimensional web slenderness ratio for shear buckling stress,

$$\lambda_w = \sqrt{[f_{yw}/(\sqrt{3}\tau_{cr})]} = \sqrt{\frac{250}{(\sqrt{3} \times 96.7)}} = 1.22 > 1.20$$

Shear stress corresponding to buckling $\tau_b = f_{yw}/(\sqrt{3}\lambda_w^2)$

$$\tau_b = 250/(\sqrt{3} \times 1.22^2) = 96.97 \text{ N/mm}^2$$

Shear force corresponding to web buckling,

$$V_{cr} = dt_w\tau_b$$
$$= 1200 \times 12 \times 96.97/10^3 = 1396.3 \text{ kN} > 877.5 \text{ kN}$$

Hence the shear strength is greater than the applied shear.

(3) Lateral torsional buckling
As the compression flange is restrained, there is no need to check for lateral torsional buckling.

(4) Design of load carrying stiffeners
(i) At the position of applied wheel load:
   As per clause 8.7.4 of the code, the local capacity of web

$$F_w = (b_1 + n_2)t_w \frac{f_{yw}}{\gamma_{m0}}$$

$$n_2 = 2 \times 35 \times 2.5 = 175 \text{ mm}$$
$$b_1 = 0$$

$$F_w = [0 + 175] \times 12 \times \frac{250}{1.10 \times 10^3} = 477.2 \text{ kN} > 225 \text{ kN}$$

The buckling resistance depends on the slenderness ratio of the web.
According to clause 8.7.1.5 of the code,

$$\lambda = 2.5 \frac{d}{t} = 2.5 \times \frac{1200}{12} = 250$$
$$n_1 = D/2 = 1270/2 = 635$$

From Table 9c of the code, $f_{cd} = 24.3 \text{N/mm}^2$
∴ Buckling resistance $= (0 + 2 \times 635) \times 12 \times 24.3/10^3 = 370.3 \text{ kN} > 225 \text{ kN}$
No stiffeners are necessary to prevent local buckling failure of the web at the point load position.

(ii) At the supports:
   The reaction is 877.5 kN, which is much greater than the point load of 225 kN.
   Hence stiffeners are necessary for both the supports.
   Try a pair of 220 × 12 mm flats.
   As per clause 8.7.1.2 of the code,

   Maximum outstand of stiffener $= 20t_q\varepsilon = 20 \times 12 \times 1 = 240 \text{ mm} > 220 \text{ mm}$
   $$= 14 \times 12 \times 1 = 168 \text{ mm} < 220 \text{ mm}$$

This means that the core area of the stiffener should be reduced to 2 × 168 mm × 12 mm.

(a) Check stiffener for buckling
As per clause 8.7.1.5, the effective section is the core area of the stiffener together with an effective length of web, on either side of the centre line of the stiffener of $20 \times t_w$. At the support, assuming that the web is available only on one side of the edge stiffener,
   Local buckling resistance,
   Area $= 2 \times 168 \times 12 + 20 \times 12 \times 12 = 6912 \text{ mm}^2$

$$I_x = \frac{12 \times 336^3}{12} + \frac{(20 \times 12)12^3}{12} = 37.967 \times 10^6 \text{ mm}^4$$

$$r = \sqrt{I_x/A} = \sqrt{\frac{37.967 \times 10^6}{6912}} = 74.11 \text{ mm}$$

$$\lambda = 0.7 \times \frac{1200}{74.11} = 11.33$$

From Table 9c of the code, $f_{cd} = 226$ N/mm^2

Buckling resistance $= 226 \times 6912/10^3 = 1562.1$ kN > 877.5 kN

Hence the stiffener is safe.

Provide a pair of 220×12 mm stiffener at both the ends of the girder. Now calculate the bearing capacity of the end stiffener.

$$F_{psd} = A_q f_{yq}/(0.8 \times \gamma_{m0}) = 2(220 - 15) \times 12 \times 250/[(0.8 \times 1.1) \times 1000]$$
$$= 1397.7 > 877.5 \text{ kN}$$

Hence the stiffener is safe.

In the preceding equation, A_q is the area of the stiffener in contact with the flanges. Normally, the flanges and web will be welded together, before the stiffeners are fitted. This means that the inside corners of the stiffener need to be coped/chamfered at the junction of the web and flanges, so that they will not foul with the web/flange weld. The chamfered width is taken as 15 mm.

(b) Torsional restraint provided by end stiffener

However, it might be deemed necessary that the ends of the plate girder are torsionally restrained during transportation and erection. This can be accomplished by checking the second moment of area of the end bearing stiffeners at the supports as per clause 8.7.9 of the code.

$$I_s \geq 0.34 \alpha_s D^3 t_{cf}$$

Assuming that there is no restraint to the compression flange for the above situation,

$$r_y \text{ of the girder} = \sqrt{\frac{I_y}{A}}$$

$$I_y = \frac{2t_f b_f^3}{12} + \frac{dt_w^3}{12} = 2 \times 35 \times 450^3/12 + 1200 \times 12^3/12$$
$$= 531.73 \times 10^6 \text{ mm}^4$$

$$A = 2 \times 450 \times 35 + 1200 \times 12 = 45900 \text{ mm}^2$$

$$r_y = \sqrt{(531.73 \times 10^6/45900)} = 107.6 \text{ mm}$$

$$\lambda = 20 \times 1000/107.6 = 185.8$$

Hence,

$$\alpha_s = 30/\lambda^2 = 30/185.8^2 = 8.69 \times 10^{-4}$$
$$I_s \geq 0.34 \times 8.69 \times 10^{-4} \times 1270^3 \times 35 = 21.183 \times 10^6 \text{ mm}^4$$

Now $I_s = 12 \times (2 \times 220)^3/12 = 85.184 \times 10^6$ mm$^4 > 21.183 \times 10^6$ mm^4

Hence the provided stiffener has the necessary torsional restraint.

(c) Weld at web flange junction

Assuming fillet weld on each side of the web,

$$q_w = VA_f \bar{y}/2I_z$$
$$I_z = b_f D^3/12 - (b_f - t_w)d^3/12$$
$$= 450 \times 1270^3/12 - (450 - 12)1200^3/12$$
$$= 13742.3 \times 10^6 \text{ mm}^4$$

$$q_w = 877.5(450 \times 35) \times 635/(2 \times 13742.3 \times 10^6)$$

$$= 0.319 \text{ kN/mm}$$

From Appendix D, provide 4 mm fillet weld (0.442 kN/mm)

(d) Weld for end stiffener

The minimum weld size required for connecting the stiffeners to the web, assuming a weld on each side of the stiffener is (clause 8.7.2.6 of the code)

$$q_1 = t_w^2/5b_s = 12^2/(5 \times 220) = 0.13 \text{ kN/mm}$$

when the stiffener also is subjected to external loading, then shear due to such loading must be added to the above value. The stiffeners have to resist the difference between the applied load and the minimum load that can be carried safely by the unstiffened web (370.3 kN in this case).

Length of weld $= 1200 - 2 \times 15 = 1170$ mm

$$q_2 = (877.5 - 370.3)/1170 = 0.43 \text{ kN/mm}$$

$$q_w = q_1 + q_2 = 0.13 + 0.43 = 0.56 \text{ kN/mm}$$

Force on each weld $= 0.56/2 = 0.28$ kN/mm

From Appendix D, provide a pair of 4 mm fillet welds (0.442 kN/mm).

Example 7.6 *Redesign the plate girder in Example 7.5 with intermediate stiffeners and not using tension field action.*

Solution

From Example 7.5,

$$M_z = 4275 \text{ kN m}$$

$$F_z = 877.5 \text{ kN m}$$

Assume a web plate of size 1400×8 mm and a flange plate of size 450×35 mm for each of the flanges. Since the flanges are of the same size as in Example 7.5, the calculations for moment capacity check are not repeated here.

(1) Check for shear

According to clause 8.4.2.2 (a) of the code,

Nominal shear strength $V_n = V_{cr}$

Assume $c/d = 1.4$; Hence stiffener spacing $= 1.4 \times 1400 = 1960$ mm

Adopt a stiffener spacing of 2000 mm. Provide 10 panels @ 2000 mm c/c (see Fig. 7.24).

Panel AB is the most critical panel for maximum shear. So design checks for the web are made for panel AB only. For the panel AB,

$$\frac{d}{t} = \frac{1400}{8} = 175$$

Since c (2000 mm) $> d$ (1400 mm) and $d/t_w < 200$, the girder satisfies serviceability requirements.

$$\frac{c}{d} = \frac{2000}{1400} = 1.4286$$

$$d/t_w = 175 < 345\varepsilon_f$$

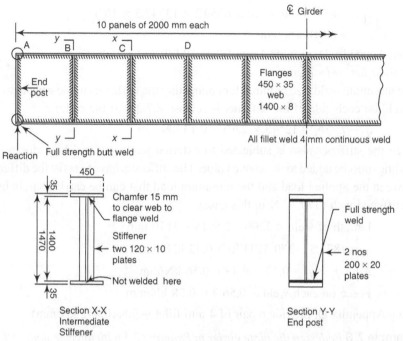

Fig. 7.24

Hence as per clause 8.6.1.2 buckling of compression flange into the web will be avoided.

For $\dfrac{c}{d} > 1.0$; $k_v = 5.35 + \dfrac{4.0}{\left(\dfrac{c}{d}\right)^2} = 5.35 + \dfrac{4.0}{(1.4286)^2} = 7.3099$

with $\mu = 0.3$
Average shear stress

$$q_v = \frac{877.5 \times 10^3}{1400 \times 8} = 78.35 \text{ N/mm}^2$$

The elastic critical shear stress of the web

$$\tau_{cr} = \frac{k_v \pi^2 E}{12(1 - \mu^2)\left(\dfrac{d}{t_w}\right)^2} = \frac{7.3099 \times \pi^2 \times 2.0 \times 10^5}{12(1 - 0.3^2)\left(\dfrac{1400}{8}\right)^2} = 43.15 \text{ N/mm}^2$$

The non-dimensional web slenderness ratio for shear buckling stress,

$$\lambda_w = \sqrt{\frac{250}{(\sqrt{3}\,\tau_{cr,\,e}}} = \sqrt{\frac{250}{(\sqrt{3} \times 45.15)}} = 1.788$$

when $\lambda_w > 1.2$

$$\tau_b = \frac{f_{yw}}{(\sqrt{3\lambda_w^2})} = \frac{250}{(\sqrt{3} \times 1.788^2)} = 45.15 \text{ N/mm}^2$$

Note that the values of τ_{cr} and τ_b are nearly the same in this case. Shear force corresponding to buckling

$$V_{cr} = dt_w \tau_b = 1400 \times 8 \times 45.15/1000 = 505.68 < 877.5 \text{ kN}$$

As per clause 8.7.1.1(a) of the code, provide intermediate stiffeners to improve the buckling strength of the slender web due to shear.

(2) *Check for shear capacity of the end panel*

According to clause 8.5.1 of the code, for end panel design (without using tension field action), we should use clause 8.5.3,

$$H_q = 1.25V_{dp}\left(1 - \frac{V_{cr}}{V_{dp}}\right)^{1/2}$$

where

$$V_{dp} = dtf_{yw}/\sqrt{3} = \frac{(1400 \times 8)250}{\sqrt{3}} = 1616.58 \text{ kN}$$

$V_{cr} = 505.68 \text{ kN}$

$\therefore$ Longitudinal shear

$$H_q = 1.25 \times 1616.58\left(1 - \frac{505.68}{1616.58}\right)^{1/2} = 1675.12 \text{ kN}$$

$$R_{tf} = \frac{H_q}{2} = \frac{1675.12}{2} = 837.56 \text{ kN}$$

$$A_v = t_w d = 8 \times 1400 = 11200 \text{ mm}^2$$

$$V_n = \frac{A_v f_{yw}}{\sqrt{3}\gamma_{m0}} = \frac{250}{\sqrt{3} \times 1.10} \times \frac{11200}{1000} = 1469.6 \text{ kN} > 837.56 \text{ kN}$$

The end panel is safe to carry the shear due to anchoring forces. Check for the moment capacity of end panel:

$$M_{tf} = \frac{H_q d}{10} = \frac{1675.12 \times 1400 \times 10^3}{10 \times 10^6} = 234.5 \text{ kN m}$$

$$y = \frac{c}{2} = \frac{2000}{2} = 1000 \text{ mm}$$

$$I = \frac{1}{12}t_w \times c^3 = \frac{1}{12} \times 8 \times 2000^3 = 533.33 \times 10^7 \text{ mm}^4$$

$$M_q = \frac{I}{y}\frac{f_y}{\gamma_{m0}} = \frac{533.33 \times 10^7}{1000 \times 10^6} \times \frac{250}{1.10} = 1212.12 \text{ kN m} > M_{tf} = 234.5 \text{ kN m}$$

$M_{tf} < M_q$

Hence, the end panel can carry the bending moment due to anchor forces.

(3) Design of stiffeners
Load bearing stiffener at point A (see Fig.7.43),
    Reaction at A = 877.5 kN

Force F_m due to $M_{tf} = \dfrac{M_{tf}}{c} = \dfrac{234.5 \times 10^6}{2000 \times 10^3} = 117.25$ kN

Total compression $F_c = 877.5 + 117.25 = 994.75$ kN
According to clause 8.7.5.2 of the code,

Area of stiffener $A_q > \left(\dfrac{0.8 \times F_c \times \gamma_{m0}}{f_{yq}} \right)$

$A_q > \left(\dfrac{0.8 \times 994.75 \times 1.1 \times 10^3}{250} \right)$ or $A_q > 3501$ mm^2

Provide stiffener of two flats of size 200 × 20 mm.
    Area = 8000 mm^2 > 3501 mm^2
(a) Check for outstand: (clause 8.7.1.2 of the code)
    Outstand b_s = 200 m < $20 t_q \varepsilon$ = 20 × 20 × 1 = 400 mm
    $14 t_q \varepsilon$ = 14 × 20 × 1 = 280 mm
    b_s = 200 mm < $14 t_q \varepsilon$
Hence, the criterion for the outstand has been satisfied.
(b) Buckling check: (clause 8.7.1.5 of the code)
Neglecting the buckling resistance of the web for simplicity,

$I_x = \left(\dfrac{20 \times 408^3}{12} \right) - \left(\dfrac{1}{12} \times 20 \times 8^3 \right)$

    = 113.19 × 10^6 mm^4
    Effective area = 200 × 20 × 2 = 8000 mm^2

Radius of gyration = $\sqrt{\dfrac{I_x}{A}} = \sqrt{\dfrac{113.19 \times 10^6}{8000}}$ = 118.951

Flange is restrained against rotation and lateral deflection.
    L_e = 0.7L = 0.7 × 1400 = 980 mm

$\lambda = \dfrac{L_e}{r_x} = \dfrac{980}{118.951}$ = 8.238 < 10

From Table 9c, for f_y = 250 N/mm^2
    f_{cd} = 227 N/mm^2
Buckling resistance of the stiffener
    $P_d = f_{cd} A_e$ = 227 × 2 × 200 × 20 =1816 kN > 994.75 kN
Hence, the stiffener is safe against buckling.
(c) Check stiffener as load bearing stiffener
Assume stiff bearing length,
    b_1 = 0
    n_2 = 2.5 × 35 = 87.5 mm

According to clause 8.7.4 of the code,

$$\text{Local capacity of web } F_w = (b_1 + n_2)t_w \frac{f_{yw}}{\gamma_{m0}}$$

$$F_w = (87.5) \times 8 \times \frac{250}{1.10 \times 10^3} = 159.09 \text{ kN}$$

Bearing stiffener is designed for $(F_c - F_w) = (994.75 - 159.09) = 835.66$ kN
Bearing capacity of stiffener alone

$$= \frac{250}{1.10} \times \frac{200 \times 20 \times 2}{1000} = 1818.18 \text{ kN} > 835.66 \text{ kN}$$

Hence the stiffener is safe as a load bearing stiffener.
At the location of concentrated loads local capacity of the web
$$= (87.5 \times 2) \times 8 \times 250/(1.10 \times 10^3) = 318.16 \text{ kN} > 225 \text{ kN}$$
Hence, the stiffener is safe.
(4) Design of intermediate stiffener at 'B' (clause 8.7.2.4 of the code)
Stiffener B is the most critical intermediate stiffener.
(a) Minimum Stiffeners

$$\frac{c}{d} = \frac{2000}{1400} = 1.429 > \sqrt{2}$$

$$I_s \geq 0.75 dt_w^3$$
$$I_s \geq 0.75 \times 1400 \times 8^3 = 0.5376 \times 10^6 \text{ mm}^4$$
Try intermediate stiffener of two flats of 120×10 mm.

$$\text{Provided } I_s = \left(\frac{10 \times 248^3}{12}\right) - \left(\frac{10 \times 8^3}{12}\right) = 12.71 \times 10^6 \text{ mm}^4$$

Hence, the stiffeners have more than the minimum required stiffeness.
(b) Check for outstand (clause 8.7.1.2):
Outstand of the stiffener $= b_s = 120$ mm
$14 t_q \varepsilon = 14 \times 10 = 140$ mm
$b_s < 14 t_q \varepsilon$
Hence the outstand provisions of the code are satisfied.
(c) Buckling check (clause 8.7.2.5):

$$F_q = \left[\frac{V - V_{cr}}{\gamma_{m0}}\right]$$

where V is the factored shear force and V_{cr} is the shear buckling resistance of the web panel designed without using tension field action $= 505.68$ kN
Shear force @ B,
$$V_B = 877.5 - 45 \times 2 = 787.5 \text{ kN}$$

$$F_q = \frac{(787.5 - 505.68)}{1.10} = 256.2 \text{ kN}$$

Buckling resistance of intermediate stiffener at B (clause 8.7.1.5):

As per clause 8.7.1.5, an effective length of web equal to $20 \times t_w$ on each side of the centre line of stiffener can be considered along with the stiffener.

$20t_w = 20 \times 8 = 160$ mm

$$I_x = \left(\frac{1}{12} \times 10 \times 248^3\right) + \left(\frac{320 \times 8^3}{12}\right) - \left(\frac{10 \times 8^3}{12}\right) = 12.724 \times 10^6 \text{ mm}^4$$

Area $= (240 \times 10) + (320 \times 8) = 4960$ mm^2

$$r_x = \sqrt{\frac{12.724 \times 10^6}{4960}} = 50.649 \text{ mm}$$

$L_e = 0.7L = 0.7 \times 1400 = 980$ mm

$$\lambda = \frac{L_e}{r_x} = \frac{980}{50.649} = 19.349$$

From Table 9c of the code, for $f_y = 250$ MPa (buckling curve c) and $\lambda = 19.349$,

$\quad f_{cd} = 224.195$ N/mm^2

$\quad$ Buckling resistance $= 224.195 \times 4960 = 1112.01$ kN > 256.2 kN

$\quad$ Hence, the stiffener is safe against buckling.

According to clause 8.7.2.5 of the code:

Intermediate stiffener at D subjected to external load should satisfy the following interaction equation.

$$\left(\frac{F_q - F_x}{F_{qd}}\right) + \left(\frac{F_x}{F_{xd}}\right) + \left(\frac{M_q}{M_{yq}}\right) \leq 1.0$$

where F_q is the stiffener force $= 256.2$ kN, F_{qd} is the design resistance of an intermediate web stiffener corresponding to buckling about an axis parallel to web $= 1112.01$ kN, F_x is the external load or reaction at the stiffener $= 225$ kN, F_{xd} is the design resistance of a load carrying stiffener corresponding to buckling about axis parallel to web at D $= 1112.01$ kN, and M_q is the moment on the stiffener due to eccentrically applied load $= 0$.

$$(F_q - F_x) = (256.2 - 225) = 31.2 \text{ kN}$$

Thus, the above check reduces to

$$\frac{31.2}{1112.01} + \left(\frac{225}{1112.01}\right) = 0.23 < 1.0$$

Hence, the stiffener at D is safe.

Note Intermediate stiffener should extend to the compression flange, but not necessarily be connected to it. Stiffeners not subjected to external load or moment can be terminated at a distance of about $4t$ ($4 \times 10 = 40$ mm) from the tension flange. However in our case, since we have a moving load, we may connect the intermediate stiffener to the top (compression) flange and just extend it up to the bottom flange. The calculation for weld at web-flange junction and weld for end and intermediate stiffener are similar to those given in Example 7.5.

In the preceding designs, it has been assumed that adequate lateral bracing will be provided for the compression flange of the plate girder. The calculations for location, size, and connection of lateral bracing are not included here (see Section 6.4.1 for the details). Normally, these bracings will be attached to the intermediate stiffeners by bolting at site. In this case, a weld between the stiffener and the compression flange would be provided to transfer the force in the bracing to the girder (Kulak & Grondin 2002). It is also usual to reduce the cross section of a welded girder at least once as the moment decreases. A transition as shown in Fig. 11.18 will be adopted which may allow for the gradual flow of stress from one section to another. In practice, the design process would be repeated for several assumed cross sections and the best cross section, in terms of material, fabrication, and erection costs, will be chosen.

Summary

A plate girder is a large beam built either from plates only or from plates and angles to achieve a more efficient use of material than is possible with the use of rolled beams, which have a limitation in depth. Plate girders may be built using bolts or welds. The advent of welding allowed designers the freedom to tailor-make a girder to suit their requirements. Now, the flanges and the web are connected together by fillet welds only, using semi- or fully automatic welding procedures. Where a splice is required, owing to the maximum length of plate rolled or because there is a change in plate thickness, a full strength butt weld is employed.

Several long span plate girders have been built all over the world. Box girders and hybrid girders are being increasingly used for long spans and heavy loads. Box girders, due to their large torsional rigidity, compared to plate girders, are adopted for curved bridges. In hybrid girders, different grades of steel are used for flanges (high strength) and webs (low strength) to obtain an efficient cross section. Sometimes, notch ductile steel is used for the tension flange of a road bridge to eliminate the possibility of low cycle brittle fracture. Since plate girders are quite deep, the steel need to be distributed economically between the flanges and the web so as to minimize the weight of the girder in a given situation.

Since thin webs are often adopted, the usual assumption made for small and medium plate girders is that the flanges resist most of the bending moment and the web carries the shear force. The flanges are assumed to be plastic, compact, or semi-compact. It is also possible to design assuming that that the bending moment and axial force are resisted by the whole section and the web resists the combined shear and normal stresses. The design of flanges in plate girders is similar to the design of flanges of rolled sections.

The webs are often thin and are likely to buckle under compression or shear. To avoid buckling of web due to shear and bending, and buckling or crippling of web at points of concentrated loads, the web has to be stiffened by intermediate stiffeners, horizontal stiffeners, and bearing stiffeners and their combinations. Experiments conducted in the past have shown that the web will be capable of withstanding loads in excess of the elastic buckling load, provided sufficiently heavy stiffeners are employed. This occurs as a result of *tension field action* in which the diagonal web tensile stresses act with the transverse stiffeners and the flanges to transfer the additional load by means of a truss type of action. Ultimate load is reached when the tension field yields. The code has adopted the tension field action as suggested by Rockey, Evans, and Porter of England. Various types of transverse stiffeners and longitudinal stiffeners have also been considered in depth. The provisions for the design of plate girders, as given in IS 800 : 2007 have been discussed at great length.

Plate girders with corrugated webs are being increasingly used, since they provide an aesthetic appearance as well as increase the shear strength of the web, without the use of stiffeners. Steel plate shear walls (SPSW) are used in several countries owing to their advantages over reinforced concrete shear walls and their ability to resist cyclic loading. The steps involved in the design of plate girders and some thumb rules for the preliminary proportioning of plate girders has also been discussed. The examples provided illustrate the use of equations given in this chapter.

Exercises

1. Determine the buckling resistance moment for a welded plate girder consisting of 400×25 mm flange plates and a 1000×10 mm web plate in grade 410 steel. Assume a laterally braced span of 5.5 m.

2. Compute the design flexural strength of a plate girder which consists of a 8×1200 mm web and 30×450 mm flanges. Use grade 410 steel and assume that the compression flange is continuously supported.

3. A welded plate girder is fabricated from two 600×35 mm flange plates and 1300×12 mm web plate of grade 410 steel. What is the moment capacity of the girder? If the section is required to carry full plastic moment, what would be the changes required in the plate thickness?

4. Determine the shear buckling resistance of the plate girder web given in Exercise 7.1, if intermediate transverse stiffeners are spaced at 1600 mm.

5. A plate girder of grade 410 steel has a 12×1500 mm web and 35×550 mm flanges.
   (a) Determine the design flexural strength, if the compression flange is continuously laterally supported.
   (b) Determine the flexural strength, if the unrestrained length is 10 m.
   (c) What is the design shear strength if no intermediate stiffeners are used.
   (d) Determine the shear strength, if the stiffeners are provided with spacing of 1.5 m.
   (e) Find the shear strength of the end panel, if the first intermediate stiffener is provided at 2.0 m from the end support.

6. Check the capacity of a pair of intermediate transverse web stiffeners of size 150×12 mm for the plate girder mentioned in Exercise 7.3.

7. A plate girder web is to be fabricated by a plate 1200 mm deep and 12 mm thick. The girder should be capable of carrying a shear load of 1400 kN without using tension field action. Assuming grade 410 steel, determine the spacing of vertical stiffeners required. What would be the increase in load carrying capacity, if tension field action is permitted?

8. A plate girder of grade 410 steel is composed of a 10×2000 mm web and 32×500 mm flanges. The girder span is 15 m (see Fig. 7.25). Stiffeners are placed at 1 m, 3 m, and 5 m from both ends. Determine the shear strength of each of the panels.

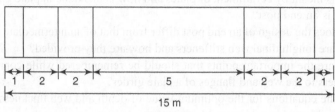

Fig. 7.25

9. Check the adequacy of a pair of load bearing stiffeners 100×12 mm, which are placed at the support of the plate girder mentioned in Exercise 7.2. Assume that the flanges of the girder are restrained by other structural elements against rotation. The girder is supported on a stiff bearing 250-mm long, the end panel width is 1000 mm and the reaction(factored) is 800 kN.

10. Design an 18-m long simply supported welded plate girder carrying a uniformly distributed load of 50 kN/m excluding self weight, and two concentrated loads of 350 kN each at quarter points of the span. Assume that the girder is laterally supported throughout.

11. A 20-m long plate girder has to support a UDL and concentrated loads at one-third points. The uniform load consists of 18 kN/m dead load and 30 kN/m live load. Each concentrated load consists of a 125 kN dead load and a 225 kN live load. There is lateral support at the ends and at the points of concentrated load. Using grade 410 steel determine the following.
    (a) Design the cross section.
    (b) Determine the location and size of intermediate stiffeners.
    (c) Design suitable bearing stiffeners at the supports and beneath the loading points.
    (d) Design welds for all the elements.

Review Questions

1. What is the difference between a beam and a plate girder?
2. Where are plate girders used?
3. What are the main characteristics of a plate girder?
4. Sketch the different types of cross sections used as plate and box girders.
5. Why have bolted and riveted plate girders become obsolete?
6. What are the different modes of failures of a plate girder?
7. List the different elements of a welded plate girder.
8. Explain the tension field action that is developed in the thin webs of plate girders?
9. What are the various types of stiffeners?
10. State the minimum web thickness provisions of a IS 800 : 2007.
11. What are the design concepts of a plate girder?

12. When do we need to consider the shear buckling of the web?
13. How does plate girder derive post-buckling strength?
14. When do we provide splicing of webs and flanges?
15. What is meant by curtailment of plates and how is this done in plate girders?
16. What is an end post?
17. How does the design of an end post differ from that of an intermediate stiffener?
18. What are longitudinal web stiffeners and how are they provided?
19. What are the important points that should be remembered while connecting the stiffeners to the web and flanges of a plate girder?
20. State the equations for the optimum value of depth and web thickness of a plate girder subjected to a moment *M*.
21. Why/where are bearing stiffeners provided?
    (a) At the end of the plate girder
    (b) At the ends of the plate girder and on both faces of the web
    (c) At the ends of the plate girder and only on one face of the web
    (d) Where concentrated loads are acting and on both faces of the web
22. Why/where are intermediate transverse stiffeners provided?
    (a) Where concentrated loads are acting
    (b) In order to reduce the panel aspect ratio
    (c) For aesthetic purposes
    (d) On one face of the web only for aesthetic reasons
23. The portion of the web that acts integrally with the vertical stiffener in resisting the load is taken as
    (a) 20 times the thickness of the web on either side of the stiffener
    (b) 40 times the thickness of the web on either side of the stiffener
    (c) 12.5 times the thickness of the web on either side of the stiffener
    (d) None of the above
24. What are the steps involved in the design of plate girders?

8

Design of Gantry Girders

Introduction

In mills and heavy industrial buildings such as factories and workshops, gantry girders supported by columns and carrying cranes are used to handle and transport heavy goods, equipment, etc. There are several types of cranes; overhead travelling, under-slung, jib, gantry, and monorail are among the most common. A building may have one or several of these, either singly or in combinations. Hand-operated overhead cranes have lifting capacities of up to 50 kN and electrically operated overhead travelling cranes, called EOT cranes, can have capacities in the range of 10–3000 kN. The overhead travelling crane runway system consists of the following components (see Figs 8.1 and 8.2):

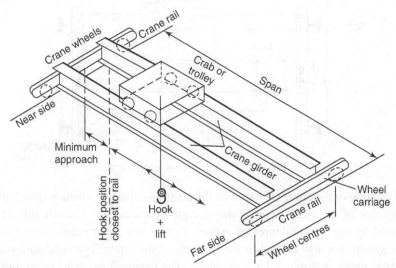

Fig. 8.1 Components of an overhead crane

1. The crane, comprising the crane girder (crane frame), crab or trolley, hoist, power transmitting devices, and a cab which houses the controls and operator
2. The crane rails and their attachments
3. The gantry girder
4. The gantry girder supporting columns or brackets
5. The crane stops

The loads are lifted using a hook and moved longitudinally and transversely anywhere in the building through the movement of a crab car or trolley on the crane frame (girder) or truss and the crane wheels on the crane rails (see Figs 8.1 and 8.2). The gantry girder is supported by brackets attached to the main columns of the building or by stepped columns.

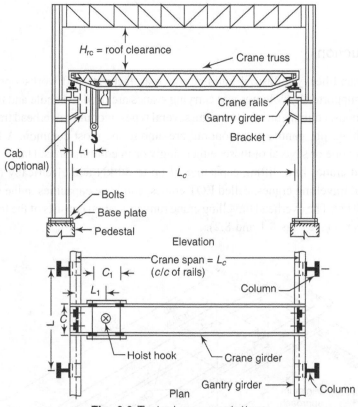

Fig. 8.2 Typical crane notations

The rating of the cranes is based on the hoisting capacity, which in turn depends on the size of the crab, wheel spacing, roof clearance, etc. Usually this data is supplied by the manufacturer of cranes. Power and other requirements of the crane system are also related to the capacity of the crane. Typical data for some cranes is given in Table 8.1 for guidance. The designer needs to check out this data from the manufacturer before carrying out the design of gantry girders.

Table 8.1 Typical data for cranes (see Fig. 8.2)

Capacity of crane (kN)	Auxiliary weight (kN)	Span L_c (m)	Wheel base c (m)	Minimum hook distance of main load, L_1 (m)	Vertical clearance (m)	Weight of crane bridge (kN)	Weight of trolley/crab (kN)	Crane rail Head width (mm)	Crane rail Weight (kN/m)	Base width (mm)
50	—	10.5–22.5	3.0–4.8	0.65–1.00	1.83	50–150	15	CR50	298	90
100	30	10.5–22.5	3.2–4.8	0.65–1.10	1.83	80–210	35	CR50	298	90
150	30	10.5–31.5	3.2–5.3	0.80–1.10	2.13	210–250	60	CR60	400	105
200	50	10.5–31.5	3.5–5.3	0.80–1.10	2.13	160–275	75	CR60	400	105
250	50	10.5–31.5	3.5–5.3	0.80–1.10	2.44	275–320	85	CR80	642	130
300	80	10.5–31.5	3.8–5.3	0.80–1.15	2.44	300–360	100	CR80	642	130
400	80	10.5–31.5	3.8–5.3	0.85–1.15	2.74	350–400	120	CR80	642	130
500	125	10.5–31.5	4.0–5.3	0.85–1.20	2.74	400–470	135	CR100	890	150
600	125	10.5–31.5	4.5–5.3	1.00–1.20	2.74	600–750	250	CR120	1180	170

Note: 1. Exact crane data must be obtained from the manufacturer.
2. The auxiliary load need not be considered in the design of the girder.

While deciding on the choice of the crane, one must consider the load capacity, space limitations, and the class of service required. While designing the crane supporting structures, the engineer should take into account these requirements and other factors such as the likely future changes in the load capacity, addition of other cranes, various load combinations, and possible future extensions. Few other structures suffer such an extreme range of stresses and as high an incidence of maximum loadings and fatigue as crane runways, and this must also be considered by the engineer. In addition, one must be aware of the infinite variety of abuses inflicted on crane systems, such as hoisting loads which exceed the crane capacity, swinging loads as a pendulum, dragging loads laterally or longitudinally, and ramming the crane against the crane stops at an excessive speed. The crane is often one of the most important parts of an industrial operation, so any significant 'downtime' required for repairs and maintenance of the same can be disastrous to the owner.

Cranes may be classified on the basis of the load carrying capacity (load lifted), the height to which the load is lifted, and the frequency of lifting the loads. For example, the Crane Manufacturers Association of America (CMAA) have classified the cranes as follows (Ricker 1982):

1. Class A1 (standby service)
2. Class A2 (infrequent use)
3. Class B (light service)
4. Class C (moderate service)
5. Class D (heavy duty)
6. Class E (severe duty cycle service)
7. Class F (Steel Mill)

Table 8.2 gives the representative crane speeds.

Table 8.2 Typical speeds of overhead cranes (Weaver 1979)

Capacity (kN)	Slow (m/min)	Medium (m/min)	Fast (m/min)
10	60	90	120
50	60	75	90
100	30	45	60
150	30	38	45
200	30	38	45
250	30	38	45

It should be remembered that the design aspects for light cranes may be different from those for heavy cranes: cranes with long spans (over 15 m) should be treated differently than those with shorter spans; fast, heavy service cranes require special consideration not required for slower, lighter cranes.

8.1 Loading Considerations

The following important quantities need to be considered by the designer while estimating the loads on the gantry girder (see Fig. 8.3).

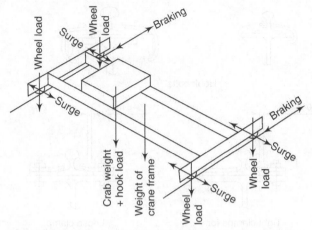

Fig. 8.3 Loads to be considered on gantry girders

Weight of the trolley or crab car (W_t) Since the trolley moves on the crane girder (see Figs 8.1 and 8.3) along the span of the truss, its weight is transferred to the crane wheels as the axle load and finally to the gantry girder. The load transferred to the gantry will be maximum when the trolley wheels are closest to the gantry girder. The wheel load that is transferred from the trolley to the gantry girder is

$$W_1 = [W_t(L_c - L_1)]/(2L_c) \tag{8.1}$$

where W_1 is load of each wheel on the gantry girder (note that there are 2 wheels), W_t is the weight of the trolley or crab car, L_c is the distance between the gantry girders (span of truss/crane span), and L_1 is the distance between the CGs of the trolley and gantry.

Weight of the crane girder (W_c) The crane manufacturers also supply a pair of crane girders, which may be in the form of open web trusses or solid I-beams on which the trolley moves. These I-beams will be mounted on a crane rail using wheel carriages having four wheels, two wheels on either end (see Figs 8.1–8.3). These wheels move on rails mounted on the gantry girders. The crane rails are attached to the gantry girders with bolted clamps or hook bolts spaced about 500–1000 mm apart, depending on the lifting capacity of the crane. Rails are generally not welded to the gantry girder because of the difficulty in the readjustment of the alignment and replacement of defective or worn out rails. The hook bolts (used with slow-moving cranes having a capacity of 50 kN and bridge span of 15 m) and clamps are shown in Fig. 8.4. More details about rails and the methods of attaching them to gantry girders may be found in Ricker (1982). The weight of the crane girder is equally distributed on all the four wheels. Hence, the weight of the crane girder on each wheel is

$$W_2 = W_c/4 \tag{8.2}$$

where W_2 is the load due to the weight of the crane truss/girder and W_c is the weight of the crane truss/girder.

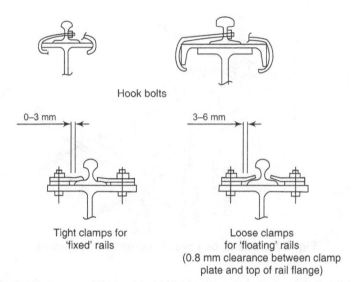

Hook bolts

0–3 mm

3–6 mm

Tight clamps for
'fixed' rails

Loose clamps
for 'floating' rails
(0.8 mm clearance between clamp
plate and top of rail flange)

Fig. 8.4 Methods of attaching rails to gantry girders (Ricker 1982)

Impact loads As the load is lifted using the crane hook and moved from one place to another, and released at the required place, an impact is felt on the gantry girder. It is due to the sudden application of brakes (frictional force), moving loaded crane's acceleration, retardation, vibration, or possible slip of slings, etc. To account for these, suitable impact factors should be applied to the loads. As per IS 875 and IS 807, additional impact loads as listed in Table 8.3 should be considered.

Table 8.3 Additional impact loads on cranes

Type of load	Impact allowance (percentage)
Vertical forces transferred to wheels	
(a) For EOT cranes	25% of the maximum static wheel load
(b) For hand-operated cranes	10% of the maximum static wheel load
Horizontal forces transverse to the rails:	
(a) For EOT cranes	10% of the weight of the crab and the weight lifted on the crane
(b) For hand-operated cranes	5% of the weight of the crab and the weight lifted on the crane
Horizontal forces along the rails	5% of the static wheel loads for EOT or hand-operated cranes

Lateral load (surge load) As the crane moves with the load, a lateral load (transverse to the rail) is developed, as shown in Fig. 8.5, due to the application of brakes or the sudden acceleration of the trolley. Suppose that the total weight including the lifted weight and the trolley weight is W, the coefficient of friction is 0.1, the number of wheels is 4, and number of breaking wheels is 2. Then,

$$\text{Horizontal force} = W \times 0.1 \times (2/4) = 0.05W.$$

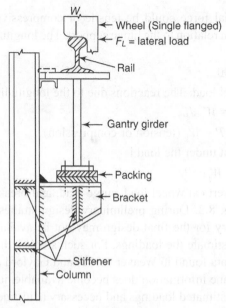

Fig. 8.5 Application of lateral load and typical bracket

However, the Indian code IS 875 recommends $10\%W$ for EOT cranes (see Table 8.3). These horizontal loads are also called *surge loads*. In the design of the girder, it is usually assumed that the lateral or longitudinal surge is resisted by the compression flange alone. Note that it is customary to assume that the entire horizontal force acts on one gantry girder at the position of the wheels. But if the wheels have guides on both sides of the flange of the wheel, the force is resisted by the two opposite gantry girders and only half the horizontal load as specified in Table 8.3 acts on one girder. Thus,

$$W_L = 10/100(W_t + W_k) \text{ for EOT crane} \tag{8.3a}$$

$$W_L = 5/100(W_t + W_k) \text{ for hand-operated crane} \tag{8.3b}$$

where W_k is the hook load (the maximum capacity of the crane is denoted by this load) and W_t is the weight of the trolley.

Longitudinal load (drag load) As the crane moves longitudinally, loads parallel to the rails are caused due to the braking (stopping) or acceleration and swing (starting) of the crane. This load is called the longitudinal load or *drag load* and is transferred at the rail level. Figure 8.6 illustrates the action of the longitudinal load and its reactions. Due to the vertical loads, the reactions R_{a1} and R_{b1} are developed. These reactions can be calculated easily.

Due to the braking action, a longitudinal load is developed at the top of the rails, which acts at a height e from the centre of gravity of the beam. This eccentric load, in turn, generates equal and opposite reactions. Thus, the gantry girder is subjected to bending moments in addition to the

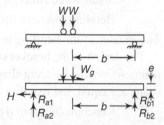

Fig. 8.6 Braking force

axial force. The axial force could be tensile or compressive depending on the direction of motion in relation to the hinge support. The longitudinal load per wheel as per Table 8.3 is

$$W_g = 5W/100 \tag{8.4}$$

where W is the wheel load. The reactions due to the longitudinal loads are

$$R_{b2} = -R_{a2} = W_g e/L \tag{8.5}$$

$$\text{Axial force } P = W_g \text{ (tension or compression)} \tag{8.6}$$

The bending moment under the load is

$$M = R_{b2}b = W_g eb/L \tag{8.7}$$

All these loads (vertical wheel load, surge load, and breaking load are shown schematically in Fig. 8.3. During preliminary design studies, the specific crane information necessary for the final design may not be available, and it becomes often necessary to estimate the loadings. For such studies, the information given in Table 8.1 or the data found in Weaver (1979) and Gaylord et al. (1996) may be useful. When the crane information does become available, it should be carefully compared with the estimated loadings and necessary adjustments should be made to the preliminary design.

8.2 Maximum Load Effects

Moving loads, such as crane wheels, result in bending moments and shear forces, which vary as the load travels along the supporting girder. In simply supported beams the maximum shear force will occur immediately adjacent to the support, while the maximum bending moment will occur in the region of mid-span. In general, influence lines should be used to find the load positions that produce maximum values of shear forces and bending moments (see Marshall & Nelson 1990; Coates et al. 1988 for a discussion and calculation for influence lines).

For a simply supported beam with two moving loads, the load positions which produce maximum shear force and bending moments are shown in Fig. 8.7. (The wheel load bending moment is maximum when the two loads are in such a position that the centres of gravity of the wheel loads and one of the wheel loads are equidistant from the centre of gravity of the girder. The shear due to the wheel load is maximum when one of the wheels is at the support.) From these, the maximum shear force and bending moments are (Marshall & Nelson 1990)

$$\text{Shear force (max)} = W(2 - c/L) \tag{8.8}$$

$$\text{Bending moment (max)} = WL/4 \text{ or } 2W(L/2 - c/4)^2/L \tag{8.9}$$

The greater of the bending moment values should be used.

The design of the bracket supporting a crane girder uses the values of maximum reaction from the adjacent simply supported beams as in Fig. 8.7. When the adjacent spans are equal, the reaction is equal to the shear force, i.e,

$$\text{Reaction (max)} = W(2 - c/L) \tag{8.10}$$

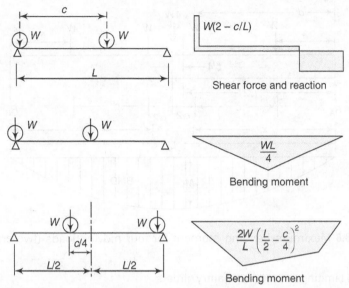

Fig. 8.7 Maximum BM, SF, and reaction for two moving loads

Note that in all the cases the effect of the self-weight of the gantry girder must also be considered as a uniformly distributed load.

The deflection at the mid-span due to placing the two loads symmetric with respect to it is given by

$$\Delta_c = W_c L^3 [(3a/4L) - (a^3/L^3)]/(6EI) \tag{8.11}$$

where $a = (L - c)/2$, E is Young's modulus of rigidity, and I is the moment of inertia of the cross section.

When two cranes are operating in tandem on the same span, these have to be located closest to each other towards the mid-span to produce maximum bending moment. In such a situation, there will be four wheels, two wheels from each crane. These four wheels must be adjusted on the span such that the centroid of the load and the nearest wheel load to the CG from the centre line of the beam are at the same distance (see Fig. 8.8). Let x be the distance from left to the load under which maximum bending moment occurs.

$$x = (L/2) - (b/4) \tag{8.12}$$

The reaction and bending moment are

$$R_A = 4W(x/L) \tag{8.13}$$

$$
\begin{aligned}
M_1 &= R_A x - W_c \\
&= (4Wx^2/L) - W_c \\
&= W/L[4x^2 - cL]
\end{aligned} \tag{8.14}
$$

The deflection at mid span may be obtained approximately as (see Fig. 8.8)

$$\Delta = W/(24EI)[a(3L^2 - 4a^2) + b(3L^2 - 4b^2)] \tag{8.15}$$

These deflections at working loads should be less than the allowable deflections as specified in Table 8.4 (see Table 6 of IS 800 : 2007).

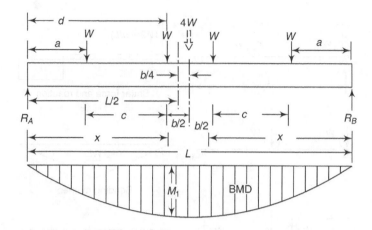

Fig. 8.8 Maximum bending moment for four moving loads (two cranes)

Table 8.4 Limiting deflection of gantry girders

Category	Maximum deflection
Vertical deflection	
Manually operated cranes	$L/500$
EOT cranes with a capacity of less than 500 kN	$L/750$
EOT cranes with a capacity of greater than 500 kN	$L/1000$
Lateral deflection	
Relative between rails	10 mm or $L/400$

8.3 Fatigue Effects

Gantry girders are subjected to fatigue effects due to the moving loads. Normally, light- and medium-duty cranes are not checked for fatigue effects if the number of cycles of load is less than 5×10^6. For heavy-duty cranes, the gantry girders are to be checked for fatigue loads (see also IS 1024 and IS 807). See section 13 of the code for design provisions for fatigue effects. Note that fatigue strength is checked at working loads!

8.4 Selection of Gantry Girder

The gantry girder is subjected to vertical loading including impact loading, lateral loading, and longitudinal loading from traction, braking, and impact on crane stops. In addition, the crane beams must withstand local buckling under wheel loads and at the bottom flange over the column (in the common case where the beam bears on a column cap plate). Figure 8.9 illustrates typical profiles used for gantry girders.

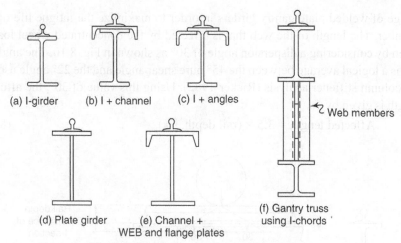

(a) I-girder   (b) I + channel   (c) I + angles

Web members

(d) Plate girder   (e) Channel +   (f) Gantry truss
WEB and flange plates   using I-chords

Fig. 8.9 Profiles used for gantry girders

- Rolled beams with or without plates, channels or angles (for spans up to 8 m and usually for 50 kN cranes), as shown in Figs 8.9(a–c).
- Plate girders (for spans from 6 m to 10 m) [Fig. 8.9(d)]
- Plate girder with channels, angles, etc. (for spans more than 10 m) [Fig. 8.9(e)].
- Box girders with angles (for spans more than 12 m)
- Crane truss using I-sections as chord members and angles as web members [Fig. 8.9(f)]

Single-span gantry girders are desirable. Two-span gantry girders can result in uplift in columns at certain loading positions, and differential settlement of columns may result in undesirable additional stresses. Moreover, making reinforcement or replacement of worn out gantry girders is more complicated and expensive in two-span gantry girders. Hence they are not adopted in practice.

Abrupt changes in cross sections of gantry girders should be avoided. Gantry girders or trusses approximately 20 m or more in length should be cambered for deflection due to the dead load plus one-half the live load without impact. (Brockenbrough & Merritt 1999). Cantilevered gantry girders should be avoided. If high-strength steel is used, the deflection should be checked, since the resulting section will be considerably small.

The major cause of problems in crane runs is the deflection of gantry girders and the accompanying end rotation. Stretching of rails, opening of splice joints, column bending, skewing of the crane girders, and undulating crane motion are among the problems created by excessive deflection. Hence it is important to limit the vertical deflection of the gantry girder as per the limits given in Table 8.4. In general, it is a good idea to keep the spans as short as possible and beam depths as large as possible.

Vertical loads are delivered to the gantry girder via the crane rail. (The depth of crane rails varies from 70 mm to 150 mm and their bottom width from 70 mm to 150 mm.) The beam must be capable of withstanding this localized load. It is recommended to use a full penetration groove weld between the web and the top flange of welded plate gantry girders in order to maximize the fatigue life of the member. The length of the web that is affected by the concentrated wheel load is taken by considering a dispersion angle of 30° as shown in Fig. 8.10. The angle of 30° is a logical average between the 45° pure shear angle and the 22° angle used in the column stiffener analysis (Ricker 1982). Using this value of 30°, the affected length is given by

$$\text{Affected length} = 3.5 \times (\text{rail depth} + k) \tag{8.16}$$

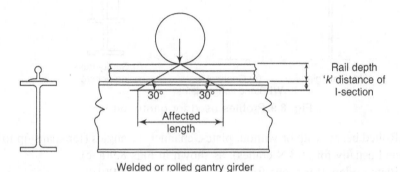

Welded or rolled gantry girder

Fig. 8.10 Angle of dispersion for the concentrated load (Ricker 1982)

In the case of a plate girder, it becomes

$$\text{Affected length} = 3.5 \times (\text{rail depth} + \text{flange thickness}) \tag{8.17}$$

However, note that a dispersion angle of 45 degrees has been assumed in the Indian code and the dispersion is taken upto the mid-depth of the girder (see Section 6.10). It has been found that web crushing is not critical in most of the cases (Ricker 1982).

The effects of an off-centre crane rail must be considered. Excessive rail eccentricity must be avoided, because it causes local flange bending and subjects the gantry girder to torsional moments. A limit of $0.75t_w$ is often specified for this eccentricity (see Fig. 8.11) for both wide flange beams and plate girders (Ricker 1982). Intermediate stiffeners may be welded to the underside of the top flange and down the web with a continuous weld to counteract the effect of rail eccentricity.

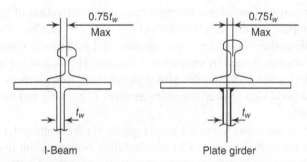

Fig. 8.11 Maximum allowable eccentricity for rails

While designing the gantry girder to resist lateral loading, the strength of the top flange alone should be considered. If this strength is inadequate, it may be reinforced by adding a channel, plate or angles, or by making a horizontal truss or girder in the case of large lateral loads (see Fig. 8.9). These reinforcing members are often attached by welding. Due to the fatigue factor associated with

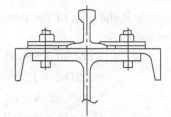

Fig. 8.12 Clamping of rails with bolts

intermittent welds, it is advisable to use continuous welds. In designing gantry girders which require channels, plates, or angles to resist lateral loads, a simplified design method which considers that the beam resists the vertical forces and the lateral forces are resisted only by the channel (or plate or angle) may be used; but this will result in uneconomical designs. Hence most designers assume that the lateral load is resisted by the channel (or plates or angles) plus the top flange of the beam and that the vertical load is resisted by both the beam and the channel (or plate or angle). If clamps are used to fasten the rails above the girder, it is necessary to select member sizes which will accept the required hole spacing (see Fig. 8.12).

8.4.1 Section Properties

With reference to Fig. 8.13, if A_p is the area of the top plate and A_s is the area of the rolled I-section, then the total area is given by

$$A = A_s + A_p$$

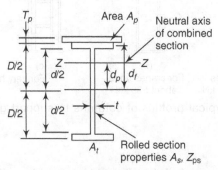

Fig. 8.13 Rolled I-beam with top plate

To find out the plastic section modulus, the neutral axis should be located at a location that divides the total area into two equal parts. Thus,

$$A_s/2 + d_p t = A_s/2 - d_p t + A_p$$

Hence $d_p = A_p/2t$. Ignoring the effect of the root fillets associated with rolled I-sections,

$$Z_{ps} = 2A_f d_f + t d^2/4 \qquad (8.18)$$

where A_f is the area of the flange and d_f is the distance of the flange from the neutral axis:

$$Z_{pz} = A_f(d_f - d_p) + A_f(d_f + d_p) + t(d/2 - d_p)^2/2 + t(d/2 + d_p)^2/2$$
$$+ A_p(D/2 + T_p/2 - d_p)$$
$$= 2A_f d_f + td^2/4 + td_p^2 + A_p(D/2 + T_p/2 - d_p)$$
$$= Z_{ps} + td_p^2 + A_p(D/2 + T_p/2 - d_p) \qquad (8.19)$$

Note that the above formula is applicable only if the neutral axis of the combined section lies within the web depth. If it lies within the flange of the section, the section properties should be determined from first principles.

8.4.2 Columns

Figure 8.14 shows the various gantry girder column profiles. If the gantry girder is supported on a bracket attached to a column, then the impact must be considered in the design of the bracket. Slots are provided in the bracket seat plate for lateral adjustment. Stiffeners are placed at the end of the beam to prevent web buckling. The bolts connecting the beam to the bracket must be strong enough to resist the longitudinal forces.

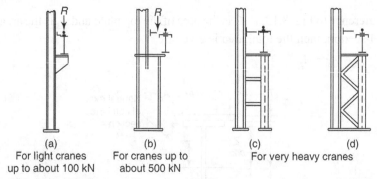

| (a) | (b) | (c) | (d) |
| For light cranes up to about 100 kN | For cranes up to about 500 kN | For very heavy cranes | |

Fig. 8.14 Typical profiles of columns to support gantry girders

8.4.3 Bracings

Columns supporting gantry girders should be braced laterally and longitudinally. The simplest and most effective bracing is the X-bracing system. It is better to limit the L/r ratio of such bracings to 200, due to the abrupt reversal of stresses in crane runways. Bracing members should be made of double angles, wide flange beams, tubes, or pipe sections; these should never be made of rods. Single-angle bracings may be used on light cranes. Bracings should never be connected directly to the underside of gantry girders.

It is better to locate the braces near the centre of the runway, since it will allow thermal expansion and contraction to advance or retreat from a centrally 'anchored' area of the runway towards the ends. Knee braces should never be used, since these are the source of many crane run problems (causing undesirable restraint, column bending, and secondary stresses).

8.5 Design of Gantry Girder

The design of the gantry girder subjected to lateral loads is a trial-and-error process. As already mentioned, it is assumed that the lateral load is resisted entirely by the compression top flange of the beam and any reinforcing plates, channels, etc. and that the vertical load is resisted by the combined beam. The various steps involved in the design are given below.

1. The first step is to find the maximum wheel load. As discussed in Section 8.1, this load is maximum when the trolley is closest to the gantry girder. It can be calculated by using Eqn (8.1) and increased for the impact as specified in Table 8.3.

2. The maximum bending moment in the gantry girder due to vertical loads needs to be computed. This consists of the bending moment due to the maximum wheel loads (including impact) and the bending moment due to the dead load of the girder and rails. Equation (8.9) gives the maximum bending moment due to wheel loads when only one crane is running over the girder and Eqn (8.14) gives the maximum bending moment when two cranes are placed on one gantry girder. The bending moment due to dead loads is maximum at the centre of the girder, whereas the bending moment due to the wheel load is maximum below one of the wheels. However, for simplifying the calculations, the maximum bending moment due to the dead load is directly added to the maximum wheel load moment.

3. Next the maximum shear force is calculated. This consists of the shear force due to wheel loads and dead loads from the gantry girder and rails. The shear force due to wheel loads can be calculated using either Eqn (8.8) or (8.13) depending on whether one or two cranes are operating on the gantry girder. Generally an I-section with a channel section is chosen, though an I-section with a plate at the top flange may be used for light cranes. When the gantry is not laterally supported, the following may be used to select a trail section:

$$Z_p = M_u/f_y \qquad (8.20)$$

$$Z_p \text{ (trial)} = kZ_p \quad (k = 1.40\text{--}1.50)$$

Generally, the economic depth of a gantry girder is about (1/12)th of the span. The width of the flange is chosen to be between (1/40) and (1/30)th of the span to prevent excessive lateral deflection.

4. The plastic section modulus of the assumed combined section is found out by considering a neutral axis which divides the area in two equal parts, at distance y to the area centroid from the neutral axis. Thus

$$M_p = 2f_y A/2\bar{y} = A\bar{y}f_y \qquad (8.21)$$

where $A\bar{y}$ is equal to the plastic modulus Z_p.

5. When lateral support is provided at the compression (top) flange, the chosen section should be checked for the moment capacity of the whole section (clause 8.2.1.2 of IS 800):

$$M_{dz} = \beta_b Z_p f_y / \gamma_{m0} \leq 1.2 Z_e f_y / \gamma_{m0} \qquad (8.22)$$

The above value should be greater than the applied bending moment. The top flange should be checked for bending in both the axes using the interaction equation

$$(M_y / M_{ndy}) + (M_z / M_{ndz}) \leq 1.0 \qquad (8.23)$$

6. If the top (compression) flange is not supported, then the buckling resistance is to be checked in the same way as in step 4 but replacing f_y with the design bending compressive stress f_{bd} (calculated using Section 8.2.2 of the code).

7. At points of concentrated load (wheel load or reactions) the web of the girder must be checked for local buckling and, if necessary, load-carrying stiffeners must be introduced to prevent local buckling of the web.

8. At points of concentrated load (wheel load or reactions) the web of the girder must be checked for local crushing. If necessary, bearing stiffeners should be introduced to prevent local crushing of the web.

9. The maximum deflection under working loads has to be checked.

Examples

Example 8.1 *Determine the moments and forces due to the vertical and horizontal loads acting on a simply-supported gantry girder given the following data.*
1. *Simply-supported span = 6 m*
2. *Crane's wheel centres = 3.6 m*
3. *Self-weight of the girder (say) = 1.5 kN/m*
4. *Maximum crane wheel load (static) = 220 kN*
5. *Weight of crab/trolley = 60 kN*
6. *Maximum hook load = 200 kN*
Calculate also the serviceability deflection (working load).

Solution
1. *Moments and forces due to self-weight*
Factored self-weight $W_d = 1.5 \times 1.5 \times 6 = 13.5$ kN
Ultimate mid-span BM, $M_1 = W_d L/8 = 13.5 \times 6/8 = 10.125$ kN m
Ultimate reaction $= R_{A1} = R_{B1} = W/2 = 13.5/2 = 6.75$ kN
2. *Moments and forces due to the vertical wheel load*
Wheel load (including γ_f and 25% impact)
$$W_c = 1.5 \times 1.25 \times 220 = 412.5 \text{ kN}$$
Ultimate BM under wheel (case 1) as per Eqn (8.9)
$$= 2W_c(L/2 - c/4)^2/L$$
$$= 2 \times 412.5(6/2 - 3.6/4)^2/6 = 606.375 \text{ kN m}$$

Ultimate BM under wheel (case 2)
$$= W_c L/4 = 412.5 \times 6/4 = 618.75 \text{ kN m}$$
Hence, the maximum ultimate BM,
$$M_2 = 618.75 \text{ kN m}$$
Ultimate reaction [Eqn (8.10)]
$$R_{A2} = W_c(2 - c/L) = 412.5(2 - 3.6/6.0) = 577.5 \text{ kN}$$
3. *Moment and forces due to horizontal wheel loads*
Horizontal surge load (including γ_f)
$$W_{hc} = 1.5 \times 0.10(200 + 60) = 39 \text{ kN}$$
This is divided among the 4 wheels (assuming double-flanged wheels).
Horizontal wheel load
$$W_{hc} = 39/4 = 9.75 \text{ kN}$$
Using calculations similar to those for vertical moments and forces, ultimate horizontal BM (case 2)
$$= 2W_{hc}L/4 = 9.75 \times 6.0/4 = 14.625 \text{ kNm}$$
Ultimate horizontal BM (case 1)
$$= 2W_{hc}(L/2 - c/4)^2/L$$
$$= 2 \times 9.75(6/2 - 3.6/4)^2/6 = 14.33 \text{ kN m}$$
4. *BM and reaction due to drag force*
Assuming that e is 0.15 m and the depth of the girder is 0.6 m,
$$R_{A3} = W_g e/L = 1.5 \times (0.05 \times 220 \times 1.25)(0.3 + 0.15)/6 = 1.55 \text{ kN}$$
Ultimate BM due to drag force
$$M_3 = R_a(L/2 - c/4) = 1.55(3 - 0.9) = 3.255 \text{ kNm}$$
Hence, the maximum ultimate design BM (vertical)
$$= M_1 + M_2 + M_3 = 10.125 + 618.75 + 3.255 = 632.13 \text{ kNm}$$
Maximum design reaction (vertical)
$$R_{A1} + R_{A2} + R_{A3} = 6.75 + 577.5 + 1.55 = 585.8 \text{ kN}$$
5. Serviceability deflection due to vertical wheel load excluding impact
$$W_c = 220 \text{ kN}$$
$$\Delta_c = W_c L^3[(3a/4L) - (a^3/L^3)]/(6EI) \text{ with } a = (L - C)/2$$
Assuming ISMB 600 with $I_z = 91,800 \times 10^4 \text{ mm}^4$
$$a = (6000 - 3600)/2 = 1200 \text{ mm}$$

$$\Delta_c = \frac{220 \times 1000 \times 6000^3 \left[\dfrac{3 \times 1200}{(4 \times 6000)} - \dfrac{1200^3}{6000^3} \right]}{[6 \times 2 \times 10^5 \times 91,800 \times 10^4]}$$

$$= 6.125 \text{ mm} < L/750 = 8 \text{ mm}$$

Example 8.2 *Design a gantry girder, without lateral restraint along its span, to be used in an industrial building carrying an overhead travelling crane for the following data (see Fig. 8.15).*
Centre-to-centre distance between columns (i.e., span of the gantry girder) = 7.5 m
Crane capacity = 200 kN
Self-weight of the crane girder excluding trolley = 200 kN
Self-weight of trolley, electrical motor, hook, etc. = 40 kN

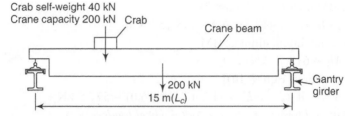

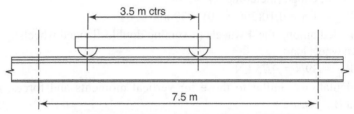

Fig. 8.15 Design data for Example 8.2

Minimum hook approach = 1.2 m
Distance between wheel centres = 3.5 m
Centre-to-centre distance between gantry rails (i.e., span of the crane) = 15 m
Self-weight of the rail section = 300 N/m
Yield stress of steel = 250 MPa

Solution
1. Load and bending moment calculations
(a) Load
(i) Vertical loading
Calculation of maximum static wheel load
Maximum static wheel load due to the weight of the crane = 200/4 = 50 kN
Maximum static wheel load due to crane load

$$W_1 = [W_t(L_c - L_1)]/(2L_c) = (200 + 40)(15 - 1.2)/(2 \times 15) = 110.4 \text{ kN}$$

Total load due to the weight of the crane and the crane load = 50 + 110.4 = 160.4 kN

To allow for impact, etc., this load should be multiplied by 25% (see Table 8.3).

Design load = 160.4 × 1.25 = 200.5 kN

∴ Factored wheel load on each wheel,

$$W_c = 200.5 \times 1.5 = 300.75 \text{ kN}$$

(ii) Lateral (horizontal) surge load

Lateral load (per wheel) = 10%(hook + crab load)/4 = 0.1 × (200 + 40)/4 = 6 kN

Factored lateral load = 1.5 × 6 = 9 kN

Note However, as per IS 875, 10% of the weight acting on one side of the rail track has to be taken for EOT cranes. Hence, the factored lateral load (per wheel) should be taken as 18kN.

(iii) Longitudinal (horizontal) braking load

Horizontal force along rails (Table 8.3) = 5% of wheel load

= $0.05 \times 200.5 = 10.025$ kN

Factored load $P_g = 1.5 \times 10.025 = 15.04$ kN

(b) Maximum bending moment

(i) Vertical maximum bending moment

Without considering the self-weight,

$$M_1 = W_c L/4 = 300.75 \times 7.5/4 = 563.90 \text{ kNm}$$
$$M_2 = 2W_c(L/2 - c/4)^2/L$$
$$= 2 \times 300.75(7.5/2 - 3.5/4)^2/7.5$$
$$= 662.90 \text{ kNm}$$

Hence $M = 662.90$ kNm.

Assume that the self-weight of the gantry girder is 1.6 kN/m.

Total dead load = 1600 + 300 (self-weight of rail) = 1.9 kN/m

Factored DL = $(1.9 \times 1.5) = 2.85$ kN/m

$$\text{BM due to dead load} = wl^2/8 = \left(2.85 \times \frac{7.5^2}{8} \right)$$

$$= 20.04 \text{ kNm}$$

(ii) Horizontal bending moment

Moment due to surge load = $2 \times 9(7.5/2 - 3.5/4)^2/7.5$

$$M_y = 19.84 \text{ kNm}$$

(iii) Bending moment due to drag (assuming the rail height as 0.15 m and the depth of girder as 0.6 m)

Reaction due to drag force = $P_g e/L = 15.04(0.3 + 0.15)/7.5 = 0.903$

$$M_3 = R(L/2 - c/4) = 0.903(7.5/2 - 3.5/4) = 2.59 \text{ kNm}$$

Total design bending moment $M_z = 662.9 + 20.04 + 2.59 = 685.53$ kNm

(c) Shear force

For maximum shear force, one of the wheel loads should be very close to the supports.

(i) Vertical shear force

Shear force due to wheel load

$$W_L(2 - c/L) = 300.75(2 - 3.5/7.5) = 461.15 \text{ kN}$$

$$\text{Shear force due to DL} = \frac{wl}{2} = 2.85 \times \frac{7.5}{2} = 10.69 \text{ kN}$$

Maximum ultimate shear force

$$V_z = 10.94 + 461.15 = 472.09 \text{ kN}$$

(ii) Lateral shear force due to surge load $V_y = 9(2 - 3.5/7.5) = 13.8$ kN

Reactions due to drag force = 0.903 kN

Maximum ultimate reaction

$$R_Z = 472.09 \text{ kN} + 0.903$$
$$= 472.99 \text{ or } 473 \text{ kN}$$

2. Preliminary selection of the girder

Since $L/12 = 7500/12 = 625$, we choose the depth as 600 mm. Therefore,

Approximate width of beam $= L/30 = 250$ mm

Since deflection governs the design, choose I, using the deflection limit of $L/750$,

$$I = \frac{15.6W(L-c)}{LE}[2L^2 + 2Lc - c^2]$$

$$= 15.6 \times 200.5 \times 10^3 \times (7500 - 3500)$$

$$\times [2 \times 7500^2 + 2 \times 7500 \times 3500 - 3500^2]/(2 \times 10^5 \times 7500)$$

$$= 1.274 \times 10^9 \text{ mm}^4$$

Required $Z_p = 1.4 \times M/f_y = 1.4 \times 685.53 \times 10^6/250 = 3.83 \times 10^6$ mm^3

Choose ISMB 600 @ 1230 N/m with a channel ISMC 300 @ 363 N/m @ the top (see Fig. 8.16).

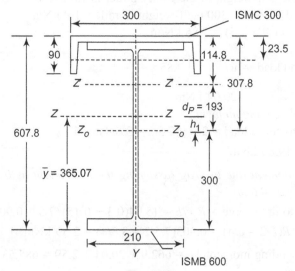

Fig. 8.16

a. Properties of the sections

ISMB 600 @ 1.23 kN/m	ISMC 300 @ 0.363 kN/m
$A = 15600$ mm^2	$A = 4630$ mm^2
$t_f = 20.3$ mm	$t_f = 13.6$ mm
$t_w = 12$ mm	$t_w = 7.8$ mm
$B = 210$ mm	$B = 90$ mm
$I_{zz} = 91800 \times 10^4$ mm^4	$I_{zz} = 6420 \times 10^4$ mm^4
$I_{yy} = 2650 \times 10^4$ mm^4	$I_{yy} = 313 \times 10^4$ mm^4
$R = 20$ mm	$C_y = 23.5$ mm

(i) Elastic properties of the combined section

Total area $A = A_B + A_{ch} = 15{,}600 + 4630 = 20{,}230$ mm^2

The distance of NA of the built-up section from the extreme fibre of tension flange,

$$\bar{y} = [15,600 \times 600/2 + 4630 \times (600 + 7.8 - 23.5)]/20,230 = 365.07 \text{ mm}$$

$$h_1 = \bar{y} - h_B/2 = 365.07 - 600/2 = 65.07 \text{ mm}$$

$$h_2 = (h_B + t_{ch}) - \bar{y} - C_y = (600 + 7.8) - 365.07 - 23.5 = 219.23 \text{ mm}$$

$$h_3 = 607.8 - 365.07 - 7.8 = 234.93 \text{ mm}$$

$$I_Z = I_{ZB} + A_B h_1^2 + (I_y)_{ch} + A_{ch} \times h_2^2$$
$$= 91,800 \times 10^4 + 15,600 \times 65.07^2 + 313 \times 10^4 + 4630 \times 219.23^2$$
$$= 1.2097 \times 10^9 \text{ mm}^4 \approx 1.274 \times 10^9 \text{ (required for deflection control)}$$

$$Z_{Zb} = 1.2097 \times 10^9/365.07 = 3.31 \times 10^6 \text{ mm}^3$$

$$Z_{Zt} = 1.2097 \times 10^9/242.73 = 4.98 \times 10^6 \text{ mm}^3$$

I_{yy} combined $= 2650 \times 10^4 + 6420 \times 10^4 = 9070 \times 10^4 \text{ mm}^4$

I_y for tension flange about the y-y axis

$I_{tf} = 20.3 \times 210^3/12 = 1566.6 \times 10^4 \text{ mm}^4$

For compression flange about the y-y axis

$I_{cf} = 1566.6 \times 10^4 + 6420 \times 10^4 = 7986.6 \times 10^4 \text{ mm}^4$

Z_y (for top flange alone) $= 7986.6 \times 10^4/150 = 532,443 \text{ mm}^3$

b. Calculation of plastic modulus (see Section 8.4.1 and Fig. 8.16)

The plastic neutral axis divides the area into two equal areas, i.e., 10,115 mm^2.

$$d_p = 4630/(2 \times 12) = 193 \text{ mm}$$

Ignoring fillets, the plastic section modulus below the equal-area axis is

$$\Sigma A\bar{y} = 20.3 \times 210 \times (493 - 20.3/2) + (493 - 20.3) \times 12 \times (493 - 20.3)/2$$
$$= 3399.1 \times 10^3 \text{ mm}^3$$

Above the equal-area axis

$$\Sigma A\bar{y} = 4630 \times (114.8 - 23.5) + 210 \times 20.3 \times (114.8 - 7.8 - 10.15)$$
$$+ 86.7 \times 12 \times 86.7/2 = 880.692 \times 10^3 \text{ mm}^3$$

$$Z_{pz} = 3399.1 \times 10^3 + 880.692 \times 10^3 = 4279.792 \times 10^3 \text{ mm}^3$$

For the top flange only

$$Z_{py} = 20.3 \times 210^2/4 + (300 - 2 \times 13.6)^2 \times 7.8/4 + 2 \times 90 \times 13.6$$
$$\times (150 - 13.6/2) = 719,479.8 \text{ mm}^3$$

3. Check for moment capacity

Check for plastic section

b/t of the flange of the I-beam $= [(210 - 12)/2]/20.3 = 4.87 < 9.4$

b/t of the flange of the channel $= (90 - 7.8)/13.6 = 6.04 < 9.4$

d/t of the web of the I-section $= (600 - 2 \times 20.3)/12 = 46.6 < 84$

Hence the section is plastic.

a. Local moment capacity

$$1.2Z_e f_y/1.1 = 1.2 \times 3.31 \times 10^6 \times (250/1.1) \times 10^{-6} = 902.72 \text{ kNm}$$

$$M_{dz} = f_y Z_p/1.1 = (250/1.1) \times 4279.792 \times 10^{-3} = 972.68 \text{ kNm} > 902.72 \text{ kNm}$$

Hence take $M_{dz} = 902.72$ kNm

$$M_{dy} = (f_y/1.1) \times Z_p(\text{top flange}) = (250/1.1) \times 719,479.8 \times 10^{-6} = 163 \text{ kNm}$$

$$1.2Z_yf_y/1.1 = 1.2 \times 532,443 \times (250/1.1) \times 10^{-6} = 145.2 \text{ kNm} < 163 \text{ kNm}$$

Hence take $M_{dy} = 145.2$ kNm.

b. Combined local capacity check

$$685.53/902.72 + 19.84/145.2 = 0.759 + 0.137 = 0.896 < 1$$

Hence the chosen section is the right choice.

4. Check for buckling resistance

As per IS 800 (Clause 8.2.2), the design bending strength

$$M_d = \beta_b Z_p f_{bd}$$

We have

$$\beta_b = 1.0$$

$$h = 600 + 7.8 = 607.8 \text{ mm}$$

$$KL = 7500 \text{ mm}$$

$$E = 2 \times 10^5 \text{ N/mm}^2$$

$$t_f = 20.3 + 7.8 = 28.1 \text{ mm}$$

$$r_y = \sqrt{I_{yy}/A}$$

$$I_{yy} = (2650 \times 10^4) + (6420 \times 10^4) = 9070 \times 10^4 \text{ mm}^4$$

$$A = 15,600 + 4630 = 20,230 \text{ mm}^2$$

$$r_y = \sqrt{\frac{9070 \times 10^4}{20,230}} = 66.96 \text{ mm}$$

According to clause 8.2.2.1 of (IS 800), elastic lateral buckling moment

$$M_{cr} = C_1 \frac{\pi^2 E I_y h}{2(KL)^2} \left[1 + \frac{1}{20} \left[\frac{KL/r_y}{h/t_f} \right]^2 \right]^{0.5},$$

$$C_1 = 1.132 \text{ (from Table 42 of IS 800 : 2007)}$$

Note The value of C_1 will be in between 1.046 and 1.132 as both udl and concentrated loads are applied.

$$M_{cr} = 1.132 \times \left(\frac{\pi^2 \times 2 \times 10^5 \times 9070 \times 10^4 \times 607.8}{2 \times 7500^2} \right)$$

$$\left[1 + \frac{1}{20} \left(\frac{7500/66.96}{607.8/28.1} \right)^2 \right]^{0.5} = 1675.22 \times 10^6 \text{ N/mm}$$

Note The above formula is valid for I-sections only. For more accurate values of M_{cr} for a compound section, the formula given in E1.2 (Annex E) of the code may be used.

Non-dimensional slenderness ratio

$$\lambda_{LTZ} = \sqrt{\frac{\beta_b Z_{pz} f_y}{M_{cr}}}$$

$$= \sqrt{\frac{1.0 \times 4279.792 \times 10^3 \times 250}{1675.22 \times 10^6}} = 0.7992$$

Along the z-direction

$$\phi_{LTZ} = 0.5[1 + \alpha_{LT}(\lambda_{LTZ} - 0.2) + \lambda_{LTZ}^2]$$

$$= 0.5[1 + 0.21(0.7992 - 0.2) + 0.7992^2] = 0.882$$

Note Since the channel will normally be connected by intermittent welds to the I-section, α_{LT} value has been taken as 0.21. If heavy welding is involved, take α_{LT} as 0.49.

$$\chi_{LTZ} = \frac{1}{[\phi_{LTZ} + (\phi_{LTZ}^2 - \lambda_{LTZ}^2)^{0.5}]} \leq 1.0$$

$$= \frac{1}{0.882 + (0.882^2 - 0.7992^2)^{0.5}}$$

$$= 0.7967$$

$$f_{bd} = f_y \chi_{LT}/\gamma_{m0}$$

$$\gamma_{m0} = 1.10 \text{ (from Table 5 of the code)}$$

$$\therefore \quad f_{bd} = 0.7967 \times 250/1.1 = 181.07 \text{ N/mm}^2$$

$$\therefore \quad M_{dz} = \beta_b Z_{pz} f_{bd}$$

$$= 1.0 \times 181.07 \times 4279.792 \times 10^{-3}$$

$$M_{dz} = 774.9 \text{ kNm} > 685.53 \text{ kNm}$$

Thus the beam is satisfactory under vertical loading. Now it is necessary to check it under biaxial bending.

For top flange only

$$M_{dy} = (f_y/1.1) \times Z_{yt}$$

$$= (250/1.1) \times 719,479 \times 10^{-6} = 163.5 \text{ kNm}$$

$$> \frac{1.2 \times 532,433 \times 250}{1.1 \times 10^6} = 145.2 \text{ kNm}$$

Hence $M_{dy} = 145.2$ kNm

a. Check for biaxial bending

In order to check for biaxial bending, we substitute the terms with their values in the following equation:

$$\frac{M_z}{M_{dz}} + \frac{M_y}{M_{dy}} \not> 1.0$$

We have

$$\frac{685.53}{774.9} + \frac{19.84}{145.2} = 0.884 + 0.137 = 1.021 \approx 1.0$$

Slightly bigger size of top channel may be selected.

5. Check for shear capacity

For vertical load,

$$V_z = 472.09 \text{ kN}$$

Shear capacity $= A_v f_{yw}/(\sqrt{3} \times 1.10)$

$$= (600 \times 12) \times 250/(\sqrt{3} \times 1.10) \times 10^{-3}$$

$$= 944.75 \text{ kN} > 472.09 \text{ kN}$$

The maximum shear force is 472.09 kN, which is less than 0.6 times the shear capacity, i.e.,

$$0.6 \times 944.75 = 566.85 \text{ kN}$$

Hence it is safe in vertical shear and there is no reduction in the moment capacity.

a. Weld design

The required shear capacity of the weld is given by

$$q = VA\bar{y}/I_Z$$
$$\bar{y} = h_3 = 234.93$$
$$A = 4630 \text{ mm}^2, \quad V = 472.09 \text{ kN}$$
$$I_Z = 1.2097 \times 10^9 \text{ mm}^4$$
$$q = 472.09 \times 10^3 \times 4630 \times 234.93/(1.2097 \times 10^9)$$
$$= 424 \text{ N/mm}$$

This shear is taken by the welds. Hence use a minimum weld of 4 mm (442 N/mm per weld) connecting the channel to the top flange of the I-beam.

For lateral shear force,

$$F_y = 13.8 \text{ kN}$$

Shear capacity $V_{ny} = A_v f_{yw}/(\sqrt{3} \times 1.10)$

$$= 250/(\sqrt{3} \times 1.10) \times (210 \times 20.3 + 300 \times 7.8) \times 10^{-3}$$

$$= 866.41 \text{ kN} > 13.8 \text{ kN}$$

Hence it is safe for resisting lateral shear.

6. Web buckling

At points of concentrated loads (wheel loads or reaction) the web of the girder must be checked for local buckling (see Section 6.10)

The dispersion length under wheel (assuming it to be 150 mm and assuming an angle of dispersion of 45°). Also assume stiff bearing length at support = 150 mm.

$$b_1 = 150 \text{ mm}$$
$$n_1 = 600/2 + 7.8 = 307.8 \text{ mm}$$

Web slenderness $\lambda = 2.5d/t$

$$= 2.5 \times [600 - 2(20.3 + 20)]/12$$
$$= 108.2$$

Stress reduction factor (from Table 8 of IS 800 : 2007) = 0.426

$$f_{cd} = 0.426 \times 250/1.1 = 96.8 \text{ MPa}$$

Buckling resistance $= (b_1 + n_1)t f_{cd}$

$$= (150 + 307.8)12 \times 96.8 \times 10^{-3}$$
$$= 531.78 \text{ kN}$$

Maximum reaction = 473 kN < 531.78 kN

Hence buckling resistance is satisfactory.

7. Web bearing (see Section 6.10)

Load dispersion at support with 1: 2.5 dispersion

Minimum stiff bearing $= R_x/(tf_{yw}/1.1) - n_2$

$$n_2 = (20.3 + 20) \times 2.5 = 100.75 \text{ mm}$$
$$R_x = 473 \text{ kN (support reaction)}$$
$$b_1 = 473 \times 10^3/(12 \times 250/1.1) - 100.75 = 72.68 \text{ mm}$$

Web bearing at support requires a minimum stiff bearing of 73 mm < 150 mm.

8. Check for deflection at working load

Serviceability vertical wheel load excluding impact = 160.4 kN

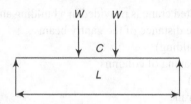

Deflection at mid-span

$$\Delta = WL^3[(3a/4L) - (a^3/L^3)/(6EI)$$

Where

$$a = (L - c)/2 = (7500 - 3500)/2 = 2000$$

(i) Vertical

Combined $I_{zz} = 1.2097 \times 10^9$ mm^4

$$\Delta = \frac{160.4 \times 10^3 \times 7500^3}{6 \times 2 \times 10^5 \times 1.2097 \times 10^9} [3 \times 2000/(4 \times 7500) - 2000^3/7500^3]$$

$$= 8.43 \text{ mm} < L/750 = 10 \text{ mm}$$

(ii) Lateral

Only the compound top flange will be assumed to resist the applied surge load as in the bending check.

$$I = (I_z)_{ch} + I_F = 7986.6 \times 10^4 \text{ mm}^4$$

$$\Delta = \frac{6 \times 10^3 \times 7500^3}{6 \times 2 \times 10^5 \times 7986.6 \times 10^4} [3 \times 2000/(4 \times 7500) - 2000^3/7500^3]$$

$$= 4.78 < 10 \text{ mm (Table 6 of IS 800 : 2007)}$$

Summary

Cranes are often employed in industrial buildings to move stock, finished goods, or new materials from one place to another for processing. These cranes are supported on gantry girders, which are supported on separate columns or brackets attached to steel columns. The design of gantry girders is often complicated since they support moving load and are, hence, subjected to fatigue. Before attempting to design a gantry girder, the engineer should know the loading and clearance details from the crane manufacturer. In this chapter, some guidelines have been provided in the form of tables to do preliminary calculations and arrive at the size of gantry girders. These calculations should be checked, during the final design, with the exact details obtained from the manufacturers. The various loadings that may occur due to the moving loads have been discussed and expressions for calculating the maximum bending moment, shear, and deflection of gantry girders provided. The various factors that may affect the choice of girders are discussed. The steps in the design of gantry girders have been explained with the use of practical examples.

Exercises

1. A 50 kN hand-operated crane is provided in a building and has the following data:

Centre-to-centre distance of the gantry beam	16 m
(width of the building)	
Longitudinal spacing of columns	7.5 m
(span of gantry)	
Weight of the crane	40 kN
Wheel spacing	3 m
Weight of the crab	10 kN
Minimum hook approach	1 m
Yield stress of steel	250 MPa

   Design a simply supported gantry girder assuming lateral support to it.

2. Design a gantry girder, without lateral restraint along its span, to be used in an industrial building carrying overhead travelling crane for the following data:

Centre-to-centre distance between columns	6 m
(span of the gantry girder)	
Crane capacity	50 kN
Self-weight of the crane girder excluding trolley	40 kN
Self-weight of the trolley, electric motor, hook, etc.	10 kN
Minimum hook approach	1.0 m
Wheel centres	3 m
Centre-to-centre distance between gantry rails	12 m
(span of crane)	
Self-weight of rail section	100 N/m
Yield stress of steel	250 N/mm^2

3. Design a simply supported crane girder to carry an electric overhead travelling crane for the following data:

Crane capacity	300 kN
Weight of the crane and crab	300 kN
Weight of the crane	200 kN
Minimum hook approach	1.2 m
Centre-to-centre distance between wheels	3.2 m
Span of the gantry girder	5 m
Centre-to-centre distance between gantries	15 m
Weight of rail	300 N/m
Height of rails	75 mm
Yield stress of steel	250 MPa

Review Questions

1. What is the main purpose of a gantry girder?
2. What are the components of a crane runway system?
3. What are the requirements to be considered by the designer while selecting a crane and designing a crane supporting structure?
4. List the loads that should be considered while designing a gantry girder.
5. The impact allowance in percentage to be applied to the vertical forces transferred to the wheels of an EOT crane is
   (a) 10% of the maximum static wheel load
   (b) 10% of the weight of the crab and the weight lifted on the crane
   (c) 25% of the maximum static wheel load
6. The impact allowance for horizontal force transverse to the rail (surge load) for an EOT crane is
   (a) 25% of the weight of the crab and the weight lifted on the crane
   (b) 10% of the static wheel load
   (c) 10% of the weight of the crab and the weight lifted on the crane
7. The impact allowance for drag forces along the rail for an EOT crane is
   (a) 10% of the static wheel load
   (b) 5% of the static wheel load
   (c) 10% of the weight of crab and the weight lifted on the crane.
   (d) 5% of the weight of the crab and the weight lifted on the crane.
8. What is the difference between surge load and drag load of cranes?
9. The surge load is assumed to be resisted by the
   (a) whole cross section
   (b) compression flange alone
   (c) compression and tension flanges
   (d) cross section above the neutral axis
10. Write down the expressions for maximum shear force, bending moment, and deflection at mid-span for a simply supported beam with two moving loads, each with a value W.
11. Write down the expressions for maximum shear force, bending moment, and deflection at mid-span for a simply supported beam with four moving loads, each with a value W (two cranes running in tandem).
12. List the different profiles of cross sections which are used for gantry girders.
13. Why are simply supported girders preferred to two-span gantry girders?
14. What is the limiting vertical deflection of the gantry girder for
    (a) manually operated cranes
    (b) EOT cranes with a capacity less than 500 kN
    (c) EOT cranes with a capacity greater than 500 kN
15. List the various steps involved in the design of a gantry girder.

CHAPTER 9

Design of Beam-columns

Introduction

Most columns are subjected to bending in addition to the axial load; considerable care should be taken in a practical situation to load a column under axial load only. When significant bending is present in addition to an axial load in a member, the member is termed as a *beam-column*. The bending moments on a column may be due to any of the following effects.

(a) *Eccentricity of axial force*

(b) *Building frame action* In a multi-storey building, usually columns support beams which have similar identical connection eccentricities at each floor level. In a rigid frame building construction, the columns carry the building load axially as well as end moments from the girders that frame into them. Most building frames are braced against sway by bracings or core walls, but the horizontal wind forces have to be resisted by bending actions in the columns. Due to the wind forces, the columns may bend in double-curvature bending (similar to the shape of the letter 'S') in contrast to the single curvature bending of columns due to moment, created by gravity loads. Single curvature bending is often the most critical design condition than reverse or double curvature bending. These are shown schematically in Fig. 9.1. Wind loads can also produce lateral loading on a column, giving beam-type bending moment distribution.

(c) *Portal or gable frame action* Another common example of a column with bending moments occurs in a portal frame where the columns and rafters are subjected to relatively light axial loads combined with bending.

(d) *Load from brackets* In industrial buildings, column brackets may be used to carry gantry girders on which the cranes move. The resulting eccentricity produces bending moments in addition to the axial loads in the columns. In this case, the column moment is not at the column ends {see Fig. 9.1(d)}.

(e) *Transverse loads* As already discussed, wind pressure on long vertical members may produce bending moments. Similarly earthquakes also produce bending in the columns. Purlins placed between panel joints of a rafter of roof trusses (in order to reduce the size of purlin or to accommodate maximum size of roofing sheets) will produce bending in rafters (see also Chapter 12).

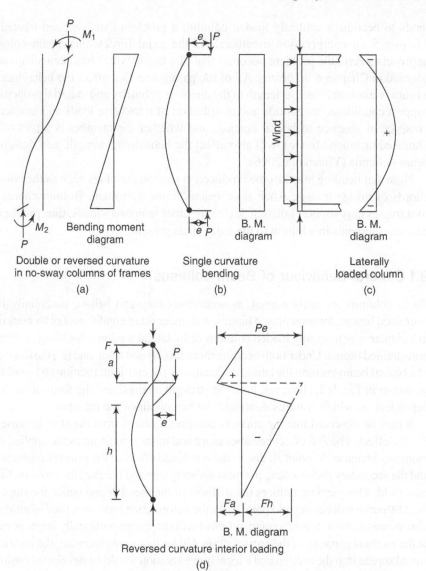

Fig. 9.1 Single and double curvature bending in beam-columns

(f) *Fixed base condition* If the base of the column is fixed due to piles, rafts, or grillage foundation, bending moments will be present at the base of the columns, even though their top ends may be hinged.

Beam columns in steel structures are often subjected to biaxial bending moments, acting in two principal planes, due to the space action of the framing system. The column cross section is usually oriented in such a way to resist significant bending about the major axis of the member. When I-shaped cross sections are used for the columns, the minor axis bending may also become significant, since the minor axis bending resistance of I-section is small compared to the major axis bending resistance.

Thus, in general, beam-columns are subjected to axial forces and bending moments. As the bending moment on a beam-column approaches zero, the member tends to become a centrally loaded column, a problem that has been treated in Chapter 5 on compression members. As the axial forces on a beam-column approaches zero, the problem becomes that of a beam, which has been adequately covered in Chapter 6 on beams. All of the parameters that affect the behaviour of a beam or a column (such as length of the member, geometry and material properties, support conditions, magnitude and distribution of transverse loads and moments, presence or absence of lateral bracing, and whether the member is a part of an unbraced or braced frame), will also affect the behaviour, strength, and design of beam-columns (Vinnakota 2006).

Note that bending may also be produced in tension members such as the bottom chords of bridge trusses, when floor beams frame into them. Bottom chords of roof trusses may support hoisting devices or other temporary loads, thus producing bending moments in addition to the axial loads present.

9.1 General Behaviour of Beam-columns

Beam columns are aptly named, as sometimes they can behave essentially like restrained beams, forming plastic hinges, and under other conditions fail by buckling in a similar way to axially loaded columns or by lateral torsional buckling similar to unrestrained beams. Under both bending moment (M) and axial load (P) the response of a typical beam-column for lateral deflection (δ) or end-joint rotation (θ) would be as shown in Fig. 9.2. However, both the strength attained and the form of curve is dependent on which features dominate the behaviour of the member.

It may be observed that the curve is non-linear almost from the start because of the P-δ effect. The P-δ effect becomes more and more significant as the applied end moments increase. At point A, due to the combined effect of the primary moment M and the secondary P-δ moment, the most severely stressed fibres of the cross sections may yield. This yielding reduces the stiffness of the member and hence the slope of the M-θ curve reduces beyond point A. As the deformation increases, the P-δ moment also increases. Now, this secondary moment will share a proportionally larger portion of the moment capacity of the cross section. Under increasing moment, the plasticity would spread into the section and a local hinge rotation would be developed (point B on the M-θ curve). The hinge would now spread a short distance down the column, which causes the slight downward slope on the M–θ curve and at point C, the moment carrying capacity of the cross section is finally exhausted.

In the preceding discussion, we assumed that other forms of failure do not occur before the formation of a plastic hinge. However, if the member is slender and the cross section is torsionally weak (e.g., open cross sections), lateral torsional buckling may occur. Lateral torsional buckling may occur in the elastic range (curve 1 shown in Fig. 9.2) or in the inelastic range (curve 2 shown in Fig. 9.2) depending on the slenderness of the member. As discussed in Chapter 6, member with a high slenderness ratio will experience elastic lateral torsional buckling, whereas a member with an intermediate slenderness ratio will experience inelastic lateral torsional buckling. Lateral torsional buckling will not occur if the slenderness ratio of the member is low or if the member is bent about the minor principal axis of the cross section. Similarly in members with cross sections having axis symmetry (e.g., circular sections) or equal moment of inertia about both principal axes (e.g., square box sections), lateral torsional buckling will not occur, regardless of the slenderness ratio.

①, ②, ③, *beam-column*   ⑤ *Elastic–plastic beam P = 0*   ⑦ *Elastic-buckling column M = 0*
④ *Linear elastic beam P = 0*   ⑥ *Rigid-plastic beam P = 0*   ⑧ *Elastic interaction between*
                                                            bending and buckling

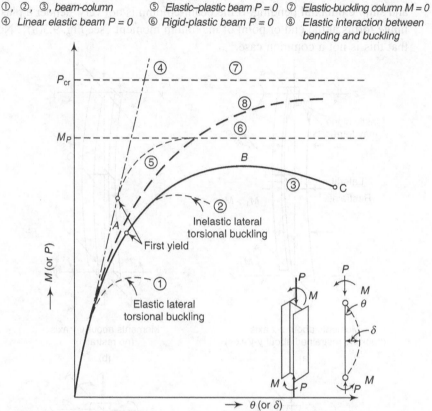

Fig. 9.2 Behaviour of a beam column compared with beams and columns

Another form of failure which may occur in the member is local buckling of component elements of the cross section. As seen in Chapter 6, the element with high width-to-thickness ratio is susceptible to local buckling. Like lateral torsional buckling, local buckling may occur in the elastic or inelastic range. The effect of both lateral torsional buckling and local buckling is to reduce the load carrying capacity of the cross section. Local buckling may be prevented by limiting the width-to-thickness ratios as specified in Table 4.3. When it is not possible to limit the width-to-thickness ratios local buckling may be accounted for in design by using a reduced width for the buckled element (see Example 5.1b).

Based on the earlier discussions, the behaviour of the beam-column may be classified into the following five cases (MacGinley & Ang 1992):

1. A short beam-column subjected to axial load and uniaxial bending about either axis or biaxial bending. Failure occurs when the plastic capacity of the section is reached, with the limitations set in the second case.

2. A slender beam-column subjected to axial load and uniaxial bending about the major axis z-z. If the beam-column is supported laterally against buckling about the minor axis y-y out of the plane bending, the beam-column fails by buckling about the z-z axis. It represents an interaction between column buckling and simple uniaxial bending. If the beam-column is not very slender a plastic hinge forms at the end or point of maximum moment {see Fig. 9.3(a)}. Note that this is not a common case.

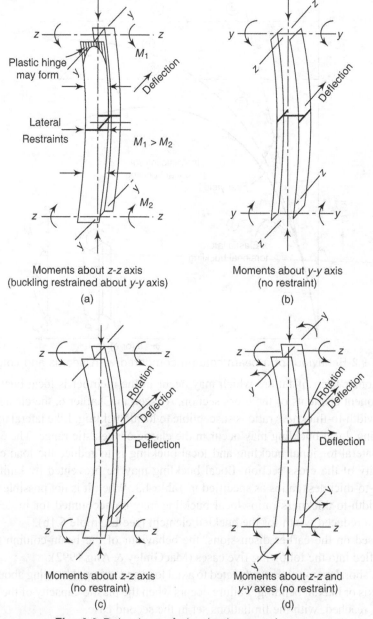

Moments about z-z axis
(buckling restrained about y-y axis)

(a)

Moments about y-y axis
(no restraint)

(b)

Moments about z-z axis
(no restraint)

(c)

Moments about z-z and
y-y axes (no restraint)

(d)

Fig. 9.3 Behaviour of slender beam-columns

3. A slender beam-column subjected to axial load and uniaxial bending about the minor axis y-y. Now there is no need for lateral support and no buckling out of the plane of bending. The beam-column fails by buckling about the y-y axis. This also represents an interaction between column buckling and simple uniaxial bending. At very low axial loads, the beam-column will attain the bending capacity about y-y axis {see Fig. 9.3(b)}.

4. A slender beam-column subjected to axial load and bending about the major axis z-z, and not restrained out of the plane of bending. The beam-column fails due to a combination of column buckling about the y-y axis and lateral torsional buckling. The beam-column fails by twisting as well as deflecting in the z-z and y-y planes {see Fig. 9.3(c)}. Thus it represents an interaction between column buckling and beam buckling.

5. A slender beam-column subjected to axial load and biaxial bending and not having any lateral support. The ultimate behaviour of the beam-column is complicated by the effect of plastification, moment magnification, and lateral torsional buckling. The failure will be similar to the fourth case but minor axis buckling will dominate. This is the general loading case {see Fig. 9.3(d)}.

9.2 Equivalent Moment Factor C_m

Member sizes of beam-columns are generally based on the magnitude of the maximum moment and the location of this maximum moment is not important in the design. Hence, the concept of equivalent moment, schematically shown in Fig. 9.4 is usually adopted in design specifications. Thus, it is assumed that the maximum second-order moment of a beam-column subjected to an axial load P and end moments M_A and M_B with $|M_B| > |M_A|$ as shown in Fig. 9.4(a) is numerically equal to the maximum second-order moment of the same member under axial load P and a pair of equal and opposite moments M_{eq}, as shown in Fig. 9.4(b). Note that the axial load P and second-order moment M_{max} are the same for both the members. The value of M_{eq} is given by (Salvadori 1956; Vinnakota 2006)

$$M_{eq} = C_m |M_B| \qquad (9.1)$$

where C_m is called the *equivalent moment factor* or *moment reduction factor*. This factor is a function of the moment ratio $\psi = M_A / M_B$ and also the axial load ratio P/P_{cr}. Many simplified expressions for C_m have been proposed, as a function of the moment ratio only. The following expression given by Austin (1961) has been adopted in the ANSI/AISC code for beam-columns subjected to end moments (Chen & Zhou 1987; Galambos 1998)

$$C_m = 0.6 + 0.4\,\psi \geq 0.4 \text{ where } -1.0 \leq \psi \leq 1.0 \qquad (9.2)$$

For beam-columns with transverse loadings, the second-order moments can be approximated by (ANSI/AISC: 360-05)

$$C_m = 1 + \beta(\alpha P_u / P_{cr}) \qquad (9.3)$$

where $\beta = \pi^2 \delta_o EI / ML^2 - 1$, $\alpha = 1.0$, δ_o is the maximum deflection due to transverse loading (in mm), M is the maximum first-order moment within the member due to transverse loading (N mm).

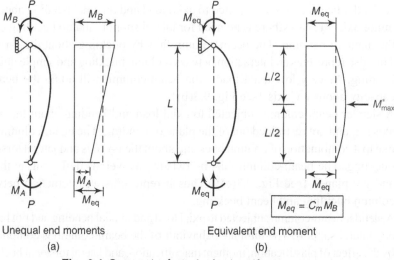

Unequal end moments
(a)

Equivalent end moment
(b)

Fig. 9.4 Concept of equivalent uniform moment

Thus, for a pin-ended beam-column of length L with a uniformly distributed load W,

$$M = WL/8, \; \delta = 5WL^3/384EI, \; P_{cr} = \pi^2 EI/L^2$$

$$\beta = [(\pi^2 EI)/L^2] \, [(5WL^3)/(384EI)] \, (8/WL) - 1 = 1.028 - 1 = 0.028$$

Hence $C_m = 1 + 0.028P/P_{cr}$.

Similarly for a beam-column of length L with a transverse load P at mid length,

$$M = PL/4; \; \delta = PL^3/48EI; \; P_{cr} = \pi^2 EI/L^2$$

Thus $\beta = [(\pi^2 EI)/L^2] \, (PL^3/48EI) \, (4/PL) - 1 = 0.822 - 1 = -0.178$
and

$$C_m = 1 - 0.178 \; P/P_{cr}$$

Note that we must consider signs when using the ratio $\psi = M_A/M_B$, with M_A being the smaller of the two end moment values. The correct value of C_m will be close to 1.0 for all the cases and hence the American code (AISC 360-05) recommends that C_m may be conservatively taken as 1.0 for members with transverse loading.

9.3 Nominal Strength—Instability in the Plane of Bending

The basic strength of beam-columns, where lateral-torsional buckling and local buckling are adequately prevented, and bending is about one axis, will be achieved when instability occurs in the plane of bending (without twisting). The elastic differential equation solution shows that the axial compression effect and the bending moment effect cannot be determined separately and combined by super-position. The relationship is also non-linear.

Residual stresses, which cause premature yielding and consequently a premature reduction in stiffness, may reduce both the initial yield and the maximum strength of beam-columns. This is similar to their effect on axially loaded compression members. Similarly, an initial crookedness, which increases the secondary bending moment caused by the axial load, reduces both the initial yield load and the maximum strength. Moreover, when a beam-column with initial imperfections (displacements u_i, v_i, and initial twist ϕ_i) is loaded by the axial force P and major axis moment M_z, the member exhibits a non-bifurcation type of instability, in which the deformations increase (from u_i, v_i, and ϕ_i), until a maximum moment is reached, beyond which static equilibrium can only be sustained by decreasing the moment. The maximum strength based on the spatial behaviour of such initially crooked beam-columns $M_{cs,max}$ could be lower than the lateral-buckling load M_{cr} of the corresponding initially straight beam-column as shown in Fig. 9.5 (Vinnakota 2006).

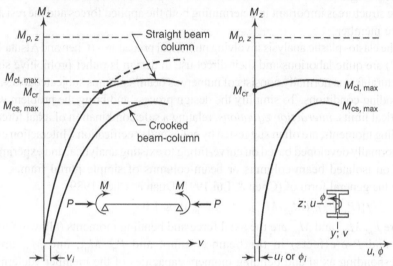

Fig. 9.5 Inelastic lateral-torsional buckling of beam-columns

The inelastic analysis to determine the strength interaction between axial compression P and bending moment M for a beam column is complicated. To trace a load deflection curve similar to that shown in Fig. 9.2, one should use some type of approximation or numerical technique. This is because the differential equations governing the inelastic behaviour of a beam-column are highly non-linear even for the simplest loading case (Chen & Atsuta 1976, 1977).

The analysis of the inelastic behaviour normally proceeds in two steps.
1. Cross section analysis
2. Member analysis
In the cross section analysis, the behaviour of a cross section subjected to the combined action of axial force and bending moment is investigated. The result is expressed as a set of M–θ–P (moment–curvature–axial compression) relationship. Once M–θ–P relationship is established, member analysis can be done.

In the member analysis, the member is divided into a number of segments whereby equilibrium and compatibility conditions along the length of the member at each division point are enforced for a given set of loadings or deflections. The analysis thus consists of finding successive solutions as the applied load or deflection of the member is increased in steps. When enough of these analyses have been performed, the load-deflection relationship of the beam-column can be traced on a pointwise basis.

9.3.1 Nominal Strength-Interaction Equations

As observed from the earlier discussions the behaviour of beam-columns is affected by a number of parameters. Moreover a real beam-column may receive end moments and axial load from its connections to other members of a structure, such as a rigid frame. Hence the relation of the beam-column to the other elements of the structure is important in determining both the applied forces and the resistance of the member.

The elasto–plastic analysis involving numerical procedures (Chen and Atsuta 1976, 1977) are quite laborious and their direct use in design is rather prohibitive since a structural frame normally consists of numerous beam-columns subjected to a variety of loading conditions. To simplify the design process and bring the problem within practical limits, *interaction equations*, relating a safe combination of axial force and bending moments, are often suggested by codes and specifications. Interaction curves are normally developed based on curve-fitting to existing analytical and experimental data on isolated beam-columns or beam-columns of simple portal frames. They have the general form of (Chen & Lui 1991; Duan & Chen 1989)

$$f\{(P_u/P_n),\ (M_{uz}/M_{nz}),\ (M_{uy}/M_{ny})\} \le 1.0 \tag{9.4}$$

where P_u, M_{uz}, and M_{uy} are the axial force and bending moments (allowing for the $P\text{-}\Delta$ and $P\text{-}\delta$ effects) in the beam-column, and P_n, M_{nz}, and M_{ny} are the corresponding axial and bending moment capacities of the member {determined as discussed in Chapter 5 (Section 5.6.1) on compression members and Chapter 6 (Sections 6.6 and 6.7) on beams}.

The three-dimensional graphical representation of Eqn (9.4) is shown in Fig. 9.6. In this figure each axis represents the capacity of the member when it is subjected to loading of one type only, while the curves represent the combination of two types of loading. The surface formed by connecting the three curves represents the interaction of axial load and biaxial bending. It is this interaction surface that is of interest to the designer.

The end points of the curves shown in Fig. 9.6 are dependent on the capacities of the members described for columns (Chapter 5) and beams (Chapter 6). The shapes of these curves between these end points will depend on (a) the cross-sectional shape and the beam-column imperfections, (b) the variation of moments along the beam-column, and (c) the end restraint conditions of the beam-column. All these variables can only be dealt with on an approximate basis and hence various formulae are given in the codes, which attempt to allow for the effects mentioned earlier.

The basic form of the three-dimensional interaction equation is

$$(P/P_n) + (M_z/M_{nz}) + (M_y/M_{ny}) \le 1.0 \tag{9.5}$$

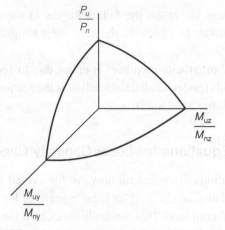

Fig. 9.6 Ultimate interaction surface for beam-columns

This interaction equation results in a straight-line representation of the interaction between any two components shown in Fig. 9.7. This simplified interaction equation gives a conservative design.

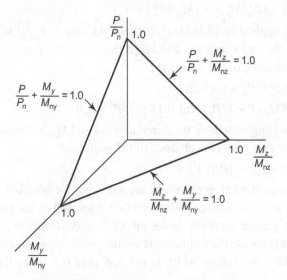

Fig. 9.7 Simplified interaction surface

The behaviour of beam-columns subject to bending moment about minor axis is similar to that subjected to major axis bending but for the following differences.
- In the case of slender members under small axial load, there is very little reduction of moment capacity below M_p, since lateral torsional buckling is not a problem in weak axis bending.
- The moment magnification is larger in the case of beam-columns bending about their weak axis.

- As the slenderness increases, the failure curves in the P/P_n, y-y axis plane change from convex to concave, showing increasing dominance of minor axis buckling.
- The failure of short/stocky members is either due to section strength being reached at the ends (under small axial load) or at the section of larger magnified moment (under large axial load).

9.4 Interaction Equations for Local Capacity Check

The interaction equations discussed till now are for overall buckling check. The beam-column should also be checked for local capacity at the point of the greatest bending moment and axial load. This is usually checked at the ends of the member, but it could be within the length of the beam-column, if lateral loads are also applied. The capacity in these cases is controlled by yielding or local buckling (if it is not prevented by limiting the width-to-thickness ratios specified in the codes). The linear interaction equation for semi-compact and slender cross section is given by

$$(P/P_y) + (M_z/M_{pz}) + (M_y/M_{py}) \leq 1 \tag{9.6}$$

where P is the applied axial load, P_y is the yield load $= A_g f_y$, M_z is the applied moment about the major axis z-z, and M_{pz} is the moment capacity about the major axis z-z in the absence of the axial load,

$$= M_{pz} \text{ if } P/P_y \leq 0.15 \text{ and}$$
$$= 1.18 M_{pz}(1 - P/P_y) \text{ for } 0.15 \leq P/P_y < 1.0$$

M_y is the applied moment about the minor axis y-y and M_{py} is the moment capacity about the minor axis y-y in the absence of the axial load

$$= 1.19 M_{py}[1 - (P/P_y)^2] \leq 1.0.$$

More accurate interaction equations are available for compact cross sections, which are based on the convex failure surface discussed in the previous section, which result in greater economy in design. Chen and Atsuta (1977) and Tebedge and Chen (1974) provide the following non-linear interaction equation for compact I-shapes in which the flange width is not less than 0.5 times the depth of the section

$$(M_z/M_{pz})^\alpha + (M_y/M_{py})^\alpha \leq 1 \tag{9.7a}$$

in which M_{pz}, M_{py}, M_z, and M_y are as defined earlier.
The value of the exponent is given by

$$\alpha = 1.6 - [(P/P_y)/\{2\ln(P/P_y)\}] \text{ for } 0.5 \leq b_f/d \leq 1.0 \tag{9.7b}$$

where ln is the natural logarithm, b_f is flange width (in mm), and d is the member depth (in mm).

A comprehensive assessment of the accuracy of the non-linear interaction equations in predicting the load carrying capacities of biaxially loaded I-sections has been made by Pillai (1981), who found that these equations predict the capacity reasonably well compared to the experimental results.

Interaction equations for a number of sections, including circular tubes, box sections, and unsymmetrical sections such as angles are available in Chen and Lui (1971), Chen and Atsuta (1977), and Shanmugam et al. (1993).

9.5 Code Design Procedures

Modern structural design specifications around the world have retained the generic form of the interaction formula given in Eqn (9.4). In every specification the moment M is always specified as the second-order (amplified) moment obtained either from a second-order structural analysis, where equilibrium is formulated in the deformed configuration of the structure, or from an approximation of using the moments from a first-order elastic analysis, which is then multiplied by an amplification factor. Depending upon whether P_{cr}, the elastic critical load, is evaluated for the member length or the storey effective length, the amplification factor accounts empirically for the member or the frame stability (Galambos 1998). This versatility of the interaction equations approach makes it very useful in design.

Most limit-states design codes use a set of load and resistance factors that are based on probabilistic principles (Bjorhovde et al. 1978).

9.5.1 Indian Code (IS 800 2007) Provisions

The Indian code (IS 800 2007) provisions are based on the Eurocode provisions and the code requires the following two checks to be performed
(a) Local capacity check and
(b) Overall buckling check

Local capacity check For beam-columns subjected to combined axial force (tension or compression) and bending moment, the following interaction equation should be satisfied.

$$(M_y/M_{ndy})^{\alpha 1} + (M_z/M_{ndz})^{\alpha 2} \leq 1.0 \tag{9.8}$$

where M_y and M_z are the factored applied moments about the minor and major axis of the cross section, respectively and M_{ndy} and M_{ndz} are the design reduced flexural strength under combined axial force and the respective uniaxial moment acting alone. The approximate value of M_{ndy} and M_{ndz} for plastic and compact I-section is given in Table 9.1.

The constants α_1 and α_2 are taken as 5n and 2 respectively for I or channel sections.

For semi-compact sections, without bolt holes, the code (IS 800 : 2007) suggests the following linear equation, when shear force is low

$$(N/N_d) + (M_y/M_{dy}) + (M_z/M_{dz}) \leq 1.0 \tag{9.9}$$

Table 9.1   Approximate value of reduced flexural strength for plastic and compact sections

Element	Equation
Welded I- or H-sections	$M_{ndz} = M_{dz}(1-n)/(1-0.5a) \le M_{dz}$
	$M_{ndy} = M_{dy}\left[1 - \left(\dfrac{n-a}{1-a}\right)^2\right] \le M_{dy}$
	$a = (A - 2\,bt_f)/A \le 0.5$
Rolled I- or H-sections without bolt holes	$M_{ndz} = 1.11 M_{dz}(1-n) \le M_{dz}$
	$M_{ndy} = M_{dy}$ for $n \le 0.2$
	$M_{ndy} = 1.56\, M_{dy}(1-n)(n+0.6)$

$n = N/N_d$
N = Factored applied axial force (Tension T or compression P)
N_d = Design strength in tension T_d, or compression P(see Chapter 5)
   $= A_g f_y/\gamma_{m0}$ where $\gamma_{m0} = 1.1$
M_{dy}, M_{dz} = Design strength under corresponding moment acting alone (see Chapter 6)
A_g = Gross area of cross section

Overall buckling check   The interaction equation for overall buckling check is given by the code as

$$(P/P_{dy}) + (K_y C_{my} M_y/M_{dy}) + (K_{LT} M_z/M_{dz}) \le 1.0 \qquad (9.10a)$$

$$(P/P_{dz}) + (0.6 K_y C_{my} M_y/M_{dy}) + (K_z C_{mz} M_z/M_{dz}) \le 1.0 \qquad (9.10b)$$

where

C_{my}, C_{mz} = Equivalent uniform moment factor obtained from Table 9.2, which depends on the shape of the bending moment diagram between lateral bracing points in the appropriate plane of bending

P = Factored applied axial compressive load

P_{dy}, P_{dz} = Design compressive strength under axial compression as governed by buckling about minor and major axis respectively (see Chapter 5, Section 5.6.1)

M_y, M_z = Maximum factored applied bending moments about minor and major axis of the member, respectively

M_{dy}, M_{dz} = Design bending strength about minor and major axis considering laterally unsupported length of the cross-section (see Chapter 6, Section 6.7).

K_y, K_z, K_{LT} = Moment amplification factors as defined below

$$K_y = 1 + (\lambda_y - 0.2)n_y \le 1 + 0.8n_y \qquad (9.10c)$$

$$K_z = 1 + (\lambda_z - 0.2)n_z \le 1 + 0.8n_z \qquad (9.10d)$$

$$K_{LT} = 1 - \frac{0.1\lambda_{LT}n_y}{(C_{mLT} - 0.25)} \ge 1 - \frac{0.1n_y}{(C_{mLT} - 0.25)} \qquad (9.10e)$$

where

n_y, n_z = Ratio of actual applied axial-force to the design axial strength for buckling about minor and major axis respectively = (P/P_{dy}) or (P/P_{dz})

C_{mLT} = Equivalent uniform moment factor for lateral-torsional buckling as per Table 9.2, which depends on the shape of the bending moment diagram between lateral bracing points

λ_y, λ_z = Non-dimensional slenderness ratio about the minor and major axis respectively, for example, $\lambda_y = (f_y/f_{cr})^{0.5}$, where $f_{cr} = \pi^2 E/(KL/r)^2$ (see Section 5.6.1)

λ_{LT} = Non-dimensional slenderness ratio in lateral buckling = $(f_y/f_{cr,b})^{0.5}$ and $f_{cr,b}$ is the extreme fibre bending compressive stress corresponding to elastic lateral buckling moment which may be determined as per Table 14 of the code.

The above Indian code provisions are based on the Eurocode 3 provisions.

Table 9.2 Equivalent uniform moment factor (Greiner & Lindner 2006)

Bending Moment Diagram	Range	C_{my}, C_{mz}, C_{mLT}	
		Uniform Loading	Concentrated Load
M ⟍ ψM	$-1 \leq \Psi \leq 1$	$0.6 + 0.4\Psi \geq 0.4$	
M_h ⟍ ψM_h ↕M_S $\alpha_s = M_s/M_h$	$0 \leq \alpha_s \leq 1 \quad -1 \leq \Psi \leq 1$	$0.2 + 0.8\alpha_s \geq 0.4$	$0.2 + 0.8\alpha_s \geq 0.4$
	$-1 \leq \alpha_s \leq 0 \quad 0 \leq \Psi \leq 1$	$0.1 - 0.8\alpha_s \geq 0.4$	$-0.8\alpha_s \geq 0.4$
	$-1 \leq \Psi \leq 0$	$0.1(1 - \Psi) - 0.8\alpha_s \geq 0.4$	$-0.2\Psi - 0.8\alpha_s \geq 0.4$
M_h ⟍ ψM_h ↕M_S $\alpha_h = M_h/M_s$	$0 \leq \alpha_h \leq 1 \quad -1 \leq \Psi \leq 1$	$0.95 + 0.05\alpha_h$	$0.90 + 0.10\alpha_h$
	$-1 \leq \alpha_h \leq 0 \quad 0 \leq \Psi \leq 1$	$0.95 + 0.05\alpha_h$	$0.90 + 0.10\alpha_h$
	$-1 \leq \Psi \leq 0$	$0.95 + 0.05\alpha_h (1 + 2\psi)$	$0.90 + 0.10\alpha_h (1 + 2\psi)$

For members with sway buckling mode the equivalent uniform moment factor $C_{my} = C_{mz} = 0.90$

C_{my}, C_{mz}, and C_{mLT} shall be obtained according to the bending moment diagram between the relevant braced points as below:

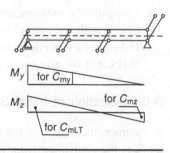

M_y for C_{my}

M_z for C_{mz}

for C_{mLT}

(Contd)

(Contd)

Moment factor	Bending axis	Points braced in direction
C_{my}	z-z	y-y
C_{mz}	y-y	z-z
C_{mLT}	z-z	z-z

9.6 Design of Beam-columns

The design of beam columns involves a trial-and-error procedure. A trail section is selected by some process and is then checked with the appropriate interaction formula. If the section does not satisfy the equation (LHS > 1.0) or if it is too much on the safer side, indicated by LHS much less than 1.0 (that is, if it is over designed), a different section is selected and the calculations are repeated till a satisfactory section is found. Thus, the different steps involved in the design of beam columns are as follows.

1. Determine the factored loads and moments acting on the beam-column using a first-order elastic analysis (though a second-order analysis is recommended by most of the codes)
2. Choose an initial section and calculate the necessary section properties.
3. Classify the cross section (plastic, compact, or semi-compact) as per clause 3.7 of the code.
4. Find out the bending strength of the cross section about the major and minor axis of the member (clause 8.2.1.2).
5. (a) Determine the shear resistance of the cross section (clause 8.4.1). When the design shear force exceeds $0.6V_d$, then the design bending strength must be reduced as given in clause 9.2.2 of the code.
   (b) Check whether shear buckling has to be taken into account (clause 8.4.2).
6. Calculate the reduced plastic flexural strength (clause 9.3.1.2), if the section is plastic or compact.
7. Check the interaction equation for cross-section resistance for biaxial bending (clause 9.3.1.1 for plastic and compact section and clause 9.3.1.3 for semi-compact section). If not satisfied go to step 2.
8. Calculate the design compressive strength P_{dz} and P_{dy} (clause 7.1.2) due to axial force.
9. Calculate the design bending strength governed by lateral-torsional buckling (clause 8.2.2).
10. Calculate the moment amplification factors (clause 9.3.2.2).
11. Check with the interaction equation for buckling resistance (clause 9.3.2.2). If the interaction equation is not satisfied (LHS > 1.0) or when it is over design (LHS $\ll$ 1.0), go to step 2.

9.6.1 Selection of Initial Section

A common method used for selecting sections to resist both moments and axial loads is the *equivalent axial load or effective axial load method*. In this method, the axial load P and the bending moments M_z and/or M_y are replaced with fictitious concentric loads P_{eff}, equivalent to the actual design axial load plus the design moment. This fictitious load is called the equivalent axial load or the effective axial load.

Yura (1988) suggested a simpler approach for the initial sizing and suggested the following equation.

$$P_{eff} = P + 2M_z/d + 7.5M_y/b \qquad (9.11)$$

where d and b are the depth and breadth of the selected beam-column.

The equivalent axial load equations were found to yield sections that are generally on the conservative side. The equivalent axial load approaches given by Eqn (9.11) are useful in preliminary sizing of beam-columns under gravity load combinations. However, when bending moment is predominant, those equations may result in considerable error. In this case, the equivalent moment approach, rather than equivalent axial force approach is preferable. Yura (1988), proposed an equation for estimating the equivalent moment as

$$M_{eq} = M_z + P_u d/2 \qquad (9.12)$$

where d is the depth of the section. This equation is applicable for initial beam-column sizing in unbraced frames under lateral-load combinations.

9.7 Beam-columns Subjected to Tension and Bending

Bending moments may occur in tension members due to connection eccentricity, self weight of the member or transverse loads such as wind acting along the length of the member. The effect of tension load will always reduce the primary bending moment. Hence, secondary bending effects can be conservatively ignored in the design of members subjected to an axial tensile force and bending.

Local capacity check The following simplified interaction equation is specified in the code for the beam-column subjected to combined axial force and bending moment

$$(N/N_d) + (M_y/M_{dy}) + (M_z/M_{dz}) \leq 1.0 \qquad (9.13)$$

where M_{dy} and M_{dz} are the design reduced flexural strength under combined axial force and the respective uniaxial moment acting alone, M_y and M_z are the factored applied moments along minor and major axis of the cross section, respectively, N_d is the design strength in tension obtained from Section 6 of the code, and N is the factored applied axial tensile force.

Overall buckling check The code stipulates that the member should be checked for lateral-torsional buckling under reduced effective moment M_{eff} due to tension and bending. The reduced effective moment is given by the code as

$$M_{eff} = [M - \psi \, TZ_{ec}/A] \leq M_d \qquad (9.14)$$

where M and T are the factored applied moment and tension, respectively, A is the area of cross section, Z_{ec} is the elastic section modulus of the section with respect to extreme compression fibre, $\psi = 0.8$, if T and M vary independently and 1.0 otherwise, and M_d is the bending strength due to lateral-torsional buckling.

9.8 Design of Eccentrically Loaded Base Plates

The design of base plate subjected to concentric compression was considered in Section 5.13. In this section column bases that transmit forces, and moments from the steel column to the concrete foundation are considered. The forces may be axial loads, shear forces, and moment about either axis or any combinations of them. The main function of the base plate is to distribute the loads to the weaker material.

The common design deals with axial load and moment about major axis. With respect to slab and gusseted base (see Fig. 5.21), there may be two separate cases (a) compression over the whole base or compression over part of the base and tension in the *holding-down bolts* (also called *anchor bolts*). The relative values of moment and axial load determine which case will occur in a given instance. Horizontal loads are resisted by shear in the weld between column and base plates, friction and bond between the base plate and concrete, and shear in the holding down bolts. As mentioned in Chapter 5, ANSI/AISC code does not allow the anchor rods to transfer substantial shear, and suggests the use of a shear key or lug to transfer a large horizontal force from the column to the foundation (see Fig. 5.22). Note that the horizontal loads are generally small except when earthquake loads are considered.

(a)	(b)	(c)

Base plate connections: (a) welded slab base plate, (b) gusseted base plate,(c) moment resisting base plate © American Institute of Steel Construction, Inc., Reprinted with permission. All right reserved.)

9.8.1 Compression Over the Whole Base Plate

A column base and loading are shown in Fig. 9.8. If the area of moment/axial load is less than $L/6$, where L is the base length, then a positive pressure exists over the whole base and may be calculated from equilibrium alone. In this case, nominal holding down bolts (2 to 4) are provided to locate the base plate accurately.

Eccentricity of the load $e = M/P$

where P is the total load on the base plate and M_z is the major axis moment on the base. The area of the base is given by

$$A = BL$$

where B is the breadth of the base plate, and L is the length of the base plate. The maximum pressure on the concrete foundation is

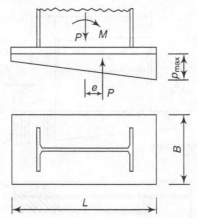

Fig. 9.8 Compression over the whole base plate

$$p_{max} = (P/A) + (M_z/Z_z) \tag{9.15}$$

where Z_z is the modulus of the base plate about Z-axis = $BL^2/6$.

When there are bi-axial moments, then

$$p_{max} = (P/A) + (M_z/Z_z) + (M_y/Z_y) \tag{9.16}$$

where M_y is the minor axis moment on the plate and Z_y is the modulus about the y-y axis = $LB^2/6$.

The maximum pressure must not exceed the bearing strength of the concrete, which is taken as $0.45f_{ck}$, where f_{ck} is the characteristic compressive strength of concrete.

In this case, the size of the base plate is established by successive trials. If the length is fixed, the breadth may be determined so that the bearing strength of the concrete is not exceeded. The weld size between base plate and the column is determined using the same requirements that were set out for the axially loaded base plate in Section 5.13. Example 9.3 shows the calculations required for a base plate with compression over the whole of the base plate.

9.8.2 Gusseted Base Plate

When gussets are provided as shown in Fig. 9.9 they support the base plate against bending and hence the thickness of the base plate could be reduced. Now, a part of the load is transmitted from the column through the gussets to the base plate.

The gussets are subjected to bending from the upward pressure under the base as shown in Fig. 9.9. The top edge of the gusset plate will be in compression and hence must be checked for buckling. To prevent this, we should choose the dimensions of outstand in such a way that they are within the limiting width-to-thickness ratios of welded sections, as given in Table 2 of the code. Thus, for the configuration shown in Fig. 9.9, the limiting width-to-thickness ratios for the gusset plate is given by

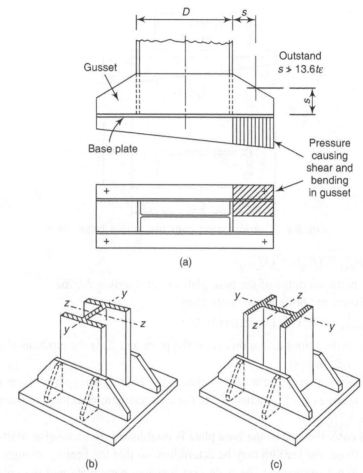

Fig. 9.9 Gusseted base plates

(a) For the portion of gusset plate welded to the flanges of the column

$$D \leq 29.3\varepsilon t \tag{9.17}$$

(b) Outstand of the gusset plate from the column or base plate

$$S \leq 13.6\ \varepsilon\ t \tag{9.18}$$

where $\varepsilon = (250/f_{yg})^{0.5}$, t is the thickness of the gusset plate, and f_{yg} is the design yield strength of the gusset plate.

The gusset plates are designed to resist shear and bending. The moment in the gusset should not exceed $f_{yg}Z_e/\gamma_{m0}$, where Z_e is the elastic modulus of the gusset and γ_{m0} is the partial safety factor for material = 1.1. An example of a gusseted base plate is provided in Example 9.4.

9.8.3 Anchor Bolts and Shear Connectors

The specification for foundation bolts is given in IS 5624-1993. Some typical shank forms of black foundation bolts as per this code are given in Fig. 9.10. The types of anchor bolts include cast-in-place anchors and post-installed anchors. Cast-in-place anchors, placed before concrete is cast, include a bolt with a curved end (used when there is no or low uplift forces) and a headed anchor. The tensile load transfer is by bonding between anchor bolt and concrete filling. Bonding between filling and concrete wall is guaranteed whenever the contact surface is sufficiently rough.

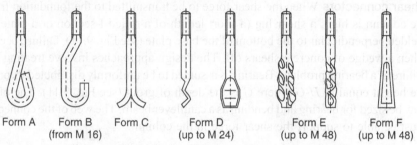

| Form A | Form B | Form C | Form D | Form E | Form F |
| (from M 16) | | | (up to M 24) | (up to M 48) | (up to M 48) |

Fig. 9.10 Typical shank forms of foundation bolts

The headed anchors transfer tensile load by mechanical bearing of the head, nut and possibly by bond between anchor shank and the surrounding concrete. As these pre-installed anchors do not allow any clearance, they need very accurate positioning.

Post-installed anchors, installed in hardened concrete, are classified according to their load-transfer mechanisms. *Adhesive or bonded anchors* transfer load through cementitious grout or chemical adhesive. Details of installation and behaviour of these types of anchors are provided by Subramanian and Cook (2002, 2004). Epoxy is the most widely used adhesive though resins such as vinyl esters, polyesters, methacrylates, and acrylics have also been used. *Mechanical anchors* transfer load by friction or bearing and include expansion anchors and undercut anchors.

Fuchs et al. (1995) suggested the concrete capacity design (CCD) approach to fastenings in concrete, which has been adopted by the ACI code (Appendix D of ACI 318-2005). According to this method the tensile capacity of anchors may be determined using the following formula (Subramanian 2000)

$$N_u = k\sqrt{f_{ck}}\,(h_{ef})^{1.5} \tag{9.19}$$

where $k = 13.5$ for post-installed anchors and 15.5 for cast-in-situ headed studs and headed anchor bolts and h_{ef} is the embedment length.

When fastenings are located so close to an edge or to an adjacent anchor, there will not be enough space for complete concrete cone to develop and hence the load-bearing capacity of the anchor has to be reduced [see Subramanian (2000) and Cook (1999) for more details of the CCD method and worked examples].

For grouted anchors bond failure at the grout-concrete interface may occur. This failure mode is best represented by a uniform bond stress model. The strength in this case is given by

$$N_{bond} = \tau_o\,\pi\,d_o h_{ef} \tag{9.20}$$

where τ_o is the grout-concrete bond strength of the product (in N/mm^2), d_o is the diameter of the hole (in mm), and h_{ef} is the embedment length.

The predicted mean strength of a headed grouted anchor is determined by the lower value of Eqn (9.19), Eqn (9.20), and the steel failure strength. Swiatek and Whitback (2004) discuss many practical problems connected with anchor rods and provide some easy solutions.

Shear connectors When the shear force to be transmitted to the foundation from the column is high, a shear lug (a short length of a rolled I-section or a plate) is welded perpendicular to the bottom of the base plate (see Fig. 9.11). Failure occurs when a wedge of concrete shears off. The design approaches involve treating the failure as a bearing problem. Bearing is assumed to be uniformly distributed through the height equal to H-G where G is the depth of grout (see Fig. 9.11). The plate may be sized for bearing and bending as a cantilever beam. The web of the connector must be able to support the shear force in the column.

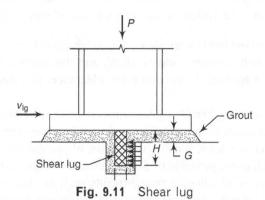

Fig. 9.11 Shear lug

The different steps involved in the design are as follows (DeWolf and Ricker 2006).

1. Determine the portion of the shear which is resisted by friction equal to μ multiplied by the factored dead load and appropriate portion of the live load, which generates the shear force. The shear force to be resisted by the lug is the difference between the factored shear force and this force. μ may be taken as 0.3 (steel friction coefficient, since steel shims are often placed under the base plate).

2. The required bearing area for the shear lug is

$$A_{lg} = V_{lg}/0.45f_{ck} \tag{9.21}$$

3. Determine the shear lug dimensions, assuming that bearing occurs on the portion of the lug below the concrete foundation.

4. The factored cantilever end moment M_{lg} acting on a unit length of the shear lug is

$$M_{lg} = (V_{lg}/W)(H + G)/2 \tag{9.22}$$

where W is the total horizontal width of the lug, H is the vertical height of the lug, and G is the thickness of grout (see Fig. 9.11)

5. The shear lug thickness is determined as follows:

$$t_{\text{lg}} = \sqrt{[4M_{\text{lg}}/(f_y/\gamma_{\text{m0}})]} \le t_b \tag{9.23}$$

where t_b is the thickness of the base plate. Example 9.5 illustrates the use of these steps.

DeWolf and Ricker (2006) provide more details about the design of base plates. Drake and Elkin (1999) present a step-by-step methodology for the design of base plates and anchor rods using factored loads and rectangular pressure distribution.

Examples

Example 9.1 *A non-sway column in a building frame with flexible joints is 4-m high and subjected to the following load and moment:*

   Factored axial load = 500 kN

   Factored moment M_z:

 at top of column = 27.0 kNm

 at bottom of column = 45.0 kNm

Design a suitable beam-column assuming $f_y = 250$ *N/mm². Take the effective length of the column as 0.8L along both the axes.*

Solution

1. Trial section

Select ISHB 200 with $A = 4750$ mm², $r_y = 45.1$ mm, $H = 200$ mm, $b = 200$ mm, and $t_f = 9.0$ mm.

   $KL = 0.8 \times 4000 = 3200$ mm

   $KL/r_y = 3200/45.1 = 70.95$

   $P_e = P + 2M_z/d = 500 + 2 \times 45000/200 = 950$ kN

From Table 10 of the code for $h/b_f < 1.2$ and $t_f < 40$ mm, using curve 'c' for minor axis buckling, we have

For $KL/r_y = 70.95$ and $f_y = 250$ MPa, from Table 9c of the code, $f_{cd} = 150$ N/mm²

Hence capacity = $150 \times 4750/1000 = 712.5$ kN < 950 kN

Hence use ISHB 250 with $A = 6500$ mm², $H = 250$ mm, and $r_y = 54.9$ mm.

   $KL/r_y = 3200/54.9 = 58.29$

   $P_{eq} = 500 + 2 \times 45,000/250 = 860$ kN

For $KL/r_y = 58.29$ and $f_y = 250$ MPa,

From Table 9c of the code, $f_{cd} = 170.56$ N/mm²

Thus capacity = $170.56 \times 6500/1000 = 1108$ kN > 860 kN

Hence, choose ISHB 250 as the trial section.

Section Properties

ISHB 250 has the following cross-sectional properties:

$H = 250$ mm	$A = 6500$ mm²	$I_z = 7740 \times 10^4$ mm⁴
$b_f = 250$ mm	$r_z = 109$ mm	$I_y = 1960 \times 10^4$ mm⁴

$t_f = 9.7$ mm $r_y = 54.9$ mm

$t_w = 6.9$ mm $Z_z = 619 \times 10^3$ mm^3

$R = 10$ mm $Z_y = 156 \times 10^3$ mm^3

$$Z_{pz} = 2b_f t_f (H - t_f)/2 + t_w(H - 2t_f)^2/4$$
$$= 2 \times 250 \times 9.7 (250 - 9.7)/2 + 6.9(250 - 2 \times 9.7)^2/4$$
$$= 674.46 \times 10^3 \text{ mm}^3$$

2. Cross-section classification

$$\varepsilon = \sqrt{(250/f_y)} = \sqrt{250/250} = 1.0$$

Outstand flanges (Table 2 of the code)

$$b/t_f = (250/2)/9.7 = 12.88 < 15.7\varepsilon$$

Hence, the flange is semi-compact.

Web

$$d = H - 2t_f - 2R = 250 - 2 \times 9.7 - 2 \times 10 = 210.6 \text{ mm}$$
$$d/t_w = 210.6/6.9 = 30.5 < 42\varepsilon$$

Hence, the cross-section is semi-compact.

3. Check for resistance of cross section to the combined effects (clause 9.3.1.3)

The interaction equation is

$$(N/N_d) + (M_z/M_{dz}) \le 1.0$$
$$N_d = A_g f_y/\gamma_{m0} = 6500 \times 250/(1.1 \times 1000) = 1477.27 \text{ kN}$$
$$M_{dz} = \beta_b Z_p f_y/\gamma_{m0}$$

where $\beta_b = Z_e/Z_p$ for a semi-compact section. Hence,

$$M_{dz} = Z_e f_y/\gamma_{m0} = 619 \times 10^3 \times 250/(1.1 \times 10^6)$$
$$= 140.68 \text{ kN m}$$

Thus,

$$(500/1477.27) + (45/140.68) = 0.338 + 0.320 = 0.658 < 1.0$$

Hence, the section is safe.

4. Member buckling resistance in compression (clause 7.1.2)

Effective length $= 0.8L = 0.8 \times 4000 = 3200$ mm

$$KL_z/r_z = 3200/109 = 29.35$$
$$KL_y/r_y = 3200/54.9 = 58.29$$

From Table 10 of the code,

$$h/b = 250/250 = 1.0 \text{ and } t_f < 40 \text{ mm}$$

Major axis buckling, use curve b

Minor axis buckling, use curve c

$$f_{cr, z} = \pi^2 \times 2 \times 10^5/(29.35)^2 = 2291.5 \text{ N/mm}^2$$

$$\lambda_z = \sqrt{(250/2291.5)} = 0.33$$

From Table 9c of the code, for $KL/r = 58.29$ and $f_y = 250$ N/mm^2,

$$f_{cd} = 170.56 \text{ N/mm}^2 \text{ and}$$

$$P_{d,y} = 170.56 \times 6500/1000 = 1108 \text{ kN} > 500 \text{ kN}.$$

From Table 9b of the code, for $KL/r = 29.35$ and $f_y = 250$ N/mm^2,

$f_{cd} = 215.6$ N/mm^2,

$P_{d,z} = 215.6 \times 6500/1000 = 1401$ kN > 500 kN.

Hence, the section is safe.

5. *Member buckling resistance in bending (clause 8.2.2)*

$M_d = \beta_b Z_p f_{bd}$

$\beta_b = Z_e/Z_p$ for semi-compact section $= 619/674.76 = 0.918$

Hence $M_d = Z_e f_{bd}$

From Table 42 of the code (assuming $k = 1$)

For $\psi = 0.75$, $C_1 = 1.141$ and for $\psi = 0.5$, $C_1 = 1.323$. Hence, for $\psi = 0.6$, $C_1 = 1.25$,

Note More refined value of $C_1 = 1.345$ may be obtained by considering $k = 0.8$ in this table.

$L_y = 4$ m, $h/t_f = 250/9.7 = 25.77$

Determination of M_{cr}

$$f_{cr,b} = C_1 [1473.5/(KL/r_y)]^2 \{1 + (1/20)[(KL/r_y)/(h/t_f)]^2\}^{0.5}$$

$$= 1.25(1473.5/58.29)^2 [1 + (1/20)(58.29/25.77)^2]^{0.5}$$

$$= 895.1 \text{ N/mm}^2$$

$M_{cr} = 895.1 \times 619 \times 10^3/10^6 = 554$ kNm

Non-dimensional lateral-torsional slenderness ratio

$$\lambda_{LT} = \sqrt{(\beta_b Z_{pz} f_y/M_{cr})}$$

$$= \sqrt{0.918 \times 674.46 \times 10^3 \times 250/(554 \times 10^6)} = 0.527$$

$\alpha_{LT} = 0.21$ for rolled sections

Reduction factor for lateral torsional buckling

$$\chi_{LT} = 1/[\phi_{LT} + (\phi_{LT}^2 - \lambda_{LT}^2)^{0.5}]$$

where $\phi_{LT} = 0.5[1 + \alpha_{LT}(\lambda_{LT} - 0.2) + \lambda_{LT}^2]$

$$= 0.5[1 + 0.21(0.527 - 0.2) + 0.527^2]$$

$$= 0.6732$$

Thus $\chi_{LT} = 1/[0.6732 + (0.6732^2 - 0.527^2)^{0.5}]$

$$= 0.9157$$

Lateral torsional buckling resistance

$$= \chi_{LT} (f_y/\gamma_{m0})Z_e = 0.9157 \times (250/1.1) \times 619 \times 10^3/10^6$$

$$= 128.82 \text{ kNm} > 45 \text{ kNm}$$

Hence, the section is safe.

6. *Member buckling resistance in combined bending and axial compression*

Determination of moment amplification factors

$$K_z = 1 + (\lambda_z - 0.2)P/P_{dz} \leq 1 + 0.8P/P_{dz}$$

$$K_z = 1 + (0.33 - 0.2)500/1401$$

$$= 1.0463 < 1 + 0.8 \times 500/1401 = 1.285$$

$$\psi_z = M_2/M_1 = 27/45 = 0.6,$$
$$C_{mz} = 0.6 + 0.4\psi$$
$$= 0.6 + 0.4 \times 0.6 = 0.84 > 0.4$$

Check with interaction formula (Clause 9.3.2.2)

$$P/P_d + [(K_z C_{mz} M_z)/M_{dz}] < 1$$

Thus, $(500/1401) + (1.0463 \times 0.84 \times 45)/128.82 = 0.357 + 0.307 = 0.664 < 1.0$

Hence section is safe against combined axial force and bending moment.

Example 9.2 *An I-section beam-column of length 4 m has to be designed as a ground floor column in a multi-storey building. The frame is moment-resisting in-plane and pinned out-of plane, with diagonal bracing provided in both directions. The column is subjected to major axis bending due to horizontal forces and minor axis bending due to eccentricity of loading from the floor beams. The design action effects for this column from a linear analysis program are as follows (see also Fig. 9.12).*

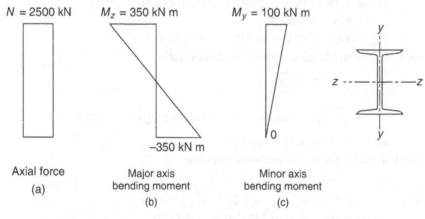

Axial force
(a)

Major axis
bending moment
(b)

Minor axis
bending moment
(c)

Fig. 9.12 Design action effects on the beam-column

$$N = 2500 \ kN$$

At the base of column: $M_z = -350 \ kNm$, $M_y = 0$
At the top of column: $M_z = 350 \ kNm$, $M_y = 100 \ kNm$
Determine whether a hot-rolled wide flange section W 310 × 310 × 226 will be suitable to resist the design action effects. Use Fe 410 grade steel.

Solution
1. Section properties
The section properties of W 310 × 310 × 226 are as follows:
$b_f = 317$ mm
$H = 348$ mm
$t_f = 35.6$ mm $\qquad I_z = 59560 \times 10^4$ mm^4
$t_w = 22.1$ mm $\qquad I_y = 18930 \times 10^4$ mm^4

$R = 15.0$ mm $\qquad\qquad r_z = 143.6$ mm

$A = 28,880$ mm^2 $\qquad\quad r_y = 81.0$ mm

$Z_{ez} = 3423 \times 10^3$ mm^3 $\qquad Z_{pz} = 3948,812$ mm^3

$Z_{ey} = 1194 \times 10^3$ mm^3 $\qquad Z_{py} = 1822,502$ mm^3

The calculations for I_t, I_w, Z_{pz}, and Z_{py} are as follows:

$$Z_{pz} = 2b_f t_f (H - t_f)/2 + t_w (H - 2t_f)^2/4$$
$$= 2 \times 317 \times 35.6 (348 - 35.6)/2 + 22.1 \times (348 - 2 \times 35.6)^2/4$$
$$= 3948,812 \text{ mm}^3$$
$$Z_{py} = 2 \times t_f b_f^2/4 + (H - 2t_f)t_w^2/4$$
$$= 2 \times 35.6 \times 317^2/4 + (348 - 2 \times 35.6)\, 22.1^2/4$$
$$= 1,822,\, 502 \text{ mm}^3$$
$$E = 2 \times 10^5 \text{ N/mm}^2$$
$$G = 76923 \text{ N/mm}^2$$

2. Cross-section classification (clause 3.7)

As the thickness is greater than 20 mm, $f_y = 240$ MPa as per Table 1 of the code. However, in the following calculations f_y has been assumed as 250 MPa.

$$\varepsilon = \sqrt{(250/f_y)} = \sqrt{(250/250)} = 1$$

Outstand flanges plastic.

$$b/t_f = (317/2)/35.6 = 4.45$$

Limit for class 1 flange $= 9.4\varepsilon = 9.4 > 4.45$

Hence flanges are class 1 (plastic)

Web

$$d = H - 2t_f - 2R = 348 - 2 \times 35.6 - 2 \times 15 = 246.8 \text{ mm}$$
$$d/t_w = 246.8/22.1 = 11.1 < 42\varepsilon = 42$$

Hence web is plastic.

The overall cross-section classification is plastic.

3. Compression resistance of the cross section

The design compression resistance of the cross section

$$N_d = A_g f_y/\gamma_{m0} = 28880 \times 250/(1.1 \times 1000) = 6563 \text{ kN} > 2500 \text{ kN}$$

Hence the design compression resistance is alright.

4. Bending resistance of the cross section (clause 8.2.1.2)

Major z-z axis

Maximum bending moment $= 350$ kN m

The design major axis bending resistance of the cross section

$$M_{dz} = \beta_b Z_p f_y/\gamma_{m0} = (1 \times 3,948,812 \times 250/1.1) \times 10^{-6}$$
$$= 897.45 \text{ kN m} > 350 \text{ kN m}$$

Minor y-y axis

Maximum bending moment $= 100$ kN m

The design minor axis bending resistance of the cross section

$$M_{dy} = (1 \times 1822,502 \times 250/1.1) \times 10^{-6} = 414.2 \text{ kN m} > 100 \text{ kN m}$$

Hence the bending resistance is fine along both major *z-z* axis and minor *y-y* axis.

5. Shear resistance of the cross-section (clause 8.4.1)

The design plastic shear resistance of the cross section

$$V_p = A_v(f_{yw}/\sqrt{3})/\gamma_{m0}$$

Load parallel to web

Maximum shear force $V = [350 - (-350)]/4.0 = 175$ kN

For a rolled section, loaded parallel to the web the shear are

$$A_v = Ht_w = 348 \times 22.1 = 7690.8 \text{ mm}^2$$

$$V_p = 7690.8 \times (250/\sqrt{3})/(1.1 \times 1000) = 1009.1 \text{ kN} > 175 \text{ kN}$$

Hence the shear resistance of the cross section is alright.

Load parallel to flanges

Maximum shear force $V = 100/4 = 25$ kN

$$A_v = 2b_f t_f = 2 \times 317 \times 35.6 = 22{,}570.4 \text{ mm}^2$$

$$V_p = 22570.4 \times (250/\sqrt{3})/(1.1 \times 1000) = 2961.5 \text{ kN} > 25 \text{ kN}$$

Hence the shear force is alright.

Shear buckling (clause 8.4.2)

Shear buckling need not be considered, provided

$d/t_w < 67\varepsilon$, for unstiffened webs

$d/t_w = 246.8/22.1 = 11.1 < 67$ ($\varepsilon = 1$)

Hence no shear buckling check is required.

6. Cross-section resistance (clause 9.3.1)

Provided the shear force is less than 60% of the design plastic shear resistance and provided shear buckling is not a concern, the cross section needs only satisfy the requirements of bending and axial force (clause 9.2.1). In this case shear force is less than 60% of design plastic shear resistance and hence the cross section needs to be checked for bending and axial force only.

Reduced plastic moment resistances

Major z-z axis

For rolled I- or H-sections,

$$M_{ndz} = 1.11M_{dz}(1 - n) \le M_{dz}$$

where $\quad n = N/N_d = 2500/6563 = 0.381$

Hence $\quad M_{ndz} = 1.11 \times 897.45 (1 - 0.381)$

$$= 616.6 \text{ kNm} > 350 \text{ kNm}$$

Minor y-y axis

For $n > 0.2$, $M_{ndy} = 1.56M_{ndy}(1 - n)(n + 0.6)$

$$= 1.56 \times 414.2(1 - 0.381)(0.381 + 0.6)$$

$$= 392.4 \text{ kNm} > 100 \text{ kNm}$$

Hence the moment resistances for major z-z axis and minor y-y axis are alright.

Cross-section check for biaxial bending (with reduced moment resistances)

$$(M_y/M_{ndy})^{\alpha_1} + (M_z/M_{ndz})^{\alpha_2} \le 1.0$$

For I- and H-sections

$\alpha_1 = 5n \geq 1$ and $\alpha_2 = 2$

$= 5 \times 0.381 = 1.905$

Thus, $(100/392.4)^{1.905} + (350/616.6)^2 = 0.396 < 1.0$

7. *Member buckling resistance in compression (clause 7.1.2)*

$N_d = A_e f_{cd}$

$f_{cd} = \{(f_y/\gamma_{m0})/(\phi + [\phi^2 - \lambda^2]^{0.5}\} \leq f_y/\gamma_{m0}$

where $\phi = 0.5[1 + \alpha(\lambda - 0.2) + \lambda^2)]$

$\lambda = \sqrt{(f_y/f_{cr})}$

$f_{cr} = \pi^2 E/(KL/r)^2$

For buckling about the major z-z axis (Table 11 of the code)

$KL_z = 0.65L = 0.65 \times 4 = 2.6$ m; $KL_z/r_z = 2600/143.6 = 18.1$

Note If the stiffness of the beams are known, the effective length should be calculated by using Wood's curves. For buckling about the minor y-y axis,

$KL_y = 1.0L = 1 \times 4 = 4$ m; $KL_y/r_y = 4000/81 = 49.38$

$f_{cr,z} = \pi^2 E/(KL_z/r_z)^2 = \pi^2 \times 2 \times 10^5/(18.1)^2 = 6025$ N/mm^2

$\lambda_z = \sqrt{(250/6025)} = 0.2037$

$f_{cr,y} = \pi^2 \times 2 \times 10^5/(49.38)^2 = 810$ N/mm^2

$\lambda_y = \sqrt{(250/810)} = 0.5557$

Selection of buckling curve and imperfection factor α

For hot rolled *H*-section (with $h/b = 348/317 = 1.09 < 1.2$ and $t_f < 100$ mm):

- For buckling about the z-z axis, use curve *b* (Table 10 of the code)
- For buckling about the y-y axis, use curve *c* (Table 10 of the code)
- For curve *b*, $\alpha = 0.34$ and for curve *c*, $\alpha = 0.49$ (Table 7 of the code)

Buckling resistance

$\phi_z = 0.5[1 + 0.34(0.2037 - 0.2) + 0.2037^2]$

$= 0.5213$

$\chi_y = 1/[0.5213 + \sqrt{(0.5213^2 - 0.2037^2)}]$

$= 0.9988$

$P_{d,z} = 0.9988 \times (250/1.1) \times 28,880/1000 = 6556$ kN > 2500 kN

Buckling resistance is, thus, fine.

Buckling resistance: Minor y-y axis

$\phi_y = 0.5[1 + 0.49(0.5557 - 0.2) + 0.5557^2] = 0.7415$

$\chi_z = 1/[0.7415 + \sqrt{(0.7415^2 - 0.5557^2)}] = 0.8113$

$P_{d,y} = 0.8113 \times (250/1.1) \times 28,880/1000 = 5325.4$ kN > 2500 kN

Hence the buckling resistance about the minor axis is fine.

8. Member buckling resistance in bending (clause 8.2.2)

The 4-m long column is unsupported along its length with no-torsional or lateral restraint. Equal and opposite design end moments of 350 kN m are applied about the major axis. Hence the full length of the column must be checked for lateral torsional buckling.

$$M = 350 \text{ kN m}$$

$$M_d = \beta_b Z_p f_{bd}$$

where $\beta_b = 1.0$ for plastic and compact sections.

Determination of M_{cr} ($L_y = 4000$ mm)

$$M_{cr} = C_1[(\pi^2 EI_y h)/2(KL_y)^2] \{1 + (1/20) [(KL_y/r_y)/(h/t_f)]^2\}^{0.5}$$

For equal and opposite end moments ($\psi = -1$) and $K = 1.0$, $C_1 = 2.752$ from Table 42 of the code, $h/t_f = 348/35.6 = 9.78$.

$$M_{cr} = 2.752[(\pi^2 \times 2.0 \times 10^5 \times 18930 \times 10^4 \times 348)/(2 \times 4000^2)]$$
$$\{1 + (1/20) [49.38/9.78]^2\}^{0.5}$$

$$= 16{,}866 \times 10^6 \text{ N mm} = 16866 \text{ kN m}$$

Non-dimensional lateral-torsional slenderness ratio,

$$\lambda_{LT} = \sqrt{(\beta_b Z_{pz} f_y / M_{cr})}$$

$$= \sqrt{[1 \times 3948{,}812 \times 250/(16866 \times 10^6)]} = 0.2419$$

$\alpha_{LT} = 0.21$ for rolled section

Reduction factor for lateral torsional buckling

$$\chi_{LT} = 1/[\phi_{LT} + (\phi_{LT}^2 - \lambda_{LT}^2)^{0.5}]$$

where $\phi_{LT} = 0.5 [1 + \alpha_{LT} (\lambda_{LT} - 0.2) + \lambda_{LT}^2]$

$$= 0.5 [1 + 0.21(0.2419 - 0.2) + 0.2419^2]$$

$$= 0.5337$$

Thus, $\chi_{LT} = 1/[0.5337 + (0.5337^2 - 0.2419^2)^{0.5}]$

$$= 0.9907$$

Lateral torsional buckling resistance

$$= \chi_{LT}(f_y/\gamma_{m0}) \beta_b Z_{pz}$$

$$= 0.9907 \times 250/1.1 \times 1 \times 3948{,}812$$

$$= 889.1 \times 10^6 \text{ N mm} = 889.1 \text{ kN m}$$

$$M/M_d = 350/889.1 = 0.394 \le 1.0$$

Hence o.k.

9. Member buckling resistance in combined bending and axial compression

Determination of moment amplification factors (clause 9.3.2.2.)

$$K_z = 1 + (\lambda_z - 0.2)P/P_{dz} \le 1 + 0.8 P/P_{dz}$$

$$K_z = 1 + (0.2037 - 0.2)2500/6556 = 1.0015 < 1 + 0.8 \times 2500/6556 = 1.305$$

$$\psi_z = M_2/M_1 = -350/350 = -1,$$

$$C_{mz} = 0.6 + 0.4\psi = 0.6 + 0.4 \times (-1) = 0.2 < 0.4; \text{ Hence } C_{mz} = 0.4$$

$$K_y = 1 + (\lambda_y - 0.2)P/P_{dy} \leq 1 + 0.8P/P_{dy}$$

$$K_y = 1 + (0.5557 - 0.2)2500/5325.4 = 1.167 < 1 + 0.8 \times 2500/5325.4 = 1.375$$

$$\psi_y = M_2/M_1 = 0,$$

$$C_{my} = 0.6 + 0.4\psi = 0.6 + 0.4 \times (0) = 0.6 > 0.4$$

$$C_{mLT} = 0.4$$

$$n_y = P/P_{dy} = 2500/5325.4 = 0.4694; \quad \lambda_{LT} = 0.2419$$

$$K_{LT} = 1 - \frac{0.1\lambda_{LT}n_y}{(C_{mLT} - 0.25)} \geq 1 - \frac{0.1n_y}{(C_{mLT} - 0.25)}$$

$$K_{LT} = 1 - 0.1 \times 0.2419 \times 0.4694/(0.4 - 0.25)$$
$$= 0.9243 \geq 1 - 0.1 \times 0.2419/(0.4 - 0.25) = 0.84$$

Check with interaction formula (clause 9.3.2.2)

$$(P/P_{dy}) + (K_y C_{my} M_y/M_{dy}) + (K_{LT}M_z/M_{dz}) \leq 1.0$$

$$(2500/5325.4) + (1.167 \times 0.6 \times 100)/414.2 + (0.9243 \times 350)/897.45$$
$$= 0.469 + 0.169 + 0.360 = 0.998 \approx 1.0$$

$$(P/P_{dz}) + (0.6K_y C_{my} M_y/M_{dy}) + (K_z C_{mz} M_z/M_{dz}) \leq 1.0$$

$$(2500/6556) + (0.6 \times 1.167 \times 0.6 \times 100)/414.2 + (1.002 \times 0.4 \times 350)/897.45$$
$$= 0.381 + 0.101 + 0.156 = 0.638 < 1.0$$

Hence the section is suitable to resist the design action effects.

Example 9.3 *Design the base plate for the column in Example 9.1 subjected to a factored moment of 45 kN m and a factored axial load of 500 kN. The column size is ISHB 250. The cube compressive strength of concrete in the foundation is f_{ck} = 25 N/mm². Use grade 410 steel.*

Solution

1. Size of the base plate

$$e = 45 \times 10^3/500 = 90 \text{ mm}$$

If the base plate is made 6e in length there will be compressive pressure over the whole of the base.

$$6e = 6 \times 90 = 540 \text{ mm}$$

The required breadth to limit the bearing pressure to $0.45f_{ck}$ (= 11.25 N/mm²) is

$$B = 2P/(L \times 0.45f_{ck}) = (2 \times 500 \times 10^3)/(540 \times 0.45 \times 25) = 164.6 \text{ mm}$$

Provide a rectangular base plate of size 540 × 400 mm. The arrangement of the base plate is shown in Fig. 9.13.

$$\text{Area} = 540 \times 400 = 216 \times 10^3 \text{ mm}^2$$

$$\text{Modulus } Z = 400 \times 540^2/6 = 19.44 \times 10^6 \text{ mm}^3$$

Maximum pressure:

$$p_{max} = 500 \times 10^3/(216 \times 10^3) + 45 \times 10^6/(19.44 \times 10^6)$$
$$= 2.31 + 2.31$$
$$= 4.62 \text{ N/mm}^2$$

$$p_{min} = 2.31 - 2.31 = 0$$

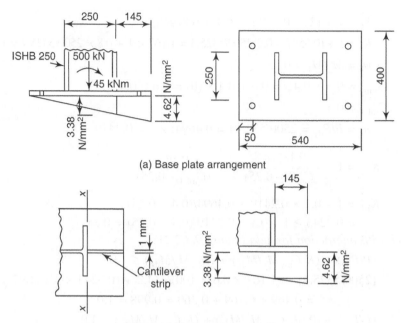

(a) Base plate arrangement

(b) Base plate design for thickness

Fig. 9.13

2. Thickness of base plate

Consider a 1-mm wide strip as shown in Fig. 9.13(b). This acts as a cantilever from the face of the column with the loading caused by pressure on the base. This method gives a conservative design for the thickness of base plate, since the plate action due to bending in two directions at right angles is not considered.

$$\text{Base pressure at section } XX = [(540 - 145)/540] \times 4.62$$
$$= 3.38 \text{ N/mm}^2$$

For the trapezoidal pressure loading on the cantilever strip as shown in the figure, the moment at XX is calculated as follows

$$M_x = (3.38 \times 145^2/2) + (4.62 - 3.38) \times 145/2 \times 2/3 \times 145$$
$$= 44.22 \times 10^3 \text{ Nmm}$$

Moment capacity of plate = $1.2 f_y Z_e / \gamma_{m0}$
where $Z_e = t^2/6$
Hence $44.22 \times 10^3 = 1.2 \times 250 \times t^2/(6 \times 1.1)$
$$= 45.45 \, t^2$$

Thickness of plate $t = \sqrt{(44.22 \times 10^3/45.45)} = 31.18$ mm

Hence, use a 32-mm thick plate.

3. Weld connecting beam-column to base plate

The base plate has been designed on the basis of linear distribution of pressure. For consistency the weld will be designed on the same basis.

Beam-column size: ISHB 250; $A = 6500$ mm^2; $Z_z = 619 \times 10^3$ mm^3

Axial stress = $500 \times 10^3/6500 = 76.92$ N/mm^2

Bending stress = $45 \times 10^6/619 \times 10^3 = 72.70$ N/mm^2

On the basis of elastic stress distribution, there is compressive stress over the whole of the base. The base plate and column are to be machined for tight contact so that the weld is required only to hold the base plate in position. Use a 6 mm continuous fillet weld around the column profile.

Example 9.4 *Redesign the base plate in Example 9.3 using gusset plates. Assume that the parts are not machined for tight contact in bearing, so the welds have to be designed to transmit the column load and bending moment to the base plate.*

Solution

Let us consider the arrangement of gusset plate as shown in Fig. 9.14.

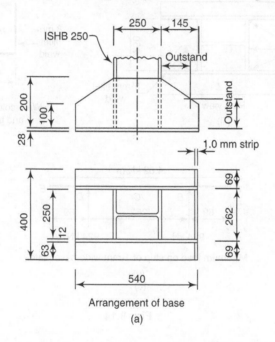

Arrangement of base

(a)

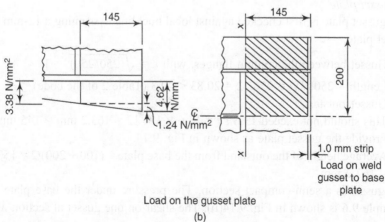

Load on the gusset plate

(b)

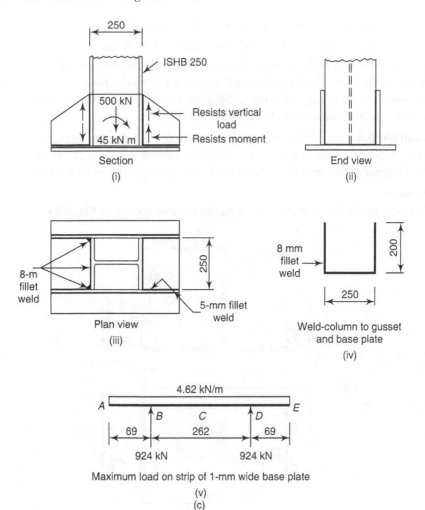

Fig. 9.14

1. Gusset plate

The gusset plate is first checked against local buckling. Assuming a 12-mm thick gusset plate,

(a) Gusset between the column flanges, with $\varepsilon = \sqrt{(250/250)} = 1$

Length = 250; $D/t = 250/12 = 20.83 < 29.3$ (Table 2 of the code)

(b) Gusset outstand

This should not exceed $13.6\,\varepsilon\,t = 13.6 \times 1 \times 12 = 163.2$ mm > 145 mm

Provide the gusset plate as shown in Fig. 9.14.

Average height of the outstand from the base plate = $(100 + 200)/2 = 150$ mm < 163.2 mm

The gusset is a semi-compact section. The pressure under the base plate from Example 9.6 is shown in Fig. 9.14(b). The shear on one gusset at section *X-X* is

$$V = (4.62 + 3.38)/2 \times 145 \times 200$$
$$= 116,000 \text{ N} = 116 \text{ kN}$$

The bending moment at X-X axis is

$$M_x = 3.38 \times 145 \times 200 \times 145/2 + 1.24 \times 145 \times 200/2 \times 2/3 \times 145$$
$$= 8.84 \times 10^6 \text{ N mm} = 8.84 \text{ kN m}$$

The shear capacity

$$V_n = V_p = A_v f_{yg}/(\sqrt{3}\ \gamma_{m0}) = 200 \times 12 \times 250/(\sqrt{3}\ \times 1.1 \times 1000)$$
$$= 314.9 \text{ kN} > 116 \text{ kN}$$
$$V = 116 \text{ kN} < 0.6 \times 314.9 = 188.95 \text{ kN}$$

The moment capacity is not reduced by the effect of shear.

$$M_g = Z_e f_y/\gamma_{m0} = (12 \times 200^2/6) \times 250/(1.1 \times 10^6)$$
$$= 18.18 \text{ kN m} > 8.84 \text{ kN m}$$

Hence the size of the gusset plate is satisfactory.

2. *Gusset plate to column weld*

The welds between the column, gussets, and base plate have to transmit all the load to the base plate. These welds are shown in Fig. 9.14(c).

Weld connecting column-gusset-base plate:

$$\text{Load per weld} = 500/2 + 45/0.25 = 430 \text{ kN}$$

Assuming an 8-mm weld,

Length of the weld {see Fig. 9.14(c)(iv)} $= 250 + 200 \times 2 - 2 \times 8 = 634$ mm

$$\text{Load per mm} = 430/634 = 0.678 \text{ kN/mm}$$

Use an 8-mm fillet weld (site weld) which has strength of 0.884 MPa (from Appendix D). The weld between one gusset plate and the base must support the maximum pressure under the base. Considering a 1-mm wide strip at the edge of the base plate, load on one weld {see Figs 9.14(a) and (b)}

$$= 4.62 \times 200/(2 \times 10^3) = 0.462 \text{ kN/mm}$$

Provide a 5-mm fillet weld with strength = 0.553 kN/mm (from Appendix D).

3. *Thickness of the base plate*

Consider a 1-mm wide strip, at the edge of the base plate, as shown in Fig. 9.14(a). It is assumed to act as a beam with overhanging ends as shown in Fig. 9.14(c)(v). The bending moments are

$$M_B = 4.62 \times 69^2/2 = 10998 \text{ N mm}$$
$$M_C = 924 \times (262/2) - 4.62 \times 200^2/2 = 28644 \text{ N mm}$$

Moment capacity of base plate $= 1.2 f_y Z_e/\gamma_{m0} = 1.2 \times (250/1.1) \times t^2/6$

Hence $28644 = 1.2 \times (250/1.1) \times t^2/6$

$$t^2 = 630.2$$

or $\qquad t = 25.1$ mm

Hence, a 28-mm thick base plate is required.

Example 9.5 *Design a shear lug for a 350 mm square base plate (thickness = 30 mm) subjected to an axial dead load of 450 kN, live load of 500 kN, and shear of 240 kN resulting from wind loading. Assume that the base plate and shear lug are of Fe 410 grade steel and the foundation is M25 grade concrete.*

Solution

Steel shims are often placed under the base plates. Hence let us assume a steel-steel friction coefficient of 0.3 (Note that IS 800 in clause 7.4.1 suggests a steel concrete friction coefficient of 0.45)

1. $V_{lg1} = 1.2(240) - 0.3[1.2 (450 + 500)] = -54$ kN

   $V_{lg2} = 1.5(240) - 0.3[1.5(450)] = 157.5$ kN

2. $A_{lg} = V/(0.45f_{ck}) = 157.5 \times 10^3/(0.45 \times 25) = 14 \times 10^3$ mm^2

3. Assume a shear lug with width $W = 200$ mm. The height of the bearing portion is

   $$H - G = 14 \times 10^3/200 = 70 \text{ mm}$$

   Assuming a grout depth of 25 mm,
   Required depth of shear lug = 70 + 25 = 95 mm

4. The cantilever end moment

   $$M_{lg} = (V/W)(H + G)/2 = (157.5 \times 10^3/200) (95 + 25)/2$$
   $$= 47250 \text{ Nmm}$$

5. The required thickness $t_{lg} = \sqrt{[4M_{lg}/(f_y/1.1)]}$

   $$= \sqrt{[4 \times 47250/(250/1.1)]}$$
   $$= 28.83 \text{ mm} < 30 \text{ mm}$$

   Use a 200-mm wide, 95-mm high, and 30-mm thick shear lug.

Summary

Bending moments will be present in many practical situations in columns, in addition to the axial loads. These bending moments may be due to (a) eccentricity of axial force, (b) building frame action, (c) portal or gable frame action, (d) load from brackets, (e) transverse loads, and (f) fixed base condition. When a member is subjected to bending moment and axial force, it is called as a beam-column. The cross section of such beam-columns is oriented in such a way to resist significant bending in the major axis of the member. In general, beam-columns may be subjected to axial forces and biaxial bending moments.

All of the parameters that affect the behaviour of a beam or a column (such as the length of the member, geometry and material properties, support conditions, magnitude and distribution of transverse loads and moments, presence or absence of lateral bracing, and whether the member is a part of an unbraced or braced frame) also affect the behaviour and strength of a beam-column. The general behaviour of beam-columns is described. Thus a beam-column may fail by local buckling, overall buckling (similar to axially loaded columns), lateral-torsional buckling (similar to beams), plastic failure (short beam-columns), and by a combination of column buckling and lateral torsional buckling. Moreover, the ultimate behaviour of beam-columns subjected to axial load and biaxial bending moments is complicated by the effect of plastification, moment magnification, and lateral-torsional buckling.

The inelastic analysis to determine the strength of beam-column is complicated and is carried out in two steps: cross-section analysis and member analysis. These methods require extensive numerical analysis procedures and hence are not suitable for design office use.

Hence codes and specifications of many countries suggest the use of interaction equations, which are developed based on curve-fitting the existing analytical and experimental data on isolated beam-columns or beam-columns of simple portal frames.

The generic form of the interaction equations have been retained in the codes of practices of several countries. The interaction equations given in the Indian code (which are based on the Eurocode provisions) are discussed.

The various steps involved in the design of beam-columns, are given. The equivalent axial load method is described, which will be useful while selecting the initial cross section of the iterative design process. Though the codal equations are based on the first-order elastic analysis and hence include the moment amplification term, it is possible to use a second-order analysis and get the second-order moments directly. In such a case, the amplification factors should not be used in the interaction equations. Similarly the advanced analysis methods may eliminate the determination of K factors. (It is of interest to note that an error of 20% in the determination of the effective length factor may result in an error of 50% in the load carrying capacity of a slender column.)

Bending moments may occur in members subjected to tension; however, the effect of tension load will always reduce the primary bending moments. Hence in the code, a reduced effective bending moment is specified to be used with the interaction equations.

The design of base plates subjected to bending and axial load are covered. The provision of gussets will result in reduced thickness of base plates, though increasing the fabrication cost. Such gusseted base plates are also discussed. Information on the design of anchor bolts and shear lugs (to resist heavy shear forces) are also included. All the design concepts are explained with illustrative examples.

Exercises

1. A beam-column of length 5 m is subjected to a compression of 800 kN and a major axis moment of 4.5 kN m. The weaker plane of the column is strengthened by bracing. If the effective length factor is 0.8, design the beam-column, assuming Fe 410 grade steel.

2. A wide flange W 310×310×143 beam-column has a height of 3.54 m and is pinned at both ends. Check whether it can support a design axial load of 600 kN together with a major axis bending moment of 300 kN m applied at the top of the beam-column. Assume grade Fe 410 steel.

3. Design a beam-column of length 3.75 m if it carries a compressive load of 500 kN, a major axis moment of 5 kN m and a minor axis moment of 2 kN m. Assume that the effective length factor is 1.2 and the column is free to buckle in any plane.

4. Redesign the beam-column given in Exercise 1 assuming that the axial load is tensile.

5. Design the base plate for an ISHB 300 column subjected to a factored axial load of 800 kN and a factored moment of 40 kN m in the major axis. Assume M25 concrete for the foundation and grade Fe 410 steel.

6. Redesign the base plate mentioned in Exercise 9 using gusset plates.

7. Design a base plate for a ISHB 300 column subjected to a factored axial load of 300 kN, bending moment of 60 kN m and a shear force of 300 kN. Assume that the base plate and shear lug are of Fe 410 grade steel and the foundation is M25 concrete.

Review Questions

1. What are beam-columns?
2. How are bending moments introduced in columns?
3. When bending moments are acting on a column in addition to axial loads, how are the columns oriented?
4. What are the parameters that affect the behaviour of beam-columns?
5. Describe the general behaviour of a beam-column.
6. Slender beam-columns may fail by
   (a) lateral torsional buckling
   (b) buckling similar to columns
   (c) local buckling
   (d) combination of (a) and (b)
   (e) combination of (a), (b), and (c).
7. Under which five cases can the behaviour of a beam-column be classified?
8. What is equivalent moment factor?
9. What is the general form of the interaction equation?
10. Identify the difference in behaviour of beam-columns subject to bending moment about minor axis compared to applied bending moment in major axis.
11. What is the form of interaction equation for biaxially loaded beam-column (state the linear format only)?
12. Interaction equations are specified in the codes for
    (a) overall buckling check
    (b) local capacity check
    (c) both (a) and (b)
13. State the general form of interaction equation specified in the code (IS 800) for
    (a) overall buckling check
    (b) local capacity check

14. What are the different steps to be followed while designing a beam column?
15. Describe the equivalent axial load method for selecting the initial section of a beam-column.
16. When equivalent moment method has to be used to select the initial section of a beam-column?
17. State the interaction equation used for the local capacity check of beam-columns subject to axial tension and bending moment.
18. What is the purpose of anchor bolts in a base plate having compression over the whole area?
19. What are the criteria for design in the case of base plate having compression over the whole area?
20. How are the gusset plates sized in the gusseted base plate?
21. How do the headed anchors transfer tensile load to the foundation?
22. State the advantage of using post-installed anchors.
23. List the different types of post-installed anchors.
24. What is a shear lug? When is it used?
25. What is the suggested value of μ for base plates? Why is this value suggested?
26. List the different design steps for sizing the shear lug.

CHAPTER 10

Bolted Connections

Introduction

Any steel structure is an assemblage of different members such as beam, columns, and tension members, which are fastened or connected to one another, usually at the member ends. Many members in a steel structure may themselves be made of different components such as plates, angles, I-beams, or channels. These different components have to be connected properly by means of *fasteners*, so that they will act together as a single composite unit. Connections between different members of a steel framework not only facilitate the flow of forces and moments from one member to another but also allow the transfer of forces up to the foundation level. It is desirable to avoid connection failure before member failure due to the following reasons.

(a) A connection failure may lead to a catastrophic failure of the whole structure.

(b) Normally, a connection failure is not as ductile as that of a steel member failure.

(c) For achieving an economical design, it is important that connectors develop full or a little extra strength of the members it is joining.

Connection failure may be avoided by adopting a higher safety factor for the joints than the members.

The basic goal of connection design is to produce a joint that is safe, economical, and simple (so that it can be manufactured and assembled at site without any difficulty). It is also important to standardize the connections in a structure and to detail it in such a way that it allows sufficient clearance and adjustment to accommodate any lack of fit, resists corrosion, is easy to maintain, and provides reasonable appearance.

Connections (or structural joints) may be classified according to the following parameters:

(a) Method of fastening such as rivets, bolts, and welding—connections using bolts are further classified as *bearing* or *friction type connections*

(b) Connection rigidity—simple, rigid (so that the forces produced in the members may be obtained by using an indeterminate structural analysis), or semi-rigid

(c) Joint resistance—bearing connections and friction connections (these are explained in subsequent sections)
(d) Fabrication location—shop or field connections
(e) Joint location—beam column, beam-to-beam, and column-to-foundation
(f) Connection geometry—single web angle, single plate, double web angle, top and seat angles (with and without stiffeners), end plates, header plate, welded connections using plates and angles, etc.
(g) Type of force transferred across the structural connection—shear connections, shear and moment connection or simply moment connection, tension or compression, and tension or compression with shear.

Structural connections transmit forces which result in linear and rotational movements. The linear movements at a joint are generally small but the rotational movement depends on the stiffness of the type of connection.

According to the IS code, based on connection rigidity, the joints can be defined as follows:

Rigid Rigid connections develop the full moment capacity of connecting members and retain the original angle between the members under any joint rotation, that is, rotational movement of the joint will be very small on these connections. Examples of rigid connections are shown in Fig. 10.1.

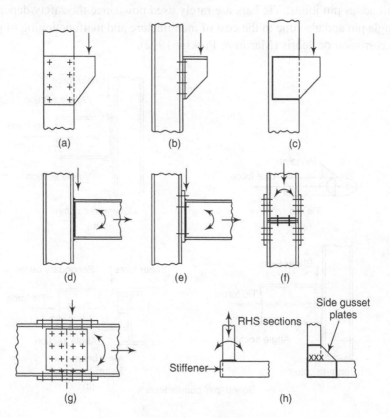

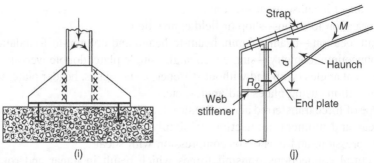

Fig. 10.1 Examples of 'rigid' connections (Martin & Purkiss 1992)

Simple In simple connections no moment transfer is assumed between the connected parts and hence are assumed as hinged (pinned). The rotational movement of the joint will be large in this case. Actually, a small amount of moment will be developed but is normally ignored in the design. Any joint eccentricity less than about 60 mm is neglected. Examples of hinged (pinned) connections are shown in Fig. 10.2. Some simple connections, for example, tie bars, are connected by real pins as shown in Fig. 10.2(a). If the pins are not corroded or blocked with debris, they will act as pin joints. Tie bars are rarely used now since the safety depends on a single pin and also due to the cost of manufacture and malfunctioning of pins due to corrosion or debris (Martin & Purkiss 1992).

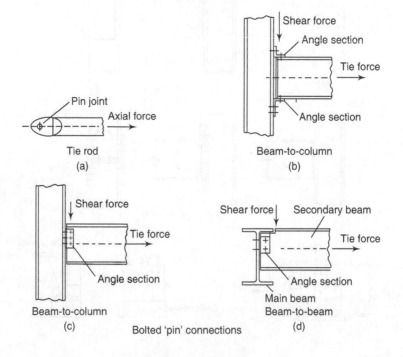

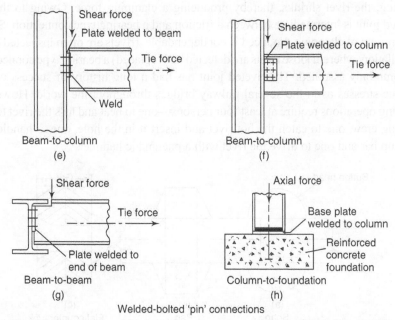

Beam-to-column
(e)

Beam-to-column
(f)

Beam-to-beam
(g)

Column-to-foundation
(h)

Welded-bolted 'pin' connections

Fig. 10.2 Examples of 'pinned' connections (Martin & Purkiss 1992)

Semi-rigid Semi-rigid connections may not have sufficient rigidity to hold the original angles between the members and develop less than the full moment capacity of the connected members. The design of these connections requires determining the amount of moment capacity (or moment–rotation relationship of the connection) based on test results or rational methods (say 20%, 30%, or 75% of moment capacity).

In reality, all the connections will be semi-rigid. However, for convenience we assume some of them as rigid and some as hinged. We will discuss bolted connections in this chapter. Welded connections are discussed in the next chapter.

10.1 Rivets and Riveted Connections

For many years rivets were the sole practical means of producing safe and serviceable steel connections. A rivet is made up of a round ductile steel bar (mild or high tensile steel as per IS 1929 and IS 2155) called *shank*, with a head at one end (see Fig. 10.3). The head may have different shapes as shown in Fig. 10.3. The snap and pan heads form a projection beyond the plate face, whereas the counter sunk head will be flush with the surface of the plate face.

The length of the rivet to be selected should be longer than the grip of the rivet (see Fig. 10.3), sufficient to form the second head. The installation of a rivet requires the heating of the rivet to a cherry red colour (approximately 980°C), inserting it into an oversize hole (approximately 1.5 mm more than the size of the rivet), applying pressure to the preformed head while at the same time squeezing the plain end of the rivet using a pneumatic driver to form a round head. During this process, the shank of the rivet completely or nearly fills the hole into which it had been inserted. Upon cooling, the rivet shrinks, thereby producing a clamping force. Owing to this, a riveted joint is intermediate between a friction and a bearing type connection. Since the amount of clamping produced is not dependable, (rivets are often inspected after installation, wherein loose rivets are detected and replaced) a bearing type connection is commonly assumed. The riveted joint has had a long history of success under fatigue stresses as in the several railway bridges throughout the world. However, riveting operations require at least four persons—one to heat and toss the rivet to the driving crew, one to catch the hot rivet and insert it in the hole, one to handle the backup bar and one to drive the rivet with a pneumatic hammer.

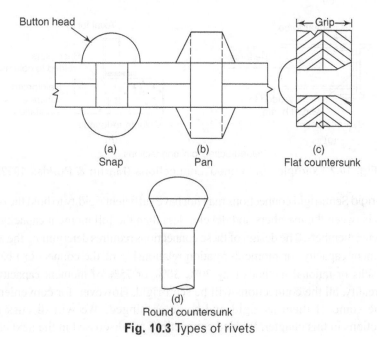

Button head

|←—Grip—→|

(a)
Snap

(b)
Pan

(c)
Flat countersunk

(d)
Round countersunk

Fig. 10.3 Types of rivets

Riveting is no longer used in engineering structures for the following reasons:
(a) The necessity of pre-heating the rivets prior to driving
(b) The labour costs associated with large riveting crews
(c) The cost involved in careful inspection and removal of poorly installed rivets
(d) The high level of noise associated with driving rivets

Readers should be aware that the design of riveted connection is similar to the design of bolted connection, except that the diameter of the rivet is taken as the diameter of the hole (diameter of rivet + clearance) in place of the nominal diameter of the bolt.

10.2 Bolted Connections

There are several types of bolts used to connect structural members. Some of them are listed as follows:
- Unfinished bolts or black bolts or C grade bolts (IS 1363 : 2002)

- Turned bolts
  - Precision bolts or A grade bolts (IS 1364 : 2002)
  - Semi-precision bolts or B grade bolts (IS 1364 : 2002)
- Ribbed bolts
- High strength bolts (IS 3757 : 1985 and IS 4000 : 1992)

10.2.1 Black Bolts

Black bolts are also referred to as ordinary, unfinished, rough, or common bolts. They are the least expensive bolts. However, they may not produce the least expensive connection since the connection may require a large number of such bolts. They are primarily used in light structures under static loads such as small trusses, purlins, girts, bracings, and platforms. They are also used as temporary fasteners during erection where HSFG bolts or welding are used as permanent fasteners. They are not recommended for connections subjected to impact, fatigue, or dynamic loads. These bolts are made from mild steel rods with a square or hexagonal head and nuts as shown in Fig. 10.4, conforming to IS 1363.

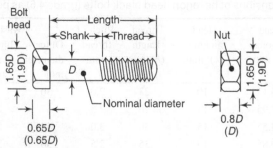

Fig. 10.4 Hexagonal head black bolt and nut. Figures in brackets are for high-strength bolts and nuts.

In steel construction, generally, bolts of property class 4.6 are used. In property class 4.6, the number 4 indicates $1/100^{th}$ of the nominal ultimate tensile strength in N/mm^2 and the number 6 indicates the ratio of yield stress to ultimate stress, expressed as a percentage. Thus, the ultimate tensile strength of a class 4.6 bolt is 400 N/mm^2 and yield strength is 0.6 times 400, which is 240 $N/mm.^2$ The tensile properties of commonly used fasteners are listed in Table 10.1. Due to the high percentage elongation of these bolts, they are more ductile. For bolts of property class 4.6, nuts of property class 4 are used and for bolts of property class 8.8, nuts of property class 8 or 10 are used.

Though square heads cost less, hexagonal heads give better appearance, are easier to hold by wrenches, and require less turning space. Most of the connections with black bolts are made by inserting them in clearance holes of about 1.5 mm to 2 mm more than the bolt diameter and by tightening them with nuts. They are produced in metric sizes ranging from 5–36 mm and designated as M5 to M36. In structural steelwork, M16, M20, M24, and M30 bolts are often used. The ratio of net tensile area at threads to nominal plain shank area of the bolt is 0.78 as per IS

1367 (Part 1). The other dimensions of commonly used bolts are as given in Table 10.2. These dimensions are so chosen that the bolt head does not fail unless the shank fails.

Table 10.1 Tensile properties of fasteners used in steel construction

Specification	Grade/ classification	Yield stress, MPa (Min)	Ultimate tensile stress, MPa (Min)	Elongation percentage (Min)
	4.6	240	400	22
IS 1367 (Part 3)	4.8	320	420	14
(ISO 898)	5.6	300	500	20
Specifications of	5.8	400	520	10
fasteners-threaded				
steel for technical	8.8 (d < 16 mm)	640	800	12
supply conditions				
	10.9	940	1040	9

Table 10.2 Dimensions of hexagon head black bolts (grade 4.6) as per IS 1364 (Part 1)

Bolt size (d), mm	Head diagonal (e), mm	Head thickness (k), mm	Thread* length (b), mm	Pitch of thread, mm	Washer (IS 5370) Outer diameter, mm	Inner diameter, mm	Thickness, mm
16	26.17	10	23	2.0	30	18	3
20	32.95	13	26	2.5	37	22	3
24	39.55	15	30	3.0	44	26	4
30	50.85	19	35	3.5	56	33	4

*For length $l \le 125$ mm. For $125 < l \le 200$, b is 6 mm more and for $l > 200$, b is 19 mm more.

10.2.2 Turned Bolts (Close Tolerance Bolts)

These are similar to unfinished bolts, with the difference that the shanks of these bolts are formed from a hexagonal rod. The surface of these bolts are prepared and machined carefully to fit in the hole. Tolerances allowed are about 0.15 mm to 0.5 mm. Since the tolerance available is small, these bolts are expensive. The small tolerance necessitates the use of special methods to ensure that all the holes align correctly. These bolts (precision and semi-precision) are used when no slippage is permitted between connected parts and where accurate alignment of components is required. They are mainly used in special jobs (in some machines and where there are dynamic loads).

10.2.3 High-Strength Bolts

High-strength bolts are made from bars of medium carbon steel. The bolts of property class 8.8 and 10.9 are commonly used in steel construction. These bolts should conform to IS 3757 and their tensile properties are given in Table 10.1. As discussed in Chapter 1, their high strength is achieved through quenching and tempering process or by alloying steel. Hence, they are less ductile than black bolts. The material of the bolts do not have a well-defined yield point. Instead of using yield stress, a so-called *proof load* is used. The proof load is obtained by multiplying the tensile stress area (may be taken as the area corresponding to root diameter at the thread and is approximately equal to 0.8 times the shank area of bolt) with the proof stress. In IS 800, the proof stress is taken as 0.7 times the ultimate tensile stress of the bolt. (In other codes such as the American code, the proof stress is taken as the yield stress, established by the 0.2% offset strain.) This bolt tension $0.7f_u$ gives adequate reserve strength, should this bolt be somewhat over stressed (e.g., 3/4 turn instead of 1/2 turn in the turn-of-the-nut method). Note that grade 10.9 bolts have lower ductility than grade 8.8 bolts and the margin between the 0.2% yield strength and the ultimate strength is also lower.

The HSFG bolt, nut, and washer dimensions are shown in Table 10.3 (also see Fig. 10.4 for approximate bolt dimensions). Bolts of sizes M16, M20, M24, and M30 are commonly used in practice. These bolts are identified by the manufacturer's identification symbol and the property class identification symbol 8S, 8.8S, 10S, or 10.9S, which will be embossed on the heads of these bolts. Since, these bolts have a tensile strength much higher than the ordinary black bolts, the number of bolts required at a joint is considerably reduced. The vibration and impact resistance of the joint are also improved considerably.

Table 10.3 High-strength friction grip bolts as per IS 3757 : 1985

Diameter, d mm	M16	M20	M24	M30
Head diagonal, e mm	29.56	37.29	45.20	55.37
Head thickness, k mm	10	12.5	15	18.7
Nut thickness, mm	13	16	19	24
Washer outer diameter,				
*D mm	30	37	44	56
Washer thickness, heavy,				
mm	4	4	4	5
Thread length, **b mm				
<100	31	36	41	49
>100	38	43	48	56

* The outside diameter of a washer is an important dimension when detailing, for example, to avoid overlapping an adjacent weld.

**The thread length depends on the length of the bolt, which is calculated as grip length plus the allowance for grip as given in Table 10.4.

The percentage elongation of 12% at failure of these bolts is less than the black bolts, but is still acceptable for design purposes. Special techniques (see Section 10.2.4) are used for tightening the nuts to induce a specified initial tension in the bolt, which causes sufficient friction between the faying faces. These bolts with induced initial tension are called *High-Strength Friction Grip (HSFG) bolts*. Due to this friction, the slip in the joint (which is associated with black bolts) is eliminated and hence the joints with HSFG bolts are called non-slip connections or *friction type connections* (as opposed to the bearing type connections of ordinary black bolts). The induced initial tension in the bolt is called the *proof-load* of the bolt and the coefficient of friction between the bolt head and the faying surfaces is called the *slip factor*. The bolts of property class 8.8 can be hot-dip galvanized {as per IS 1367 (Part 13)} whereas class 10.9 bolts should not be hot-dip galvanized since this may cause hydrogen embrittlement (IS 3757).

Table 10.4 Allowance for Bolt Length

Nominal size of the bolt	Allowance for Grip*, in mm
16	26
20	31
24	36
30	42
36	48

*The allowance includes thickness of one nut and one washer only. If additional washers are used or where threads are excluded (in bearing type joints) from shear plane, high allowance may be required.

10.2.4 Bolt Tightening Methods

When slip resistant connections are not required and when bolts are not subjected to tension, high-strength bolts are tightened to a 'snug-tight' condition to ensure that the load transmitting plies are brought into effective contact (this may be achieved as a result of a few impacts of an impact wrench or the full effort of a man using an ordinary spud wrench).

However, the reliability of HSFG bolts in a non-slip or friction type connection depends on the method of tightening of the bolt, which will ensure whether the required proof load (pre-tension) is obtained. The three methods that may be used in practice are *the turn-of-the-nut tightening* (part-turn method), *direct tension indicator tightening, and calibrated wrench tightening* (torque control method). Only turn-of-the-nut tightening method is described below. For other methods, refer Owens and Cheal (1989) and Struik et al. (1973).

Turn-of-the-nut tightening method Turn-of-the-nut tightening method, also known as part-turn method, is the simplest and most common method. Developed in the 1950s and 1960s, the specified pre-tension in the bolt is considered to be obtained by a specified rotation of the nut from the 'snug-tight' condition. In this method, after the bolts are snug-tight, permanent marks (these permanent marks may be used in a subsequent inspection) are made on bolts and nuts to identify the relative position of the bolt and nut and to control the final nut rotation. Each nut is then tightened by a specified turn of the nut from the snug-tight position depending on the length of the bolt as prescribed in IS 4000 (see Table 10.5 which also gives the minimum tension which should be available in the bolt after tightening).

Table 10.5 Minimum Bolt Tension and Nut Rotation from Snug-Tight Condition (IS 4000 : 1992).

Nominal size of bolt	Length of bolt,* mm		Minimum bolt tension in kN for bolts of property class	
	Nut rotation 1/2 turn	Nut rotation 3/4 turn	8.8	10.9
M16	≤ 120	>120 ≤ 240	94.5	130
M20	≤ 120	>120 ≤ 240	147	203
M24	≤ 160	>160 ≤ 350	212	293
M30	≤ 160	>160 ≤ 350	337	466

*Length is measured from the underside of the head to the extreme end of the shank.

Whatever be the tightening method, the installation must begin at the most rigid part of the connection and progress systematically towards the least rigid areas. Similarly, where there are more than four bolts in a group, the bolts should be tightened in a staggered manner, working from the centre of the joint outward. It has been observed that the behaviour of galvanized bolts may differ from the behaviour of normal, uncoated high-strength bolts.

Since the turn-of-the-nut method often induces a bolt tension that may exceed the elastic limit of the threaded portion, repeated tightening of high-strength bolts may be undesirable.

10.2.5 Advantages of Bolted Connections

The black bolts offer the following advantages over riveted or welded connections:
(a) Use of unskilled labour and simple tools
(b) Noiseless and quick fabrication
(c) No special equipment/process needed for installation
(d) Fast progress of work
(e) Accommodates minor discrepancies in dimensions
(f) The connection supports loads as soon as the bolts are tightened (in welds and rivets, cooling period is involved).

The main drawback of the black bolt is the slip of the joint when subjected to loading. When large forces are to be resisted, the space required for the joint is extensive. Also, precautions such as the provision of special locking devices or the use of pre-loaded high-strength bolts are required in situations involving fluctuating loads.

Though the material cost of HSFG bolts are about 50% higher than black bolts and require special workmanship for installation, they provide the following advantages.
(a) HSFG bolts do not allow any slip between the elements connected, especially in close tolerance holes (see Fig. 10.5), thus providing rigid connections.
(b) Due to the clamping action, load is transmitted by friction only and the bolts are not subjected to shear and bearing.
(c) Due to the smaller number of bolts, the gusset plate sizes are reduced.
(d) Deformation is minimized.

(e) Since HSFG bolts under working loads do not rely on resistance from bearing, holes larger than usual can be provided to ease erection and take care of lack of fit. Thus the holes may be standard, extra large, or short/long slotted. However, the type of hole will govern the strength of the connection.

(f) Noiseless fabrication, since the bolts are tightened with wrenches.

(g) The possibility of failure at the net section under the working loads is eliminated.

(h) Since the loads causing fatigue will be within proof load, the nuts are prevented from loosening and the fatigue strength of the joint will be greater and better than welded and riveted joints. Moreover, since the load is transferred by friction, there is no stress concentration in the holes.

(i) Unlike riveted joints, few persons are required for making the connections.

(j) No heating is required and no danger of tossing of bolt. Thus, the safety of the workers is enhanced.

(k) Alterations, if any (e.g. replacement of the defective bolt) are done easily than in welded or riveted connections.

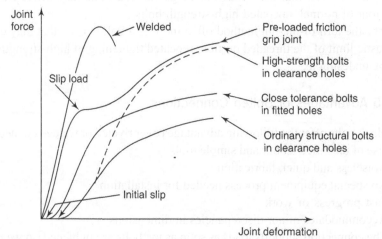

Fig. 10.5 Typical load–deformation behaviour of different types of fasteners.

However, bolting usually involves a significant fabrication effort to produce the bolt holes and associated plates or cleats. In addition, special procedures are required to ensure that the clamping actions required for pre-loaded friction-grip joints are achieved. The connections with HSFG bolts may not be as rigid as a welded connection.

10.2.6 Bolt Holes

Bolt holes are usually drilled. Punched holes (punched full size or punched undersize and reamed) are preferred by steel fabricators because it is simple and saves time and cost. However, punching can reduce ductility and toughness and may lead to brittle fracture. Hence, punched holes should not be used where plastic tensile straining can occur (Owens et al. 1981). IS 800 allows punched holes only in materials whose yield stress f_y does not exceed 360 MPa and where thickness does not exceed $(5600/f_y)$ mm. It also disallows punched holes in cyclically loaded details. Holes should not be formed by gas cutting, since they affect the local properties of steel, though plasma cutting is allowed in the code for statically loaded members (clause 17.2.4.5).

Bolt holes are made larger than the bolt diameter to facilitate erection and to allow for inaccuracies in fabrication. Table 10.6 shows the standard values of holes for different bolt sizes (the clearance is 1.0 mm for bolts less than 14 mm and 2 mm for bolts between 16 mm and 24 mm and 3 mm for bolts exceeding 24 mm).

Table 10.6 Bolt diameter, pitch, and edge distances as per IS 800

Nominal diameter of bolt, d, mm	12	14	16	18	20	22	24	Above 24
Diameter of hole, d_R, mm	13.0	15.0	18.0	20.0	22.0	24.0	26.0	Bolt diameter + 3 mm
Minimum edge distance,* e_b, mm								
(a) for sheared or rough edge	22	26	31	34	37	41	44	1.7 × hole diameter
(b) for rolled, sawn, or planed edge	18	23	27	30	33	36	39	1.5 × hole diameter

*The edge distances in this table, which are for standard holes, must be increased if oversize or slotted holes are used.

Max. edge distance = $12t\varepsilon$ where $\varepsilon = (250/f_y)^{0.5}$

Pitch (min.)	2.5 × nominal diameter of bolt
Pitch (max.)	32t or 300 mm
(a) parts in tension	16t or 200 mm, whichever is less
(b) parts in compression	12t or 200 mm, whichever is less
(c) tacking fasteners	⎡ 32t or 300 mm, whichever is less
	⎣ 16t or 200 mm, whichever is less for plates exposed to weather

where t is the thickness of the thinner outside plate or angle.

Oversize holes {should not exceed $1.25d$ or $(d + 8)$ mm in diameter, where d is the nominal bolt diameter in mm} and slotted holes are allowable but should not be used often. A slotted hole should not exceed the appropriate hole size in width and $1.33d$ in length, for short slotted hole and $2.5d$ in length, for long slotted hole. Slotted holes are used to accommodate movements in a structure. However, if holes are longer than $2.5d$, shear transfer in the direction of the slot is not admissible even in a friction type connection (see also Section 10.2.1 of code).

Bolt holes reduce the gross cross-sectional area of sections (plates, angles, etc.). The net value is used in the calculations, when the element is in tension (see Chapter 3). As already discussed, bolt holes produce stress concentration, but this is offset by the fact that yield at highly stressed cross section will work-harden, before fracture, resulting in the yield of adjacent cross section also. Whereas, if the member is in compression, then the gross cross section of the member is used in the calculation, because at yield the bolt hole deforms, transferring part of the load to the shank of the bolt or be resisted by bearing.

10.2.7 Spacing and Edge Distance of Bolt Holes

The centre-to-centre distance between individual fasteners in a line, in the direction of load/stress is called the *pitch*. The distance between any two consecutive fasteners in a zigzag pattern of bolts, measured parallel to the direction of load/stress is called a *staggered pitch*. A minimum spacing of 2.5 times the nominal diameter of the fastener is specified in the code to ensure that there is sufficient space to tighten the bolts, prevent overlapping of the washers, and provide adequate resistance to tear-out of the bolts. It also limits any adverse interaction between high bearing stresses due to neighbouring bolts. Similarly, the code specifies maximum pitch values, as given in Table 10.6. These values are specified to prevent buckling of plates in compression between the bolts, to ensure that the bolts act together as a group to resist the loads and to avoid corrosion by ensuring adequate bridging of the paint film between plates. The spacing between adjacent parallel line of fasteners, transverse to the direction of load/stress is called *gauge distance*. The gauge distance as specified in SP-1, published by the Bureau of Indian Standards, is given in Appendix D.

The distance from the centre of a fastener hole to the edge of an element (measured at right angles to the direction of the load) is called the *end* or *edge distance*. The edge distance should be sufficient for bearing capacity and to provide space for the bolt head, washer, and nut. Hence, minimum edge distances are specified in the code and are given in Table 10.6. The maximum edge distance to the nearest line of fasteners from an edge of any unstiffened part should not exceed $12t\varepsilon$ where $\varepsilon = (250/f_y)^{0.5}$ and t is the thickness of the thinner outer plate. (This rule is not applicable to fasteners interconnecting the components of back-to-back tension members.) In corrosive environment, the maximum edge distance should not exceed 40 mm plus $4t$.

10.3 Behaviour of Bolted Joints

Loads are transferred from one member to another by means of the connections between them. A few typical bolted connections are given in Fig. 10.6.

The possible 'limit states' or failure modes that may control the strength of a bolted connection are shown in Fig. 10.7. Thus, any joint may fail in any one of the following modes:
- Shear failure of bolt
- Shear failure of plate
- Bearing failure of bolt
- Bearing failure of plate
- Tensile failure of bolts
- Bending of bolts
- Tensile failure of plate

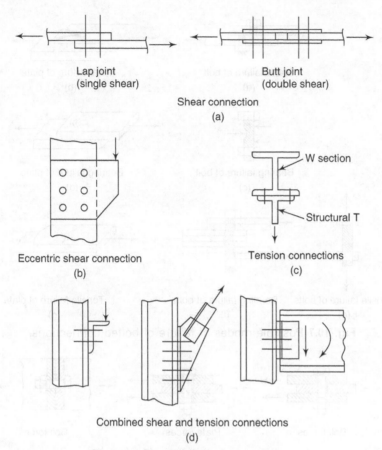

Lap joint
(single shear)

Butt joint
(double shear)

Shear connection
(a)

Eccentric shear connection
(b)

W section

Structural T

Tension connections
(c)

Combined shear and tension connections
(d)

Fig. 10.6 Typical bolted connections

In bearing type connections using black bolts or high-strength bolts, as soon as the applied load overcomes the very small amount of friction at the interface, slip will occur and the force is transferred from one element to another element through bearing of bolts (see Fig. 10.5).

Once the bolts are in bearing, the connection will behave linearly, until yielding takes place at one or more of the following positions (Owens & Cheal, 1989):

- At the net section of the plate(s), under combined tension and flexure
- On the bolt shear plane(s)
- In bearing between the bolt and the side of the hole

The forces acting on the bolt are shown in Fig. 10.8(a).

The response of the connection becomes non-linear after yielding takes place, as plasticity spreads in the presence of strain hardening and failure takes place at one of the critical sections/locations listed above (see Fig. 10.7). The mode of failure and the point of initiation of yielding depends upon the proportions and relative material strength of the components.

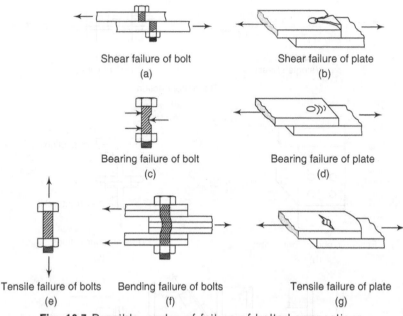

Shear failure of bolt
(a)

Shear failure of plate
(b)

Bearing failure of bolt
(c)

Bearing failure of plate
(d)

Tensile failure of bolts
(e)

Bending failure of bolts
(f)

Tensile failure of plate
(g)

Fig. 10.7 Possible modes of failure of bolted connections.

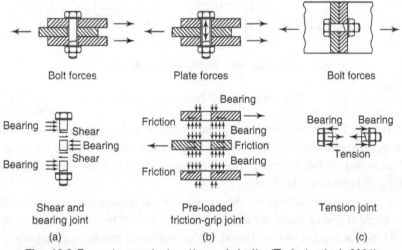

Bolt forces

Plate forces

Bolt forces

Bearing
Shear
Bearing
Shear
Bearing

Friction
Bearing
Friction
Bearing
Friction

Bearing
Bearing
Bearing
Bearing
Tension

Shear and
bearing joint
(a)

Pre-loaded
friction-grip joint
(b)

Tension joint
(c)

Fig. 10.8 Force transmission through bolts (Trahair et al. 2001)

In a multibolt connection, the behaviour is similar except that the more highly loaded bolt starts to yield first, and the connection will become less stiff. At a later stage, due to redistribution of forces, each bolt is loaded to its maximum capacity. However, it is generally assumed that equal size bolts share equally in transferring the external force as shown in Fig. 10.9(b), even during service loads. However, in a long bolted connection, the shear force is not evenly distributed among the bolts, and consequently the bolts at the end of a joint resist the highest amount of shear force, as shown in Fig. 10.9(c). In such joints, the end bolt forces may be so high that it may lead to a progressive joint failure called 'unbuttoning'. If the joint is short, the forces in the bolts will be redistributed by plastic action, and hence the bolts will share the shear force equally.

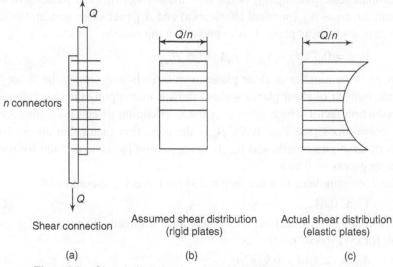

Shear connection | Assumed shear distribution (rigid plates) | Actual shear distribution (elastic plates)

(a)   (b)   (c)

Fig. 10.9 Shear distribution in a long bolted connection

Shear and bearing connections using close tolerance bolts in fitted holes behave in a similar manner to connections with clearance holes, except that the bolt slip is considerably small (see Fig. 10.5). As mentioned earlier, close tolerance bolts are rarely used.

In the case of HSFG bolts, the slip in the bolt will not occur immediately but at a load which overcomes the frictional resistance provided by the pre-load of the bolt (see Fig. 10.5). After slip occurs, the behaviour of the bolt is similar to the normal bolts. In this case also, it is commonly assumed that equal size bolts share the loads equally in transferring the external force.

The flexibility of a connection is determined from the sum of the flexibilities of its different components (bolts, plates, and cleats used in the connection). Note that plates (gusset plates) are comparatively stiff when loaded in their own plane and are considered as rigid. However, when bent out of their plane, they are comparatively flexible. The overall behaviour of any connection should be carefully assessed by determining the force flow through the connection and by synthesizing the responses of the elements to their individual loads (Trahair et al. 2001).

10.4 Design Strength of Ordinary Black Bolts

Expressions for design strength of ordinary black bolts subjected to shear, tension, and bearing forces are given in this section. In addition, when bolts are subjected to tension, there may be additional forces due to flexibility of connections, which are called prying forces. Methods to calculate prying forces and interaction equation for bolts subjected to combined shear and tension forces are also covered. Tension capacity of plates and efficiency of joints are also discussed.

10.4.1 Bearing Bolts in Shear

The nominal capacity of a bolt in shear V_{nsb} [Figs 10.8(a) and 10.7(a)] depends on the ultimate tensile strength f_u of the bolt, the number of shear planes n ($n = n_n + n_s$), and the areas A_{sb} (nominal shank area) and A_{nb} (net shear area at threads) of the bolt in each shear plane. It is expressed in the code as

$$V_{nsb} = 0.577 f_u (n_n A_{nb} + n_s A_{sb}) \beta_{lj} \beta_{lg} \beta_{pkg} \qquad (10.1)$$

where n_n is the number of shear planes with threads intercepting the shear plane, n_s is the number of shear planes without threads intercepting the shear plane, β_{lj} is the reduction factor which allows for the overloading of end bolts that occur in long connections (see Fig. 10.9), β_{lg} is the reduction factor that allows for the effect of large grip length, and β_{pkg} is the reduction factor to account for packing plates in excess of 6 mm.

The code stipulates that the factored shear force V_{sb} should satisfy

$$V_{sb} \leq 0.8 V_{nsb} \qquad (10.2)$$

When the net tensile stress area (shear area at threads) is not given, it may be taken, for ISO thread profile, as

$$A_{nb} = (\pi/4)(d - 0.9382p)^2 \qquad (10.3)$$

where d is the shank or nominal diameter of bolt in mm and p is the pitch of the thread in mm. The net tensile stress area will be approximately 78 – 80% of the gross area.

Reduction factor for long joints When the joint length l_j, of a splice or end connection, in tension or compression, exceeds $15d$ in the direction of load (the joint length is taken as the distance between the first and last rows of the bolts in a joint, measured in the direction of the load transfer), the nominal shear capacity V_{nsb} is multiplied by a reduction factor β_{lj} as shown in Eqn (10.1). This reduction factor is given by

$$\beta_{lj} = 1.075 - l_j/(200\ d) \quad \text{for } 0.75 \leq \beta_{lj} \leq 1.0 \qquad (10.4)$$

This reduction factor should not be applied when the distribution of shear over the length of joint is uniform as in the connection of web of a section to the flanges.

Reduction factor for large grip lengths When the total thickness of connected plates or plies (grip length) l_g exceeds five times the nominal diameter of the bolts, the nominal shear capacity V_{nsb} is multiplied by a reduction factor β_{lg} as shown in Eqn (10.1). This reduction factor is given by

$$\beta_{lg} = 8d/(3d + l_g) \qquad (10.5)$$

The value of β_{lg} calculated using Eqn (10.5) should not be more than β_{lj} given in Eqn (10.4) and the grip length l_g is also restricted to $8d$ by the code.

The reason for the above reduction in strength is that as the grip length increases, the bolt is subjected to greater bending moments due to the shear forces acting on them [see Fig. 10.7(f)].

Reduction factor for packing plates Similar to the grip length, the thickness of packing plates also influence the nominal shear capacity V_{nsb}. Thus, when the packing plates are greater than 6 mm, the shear capacity is multiplied by the reduction factor β_{pkg}. This reduction factor is given by

$$\beta_{pkg} = 1 - 0.0125t_{pkg} \tag{10.6}$$

where t_{pkg} is the thickness of the thicker packing plate in mm.

10.4.2 Bolts in Tension

The nominal capacity of a bolt in tension T_{nb} [Figs 10.8(c) and 10.7(c)] depends on the ultimate tensile strength f_{ub} of the bolt and net tensile stress area A_n (taken as the area at the bottom of the threads) of the bolt, and is given by

$$T_{nb} = 0.90f_{ub}A_n < 1.136\,f_{yb}A_{sb} \tag{10.7}$$

where A_{sb} is the shank area of the bolt, and f_{yb} is the yield stress of the bolt. IS 800 stipulates that the factored tension force T_b should satisfy

$$T_b \leq 0.8\,T_{nb} \tag{10.8}$$

If any of the connecting plates is sufficiently flexible, then additional prying forces may be induced in the bolt (see Section 10.4.4 for the details).

10.4.3 Bolts in Bearing

When an ordinary bolt is subjected to shear forces, it comes into contact with the plates, after the slip occurs. The bearing limit state relates to deformation around a bolt hole, as shown in Fig. 10.7(d) (bearing failure for a large end distance). A shear tear-out failure (also called end bearing failure) as shown in Fig. 10.7(b) occurs when the end distance is small. Bearing failure in bolts {see Fig. 10.7(c)} is possible only by using low strength bolt with very high grade plates, which will not occur in practice.

The nominal bearing strength of the bolt V_{npb} is given by

$$V_{npb} = 2.5k_b dt f_u \tag{10.9}$$

where f_u is the ultimate tensile stress of the plate in MPa, d is the nominal diameter of the bolt in mm, and t is the summation of the thicknesses of the connected plates experiencing bearing stress in the same direction. (If the bolts are countersunk, then t is equal to the thickness of the plate minus one half of the depth of counter sinking.)

k_b is smaller of $e/(3d_h)$, $p/(3d_h) - 0.25$, f_{ub}/f_u and 1.0 where f_{ub} is the ultimate tensile stress of the bolt, e is the end distance, p is the pitch of the fastener along bearing direction, and d_h is the diameter of bolt hole. V_{npb} should be multiplied by a factor 0.7 for over size or short slotted holes and by 0.5 for long slotted holes. The factor k_b takes care of inadequate edge distance or pitch and also prevents bearing failure of bolts. If we adopt a minimum edge distance of 1.5 × bolt hole diameter and a minimum pitch of 2.5 × diameter of bolt, k_b will be approximately 0.50.

The code stipulates that the bolt bearing on any plate subjected to a factored shear force V_{sb}, should satisfy

$$V_{sb} \leq 0.8 \, V_{npb} \tag{10.10}$$

Both bolts and plates are subject to significant triaxial containment. Due to this, bearing behaviour of plates is influenced by the proximity of neighbouring holes or boundary (edge distance). Away from holes or boundaries, significant hole elongations commence at a nominal stress of $2f_u$ but failure will occur only at about $3f_u$. Though the presence of threads in the bearing zone increase the flexibility, they do not reduce the bearing strength. Similarly, bolt material often sustains bearing stresses in excess of twice the ultimate tensile . It is not generally necessary to consider bolt bearing in design (Owens & Cheal 1989). Bearing in the thinner plate will control for plate thicknesses up to about one half of the bolt diameter.

Equations (10.1), (10.7), and (10.9), which express the design shear, tensile strength, and bearing strength of a bolt, respectively, can be presented in the form of tables to avoid repeating these calculations. Tables are presented in Appendix D which will aid the designer while designing joints using ordinary bolts.

10.4.4 Prying Forces

Moment resisting beam-to-column connections often contain regions in which the bolts will be required to transfer load by direct tension, such as the upper bolts in the end plate connection as shown in Fig. 10.10. In the design of such connections, we should consider an additional force induced in the bolts as a result of so-called 'prying action' (Douty & McGuire 1965, Agerskov 1979, Holmes & Martin 1983, Subramanian 1984). These additional prying forces induced in the bolts are mainly due to the flexibility of connected plates (see Fig. 10.11). Thus, in a simple T-stub connection as shown in Fig. 10.11, the prying force will develop only when the ends of the flanges are in contact due to the external load, as shown in Figs 10.11(b) and (c). The plastic hinges do not always form before bolt failure. The development of prying force as the external load is raised from zero to the maximum in a T-stub connection as shown in Fig. 10.12.

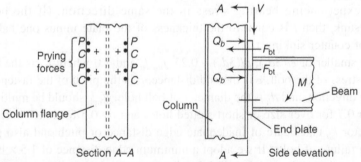

Fig. 10.10 Prying forces in a beam-to-column connection

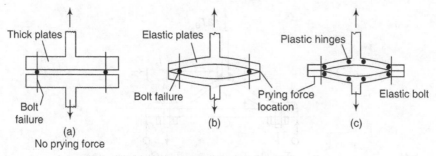

Fig. 10.11 Failure modes due to prying forces

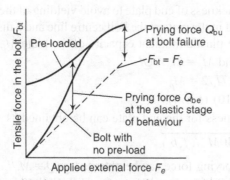

Fig. 10.12 Relationship between external force and bolt force for a T-stub

Several researchers have studied this problem and proposed equations to calculate the prying force developed in the bolt (Astanesh 1985; Kulak et al. 1987; Owens & Cheal 1989; Thornton 1985, 1992). IS 800 has adopted the equation proposed by Owens & Cheal (1989) and the additional force Q in the bolt due to prying action (see Fig. 10.13)

$$Q = \{l_v/2l_e\}\ [T_e - \beta\gamma f_o\ b_e t^4/(27l_e l_v^2)] \qquad (10.11)$$

where l_v is the distance from the bolt centre line to the toe of the fillet weld or to half the root radius of a rolled section in mm and l_e is the distance between prying force and bolt centre line in mm.

This distance is taken as the minimum of either the end distance or the value given by

$$l_e = 1.1t\sqrt{(\beta f_o/f_y)} \qquad (10.12)$$

where $\beta = 2$ for non-tensioned bolt and 1 for pre-tensioned bolt, $\gamma = 1.5$, b_e is the effective width of flange per pair of bolts in mm, f_o is the proof stress which must be consistent with the units of T_e (kN or kN/mm^2), and t is the thickness of the end plate in mm.

The second term in Eqn (10.11) is usually relatively small and hence may be neglected to yield the formula

$$Q = T_e l_v/(2l_e) \qquad (10.13)$$

This formula is obtained if plastic hinges are assumed at the bolt line and the root, that is, when minimum flange thickness is used in design.

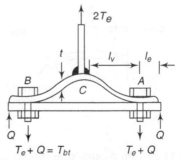

Fig. 10.13 Forces acting in the elastic stage for prying force theory

The maximum thickness of end plate to avoid yielding of the plate is obtained by equating the moment in the plate at the bolt centre line and a distance l_v from it [see Fig. 10.11(c)] to the plastic moment capacity of the plate M_p. Thus we have,

$$M_A = Ql_e \text{ and } M_c = Tl_v - Ql_e \tag{10.14}$$
$$M_A = M_c = Tl_v/2 = M_p \tag{10.15}$$

Taking $M_p = (f_y/1.10)(b_e t^2/4)$ (10.16)

The minimum thickness for the end plate can be obtained as

$$t_{min} = \sqrt{4.40 M_p/(f_y b_e)} \tag{10.17}$$

The corresponding prying force can be obtained as $Q = M_p/l_e$. If the total force in the bolt $(T + Q)$ exceeds the tensile capacity of the bolt, then the thickness of the end plate has to be increased. Example 10.2 illustrates the effect of prying forces.

10.4.5 Bolts with Shear and Tension

Where bolts are subjected to both tension and shear, as in the connections shown in Fig. 10.14, then their combined effect may be conveniently assessed from a suitable interaction diagram. Tests on bolts under shear and tension showed elliptical or circular interaction curves for the ultimate strength of bolts (Chesson et al. 1965, Khalil & Ho 1979). The following equation for the circular interaction curve has been proposed in the code.

$$(V/V_{sd})^2 + (T_e/T_{nd})^2 \le 1.0 \tag{10.18}$$

where V is the applied factored shear, V_{sd} is the design shear capacity, T_e is the externally applied factored tension, and T_{nd} is the design tension capacity.

10.4.6 Efficiency of a Joint

Holes are drilled in the plates for the connection with bolts, hence the original strength of the full section is reduced. The joint which causes minimum reduction in strength is said to be more efficient. Thus, for better efficiency, a section should have the least number of holes at the critical section. The efficiency, expressed in percentage, is the ratio of the actual strength of the connection to the gross strength of the connected members. It can also be expressed as

Efficiency = (Strength of joint per pitch length/Strength of solid plate per pitch length) × 100 (10.19)

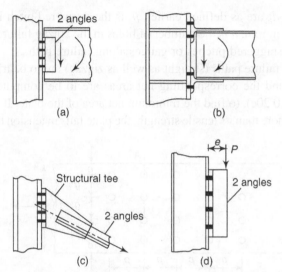

Fig. 10.14 Typical combined shear and tension connections.

10.4.7 Tension Capacity of Plate

The plate in a joint may fail in tension through the weakest section due to the holes. The holes may be arranged in the longitudinal direction of the plate, so that the number of holes is equal in all the rows across the width [see Fig. 10.15(a)], or staggered so that the number of holes across the width is reduced. In the first case, the plate will fail across the weakest section, whereas in the second case the failure is along a zigzag pattern. The tension capacity T_{nd} of the plate is expressed as

$$T_{nd} = 0.72 f_u A_n \qquad (10.20\text{a})$$

where f_u is the ultimate stress of material in MPa, and A_n is the net effective area of the plate in mm^2.

Thus, the load carrying capacity of the plate depends on the net effective area of the plate, which in turn depends on the arrangement of the holes. If the holes are not-staggered, the net area A_n can be easily computed as

$$A_n = (b - n d_h) t \qquad (10.20\text{b})$$

where b is the width of the plate in mm, n is the number of holes along the width b, perpendicular to the direction of load, d_h is the diameter of the hole in mm, and t is the thickness of the plate in mm.

Based on experimental evidence, a simplified empirical relationship has been proposed by Cochrane (1922) for staggered rows of holes [see Fig. 10.15 (b)] as

$$A_n = \left[b - n d_h + \sum_{i=1}^{m} p_{si}^2/(4 g_i) \right] t \qquad (10.20\text{c})$$

where A_n, b, t, d_h are as defined earlier, p_s is the staggered pitch in mm, g is the gauge distance in mm, n is the number of holes in the zigzag failure path, and m is the number of staggered pitches or gauges along failure path.

All possible failure paths (straight as well as zigzag) are to be tried as shown in Fig. 10.15(b) and the corresponding net areas are to be computed as per Eqns (10.20b) and (10.20c), to find the minimum net area of the plate. If the tensile load on the plate is more than its tensile strength, the plate fails in tension through rupture.

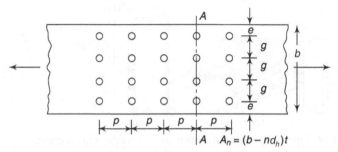

$$A_n = (b - nd_h)t$$

(a) Chain of holes in rows

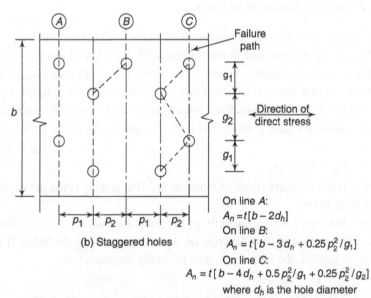

On line A:
$$A_n = t[b - 2d_h]$$
On line B:
$$A_n = t[b - 3d_h + 0.25 p_2^2/g_1]$$
On line C:
$$A_n = t[b - 4d_h + 0.5 p_2^2/g_1 + 0.25 p_2^2/g_2]$$
where d_h is the hole diameter

(b) Staggered holes

Fig. 10.15 Tension capacity of plates

10.5 Design Strength of High Strength Friction Grip Bolts

As we have seen already, HSFG bolts are used when forces are large, where space for the connection is limited, where erection cost can be reduced by using fewer bolts or where the structures are subjected to dynamic loads. Thus, they provide 'rigid' fatigue resistant joints.

It may be noted that HSFG bolts may be subdivided into parallel shank and waisted shank types. A parallel shank bolt, which is the most commonly used (and discussed in this section), is designed not to slip at serviceability load; but slips into bearing at ultimate load. Thus, only when the externally applied load exceeds the frictional resistance between the plates, the plates slip and the bolts bear against the bolt holes. A waisted shank bolt has higher strength and is designed not to slip both at service and ultimate load. Hence, waisted shank HSFG bolts are more rigid at ultimate load and need not be checked for bearing or long joint capacity (BS 5950, Martin & Purkiss 1992).

10.5.1 Slip Resistance

As mentioned earlier, the initial pretension in bolt develops clamping forces at the interface of elements being joined [see Fig. 10.8(b)]. The frictional resistance to slip between the plate surfaces subjected to clamping force, opposes slip due to externally applied shear.

The design slip resistance or nominal shear capacity of a bolt V_{nsf} of the parallel shank and waisted shank friction grip bolts is given by the code as

$$V_{nsf} = \mu_f n_e K_h F_o \qquad (10.21)$$

where μ_f is the coefficient of friction (called as slip factor) as specified in Table 10.7 ($\mu_f \leq 0.55$); n_e is the number of effective interfaces offering frictional resistance to slip; $K_h = 1.0$ for fasteners in clearance holes, 0.85 for fasteners in oversized and short slotted holes and for fasteners in long slotted holes loaded perpendicular to the slot, and 0.7 for fasteners in long slotted holes loaded parallel to the slot; F_o is the minimum bolt tension (proof load) at installation and may be taken as $A_{nb}f_o$, A_{nb} is the net area of the bolt at the threads; f_o is the proof stress, taken as $0.7f_{ub}$; and f_{ub} is the ultimate tensile stress of bolt.

Table 10.7 Typical average values for coefficient of friction (μ_f)

Treatment of surface	Coefficient of friction (μ_f)	Treatment of surface	Coefficient of friction (μ_f)
Surfaces not treated	0.20	Surfaces blasted with shot or grit and painted with ethylzinc silicate coat (thickness 60–80 μm)	0.30
Surfaces blasted with shot or grit with any loose rust removed, no pitting	0.50	Surfaces blasted with shot or grit and painted with alkalizinc silicate coat (thickness 60–80 μm)	0.30
Surfaces blasted with shot or grit and hot-dip galvanized or red lead painted surface	0.10	Surface blasted with shot or grit and spray-metallized with aluminium (thickness > 50 μm)	0.50
Surfaces blasted with shot or grit and spray-metallized with zinc (thickness 50–70 μm)	0.25	Clean mill scale	0.33

IS 800 stipulates that for a bolt subjected only to a factored design force V_{sf}, in the interface of connections at which slip cannot be tolerated, should satisfy the following

$$V_{sf} \le V_{nsf}/\gamma_{mf} \qquad (10.22)$$

where $\gamma_{mf} = 1.10$ if slip resistance is designed at service load and $\gamma_{mf} = 1.25$ if slip resistance is designed at ultimate load.

It may be noted that the resistance of a friction grip connection to slip in service is a serviceability criterion, but for ease of use it is presented in the code in a modified form, suitable for checking under factored loads.

10.5.2 Long Joints

Similar to black bolts, the design slip resistance V_{nsf} for parallel shank friction grip bolts is reduced for long joints by a factor β_{lj} given by Eqn (10.4)

It should be understood that overcoming slip does not imply that a failure mode has been reached. However, when connections are subjected to stress reversal, there is great concern regarding any slip at service load. Repeated loading may introduce fatigue concerns, if slip is occurring, especially when oversized or slotted holes are used. Generally, slip resistance governs the number of bolts used in slip-critical connections, rather than strength in bearing or shear. However, the bearing related equations for spacing of fasteners and end distance will result in smaller spacings and end distances.

10.5.3 Bearing Resistance

As a parallel shank friction grip bolt slips into bearing at ultimate limit state when subject to shear forces, the bearing stresses between the bolt and the plate need to be checked. The bolt may deform due to high local bearing stresses between the bolt and plate and the design bearing capacity of the bolt V_{npb} is obtained by using Eqn (10.9). The code stipulates that the factored shear force V_{sf} should satisfy Eqn (10.10).

Note that while checking black bolts, the ultimate tensile capacity of the bolt or the plate, whichever is smaller is used. Since the bearing strength of HSFG bolts will be greater than the plates, no check on bearing strength of bolt is necessary.

An alternative mode of failure is that of the bolt shearing through the end of the plate as shown in Fig. 10.7(b). This may be controlled by specifying the end distances and pitches. The block shear resistance of the edge distance due to bearing force should also be checked for the connection (see Section 2.5.3 and by the k_b factor of Eqn (10.9) for details).

10.5.4 Tension Resistance

The design tensile strength of parallel shank and waisted shank friction grip bolts is similar to that of black bolts (see Section 10.4.2) and is given by Eqn (10.7).

As per the code, the HSFG bolt subjected to a factored tension force T_b should satisfy Eqn (10.8).

The effect of the prying force Q has been shown in Fig. 10.12. When the external load is applied, part of the load (approximately 10%) of the load is equilibrated by the increase in bolt force. The balance of the force is equilibrated by the reduction in contact between the plates. This process continues and the contact between the plates is maintained until the contact force due to pre-tensioning is reduced to zero by the externally applied load. Normally, the design is done such that the externally applied tension does not exceed this level. After the external force exceeds this level, the behaviour of the bolt under tension is exactly similar to that of a bearing type of bolt. Eqns (10.11), (10.12), and (10.17) may be used in the calculation of HSFG bolts subjected to prying forces.

10.5.5 Combined Shear and Tension

The interaction curve suggested for combined shear and tension for HSFG bolts (for which slip in the serviceability limit state is limited) is similar to that of black bolts and is given below

$$(V_{sf}/V_{sdf})^2 + (T_f/T_{ndf})^2 \leq 1.0 \tag{10.23}$$

where V_{sf} is the applied shear at service load, V_{sdf} is the design shear strength, T_f is the externally applied tension at service load, and T_{ndf} is the design tension strength.

Since slip resistance is a service load consideration, the numerator terms of Eqn (10.23) are service loads T and V (tension and shear per bolt). It should be observed that any external tension will produce a corresponding reduction in the clamping force between the plies. Until the external load on a bolt exceeds the pre-compression force between the pieces, the tension force in the bolt will not change significantly from its initial tension (see Fig. 10.12).

The design shear, bearing, and tensile resistance of HSFG bolts can be presented in the form of a table, as shown in Appendix D, to avoid repeated calculations. Figure 10.16 shows a bracing member connected to the other members of a bridge structure using guest plates and HSFG bolts.

10.5.6 Block Shear Failure

As a result of some research work carried out in USA, it was found that angle, gusset plate, and coped beams connections may fail as a result of block shear (Kulak & Grondin 2000). Failure occurs in shear at a row of bolt holes parallel to the applied loads, accompanied by tensile rupture along a perpendicular face. This type of failure results in a block of material being torn out by the applied shear force as shown in Fig. 10.17. The block shear strength T_{db} of a connection is taken as the smaller of

$$T_{db1} = [0.525A_{vg}f_y + 0.72f_u A_{tn}] \tag{10.24a}$$

and

$$T_{db2} = [0.416f_u A_{vn} + 0.909\, f_y A_{tg}] \tag{10.24b}$$

484   *Steel Structures: Design and Practice*

Fig. 10.16 Example of a connection using HSFG bolts

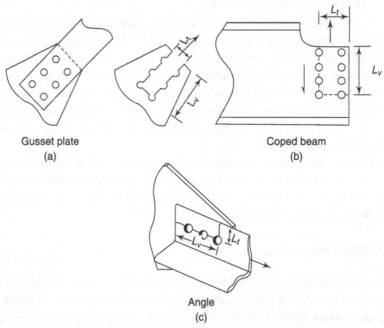

Gusset plate	Coped beam
(a)	(b)

Angle

(c)

Fig. 10.17 Examples of block shear failure

where A_{vg} and A_{vn} are the minimum gross and net area, respectively, in shear along a line of transmitted force (along L_v in Fig. 10.17); A_{tg} and A_{tn} are the minimum gross and net area, respectively, in tension from the hole to the toe of the angle or next last row of bolt in gusset plates (along L_t in Fig. 10.17); and f_u and f_y are the ultimate and yield stress of the material, respectively.

10.6 Simple Connections

In many cases, a connection is required to transmit a force only and there may not be any moment acting on the group of connectors, even though the connection may be capable of transmitting some amount of moment. Such a connection is referred to as a *simple, force, pinned*, or *flexible connection*.

As already shown in Fig. 10.8, two types of load transfers occur in these connections. In the first, the force acts in the connection plane (formed by the interface between the two connected plates) and the fasteners between these plates act in shear [Fig. 10.8(a)]. In the second type, the force acts out of the plane of the connection and the fasteners act in tension as shown in Fig. 10.8(c). In practice there will always be some eccentricity and the moment due to this small eccentricity is ignored. The different types of simple connections found in steel structures may be classified as follows:

- Lap and butt joints
- Truss joint connections
- Connections at beam column junctions
  - Seat angle connection
  - Web angle connection
  - Stiffened seat angle connection
  - Header plate connection
- Tension and flange splices

Let us now discuss these connections briefly.

10.6.1 Lap and Butt Joints

Lap and butt joints are often used to connect plates or members composed of plate elements. Though lap joints are the simplest, they result in eccentricity of the applied loads. Butt joints on the other hand eliminate eccentricity at the connection.

10.6.1.1 Lap joints

When two members which are to be connected are simply overlapped and connected together by means of bolts or welds, the joint is called a lap joint [see Figs 10.18(a)–(d)]. A single bolted lap joint and a double bolted lap joint are shown in Figs 10.18(b) and 10.18(c), respectively. The drawback of such a lap joint is that the centre of gravity of load in one member and the centre of gravity of load in the second member do not coincide and hence an eccentricity as shown in Fig. 10.18(d) is created. Due to this a couple $P \times e$ is formed, which causes undesirable bending in the connection leading to failure of bolts in tension. To minimize the effect of bending in a lap joint, at least two bolts in a line should be provided. Moreover, due to the eccentricity, the stresses are distributed unevenly across the contact area between the bolts and members to be connected. Hence, the use of lap joints is not often recommended. The design of lap joint is illustrated in Examples 10.1, 10.3, 10.4, and 10.5.

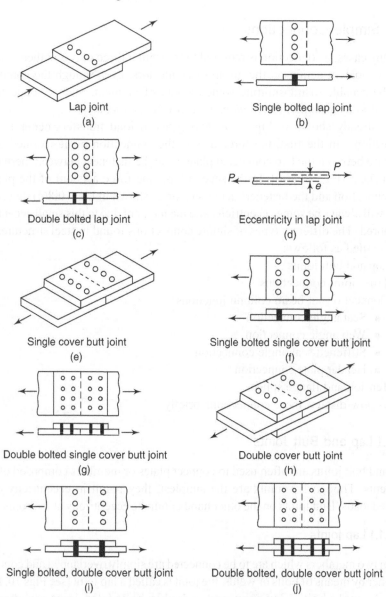

Lap joint
(a)

Single bolted lap joint
(b)

Double bolted lap joint
(c)

Eccentricity in lap joint
(d)

Single cover butt joint
(e)

Single bolted single cover butt joint
(f)

Double bolted single cover butt joint
(g)

Double cover butt joint
(h)

Single bolted, double cover butt joint
(i)

Double bolted, double cover butt joint
(j)

Fig. 10.18 Lap and butt joints

10.6.1.2 Butt Joints

In butt joints, the members to be connected are placed against each other and are bolted or welded together through the use of additional plates, called *cover plates*. The cover plates may be provided on either one or both sides of the connection as shown in Figs 10.18(e)–(j). If the cover plate is provided on one side of the joint only it is called as a *single cover butt joint* [see Figs 10.18(e)–(g)] and when provided on both sides of the joint, it is called as a *double cover butt joint* [see Figs

10.18(i) and (j)]. As in the case of lap joints, in the case of single cover butt joint also, some eccentricity is involved. Double cover butt joints are preferred due to the following reasons.

(a) In double cover butt joints, the eccentricity of load is eliminated and hence there is no bending in the bolts.

(b) In double cover butt joints, the total shear force to be transmitted is split into two parts and hence the bolts are in double shear. Whereas in the case of lap joint or single cover butt joint, the force acts in one plane of the bolt and hence the bolts are said to be in single shear. Thus, the shear carrying capacity of bolts in a double cover butt joint is double that of a bolt in a lap joint or single cover butt joint. The design of butt joint is illustrated in Examples 10.6 – 10.8.

10.6.2 Truss Joint Connections

Even though truss joints are assumed to be hinged, the detailing using gusset plates and multiple fasteners or welding does not represent hinged condition. However, in practice the secondary moment associated with such a joint is disregarded, unless the loading is cyclic.

Typical truss joints are shown in Fig. 10.19, which use gusset plates in order to accommodate the bolts. These gusset plates are connected to the members of the truss by means of bolts or welds using lap or butt joints depending on whether one or two angles are used for the member. Note that in Fig. 10.19(c), end plates are used to facilitate the connection of shop welded parts of a truss at the site using bolts. As stated earlier, the lateral dimensions of a gusset plate are determined principally by the fastener requirements of the members. However, the determination of the thickness of the gusset plate poses problems. The gusset plates are not easy to analyse because of their odd shape and the discrete introduction of forces through fasteners. Moreover, the elementary formula based on beam theory is valid only for beams whose span is more than twice the depth and at cross sections not closer to concentrated loads than about half the depth. The ordinary gusset plates fall considerably short of these requirements and hence the results obtained by the application of beam formulae are questionable and may be misleading (Gaylord et al. 1992).

Based on the experimental study on gusset plates commonly found in Warren type bridges, Whitmore (1952) proposed that the maximum direct stress in a gusset plate from an individual member could be estimated adequately by ensuring that the member force is distributed uniformly over an effective area given by 30° dispersion from the outer row of fasteners as shown in Fig. 10.20(a) [similar dispersion can be assumed for welded connections and are shown in Figs 10.20(b) and 10.20(c)]. Whitmore's proposal is valid for stresses in gusset plates loaded in the elastic range.

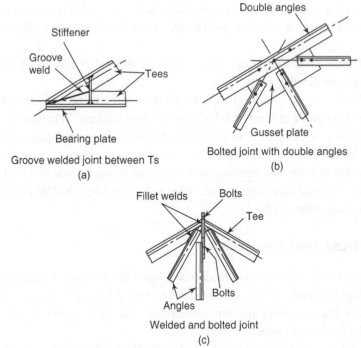

Fig. 10.19 Typical joints in trusses

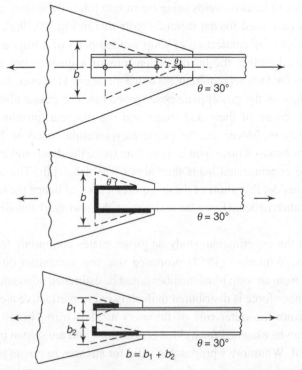

Fig. 10.20 Effective width of gusset plates as per Whitmore (1952)

The behaviour of gusset plates connections in the inelastic range was studied by Hardash & Bjorhovde (1985), who proposed a block shear model [see Eqn (10.24)] to predict the ultimate capacity of gusset plate connections in tension. Compressive stresses may develop parallel to and at the edge of the gusset plates. Therefore, the width of the top edge must not be too large, compared with the thickness, or the plate may buckle (Gaylord et al. 1992; Cheng et al. 2000). This kind of buckling, called *local buckling*, may be prevented if the unsupported edge of a gusset plate is restricted to 42ε times the thickness (Gaylord et al. 1992), where $\varepsilon = (250/f_y)^{0.5}$.

Usually, a gusset plate thickness slightly higher than the thickness of the angle is chosen to ensure safety.

Trusses may be single plane trusses in which the members are connected on the same side of the gusset plate or double plane trusses in which the members are connected to both sides of the gusset plate.

The centre line of all the members connected at a joint, should meet at a point as shown in Fig. 10.19 and 10.21(a), in order to avoid eccentricity. If the bolt centre lines do not intersect, then eccentricity should be taken into account in the calculations. In some cases, eccentricities may be unavoidable [see Fig. 10.21(b)], in which case, the gusset plates should be designed to be strong enough to resist and transmit the force without buckling. The bolts should also be designed to resist the moments due to in-plane eccentricities. If out-of-plane instability is encountered, splice plate as shown in Fig. 10.21(a) could be used. In truss connections, the clearance of the members is usually kept at 2 to 5 mm. Example 10.12 shows the calculations required for the design of a truss joint.

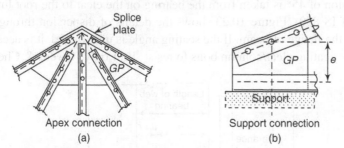

Apex connection   Support connection
(a)   (b)

Fig. 10.21 Apex and support connections in a truss

10.6.3 Clip and Seating Angle Connections

Clip and seating angle connection also called *angle seat connection* or *seat angle connection* [Figs 10.2(b) and 10.22] transfers reaction from the beam to the column through the angle seat. The top cleat (called clip angle) is provided for lateral or torsional restraint (support) to the top flange of the beam and is bolted to the top flange. This is important for lateral stability. The angle seat may be bolted or fillet welded to the column and the angles are assumed to be rigid compared to the bolts.

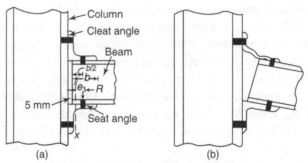

Fig. 10.22 Angle seat connections

The beam reaction is transferred by bearing, shear, and bending of the horizontal leg of the bottom angle, by vertical shear through the fasteners, and by horizontal force in the fasteners between the vertical leg and the column. The beam is designed as simply supported and the column is designed for the eccentric beam reaction. The clearance between the end of the beam and flanges of the column should be minimum (between 2 and 5 mm) so that maximum bearing length is available. During erection, the beam is placed over the bottom cleat and the top cleat is bolted. This connection, because of the close depth tolerances required at the top and bottom of the beam, results in some difficulties during erection. It is important to include tolerance in the design of seating cleats, since small dimensional changes can have a large effect on the capacity.

Minimum length of bearing at the edge of root radius

$$= \text{Reaction}/(\text{web thickness} \times \text{design strength of web}) \qquad (10.25)$$

A dispersion of 45° is taken from the bearing on the cleat to the root line (clause 8.7.1.3 of IS 800). Figure 10.23 shows the details of dispersion through seating cleat and the flange of beam. If the seating angle is unstiffened, it is acceptable to design the seating cleat/column bolts to resist shear only (Owens & Cheal 1989).

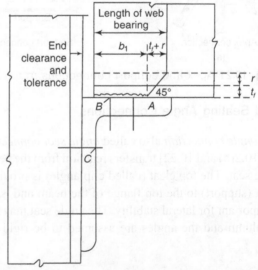

Fig. 10.23 Details of dispersion through seating angle and flange of beam

The outstanding leg of the seat angle must be stiffened when the reaction from the beam is too large or when the seating leg is not able to provide the bearing area. The beam web may also need to be stiffened to resist shear and bearing (see Chapter 6). When large reactions from the beam occur, more number of bolts may have to be provided to connect the seat angle to the column (normally four bolts are provided). In such a case, an additional angle called stiffener angle is provided. Packings equal to the thickness of seat angle are used between the column flange and the stiffener angle (see Fig. 10.24). In general, a pair of stiffener angles is provided and the outstanding legs are tack bolted. Note that when the seating angle is stiffened, the horizontal leg of the angle can no longer flex, and the bolts connecting the column flange to the stiffener angles should be designed to resist the corresponding bending moment (arising due to the eccentricity between the centre of the bearing length and the column face) as well as shear. The eccentricity should be taken as the outstand of the horizontal leg.

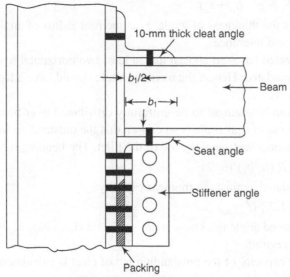

Fig. 10.24 Stiffened seat connection

The clip angle at the top of the beam can only resist shear by flexure of the legs [see Fig. 10.22(b)]. This is much more flexible than the bottom seating angle and hence should not be assumed to contribute to the shear resistance of the connection.

10.6.4 Design of Un-stiffened and Stiffened Seat Connection

The various steps involved in the design of unstiffened and stiffened seat connections are explained in this section.

Unstiffened seat connection The design of unstiffened seating angle connection, consists of the following steps:
1. The selection of seat angle having a length equal to the 'width of the beam.

2. The length of the outstanding leg of the seat angle is calculated on the basis of web crippling of the beam. The seat leg length is kept more than the calculated bearing length given by

$$b = [R/t_w \ (f_{yw}/\gamma_{m0})] \tag{10.26}$$

where R is the reaction from the beam, t_w is the thickness of the web of the beam, f_{yw} is the yield strength of the web of the beam, and γ_{m0} is the partial safety factor for material = 1.10.

3. A dispersion of 45° is taken from the bearing on the cleat to the root line [Note: The dispersion at a slope of 1 in 2.5 as specified in IS 800 clause 8.7.4 is not applicable for this purpose (Owens & Cheal 1989)]. Thus, the length of bearing on the cleat

$$b_1 = b - (t_f + r_b) \tag{10.27}$$

where t_f is the thickness of beam flange and r_b is the root radius of beam flange. The distance of end bearing on cleat to root angle

$$b_2 = b_1 + g - (t_a + r_a) \tag{10.28}$$

where t_a is the thickness of angle, r_a is the root radius of angle, and g is the clearance and tolerance.

4. The connected leg is so chosen that at least two horizontal rows of bolts can be accommodated. Hence, the assumed angle should have a leg length of 100 mm or more.

5. The reaction is assumed to be uniformly distributed over bearing length b_1. The thickness of seat angle is so chosen that the outstanding leg does not fail in bending on a section at the toe of the fillet. The bending moment

$$M_u = R \ (b_2/b_1) \ (b_2/2) \tag{10.29}$$

This is equated against the moment capacity

$$M_d = 1.2Z \ (f_y/\gamma_{m0}) \tag{10.30}$$

If the assumed angle thickness is not sufficient (i.e. when $M_d < M_u$), then the section is revised.

6. The shear capacity of the outstanding leg of cleat is calculated as

$$V_{dp} = wtf_y/(\sqrt{3}\gamma_{m0}) \tag{10.31}$$

The above value should be more than the reaction from the beam.

7. The number of bolts is calculated as below

$$n = \text{Reaction/Strength of bolt} \tag{10.32}$$

8. A cleat angle of nominal size is provided on the top flange of the beam, connected by two bolts on each of its legs.

The design of a seat angle connection is presented in Example 10.13.

Stiffened Seat Connection

As discussed in Section 10.6.3, the behaviour of a stiffened seat connection is different than that of an unstiffened seat connection. Hence, the design of a stiffened seat connection consists of the following steps.

1. Assume the size of seat angle on the basis of bearing length as described in unstiffened seat connection.

2. The outstanding leg of the stiffener angle must provide the required bearing area. The outstanding leg must not exceed 14ε times the thickness, where $\varepsilon = (250/f_y)^{0.5}$ to avoid local buckling (clause 8.7.1.2). The required bearing area is calculated as

$$A_{br} = R/(f_y/\gamma_{m0}) \tag{10.33}$$

where R is the reaction from beam, f_y is the yield strength, and γ_{m0} is the partial safety factor for material, taken as 1.10. The thickness of the stiffener angle should be more than the thickness of the web of the beam being supported.

3. Due to the stiffener, the seat angle is not flexible and the bolts in the connecting leg are subjected to moments in addition to shear. This connection behaves like a bracket connection (see Section 10.7). It can be assumed that the reaction from the beam acts at the middle of the outstanding leg of the angle.

4. From step 3, the eccentricity, the bending moment, and the tension acting in the critical bolts are computed, similar to the bracket connection.

5. Using the interaction formula [Eqn (10.18)], the critical bolt is checked.

6. A nominal angle is provided at the top of the beam and is connected with two nominal size bolts on each leg of the cleat angle.

The calculations for a typical stiffened seat connection are provided in Example 10.14.

10.6.5 Web Angle Connection

A *web cleat connection*, also known as *web cleat* or *angle cleat connection* [see Fig. 10.2(c) and 10.2 (d)] is used to transfer beam reaction through web angles either to the flange or to the web of the supporting member. One or two angle cleats may be used, and bolted to the beam web and to the supporting members. Double web cleat connection is preferred over single sided web cleat connection. The flanges may be notched or coped [see Fig. 10.2(d)], if required. The clearance between the beam and the supporting member is kept at 2-5 mm to provide sufficient end distance in the web. This connection also has very little moment capacity, as there is significant flexibility in the angle legs connected to the supporting member and the bolted connection to the beam web (see Fig. 10.25).

The beam reaction is transferred by shear and bearing from the web of the beam to the web bolts and to the angle cleats. These are then transferred by the cleat angle to the bolts at the junction of supporting member, and then to the supporting member mainly by shear and also by tension and compression. The beam is designed as a simply supported beam and the supporting member is designed for the eccentric beam reaction.

The length of the web angle is decided based on the number of bolts and the pitch and is normally not less than 0.6 times the depth of the beam. (The maximum length is the clear depth of the web between the fillets, which may be approximately taken as 0.75 times the depth of the beam.) If the depth of the web cleat is less than about 0.6 times the depth of beam web, then the bolts may be designed for shear forces only. Otherwise, assuming pure shear transfer at the column face, the bolts connecting the cleats to the beam web should be designed for the moment due to eccentricity. Normally, the connection is made in the compression zone of the beam in order to provide lateral restraint to the compression flange. A minimum thickness of 8 mm is chosen for beams having a depth of up to 450 mm. Note that thicker angles will decrease the flexibility of the joint and may introduce end moments. The joint may be made flexible by making the gauge distance [see Fig. 10.25(c)] on the column flange as large as possible, which is usually kept in the range of 100 mm to 140 mm. The web angle connection may be combined with seat angle connection, when the reaction from the beam is heavy.

It is preferable to use friction grip fasteners for the cleat-to-column connection. If ordinary bolts are used, the combined effect of the eccentric load and the clearances in the bolt holes could lead to an unacceptable twist at the end of the beam (Owens & Cheal 1989). The design calculations for a typical web angle conncetion are shown in Example 10.15.

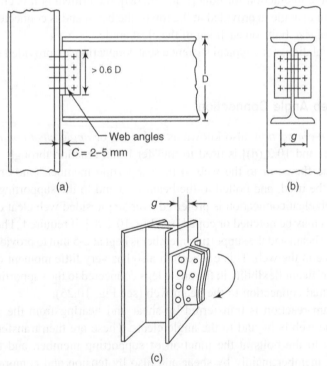

Fig. 10.25 Web angle connection and flexibility of web angle

10.6.6 Flexible End Plate Connection

Flexible end plate connection, also called as *header plate connection* [Fig. 10.2 (e)] consists of an end plate connected to the web of the beam, at the beam ends, by fillet welds. This plate is, in turn, connected to the column flange or web by means of bolts. This connection behaviour is similar to the legs of web angles connected to the column flange. However, it is difficult to ensure that the connection has sufficient flexibility/ductility to accommodate the end rotation of the beam. In practice, the flexibility may be achieved by limiting the thickness of the plate ($t < d/3$ for grade 4.6 bolts and $t < d/2$ for grade 8.8 bolts) and by positioning the bolts not too close to the web and flange of the beam, and by suitably selecting pitch and gauge length of bolts. A empirical thickness of 8 mm is used for beams having a depth of 450 mm and 10 mm for beams having greater heights. The end plate depth is kept to a minimum to reduce end moments. The length 'a' in Fig. 10.26

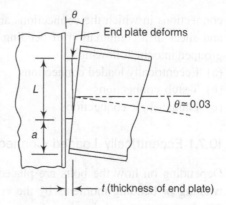

Fig. 10.26 Header plate connection (connecting bolts are not shown for clarity)

is kept less than $30t$ (where t is the thickness of end plate) so that the web of the beam does not come into contact with the column. The flanges of the beam may be notched or coped, if required. In these cases, the length of the cut portion is limited to $20t_w$ to prevent buckling of the web, where t_w is the thickness of the web.

The beam is designed for zero end moment, while the end plate augments the web shear and bearing capacity of the beams. The supporting member (e.g., column) has to be designed for the eccentric beam reaction. The reaction from the beam is transferred by weld shear to the end plate, by shear and bearing to the bolts, and by shear and bearing to the supporting member.

The weld is always referred by its leg size and the design strength of a shop-site welded fillet weld may be found from Appendix D. If V is the reaction, L_w is the length of weld, s is the size of weld, and f_{wd} is the design strength in weld material, then the equation for the size of the weld is given by

$$s = V/(L_w f_{wd}) \geq 6 \text{ mm} \qquad (10.34)$$

A typical design of flexible end plate connection is shown in Example 10.16.

10.7 Moment Resistant Connections

As discussed already simple connections are assumed to transfer only shear at some nominal eccentricity. Hence, they are used only in non-sway frames where the lateral load resistance is provided by bracings or shear walls. Whereas *moment resistant connection* is capable of transferring moment, axial force, and shear from one member to another. These kinds of connections are used in framed structures, where the joints are considered rigid. Thus the connections, where the members of a frame meet, have sufficient rigidity to prevent rotation of individual members. Another kind of situation is one in which the eccentricity of load from the centroid of the fastener group may cause forces and moments on the joint. These kinds of connections in which the connections are designed to transfer bending moments and shear or a combination of bending moment, shear, and axial force may be grouped into the following:

(a) Eccentrically loaded connections
(b) T-stub connections
(c) Flange angle connections

10.7.1 Eccentrically Loaded Connections

Depending on how the bolts are placed in a connection, they are subjected to twisting or bending moments by the eccentrically applied loads. The effect of twisting moment and the shear force on the bolt group is to cause shear forces along two directions of the bolt, whereas the bending moment and shear cause tension and shear in the bolts. Hence, they are considered separately in the following sections.

10.7.1.1 Eccentric load causing twisting moments (type I connections)

Generally, when eccentricity of a load on a bolt group is less than approximately 60 mm, it is neglected. Joints such as the simple frame connection of Fig. 10.2(c) or 10.2(d), which is widely used, belong to this category (Bowles 1980). However, in some connections, such as the bracket connection as shown in Fig. 10.1(a), the eccentricity is too large and hence cannot be neglected. In the framed beam connections, such as those shown in Fig. 10.1(b) and (e), also the resulting eccentricity may be too large to be neglected and hence taken into account in the design.

When bolt groups are subjected to shear and moment in their shear plane, the load that is eccentric with respect to the centroid of the bolt group can be replaced with a force acting through the centroid of the bolt group and a moment, with the magnitude $M = Pe$ (where e is the eccentricity of the load), as shown in Fig. 10.27. Since both the moment and the concentric load result in shear effects in the bolts of the group, it is termed as *eccentric shear*. These eccentric shear forces tend to rotate the connections (i.e., a twisting moment is applied on the bolts).

Essentially two approaches are available to the designer (a) traditional elastic (vector) analysis and (b) strength (plastic) analysis.

Elastic (Vector) Analysis This method of analysis assumes the following:

(a) Deformation of the connected parts may be ignored
(b) The relative movement of the connected parts are considered as the relative rigid body rotation of the two parts about some centre of rotation
(c) There is friction between the 'rigid' plates and the elastic fasteners
(d) This movement defines the deformations of the individual bolts, which are tangential to this centre of rotation
(e) This deformation induces reactive bolt forces, which are also tangential to the centre of rotation

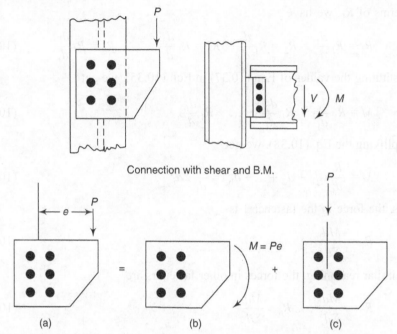

Fig. 10.27 Replacement of eccentric shear with a moment and direct shear.

This elastic method of analysis has been used in the past several years, since it makes use of simple mechanics of materials concepts and has been found to yield conservative results.

Consider a connection acted upon by a moment M as shown in Fig. 10.28. Neglecting friction between the plates for rotational equilibrium, the moment equals the sum of the forces multiplied by their respective distances to the centroid of the bolt group. Thus,

$$M = R_1d_1 + R_2d_2 + \ldots + R_6d_6 = \Sigma Rd \qquad (10.35)$$

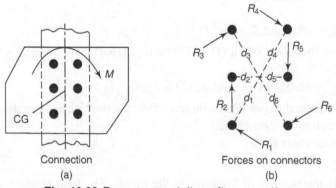

Fig. 10.28 Pure moment (type I) connection

If we assume that the value of R_i is proportional to the distance from the centroid of the fastener group, we have

$$R_1/d_1 = R_2/d_2 = \ldots = R_6/d_6 \qquad (10.36)$$

In terms of R_1, we have

$$R_1 = R_1 \frac{d_1}{d_1} \quad R_2 = R_1 \frac{d_2}{d_1} \quad R_3 = R_1 \frac{d_3}{d_1} \quad ... \quad R_6 = R_1 \frac{d_6}{d_1} \tag{10.37}$$

Substituting the values of Eqn (10.37) in Eqn (10.35), we get

$$M = R_1 \frac{d_1^2}{d_1} + R_1 \frac{d_2^2}{d_1} + ... + R_1 \frac{d_6^2}{d_1} \tag{10.38}$$

Simplifying the Eq. (10.38), we get

$$M = \frac{R_1}{d_1}(d_1^2 + d_2^2 + ... + d_6^2) = \frac{R_1}{d_1} \Sigma d_i^2 \tag{10.39}$$

Thus, the force in the fastener 1 is

$$R_1 = \frac{Md_1}{\Sigma d_i^2} \tag{10.40}$$

By similar reasoning, the forces in other fasteners are

$$R_2 = \frac{Md_2}{\Sigma d_i^2}, ..., R_6 = \frac{Md_6}{\Sigma d_i^2} \tag{10.41}$$

Or, in general

$$R = \frac{Md_i}{\Sigma d_i^2} = \frac{P_e d_i}{\Sigma d_i^2} \tag{10.42}$$

where R is the force on the fastener at the distance d_i from the centre of rotation. Note that if we want to obtain stress, we may divide R by A_b and the denominator in Eqn (10.41) will now become $\Sigma A d_i^2$, which is the polar moment of inertia J about the centre of rotation for the cross-sectional areas of the bolts in the group.

It is usually convenient to work with horizontal and vertical components of R- R_x and R_y respectively, which are associated with the horizontal and vertical distances x and y of the distance d.

Thus,

$$R_x = yR/d \text{ and } R_y = xR/d \tag{10.43}$$

Substituting the values of Eqn (10.43) into Eqn (10.42) and using $d^2 = x^2 + y^2$, we get

$$R_x = My/\Sigma(x^2 + y^2) \text{ and } R_y = Mx/\Sigma(x^2 + y^2) \tag{10.44}$$

Note that the x and y coordinates should reflect the positive and negative values of the bolt location as appropriate.

The direct shear force

$$R_v = P/n$$

where n is the number of fasteners in the group. The total resultant force R is given by

$$R = \sqrt{(R_y + R_v)^2 + R_x^2} \tag{10.45}$$

If there is horizontal force (H) and vertical force (V) acting eccentrically to the centroid, (with eccentricities e_v and e_h respectively) of the bolt group, then Eqn (10.45) can be extended as

$$R_x = H/n + (Ve_h + He_v)y/\Sigma(x^2 + y^2) \tag{10.46}$$
$$R_y = V/n + (Ve_h + He_v)x/\Sigma(x^2 + y^2) \tag{10.47}$$

This method is used in practice for ordinary bolts, high-strength friction grip bolts, and welds.

A critical bolt is the one which is subjected to the maximum resultant force. A close examination reveals that the bolt which is farthest from the centre of gravity of the bolt group and nearest to the applied load line is the most critical.

If the bolt group is subjected to applied moment or torque, the number of required bolts can be found out by using (Shedd 1934; Salmon & Johnson 1996)

$$n = \sqrt{(6M/n \, pV_{sd})} \tag{10.48}$$

where p is the pitch, M is the applied moment, V_{sd} is the design shear strength of a single bolt, and n' is the number of rows of bolts (e.g., if two vertical rows of bolts are provided, use $n' = 2$). Since the connection is subjected to shear and moment, the number of bolts found out from Eqn (10.48), may be revised, if required. Examples 10.17 to 10.19 show the design of bracket connections.

Strength (Plastic) Analysis This method, also called *ultimate strength analysis* is considered most rational. In this analysis, it is assumed that the eccentrically loaded fastener group rotates about an *instantaneous centre of rotation*. It is also assumed that the deformation at each fastener is proportional to its distance from the centre of rotation.

The procedure starts with the assumption of a location for the instantaneous centre of rotation of the fastener group. A check of the equation of equilibrium is done to find out whether or not the trial location of the instantaneous centre of rotation is the correct one. If the equations are not satisfied, a new location of the instantaneous centre is chosen and the process is repeated.

A comparison of the values obtained by the elastic (vector) analysis and the strength (plastic) analysis shows that the values obtained by the elastic analysis are conservative.

For the design of bolt groups subjected to torque, the following procedure may be adopted.

1. The number and diameter of bolts for the connection are assumed based on Eqn (10.48) and these are placed in two or more vertical rows at a suitable pitch and edge distance.
2. The resultant force on the critical bolt is worked out by using Eqn (10.45).
3. The design strength of bolt is computed. It should be more than the resultant force on the bolt. Otherwise, the number of bolts are increased and steps 1 to 3 are repeated.

10.7.1.2 Eccentric load causing bending moments (type II connections)

In connections where the brackets are placed in the plane of the column web (Figs 10.1(b), 10.1(c), and 10.29), bolt groups are subject to loading eccentric to the shear plane. That is, the line of action of load does not lie in the plane of the group of bolts and the line of rotation does not pass through the centre of gravity of the bolt group. Thus, the bolts are subjected to direct shear along with tension due to the moment. Extra angle seats are often provided to the bracket at the point where the reactions are transferred.

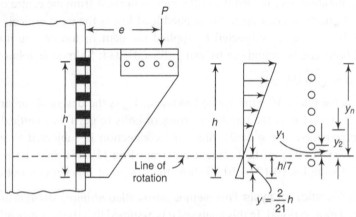

Fig. 10.29 Bracket type II connection

There are different approaches and assumptions to the distribution of shear between the bolts. The more conservative one is to assume all the bolts to carry equal shear. On this basis, the critical bolts are always those at the top of the connection, and those must be checked under combined shear and tension. Thus, according to this approach the shear in bolt is given by

$$V = P/n \qquad (10.49)$$

where P is applied load and n is the total number of bolts in the bolt group. It would be more optimistic to assume that all bolts can be working at their design capacity under varying ratios of tension and shear. On this basis the residual shear capacities of all the bolts are determined, taking account of coincident bolt tension. Adequacy in shear is then checked by ensuring that the sum of those residual shear capacities is greater than applied shear. However, Owens & Cheal (1989) suggest that without matched drilling, the usual situation for general fabrication, the more conservative approach of an equal distribution of bolt shears, should be used. When HSFG bolts are used or matched drilling is ensured the latter method could be adopted. Hence, we will adopt the conservative approach only.

To estimate the amount of tension in the bolts, the position of neutral axis (line of rotation) has to be located. The bracket section below the line of rotation presses against the flange of the column and hence stiffeners are often provided to the column web opposite the lower bracket flanges, especially in connection involving end plates (see Fig. 10.30). In such cases, it is logical to assume that the compression acts at the mid depth of the stiffener as shown in Fig. 10.30 (which is referred to as 'hard spot' on the load path).

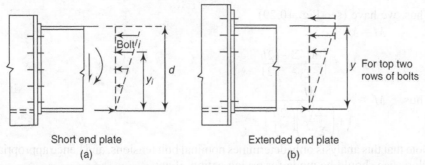

Short end plate   Extended end plate
(a)   (b)

Fig. 10.30 Bolt and compressive force distribution for end-plate connection

Without such a well-defined load path (hard spot), it is difficult to define the position of neutral axis, since it would be a function of stiffness of both tensile and compressive regions. In the case of end plate connections, the deformation of the column flange and end plate flexure would have significant influence on the behaviour. It is generally acceptable to assume a neutral axis position that is one-sixth to one-seventh of the depth of the connection from the base of the bracket (Owens & Cheal 1989). The depth is measured from the bottom of the bracket to the topmost bolt in the connection (see Fig. 10.29). The bolt above the neutral axis will be in tension and in direct shear. The necessary compression to balance this tension is assumed to be provided by the bracket below the neutral axis. Note that where the connection is required to develop a moment which is about 80% or more than the yield moment of the beam, the above procedure will lead to flange forces that are greater than their design capacity. In such circumstances, the 'load path' method, described by Owens & Cheal (1989), has to be adopted, mobilizing the bending capacity of the web directly.

The distribution of nominal bolt tensile forces is generally assumed to be a linear function of distance from the neutral axis, where extended end plates are used [see Fig. 10.30(b)]. It has been found that the top portion of the plate behaves as a T-stub that is symmetric about the tension flange. In such cases, the forces in the top two rows of bolts may be assumed to be constant and are calculated based on the distance from the top flange centroid to the neutral axis or line of action of compressive force, as appropriate (Owens & Cheal 1989).

Based on the preceding assumptions, the nominal tension in the ith bolt T_i is given by
$$T_i = ky_i \tag{10.50}$$
where k is some elastic constant and y_i is the distance from the centre of rotation.
For moment equilibrium,
$$M^* = \Sigma T_i y_i = k\Sigma y_i^2 \tag{10.51}$$
where y_i is the lever arm of the ith bolt. Substituting k in Eqn (10.50) yields
$$T_i = M^* y_i / \Sigma y_i^2 \tag{10.52}$$
Thus, the tensile force in the extreme critical bolt is
$$T_e = M^* y_n / \Sigma y_i^2 \tag{10.53}$$
From equilibrium considerations, we have
Total tensile force = Total compressive force
Total compressive force, $C = M'\Sigma y/\Sigma y^2$
External moment = Moment resisted by bolts in tension + moment of the compressive force

Thus, we have (see Fig. 10.29)

$$M = M' + c\,\bar{y}$$

$$= M' + \left(\frac{M'\Sigma y}{\Sigma y^2}\right)\frac{2h}{21}$$

Thus, $M' = \dfrac{M}{1 + \left(\dfrac{\Sigma y}{\Sigma y^2}\right)\left(\dfrac{2h}{21}\right)}$

Note that this analysis only determines nominal bolt tensions, and hence appropriate allowance should be made for prying action, if any.

The design of the bracket connection as shown in Fig. 10.29 involves the following steps:

1. Assume the diameter, pitch, and edge distance for bolts
2. Compute the design strength of the bolt in shear and tension
3. Compute the required number of bolts using Eqn (10.48)
4. Compute the shear force V in the extreme critical bolt
5. Calculate the tensile force T_i in the extreme critical bolt using Eqn (10.52)
6. Check the connection for combined tension and shear by using Eqn (10.18), i.e.,

$$(V/V_{sd})^2 + (T_e/T_{nd})^2 \leq 1.0.$$

The various steps involved in the design of a bracket (Type II) connections are shown in Example 10.20.

10.7.2 Bolted Moment End Plate Connection

A bolted moment end plate connection [Fig. 10.1(e)] is used to transfer moment, axial force, and shear force from one member to another. The end plate is fillet welded to the web and flange of beam and bolted to the column flange or web. Due to the concentration of forces in the beam, the web of columns may require additional strengthening in the form of web stiffeners, diagonal stiffeners, or web plate (see Fig. 10.31). The end plates may be extended beyond the face of the beam to carry more moments [see Fig. 10.31(b)]. In the case of deep beams connected to slender columns, haunched connections as shown in Fig. 10.31(c) may be adopted. As discussed in the previous section, by using the extended end plate connections, it is possible to transfer about 80% of the yield moment capacity of the beam. However, this connection is less stiff than welded moment connection, due to the flexure of end plate. However, they provide ease of use, since the end plates may be welded to the beams in shop and bolted at site to the column. They provide economic alternatives to connections using angle or T-sections, which require large number of bolts. (Note: The cost of bolts are expensive than steel sections.)

In these connections, the bending moment, axial force, and shear are transferred by tension and compression or shear through the flange welds and by shear through the web welds to the end plate. Then they are transferred from the end plates to the bolts by bending and shear. Finally, they are transferred to the supporting member by bolt shear and tension and also by horizontal plate reaction (Trahair et al. 2001).

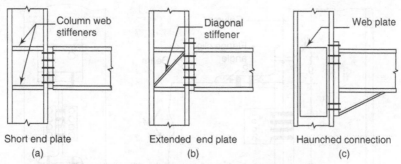

Short end plate   Extended end plate   Haunched connection
(a)   (b)   (c)

Fig. 10.31 Rigid beam-to-column connections

Due to the bending of the end plate, the bolts in the extended region of the end plate are subjected to prying actions, and hence must be considered on the design. Increases in bolt load of up to 30% have been observed.

In the design the following steps are adopted.

1. The dimensions (length and breadth) of the end plate are selected based on the dimensions of the beam and column flange. The layout of the bolts are also fixed, by providing nominal minimum distance (of 50–60 mm) from the flanges and web of the beam, so that sufficient clearances are available for the use of power tools to tighten HSFG bolts. Six bolts are sufficient to carry the loads in most of the rolled beams.
2. The connection is assumed to 'pivot' about the 'hot spot' (at the centre of the bottom flange thickness) and the loads in the bolts are assumed to be proportional to their distances from the centre of the bottom flange. It is assumed that the bolts in the cantilever end of the plate and the bolts near the top flange of beam carry equal loads.
3. Tension and shear capacity of the bolt are computed. An approximate value of the prying force is assumed: Q = tension capacity – bolt load. Using the equations for calculating the prying forces (Section 10.4.4), the moment at the toe of the weld is calculated. Equating it with the moment capacity of the plate, the thickness of end plate may be obtained. Using this thickness, the prying force is computed and checked with the assumed prying force. If it is more than the assumed prying force, the thickness has to be changed.
4. A check is made for combined shear and tension in the bolt.
5. The capacity of the end plate to carry the reaction due to bolt forces has to be checked.
6. The welds connecting the beam to end plate are to be checked.
7. The capacity of column web to support the flange reaction has to be checked. If the column web is not able to support the reaction, the web has to be strengthened by means of web plates, web stiffeners, etc.

Example 10.20 shows the calculations involved in the preceding steps.

10.7.3 Flange Angle Connections and Split-Beam T-stub Connections

If we use an additional pair of angles in the angle seat connections to connect the web of the beam to the flange of the column, as shown in Fig. 10.32 the connection can be designed to transfer small end moments in addition to large end shears. This kind of connection is also called *clip angle connection* or *light moment connection*.

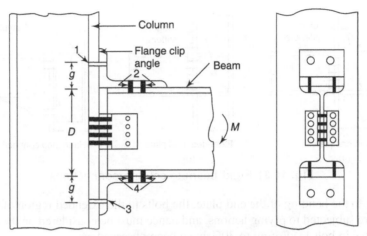

Fig. 10.32 Flange-angle connection

As shown in Fig. 10.32, this connection consists of four angles (two pairs). One pair of angle is used to connect the web of the beam with the column flange (one angle on each side of the web) and the other pair of angles (called *clip angles*) is placed one on top of the beam flange and the other below the bottom flange of the beam. If a clockwise end moment acts on the connection, the bolts marked as 1 in the figure are subjected to tension and those marked as 2 and 4 are subjected to shear. Similarly, an anticlockwise moment will produce tension in bolts 3 and shear in bolts 2 and 4. The end shear force in the beam is assumed to be transferred by the web angle connection. The force acting on the bolts (at the top and bottom clip angles) is the moment divided by the lever arm of the bolts $[M/(D + 2g)]$, where g is the length as shown in Fig. 10.32. In the simple beam column connection designed for shear only, the connecting angles at the bottom and top were flexible (so that they can deform sufficiently) and hence permit the rotation necessary for simple support condition at the ends of the beam. Unless this condition is fulfilled, end moments will be induced, which will overstress the connecting bolts. On the other hand, if the connection is considered rigid, the deformation of the connecting angle should be nil or very small. Hence, it is necessary to analyse the forces and deformation in the top and bottom angles of clip-angle connection. The flange angles are deformed due to eccentric forces and moments. The thickness of the angles should be adequate to resist the internal moments imposed on them, as shown here.

Consider the free body diagram of legs AB and BC (Fig. 10.33).

For leg BC, $\qquad \theta_B = M_B g_2/4EI$ $\hfill$ (10.54a)
For leg AB, $\qquad \theta_B = P g_1^2/2EI - M_B g_1/EI$ $\hfill$ (10.54b)
where P is the tensile force in the bolt, g_1 and g_2 are gauge lengths, E is Young's modulus, and I is the moment of inertia of the leg of the angle.

Equating Eqns (10.54a) and (10.54b), we get
$$M_B = 2P g_1^2/(4g_1 + g_2)$$

Since,
$$Pg_1 = M_B + M_A$$
$$M_A = Pg_1 (2g_1 + g_2)/(4g_1 + g_2)$$

If we consider an equal angle, $g_1 = g_2 = g$, we get
$$M_A = 0.6Pg \hfill (10.55a)$$

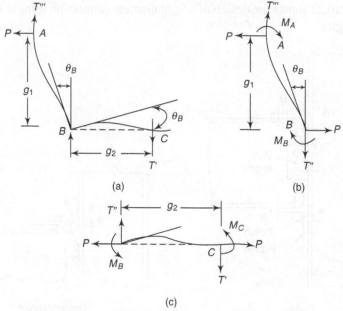

Fig. 10.33 Deformation and forces in clip angles

$$M_B = 0.4Pg \qquad (10.55b)$$
$$M_C = 0.2\ Pg \qquad (10.55c)$$

Hence, the maximum moment in leg AB is $0.6Pg$, where g is the gauge length for the angle if only two bolts are used in the connecting leg of the column. Thus, the angle should be designed to have sufficient thickness to resist the maximum moment.

The deflection is given by Arya & Ajmani (1989)

$$\Delta = 0.133Pg^3/EI \qquad (10.56)$$

Example 10.21 shows the calculations required for the design of a light moment connection using clip and flange angles.

Instead of an angle, a T-stub cut from an I-section can be used to clip the top flange of an I-beam as shown in Fig. 10.34, the deformation of the T is symmetrical with respect to point B so that the tangent at B remains vertical [Fig. 10.34(b)]. Then if P is the tensile force in the bolt, the moments in the T will be

$$M_A = M_B = 0.5Pg \qquad (10.57)$$

In *split-beam T-connections* (also called *T-stub connection*), the prying action should be considered and the bolts connecting the T with the column flange should be capable of resisting the additional tension due to prying action.

Clip angle connection has the following limitations: (a) Only two bolts can be provided in one gauge line to connect the outstanding leg with the column flange. If more bolts are provided in one row (gauge line), the bolts will be subjected to non-uniform forces [see Fig. 10.35(a)] and (b). If more than one gauge line is provided, then the bolts will be subjected to non-uniform forces [see Fig. 10.35(b)]. Due to these, the number of bolts to connect the clip angle to the column flange is limited to two only and hence the connection can transfer only limited moments.

Example 10.22 shows the design of a light moment connection using a T-stub and flange angles.

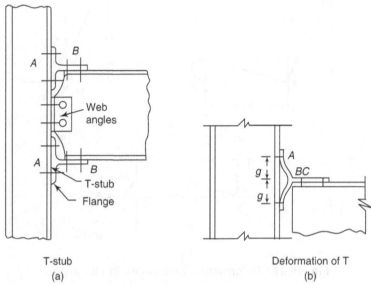

T-stub
(a)

Deformation of T
(b)

Fig. 10.34 T-stub connection

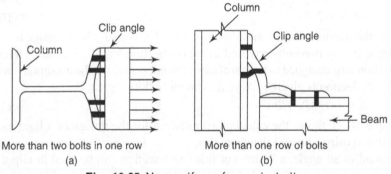

More than two bolts in one row
(a)

More than one row of bolts
(b)

Fig. 10.35 Non-uniform forces in bolts

Bracket connection If the lever arm is to be extended to accommodate more bolts, a bracket type connection is sometimes used (Fig. 10.36). These bracket type connections are more rigid compared to any type of moment connection discussed so far. But the fabrication cost is very high and hence they are not adopted in practice, except in some special cases.

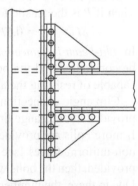

10.8 Beam-to-Beam Connections

The simple connections such as clip and seating angle connection, web angle connection, and flexible end plate

Fig. 10.36 Bracket type connection

connections, discussed in the earlier sections for connecting beam-to-columns, can be adopted for beam-to-beam connections also. If the supporting beam and the supporting girder are required to have their top flange at the same level (for fixing floor chequred plates or floor slabs), the top flange of the beam is cut away (coped) to provide space for the supporting girder.

In practice most of the secondary beams are connected to main beams by web cleats and bolts [see Fig. 10.37(a)] since the web of the main beam may not be strong enough to support a seating angle. (Moreover, the seat connection requires more space in the vertical direction.) When the beam ends are coped, the web of the secondary beam has to be checked for block shear. In addition, the loss of top flange will also reduce the bending strength. To reduce the effect of notched end and to increase the robustness of the element, stiffeners may be welded just below the coping [see Figs 10.37(c) and (d)], which are required only on one side of the web. If two secondary beams on either side of the main supporting girder are on the same grid line, same bolts are used to connect both beams to the supporting girder (see Fig. 10.37). The arrangement used for bolts and cleats, is suitable where larger (left hand) beam is erected before the smaller one [Fig. 10.37(c)]. The top bolts through main girder webs will be in double shear after the secondary beam is attached. The larger beam can be held in position by the bottom pair of bolts while the smaller one is erected. If the smaller beams are to be erected first, they are propped and supported in some way until the larger one is erected.

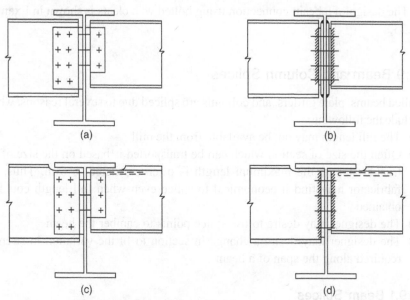

(a)

(b)

(c)

(d)

Fig. 10.37 Simple beam-to-beam connections

Sometimes rigid connections may be provided for moment continuity between secondary beams, which may result in economy in beam section. Rigid connections are also necessary if a cantilever beam is supported by a girder. However, in this case, the end of the cantilever will transfer shear force and bending moment—this bending moment will be transferred to the primary beam as a torsion. Hence, the primary beam has to be designed for bending, shear, and torsion. If the primary beam is torsionally flexible, then torsion transferred to it may be neglected. Figure 10.38 shows some typical details of moment resisting beam connections. To ease erection, packing plates are used while connecting short, wide secondary beams (Owens & Cheal 1989). Note that when top flanges are required to be at the same level, a combined splice plate/end plate connection with coped ends, as shown in Fig. 10.38(b) is used. For avoiding local rotation, HSFG bolts may be utilized in these connections. More details on beam-to-beam connections may be found in Owens & Cheal (1989).

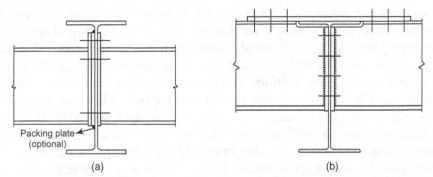

(a)   (b)

Fig. 10.38 Moment resisting beam-to-beam connections

The design of a beam connection using bolted web cleats is shown in Example 10.23.

10.9 Beam and Column Splices

Rolled beams, plate girders, and columns are spliced due to several reasons, which include the following:
(a) The full length may not be available from the mill
(b) Often the size of section which can be transported is based on the size of the trucks. Hence, the maximum length is often restricted to 6 m. Thus, the fabricator may find it economical to splice even when full length could be obtained
(c) The designer may desire to use splice points to camber the beam
(d) The designer may desire a change in section to fit the variation in strength required along the span of a beam

10.9.1 Beam Splices

Many field splices for beam and plate girders use laping splice plates and high-strength bolts as connectors (see Fig. 10.39). Splices are designed for the moment M and the shear V occurring at the spliced section, or they are designed for some specification prescribed higher values. On each side of the joint, one must provide enough bolts to transfer the shear and bending moment.

Fig. 10.39 Beam splice in a bridge structure

For beam splices, each element of the splice is designed to do the work the sections underlying the splice plates could do, if uncut. Plates on the flange should be designed to do the work of the flange (area of splice plate should not be less than 5% in excess of the area of flange element splices) and plates over the web should be designed to do the work of the web. The simplified partition of resisting all the moments by the flange splices and all shear by the web splices should not be used for deep plate girders with slender webs. These web splices should be designed to resist both shear and its share of the moment. Also double row of bolts should be used on each side of such web splices [where tension field action (see Chapter 7 on plate girders) is predominant]. In web splices, the moment about the centroid of the bolt group on either side of the splice should be designed for moment due to eccentricity. IS code stipulates that in no case should the strength of the spliced portion be less than 50% of the effective strength of the material spliced.

In accordance with the principles used for designing bolted connections, the forces designed should be those acting at the centre of gravity of the bolt group (Salmon & Johnson 1996). Most splices are located near a point of effective restraint; if possible, splices should not be located in critical sections. Thus, they are often located where either the shear or the bending moment is low, that is, at locations where minimum strength is required.

Figure 10.40 illustrates two basic forms of beam splice, both of which may have several variants (Nethercot 2001). Flush end plates are used when the bending moments to be resisted are modest. Singly or doubly extended end plates are used when resisting high moments of one sign or full reversal, respectively. When end plate thickness is greater than 25 mm, the susceptibility of lamellar tearing should be checked.

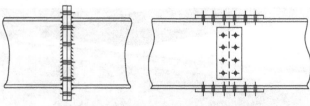

Fig. 10.40 Types of beam splices

Reasonable care should be given to the flatness of the end plate, which will tend to distort in the presence of heavy welding on one face. When a change in size between the two sections of the beam occurs at an end plate splice, it can be easily accommodated by introducing longitudinal stiffeners to the larger beam as shown in Fig. 10.41. The flange cover plates and bolts (after deducting the bolt hole area) in Fig. 10.40(b), should be designed to take the moment, while the web plates and bolts are designed to resist the shear force plus the secondary moment due to eccentricity. For large beams and plate girders, flange plates are provided on both faces (top and bottom) of the beam flanges. The use of HSFG bolts will result in reasonable number of bolts, thus reducing the splice length.

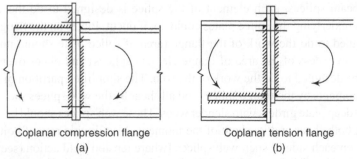

Coplanar compression flange   Coplanar tension flange
(a)   (b)

Fig. 10.41 End plate connections between beams of different sizes

If there is a small difference in the height of the two sections of beams that are being jointed, the first line of bolts near the junction in the cover plate may be ignored when calculating the slip resistance of the connection, using HSFG bolts (Owens & Cheal 1989). If the difference in thickness is relatively large, shims or packing plates may be used. The design of a bolted beam splice is worked out in Example 10.24.

10.9.2 Column Splices

The connections as discussed for beams may be adopted for column splices also. Design of column splices does not pose much difficulty, since the high tensile forces that caused problems in the case of beam splices are usually absent. Although such splices provide an opportunity to change the size of columns in the upper floors, the same size is continued in most of the cases due to economic and practical reasons. With modern steel rolling practices it is also possible for retaining the common outside dimensions over the full column height and reduce only the thickness of flanges/web in the upper floors.

When the columns carry predominantly axial forces, either direct end bearing arrangement or leaving a gap between the ends may be used (see Fig. 10.42). In the first case the load is transferred through the contact area (the ends of both columns are machined or cut precisely) and the splices are designed only to resist any accidental tension due to some uplift loading or internal explosion in the building. Whereas in the second case [Fig. 10.42(c)], the whole load is transmitted by means of the splice plates. In these cases it is preferable to use HSFG bolts in the connection. Changes in the size of the columns can be accommodated by using packing plates as shown in Figs 10.42(b) and 10.42(c).

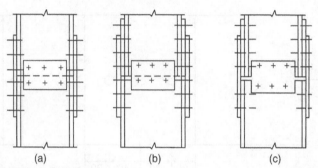

Fig. 10.42 Bolted column splices (a) same size columns, (b) small change in size and (c) different size columns with gap between ends

Similar to beam end plate splices, end plate splices can be used for columns as shown in Fig. 10.43, which provide for load reversals in columns in a convenient way. Short end plates are used for moderate moments and extended end plates for heavy moments. It is normal practice to position splices just above floor level so that the effects of flexing of the column may be neglected. However, in regions of seismic activity, the splices should be placed near mid-height of the columns, where bending moments will be minimum. The design of end plate connections can be treated in a similar way to end plate beam splices. Changes in column sizes can be easily tackled with the use of appropriate stiffeners, as shown in Fig. 10.43(c). IS code also stipulates that the centroidal axis of the splices should coincide with the centroidal axis of members jointed; if this condition is not satisfied, the design should cater for the resulting stress considering eccentricity. Examples 10.25 and 10.26 show the calculations required for the design of column splices.

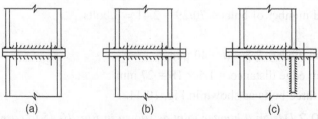

Fig. 10.43 Column splices using end plates (a) same size columns, (b) same size columns with high moment, and (c) different size columns with high moment

Examples

Example 10.1 *Design a lap joint between two plates as shown in Fig. 10.44 so as to transmit a factored load of 70 kN using M16 bolts of grade 4.6 and grade 410 plates.*

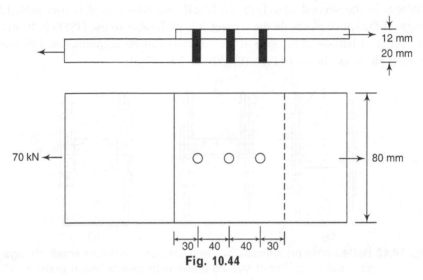

12 mm

20 mm

70 kN ←

80 mm

30 40 40 30

Fig. 10.44

Solution
Strength calculation

Nominal diameter of the bolt $d = 16$ mm

Hole diameter $= 16 + 2 = 18$ mm

Bolts are in single shear and hence shear capacity of the bolt

$= (f_u/\sqrt{3}) (n_n A_{nb} + n_s A_{sb})/\gamma_{mb} = (400/\sqrt{3}) (1 \times 157)/1.25 = 29.0$ kN

Since the top plate is only 12 mm, it is assumed that the shear plane is through the threaded portion and hence n_s is taken as zero.

Bearing capacity of the thinner plate

$= 2.5 k_b dt f_u/\gamma_{mb} = (2.5 \times 0.49 \times 410/1.25) \times 16 \times 12 = 77.15$ N

k_b is smaller of $30/(3 \times 18)$, $40/(3 \times 18) - 0.25$, $400/410$, 1.0.

Hence, $k_b = 0.49$,

Bolt value $= 29$ kN

Required number of bolts $= 70/29 = 2.41 \approx 3$ bolts

Detailing

Minimum pitch $= 2.5 \times d = 40$ mm

Minimum edge distance $= 1.5 \times 18 = 27$ mm

Provide three bolts as shown in Fig. 10.44.

Example 10.2 *Design a hanger joint as shown in Fig. 10.45 to carry a factored load of 300 kN. Use an end plate of size 250 mm × 150 mm and appropriate thickness, M24 HSFG bolts (2nos) and Fe 410 steel for end plate ($f_y = 250$ MPa).*

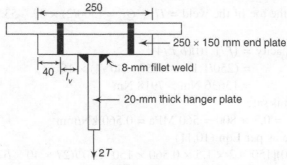

Fig. 10.45

Solution

Assuming an 8-mm fillet weld between the hanger and the end plate and an end distance of 40 mm, the distance from the centre line of the bolt to the toe of the fillet weld,

$$l_v = (250/2) - 10 - 8 - 40 = 67 \text{ mm}$$

For minimum thickness design,

$$M = Tl_v/2 = (300/2) \times 67/2 = 5025 \text{ kNmm}$$

and

$$t_{min} = \sqrt{[(4.4 \times 5025 \times 10^3)/(250 \times 150)]} = 24.28 \text{ mm, say 25 mm.}$$

Check for prying forces:

Distance l_e from the centre line of the bolt to the point where prying force acts is the minimum of edge distance (40 mm) or

$$L_e = 1.1t\sqrt{\beta f_o/f_y} = 1.1 \times 25\sqrt{(2 \times 560/250)} = 58 \text{ mm}$$

Note proof stress of the $M24$ bolt = $0.7 \times 800 = 560$ MPa

Hence, adopt

$$l_e = 40 \text{ mm}$$

Prying force = $M/l_e = 5025/40 = 125.6$ kN

Bolt load = $300/2 + 125.6 = 276$ kN

Tension capacity of $M24$ bolt = $0.9f_u A_{nb}/1.25$

$$= 0.9 \times 800 \times 353/1.25 = 203.3 \text{ kN} < 276 \text{ kN}$$

Hence the joint is unsafe.

To reduce prying force and hence bolt load, use a thicker plate

Assuming a 40-mm thick plate,

$$l_e = 1.1 \times 40\sqrt{(2 \times 560/250)} = 93 \text{ mm}$$

Hence, use

$$l_e = 40 \text{ mm (edge distance)}$$

Allowable prying force

$$Q = 203.3 - (300/2) = 53.3 \text{ kN}$$

Moment at the toe of the weld = $Tl_v - Ql_e$ = (300/2) × 67 – 53.3 × 40 = 7918 Nm

Moment capacity = $(f_y/1.10)(b_e t^2/4)$
$$= (250/1.10) × (150 × 40^2/4) × 10^{-3}$$
$$= 13636 \text{ Nm} > 7918 \text{ Nm}$$

Hence the joint is safe

Proof stress = 0.7 × 800 = 560 MPa = 0.560 kN/mm²

Prying force as per Eqn (10.11)

= 67/(2 × 40)[150 – 2 × 1.5 × 0.560 × 150 × 40⁴/(27 × 40 × 67²)]

= 0.8375[150 – 133] = 14.24 kN < 53.3 kN

Hence the joint is safe.

Thus, it is observed that an end plate thickness of 40 mm is required to avoid the development of significant prying forces in the bolts. Note that the second term in the above calculation is not small. It will be small only when the thickness of the plate is small.

Example 10.3 *A member of a truss consists of two angles ISA 75 × 75 × 6 placed back to back. It carries an ultimate tensile load of 150 kN and is connected to a gusset plate 8-mm thick placed in between the two connected legs. Determine the number of 16-mm-diameter 4.6 grade ordinary bolts required for the joint. Assume f_u of plate as 410 MPa.*

Solution

The arrangement of joint is as shown in Fig. 10.46. The bolts are in double shear. They bear against 8-mm gusset and two 6-mm angles, the former controlling the value in bearing as shown below. From Appendix D,

Strength in double shear for 16-mm-diameter 4.6 grade bolts = 2 × 29 = 58 kN

Strength in bearing on 8-mm plate = (2.5 × 0.49 × 16 × 8 × 410)/(1.25 × 1000)
$$= 51.4 \text{ kN}$$

(Note: with e = 30 and p = 40, k_b = 0.49 from Example 10.1)

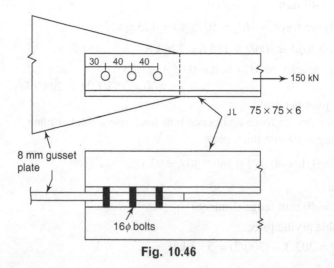

Fig. 10.46

Strength in bearing on two 6-mm thick angles

$$= [2.5 \times 0.49 \times (2 \times 6) \times 16 \times 410]/(1.25 \times 1000) = 77.0 \text{ kN}$$

Thus,

Strength of the bolt = 51.4 kN

Required number of bolts = 150/51.4 = 2.92

Therefore, provide three bolts as shown in Fig. 10.46.

Example 10.4 *The plates of a 6-mm thick tank are connected by a single bolted lap joint with 20-mm diameter bolts at 60-mm pitch. Calculate the efficiency of the joint. Take f_u of plate as 410 MPa and assume 4.6 grade bolts.*

Solution

Strength of the bolt in single shear (from Appendix D) = $(245 \times 185)/1000 = 45.3$ kN

Strength of the bolt in bearing on a 6-mm plate

$$= 2.5 \times 0.53 \times (410/1.25) \ 6 \times 20/1000 = 52.15 \text{ kN}$$

Note: Assuming $e = 35$ mm, k_b is the least of $35/(3 \times 22)$, $60/(3 \times 22) - 0.25$, $400/410$ and 1.0. Thus, $k_b = 0.53$

Strength of the net section per pitch length (see Chapter 3) = $0.9A_n f_u/\gamma_{ml}$

$$= 0.9(60 - 22) \times 6 \times 410/(1.25 \times 1000) = 67.3 \text{ kN}$$

Strength of the bolt in single shear is 45.3 kN, being the smallest of 45.3, 52.15, and 67.3, determines the strength of the joint. Thus,

Strength of the joint per pitch length = 45.3 kN

Original strength of the plate per pitch length = $0.9f_u A/\gamma_{ml}$

$$= 0.9 \times 410 \times 60 \times 6/(1.25 \times 1000) = 106.3 \text{ kN}$$

Therefore,

Joint efficiency = $(45.3/106.3) \times 100 = 42.6\%$

Note that the strength of bolt per pitch length is equal to the strength of plate in tearing per pitch length.

Thus, for best design

$$45.3 = [0.9 \times (p - 22) \times 6 \times 410]/(1.25 \times 1000)$$
$$56.625 \times 10^3 = 2214(p - 22)$$

or

$$p = 47 \text{ mm} < 2.5\phi = 50 \text{ mm}$$

Hence, provide $p = 50$ mm

Now,

Original strength of the plate = $(0.9 \times 410 \times 6 \times 50)/(1.25 \times 1000) = 88.56$ kN

Efficiency = $(45.3/88.56) \times 100 = 51.2\%$

Note that we have adopted a bolt diameter of 20 mm for a 6-mm plate. As per Unwin's formula $\left(\phi = 6.01\sqrt{t}\right)$, we can use a 16-mm bolt. If we rework with a 16-mm bolt, and spacing of 40 mm

Bolt strength in single shear = 29 kN

Original strength of the plate = $(0.9 \times 410 \times 6 \times 40)/(1.25 \times 1000) = 70.85$ kN

Therefore,

Efficiency = $(29/70.85) \times 100 = 40.9\%$

Thus, the use of a bolt with a large diameter slightly increases the efficiency.

Example 10.5 *Calculate the efficiency of a zigzag double bolted lap joint as shown in Fig. 10.47. Assume Fe 410 grade plate and grade 4.6 bolts of diameter 20 mm and 8-mm thick plates.*

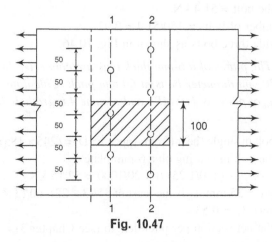

Fig. 10.47

Solution
Strength of a 20-mm-diameter bolt
(i) in single shear (from Appendix D) = 45.3 kN
(ii) in bearing on a 8-mm plate (with k_b = 0.5) = (2.5 × 0.5 × 20 × 8 × 410)/(1.25 × 1000) = 65.6 kN

Consider the 100-mm length of the joint shaded as shown in Fig. 10.47, which represents typical conditions. Line 1-1 has one bolt hole 22 mm in diameter. Line 2-2 also has one bolt hole, but if plate-*A* tears along 2-2, the bolt line 1-1 must fall with it. Hence, the resistance of plate-*A* in tearing along 2-2 = Strength of the net section of plate 2-2 plus one bolt strength.

Hence for plate-*A*, net section 1-1 is critical. Here,
Net strength of the plate = [0.9(100 – 22) × 8 × 410]/(1.25 × 1000) =184 kN
Strength of two bolts = 2 × 45.3 = 90.6 kN
Therefore,
Strength of the joint = 90.6 kN (smaller of 184 kN and 90.6 kN)
Gross strength of the plate = (0.9 × 100 × 8 × 410)/(1.25 × 1000) = 236 kN
Efficiency = (90.6/236) × 100 = 38.4%

Example 10.6 *A single-bolted double-cover butt joint is used to connect two plates 6-mm thick (see Fig. 10.48). Assuming the bolts of 20-mm diameter at 60-mm pitch, calculate the efficiency of the joint. Use 410 MPa plates and 4.6 grade bolts.*

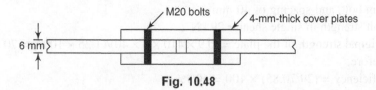

Fig. 10.48

Solution
Strength of a 20-mm-diameter bolt in double shear (from Appendix D)
$$= 45.3 \times 2 = 90.6 \text{ kN}$$
Thickness of the cover plate $> 5/8t = 5/8 \times 6 = 3.75$ mm
Therefore, provide 4-mm thick cover plates.

The thickness to be considered in the calculations are the thickness of the main plate or the sum of the cover plates, whichever is less. Hence,
Strength of the bolt in bearing $= (2.5 \times 0.5 \times 20 \times 6 \times 400)/(1.25 \times 1000) = 48$ kN (Assuming $k_b = 0.5$)
Strength of the plate in tearing $= 0.9(p - d) \times tf_u/1.25$
$$= 0.9(60 - 22) \times 6 \times 410/(1.25 \times 1000)$$
$$= 67.3 \text{ kN}$$
The strength of the plate in bearing is the minimum. Hence,
Strength of the joint = 48 kN
Strength of the solid plate per pitch length $= (0.9 \times 60 \times 6 \times 410)/(1.25 \times 1000)$
$$= 106.3 \text{ kN}$$
Joint efficiency $= (48/106.3) \times 100 = 45.17\%$

Example 10.7 *Two plates 10-mm and 18-mm thick are to be jointed by a double-cover butt joint. Assuming cover plates of 8-mm thickness, design the joint to transmit a factored load of 500 kN. Assume Fe 410 plate and grade 4.6 bolt.*

Solution
Assume 20-mm diameter bolts.
Bolt strength in double shear (from Appendix D) $= 2 \times 45.3 = 90.6$ kN
Since the joint has a packing plate greater than 6 mm (see Fig. 10.49), the bolt strength has to be reduced by
$$\beta_{pkg} = (1 - 0.0125 t_{pkg})$$
$$= (1 - 0.0125 \times 8) = 0.9$$

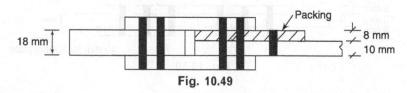

Fig. 10.49

Hence,
Bolt strength in shear $= 90.6 \times 0.9 = 81.54$ kN
Bolt strength in bearing $= (2.5 \times 0.5 \times 20 \times 10 \times 410/(1.25 \times 1000) = 82$ kN
Hence,
Bolt strength = 81.54 kN
Required number of bolts $= 500/81.54 = 6.13$
Provide eight bolts, four bolts in two rows.
Strength of bolt/pitch length $= 2 \times 81.54 = 163.08$ kN

Equating it to the strength/pitch length of the plate in tearing

$$163.08 = 0.9 \times (p - 22) \times 10 \times 410/(1.25 \times 1000)$$
$$203.85 \times 10^3 = 3690(p - 22)$$
$$p = 77.24 \text{ mm} > 2.5 \times 20 = 50 \text{ mm}$$

Hence, provide $p = 75$ mm. Provide additional two bolts of 20-mm diameter on the packing plate as shown in Fig. 10.49.

Example 10.8 *Design a butt joint to connect two plates 175 × 10 mm (Fe 410 grade) using M20 bolts. Arrange the bolts to give maximum efficiency.*

Solution
Thickness of the cover plate = $(5/8) \times 10 = 6.25$ mm
Provide double cover plates each having a thickness of 8 mm.
  The tensile force the main plate can carry (assuming two bolts in a line)

$$= 0.9 \times (175 - 2 \times 22) \times 10 \times 410/(1.25 \times 1000) = 386.7 \text{ kN}$$

  Shear strength of the bolt = $2 \times 45.3 = 90.6$ kN
  Bearing strength of the bolt (with $k_b = 0.5$) = $(2.5 \times 0.5 \times 20 \times 10 \times 410)/(1.25 \times 1000) = 82$ kN
Hence,
  Strength of the bolt = 82 kN
  Required number of bolts = 386.7/82 = 4.7
Hence, provide five bolts of 20-mm-diameter as shown in Fig. 10.50.

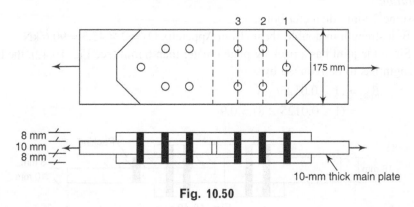

Fig. 10.50

Example 10.9 *Determine the adequacy of the fasteners in Fig. 10.51, when 20-mm-diameter grade 4.6 bolts are used. Assume that the strength of the column flange and the structural T (ST) sections do not govern the design. Neglect prying action.*

Solution
  Tension component $P_{ux} = 250 \times 4/5 = 200$ kN
  Shear component $P_{uy} = 250 \times 3/5 = 150$ kN
The factored loads T_u and V_u per bolt
    Tension $T_u = 200/6 = 33.33$ kN/bolt
    Shear $V_u = 150/6 = 25$ kN/bolt

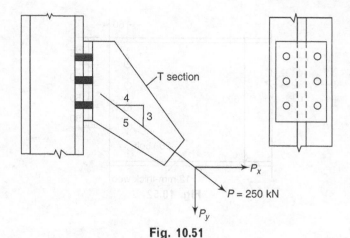

Fig. 10.51

For a 20-mm-diameter bolt,
Strength in single shear (from Appendix D) = 45.3 kN
Strength in tension (from Appendix D) = 68.5 kN
As per Eqn (10.18),

$$(V/V_{sd})^2 + (T_e/T_{nd})^2 \leq 1.0$$
$$(25/45.3)^2 + (33.33/68.5)^2 = 0.305 + 0.238 = 0.54 < 1.0$$

Hence, the six grade 4.6 bolts of 20-mm-diameter are sufficient to carry the load of 250 kN applied at the joint.

Example 10.10 *Repeat Example 10.9 assuming the use of M20 parallel-shank HSFG grade 8.8 bolts in clearance holes and a factored load of 500 kN.*

Solution
Taking $\mu = 0.48$,
Tensile capacity of one bolt (Appendix D) = 141 kN
Slip resistance of one bolt = 52.6
Tensile load per bolt = 400/6 = 66.67 kN
Shear load per bolt = 300/6 = 50 kN
Checking the interaction Eqn (10.23),

$$(50/52.6)^2 + (66.67/141)^2 = 0.903 + 0.224 = 1.13 > 1.0.$$ Hence unsafe.

Note that if the connection is only required to function as non-slip under service loads, the interaction equation has to be checked at service stage as follows:
Tensile load per load (assuming load factor of 1.5) = 400/(1.5 × 6) = 44.44 kN
Shear load per bolt = 300/(1.5 × 6) = 33.33 kN
Thus, at service load,

$$(33.33/52.6)^2 + (44.44/141)^2 = 0.402 + 0.100 = 0.502 < 1.00$$

Hence the connection is safe.

Example 10.11 *An ISMB 600 is connected to a column by web cleats with a single row of bolts. If the reaction is 350 kN and there are four 20-mm-diameter bolts through the web, as in Fig. 10.52, check if the section is adequate for block shear failure.*

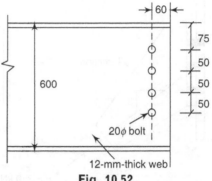

Fig. 10.52

Solution

Block shear capacity

$$T_{db1} = A_{vg} f_y/(\sqrt{3} \times 1.1) + 0.9 f_u A_{tn}/1.25$$

$$T_{db2} = 0.9 f_u A_{vn}/(\sqrt{3} \times 1.25) + f_y A_{tg}/1.1$$

Design strength of the web = 250 N/mm^2

Plate or web thickness = 12 mm

Net length of shear face = $(3 \times 50 + 75) - 3.5 \times 22 = 148$ mm

Net length of tension face = $60 - 0.5 \times 22 = 49$ mm

$$A_{vg} = 12 \times (3 \times 50 + 75) = 2700 \text{ mm}^2$$

$$A_{vn} = 12 \times 148 = 1776 \text{ mm}^2$$

$$A_{tg} = 12 \times 60 = 720 \text{ mm}^2$$

$$A_{tn} = 12 \times 49 = 588 \text{ mm}^2$$

Therefore,

$$T_{db1} = [2700 \times 250/(\sqrt{3} \times 1.1) + 0.9 \times 410 \times 588/1.25]/1000$$

$$= 354.283 + 173.58 = 527.86 \text{ kN}$$

or

$$T_{db2} = [0.9 \times 410 \times 1776/(\sqrt{3} \times 1.25) + 250 \times 720/1.1]/1000$$

$$= 302.69 + 163.636 = 466.32 \text{ kN}$$

Therefore,

$$T_{db} = 466.32 \text{ kN}$$

The value of $T_{db} = 466.32$ kN is much higher than the applied reaction of 350 kN and hence there will not be any block shear failure in this case.

Example 10.12 *Design a connection of a truss joint as shown in Fig. 10.53, using M16 black bolts of property class 4.6 and grade 410 steel. Assume that the members shown are capable of resisting the loads.*

Solution

Assume 8-mm thick gusset. The gusset plate is sandwiched between the angles and hence the bolts will be in double shear.

For 16-mm diameter property class 4.6 bolt, from Appendix D

Strength in double shear = $29.0 \times 2 = 58$ kN; $k_b = 40/(3 \times 18) - 0.25 = 0.49$

Strength in bearing = $2.5 \times 0.49 \times 16 \times 8 \times 410/(1.25 \times 1000) = 51.4$ kN

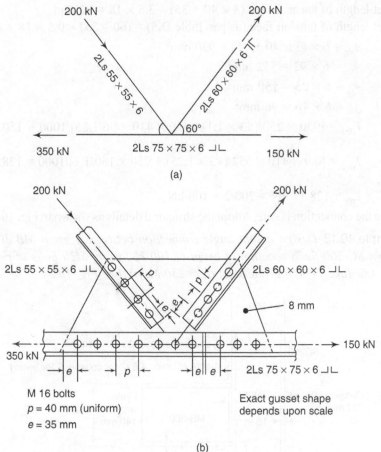

Fig. 10.53

Hence,

    Strength of bolt = 51.4 kN

    Bolts for 200 kN = 200/51.4 = 3.9 (hence provide four bolts)

    Bolts for 150 kN = 150/51.4 = 2.9 (hence provide three bolts)

    Bolts for 350 kN = 350/51.4 = 6.8 (hence provide seven bolts)

    Provide edge distance = 2 × 16 = 32, say 35 mm

    Pitch = 2.5 × 16 = 40 mm

Check for gusset plate

    Distance from first bolt to last bolt in member carrying 200 kN

        = 3p = 3 × 40 = 120 mm

    'Whitmore' effective width = 2 tan 30° × 120 = 138.56 mm

Capacity of plate = 0.9 × 138.56 × 8 × 410/(1.25 × 1000)

        = 327.22 kN > 200 kN

Hence the connection is safe. Note that the connection should be checked for block shear failure as shown in Example 10.11. The required calculations for 60 × 60 × 6 angle are shown as follows:

Net length of shear face = $(3 \times 40 + 35) - 3.5 \times 18 = 92$ mm

Net length of tension face (as per Table D.5) = $(60 - 35) - 0.5 \times 18 = 16$ mm

$$A_{vg} = 6 \times (3 \times 40 + 35) = 930 \text{ mm}^2$$
$$A_{vn} = 6 \times 92 = 552 \text{ mm}^2$$
$$A_{tg} = 6 \times 25 = 150 \text{ mm}^2$$
$$A_{tn} = 6 \times 16 = 96 \text{ mm}^2$$
$$T_{db1} = [930 \times 250/(\sqrt{3} \times 1.1) + 0.9 \times 410 \times 96/1.25]/1000 = 150.37 \text{ kN}$$

and

$$T_{db2} = [0.9 \times 410 \times 552/(\sqrt{3} \times 1.25) + 250 \times 150/1.1]/1000 = 128.17 \text{ kN}$$

Thus,

$$T_{db} = 128.17 \text{ kN} > 200/2 = 100 \text{ kN}$$

Hence the connection is safe. Adopt the structural details as shown in Fig. 10.53(b).

Example 10.13 *Design a seat angle connection between a beam MB 300 and column SC 200 for a reaction of beam of 100 kN, using M20 bolts of property class 4.6. Take Fe 410 grade steel (f_y = 250 MPa). See Fig. 10.54.*

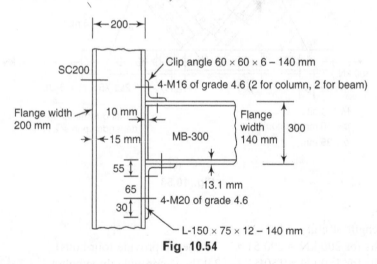

Fig. 10.54

Solution

Assuming $150 \times 75 \times 12$ angle

For M20 bolt,

Thickness of column flange (ISSC 200) = 15 mm

Strength in single shear (from Appendix D) = 45.3 kN

Strength in bearing = $2.5 \times 0.5 \times 20 \times 12 \times 400/(1.25 \times 1000) = 96$ kN

Hence,

Strength of bolt = 45.3 kN

Required number of bolts = $100/45.3 = 2.21$

Provide four M20 bolts.

Seating angle

Width of MB 300 = 140 mm

Hence,

Length of angle = 140 mm

Length of bearing required at root line of beam

$$= R/(t_w f_{yw}/\gamma_{m0})$$
$$= (100 \times 1000)/(7.7 \times 250/1.1) = 57.15 \text{ mm}$$

Assume end clearance of beam from face of column as 5 mm and tolerance of 5 mm.

Required length of outstanding leg = 57.15 + 10 = 67.15 < 75 mm. Hence, the length is as required.

Length of bearing on cleat = b_1 = 57.15 − $(T + R)$
$$= 57.15 - (13.1 + 14) = 30.05 \text{ mm}$$

For 150 × 75 × 12 angle, distance from the end of bearing on cleat to root angle (A to B) in Fig. 10.23,

$$b_2 = b_1 + 5 + 5 - (t + r_a) \text{ of angle}$$
$$= 30.05 + 10 - (12 + 10) = 18.05 \text{ mm}$$

Assume uniformly distributed load over bearing length b_1.

Moment at root of angle (point B) due to load to right of B
$$= (100 \times 18.05/30.05) \times (18.05/2) = 542 \text{ kN mm}$$

Moment capacity = $1.2 Z f_y/\gamma_{m0}$
$$= 1.2 \times (250/1.1) \times 140 \times 12^2/6 \times 10^{-3}$$
$$= 916 \text{ kN mm} < 542 \text{ kN mm}$$

Hence, the connection is safe.

Therefore, provide 150 × 75 × 12 mm seating angle

Shear capacity of the outstanding leg of cleat
$$= wtf_y/(\sqrt{3} \times 1.10) = 140 \times 12 \times 250/(\sqrt{3} \times 1.10 \times 1000)$$
$$= 220 \text{ kN} > 100 \text{ kN}$$

Hence, the shear capacity is as required.

Shear strength of beam
$$V_d = A_v f_{yw}/(\sqrt{3} \times 1.10) = (300 \times 7.7) \times 250/(\sqrt{3} \times 1.10 \times 1000)$$
$$= 303 \text{ kN} > 100 \text{ kN}$$

Hence, web does not need any stiffener at support.

Hence, provide a seating angle of 150 × 75 × 12 mm with four M20 bolts and provide a top clip angle of 60 × 60 × 6 with four M16 bolts of grade 4.6 as shown in Fig. 10.54.

Example 10.14 *Design a stiffened seat angle for a reaction of 250kN from a beam of ISMB 400 using M20 bolts of grade 4.6. The beam has to be connected to ISSC 200 column. Assume Fe 410 grade steel (f_y = 250 MPa).*

Solution

Length of bearing required at root line of beam
$$= R/(t_w f_{yw}/\gamma_{m0})$$
$$= (250 \times 1000)/(8.9 \times 250/1.1) = 124 \text{ mm}$$

Assuming a clearance including tolerance of 6 mm,

Required length of the outstanding leg = 124 + 6 = 130 mm

Width of the beam = 140 mm

Provide a seat angle of 130 × 130 × 8 mm of length 140 mm connected to the beam by two M20 bolts.

Stiffener angle

Bearing area required for stiffener angle = (250 × 1000)/(250/1.10) = 1100 mm^2

Select two angles ISA (80 × 80 × 8) with Area = 1220 mm^2

Length of outstanding leg = 80 – 8 = 72 mm

Thickness of the angle t = 8 mm

B/t = 72/8 = 9 < 14. Hence the thickness is fine. (clause 8.7.1.2 of IS code)

Distance of end reaction from column flange, e_x = 130/2 = 65 mm

Provide 20-mm-diameter grade 4.6 bolts at a pitch of 55 mm.

Strength of bolt in single shear (from Appendix D) = 45.3 kN

Strength of bolt in bearing = 2.5 × 0.5 × 20 × 8 × 410/(1.25 × 1000) = 65.6 kN

Hence,

Strength of bolt = 45.3 kN

Let us provide bolts in two rows with a pitch of 55 m [see Fig. 10.55]

$$n = \sqrt{[6M/(pn'V_{db})]}$$

$$n = \sqrt{[(6 \times 250 \times 65)/(55 \times 2 \times 45.3)]}$$
$$= 4.42$$

Provide five bolts in each row with an edge distance of 45 mm.

Depth of stiffener angle = 4 × 55 + 2 × 45 = 310 mm

h = 310 – 45 = 265 mm

The neutral axis lies at $h/7$ = 265/7 = 37.86 mm

Shear force in each bolt = 250/10 = 25 kN

The critical bolt will be at the top of the connection.

$$\Sigma y = 2[(45 - 37.86) + (100 - 37.86) + (155 - 37.86) + (210 - 37.86)$$
$$+ (265 - 37.86)] = 1171.4 \text{ mm}$$
$$\Sigma y^2 = 2[7.14^2 + 62.14^2 + 117.14^2 + 172.14^2 + 227.14^2]$$
$$= 197,717.8 \text{ mm}^2$$
$$M^1 = M/[1 + (2h/21)(\Sigma y/\Sigma y^2)]$$
$$= 250 \times 10^3 \times 65/[1 + (2 \times 265/21)(1171.4/197,717.8)]$$
$$= 16.25 \times 10^6/1.15 = 14,136 \times 10^3 \text{ Nmm}$$

Tensile force = $M^1 y_n/\Sigma y^2$
$$= 14,136 \times 10^3 \times 227.14/197,717.8 = 16,240 \text{ N}$$

Tensile strength of 20-mm bolt (Appendix D) = 68.5 kN

As per clause 10.3.6 of IS code,

$$(25/45.3)^2 + (16.24/68.5)^2 = 0.305 + 0.056 = 0.361 < 1.0$$

Hence the connection is safe.

Provide a top clip angle of 60 × 60 × 6 mm with four M16 bolts and a packing plate of size 185 × 160 × 8 mm as shown in Fig. 10.55.

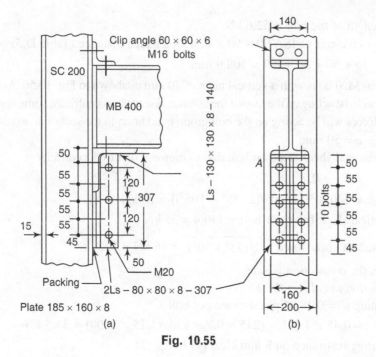

Fig. 10.55

Example 10.15 *Design a bolted web angle connection for a ISMB 400 beam, to carry a reaction of 140 kN due to factored loads. The connection is to the flange of a column ISSC 200 in grade Fe 410 steel (see Fig. 10.56).*

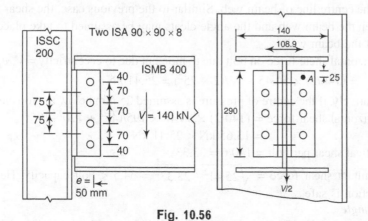

Fig. 10.56

Solution

For IS MB 400, b_f = 140 mm, t_w = 8.9 mm

Using M20 HSFG bolts of grade 8.8

    Shear capacity of M20 bolt in single shear (from Appendix D) = 52.6 kN

    Bearing capacity on web of beam, assuming k_b = 0.5 (web thickness of ISMB 400 = 8.9 mm)

$$= 2.5 \times 0.5 \times 20 \times 8.9 \times 410/(1.25 \times 1000) = 72.98 \text{ kN}$$

Strength of the bolt = 52.6 kN

Using web cleats of ISA 90 × 90 × 8, with a gauge distance (Table D.5)

$$g = 50 + 50 + 8.9 = 108.9 \text{ mm}$$

Use four M20 bolts with a vertical pitch of 70 mm as shown in Fig. 10.56. Assuming the shear to be acting on the face of the column, due to the eccentricity, some horizontal shear forces will be acting on the bolt group in addition to the shear due to reaction.

$$e = 50 \text{ mm}$$

Horizontal shear force on bolt due to moment due to eccentricity

$$= V_x e_x r_i / \Sigma r_i^2$$
$$= 140 \times 50 \times 105/[2 (35^2 + 105^2)] = 30 \text{ kN}$$

Vertical shear force per bolt = 140/4 = 35 kN

Resultant shear force = $\sqrt{(35^2 + 30^2)}$ = 46 kN < 52.6 kN

Hence, the connection is safe.

Connection to column flange

Assuming μ = 0.48, slip resistance per bolt

$$= 0.48 \times 1 \times 1 \times (245 \times 0.7 \times 830)/(1.25 \times 1000) = 54.5 \text{ kN}$$

Bearing resistance on 8 mm cleat per bolt

$$= 2.5 \times 0.5 \times 8 \times 20 \times 410/(1.25 \times 1000) = 65.6 \text{ kN}$$

Hence,

Bolt strength = 54.5 kN

Try six bolts as shown in Fig. 10.56 with a vertical pitch of 75 mm, 54.45 mm from the centre line of beam web. Similar to the previous case, the shear transfer between the beam web and the angle cleats may be assumed to take place on the face of the beam web.

Horizontal shear force on bolt due to moment due to eccentricity = $V_x e_x r_i / \Sigma r_i^2$

$$= (140/2) \times 54.45 \times 75/(2 \times 75^2) = 25.41 \text{ kN}$$

Alternatively, if the centre of pressure is assumed 25 mm below the top cleat, then

Horizontal shear force = $(140/2) \times 54.45 \times 195/(45^2 + 120^2 + 195^2)$
$$= 13.65 \text{ kN} < 25.41 \text{ kN}$$

Vertical shear per bolt = 140/6 = 23.33

Resultant shear force = $\sqrt{25.41^2 + 23.33^2}$ = 34.5 < bolt capacity. Hence the connection is safe.

Cleat angle

Provide two angles ISA 90 × 90 × 8 of length 290 mm. Check bending at bolt line of connection to column flange.

Bending moment = 140/2 × 54.45 = 3811.5 N m

Moment capacity = $1.2 f_y Z / \gamma_{m0}$ = 1.2 × 250 × (8 × 290²/6) × 10⁻³/1.10 = 33640 N m > 3811.5 N m

Hence, the chosen cleat angle size is as required.

Shear strength of single cleat = $f_y A_v / (1.1 \times \sqrt{3})$

$$= 250 \times (290 \times 8)/(1.1 \times \sqrt{3})$$

$$= 304 \text{ kN} > N/2 = 70 \text{ kN}$$

Local shear strength of beam

Shear capacity $= f_y A_v/(1.1 \times \sqrt{3}) = 250 \times (400 \times 8.9)/(1.1 \times \sqrt{3})$
$$= 467 \text{ kN} > 140 \text{ kN}$$

Example 10.16 *Design a header plate connection for a ISMB 400 beam to carry a reaction of 140 kN due to factored loads. The connection is to the flange of a ISSC 200 column. Use Fe 410 grade steel (f_y = 250 MPa) and M20 bolts of grade 4.6.*

Solution

Bolted connection to column

Assume the thickness of the end plate as 6 mm.

Shear capacity of M20 bolt in single shear (Appendix D) = 45.3 kN

Bearing capacity of bolt on 6-mm end plate (with k_b = 0.5)
$$= (2.5 \times 0.5 \times 20 \times 6 \times 410)/(1.25 \times 1000) = 49.2 \text{ kN}$$

Strength of bolt = 45.3 kN

Required number of bolts = 140/45.3 = 3.09

Provide four M20 bolts of grade 4.6.

End plate

Length of end plate = 60 + 2 × 45 = 150 mm

Provide a = 150 < 30 × 6 = 180 mm (see section 10.6.6)

Use a 160 × 150 × 6-mm end plate

Length of the fillet weld connecting end plate to beam web = 150 – 2 × 6 = 138

Assuming site welded (see Chapter 11)

Size of the weld = $(140 \times 10^3)/(2 \times 138 \times 158) = 3.21$

Provide a 6-mm fillet weld.

Shear stress on the web of the beam at the end plate

$$= (140 \times 10^3)/(8.9 \times 150) = 104.9 \text{ MPa} < 250/(\sqrt{3} \times 1.10) = 131.2 \text{ MPa}$$

Hence, the connection is safe.

Adopt the details as shown in Fig. 10.57.

End plate shear strength $= f_y A_v/(1.1 \times \sqrt{3}) = 0.525 \times 250 \times 636 \times 10^{-3} =$ 83.5 kN < 140 kN

Note: $A_v = t_p(l_p - 2d_h) = 6 \times (150 - 2 \times 22) = 636 \text{ mm}^2$

Hence increase thickness to 10 mm

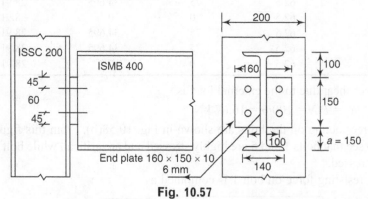

Fig. 10.57

adequate for the given load in a bearing type connection assuming threads in the shear plane?

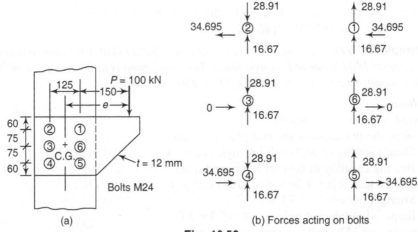

(a)

(b) Forces acting on bolts

Fig. 10.58

Solution

Since the bolt pattern is symmetrical (as in most practical problems), the centroid is readily located and marked as *CG* in Fig. 10.58.

$$x = 125/2 = 62.5 \text{ mm}$$
$$\Sigma x^2 = 6 \times 62.5^2 = 23{,}437.5 \text{ mm}^2$$
$$\Sigma y^2 = 2(2 \times 75^2) = 22{,}500 \text{ mm}^2$$
$$\Sigma(x^2 + y^2) = 45{,}937.5 \text{ mm}^2$$
$$e = 150 + 125/2 = 212.5 \text{ mm}$$

Hence,

$$R_{x_i} = [212.5(100)y_i]/45{,}937.5 = 0.4626y_i$$
$$R_{y_i} = 0.4626x_i$$

The forces acting on the bolts are calculated as shown in the table below.

Bolt	x (mm)	y (mm)	R_x (kN)	R_y (kN)
1	62.5	75	34.695	28.91
2	62.5	75	34.695	28.91
3	62.5	0	0	28.91
4	62.5	75	34.695	28.91
5	62.5	75	34.695	28.91
6	62.5	0	0	28.91

Direct shear due to the external load is

$$R_v = P/N = 100/6 = 16.67 \text{ kN}$$

The forces acting on the bolts are shown in Fig. 10.58(b). From this figure, it is easy to see that bolts 1 and 5 are highly stressed and are critical, while bolt 3 is the least stressed.

The resisting force on bolt 1 is computed as

$$R = \sqrt{(28.91 + 16.67)^2 + 34.695^2} = \sqrt{3281.28} = 57.28 \text{ kN}$$

$$R = \sqrt{(28.91+16.67)^2 + 34.695^2} = \sqrt{3281.28} = 57.28 \text{ kN}$$

$$f_v = \text{Stress in the bolt} = 57.28 \times 1000/353 = 162.3 \text{ MPa} < 400/(\sqrt{3} \times 1.25)$$
$$= 185 \text{ MPa}$$

Check for plate bearing:

$f_b = 57.28/(0.024 \times 12) = 198.88$ MPa < 410 MPa, Hence ok.

Check for possible tension rupture of the plate along the forward bolt line:

Moment of inertia $I = 12(2 \times 60 + 2 \times 75)^3/12 - 2(12 \times 24)(75)^2 = 16,443,000$ mm^4

Section modulus $\quad Z = 16,443,000 \times (2/270) = 121,800$ mm^3

Bending moment at forward bolt line $= 100 \times 150 = 15,000$ kNmm

$\quad f_b = M/Z = 15,000 \times 1,000/121,800$
$\quad\quad = 123.2$ MPa $< 250/1.1$ MPa $= 227$ MPa

Check plate buckling $b/t = 150/12 = 12.5 < 29.3$ (Table 2 of IS 800)

Shear strength of plate $= f_y A_v/(1.1 \times \sqrt{3}) = 250 \times (270 - 3 \times 27) \times 12/(1.1 \times \sqrt{3})$
$\quad\quad = 297$ kN < 140 kN

Thus, the joint is adequate for bolt shear, plate bearing on bolt, and bending.

Example 10.18 *Determine the size of the bolts required to connect the bracket (supporting a crane girder) to the column shown in Fig. 10.59.*

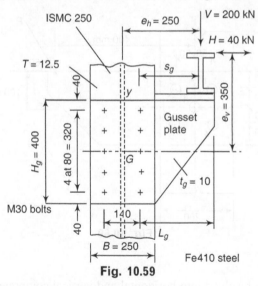

Fig. 10.59

Solution

$\quad x = 140/2 = 70$ mm
$\quad \Sigma x^2 = 10 \times 70^2 = 49,000$ mm^4
$\quad \Sigma y^2 = 4(80^2 + 160^2) = 128,000$ mm^4
$\quad \Sigma(x^2 + y^2) = 177,000$ mm^4
$\quad e_h = 250$ mm; $\quad e_v = 350$ mm

The maximum vector force in the direction of *y-y* axis on a bolt farthest from the centroid of the bolt group.

$\quad R_y = V/n + [(Ve_h + He_v) x_n/\Sigma(x^2 + y^2)]$
$\quad\quad = 200/10 + [(200 \times 250 + 40 \times 350)70/177,000] = 45.3$ kN

The maximum vector force in the direction of x-x axis on the same bolt

$$R_x = H/n + [(Ve_h + He_v) y_n/\Sigma(x^2 + y^2)]$$
$$= 40/10 + [(200 \times 250 + 40 \times 350)160]/177,000 = 61.8 \text{ kN}$$

Resultant vector force on the bolt

$$R = \sqrt{45.3^2 + 61.8^2} = 76.62 \text{ kN}$$

Assuming ten 24-mm-diameter grade 4.6 bolts, stress in the bolt

$$f_v = 76.62 \times 1000/353 = 217 \text{ MPa} > 185 \text{ MPa}$$

Hence, provide ten 30-mm-diameter grade 4.6 bolts, then stress in the bolt,

$$f_v = 76.62 \times 1000/561 = 136.5 \text{ MPa} < 185 \text{ MPa}$$

Note that some codes restrict the bolt diameter to 24 mm for a flange width of 250 mm. As in the previous example, checks should be made for plate bearing, tension rupture of the plate, and plate buckling. This is left as an exercise for the reader.

Example 10.19 *Find the maximum load inclined at 60° to the horizontal, which the bracket shown in Fig. 10.60 can transmit if five grade 8.8 bolts with a diameter of 20 mm are used and plates connected are 10-mm thick. Determine the load (i) if the joint is considered a slip joint and (ii) if the joint is considered as a non-slip joint.*

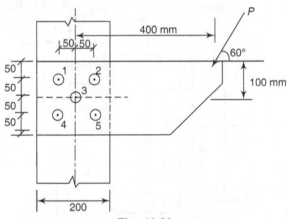

Fig. 10.60

Solution
The force can be resolved into horizontal and vertical components as

$$P_h = P \cos 60° = 0.5P$$
$$P_v = P \sin 60° = 0.866P$$

Bolt 5 will have the maximum stress (this can be verified by calculating the forces in various bolts in a tabular form as shown in Example 10.17)

Force in bolt 5 in y-y direction

$$\Sigma(x^2 + y^2) = 4(50^2 + 50^2) = 20,000 \text{ mm}^4$$
$$R_y = 0.866P/5 + [(0.866P \times 400 - 0.5P \times 100) \times 50/20,000] = 0.9142P$$

$$R_x = 0.5P/5 + [(0.866P \times 400 - 0.5P \times 100) \times 50/20,000] = 0.841P$$

Resultant force in the bolt $= P\sqrt{0.9142^2 + 0.841^2} = 1.242P$

(i) Joint is considered as slip joint

Strength of a 20-mm-diameter bolt (grade 8.8) from Appendix D:

Bearing $= 410 \times 10 \times 20/1000 = 82$ kN

Shear $= 370 \times 245/1000 = 90.65$

Therefore,

Bolt strength $= 82$ kN

Equating bolt strength with the maximum force, we get

$1.242P = 82$ kN

or

$P = 82/1.242 = 66$ kN

(ii) Joint is considered as non-slip joint

Slip resistance of a 20-mm-diameter bolt from Appendix D $= 52.6$ kN

Therefore,

$1.242P = 52.6$ kN

Thus,

$P = 42.35$ kN

Hence, if the joint is a non-slip joint, the load carrying capacity of the bolt group is reduced by 36% (from 66 kN to 42.35 kN), for the same grade and diameter of the bolt.

Example 10.20 *Design a bolted end plate connection between an ISMB 300 beam and an ISHB 200 column, to transfer a vertical factored shear of 120 kN and a factored hogging bending moment of 120 kNm. Use HSFG bolts of diameter 20 mm. The connection is as shown in Fig. 10.61.*

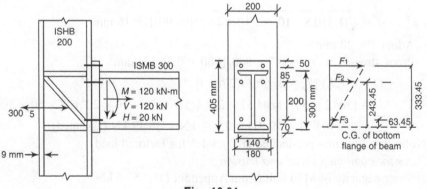

Fig. 10.61

Solution

For ISMB 300, $b_f = 140$ mm, $t_f = 13.1$ mm, $t_w = 7.7$ mm

Bolt forces

By taking moment about the centre of the bottom flange

$$120 \times 10^3 + 20 \times (300/2 - 13.1/2) = (2F_1 + 2F_2) \times (300 - 13.1)$$
$$+ 2F_3 \times 63.45$$

Rewritting in terms of F_1, we get

$$122,869 = (2F_1/286.9) [2 \times 286.9^2 + 63.45^2]$$
$$122,869 = 1,175.67F_1$$
$$F_2 = F_1 = 104.51 \text{ kN}$$
$$F_3 = (63.45/286.9)F_1 = 23.11 \text{ kN}$$

Reaction at bottom flange

$$F_c = 2(104.5 + 104.51 + 23.11) - 20$$
$$= 444.26 \text{ kN}$$

Capacity of beam flange $= (f_y/\gamma_{m0})A$

$$= (250/1.1) \times 13.1 \times 140 \times 10^{-3} = 416.8 \text{ kN}$$

This is only 6.2% less than F_c. Hence the connection is safe.

End plate and bolts

Consider the portion of end plate above top flange (assume 10-mm fillet welds to flange and 180-mm wide end plate)

Distance from the centre line of the bolt to the toe of the fillet weld

$$l_v = 40 - 10 = 30 \text{ mm}$$

Adopted end distance $l_e = 50$ mm

Effective length of end plate per bolt $= 180/2 = 90$ mm

Tension capacity of M20 bolt (Appendix D) = 141 kN

Allowable prying load $Q = 141 - 104.51 = 36.49$ kN

Moment at the toe of the weld $= Tl_v - Ql_e$

$$= 104.51 \times 30 - 36.49 \times 50 = 1,310.8 \text{ Nm}$$

Moment capacity of the plate $= (f_y/1.10)(wT^2/4)$

Therefore,

$$T = \sqrt{[1,310.8 \times 10^3 \times 1.10 \times 4/(250 \times 90)]} = 16 \text{ mm}$$

Adopt $T = 20$ mm

Proof stress $= 0.7f_{ub} = 0.7 \times 800/1000 = 0.56$ kN/mm^2

$$Q = (l_v/2l_e) [T_e - \beta\gamma f_o b_e t^4/(27l_e l_v^2)]$$
$$= [30/(2 \times 50)] [104.51 - 2 \times 1.5 \times 0.56 \times 90 \times 20^4/(27 \times 50 \times 30^2)]$$
$$= 0.3(104.51 - 19.91) = 25.38 \text{ kN} < 36.49 \text{ kN (assumed)}$$

(Note: $\beta = 2$ for non-pre-loaded and $\gamma = 1.5$ for factored load)

Check for combined shear and tension

Shear capacity of M20 bolt (from Appendix D) = 52.6 kN

Shear per bolt $= 120/6 = 20$ kN

Tensile capacity of the bolt = 141 kN

$$(20/52.6)^2 + [(104.51 + 25.38)/141]^2 = 0.145 + 0.848 = 0.993 < 1.00$$

Hence the connection is safe.

In addition to the above checks, the welds connecting flange to end plate and beam web to end plate should be checked. The column flange should also be checked. Usually, stiffeners are provided in the web of the column where the beam flange is supported as shown in Fig. 10.61. These calculations are shown in the next chapter on welded connections.

Example 10.21 *Design a flange angle connection using M16 bolts of grade 4.6 to transfer a factored moment of 12kNm and a shear of 150kN from a beam of ISMB 350 to a column of ISHB 300.*

Solution

For M16 bolts (from Appendix D):

   Strength of a bolt in single shear = 29 kN

   Strength in double shear = 29 × 2 = 58 kN

Bearing strength on web of thickness 8.1 mm (ISMB 350), assuming $k_b = 0.5$

$$= 2.5 \times 0.5 \times 16 \times 8.1 \times 410/(1.25 \times 1000) = 53.1 \text{ kN}$$

   Tensile strength of bolt (from Table D.3 of Appendix D) = 43.8 kN

Flange clip angles

Provide two flange clip angles of size ISA 125 × 95 × 12 with 95-mm leg, connected to the column flange.

   Gauge distance for 95-mm leg (from Table D.5) = 55 mm

   Lever arm = 350 + 2 × 55 = 460 mm

Providing two bolts to connect flange clip angle with column flange,

   Moment capacity = 2 × 42.7 × 0.46 = 39.28 > 12 kNm (applied moment)

   Force on bolts connected to beam flange= Moment/depth of beam

$$= 12/0.35 = 34.29 \text{ kN}$$

   Required number of bolts = 34.29/29 = 1.18

Therefore, provide four bolts in 2 rows.

Flange angle thickness

   Pull P in the bolt connecting column flange with clip angle

$$= 12 \times 10^3/460 = 26.09 \text{ kN}$$

$$M_A = 0.6 \times 26.09 \times (55 - 12/2) = 767 \text{ kNm}$$

   Moment capacity of the 12-mm thick leg of the angle = $1.2 \times (f_y/\gamma_{m0})Z$

$$= 1.2 \times (250/1.1) \times (140 \times 12^2/6) \times 10^{-3}$$

$$= 916.4 \text{ kNm} > 767 \text{ kNm. Hence, the thickness provided is adequate.}$$

Web clip angle

The bolts will be in double shear (connected to web of beam)

   Required number of bolts = 150/58 = 2.58

Provide 3 bolts of M16 at a pitch of 50 mm and edge distance of 40 mm. Use thickness of web angle as 8 mm (depth of beam < 400 mm).

   Length of web angle = 2 × (50 + 40) = 180 mm

Provide 180-mm long angle of size 75 × 75 × 8 mm. The bolts connecting the web angle to the column flange will be in single shear. Hence, provide six M16 bolts.

Check for prying force

$$l_v = \text{edge distance} = 40 \text{ mm}$$

$$\gamma = 1.5 \text{ (factored load)}$$

$$\beta = 2 \text{ (non-pre-loaded)}$$

$$f_o = 0.7f_{ub} = 0.7 \times 400 = 280 \text{ MPa}$$

$$Q = [l_v/(2l_e)] \, [T_e - \beta \gamma f_o b_e t^4/(27 l_e l_v^2)]$$
$$= [55/(2 \times 40)] \, [26.09 - 2 \times 1.5 \times 0.28 \times 140 \times 12^4/(27 \times 40 \times 51.5^2)]$$
$$= 15.45 \text{ kN}$$

Total load in the bolts = 26.09 + 15.45 = 41.54 kN

Capacity of two M16 bolts in tension = $2 \times 43.8 = 87.6$ kN > 41.54 kN
Hence the connection is safe.

Shear strength of web angle = $2 \times (180 - 3 \times 18) \times 8 \times 250/(\sqrt{3} \times 1.1)$
$$= 264 \text{ kN} > 150 \text{ kN}$$

The designed details are shown in Fig. 10.62.

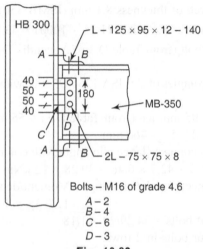

Fig. 10.62

Example 10.22 *Design a T-stub (split beam T) connection to transfer a factored shear of 150 kN and a factored moment of 50 kNm from the end of a beam ISMB 350 to a column of ISHB 300 using 16 mm bolts.*

Solution

Try M16 bolts. For M16 bolts (from the previous example):

Strength in single shear = 29 kN
Strength in double shear = 58 kN
Bearing strength (against 8.1 thick plate) = 53.1 kN
Tensile strength = 43.8 kN

Web angle connection

Provide similar arrangement as in the previous example, since the shear force is same.

Thus provide 75 × 75 × 8 mm angle with six M16 bolts connecting to the flanges of the column and three M16 bolts connecting to the beam web. Use a T-stub 220 mm long cut from ISMB 450.

Flange width of ISMB 450 = 150 mm

Flange thickness = 17.4 mm

Web thickness = 9.4 mm

Root radius = 15 mm

Split beam-to-beam top flange connection

Using 20 mm bolts

Flange force, $P = 50 \times 10^3/350 = 142.86$ kN

Required number of bolts = 142.86/45.3 = 3.2

Provide four bolts in two rows.

Thickness of T-web (adopting a width of 140 mm)

$t = P/(b \times f_y/\gamma_{m0})$

$= 142.86 \times 1000/[(140 - 2 \times 18) \times (250/1.10)]$

$= 6$ mm < 9.4 mm. Hence the thickness provided is adequate.

Split beam-to-column flange connection

l_e = minimum edge distance = $1.5 \times 16 = 24$ mm

Provide an edge distance of 30 mm

a = distance from fillet line to bolt = $[(150/2) - (9.4/2) - 15 - 30]$

$= 25.3$ mm

Moment in T-stub = $0.5P \times a$

$= 0.5 \times 142.86 \times 25.3$

$= 1807.2$ kNmm

Moment capacity of T-stub flange

$= 1.2 \times (250/1.10) \times 140 \times 17.4^2/6 \times 10^{-3}$

$= 1926.65 > 1807.2$ kNmm

Hence, thickness provided is as required. The force in a row of bolts connecting the T-stub to column

$= P/2 = 142.86/2 = 71.43$ kN

The tensile force in 16-mm-diameter bolts = $43.8 \times 2 = 87.6$ kN > 71.43 kN However, provide four 20-mm-diameter bolts, in two rows with a pitch of 60 mm and an edge distance of 40 mm.

Tension capacity of four M20 bolts = $4 \times 68.5 = 274$ kN > 142.86 kN

Hence ok.

Tensile force in bolt due to the applied moment

$= 142.86/4 = 35.72$ kN

Check for prying force

$l_e = 30$ mm (provided)

$l_v = [(150/2) - (9.4/2) - 30] = 40.3$ mm

$\gamma = 1.5$ (factored load)

$\beta = 2$ (non-pre-loaded)

$f_o = 0.7f_{ub} = 0.7 \times 400 = 280$ MPa

$Q = [40.3/(2 \times 30)][35.72 - 2 \times 1.5 \times 0.28 \times 140 \times 17.4^4/(27 \times 30$

$\times 40.3^2)] = 18.48$ kN

Total tensile load in bolt = 35.72 + 18.48 = 54.2 kN < 68.5 kN

Hence the bolts are adequate. The details of the connection are given in Fig. 10.63.

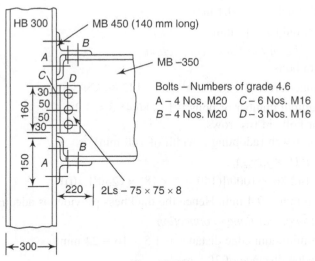

Fig. 10.63

Example 10.23 *Design a bolted web cleat connection for an ISMB 600 and two coped beams of size ISMB 400 (300 kN reaction due to factored loads) and ISMB 250 (75 kN reaction due to factored loads) using grade 8.8 bolts of 20 mm diameter (see Fig. 10.64).*

Solution

For M20 grade 8.8 bolts:

Shear capacity in single shear = $245 \times 800/(\sqrt{3} \times 1.25 \times 1000) = 90.5$ kN

Shear capacity in double shear = $2 \times 90.5 = 181$ kN

Left hand side beam ISMB 400

Web thickness = 8.9 mm

Bearing capacity of bolt against the web, with $k_b = 0.6$

$\qquad = 2.5 \times 0.6 \times 20 \times 8.9 \times 410/(1.25 \times 1000) = 87.6$ kN

Try five bolts at 60 mm spacing and 40-mm edge distance. Horizontal shear force on bolt due to eccentricity

$\qquad = 300 \times (50 + 0.5 \times 12) \times 120/[2(60^2 + 120^2)] = 56$ kN

Vertical shear force per bolt = $300/5 = 60$ kN

Resultant shear force = $\sqrt{(56^2 + 60^2)}$ = 82.07 kN > bolt capacity. Hence the connection is safe.

Connection to web of supporting beam

Assuming a cleat of 2L ISA 90 × 90 × 8 mm

Bearing capacity on 8 mm cleat = $2.5 \times 0.6 \times 20 \times 8 \times 410/(1.25 \times 1000) = 78.72$ kN

Capacity of 78.72 kN governs

Try eight bolts, two columns at a horizontal gauge length of 120 mm with a vertical pitch of 50 mm for the first three bolts and 140 mm for the last bolt with an edge distance of 40 mm in the cleat. Assuming centre of pressure at 25 mm below top of cleat,

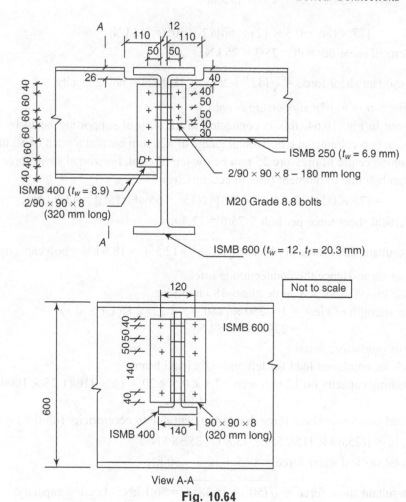

View A-A

Fig. 10.64

Horizontal shear force on bottom bolt due to moment due to eccentricity

$$= [300 \times 0.5(120 - 8.9) \times 255]/[2(15^2 + 65^2 + 115^2 + 255^2)] = 25.69 \text{ kN}$$

Vertical shear force per bolt = 300/8 = 37.5 kN

Resultant shear force = $\sqrt{(37.5^2 + 25.69^2)}$ = 45.46 kN < bolt capacity

Hence the connection is ok.

Use two 90 × 90 × 8 angle cleats, 320 mm long.

Shear strength of cleat = $f_y A_v/(1.1 \times \sqrt{3})$

$$= 2 \times 250 \times (320 - 5 \times 22) \times 8/(1.1 \times \sqrt{3})$$

$$= 440 \text{ kN} > 300 \text{ kN}$$

Right hand side beam

ISMB 250 with t_w = 6.9 mm

Bearing capacity of web of beam = $2.5 \times 0.6 \times 20 \times 6.9 \times 410/(1.25 \times 1000) = 56.58$ kN
Try three bolts at 50-mm vertical pitch and 40-mm edge distance. Horizontal shear force on bolt due to moment (due to eccentricity)

$$= [75 \times (50 + 0.5 \times 12) \times 50]/(2 \times 50^2) = 42 \text{ kN}$$

Vertical shear per bolt = 75/3 = 25 kN

Resultant shear force = $\sqrt{(42^2 + 25^2)}$ = 48.9 kN < bolt capacity

Connection to web of supporting beam

As seen in Fig. 10.64, this is connected to the web of supporting beam by six bolts, in two columns at a horizontal gauge of 120 mm c/c and a pitch of 50 mm. Assuming centre of pressure 25 mm below top of cleat, horizontal shear force on bottom bolt due to moment (due to eccentricity)

$$= [75 \times 0.5(120 - 6.9) \times 115]/[2(15^2 + 65^2 + 115^2)] = 13.8 \text{ kN}$$

Vertical shear force per bolt = 75/6 = 12.5 kN

Resultant shear force per bolt = $\sqrt{(13.8^2 + 12.5^2)}$ = 18.6 kN < bolt capacity in double shear. Hence the connection is safe.

Use two 90 × 90 × 8 angle cleats 180 mm long.

Shear strength of cleat = 2 × 250 × (180 − 3 × 22) × 8/(1.1 × $\sqrt{3}$)
$$= 239 \text{ kN} > 75 \text{ kN}$$

Web of supporting beam

Check for combined load for left and right hand beam.

Bearing capacity on 12 mm web= 2.5 × 0.5 × 20 × 12 × 410/(1.25 × 1000)
$$= 98.4 \text{ kN}$$

Total horizontal shear force due to moment due to eccentricity (on third row)

$$= [(25.69 \times 115/255] + 13.80 = 25.38 \text{ kN}$$

Total vertical shear force = 37.5 + 12.5 = 50 kN

Resultant shear force = $\sqrt{(50^2 + 25.38^2)}$ = 56.1 kN < bearing capacity.

Hence, the connection is safe.

Left hand beam

At the end of the notch, depth of section = 400 − 26 = 374 mm

Shear capacity = (374 × 8.9) × (250/√3)/(1.1 × 1000)
$$= 436.76 \text{ kN} > \text{shear force} = 300 \text{ kN}$$

Shear capacity through bolt holes

Block shear capacity:

$$T_{db1} = [250/(\sqrt{3} \times 1.1) \times 8.9 \times (54 + 4 \times 60) + (0.9 \times 410/1.25) \times 8.9$$
$$\times (50 - 22/2)]/1000 = 445.8 \text{ kN}$$

$$T_{db2} = [(0.9 \times 410/(\sqrt{3} \times 1.25) \times 8.9 \times (54 + 4 \times 60 - 4.5 \times 22) + 50 \times 8.9$$
$$\times (250/1.10)]/1000 = 396.92 \text{ kN}$$

Hence,

$$T_{db} = 396.92 \text{ kN}$$

Shear force = 300 kN < shear capacity. Hence the connection is safe.

Right hand side beam

At the end of the notch, depth of section = 250 − 26 = 224

Shear capacity = $(224) \times 6.9 \times (250/\sqrt{3})/(1.1 \times 1000)$
　　　　　　 = 202.80 kN > shear force = 75 kN

Block shear capacity of web

$T_{db1} = [250/(\sqrt{3} \times 1.1) \times 6.9 \times (54 + 2 \times 50) + (0.9 \times 410/1.25)$
　　　　　 $\times 6.9 (50 - 22/2)]/1000 = 218.86$ kN

$T_{db2} = [0.9 \times 410/(\sqrt{3} \times 1.25) \times 6.9 \times (54 + 2 \times 50 - 2.5 \times 22) + 50 \times 6.9$
　　　　　 $\times (250/1.1)]/1000 = 194.83$ kN

$T_{db} = 194.83$ kN

Shear force = 75 kN < Shear capacity
Hence, the connection is safe.

Example 10.24 *Design a bolted splice for an ISMB 400 section to transfer a factored bending moment of 120 kNm and a factored shear of 80 kN. Assume that the flange splice carries all the moment and that the web splice carries only the shear (see Fig. 10.65).*

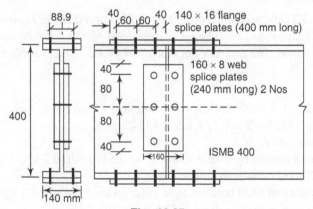

Fig. 10.65

Solution
For ISMB 400, $t_f = 16.0$ mm and $t_w = 8.9$ mm
Flange splice
　　Force in the flanges = $BM/(D - t_f) = 120 \times 10^3/(400 - 16) = 312.5$ kN
Assuming M20 grade 8.8 HSFG bolts,
　　Slip resistance per bolt (from Appendix D) = 52.6 kN
　　Bearing resistance on flange = $2.5 \times 0.5 \times 20 \times 16 \times 410/(1.25 \times 1000) = 131.2$ kN
　　Bolt value = 52.6 kN
Use three rows of two bolts at a pitch of 60 mm.
　　Net area of flange = $(140 - 2 \times 22) \times 16 = 1536$ mm^2
　　Flange capacity = $(250/1.1) \times 1536/1000 = 349$ kN > flange force = 312.5 kN
Hence, the connection is safe.
　　Use a 140 mm wide splice plate.

Required thickness of plate $= 312.5 \times 10^3/[250(140 - 2 \times 22)/1.10]$
$$= 14.32 \text{ mm}$$
Use two flange splice plates of size $400 \times 140 \times 16$ mm
Web splice
For M20 HSFG bolt in grade 8.8 in double shear
    Slip resistance per bolt (from Appendix D) $= 2 \times 52.6 = 105.2$ kN
Try 8-mm thick web splice plate on both sides of the web.
    Bearing resistance per bolt $= 2.5 \times 0.5 \times 20 \times 8 \times 410/(1.25 \times 1000) = 65.6$ kN
    Bolt value $= 65.6$ kN
Try three bolts at a vertical pitch of 80 mm and 40 mm from the centre of the joint
    Horizontal shear force on bolt due to moment (due to eccentricity)
$$= 80 \times 40 \times 80/(2 \times 80^2) = 20 \text{ kN}$$
Vertical shear force per bolt $= 80/3 = 26.67$ kN

Resultant shear force $= \sqrt{(26.67^2 + 20^2)} = 33.34$ kN < 65.6 kN
Use two web splice plates of size $240 \times 160 \times 8$ mm. The designed splice details
are shown in Fig. 10.65.

Example 10.25 *Design a bolted cover plate splice for a ISHB 225 column connected
to an ISHB 225, to transfer a factored axial load of 450 kN, both columns are of
grade Fe 410 steel. The ends are not machined for full contact in bearing (see Fig.
10.66).*

Solution
For ISHB 225, $A = 5490$ mm^2, $t_f = 9.1$ mm, and $t_w = 6.5$ mm
Area of web $= (225 - 2 \times 9.1) \times 6.5 = 1344.2$ mm^2
Design of web splice
Portion of load carried by web $= 450 \times 1344.2/5490 = 110.2$ kN
Use two M20 HSFG bolts in double shear.
    Slip resistance of M20 bolt in single shear (from Appendix D) $= 52.6$ mm
    Slip resistance in double shear $= 52.6 \times 2 = 105.2$ kN
    Bearing resistance (with $k_b = 0.6$) $= 2.5 \times 0.6 \times 20 \times 6.5 \times 410/(1.25 \times 1000) = 64$ kN
    Bolt value $= 64$ kN
    Shear force per bolt $= 110.2/2 = 55.1$ kN < 64 kN
Provide an edge distance of 45 mm and provide two $150 \times 150 \times 8$ web splice.
Flange splice
    Portion of load carried by each flange $= (450 - 110.2)/2 = 169.9$ kN
For four M20 HSFG bolts of grade 8.8 in two rows in single shear.
    Total slip resistance $= 4 \times 52.6 = 210.4$ kN
    Total bearing resistance (with $k_b = 0.5$) $= 4 \times 2.5 \times 0.5 \times 20 \times 9.1 \times 410/(1.25 \times 1000)$
$$= 298.5 \text{ kN}$$
    Total bolt strength $= 210.4$ kN > 169.9 kN
Hence the connection is safe.
    Provide an edge distance of 45 mm and a vertical pitch of 60 mm. Provide a
225-mm wide splice plate.
    Thickness of flange splice plate $= 169.9 \times 10^3/[(250/1.1) \times (225 - 2 \times 22)]$
$$= 4.13 \text{ mm}$$

However, provide 10-mm thick flange plate (thickness should be equal to the flange thickness of the column).

Adopt two splice plate of size $270 \times 225 \times 10$ mm. The designed splice details are shown in Fig. 10.66.

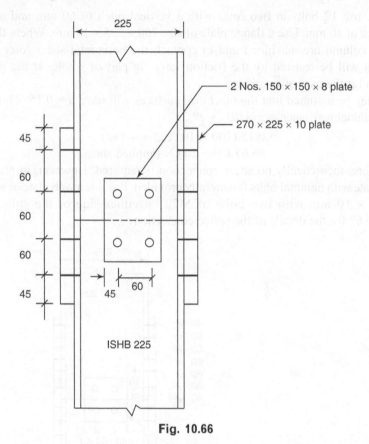

Fig. 10.66

Example 10.26 *Check the ability of the column splice illustrated in Fig. 10.67 to transfer a combination of the following: compression of 400 kN, moment of 100 kNm, and shear of 20 kN. Assume that the splice is designed for direct bearing and that M20 bolts are used. The steel is Fe 410 and the bolts are grade 8.8.*

Solution

The bending moment is assumed to be carried by the flange cover plates.

The force in the cover plate = $\pm 400/2 + 100 \times 10^3/225$

$$= 644.44 \text{ kN and } 244.44 \text{ kN}$$

Adopt a cover plate of width = 225 mm

Thickness of the flange plate, (assuming the width of cover plate as 225 mm)

$$T_p = [M \pm P(D_u/2)]/[(B_p - 2d_b)\ (f_y/\gamma_{m0})]$$

$$= 644.44 \times 10^3/[(225 - 2 \times 22) \times (250/1.1)]$$

$$= 15.67 \text{ mm}$$

Provide a 16 mm thick plate.

Number of M20 bolts (slip resistance V_{sb} for an M20 bolt = 52.6 kN as per Appendix D)

$$n_b = 644.44/52.6$$
$$= 12$$

Hence, use 12 bolts in two rows with a vertical pitch of 60 mm and an edge distance of 40 mm. Use a flange plate of size $760 \times 225 \times 16$ mm. Where the ends of the column are machined and in contact, the horizontal shear force on the column will be resisted by the friction force, in part of whole, at the point of contact (at the axis of rotation).

It may be assumed that the machined surfaces will have $\mu = 0.15$. Thus,

Frictional resistance = $\mu(M/D_u + P/2)$

$$= 0.15\ (100 \times 10^3/225 + 400/2)$$
$$= 96.67\ \text{kN} > 20\ \text{kN (applied shear)}$$

Therefore, theoretically no shear connection is required. However, in practice a web plate with nominal bolts is generally provided. Provide a web plate of size 160 × 140 × 10 mm with two bolts of M20 on either side of the splice. (see Fig. 10.67 for the details of the splice connection)

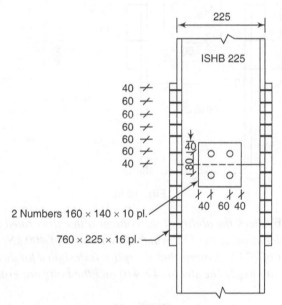

Fig. 10.67

Summary

Different components in a steel structure such as beams, columns, and tension members have to be connected by means of fasteners, so that all these members will act together to resist the applied forces. Riveted, bolted, and welded connections are normally used in steel structures. Out of these, the riveted connections have become obsolete, due to the difficulties encountered while making these connections.

Hence, in this chapter, the various bolts and their tightening methods have been discussed. The design expressions for black and HSFG bolts were presented. The various types of simple and moment resistant (rigid) connections have been covered along with their design methods. Due to the variety of possible connection configurations, the variability of behaviour, and the practical limitations in fabrication and erection, the design methods presented are based on the rational philosophy of simple analysis, using higher load factors of safety, rather than on complex design procedures. Several examples have been worked out to explain the concepts presented. The various types of welded connections are discussed in the next chapter.

Exercises

1. Design a lap joint between two plates of size 100×16 mm thick and 100×10 mm thick so as to transmit a factored load of 100 kN using a single row of M16 bolts of grade 4.6 and grade 410 plates.
2. Design a hanger joint as shown in Fig. 10.45 to carry a factored load of 200 kN. Use end plate of size 250 mm × 100 mm and appropriate thickness, two M20 HSFG bolts and Fe 410 steel for end plate ($f_y = 250$ MPa). Assume 8-mm fillet weld between the hanger and the end plate.
3. A member of a truss consists of two angles ISA $65 \times 65 \times 6$ placed back to back. It carries an ultimate tensile load of 125 kN and is connected to a gusset plate 8-mm thick placed in between the two connected legs. Determine the number of 16-mm-diameter grade 4.6 ordinary bolts required for the joint. Assume f_u of plate as 410 MPa.
4. The plates of a tank 8-mm thick are connected by a single bolted lap joint with 20-mm-diameter bolts at 50-mm pitch. Calculate the efficiency of the joint. Take f_u of plate as 410 MPa and assume grade 4.6 bolts.
5. A single bolted double cover butt joint is used to connect two plates 8-mm thick (see Fig. 10.48). Assuming the bolts of 20-mm diameter at 50-mm pitch, calculate the efficiency of the joint. Use 410 MPa plates and grade 4.6 bolts.
6. Two plates 10-mm and 14-mm thick are to be jointed by double cover butt joint. Assuming cover plates of 8-mm thickness, design the joint to transmit factored load of 300 kN. Assume Fe 410 plate and 16-mm-diameter grade 4.6 bolt.
7. Design a butt joint to connect two plates 200×8 mm (Fe 410 grade) using M16 bolts. Arrange the bolts to give maximum efficiency.
8. Determine the adequacy of the fasteners in Fig. 10.51, to carry a factored load of 160 kN when 16-mm-diameter grade 4.6 bolts are used. Assume that the strength of the column flange and the structural T (ST) sections do not govern the design. Neglect prying action.
9. Repeat the above exercise assuming the use of M16 parallels-shank HSFG 8.8 grade bolts in clearance holes and a load of 320 kN.
10. An ISMB 400 is connected to a column by web cleats with a single row of bolts. If the reaction is 220 kN and there are four 16-mm-diameter bolts through the web, as in Fig. 10.52, check if the section is adequate for block shear failure.

11. Design a connection of a truss joint as shown below, using M16 black bolts of property class 4.6 and grade 410 steel. Assume that the members shown are capable of resisting the loads and use 12 mm thick gusset plates.

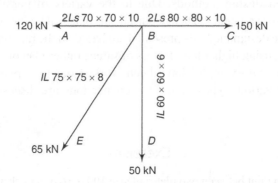

12. Design a seat angle connection between a beam MB 250 and column HB 200 for a reaction of beam 90 kN, using M16 bolts of property class 4.6. Take Fe 410 grade steel with $f_y = 250$ MPa. (see Fig. 10.54 for the arrangement of the connection)
13. Design a stiffened seat angle for a reaction of 160 kN from beam of ISMB 300 using M16 bolts of grade 4.6. The beam has to be connected to ISHC 200 column. Assume Fe 410 grade steel with $f_y = 250$ MPa (see Fig. 10.55 for the arrangement of connection).
14. Design a bolted web angle connection for a ISMB 350 beam, to carry a reaction of 90 kN due to factored loads. The connection is to the flange of a column ISHB 200 in grade Fe 410 steel. (see Fig. 10.56 for the arrangement use *M*16 bolts).
15. Design a header plate connection for a ISMB 300 beam to carry a reaction of 90 kN due to factored loads. The connection is to the flange of a ISHB 200 column. Use Fe 410 grade steel ($f_y = 250$ MPa) and *M*16 bolts of grade 4.6.
16. Given the bracket connection shown in Fig. 10.58. With 16-mm-diameter grade 4.6 bolts and plate of Fe 410 steel, is the bolt pattern and plate adequate for supporting the given load of 50 kN in a bearing type connection assuming threads in the shear plane?
17. Determine the size of the bolts required to connect the bracket (supporting a crane girder) to the ISHB column 200 shown in Fig. 10.59, with $V = 100$ kN, H = 25 kN, $e_v = 250$ mm, $e_h = 200$ mm, c/c spacing of bolts = 100 mm, and M24 grade 4.6 bolts.
18. Find the maximum load inclined at 45° to the horizontal, which the bracket shown in Fig. 10.60 can transmit if 5–16 mm diameter grade 8.8 bolts are used and plates connected are 10 mm thick. Determine the load (i) if the joint is considered a slip joint and (ii) if the joint is considered as a non-slip joint.
19. Design a bolted end plate connection between an ISMB 250 beam and an ISHB 200 column, to transfer a vertical factored shear of 90 kN and a factored hogging bending moment of 80 kNm. Use HSFG bolts of diameter 20 mm. The arrangement of connection is as shown in Fig. 10.61.
20. Design a flange angle connection using M20 bolts of grade 4.6 to transfer a moment of 10 kNm and a shear of 200kN from a beam of ISMB 350 to a column of ISHB 300 (see Fig. 10.62 for the arrangement of connection).

21. Design a T-stub connection to transfer a shear of 100 kN and a moment of 30 kNm from the end of a beam MB 250 to a column of ISHB 300 (See Fig. 10.63 for the arrangement of connection).
22. Design a bolted web cleat connection for an ISMB 400 and a coped beam of size ISMB 300 (180 kN reaction due to factored loads) using bolts of 16-mm diameter and grade 8.8.
23. Design a bolted splice for an ISMB 300 section to transfer a factored bending moment of 100 kNm and a factored shear of 60 kN. Assume that the flange splice carries all the moment and that the web splice carries only the shear. Use M16 bolts.
24. Design a bolted cover plate splice for a ISHB 200 column connected to an ISHB 200, to transfer a factored axial load of 370 kN, both columns are of grade Fe 410 steel. The ends are not machined for full contact in bearing.

Review Questions

1. What are the factors that influence the behaviour of joints or connections?
2. What is the basic goal of a connection design?
3. How are connections classified?
4. Write short notes on rigid, simple, and semi-rigid joints.
5. Describe the installation procedure of a rivet.
6. What are the reasons for riveting to become obsolete?
7. List some of the bolts that are used in structural connections.
8. Under what circumstances are black bolts not recommended?
9. In class 4.6 bolts, what do the numbers 4 and 6 indicate?
10. What are the normal bolt sizes used in structural applications?
11. The diameter of the bolt hole can exceed the diameter of the bolt by about
    (a) 1.5 mm to 2 mm (b) 2 mm to 5 mm (c) 1 mm to 3 mm (d) > 5 mm
12. What is the main purpose of a washer in HSFG bolts?
13. Why are connections using HSFG bolts called slip-critical connections?
14. What is the main difference between bearing type connection and friction type connection? Which type of connection is preferable under alternating loads?
15. What are the advantages of bolted connections over riveted or welded connections?
16. What are the advantages of HSFG bolts over black bolts?
17. Why are drilled holes preferable than punched holes?
18. Where are slotted holes used?
19. What is the difference between a pitch and a staggered pitch?
20. The minimum pitch allowed in the code is
    (a) 2.5 × diameter of bolt   (b) 1.5 × diameter of bolt
    (c) 3 × diameter of bolt   (d) 1.0 × diameter of bolt
21. The maximum longitudinal pitch allowed in a bolted compression member is
    (a) 400 mm   (b) 300 mm   (c) 200 mm   (d) 100 mm
22. The maximum longitudinal pitch allowed in a bolted tension member is
    (a) 300 mm   (b) 200 mm   (c) 150 mm   (d) 400 mm
23. The maximum longitudinal pitch allowed in a bolted compression member is
    (a) 16 × thickness of plate   (b) 12 × thickness of plate
    (c) 12 × diameter of bolt   (d) 16 × diameter of bolt

24. The maximum longitudinal pitch allowed in a bolted tension member is
    (a) 16 × thickness of plate   (b) 12 × thickness of plate
    (c) 12 × diameter of bolt   (d) 16 × diameter of bolt
25. The minimum edge distance in a member with rolled edge is approximately
    (a) 1.5 × hole diameter   (b) 2 × bolt diameter
    (c) 1.7 × hole diameter   (d) 1.7 × bolt diameter
26. Why are minimum pitch values specified in the code?
27. List the failure modes that may control the strength of a bolted joint.
28. What is the maximum slip, a bolt in normal clearance hole may experience?
29. What is the major difference in behaviour between a joint with black bolts and a joint with HSFG bolts?
30. Write the expression for calculating the nominal capacity of black bolts.
31. Write the expression for reduction factors for joints with black bolts and with (a) long joints (b) large grip length (c) thick packing plates
32. Write the expression for calculating the nominal capacity of bolt in tension.
33. Write the expression for calculating the nominal bearing strength of the bolt.
34. What do you mean by prying forces?
35. Prying effect will be more pronounced in
    (a) thin plates   (b) thick plates
    (c) intermediate thickness plate   (d) large diameter bolts
36. Write the interaction equation specified by the code for black bolts.
37. Write the expression for calculating the tensile capacity of plate.
38. Define efficiency of a joint.
39. Write the empirical expression proposed by Cochrane to calculate the net area of the plate when staggered rows of holes are present.
40. Write the expression given in the code for calculating the design slip resistance of HSFG bolts.
41. Write the value of coefficient of friction (slip factor) for
    (a) surfaces blasted with shot or grit and hot-dip galvanized
    (b) sand blasted surfaces
    (c) red lead painted surfaces
    (d) surfaces blasted with shot or grit and painted with ethylzinc silicate coat of 45 μm thickness
42. Write short notes on:
    (a) Lap joint   (b) Butt joint
43. What are the advantages of butt joints over lap joints?
44. Describe the Whitmore's (30°) method of designing gusset plate?
45. What is the difference between unstiffened and stiffened seat connection?
46. Write short notes on:
    (a) Unstiffened seat angle connection   (b) Stiffened seat angle connection
    (c) Web angle connection   (d) Flexible end plate connection
47. What are the two main types of moment–resistance connections?
48. Write the expression for calculating the force R in a bolt subjected to a moment M and located at a distance d from the centre of rotation.
49. What is the expression used to find the number of bolts when a bolt group is subjected to applied moment or torque?

11

Welded Connections

Introduction

Welding is a method of connecting two pieces of metal by heating to a plastic or fluid state (with or without pressure), so that fusion occurs. Welding is one of the oldest and reliable methods of jointing. Little progress in welding technology, as is known now, was made until 1877, though welding processes such as forge welding and brazing were known for at least 3000 years. Although there had been initial work on arc welding in the 1700s using carbon electrodes powered by batteries, development work intensified between 1880 and 1900 with the availability of electric generators to replace batteries. Professor Elihu Thompson was the first to patent the first resistance groove welding machine in 1885. Charles Coffin invented the metal arc process and patented it in USA in 1892, though Zerner introduced the carbon arc welding process in 1885. The concept of coated metal electrodes, which eliminated the problems associated with the use of bare electrodes, was introduced in 1889 by A.P. Strohmeyer. The metal arc process was first used (in 1889) in Russia by using uncoated, bare electrodes (Salmon & Johnson 1996).

Oxyacetylene welding and cutting was employed after 1903, due to the development of acetylene torches by Fouche and Picard. By the early 1900s, Lincoln Electric offered the first arc welding machine and by 1912 covered electrodes were patented. During World War I (1914–1918), welding techniques were used for repairing damaged ships. During the period 1930–1950, several improvements and techniques such as the use of granular flux to protect the weld and submerged arc welding were developed.

Today there are several welding processes available to join various metals and their alloys. The types of welds and welded joints and the advantages of using welding over bolts or rivets are also discussed. The behaviour and design of various welded connections are also outlined. A brief review of the methods of joining tubular connections is given. Several examples are given to illustrate the design procedures adopted for welded connections. This chapter concludes with the recent developments in the design of joints to resist earthquake loads.

11.1 Welding Processes

Structural welding is nearly all electric; though some gas welding may also be used ['gas' denotes the use of a gas (usually acetylene/oxygen mixture) to produce a very hot flame to heat the parts and the weld filler material]. However, gas is used primarily for cutting pieces to shape. It is now possible to cut metals using mechanically controlled gas cutting equipment in fabrication shop, which results in smooth cuts similar to sawed cuts. Though gas welding is simple and inexpensive, it is slow and hence it is generally used for repair and maintenance work only.

In the most common welding processes of welding structural steel, electric energy is used as the heat source. Electric welding involves passing either direct or alternating current (mostly direct current is used) through an electrode (commonly the electrode is the anode and the operation uses 'reversed polarity'). By holding the electrode at a very short distance from the base metal, which is connected to one side of the circuit, an arc forms as the circuit is essentially 'shorted' [see Fig. 11.1(a)]. With this 'shorting' of the circuit, a very large current flow takes place, which melts the electrode's tip (at the arc) and the base metal in the vicinity of the arc. A temperature of about 3300–5000°C is produced in the arc. The electron flow making the circuit 'carries the molten electrode metal' to the base metal to build up the joint. The parameters that control the quality of weld are the electrode size and the current that produces sufficient heat to melt the base metal and minimizes electrode splatter.

The different processes of arc welding that are used in structural steel applications are as follows:

- Shielded metal arc welding (SMAW)
- Submerged arc welding (SAW)
- Gas-shielded metal arc welding (GMAW)
- Flux core arc welding (FCAW)
- Electro slag welding (ESW)
- Stud welding (SW)

Details of these processes may be found in Galvery & Mavlow (2001), Jeffus (2002) and Subramaniam (2008).

11.1.1 Shielded Metal Arc Welding (SMAW)

Shielded Metal Arc Welding (SMAW) also called 'stick' welding is a manual process and is the most common method of welding used in structural connections owing to low capital cost and flexibility. However, for long continuous welds automatic processes are preferred due to the consistent quality. The SMAW processes require the following set up (see Fig. 11.1(a)):

(a) Constant-current (CC) welding power supply
(b) Electrode holder, lead, and its terminals
(c) Ground clamp, lead, and its terminals
(d) Welding electrodes

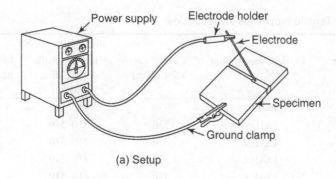

(a) Setup

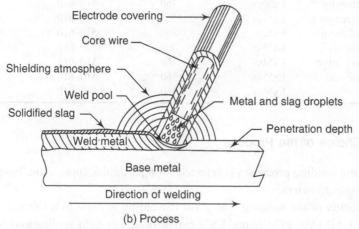

(b) Process

Fig. 11.1 Shielded metal arc welding

As stated earlier, the electrons flowing through the gap between the electrode and the metal produce an arc that furnishes the heat to melt both the electrode metal and the base metal. Temperatures within the arc exceed 3300°C. The arc heats both the electrode and the metal beneath it. Tiny globules of metal form at the tip of the electrode and are transferred to the molten weld pool on the base metal. As the electrode moves away from the molten pool, the molten mixture of electrode and base metal solidifies and the weld is completed (see Fig. 11.1b).

Generally the electrode is stronger than the parent metal. For example, an E41 electrode, which would be used to weld grade 410 steel, gives a weld deposit which has a maximum yield strength of 330 MPa with a tensile strength in the range of 410–510 MPa (see Table 11.1). For *manual metal arc welding (MMA)*, the electrodes should comply with IS 2879, IS 1395, and IS 814.

The electrodes are available in lengths of 225–450 mm, and diameters ranging from 3.2 to 6 mm. The maximum size of weld produced in one pass is about 8 mm (Bowles 1980).

Table 11.1 Tensile properties of electrodes

Specification	Grade/Classification	Yield stress, MPa (Min)	Properties Ultimate tensile stress, MPa, (Min)	Elongation percentage (Min)
	Ex40xx	330	410–510	16
	Ex41xx	330	410–510	20
	Ex42xx	330	410–510	22
IS 814 : 2004	Ex43xx	330	410–510	24
	Ex44xx	330	410–510	24
Specification for	Ex50xx	360	510–610	16
covered electrodes	Ex51xx	360	510–610	18
for manual metal	Ex52xx	360	510–610	18
arc welding of	Ex53xx	360	510–610	20
carbon and carbon	Ex54xx	360	510–610	20
manganese steel	Ex55xx	360	510–610	20
	Ex56xx	360	510–610	20

11.1.2 Choice of the Process

One of the welding processes is selected for a particular application, based on the following parameters.

(a) *Location of the welding operation* If welding is done in a fabrication shop, SAW, GMAW, FCAW, and ESW can be used. For field applications SMAW is preferred.

(b) *Accuracy of setting up* SAW, spray transfer GMAW, and ESW require accurate set-up.

(c) *Penetration of weld* Penetration of FCAW and SAW is better than SMAW.

(d) *Volume of weld to be deposited* FCAW, GMAW, and ESW have high deposition rates.

(e) *Position of welding* SAW and ESW are not suitable for overhead positions. FCAW and GMAW can be used in all positions. SMAW is probably the best for overhead works, especially at site.

(f) *Access to joint* In easily accessible joints SAW and GMAW are used. In cramped joints SMAW is used.

(g) *Steel composition* GMAW and SAW are less likely to lead to heat-affected zone (HAZ) cracking.

(h) Thickness of connecting parts.

(i) Comparative cost.

11.2 Welding Electrodes

As stated earlier, a variety of electrodes are available so that a proper match of base metal strength and metallurgical properties to the weld metal can be chosen. Only coated electrodes are used in structural welding. Welding electrodes are classified (by the American Welding Society, in cooperation with ANSI) using the following numbering system for shielded metal arc welding (SMAW):

Exxxbc

In this numbering system, E stands for electrodes. xxx stands for two or three-digit number establishing the ultimate tensile strength of the weld metal. As per IS 814 the following values are available: 40, 41, 42, 43, 44, 50, 51, 53, 54, 55, 56 kg/cm^2. The value of b indicates the suitability of welding positions, which may be flat, horizontal, vertical, and overhead, that is, b = 1 denotes suitability for all positions, 2 denotes suitability for flat positioning of work, 4 denotes suitability for flat, horizontal, overhead, and vertical down. 'c' stands for coating and operating characteristics. The value of c equal to 5, 6, 8 indicates low hydrogen.

The various grades of electrodes as per IS 814 and their tensile properties are shown in Table 11.1.

11.3 Advantages of Welding

Welding offers the following advantages over bolting or riveting.

(a) Welded connections eliminate the need for making holes in the members, except for a few employed for erection purposes. Since the holes at the ends govern the design of bolted connections (edge distance), a welded connection results in a member with a smaller gross section. This has a greater influence in the case of tension members, since the calculation of net section is eliminated.

(b) Welding offers airtight and watertight jointing of plates and hence is employed in the construction of water/oil storage tanks, ships, etc.

(c) Welded joints are economical, since they enable direct transfer of stresses between the members. Moreover, the splice plates and bolt material are eliminated. The required size of gusset plates is also smaller, because of reduced connection length. Due to the elimination of operations such as drilling and punching, welding results in less fabrication costs. In addition, due to the simple design details, time is also saved in detailing, fabrication, and field erection. Welding also requires considerably less labour for executing the work. It is estimated that the total overall savings by employing welding over bolting may be up to 15%.

(d) Welded structures are more rigid (due to the direct connection of members by welding) as compared to bolted joints. In bolted joints, the cover plates, connecting angle, etc. may deflect with the member during load transfer thus making a structure flexible. Rigid structures are always more economical than flexible structures, due to the transfer of moments from one member to another.

(e) Welded connections are usually aesthetic in appearance and appear less cluttered in contrast to bolted connections. This is evident from Fig. 11.2, which shows a bolted and welded plate girder.

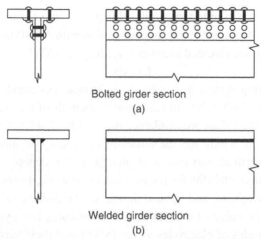

Bolted girder section
(a)

Welded girder section
(b)

Fig. 11.2 Appearance of bolted and welded plate girders

(f) Welding offers more freedom to the designer in choosing sections. The designer is not bound by the available rolled sections, but may build up any cross section, which may be economical and advantageous. Welding has resulted in the innovation of open web joists, castellated beams, tapered beams, vierendeel trusses, composite construction, tubular trusses, and offshore platforms.

(g) Welding is practicable even for complicated shapes of joints. For example, connections with tubular sections can be made easily by welding, whereas it is difficult to make them using bolting. Tubular sections are structurally economical as compression members and their use in trusses is feasible due to welding.

(h) Alterations can be made with less expense in case of welding as compared to bolting. It is also easy to correct mistakes in fabrication during erection, whereas a mismatch of holes in a bolted connection is very difficult to correct. Also members can be shortened by cutting and rejoined by suitable welding. In the same way, members can be lengthened by splicing a piece of the same cross section.

(i) A truly continuous structure is formed by the process of fusing the members together. This gives the appearance of a one-piece integrated structure. Usually, the strength of a welded joint is as strong as or stronger than the base metal, thus there are no restrictions in the placement of joints.

(j) The efficiency of a welded joint is more than a bolted joint. In fact 100% efficiency can be obtained using welding.

(k) Due to the elimination of holes, stress concentration effect is considerably less in welded connections.

(l) The process of welding is relatively silent compared to riveting and bolting (drilling holes) and requires less safety precautions.

However, welding has the following disadvantages.

(a) Welding requires highly skilled human resources.

(b) The inspection of welded joints is difficult and expensive, whereas inspection of bolted joints is simple. Moreover, non-destructive testing is required in important structures.
(c) Members jointed by welding may distort, unless proper precautions are taken. Welded joints have large residual stresses.
(d) Costly equipment is necessary to make welded connections.
(e) Welded connections are prone to cracking under fatigue loading.
(f) Proper welding may not be done in field conditions, especially in vertical and overhead positions.
(g) The possibility of brittle fracture is more in the case of welded joints than in bolted connections.
(h) The welding performed in the field is expensive than performed in the shop.
(i) Welding at the site may not be feasible due to lack of power supply.

Several factors influence the welding cost, which include the following (Salmon & Johnson 1996):
(a) Cost of preparing the edges to be welded (in case of groove welds)
(b) Amount of weld material required
(c) Ratio of the actual arc time to overall welding time
(d) The handling required (cranes and special equipment needed during erection)
(e) General over head costs
(f) Cost of pre-heating, if any

11.4 Types and Properties of Welds

The welds may be grouped into four types as follows:
(a) Groove welds
(b) Fillet welds
(c) Slot welds
(d) Plug welds

These are shown in Fig. 11.3. Each type of weld has its own advantage and may be selected depending on the situation. It has been found that fillet welds are used extensively (about 80%) followed by groove welds (15%). Slot and plug welds are used rarely (less then 5%) in structural engineering applications. Fillet welds are suitable for lap and T-joints (see Section 11.5) and groove welds are suitable for butt, corner, and edge joints. Each of these four types of welds are discussed further in the following sections.

11.4.1 Groove Welds

Groove welds are used to connect structural members that are aligned in the same plane and often used in butt joints. Groove welds may also be used in T-connections. The grooves have a slope of 30°–60°. Edge preparation becomes necessary for plates over 10-mm thick for manual arc welding, and over 16-mm thick for automatic welding. Various types of groove welds are shown in Fig. 11.4. The

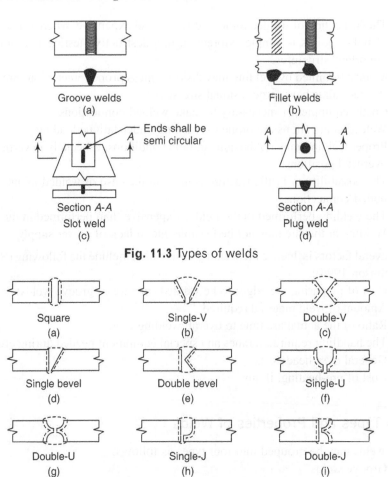

Groove welds
(a)

Fillet welds
(b)

Ends shall be semi circular

Section *A-A*
Slot weld
(c)

Section *A-A*
Plug weld
(d)

Fig. 11.3 Types of welds

Square
(a)

Single-V
(b)

Double-V
(c)

Single bevel
(d)

Double bevel
(e)

Single-U
(f)

Double-U
(g)

Single-J
(h)

Double-J
(i)

Fig. 11.4 Types of groove welds

square groove weld is used to connect plates up to 8-mm thickness. The terms that are associated with a completed groove weld are shown in Fig. 11.5. Partial penetration groove welds should not be used especially in fatigue situations.

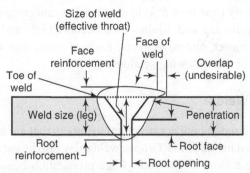

Size of weld
(effective throat)

Face of
weld

Face
reinforcement

Overlap
(undesirable)

Toe of
weld

Weld size (leg)

Penetration

Root
reinforcement

Root face

Root opening

Fig. 11.5 Terms used to describe the parts of a groove weld

To ensure full penetration and a sound weld, a back-up strip is provided at the bottom of single-V/bevel/J or U grooves. Thus, the back-up strips are commonly used when all welding is done from one side or when the root opening is excessive (see Fig. 11.6). The back-up strip introduces a crevice into the weld geometry and prevents the problem of burn-through. The back-up strip can be left in place or removed after welding the pieces.

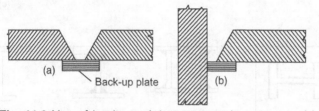

Fig. 11.6 Use of back-up plate or spacer in groove weld

For a groove weld, the root opening or gap (see Fig. 11.5), is provided for the electrode to access the base of the joint. The smaller the root opening, the greater will be the angle of the bevel (for root openings of 3 mm, 6 mm, and 9 mm, angles of 60°, 45°, and 30°, respectively, may be chosen).

The choice between single or double penetration depends on access on both sides, the thickness of the plate, the type of welding equipment, the position of the weld, and the means by which the distortion is controlled.

Since weld metal is expensive compared to the base metal, the groove is made of double-bevel or double-V for plates of thickness more than 12 mm, and made of double-U or double-J for plates of thickness more than 40 mm. For plates between 12–40 mm, single-J and single-U grooves may be used.

Since groove welds will transmit the full load of the members they join, they should have the same strength as the members they join. Hence, only full penetration groove welds are often used.

11.4.2 Fillet Welds

Fillet welds are most widely used due to their economy, ease of fabrication, and adoptability at site. They are approximately triangular in cross section and a few examples of application of fillet weld are shown in Fig. 11.7. Unlike groove welds, they require less precision in 'fitting up' two sections, due to the overlapping of pieces. Hence, they are adopted in field as well as shop welding. Since they do not require any edge preparation (edge conditions resulting from flame cutting or shear cutting procedures are generally adequate), they are cheaper than groove welds.

In connections, members generally intersect at right angles, but intersection angles between 60° and 120° can be used, provided the correct throat size is used in design calculations (see Section 11.9.2). Fillet welds are assumed to fail in shear.

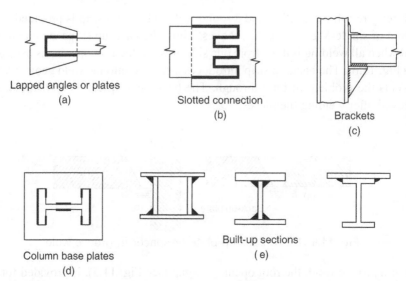

Lapped angles or plates
(a)

Slotted connection
(b)

Brackets
(c)

Column base plates
(d)

Built-up sections
(e)

Fig. 11.7 Typical uses of fillet welds

11.4.3 Slot and Plug Welds

Slot and plug welds are not used exclusively in steel construction. When it becomes impossible to use fillet welds or when the length of the fillet weld is limited, slot and plug welds are used to supplement the fillet welds. They are also assumed to fail in shear. Thus, their design strength is similar to that of fillet welds.

11.4.4 Structure and Properties of Weld Metal

The weld metal is a mixture of parent metal and steel melted from the electrode. The solidified weld metal has properties characteristic of cast steel. Hence, it has higher yield to ultimate ratio but lower ductility compared to structural steel. When the weld pool is cooling and solidifying, the parent metal along side the joint is subjected to heating and cooling cycles and the metallurgical structure of this steel in this region will be changed. This region is called the *heat-affected zone* (HAZ).

The change in structure in HAZ should be considered in the design stage by selecting a suitable Charpy V-impact value for the (see Section 1.8.5) welding electrode (its Charpy impact value should be equal to or greater than that specified for the parent metal), corrosion resistance, etc. Pre-heating of joints will also help to reduce HAZ cracks. However, pre-heating increases the cost of welding.

Charts for finding the required pre-heat temperature are provided by Blodgett (1966). In critical cases, 'pre-heat' is maintained for a considerable period of time after welding.

11.4.5 Weld Defects

The production of sound welds is governed by the type of joint, its preparation and fit-up, the root opening, etc. In addition to this, the choice of electrode, the welding position, the welding current and voltage, the arc length, and the rate of travel also affect the quality of weld (Gaylord et al. 1992). Accessibility of the welding operation is also important, since the quality of weld is determined to a considerable extent by the positioning of the electrode. Some of the common defects in the welds are as follows:

(a) Incomplete fusion
(b) Incomplete penetration
(c) Porosity
(d) Inclusion of slag
(e) Cracks
(f) Undercutting
(g) Lamellar tearing

These defects are shown schematically in Fig. 11.8. For more details about these defects and the methods to eliminate them, refer Blodgett (1966) and Jeffus (2002).

Lamellar tearing is discussed in Section 1.8.6 also.

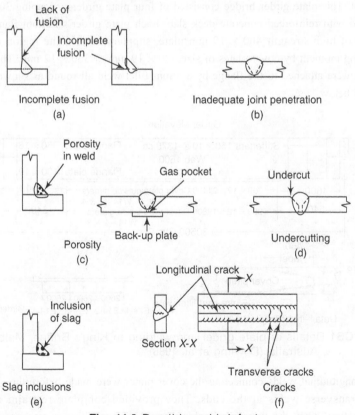

Fig. 11.8 Possible weld defects

Since a small error in a weld may lead to a catastrophic collapse, checks are to be made before, during, and after welding (Blodgett 1966).

In addition, a qualified welder, who knows the weld qualification procedures, should be employed to execute the job. Visual inspection (which is dependent on the competence of the observer) and non-destructive tests (may be employed for important structures) should be used to determine the type and distribution of weld defects (Gaylord et al. 1992). Any poor or suspicious weld should be cut and replaced. A welding gauge may be used to rapidly check the size of the fillet welds. The non-destructive tests usually employed include the following:

(a) Liquid penetrant inspection,
(b) Magnetic particle inspection,
(c) Radiographic inspection, and
(d) Ultrasonic inspection.

Failure of the King's Bridge, Australia

King's bridge in Melbourne, Australia, failed while in service on 10th July 1962 (Melbourne's winter) due to brittle fracture, when a 45-ton vehicle was passing over it. This plate girder bridge consisted of four plate girders, spanning 30 m and topped with reinforced concrete deck slab. Each plate girder's bottom flange was made of high-strength 400×19 mm plate, supplemented in the region of high bending moment by cover plates of size 300×19 mm or 360×12 mm. The cover plates were attached to the flange by a 5 mm fillet weld all round as shown in the figure below.

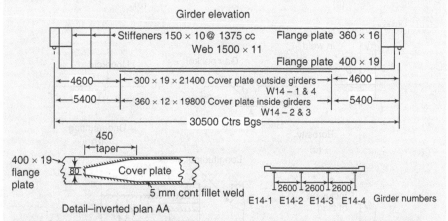

Fig. CS1 Details of plate grider and welding in King's Bridge, Melbourne, Austraila (Dowling et al. 1988)

The longitudinal welds connecting the cover plates were made before the short 80 mm transverse welds at the ends. They provided complete restraint against contraction, when the transverse welds were placed, resulting in transverse crack in flange plates. The transverse welds were made in three passes. In some instances, the cracks caused in the main flange plate by the first run were covered up by a subsequent pass. In other cases, the cracks caused by the last run were covered up with priming paint. The penetration of paint coats into the cracks showed later that the cracks passed through the full thickness of the flange even before the girders left the factory. In the span that failed, cracks existed in the main flange plate under seven of the eight transverse fillet welds. Thus, the most likely and most dangerous cracks were regularly missed by the inspectors, who however repaired several less harmful longitudinal cracks. All the seven cracks developed into complete flange failure, partly by brittle fracture and partly by fatigue, under a load that was well within the design load of the bridge.

11.5 Types of Joints

The five basic types of welded joints which can be made in four different welding positions such as flat, horizontal, vertical, and overhead are as follows (see Fig. 11.9):
(a) Butt joint
(b) Lap joint
(c) T-joint
(d) Corner joint
(e) Edge joint

These joints are discussed briefly in the following sections.

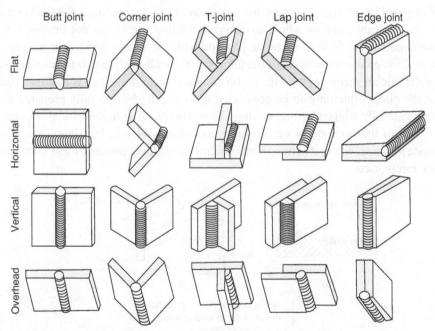

Fig. 11.9 Five basic weld joints may be made in four different welding positions

11.5.1 Butt Joints

A butt joint is used to join the ends of flat plates of nearly equal thickness. This type of joint eliminates the eccentricity developed using a lap joint (see Fig. 11.9). The butt joint obtained from a full penetration groove weld has 100% efficiency (i.e. the weld is considered as strong as the parent plate). As mentioned previously, the groove welds used in butt joints can be full penetration or partial penetration depending upon whether the penetration is complete or partial through the thickness (see Section 11.4.1 for the discussion on groove welds). Such butt joints also minimize the size of the connection and are aesthetically pleasing than lap joints. Face reinforcement (see Fig. 11.5) is the extra weld metal that makes the throat dimension greater than the thickness of the welded material. The provision of reinforcement increases the efficiency of the joint and ensures that the depth of the weld is at least equal to the thickness of the plate. Reinforcement is normally provided, since it is difficult for the welder to make the weld flush with the parent metal.

Reinforcement makes the butt joint stronger under static load and the flow of forces is generally smooth. However, when in the case of fatigue loads, stress concentration develops in the reinforcement, leading to cracking and early failure. Hence, under these circumstances, the reinforcement should be either removed by machining or kept within limits (normally within 0.75–3 mm) to avoid stress concentration. Similarly, when plates of two different thicknesses and/or widths are joined, the wider or thicker part should be reduced at the butt joint to make the width or thickness equal to the smaller part, the slope being not steeper than one in five [see Fig. 11.10(b)]. Where the reduction of the dimension of the thicker part is impracticable, and/or where dynamic/alternating forces are not involved, the weld metal shall be built up at the junction with the thicker part to dimensions at least 25% greater than those of the thinner part, or alternatively to the dimensions of the thicker member [see Fig. 11.10(c)]. Their main drawback is that the edges of the plates which are to be connected must usually be specially prepared and very carefully aligned before welding. They also result in high residual stresses.

Due to the accurate placement of parts before welding, butt joints are often made in shops, where it is possible to control the welding process. Field butt joints are rarely used.

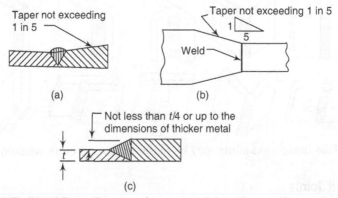

Fig.11.10 Butt welding of parts of unequal thickness and/or unequal width

11.5.2 Lap Joints

Lap joints are most commonly used because they offer ease of fitting and ease of jointing. Thus, they do not require any special preparation (even flame cut or sheared edges can be used) and can accommodate minor errors in fabrication or minor adjustment in length. Lap joints utilize *fillet welds* (see Section 11.4.2) and hence are well suited for shop as well as field welding. Some examples of lap joints are shown in Fig. 11.11. The connections using lap joints may require a small number of erection bolts, which may either be removed after welding or left in place. The additional advantage of lap joints is that plates with different thicknesses can be joined without any difficulty. Hence, it is often preferred in truss joints as shown in Fig. 11.11(c). However, the main drawback of a lap joint is that it introduces some eccentricity of loads, unless a double lap joint is used as in Fig. 11.11(e).

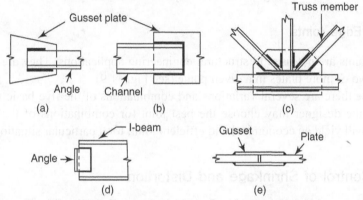

Fig. 11.11 Some examples of welded lap joints

11.5.3 Tee Joints

T-joints are often used to fabricate built-up sections such as T-shapes, I-shapes, plate girders, hangers, brackets, and stiffeners, where two plates are joined at right angles. T-joints can be made by using either fillet or groove welds. The groove weld edge shapes used on T-joints are shown in Fig. 11.12.

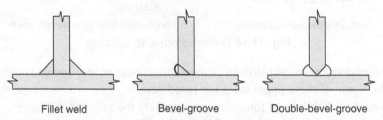

Fillet weld   Bevel-groove   Double-bevel-groove

Fig. 11.12 Fillet and groove welded T-joints (Double bevel or J groove are often used with thick plates)

11.5.4 Corner Joints

Corner joints are used to form built-up rectangular box sections, which may be used as columns or beams to resist high torsional forces. Fillet weld and a few groove weld edge shapes for corner joints are shown in Fig. 11.13.

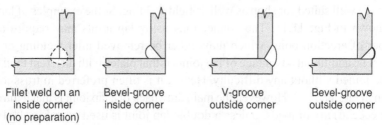

| Fillet weld on an inside corner (no preparation) | Bevel-groove inside corner | V-groove outside corner | Bevel-groove outside corner |

Fig. 11.13 Corner joint edge shapes

11.5.5 Edge Joints

Edge joints are not used in structural engineering applications. They are used to keep two or more plates in a given plane (see Fig. 11.9).

Since there are several variations and combinations of the five basic types of joints, the designer may choose the best joint (or combinations of the joints), which will yield an economical and efficient joint, for a particular situation.

11.6 Control of Shrinkage and Distortion

The molten weld bead that has been deposited starts to cool and while solidifying attempts to contract both along and transverse to its axis. This tendency to contract will induce tensile residual stresses and distortions (see Fig. 11.14).

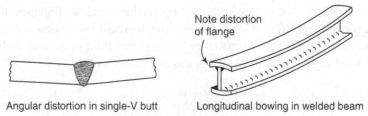

Note distortion of flange

| Angular distortion in single-V butt | Longitudinal bowing in welded beam |

Fig. 11.14 Distortion due to welding

There are several ways to minimize these distortions and are provided by Blodgett (1966). Some of these approaches are listed here.

(a) Reduce the shrinkage forces by incorporating the following:
- Use minimum weld metal; for groove welds use the minimum root opening that is necessary; do not over weld
- Use only a few passes to complete the weld
- Use proper edge preparation and fit-up
- Use intermittent welds

- Deposit the weld metal in the direction opposite to the progress of welding the joint

(b) Allow the shrinkage to occur freely as follows.

(c) Balance shrinkage forces by incorporating the following.
- Use symmetry in welding
- Use scattered and intermittent weld segments.
- Use peening (i.e. stretching the metal by a series of blows, using a hammer).
- Use clamps, jigs, etc., to force the weld metal to stretch as it cools.

In practice, more than one method may be used at the same time for a particular situation. Minimum pre-heat and *interpass temperature* (for welds requiring more than one pass of welding operation along a joint, the interpass temperature is the temperature of the deposited weld when the next pass is about to begin) are sometimes prescribed to minimize shrinkage and ensure adequate ductility (Salmon & Johnson 1996).

11.7 Weld Symbols

The standard weld symbols used on drawings for different types of welds are shown in Fig. 11.15. Symbols save a lot of space as descriptive notes can be omitted. The location and details of the weld are shown by an arrow, a horizontal line ending with a fork (see Fig. 11.16). The 'side' below the arrow is called the arrow side and the 'side' above is called the other side. A circle at the kink indicates a weld all round and a vertical line and triangular pennant at the kink shows a field weld. The weld is denoted by symbols on both 'arrow' and 'other' side, but the weld symbol on the 'arrow side' is inverted. The surface condition is shown by a convex or horizontal line (contour symbol). The size of the welds is shown near the fork on the horizontal line. The length and pitch of the weld (for intermittent welds only) are shown after the weld symbol. The use of some of these symbols is illustrated in Fig. 11.16.

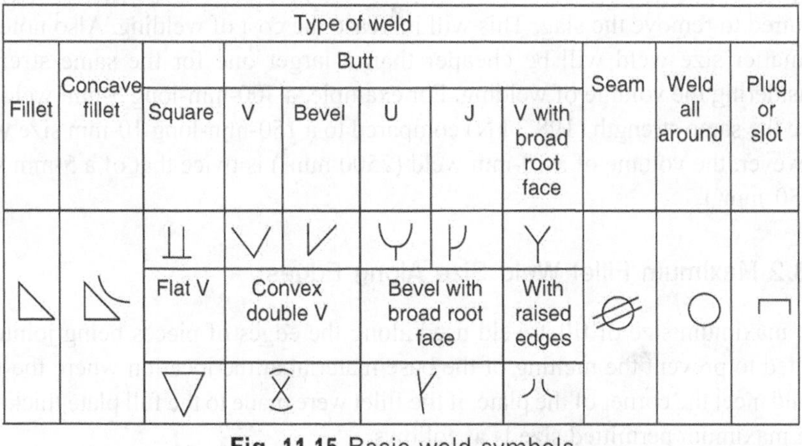

					Type of weld					
				Butt						
Fillet	Concave fillet	Square	V	Bevel	U	J	V with broad root face	Seam	Weld all around	Plug or slot
△	◺	⊥⊥ Flat V	∨ Convex double V	∨̷	⋎ Bevel with broad root face	⊍ With raised edges	⋎	⊘	○	⊓
		▽	⊗		⋎	⋏				

Fig. 11.15 Basic weld symbols

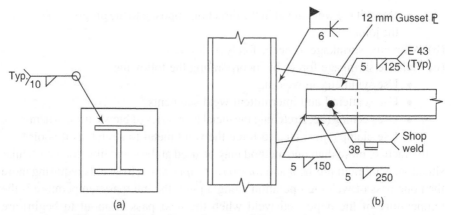

(a)   (b)

Fig. 11.16 Illusration of some of the welding symbols

11.8 Weld Specifications

Thicker plates dissipate the heat due to arc welding vertically as well as horizontally while thinner plates dissipate heat only horizontally. Thus, in thicker plates heat is removed from the welding area quickly and hence results in lack of fusion. For this reason, specifications often stipulate minimum and maximum weld sizes to achieve proper fusion of the base metal and the electrode.

11.8.1 Minimum Weld Size

To ensure fusion, minimize distortion, and to avoid the risk of cracking IS 800 : 2007 and IS 816 provide for a minimum size weld based on the thickness of the pieces being joined. The size of fillet weld should not be less than 3 mm nor more than the thickness of the thinner part joined. The minimum size of the first run or of a single run fillet weld should be as per Table 11.2. Usually the weld size closer to the minimum size is selected. Large size welds require more than one run of welding, which means that after the first run, chipping and cleaning of the weld is required to remove the slag. This will increase the cost of welding. Also note that a smaller size weld will be cheaper than a larger one for the same strength, considering the volume of welding. For example, a 300-mm-long 5-mm weld will have the same strength (198.3 kN) compared to a 150-mm-long 10-mm size weld. However, the volume of a 10-mm weld (7500 mm^3) is twice that of a 5-mm weld (3750 mm^3).

11.8.2 Maximum Fillet Weld Size Along Edges

The maximum size of fillet weld used along the edges of pieces being jointed is limited to prevent the melting of the base material at the location where the fillet would meet the corner of the plate, if the fillet were made to the full plate thickness. The maximum permitted size is as follows.

Table 11.2 Minimum size of a single run fillet weld (as per IS 800)

Thickness of thicker part		Minimum size, mm
Over, mm	Up to and including, mm	
—	10	3
10	20	5
20	32	6
32	50 (see notes below)	8 for first run, 10 for minimum size of weld

Note 1: When the minimum size is greater than the thickness of the thinner part, the minimum size should be equal to the thickness of the thinner part. Pre-heating of thicker part may be necessary.

Note 2: Where the thicker part is more than 50-mm thick, special precautions like pre-heating should be taken.

(a) Along the edge of the plate, less than 6-mm thick, the maximum size is equal to the thickness of the plate.

(b) Where the fillet weld is applied to the square edge of a plate of thickness greater than 6 mm, size of the weld should be at least 1.5 mm less than the edge thickness [see Fig. 11.17(a)]. This limit is specified such that the total strength may be developed without overstressing the adjacent metal.

(c) Where the fillet weld is applied to the rounded toe of the rolled section, the size of the weld should not exceed 3/4 of the thickness of the section at the toe [see Fig. 11.17(b)].

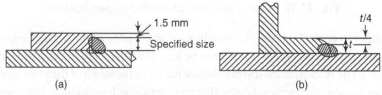

(a) (b)

Fig. 11.17 Size of fillet welds

11.8.3 Minimum Effective Length of Fillet Weld

When placing a fillet weld, though the welder tries to build up the weld to its full dimension from the beginning, there is always a slight tapering off where the weld starts and where it ends. Therefore, a minimum length of four times the size of the weld is specified [see Fig. 11.18(b)]. If this requirement is not met, the size of the weld should be one fourth of the effective length.

For the above reasons, the effective length is taken equal to its overall length minus twice the size of weld. *End returns* as shown in Fig. 11.18(d) are made equal to twice the size of the weld to relieve the high stress concentration at the ends. Most designers neglect the end returns in the effective length calculation of the welds. End returns must be provided for welded joints that are subject to eccentricity, stress reversals, or impact loads.

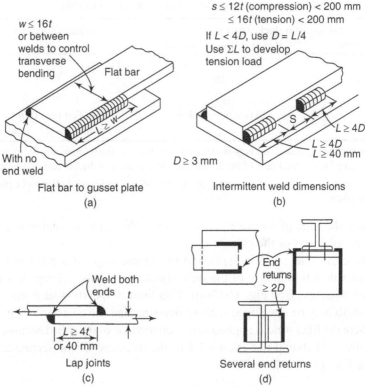

Fig. 11.18 Several welded connection specifications

In order to control the stress concentration at the edge of the plate, the length of the longitudinal (side) fillets should not be less than the width of the plate [see Fig. 11.18(a)]. The uneven stress distribution increases as the width of the plate increases. For this reason, the perpendicular distance between longitudinal fillet welds is limited to 16 times the thickness of the thinner plate jointed. If the plate is wider than this limit, slot or plug welds may be introduced, which tend to improve the distribution of stress in plate.

11.8.4 Overlap

The overlap of plates to be fillet welded in a lap joint should not be less than 4 times the thickness of the thinner part [see Fig. 11.18(c)].

11.8.5 Effective Length of Groove Welds

The effective length of groove welds in butt joints is taken as the length of continuous full size weld, but it should not be less than four times the size of the weld.

11.8.6 Effective Length of Intermittent Welds

The intermittent fillet welds should have an effective length not less than four times the weld size, with a minimum of 40 mm, as already shown in Fig. 11.18(b).

The clear spacing between the effective lengths of intermittent welds should not exceed 12 and 16 times the thickness of thinner plate jointed for compression and the tension joint respectively, and should never be more than 200 mm. The intermittent groove weld in butt joints should have an effective length of not less than four times the weld size and the longitudinal space between the effective lengths of the intermittent welds should not be more than 16 times the thickness of the thinner part joined. The IS code prohibits the use of intermittent welds in joints subjected to dynamic, repetitive, and alternate stresses.

11.8.7 Effective Area of Plug Welds

Effective area of plug welds should be taken as the nominal area of the hole in the plane of the faying surface. IS code stipulates that they should not be designed to carry any stresses.

11.9 Effective Area of Welds

The effective areas of a groove or fillet weld is the product of the effective throat dimension (t_e) multiplied by the effective length of the weld. The effective throat dimension of a groove weld or a fillet weld depends on the minimum width of expected failure plane and is explained in the next section.

11.9.1 Groove Weld

The effective throat thickness of a complete penetration groove weld is taken as the thickness of the thinner part joined [see Figs 11.19(a) and (b)]. The effective throat thickness of T- or L-joints are taken as the thickness of the abutting part. Reinforcement (see Fig. 11.5), which is provided to ensure full cross-sectional area, is not considered as part of the effective throat thickness.

The effective throat thickness of a partial penetration joint weld is taken as the minimum thickness of the weld metal common to the parts joined, excluding reinforcement (see Fig. 11.19). In unsealed single groove welds of V-, U-, J-, and bevel-types and groove welds welded from one side only, the throat thickness should be at least $7/8^{th}$ of the thickness of the thinner part joined. However, for the purpose of stress calculation, the effective throat thickness of $5/8^{th}$ thickness of the thinner member only should be used (IS 816 : 1969). The unwelded portion in incomplete penetration welds, welded from both sides, should not be greater than 0.25 times the thickness of the thinner part joined, and should be central in the depth of the weld [Fig. 11.19(d)]. In this case also, a reduced effective throat thickness of $5/8^{th}$ of the thickness of the thinner part should only be used in the calculations. Groove welds used in butt joints, where the penetration is less than those specified above, due to non-accessibility, should be considered as non-load carrying for the purposes of design calculations.

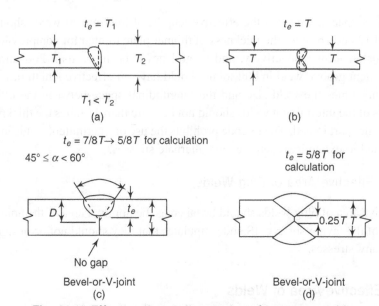

Fig. 11.19 Effective throat diamensions for groove welds

11.9.2 Fillet Weld

The effective throat dimension of a fillet weld is the shortest distance from the root of the face of the weld, as shown in Fig. 11.20. The effective throat thickness of a fillet weld should not be less than 3 mm and should not exceed $0.7a$ ($1.0a$, under special circumstances), where a is the size of the weld in mm. Thus, if the fillet weld is having unequal lengths (which is a rare situation), as shown in Fig. 11.20(b), the value of t_e should be computed from the diagrammatic shape of the weld.

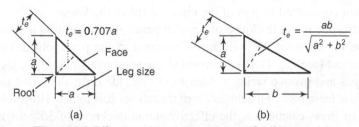

Fig. 11.20 Effective throat dimensions for fillet welds

The load–deformation relationship of fillet welds has been studied by several researchers (e.g. Butler et al. 1972; Swannel 1981; Neis 1985). Figure 11.21 shows the variations in fillet weld behaviour with the relative direction of the load vector to the weld axis (for a 8-mm fillet weld of ultimate strength 565 MPa, weld length 50 mm, plate thickness 19 mm, and ultimate strength of plate = 511 MPa). When $\theta = 0°$, the weld axis is normal to the load vector, the so-called *end fillet* (transverse fillet) situation, and the weld develops a high strength with less ductility (with deformation at rupture less than 1 mm). On the other hand, when $\theta = 90°$, the weld axis is parallel to the load vector, the *side-fillet* (longitudinal fillet) situations, and the weld shear strength is limited to about 56% of the weld metal tensile strength. However, the side fillet exhibits more ductility (rupture occurring at over 2-mm deformation). Intermediate orientations show intermediate values of both strength and ductility. Thus, the end fillet welds are 30%–40% stronger than side fillet welds.

Apart from the large difference in strength, end and side fillets also differ in both stiffness and ductility. Note that when the two types of welds are mixed, a greater share of the load is attracted to end fillets because of their high stiffness, which may result in the side fillets not developing to their full capacity. The code stipulates that the effective throat thickness in fillet welds joining faces inclined to each other should be taken as follows:

$$\text{Effective throat thickness} = K \times \text{size of weld}$$

where K is a constant depending upon the angle between fusion faces (see Fig. 11.21), as given in Table 11.3.

Table 11.3 Values of K for different angles between fusion faces (as per IS 800)

Angle between fusion faces	60°–90°	91°–100°	101°–106°	107°–113°	114°–120°
Constant K	0.70	0.65	0.60	0.55	0.50

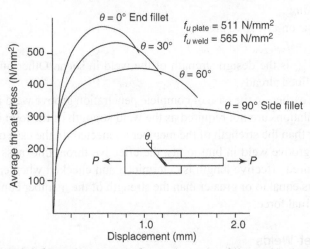

Fig. 11.21 Load-deformation curves for an 8 mm leg fillet weld at varying angles θ to load vector

11.9.4 Long Joints

If the maximum length of side weld exceeds $150t_e$, where t_e is the throat size of weld, a reduction factor as per clause 10.5.7.3 of the code has to be applied to the calculated strength.

11.10 Design of Welds

The following assumptions are usually made in the analysis of welded joints.
(a) The welds connecting the various parts are homogenous, isotropic, and elastic.
(b) The parts connected by the welds are rigid and their deformation is, therefore, neglected.
(c) Only stresses due to external forces are considered. The effects of residual stresses, stress concentrations, and the shape of the weld are neglected.

11.10.1 Groove Welds

As per IS 800 : 2007, the groove welds in butt joints will be treated as parent metal with a thickness equal to the throat thickness and the stresses shall not exceed those permitted in the parent metal.
(a) For tension or compression normal to effective area and tension and compression parallel to the axis of the weld

$$T_{dw} = f_y L_w t_e / \gamma_{mw} \qquad (11.1)$$

where T_{dw} is the design strength of the weld in tension, f_y is the smaller of yield stress of the weld and the parent metal in MPa, t_e is the effective throat thickness of the weld in mm, L_w is the effective length of the weld in mm, and γ_{mw} is the partial safety factor taken as 1.25 for shop welding and as 1.5 for site welding.
(b) For shear on effective area

$$V_{dw} = L_w t_e f_{yw} / (\sqrt{3} \times \gamma_{mw}) \qquad (11.2)$$

where V_{dw} is the design strength of the weld in shear. Other quantities have been defined already.

As stated earlier, in the case of complete penetration groove weld in butt joints, design calculations are not required as the weld strength of the joint is equal to or even greater than the strength of the member connected. In the case of incomplete penetration groove weld in butt joints, the effective throat thickness is computed and the required effective length is determined and checked whether the strength of the weld is equal to or greater than the strength of the member connected or the applied external force.

11.10.2 Fillet Welds

The actual distribution of stress in a fillet weld is very complex. A rigorous analysis of weld behaviour has not been possible so far. Multi-axial stress state, variation in yield stress, residual stresses, and strain hardening effects are some of the factors, which complicate the analysis. In many cases, it is possible to use the simplified approach of average stresses in the weld throat.

In the code, the design strength of fillet weld, f_{wd}, is given by

$$f_{wd} = f_u / (\sqrt{3} \gamma_{mw}) \qquad (11.3)$$

where f_u is the smaller of the ultimate stress of the weld and parent metal, and γ_{mw} is the partial safety factor which equals 1.25 or 1.5 depending on whether the weld is made at a shop or at the site, respectively. (Note that the weld metal always has a higher strength; hence we should use the parent metal strength only in the equation as per IS code.)

Hence as per IS 800 : 2007 the design strength is given by

$$P_{dw} = L_w t_e f_u / (\sqrt{3} \gamma_{mw})$$ (11.4a)

or

$$P_{dw} = L_w K s f_u / (\sqrt{3} \gamma_{mw})$$ (11.4b)

where P_{dw} is the design strength of the fillet weld and s is the size of the weld. Other terms have been defined earlier.

Tables have been prepared to simplify the calculation while using Eqn (11.4) and are presented in Appendix D.

11.10.2.1 Design Procedure

The design procedure is as follows.

1. Assume the size of the weld based on the thickness of the members to be joined.
2. By equating the design strength of the weld to the external factored load, the effective length of the weld to be provided is calculated. The length may be provided either as longitudinal fillet welds (parallel to the load axis) or as transverse fillet welds (perpendicular to the load axis) along with longitudinal fillet welds. It is a common practice to treat both the welds as if they are stressed equally. If the length exceeds $150 t_e$, reduce design capacity by a factor β_{lw} as per clause 10.5.7.3 of the code.
3. If only the longitudinal fillet weld is provided, a check is made to see if the length of each longitudinal fillet weld is more than the perpendicular distance between them.
4. End returns of length equal to twice the size of the weld are provided at each end of the longitudinal fillet weld.

When subjected to combined tensile and shear stress, the equivalent stress, f_e, should satisfy

$$f_e = \sqrt{(f_a^2 + 3q^2)} \le f_u / (\sqrt{3}\, \gamma_{mw})$$ (11.5)

where f_a = normal stress due to axial force or bending moment, and q = shear stress due to shear force or tension.

11.10.3 Intermittent Fillet Welds

Intermittent fillet welds are provided to transfer calculated stress across a joint, when the strength required is less than that developed by a continuous fillet weld of the smallest practical size. Such intermittent welds are often found in the connection of stiffeners to the web of plate girders. In such situations, first the fillet weld length required is computed as a continuous fillet weld. A chain of intermittent fillet welds of total length equal to the computed length, is provided as shown in Fig. 11.22. Intermittent fillet welds shown in Fig. 11.22(a) are structurally better than those shown in Fig. 11.22(b), since they reduce the distortion due to the balancing nature of the welds.

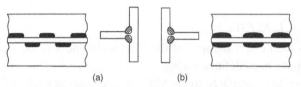

(a)   (b)
Fig. 11.22 Intermittent fillet weld

In the design of intermittent welds, the following procedure is adopted (IS 816 : 1969).
1. Assume the size of weld and compute the total length of required intermittent weld.
2. The minimum effective length (four times the size of weld or 40 mm) and clear spacing ($12t$ for compressions and $16t$ for tension and should not be less than 200 mm, where t is the thickness of the thinner plate joined) clauses of IS codes should be followed.
3. At the ends, the longitudinal intermittent fillet weld should be of length not less than the width of the member, otherwise transverse welds should be provided. If transverse welds are provided along with longitudinal intermittent welds, the total weld length at the ends should not be less than twice the width of the member.

11.11 Simple Joints

In this section we will discuss the design of some simple welded joints such as truss member connections, angle seat connections, web angle and end seat connections, and end plate connections.

 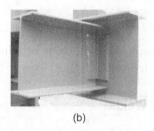 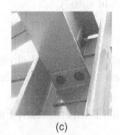

(a)   (b)   (c)

Some simple welded connections: (a) column splice with CJP grove weld, (b) welded double angle connection, (c) unstiffened seat connection © American Institute of Steel Construction, Inc., Reprinted with permission. All rights reserved.)

11.11.1 Design of Fillet Welds for Truss Members

In the design of welds connecting tension or compression members, the welds should be at least as strong as the members they connect and the connection should not result in significant eccentricity of loading. Truss members often consist of single or double angles, and occasionally T-shapes and channels. Consider the angle tension member shown in Fig. 11.23, with two longitudinal welds (on the two sides parallel to the axis of the load) and a transverse weld (perpendicular to the axis of the load). The axial force T in the member will act along the centroid of the member. The force T has to be resisted by the forces P_1, P_2, and P_3 developed by the weld lines. The forces P_1 and P_2 are assumed to act at the edges of the angle and the force P_3 at the centroid of the weld length, located at $d/2$. Taking moments about point A located at the bottom edge of the member and considering clockwise moments as positive, we get

$$\Sigma M_A = -P_1 d - P_2 d/2 + Ty = 0 \tag{11.6}$$

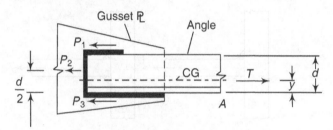

Fig. 11.23 Balancing the welds on a tension member connection.

Hence,

$$P_1 = Ty/d - P_2/2 \tag{11.7}$$

The force P_2 is equal to the resistance R_w of the weld per mm multiplied by the length L_w of the weld.

$$P_2 = R_w L_{w2} \tag{11.8}$$

Considering the horizontal equilibrium, we get

$$\Sigma F_H = T - P_1 - P_2 - P_3 = 0 \tag{11.9}$$

Solving Eqns (11.7) and (11.9) simultaneously, we get

$$P_3 = T(1 - y/d) - P_2/2 \tag{11.10}$$

Designing the connection shown in Fig. 11.23, to eliminate the eccentricity caused by the unsymmetrical welds is called *balancing the weld*. The procedure adopted for balancing the weld is as follows.

1. After selecting the proper weld size and electrode, compute the force resisted by the end weld P_2 (if any) using Eqn (11.8).
2. Compute P_1 using Eqn (11.7).
3. Compute P_3 using Eqn (11.10) or

4. Compute the lengths L_{w1} and L_{w3} on the basis of

$$L_{w1} = P_1/R_w \text{ and } L_{w3} = P_3/R_w \qquad (11.11)$$

Alternatively, the total length required to resist the load, L_w may be calculated. The length of end weld may then be subtracted from the total and the remaining length is allocated to P_1 and P_2 in inverse proportion to the distances from the centre of gravity.

Single sided welds in tension

Examples of unsatisfactory and satisfactory welds for tension connections are given in Fig. 11.24. In the examples shown in Fig. 11.24(a), the eccentricity between the line of action of the load and the throat centroid creates a moment on the weld throat. Hence, this should be avoided in practice. In the symmetric arrangements shown in Fig. 11.24(b), though there is a small variation in stress across the weld throats, with little ductility, this variation is redistributed, resulting in uniform stress fields on the weld throats.

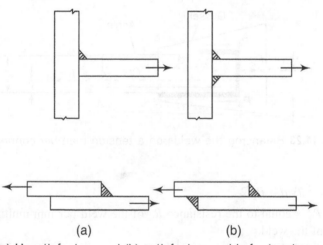

(a)   (b)

Fig. 11.24 (a) Unsatisfactory and (b) satisfactory welds for tension connections

11.11.2 Angle Seat Connections

As discussed in Section 10.6.3, a beam may be supported on a seat, either unstiffened or stiffened. In this section, the unstiffened seat connection as shown in Fig. 11.25 is discussed, where an angle is designed to carry the entire reaction. This type of connection uses a top clip angle, whose intended function is to provide lateral support to the compression flange. The seated connection is designed to transfer only the vertical reaction and should not give significant restraining moment at the end of the beam. Hence, the seat and top angle are selected in such a way that they are relatively flexible.

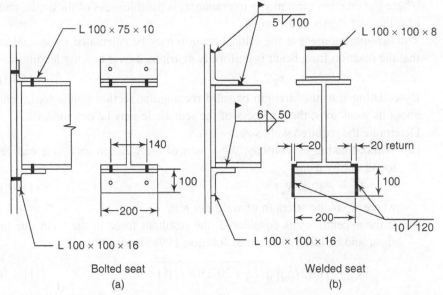

Bolted seat
(a)

Welded seat
(b)

Fig. 11.25 Welded seat angle connection (for comparison bolted seat angle connection is also shown; The angle sizes and weld sizes and weld sizes may change depending on load)

For welded seat, since the weld along the end holds the angle tight against the column, the critical section is the same (whether or not the beam is attached to the seat) as that for the previous case of bolted beam, connected to seat (see Fig. 10.23).

The design of the unstiffened angle seat involves the following steps (see Section 10.6.4 and Fig. 10.23).

1. Selection of seat angle having a length equal to width of the beam.
2. Length of the outstanding leg of the seat angle is calculated on the basis of web crippling of the beam.

$$b = R/[t_w(f_{yw}/\gamma_{m0})]$$ (11.12)

   where R is the reaction of the beam, t_w is the thickness of the web of the beam, f_{yw} is the yield strength of the web, and γ_{m0} is the partial safety factor for material = 1.10.
3. Determine the length of the bearing on cleat

$$b_1 = b - (t_f + r_b)$$ (11.13)

   where t_f is the thickness of the flange of the beam and r_b is the root radius of the beam flange.
4. Determine the distance from the end of the bearing on cleat to the root of the angle

$$b_2 = b_1 + g - (t_a + r_a)$$ (11.14)

where g = erection clearance + tolerance, t_a is the thickness of the angle, and r_a is the root radius of the angle.

5. The bending moment at the critical section may be calculated by assuming that the reaction from beam is uniformly distributed over bearing length b_2

$$M_u = R \times (b_2/b_1) \times b_2/2 \tag{11.15}$$

By equating it to the strength of solid rectangular section (angle leg), bent about its weak axis, the thickness of the seat angle may be determined.

6. Determine the required weld size.

(a) Without taking eccentricity, the length of the weld on each side can be found out by using

$$L_w = R/(2 \times R_w) \tag{11.16}$$

where R_w is the strength of weld per mm.

(b) If the eccentricity is considered, the resultant force in the weld due to shear and bending (Salmon & Johnson 1996) is given by

$$R_{res} = [R/(2L_w^2)]\sqrt{[L_w^2 + 20.25(b_2/2)^2]} \tag{11.17}$$

11.11.3 Web Angle and End Plate Connections

The field-welded shear connection using web angles is shown in Fig. 11.26. The intension of such a connection is that the angles are as flexible as possible so that the beams are capable of rotating at the ends. They are assumed to provide simply supported end condition and designed to transmit shear only. The pair of angles are shop welded to the beam and field welded or connected to the column by means of HSFG bolts at site. The angles (called clip angles) project out of the beam web by a distance of about 12 mm (called set back), so that the beam can be fitted with acceptable tolerances. When beams intersect and have the same depth, the flanges are coped (cut away) as shown in Fig. 11.26(c), resulting in some loss of shear strength. The coping of beams will result in block shear failure and are susceptible to local web buckling (Gupta 1984; Yura et al. 1982).

Erection bolts are used to erect these beams and then the angles are welded at site. These bolts are often provided at the bottom of the angle. Generally 100-mm size legs are used for connecting the beam and the leg size at the column or girder side is kept a little longer. The length of the angle is kept equal to the distance between fillets of the beam, so that sufficient length is available for welding. Usually a weld size 2–3 mm smaller than the web angle thickness is chosen.

The connection, though assumed to transfer only shear forces, due to the eccentricity of connection, is also subjected to a bending moment (Blodgett 1966). Due to the rotation effect, the field welds cause web angles to press against the beam web at the top and tear apart from the bottom, thus indicating horizontal shear in the fillet weld. It is assumed that the neutral axis is at a distance of $L/6$ from the top of the angle. The horizontal shear is taken as zero at this point and maximum at the bottom of the angle [see Fig. 11.27(b)]. Neglecting the effects of the returns at top, the horizontal component R_x can be obtained from moment equilibrium (Blodgett 1966).

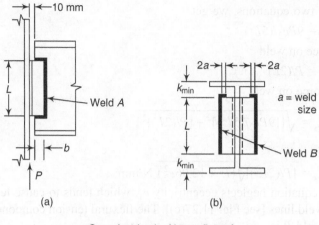

(a)

(b)

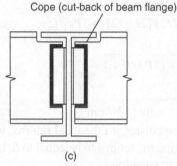

(c)

Fig. 11.26 Simple shear double-anlge connections

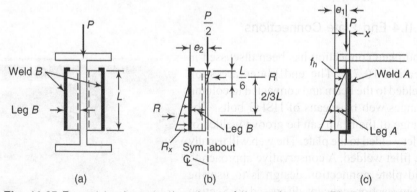

(a)

(b)

(c)

Fig. 11.27 Eccentric shear in the plane of the web as per Blodgett (1966)

Applied moment from load = Resisting moment of weld

$$(P/2)e_2 = (2/3)RL \qquad (11.18)$$

where L is the length of weld

Hence,

$$R = 0.75Pe_2/L$$

From force triangle, we get

$$R = 0.5R_x \times (5/6)L$$

From these two equations, we get

$$R_x = 9Pe_2/(5L^2) \qquad (11.19)$$

Vertical force on weld

$$R_v = P/(2L) \qquad (11.20)$$

Resultant force on weld

$$R_{res} = \sqrt{\{[9Pe_2/(5L^2)]^2 + [P/(2L)]^2\}} \qquad (11.21)$$

or

$$R_{res} = [P/(2L^2)]\sqrt{(L^2 + 12.96e_2^2)} \text{ N/mm} \qquad (11.22)$$

The above equation neglects eccentricity e_1, which tends to cause tension at the top of the weld lines [see Fig. 11.27(c)]. The flexural tension component R_x at the top of the weld B is

$$R_x = My/I = Pe_1(L/2)/[2L^3/12] = 3Pe_1/L^2 \qquad (11.23)$$

Thus,

$$R_{res} = \sqrt{[(P/2L)^2 + (3Pe_1/L^2)^2]} \qquad (11.24)$$

or

$$R_{res} = P/(2L^2)\sqrt{(L^2 + 36e_1^2)} \text{ N/mm} \qquad (11.25)$$

Note that the above equations disregard the weld returns, which have the greatest effect if L is short. Considering the returns to be equal to $L/12$, Salmon and Johnson (1996), derived the following equation

$$R_{res} = P/(2L^2)\sqrt{(L^2 + 20.25e_1^2)} \text{ N/mm} \qquad (11.26)$$

11.11.4 End Plate Connections

End plate connection has been discussed in Section 10.7.2. The end plates are shop welded to the beam and connected to column flanges/web by means of HSFG bolts. The flange of the beam can be groove welded or fillet welded to the plate. The web will usually be fillet welded. A conservative approach to end plate connection design is to use the prying action concept discussed in Section 10.4.4. The region near the tension flange of the beam is designed similar to that of a split-beam T-connection. This fastener group is designed for shear and tension, including the effect of prying action. The welds are designed for the resultant force using the elastic vector analysis as below.

$$\text{Resultant force} = \sqrt{(P/A)^2 + (My/I)^2} \leq \text{design strength of the weld} \qquad (11.27)$$

where P/A is the vertical stress due to shear force and My/I is the tension component (horizontal) due to the bending moment.

11.12 Moment Resistant Connections

When beams are connected to columns through brackets, depending on the way in which they are connected, the welds may be subjected to either twisting moment or bending moment, in addition to shear forces. Welded stiffened seat connections are also subjected to bending moment and shear forces. These connections are termed as moment resistant connections and are discussed in this section.

11.12.1 Eccentrically Loaded Connections

Loads acting eccentrically from the centroid of a weld line or weld group may cause either a twisting moment or a bending moment on the weld, depending upon the location of the welds, in addition to the direct shear forces.

Eccentric load causing twisting moment Since no initial tension is involved with welded connections, the eccentricity of loading, even though small, has to be considered. Also, there are situations where the loading of fillet welds is neither parallel not transverse to the axis of fillet welds, as shown in Fig. 11.28. Analysis of such eccentric loading is complicated since the load–deformation behaviour is a function of the angle θ between the direction of applied load and axis of the fillet weld (see Fig. 11.21).

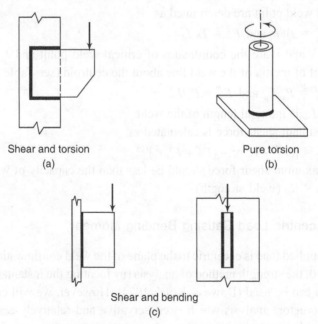

Shear and torsion
(a)

Pure torsion
(b)

Shear and bending
(c)

Fig. 11.28 Types of eccentric loading.

As discussed in Section 10.7.1, the strength of an eccentrically loaded fillet weld can also be determined by locating an instantaneous centre of rotation, using the load–deformation relationship of the fillet weld.

Here the more conservative traditional elastic vector analysis which is easier than the strength method is described. The following assumptions are made in the elastic method.

(a) Each segment of weld (of the same size) resists the concentrically applied load with an equal force.

(b) The rotation caused by the torsional moment is assumed to occur about the centroid of weld configuration.

(c) The load on a weld segment caused by the torsional moment is assumed to be proportional to the distance from the centroid of the weld configuration.

(d) The components of the forces caused by the direct load and by torsion are combined vectorially to obtain a resultant force.

The steps involved in checking the adequacy of the weld are as follows.

1. The centroid of the weld line is calculated. The twisting moment and the forces at the centroid are determined (see Fig. 11.29)

$$T = P_x e_x + P_y e_y \tag{11.28}$$

where P_x and P_y are the x and y components of the eccentric load and e_x and e_y are the eccentricities of P_x and P_y with respect to the centroid of weld line.

2. The critical weld points are located.

3. The force components due to twisting moment and maximum shear force for critical weld point are determined as

$$F_x^T = Ty/I_p \text{ and } F_y^T = Tx/I_p \tag{11.29}$$

where x and y are the coordinates of critical weld point and I_p is the polar moment of inertia of the weld line about the centroid (see Table 11.4) and

$$F_x^P = P_x/L_w \text{ and } F_y^P = P_y/L_w \tag{11.30}$$

where L_w is the total length of the weld.

4. The resultant shear force is calculated as

$$F_R = [(F_x^P + F_x^T)^2 + (F_y^P + F_y^T)^2]^{0.5} \tag{11.31}$$

5. The maximum shear force should be less than the capacity of weld

$$F_R < R_w \text{ (weld strength)} \tag{11.32}$$

11.12.2 Eccentric Load Causing Bending Moment

When the applied load is eccentric to the plane of the weld configuration, as shown in Fig. 11.30, the strength method of analysis (by locating the instantaneous center of rotation) can be used (Dawe & Kulak 1974). However, we will consider only the elastic (vector) analysis which is conservative and relatively easy to use for loading resulting in shear and tension.

Table 11.4 Properties of welds treated as lines

Section b = width; $d = L_w$ = depth		Section modulus $I_x/\bar{y}$	Polar moment of interia, I_p about centre of gravity
1.		$Z = \dfrac{d^2}{6}$	$I_p = \dfrac{d^3}{12}$
2.		$Z = \dfrac{d^2}{3}$	$I_p = \dfrac{d(3b^2 + d^2)}{6}$
3.		$Z = bd$	$I_p = \dfrac{b(3d^2 + b^2)}{6}$
4.	$\bar{y} = \dfrac{d^2}{2(b+d)}$ $\bar{x} = \dfrac{b^2}{2(b+d)}$	$Z = \dfrac{4bd + d^2}{6}$	$I_p = \dfrac{(b+d)^4 - 6b^2d^2}{12(b+d)}$
5.	$\bar{x} = \dfrac{b^2}{2b+d}$	$Z = bd + \dfrac{d^2}{6}$	$I_p = \dfrac{8b^3 + 6bd^2 + d^3}{12} - \dfrac{b^4}{2b+d}$
6.	$\bar{y} = \dfrac{d^2}{b+2d}$	$Z = \dfrac{2bd + d^2}{3}$	$I_p = \dfrac{b^3 + 6b^2d + 8d^3}{12} - \dfrac{d^4}{2d+b}$
7.		$Z = bd + \dfrac{d^2}{3}$	$I_p = \dfrac{(b+d)^3}{6}$
8.	$\bar{y} = \dfrac{d^2}{b+2d}$	$Z = \dfrac{2bd + d^2}{3}$	$I_p = \dfrac{b^3 + 8d^3}{12} - \dfrac{d^4}{b+2d}$
9.		$Z = bd + \dfrac{d^2}{3}$	$I_p = \dfrac{b^3 + 3bd^2 + d^3}{6}$
10.		$Z = \pi r^2$	$I_p = 2\pi r^3$

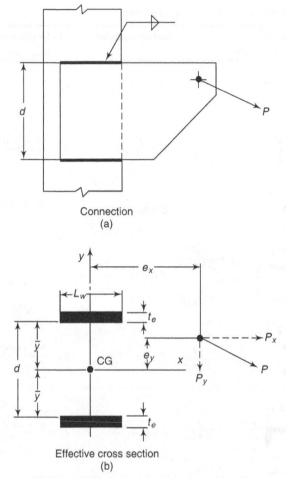

Connection
(a)

Effective cross section
(b)

Fig. 11.29 Eccentric bracket connection

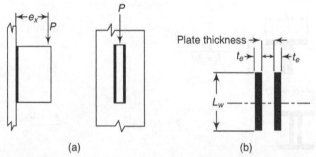

(a)   (b)

Fig. 11.30 Loads applied eccentric to the plane of weld

The effect of eccentric load on the weld group about its centroid is equivalent to a bending moment and direct force at the centroid. The bending moments (Pe) cause bending tensile and compressive stresses (i.e. normal stresses) at the throat section of the fillet weld, while the direct forces cause direct or shear stresses. Either a groove weld or a fillet weld can be used in such connections. Thus, the welds must carry the loads in the same manner as the members being connected carry them. The stresses are shown in Fig. 11.31. Thus, the direct stresses in the weld = load/effective area of weld

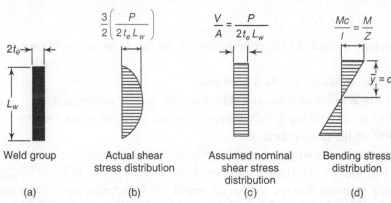

Weld group	Actual shear stress distribution	Assumed nominal shear stress distribution	Bending stress distribution
(a)	(b)	(c)	(d)

Fig. 11.31 Stresses in weld subject to eccentric load causing bending moment

(a) In the case of a fillet weld

$$\tau_{vf,\ cal} = P/(2L_w t_e) \tag{11.33}$$

(b) For groove welds

$$\tau_{vf,\ cal} = P/dt \tag{11.34}$$

where t is the thickness of plate.

The bending stress in the weld = moment/section modules

1. For fillet weld the failure is due to critical stress in the throat of the weld (at 45° to the weld leg length)
   Hence,

   $$f_b = M/Z = [Pe(L_w/2)]/[2 \times (L_w^3 t_e/12)] = 3Pe/(t_e L_w^2)$$

   The throat stress is treated as shear since a 45° line of failure is assumed. This shear stress is assumed linearly varying from zero at mid depth to the maximum value at the extreme fibres. Hence,

   $$\tau_{vf1,\ cal} = 3Pe/(t_e L_w^2) \tag{11.35}$$

2. For groove welds

   $$f_b = 6Pe/(t L_w^2) \tag{11.36}$$

   The combined stress in the fillet weld is given by the following equation

   $$f_e = \sqrt{(\tau_{vf,\ cal}^2 + \tau_{vf1,\ cal}^2)} < \text{weld strength} = f_u/(\sqrt{3}\gamma_{mw}) \tag{11.37}$$

   The combined bending and shear stress in the groove weld is checked by using the interaction formula (clause 10.5.10.1.1 of code)

$$f_e = \sqrt{(f_{b,\,\mathrm{cal}}^2 + 3\tau^2)} \tag{11.38}$$

and f_e should not exceed the values allowed for the parent metal. The code also states that the check for the combination of stresses need not be done:
(a) for side fillet welds joining cover plates and flange plates and
(b) for fillet welds, where sum of normal and shear stresses does not exceed

$$f_{\mathrm{wd}} = f_u^{\hat{}}/(\sqrt{3}\gamma_{\mathrm{mw}})$$

Similarly a check for the combination of stresses in groove welds need not be done if:
• groove welds are axially loaded and
• in single and double bevel welds, where the sum of normal and shear stresses does not exceed the design normal stress and the shear stress does not exceed 50% of the design shear stress.

Note that the locations of maximum bending and shearing stresses are not the same (see Fig. 11.31). Hence, if the welds are used as shown in Fig. 11.32, it can be safely assumed that the web welds would carry the entire shear force and the flange welds would carry the entire bending moment.

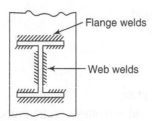

Fig. 11.32 Welding of brackets for carrying shear and moment.

11.12.2.1 Design

The design of brackets subject to combined bending and shear is done using the following steps.
1. Assume the size of the weld and compute the throat thickness, design strength, and capacity of weld (R_{nw}).
2. Calculate the depth of bracket (length of weld) using the following appropriate equations.
   (a) In the case of groove welds
$$L_w = [6M/(tf_b)]^{1/2} \tag{11.39}$$
      Where $f_b = f_y/\gamma_{m0}$ with $\gamma_{m0} = 1.10$
   (b) In the case of fillet welds
$$L_w = [6M/\{2t_e R_{\mathrm{nw}}\;(\mathrm{est.})\}]^{1/2} \tag{11.40}$$

A reduced value of R_{nw} is used to account for the direct shear effect also.
3. The direct shear stress is computed using Eqn (11.33) or (11.34), as appropriate.

4. The stress due to bending moment is computed from Eqn (11.35) or (11.36), as appropriate.
5. The equivalent stress is computed from Eqn (11.37) or (11.38), as appropriate.
6. If the equivalent stress exceeds the weld strength (fillet welds) or the design stress of the parent metal (groove weld), the length of the bracket (weld length) may be increased and the process repeated till the checks are satisfied.

11.12.3 Stiffened Beam Seat Connection

A welded stiffened seat connection for a beam is much simpler than the one with a bolted connection. It consists of two plates forming a T or a split I-section used as a seat (see Fig. 11.33). The thickness of the stem of the T should not be less than the web thickness of the beam it supports. Similarly, the thickness of seat plate should not be less than the thickness of the flange of the beam. The length of bearing is governed by the strength as well as by the web crippling requirement of the beam (as in the case of unstiffened seat angle). The depth of the stem should be short enough to avoid local buckling and it depends upon the length of the vertical weld required. The under side of the flange of the T is also welded to increase the torsional stiffness of the connection (see Fig. 11.34). The seat plate is kept wider than the flange of the beam by at least twice the size of the weld on each side of beam flange to facilitate welding. As in the unstiffened beam seat connection, a cleat angle of nominal size is welded to the top of beam in the shop and to the column at the field to provide lateral support to the beam's top flange.

Fig. 11.33 Welded stiffened seat connection using the split I-section in a car parking structure in Bethesda, USA. (The bolts are erection bolts.)

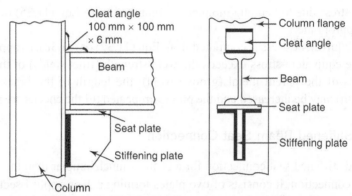

Fig. 11.34 Welded stiffened seat connection

There are two basic types of loading used on stiffened seats. The common one is shown in Fig. 11.34, where a beam web is placed directly in line with the stiffener. The other type which occurs when supporting gantry girders, is shown in Fig. 11.35, in this case the beam is oriented in such a way that the plane of the web is at 90° to the plane of the stiffener. This stiffener behaves similar to an unstiffened element under uniform compression, and local buckling may be prevented by satisfying the limiting width to thickness ratios given in Table 2 of the code.

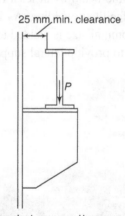

Fig. 11.35 Bracket supporting concentrated load

11.12.3.1 Design

The design of the stiffened welded seat connection is similar to that of the unstiffened welded seat connection. The steps to be followed are as follows.

1. The width of the seat angle is calculated.
2. The thickness of the seat plate is chosen as equal to the thickness of the flange plate.
3. The thickness of stiffening plate is chosen as equal to the thickness of the web of the beam.
4. The eccentricity of the load and bending moment due to it are calculated.

5. Vertical and horizontal shear stresses due to the shear force and bending moment are calculated. From these values, the resultant shear of the weld is computed as the vector sum.

6. From the above resultant shear value, the weld size is determined.

11.13 Continuous Beam-to-Column Connections

In continuous beam-to-column connections, it is assumed that there is full transfer of moment and little or no relative rotation of members within the joint. It is obvious that the connection must be of adequate strength to accomplish this. Local elements of the connection must be so proportioned that their premature failure will not result in a lowering of overall connection strength.

Rigid frame connections (except in pre-fabricated buildings) are always welded. Some of the common welded connections to column flanges are shown in Fig. 11.36. (Also see the photos given below.)

 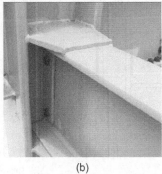

(a)   (b)

(a) Beam flange directly welded to column flange (b) welded flange plate connection to column web (© American Institute of Steel Construction, Inc., Reprinted with permission. All rights reserved.)

Most of these connections are partly shop welded and then completed at site using site welding. Splices in beams/columns are made away from this region, and at location of lower shear and moment.

For connecting the tension flange of the beam to the column, a cover plate is used Fig. 11.36(c). This plate is connected to the column flange by means of full penetration groove welds and to the beam flange by fillet welds. This welding requires a minimum gap of 4–5 mm between the plate and the column. Hence, a backing strip may be used to install the weld efficiently. The length of the plate is kept sufficiently long (about 1.5 times the width of the beam flange) so that a length at least equal to its own width is left unwelded. This unwelded length imparts ductility to the connection, since this length will yield and elongate sufficiently before the connection fails. Also the welded length is kept at least equal to the width of the plate so that full stresses are transferred to the cover plate from the flange by shear lag. (In proportioning the connection, it is usual to assume that the bending moments and thrust are carried by the flanges and the shear is carried entirely by the web.) If a long length of fillet weld is required to connect the cover plate with the beam, U-shaped slots may be cut at the ends of the plate to increase the length of the weld. It is also tapered to facilitate fillet welding.

Under heavy moments, the column flange may get deformed due to the loads acting on the beam flanges (tension or compression), and its moment carrying capacity is reduced. To control this, additional plates, called *stiffeners* are welded between column flanges opposite the beam flanges as shown in Figs 11.36(b) and (c).

Normally, in continuous construction, the beam develops tension at the top and compression at the bottom near the supports. But in some tall frames subjected to lateral forces (wind or earthquake), tension may develop in the bottom flange, due to stress reversal. In such cases, it is advisable to provide a bottom plate similar to the top plate given in Fig. 11.36(c), capable of yielding at bottom also.

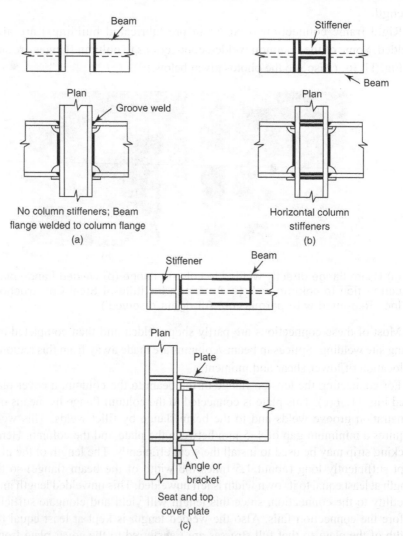

Fig. 11.36 Continuous beam-to-column connections: Beam welded to column flange

11.14 Beam and Column Splices

As already mentioned in Chapter 10, sections used for beams and columns may have to be spliced due to practical reasons. In this section, we will consider the methods of providing splices in columns and beams by welding (see also clause 10.7 of the code).

11.14.1 Beam Splices

Long span beams may require site connections between successive pieces. The maximum length that can be transported to the site is restricted by the size of available vehicles. Normally, the length that can be transported is up to 6 m. Splices are located in regions of the beams where the bending moment is low. Splicing of webs is very common in plate girders. Splicing is done to change the thickness of plates in the girder, for cambering or when the required length of plate is not available.

Beams may be spliced using groove welds and using a staggered form of arrangement as shown in Fig. 11.37. This kind of splicing requires temporary support during welding and is often expensive (though providing aesthetic appearance). In order to minimize distortion effects, flange welding is done before web welding.

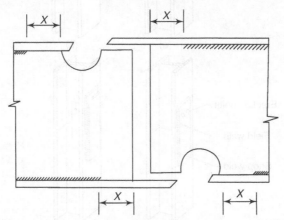

Fig. 11.37 Staggered form of arrangement for groove welded beam splice

The alignment problems of groove welded splices can be overcome by using splice plates, which can be fillet welded (see Fig. 11.38).

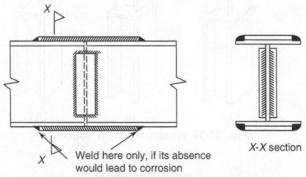

X-X section

Weld here only, if its absence would lead to corrosion

Fig. 11.38 Fillet welded beam splice

11.14.2 Column Splices

When column sections are to be spliced by welding (see Fig. 11.39), the ends are first milled for a square bearing surface. Then the splice angles are shop welded [see Figs 11.39(a)]. The outstanding legs of these angles are provided with holes for erection bolts to engage the outstanding legs of the other two angles that are shop welded to the upper column section. Since they do not project beyond the ends of the columns, they are not damaged during transit. Once these two sections are brought to the site, the splice angles are bolted and the two column sections are

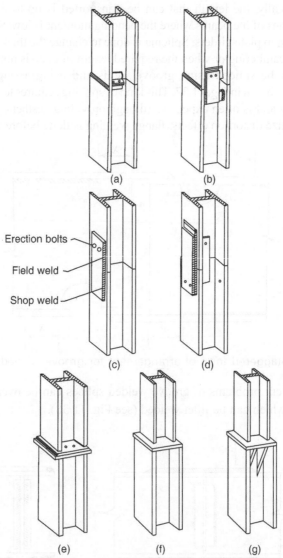

Erection bolts

Field weld

Shop weld

Fig. 11.39 Welded column splices.

welded using full penetration groove welds or partial penetration welds. Instead of splice angles, splice plates can be used [see Figs 11.39(b), (c), and (d)]. In Fig. 11.39(d), notice the filler plates, which are used when two different sizes of columns are connected. It is also possible to use welded end plates for making column splices. Both faces of these end plates are machined for bearing and normally have a thickness of more than 20 mm [see Figs 11.39(e), (f), and (g)]. When two different sizes of columns are connected, a web stiffener [Fig. 11.39(g)] is used, which assists in diffusing the load into the lower column. This web stiffener should have similar proportions to the upper column flange. Though column splices are normally placed just above floor level, in earthquake regions they must be located in regions of near zero bending moments.

Pre-qualified Seismic Moment Connections

Subsequent to the January 1994 Northridge earthquake in California and the Kobe earthquake in Japan in 1995, it was determined that some damage to moment-resisting frames occurred at the beam-column connections. Failures included the fractures of bottom beam flange-to-column flange complete-joint-penetration groove welds, which propagated into the adjacent column flange and web and into the beam bottom flange. This failure was accompanied in some instances by the secondary cracking of the beam web shear plate and the failure of the beam top flange weld. The factors that contributed to the damage include the following (FEMA 2000):
• Stress concentration at the bottom flange weld, due to the notch effect produced by the backing strips left in place,
• The use of low toughness weld metal at the beam-column connection,
• Uncontrolled deposition rates,
• The use of larger members than those previously tested,
• Lack of control of basic material properties (large variations of member strength from the prescribed values),
• Inadequate quality control during construction, and
• The tri-axial restraint existing at the center of beam flanges and at the beam-column interface, which inhibits yielding.
A multi-billion dollar research conducted over 10 years resulted in the development of current design provisions for moment-resistant frames, prescribed in AISC 341-05 (*seismic provisions for structural steel buildings, American Institute of Steel Construction*). In addition, AISC has developed another American National Standards Institute approved standard, AISC 358-05 (*pre-qualified connections for special and intermediate steel moment frames for seismic applications including Supplement No.1*), which presents materials, design, detailing, fabrication, and inspection requirements for a series of pre-qualified moment connections. AISC

updates and reissues this standard from time to time, as and when additional research results are available. The draft AISC 358-2010 contains a number of pre-qualified connections and are discussed briefly here.

Reduced beam section connection

In reduced beam section (RBS) moment connection, some portions of the beam flanges are removed in a predetermined fashion, adjacent to the beam-column connection, as shown in Fig. CS1. In such a connection, yielding and plastic hinges are forced to form away from the connection at the reduced section of the beam.

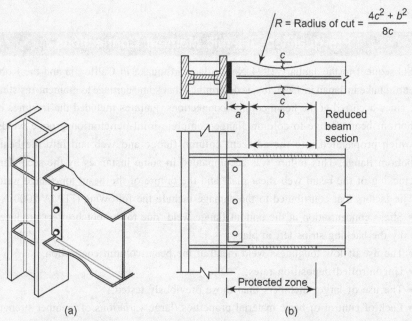

$$R = \text{Radius of cut} = \frac{4c^2 + b^2}{8c}$$

(a)   (b)

Fig. CS1 Reduced beam section (RBS) moment connection (a) connection, (b) details of reduced beam flange (Carter and Grubb, 2010)

Bolted Unstiffened and Stiffened Extended End-plate Moment Connections

Bolted end-plate connections are made by welding the beam section to an end-plate which is in turn bolted to the column flange. Three types of these connections are pre-qualified by AISC 358. It gives equations to check the various limit states of this type of connection, such as flexural yielding of the beam section or end-plate, yielding of column panel zone, shear or tension failure of the end-plate bolts, and failure of the various welded joints. These provisions are intended to ensure inelastic deformation of the connection by beam yielding.

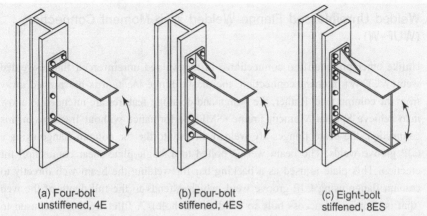

(a) Four-bolt
unstiffened, 4E

(b) Four-bolt
stiffened, 4ES

(c) Eight-bolt
stiffened, 8ES

Fig. CS2 Bolted unstiffened extended end-plate (BUEEP) and bolted stiffened extended end-plate (BSEEP) moment connections (Carter and Grubb, 2010)

Bolted Flange Plate (BFP) Moment Connection

These connections consist of plates welded to column flanges and bolted to beam flanges as shown in Fig. CS3. Identical top and bottom plates are used. Flange plates are connected to the column flange by using complete joint penetration (CJP) groove welds and beam flanges are connected to the plates by using high-strength friction grip bolts. The web of the beam is connected to the column flange using a bolted single-plate shear connection, with bolts in short-slotted holes. In this connection, yielding and plastic hinge formation are designed to occur in the beam near the end of the flange plates. The design procedure for this type of connection is more complex than other pre-qualified connections.

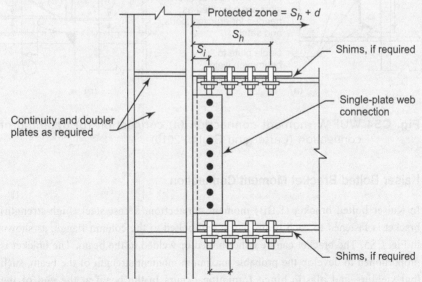

Fig. CS3 Bolted flange plate (BFP) moment connection (Carter and Grubb, 2010)

Welded Unreinforced Flange-Welded Web Moment Connection (WUF-W)

Unlike other pre-qualified connections, in the welded unreinforced flange-welded web (WUF-W) moment connection, the plastic hinge location is not moved away from the column face. Rather, the design and detailing features are intended to allow it to achieve Special Moment Frame (SMF) performance without fracture. In this connection, the beam flanges are welded directly to the the column flange using a CJP groove welds. The beam web is bolted to a single-plate shear connection for erection. This plate is used as a backing bar for welding the beam web directly to column flange using CJP groove weld, which extends to the full depth of the web (that is, from weld access hole to weld access hole). A fillet weld is also used to connect the shear plate to the beam web, as shown in Fig. CS4. A special seismic weld access hole and detailing, as shown in Fig. CS4(b), is specified for the WUF-W moment connection, to reduce stress-concentration in the region around the access hole,

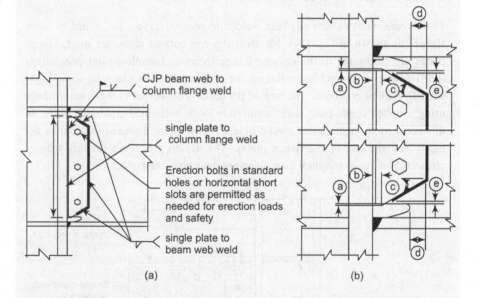

CJP beam web to column flange weld

single plate to column flange weld

Erection bolts in standard holes or horizontal short slots are permitted as needed for erection loads and safety

single plate to beam web weld

(a)

(b)

Fig. CS4 WUF-W moment connection (a) connection (b) detailing of connection (Carter and Grubb, 2010)

Kaiser Bolted Bracket Moment Connection

In Kaiser bolted bracket (KBB) moment connection, a cast steel (high-strength) bracket is fastened to each beam flange and bolted to the column flange, as shown in Fig. CS5. The bracket can be either bolted or welded to the beam. The bracket is proportioned to develop the probable maximum moment strength of the beam, such that yielding and plastic hinge formation occurs in the beam at the end of the bracket away from the column flange. This connection is designed to eliminate field welding and facilitate erection. This bracket connection is a propitiatory design of Steel Cast Connection LLC, USA. The advantage of using casting is that it will not have HAZ issues or residual stresses that would be found in welds.

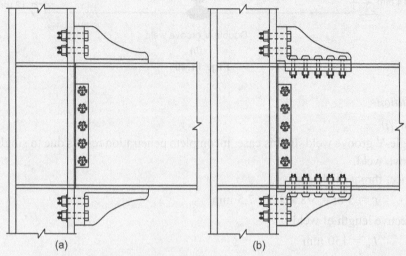

(a)   (b)

Fig. CS5 Kaiser bolted bracket moment connections (a) beam welded to bracket (b) beam bolted to bracket (Carter and Grubb)

The design procedure and detailing requirements for these connections are given in AISC 358-2010.

References

1. www.sacsteel.org/
2. Hamburger, R.O., Krawinkler, H., Malley, J.O., and Adan, S.M., *Seismic Design of steel special moment frames: A guide for practicing engineers,* NEHRP Seismic design technical brief no.2, National Institute of Standards and Technology, Gaithersburg, USA (http://www.nehrp.gov/pdf/nistgcr9-917-3.pdf)
3. Carter, C.J. and Grubb, K.A., Prequalified Moment Connections (revisited), *Modern Steel Construction,* Vol.50, No.1, Jan 2010, pp. 54–57
4. www.steelcastconnections.com

Examples

Example 11.1 *Two plates of thickness 14 mm and 12 mm are to be jointed by a groove weld as shown in Fig. 11.40. The joint is subjected to a factored tensile force of 350 kN. Assuming an effective length of 150 mm, check the safety of the joint for*

Case (i) Single-V groove weld joint

Case (ii) double-V groove weld joint

Assume that Fe 410 grade steel plates are used and that the welds are shop welded.

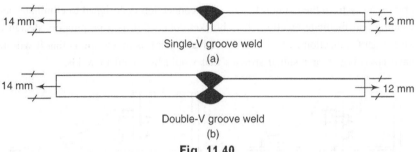

Single-V groove weld

(a)

Double-V groove weld

(b)

Fig. 11.40

Solution

Case (i)

Single-V groove weld: In this case, incomplete penetration results due to single-V groove weld.

Hence, throat thickness

$$t_e = 5/8t = 5/8 \times 12 = 7.5 \text{ mm}$$

Effective length of weld

$$L_w = 150 \text{ mm}$$

Strength of weld $= L_w t_e f_y/\gamma_{mw} = 7.5 \times 150 \times (250/1.25)/1000$
$$= 225 \text{ kN} < 350 \text{ kN}$$

Hence the joint is not safe.

Case (ii)

In the case of double-V groove weld, complete penetration takes place

Throat thickness = thickness of thinner plate = 12 mm
Strength of weld $= 12 \times 150 \times (250/1.25)/1000$
$$= 360 \text{ kN} > 350 \text{ kN}$$

Hence, the joint is safe.

Note Though the yield stress of electrodes is 330 MPa, only the yield stress in the plate is used in the calculations. Since the mechanical properties of the weld metal are always assumed to be stronger than the connecting parts, full penetration groove welds are adequately strong and hence no checking is required. The above calculations have been given just to show the strength of the welded joint.

Example 11.2 *The tie member of a truss is made of ISA 65 × 65 × 6 and is subjected to a factored tension load of 90 kN. The length of the angle is not enough to go from end to end and hence a splice has to be provided. Design a groove welded joint.*

Solution

Provide a single-V groove weld.

The effective throat thickness $= 5/8t = 5/8 \times 6 = 3.75$ mm

Perimeter length of angle available for welding $= 65 + 65 = 130$ mm

Strength of the weld $= 3.75 \times 130 \times (250/1.25)/1000$
$$= 97.5 \text{ kN} > 90 \text{ kN}$$

Area of the angle $= 744$ mm^2

Design strength of the member $= 744 \times 250/1.1$

$= 169$ kN > 90 kN

Note Though the strength of the weld is adequate, partial penetration welds should be avoided in direct tension. If not, the secondary bending stresses arising from the eccentricity of the throat area have to be considered.

Example 11.3 *Determine the effective throat dimension of a 10-mm fillet weld made by (a) shielded metal arc welding (SMAW) and (b) submerged arc welding (SAW).*

Solution

(a) Using SMAW process

Effective throat thickness

$$t_e = 0.7a = 0.7 \times 10 = 7 \text{ mm}$$

(b) Using SAW process

Using SAW process, we are in a position to get better penetration than SMAW process. Thus, as per clause 10.5.2.1 of the code

$$a = 10 + 2.4 = 12.4 \text{ mm}$$

Hence,

$$t_e = 0.7a = 0.7 \times 12.4 = 8.68 \text{ mm}$$

Example 11.4 *Determine the design shear strength of a 10-mm fillet weld produced by (a) SMAW and (b) SAW, assuming E 51 electrodes, Fe 410 material, and shop welding.*

Solution

Using the results obtained in Example 11.3, we get

(a) Using SMAW process

$$R_w = t_e(f_u/\sqrt{3})/\gamma_{mw} = 7(410/\sqrt{3})/1.25 = 1325 \text{ N/mm}$$

(b) Using SAW process

$$R_w = 8.68(410/\sqrt{3})/1.25 = 1643 \text{ N/mm}$$

Thus, the strength of the weld produced by using the SAW process is about 1.25 times higher than the strength of the weld produced by using the SMAW process.

Example 11.5 *Design a connection to joint two plates of size 250 × 12 mm of grade Fe 410, to mobilize full plate tensile strength using shop fillet welds, if (i) a lap joint is used (ii) a double cover butt joint is used.*

Solution

(i) Connection using lap joint

Plate strength $= f_y A_g / \gamma_{m0}$

$$= 250(250 \times 12)/(1.10 \times 1000) = 681.82 \text{ kN}$$

Size of the weld:

Minimum = 5 mm (Table 11.2)

Maximum = 12 – 1.5 = 10.5 (clause 10.5.8.1 of IS 800)

Though the minimum size is preferable, let us take the size of the weld as 8 mm in order to reduce the length of the connection.

Strength of the weld = $[f_u/(\sqrt{3} \times 1.25)]\ 0.7 \times$ size

$$= [410/(\sqrt{3} \times 1.25)] \times 0.7 \times 8$$
$$= 189 \times 0.7 \times 8 = 1058 \text{ N/mm}$$

Required length of weld = 681.82 × 1000/1058 = 644 mm, say 650 mm

Adopting a welding scheme as shown in Fig. 11.41(a).

Weld length available at end = 150 mm

Length of weld on one side = (650 – 150)/2 = 250 mm < 150 × 0.7 × 8 = 840 mm. Hence, long joint correction is not required.

(ii) Connection using Butt joint

Assume width of cover plate = 250 – 2 × 15 = 220 mm

Use an 8-mm thick cover plate.

Area of cover plate = 220 × 8 = 1760 mm^2

Required area of cover plate = 1.05 × 250 × 12/2 = 1575 mm^2 < 1760 mm^2

Use a 5-mm fillet weld.

Strength of the 5-mm weld = 189 × 0.7 × 5 = 661.5 N/mm

Required length of weld = 681.82 × 1000/661.5 = 1031 mm, say 1040 mm

Length of the connection = [(1040 – 2 × 220)/4] × 2 = 300 mm < 150 × 0.7 × 8, hence long joint connection not required.

Hence, provide two cover plates of size 300 × 220 mm as shown in Fig. 11.41(b).

Example 11.6 *Determine the size and length of the fillet weld for the lap joint to transmit a factored load of 120 kN shown in Fig. 11.42, assuming site welds, Fe 410 steel, and E 41 electrode. Assume width of the plate as 75 mm.*

Solution

Minimum size of weld for a 8-mm thick section = 3 mm

Maximum size of weld = 8 – 1.5 = 6.5 mm

Choose the size of weld as 6 mm.

Effective throat thickness = 0.7 × 6 = 4.2 mm

Strength of weld = 4.2 × 410/($\sqrt{3}$ × 1.5) = 662.7 N/mm

Assuming that there are only two longitudinal (side) welds,

Required length of weld = 120 × 10^3/662.7

$$= 181 \text{ mm}$$

Length to be provided on each side = 181/2 = 90.5 mm > 75 mm

Hence, provide 90.5-mm weld on each side with an end return of 2 × 6 = 12 mm.

Therefore, the overall length of the weld provided = 2 × (90.5 + 2 × 6) = 205 mm

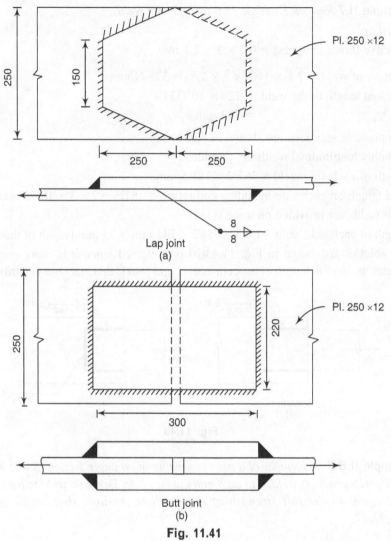

Lap joint
(a)

Butt joint
(b)

Fig. 11.41

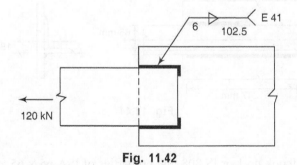

Fig. 11.42

Example 11.7 *Rework Example 11.6 using 3 mm welds.*

Solution

Effective throat thickness = $0.7 \times 3 = 2.1$ mm

Strength of weld = $2.1 \times 410/(\sqrt{3} \times 1.5) = 331$ N/mm

Required length of the weld = $120 \times 10^3/331$
$$= 363 \text{ mm}$$

Two possible solutions are shown in Fig. 11.43.

(i) If only longitudinal welds are provided,

Length of each side weld = $363/2 = 181.5$ mm

Total length on each side including end return = $181.5 + 2 \times 3 = 187.5$ mm

(ii) If welds are provided on three sides

Length of each side weld = $(363 - 75)/2 = 144$ mm > 75 mm (width of the plate)

The solution as shown in Fig. 11.43(b) is preferred since it is more compact, reduces the overall length of the connection, and provides better stress distribution.

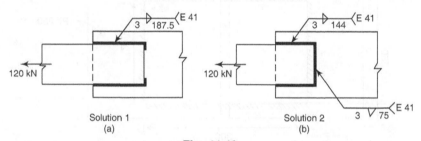

Solution 1   Solution 2
(a)   (b)

Fig. 11.43

Example 11.8 *A tie member of a truss consisting of an angle section ISA 65 × 65 × 6 of Fe 410 grade, is welded to an 8-mm gusset plate. Design a weld to transmit a load equal to the full strength of the member. Assume shop welding. See Fig. 11.44.*

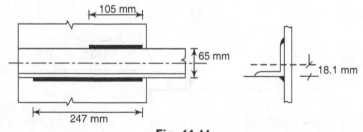

Fig. 11.44

Solution

From IS handbook no.1 or IS 808, the properties of ISA 65 × 65 × 6 are
$$A = 744 \text{ mm}^2$$
$$C_z = 18.1 \text{ mm}$$

Tensile capacity of the member = 744 × 250/1.1 = 169.1 kN
The force resisted by the weld at the lower side of the angle

P_1 = 169.1 × (65 – 18.1)/65 = 122.01 kN

Force to be resisted by the upper side of the angle

P_2 = 169.1 × 18.1/65 = 47.09 kN

Assuming a weld size of 4 mm [> 3 mm and < {3/4} × 6 = 4.5 mm]
Effective throat thickness of the weld = 0.7 × 4 = 2.8 mm
Strength of the weld = 2.8 × 410/($\sqrt{3}$ × 1.25) = 530 N/mm

L_{w1} = 122.01 × 10³/530 = 230.1 mm

Hence, provide 230.1 + (2 × 4)2 = 246.1 mm, say 247 mm length at the bottom

L_{w2} = 47.09 × 10³/530 = 88.8

Provide 88.8 + (2 × 4) × 2 = 104.8 mm, say 105 mm length at the top.
The block shear failure has been checked and it is found that the thickness of gusset plate is adequate (see Example 11.9 for the calculations involved in the block shear failure check).

Example 11.9 *Design a joint according to the instructions given in Example 11.8, if the welding is done on three sides of the angle as shown in Fig. 11.45.*

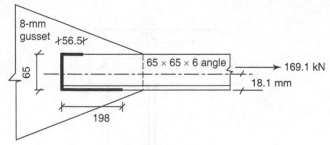

Fig. 11.45

Solution

Strength of 4-mm weld = 2.8 × 410/($\sqrt{3}$ × 1.25) = 530 N/mm

P_2 = 530 × 65/1000 = 34.45 kN
P_1 = Ty/d – P_2/2 = 169.1 × 18.1/65 – 34.45/2
           = 29.86 kN
P_3 = T – P_1 – P_2 = 169.1 – 34.45 – 29.86
           = 104.79 kN

L_{w1} = 29.86 × 1000/530 = 56.3 mm, say 56.5 mm
L_{w3} = 104.79 × 1000/530 = 197.7 mm, say 198 mm

Total length of weld = 65 + 56.5 + 198 = 319.5 mm
Note: Add twice the weld size at the ends.

Check for block-shear failure

Since the member is welded to the gusset plate, no net areas are involved and hence A_{vn} and A_{tn} in the equation for T_{db} (Section 6.4.1 of the code) should be taken to be the corresponding gross areas. Using the weldment with $L_1 = 198$ mm, $L_2 = 56.5$ mm and 65 mm at the end of the angle yields

$$T_{db1} = [8 \times (198 \times 2) \times 250/(\sqrt{3} \times 1.1) + 0.9 \times 410 \times 8 \times 65/1.25]/1000$$
$$= 569.2 \text{ kN}$$

$$T_{db2} = [0.9 \times 410 \times (198 \times 2)/(\sqrt{3} \times 1.25) + 250 \times 8 \times 65/1.1]/1000$$
$$= 658.1 \text{ kN}$$

Hence,

$$T_{db} = 569.2 \text{ kN} > 169.1 \text{ kN}$$

Hence, the thickness of gusset plate is adequate.

Note L_2 does not enter into this calculation because a shear rupture of the gusset plate along the toe of the angle runs for the full length of the contact with the toe, 198 mm, instead of only the length L_2.

Example 11.10 *Design a suitable fillet weld to connect the web plate to the flange plate and the flange plate to the cover plate of the compression flange of a plate girder as shown in Fig. 11.46. The size of the web plate is 1000 × 10 mm. The flange and cover plate are of the dimensions 300 × 16 mm and 275 × 12 mm, respectively. The maximum factored shear force V = 800 kN. Assume shop welding.*

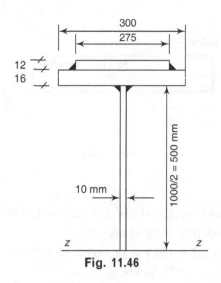

Fig. 11.46

Solution
Connection of web and flange plate
Minimum size of weld = 3 mm
Maximum size of weld = 10 − 1.5 = 8.5 mm
Adopt a 6-mm size weld

$$\tau_{vf, cal} = VA_f\bar{Y}_f/(I_{zz} \Sigma t)$$
$$t = 0.7 \times 6 = 4.2 \text{ mm}$$
$$\Sigma t = 2 \times 4.2 = 8.4 \text{ mm}$$

$$A_f \overline{Y}_f = 300 \times 16(500 + 8) + 275 \times 12(500 + 16 + 6)$$
$$= 243.84 \times 10^4 + 172.26 \times 10^4$$
$$= 416.1 \times 10^4$$
$$I_{zz} = 2 \times [275 \times 12^3/12 + 275 \times 12 \times 522^2 + 300 \times 16^3/12$$
$$+ 300 \times 16 \times 508^2] + 10 \times 1000^3/12$$
$$= 5.109 \times 10^9 \text{ mm}^4$$
$$\tau_{vf,\ cal} = 800 \times 10^3 \times 416.1 \times 10^4/(5.109 \times 10^9 \times 8.4)$$
$$= 77.56 \text{ N/mm}^2 < 410/(\sqrt{3} \times 1.25) = 189 \text{ N/mm}^2$$

Hence, the weld is safe.

Connection of flange plate to cover plate

Adopt a 6-mm fillet weld.

$$t_e = 6 \times 0.7 = 4.2 \text{ mm}$$
$$\Sigma t = 2 \times 4.2 = 8.4 \text{ mm}$$

$$A\overline{Y} = 275 \times 12 \times (500 + 16 + 6) = 172.26 \times 10^4$$
$$\tau_{vf,\ cal} = 800 \times 10^3 \times 172.26 \times 10^4/(5.109 \times 10^9 \times 8.4)$$
$$= 32.11 \text{ N/mm}^2 < 189 \text{ N/mm}^2$$

Hence the weld is safe.

Example 11.11 *Determine the service load permitted on the connection shown in Fig. 11.47. Assume field welding and Fe 410 steel.*

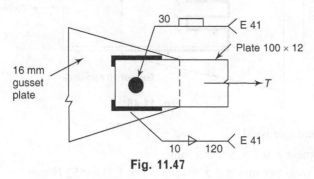

Fig. 11.47

Solution

The design strength per mm supplied by a 10-mm fillet weld,

$$R_{nw} = t_e f_y/(\sqrt{3} \times \gamma_{mw})$$
$$= 0.7 \times 10 \times 410/(\sqrt{3} \times 1.5) = 1104 \text{ N/mm}$$

Strength provided by the fillet welds $= 120 \times 1104/1000$
$$= 132.48 \text{ kN}$$

The design strength of a plug weld of diameter 30 mm
$$= \pi \times 30^2/4 \times 410/(\sqrt{3} \times 1.5 \times 1000) = 111.55 \text{ kN}$$

Design strength based on the weld $= 132.48 + 111.55 = 244.03$ kN

Tensile capacity of the plate

$$f_y A_g/1.1 = 250 \times (12 \times 100)/(1.1 \times 1000) = 272.7 \text{ kN} > 244.03 \text{ kN}$$

Service load carrying capacity

$$244.03 = 1.5T$$

Therefore,

$$T = 244.03/1.5 = 162.68 \text{ kN}$$

Example 11.12 *An ISMC 250 is used to transmit a factored force of 700 kN. The channel section is connected to a gusset plate 10 mm thick as shown in Fig. 11.48. Design a fillet weld if the overlap is limited to 300 mm. Use slot welds if required.*

Solution

The properties of ISMC 250 are

$$A = 3900 \text{ mm}^2$$
$$t_f = 14.1 \text{ mm}$$
$$t_w = 7.2 \text{ mm}$$

As slot welds are required

Maximum size of weld = 7.2 − 1.5 = 5.7 mm

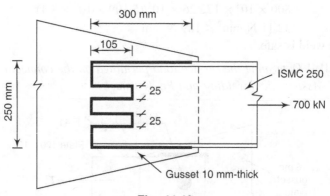

Fig. 11.48

Adopt a 5-mm size weld.

Throat thickness = 0.7 × 5 = 3.5 mm

Strength of weld per mm = 3.5 × 410/($\sqrt{3}$ × 1.5) = 552 N/mm

Required length of weld = 700 × 1000/552 = 1268 mm, say 1270 mm

The maximum length of weld that can be provided in the channel

$$= 300 \times 2 + 250 = 850 \text{ mm} < 1270 \text{ mm}$$

Hence, use two slot welds of width 25 mm ($3t$ = 3 × 7.2 = 21.6 or 25 mm, whichever is greater)

Assume the length of slot weld as x mm, then

$$1270 = 2 \times 300 + 250 + 4x$$

Hence,

$$x = 105 \text{ mm}$$

Therefore, provide 105-mm long fillet welds in slots as shown in Fig. 11.48.

Example 11.13 *Design a welded seat angle connection between a beam MB 300 and column HB 200 for a reaction of beam 100 kN, assuming Fe 410 grade steel (f_y = 250 MPa) and site welding.*

Solution

For MB 300, t_f = 13.1 mm, r_b = 14.0 mm, and t_w = 7.7 mm

1. Width of flange MB 300 = 140 mm. Since the bottom seat angle has to be welded to the flange, assume length of angle = 140 + 2 × 10 = 160 mm.
2. Length of bearing required at the root line of the beam

$$b = R/[t_w(f_{yw}/\gamma_{m0})]$$
$$= 100 \times 1000/(7.7 \times 250/1.1)$$
$$= 57.15 \text{ mm}$$

Length of bearing on cleat = $b_1 = b - (t_f + r_b)$
$$= 57.15 - (13.1 + 14.0)$$
$$= 30.05 \text{ mm}$$

Assume end clearance = 5 mm, tolerance = 3.5 mm, and therefore

$$g = 5 + 3.5 = 8.5 \text{ mm}$$

Required length of outstanding angle = 57.15 + 8.5 = 65.65

Distance from the end of bearing on cleat to the root of the angle (assuming an angle of 100 × 75 × 10 mm) with r_a = 8.5 mm

$$b_2 = b_1 + g - (t_a + r_a)$$
$$= 30.05 + 8.5 - (10 + 8.5) = 20.05$$

Assume a uniformly distributed load over the bearing length b_1. Moment at root of angle due to load,

$$M_u = 100(20.05/30.05) \times 20.05/2 = 668 \text{ Nm}$$

Moment capacity of the leg of the angle

$$= 1.2 \times (250/1.1) \times (160 \times 10^2/6) \times 10^{-3} = 727 \text{ Nm} > 668 \text{ Nm}$$

Size of weld

Maximum weld size = 10 − 1.5 = 8.5 mm

Minimum weld size = 5 mm

Use a 5-mm weld.

Strength of the 5-mm weld = (5 × 0.7) × 410/($\sqrt{3}$ × 1.5) = 552 N/mm

Length of the weld per side = 100 × 10³/(2 × 552) = 90.5 mm

Take the length of the weld per side = 90.5 + 2 × 5 = 100.5 mm, say 110 mm.

Use a clip angle of size 60 × 60 × 6 with a 3 mm weld.

The connection details are shown in Fig. 11.49.

Taking eccentricity into account (L_w = 110 − end returns = 100 mm),

$$R_{res} = R/(2L_w^2)\sqrt{[L_w^2 + 20.25 \times (b_2/2)^2]}$$
$$= 100 \times 1000/(2 \times 100^2)\sqrt{[100^2 + 20.25 \times (20.05/2)^2]}$$
$$= 549 \text{ N/mm}$$

Capacity of a 5-mm weld = 552 N/mm > 549 N/mm

Hence the connection is safe.

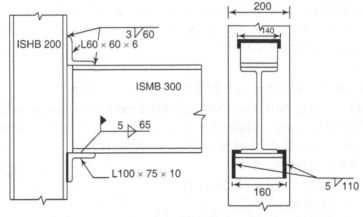

Fig. 11.49

Example 11.14 *Determine the maximum load that could be resisted by the bracket shown in Fig. 11.50, assuming the plate thickness does not affect the result and if 6-mm shop fillet welds are used.*

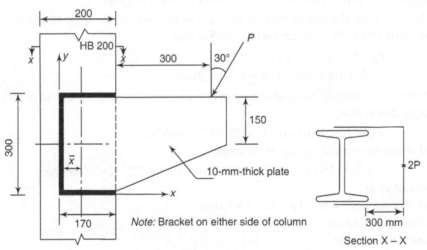

Fig. 11.50

Solution

Total length of weld = $2 \times 170 + 300 = 640$ mm

Polar moment of inertia (from Table 11.4)

$$I_p = (8b^3 + 6bd^2 + d^3)/12 - b^4/(2b + d)$$
$$= (8 \times 170^3 + 6 \times 170 \times 300^2 + 300^3)/12 - 170^4/(2 \times 170 + 300)$$
$$= 11.87 \times 10^6 \text{ mm}^4$$

C.G. of the weld line (Table 11.4)

$$\bar{Y} = 300/2 = 150 \text{ mm}$$
$$\bar{X} = b^2/(2b + d) = 170^2/(2 \times 170 + 300)$$
$$= 45.15 \text{ mm}$$

Let P be the maximum load in one bracket. The load components are

$P_x = P\sin 30° = 0.5P$ kN

$P_y = P\cos 30° = 0.866P$ kN

Eccentricities $e_x = 150$ mm

$e_y = 300 + 170 - \bar{X}$

$= 300 + 170 - 45.15$

$= 424.85$ mm

Coordinates of the critical weld point from the centroid (bottom right)

$x = 170 - 45.15 = 124.85$ mm

$y = 150$ mm

Twisting moment

$T = -0.5P \times 150 + 0.866P \times 424.85 = 292.92P$ kN mm

$F_x^P = 0.5P/640 = 0.7813 \times 10^{-3}P \leftarrow$

$F_y^P = 0.866P/640 = 1.3531 \times 10^{-3}P \downarrow$

$F_x^T = Ty/I_p = 292.92P \times 150/(11.87 \times 10^6) = 3.702 \times 10^{-3}P$

$F_y^T = Tx/I_p = 292.92P \times 124.85/(11.87 \times 10^6) = 3.081 \times 10^{-3}P$

$F_R = [(F_x^P + F_x^T)^2 + (F_y^P + F_y^T)^2]^{0.5}$

$= [(0.7813 \times 10^{-3} + 3.702 \times 10^{-3})^2 + (1.3531 \times 10^{-3} + 3.081 \times 10^{-3})^2]^{0.5}P$

$= (20.10 \times 10^{-6} + 19.66 \times 10^{-6})^{0.5}P = 6.306 \times 10^{-3}P$

Weld strength $= 6 \times 0.7 \times 410/(\sqrt{3} \times 1.25) \times 10^{-3} = 0.795$ kN/mm

Equating F_R with weld strength, we get

$P_u = 0.795/(6.306 \times 10^{-3}) = 126.1$ kN

Thus, the maximum load that can be resisted $= 126.1 \times 2 = 252.2$ kN.

Example 11.15 *A bracket plate is welded to the flange of a column ISHB 200 as shown in Fig. 11.51. Calculate the size of the weld required to support a factored load of 100 kN.*

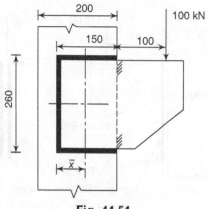

Fig. 11.51

Solution

Total length of weld (ignoring end return) = 260 + 2 × 150 = 560 mm

$\bar{X}$ = distance of centroid of weld group
$$= b^2/(2b + d) = 150^2/(2 \times 150 + 260) = 40.18 \text{ mm}$$

$\bar{Y}$ = 260/2 = 130 mm

Eccentricity e_y = 100 + 150 – 40.18 = 209.82 mm

Polar moment inertia (Table 11.4)
$$I_p = [(8b^3 + 6bd^2 + d^3)/12] - [(b^4/(2b + d)]$$
$$= (8 \times 150^3 + 6 \times 150 \times 260^2 + 260^3)/12 - 150^4/(2 \times 150 + 260)$$
$$= 7.88 \times 10^6 \text{ mm}^4$$

Coordinates of critical weld point from the centroid
$$x = 150 - \bar{X} = 150 - 40.18 = 109.82$$
$$y = 130 \text{ mm}$$

Stress due to direct shear
$$R_v = P/L_w = 100 \times 1000/560 = 178.57 \text{ N/mm}$$

From torsion T about the centroid of configuration
$$R_x = Ty/I_p = (100 \times 10^3 \times 209.82) \times 130/7.88 \times 10^6 = 346.15 \text{ N/mm}$$
$$R_y = Tx/I_p = (100 \times 10^3 \times 209.82) \times 109.82/7.88 \times 10^6 = 292.42 \text{ N/mm}$$

Resultant force $R = \sqrt{[346.15^2 + (292.42 + 178.57)^2]}$
$$= 584.5 \text{ N/mm}$$

Weld resistance = $0.7 \times a \times 410/(\sqrt{3} \times 1.25) = 132.56a$

Therefore,
$$132.56a = 584.5$$
or $\quad a = 4.4$ mm, say 5 mm

Hence, use a 5 mm E43 weld.

Example 11.16 *Determine the length of a fillet weld required to carry the load indicated by Fig. 11.52. Assume an 8-mm shop weld.*

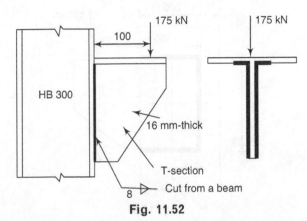

Fig. 11.52

Solution
Factored load = $1.5 \times 175 = 262.5$ kN

Required length of the weld = $L_w = \sqrt{6M/R_{nw}}$
Flange thickness of ISHB 300 = 10.6 mm
Maximum size of weld = $10.6 - 1.5 = 9.1$ mm
Minimum size of weld (as per Table 21 of code) = 5 mm
Adopt 8-mm weld

Design strength of the weld = $f_{wd} = 410/(\sqrt{3} \times 1.25) = 189$ N/mm^2
$\quad R_{nw} = 189 \times 0.7 \times 8 = 1058.4$ N/mm
$\quad M_u = 262.5 \times 100 = 26250$ kN/mm per two lines of weld
$\quad L_w = \sqrt{[(6 \times 26250 \times 1000/2)/900(\text{est})]} = 295.8$ mm
A reduced value of R_{nw} has been used to account for the direct shear effect also.
Try $\quad L_w = 300$ mm
Neglecting the end returns, stress due to direct shear
$\quad \tau_v = P/2L_w t_e$
$\quad\quad = (262.5 \times 1000)/(2 \times 300 \times 0.7 \times 8)$
$\quad\quad = 78.125$ N/mm^2
Stress due to bending
$\quad f_b = My/I = 262.5 \times 1000 \times 100 \times 150/[(2 \times (300^3/12) \times 0.7 \times 8)]$
$\quad\quad = 156.25$ N/mm^2
Resultant stress = $\sqrt{(78.125)^2 + 156.25^2)}$
$\quad\quad = 174.69$ N/mm$^2 < 189$ N/mm^2
Hence with a length of 300 mm, the fillet weld will be able to carry the required load.

Example 11.17 *Determine the size of weld for a connection as shown in Fig. 11.53. Assume site welding. The joint is subjected to a factored shear force of 180 kN and factored bending moment of 25 kNm.*

Solution
Assuming unit weld throat
$\quad$ Area = $2 \times 140 + 2 \times 300 = 880$ mm^2
$\quad I_{zz} = 2 \times 140 \times (400/2)^2 + 2 \times 300^3/12 = 15.7 \times 10^6$ mm^4
$\quad$ Z for the extreme fibre = $15.7 \times 10^6/(400/2) = 78.5 \times 10^3$ mm^3
$\quad$ Direct stress = $180/880 = 0.205$ kN/mm
$\quad$ Bending stress = $25 \times 10^3/78.5 \times 10^3 = 0.318$ kN/mm
$\quad$ Vector sum = $(0.318^2 + 0.205^2)^{0.5} = 0.378$ kN/mm
$\quad$ Weld size = $0.378 \times 10^3/[410/(\sqrt{3} \times 1.50) \times 0.7] = 3.43$ mm
Thickness of flange and web of MB 400 are 16.0 mm and 8.9 mm, respectively.
Hence,
$\quad$ Minimum thickness = 3 mm (Table 21 of code)
$\quad$ Maximum thickness = $8.9 - 1.5 = 7.4$ mm
$\quad$ Adopt a 5-mm fillet weld of grade E 43.

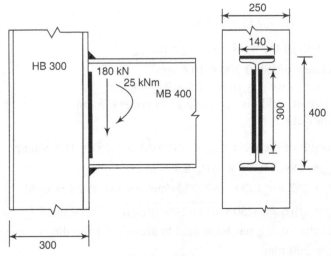

Fig. 11.53

Example 11.18 *A 10-mm thick bracket plate is used to transmit a reaction of 80 kN (factored load) at an eccentricity of 80 mm from the column flange. Design the butt weld for grade Fe 410 steel and E 43 electrode. See Fig. 11.54 for details.*

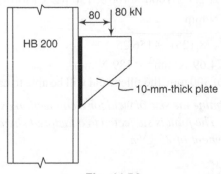

Fig. 11.54

Solution
Adopting a double-V groove weld,
Effective throat thickness = 10 mm
Depth of bracket plate from Eqn (11.39)

$$d = \sqrt{6M/(tf_b)}$$

$$= \sqrt{[6 \times 80 \times 10^3 \times 80/(10 \times 250/1.1)]}$$

$$= 130 \text{ mm}$$

Adopt a depth of 150 mm. Direct shear stress,

$$\tau_{vf} = 80 \times 10^3/(150 \times 10)$$

$$= 53.33 \text{ N/mm}^2$$

Bending stress $= f_b = 6M/td^2$

$\qquad = 6 \times 80 \times 10^3 \times 80/(10 \times 150^2)$

$\qquad = 170.67$ N/mm^2

Check

$$f_e = \sqrt{(170.67^2 + 3 \times 53.33^2)} = 194.1 \text{ N/mm}^2$$

$$< 250/1.1 = 227 \text{ N/mm}^2$$

Hence the joint is safe.

Example 11.19 *Design stiffened seat angle for a reaction of 250 kN from beam of ISMB 400. This beam has to be connected to a column of size ISHB 200. Assume Fe 410 grade steel and shop welding.*

Solution

From Example 10.14 of Chapter 10,

Length of bearing required = 124 mm

Provide a clearance of 5 mm (including tolerance).

Required width of seating plate = 124 + 5 = 129 mm

Width of flange of beam = 140 mm

Thickness of beam flange = 16 mm

Adopt a seating plate of size 130 × 160 mm and thickness 16 mm. Thickness of stiffening plate should be equal or greater than thickness of web of beam = 8.9 mm

Provide a 12-mm thick stiffening plate.

Distance of end reaction from the outer edge of the seat plate

$\qquad = 129 - 124/2 = 67$ mm

Hence,

Bending moment = 250 × 67 = 16,750 kN/mm

The vertical stem should not buckle. As per Table 2 of IS 800,

$\qquad d/t < 18.9$

Therefore,

$\qquad d < 18.9 \times 12 = 226.8$ mm

Adopt a depth of 220 mm. Taking moments from the bottom of the seat plate, C.G. of the weld group

$$\bar{y}_1 = (2 \times 220 \times 110 + 0)/[(2 \times 220 + (160 - 12)] = 82.3 \text{ mm}$$

$$\bar{y}_2 = 220 - 82.3 = 137.7 \text{ mm}$$

Assuming unit weld thickness,

I_{xx} of the weld $= [(160 - 12) \times 82.3^2 + 2 \times (220^3/12) + 2 \times 220 \times (137.7 - 110)^2]$

$\qquad = 3.114 \times 10^6$ mm^4

Vertical shear $\tau_{vfl} = 250 \times 10^3/(2 \times 220 + 148)$

$\qquad = 425.2$ N/mm

Horizontal shear $\tau_{vf2} = [16750 \times 10^3 \times 82.3/(3.114 \times 10^6)]$
$$= 442.7 \text{ N/mm}$$

Resultant shear stress/mm = Strength of weld/mm

$$\sqrt{(425.2^2 + 442.7^2)} = 0.7 \times s \times 410/(\sqrt{3} \times 1.25)$$
$$613.8 = 132.56s$$

Therefore, $s = 4.63$. Provide 5-mm weld.

Also provide a nominal size of top cleat of size $75 \times 75 \times 6$ mm and connect it by a 4-mm fillet weld as shown in Fig. 11.55.

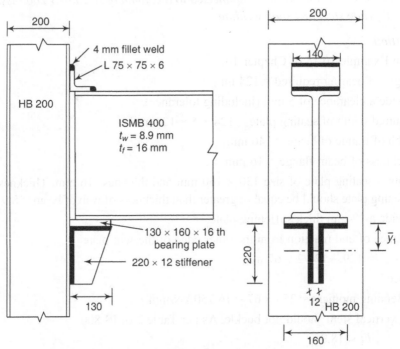

Fig. 11.55

Example 11.20 *An ISA 125 × 75 × 8 is welded with the flange of a column ISHB 300. The bracket carries a factored load of 75 kN at a distance of 38 mm as shown in Fig. 11.56. Design the connection. Use Fe 410 grade steel.*

Solution

Flange width of ISHB 300 = 250 mm

Provide welding at the top and bottom of the seating angle, which acts as a bracket. The top weld will be in tension and the bottom will be in compression, in addition to sharing the shear force. Provide 250 mm length of weld at the top and bottom of the angle, and denoting t as the effective throat thickness.

Strength of the weld = $t \times 0.7 \times 410/(\sqrt{3} \times 1.5) = 110.4t$

Shear force in 1 mm length of weld = $75,000/(2 \times 250) = 150$ N

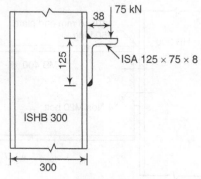

Fig. 11.56

Force in 1-mm length of weld due to bending moment (see Table 11.4)
$$= (75,000 \times 38)/(125 \times 250) = 91.2 \text{ N}$$

Resultant force $= \sqrt{(150^2 + 91.2^2)} = 175.6 \text{ N}$

Thus,
$$110.4t \geq 175.6$$

or
$$t \geq 1.59$$

Hence, a weld size of 3 mm should be provided.

Example 11.21 *Design a welded end plate connection for a ISMB 400 in grade Fe 410 steel to carry a reaction of 120 kN due to factored loads. The connection has to be made to the flanges of ISHB 300.*

Solution
For MB 400, $b_{fb} = 140$ mm, $t_{fb} = 16$ mm, and $r_b = 14$ mm, $t_{wb} = 8.9$ mm
For HB 300, $b_{fc} = 250$ mm

Bolted connection to column
For M20 grade 4.6 bolts and a 6-mm thick end plate,
Shear capacity of the bolt in single shear (from Appendix D) = 45.3 kN
Bearing capacity of the bolt on a 6-mm thick end plate (with $k_b = 0.5$)
$$= 2.5 \times 0.5 \times 20 \times 6 \times 410/1.25 = 49.2 \text{ kN}$$
Required no. of bolts = 120/45.3 = 2.64
Hence, provide four M20 as shown in Fig. 11.57

End plate
Use a 140 × 6-mm plate, 250 mm deep.
Length of fillet welds connecting the end plate to the beam web
$$= 250 - 16 - 14 = 220 \text{ mm on each side}$$
Shear on weld $= 120 \times 10^3/(2 \times 220) = 272$ N/mm
Use a 3-mm fillet weld.
Strength of the weld $= 3 \times 0.7 \times 410/(\sqrt{3} \times 1.5) = 331$ N/mm > 272 N/mm
Hence, the connection is safe.

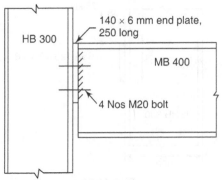

Fig. 11.57

Shear strength of end plate $= f_y A_v/(1.1 \times \sqrt{3}) = 250 \times (250 - 2 \times 22) \times 6/(\sqrt{3} \times 1.1)$
$$= 162 \text{ kN} > 120 \text{ kN}$$

Example 11.22 *Design a welded splice for an ISMB 400 section to transfer a factored bending moment of 120 kNm and a factored shear of 80 kN. Assume that the flange splice carries all the moment and that the web splice carries only the shear.*

Solution
ISMB 400
$$t_f = 16 \text{ mm}, t_w = 8.9 \text{ mm}, \text{ and } b_f = 140 \text{ mm}$$
Force in the flange $= \text{BM}/(D - t_f)$
$$= 120 \times 10^3/(400 - 16) = 312.5 \text{ kN}$$
Assuming 8-mm weld,
Capacity $= 8 \times 0.7 \times 410/(\sqrt{3} \times 1.5) = 883.7$ N/mm
Required length of the weld $= 312.5 \times 1000/883.7 = 354$ mm
Hence, adopt a cover plate of length $= 400$ mm
Use a 100-mm splice plate.
Required thickness of plate $= 312.5 \times 10^3/(250 \times 100/1.1) = 13.75$ mm
Provide two flange splice plates of size $400 \times 100 \times 16$ mm.

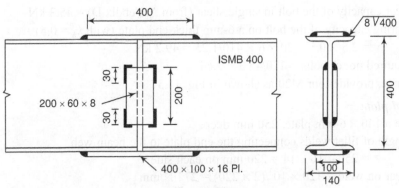

Fig. 11.58

Web splice
Try 8-mm thick web splice plates on both sides of the web
Adopt a 6-mm fillet weld,

Capacity of the weld = 662.7 N/mm
Required length of the weld = 80 × 1000/662.7 = 120 mm
Since the welds are on both sides, provide a 60-mm weld on each side and provide a web splice plate of size 200 × 60 × 8 mm.

Summary

Welding is employed in several connections due to its aesthetic appearance. Moreover by using welding any difficult geometry can be jointed, whereas bolting requires some edge distances and also prior drilling. Several welding processes are available, each having some advantages and disadvantages. Users should choose a particular process based on their requirements. Welding, due to high heating, introduces distortion of connected parts. Some techniques have been given to minimize these distortions and also to reduce welding defects.

The various types of welds (groove, fillet, slot, and plug) and welded joints (butt, lap, T, and corner) have been discussed. The specifications for welded joints are also given. Most widely used welding symbols are explained. The concepts and design procedures of simple joints such as butt, lap, truss joints, angle seat, web angle, and stiffened seat connections have been discussed. Details of various moment resistant connections such as eccentrically loaded connections (load causing twisting moment and bending moment), end plate connections, and beam-column moment joints have been covered. Since it is impossible to consider all the combinations of connections, only those that are frequently encountered in practice are discussed. Splicing of beams and columns, which may be necessary due to the difficulty of transporting long length of members, are also included. The concepts presented have been explained with several numerical examples.

Exercises

1. Two plates of thickness 12 mm and 10 mm are to be jointed by a groove weld. The joint is subjected to a factored tensile force of 275 kN. Assuming an effective length of 150 mm, check the safety of the joint for
   (i) single-V groove weld joint and
   (ii) double-V groove weld joint.
   Assume Fe 410 grade steel plates and that the welds are shop welded.
2. The tie member of a truss is made of ISA 75 × 75 × 6 and is subjected to a factored tension load of 110 kN. The length of the angle is not enough to go from end-to-end, and hence a splice has to be provided. Design a groove welded joint.
3. Determine the effective throat dimension of 8-mm fillet weld made by (a) shielded metal arc welding (SMAW) and (b) submerged arc welding (SAW).
4. Determine the design shear strength of a 8-mm fillet weld produced by (a) SMAW and (b) SAW. Assume E 51 electrodes, Fe 410 material, and shop welding.
5. Design a connection to joint two plates of size 200 × 10 mm of grade Fe 410 to mobilize full plate tensile strength using shop fillet welds, if (i) a lap joint is used, (ii) a double cover butt joint is used.

6. Determine the size and length of the fillet weld for the lap joint to transmit a factored load of 100 kN shown in Fig. 11.42, assuming site welds, Fe 410 steel, and E 41 electrode.

7. Rework the Exercise 6 using 4-mm welds.

8. A tie member of a truss consisting of an angle section ISA 75 × 75 × 8 of Fe 410 grade, is welded to a 10-mm gusset plate. Design a weld to transmit a load equal to the full strength of the member. Assume shop welding (see also Fig. 11.44).

9. Redesign the joint in the above problem, if the welding is done on three sides of the angle as shown in Fig. 11.45.

10. Design a suitable fillet weld to connect the web plate to the flange plate and flange plate to the cover plate of the compression flange of a plate girder as shown in Fig. 11.46. The size of the web plate is 800 × 8 mm. The flange and cover plate are of the size 250 × 12 mm and 200 × 10 mm, respectively. The maximum factored shear force $V = 650$ kN. Assume shop welding.

11. Determine the service load permitted on the connection similar to that shown in Fig. 11.47. The size of plate is 100 × 10 mm and the size of the gusset plate is 12 mm. Assume a plug weld of 20 mm diameter and an 8-mm fillet weld of length 100 mm on each face of the plate. Assume field welding and Fe 410 grade steel.

12. An ISMC 300 is used to transmit a factored force of 800 kN. The channel section is connected to a gusset plate 10-mm thick as shown in Fig. 11.48. Design a fillet weld if the overlap is limited to 350 mm. Use slot welds if required.

13. Design a welded seat angle connection between a beam MB 400 and column HB 200 for a reaction of beam 120 kN, assuming Fe 410 grade steel ($f_y = 250$ MPa) and site welding.

14. Determine the maximum load that could be resisted by the bracket shown in Fig. 11.50 (with the load acting at 45°), assuming the plate thickness does not affect the result and assuming 8-mm shop fillet welds are used.

15. A bracket plate is welded to the flange of a column ISHB 300 as shown in Fig. 11.51. Calculate the size of the weld required to support a factored load of 150 kN.

16. Determine the length of a fillet weld required to carry the load indicated by Fig. 11.52. Assume an 8-mm shop weld and that the load of 200 kN is acting at 120 mm eccentricity from the face of the column.

17. Determine the size of the weld for a connection as shown in Fig. 11.53 to connect a beam of MB 300 to column of HB 200. Assume site welding. The joint is subjected to a factored shear force of 120 kN and factored bending moment of 20 kNm.

18. A bracket plate 10-mm thick is used to transmit a reaction of 100 kN (factored load) at an eccentricity of 60 mm from the column flange. Design the butt weld for grade Fe 410 steel and E 43 electrode. See Fig. 11.54 for details.

19. Design a stiffened seat angle for a reaction of 200 kN from beam of ISMB 300. This beam has to be connected to a column of size ISHB 200. Assume Fe 410 grade steel and shop welding. (See Fig. 11.55 for the arrangement.)

20. An ISA 150 × 115 × 8 is welded with the flange of a column ISHB 300. The bracket carries a factored load of 100 kN at a distance of 50 mm as shown in Fig. 11.56. Design the connection using Fe 410 grade steel.

21. Design a welded header plate connection for a ISMB 300 in Fe 410 grade steel to carry a reaction of 80 kN due to factored loads. The connection has to be made to the flanges of ISHB 200.

22. Design a welded splice for an ISMB 300 section to transfer a factored bending moment of 80 kNm and a factored shear of 60 kN. Assume that the flange splice carries all the moment and that the web splice carries only the shear.

Review Questions

1. List the different processes of arc welding used in structural steel applications.
2. Write the expansions for the following abbreviations:
   (a) SMAW   (b) SAW   (c) GMAW   (d) FCAW
   (e) ESW   (f) SW
3. Why is SMAW chosen as the common method of welding in structural steel applications?
4. What are the different components required in the SMAW process?
5. What is the maximum yield stress and ultimate tensile strength of E41 electrode?
6. List the parameters based on which a particular welding process is selected.
7. In the numbering system Exxxbc, what does each component signify?
8. List some of the important advantages of welding over bolting?
9. Are there any disadvantages of using welding? What are they?
10. List the four types of welds.
11. The type of welding used for connecting the member without eccentricity is
    (a) Groove welding   (b) Fillet welding
    (c) Groove or fillet welding   (d) Slot welding
12. When are backup strips used in groove welds? Under what condition should it not be used?
13. What are the various types of groove welds?
14. What do you mean by partial penetration and full penetration groove weld? Which one is preferred and why?
15. If the thickness of the groove weld is equal to that of the main plates, then the weld strength is
    (a) Equal to that of the main members
    (b) Equal or even greater than that of the main members
    (c) Greater than that of the main member
    (d) About 95% of the main member
16. Why are fillet welds often adopted at site?
17. The members meeting at an angle can be welded using
    (a) Groove welds only   (b) Fillet welds only
    (c) Both groove or fillet welds
18. Is edge preparation necessary in fillet welds?
19. The size of a fillet weld is taken as
    (a) Side of the triangle of the fillet   (b) Throat of the fillet
    (c) Size of the plate   (d) Length of the fillet weld
20. The ratio of the effective throat of the fillet weld to its size is
    (a) 0.707   (b) Less than or equal to 0.707
    (c) Less than 1.0   (d) More than 0.707
21. Under what circumstances are slot and plug welds used?
22. List some of the common defects associated with welds.
23. What are the five basic types of welded joints?
24. What is the main drawback of a butt joint?
25. What is the type of weld used in a butt joint?
26. What are lap joints? What is their main advantage? What welds are used in lap joints?
27. How can distortions in welded joints be minimized?

28. Write short notes on weld symbols.
29. What is the advantage of using weld symbols?
30. The maximum size of a fillet weld applied to the square edge of a plate of thickness greater than 6 mm is
    (a) not more than the thickness of the plate
    (b) 1.0 mm less than the thickness of the plate
    (c) 1.5 mm less than the thickness of the plate
    (d) half the thickness of the plate
31. Minimum length of the fillet weld should be
    (a) 4 times the size of the weld   (b) 2 times the size of the weld
    (c) 3 times the size of the weld
32. What is the use of end returns? What is the minimum length of an end return?
33. In order to reduce the effect of stress concentration at the edge of a plate, the length of longitudinal (side) fillets should be
    (a) Not less than 100 mm
    (b) Not less than the width of the plate
    (c) 16 times the thickness of the thinner plate joined
34. The clear spacing between the effective lengths of the intermittent welds in plates subject to compression should be
    (a) Not less than 16 times the thickness of the thinner plate joined
    (b) Between 12 to 16 times the thickness of the thinner plate
    (c) 12 times the thickness of the thinner plate
    (d) 16 times the thickness of the thinner plate
35. Which fillet weld is stronger—side or end?
36. What are the assumptions made in the design of welded joints?
37. Write down the expression given in IS 800 to calculate the design strength of a groove weld in tension and a groove weld in shear.
38. Write down the expression given in IS 800 to calculate the design strength of fillet welds.
39. Cite a location where intermittent fillet welds are used.
40. Write short notes on
    (a) Fillet welds in truss members
    (b) Web angle and end plate connections
    (c) Angle seat connections
    (d) Moment resistant connections
    (e) Stiffened beam seat connections
    (f) Continuous beam-to-column connections
    (g) Continuous beam-to-beam connections
    (h) Beam and column splices
41. What are the factors that contributed to the damage of welded moment resistant connections during Northridge and Kobe earthquakes?

12

Design of Industrial Buildings

Introduction

High-rise steel buildings account for a very small percentage of the total number of structures that are built around the world. The majority of steel structures being built are low-rise buildings, which are generally of one storey only. Industrial buildings, a subset of low-rise buildings are normally used for steel plants, automobile industries, utility and process industries, thermal power stations, warehouses, assembly plants, storage, garages, small scale industries, etc. These buildings require large column free areas. Hence interior columns, walls, and partitions are often eliminated or kept to a minimum. Most of these buildings may require adequate head room for the use of an over head travelling crane.

The structural engineer has to consider the following points during the planning and design of industrial buildings (Fisher 1984):
(a) Selection of roofing and wall material
(b) Selection of bay width
(c) Selection of structural framing system
(d) Roof trusses
(e) Purlins, girts, and sag rods
(f) Bracing systems to resist lateral loads
(g) Gantry girders, columns, base plates, and foundations

Out of the listed points, gantry girders which support cranes have been discussed in Chapter 8 and columns and base plates have been discussed in Chapters 5 and 9. Foundations are made with reinforced concrete and are outside the scope of this book. Hence, this chapter focusses on the rest of points in the sections to follow.

12.1 Selection of Roofing and Wall Material

The type of roof deck, type of purlin used, purlin spacing, deflections of secondary structural members, roof pitch, and drainage requirements are all determined by the choice of roofing. The roof weight also affects the gravity load design of the roof system and in the case of seismic calculations, the lateral load design.

Similar considerations apply to the cladding/wall systems. In selecting the cladding/wall system, the designer should consider the following areas: (a) cost, (b) interior surface requirements, (c) aesthetic appearance (including colour), (d) acoustics and dust control, (e) maintenance, (f) ease and speed of erection, (g) insulating properties, and (h) fire resistance.

Note that *cladding* carries only its own weight and the weight of the loads imposed by wind. In the case of roofs, the sheeting supports insulation and water proofing in addition to self weight and weight of loads due to wind and/or snow. Hence, it is often termed as *roof decking*. The cladding/wall system will have an impact on the design of girts, wall bracing, eave members, and foundation.

In India, corrugated galvanized iron (GI) sheets are usually adopted as coverings for roofs and sides of industrial buildings. Now light-gauge cold-formed ribbed steel or aluminium decking (manufactured by cold drawing flat steel or aluminium strips through dies to produce the required section) is also available. Sometimes asbestos cement (AC) sheets are also provided as roof coverings owing to their superior insulating properties. Their insulating properties may be enhanced by painting them white on the top surface. These three types of sheets are discussed briefly in the following section.

12.1.1 Steel or Aluminium Decking/Cladding

The modern built-up roof system consists of three basic components: steel/aluminium deck, thermal insulation, and membrane. The structural deck transmits gravity, wind, and earthquake forces to the roof framing. Thermal insulation is used for reducing heating and cooling costs, increasing thermal comfort, and preventing condensation on interior building surfaces. The membrane is the water-proofing component of the roof systems. On sloping roofs, the insulation consists of the insulation board or glass wool. On flat roofs, insulation board, felt, and bitumen are laid over the steel decking as shown in Fig. 12.1.

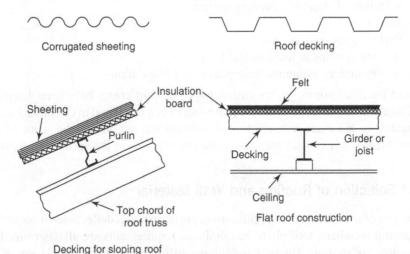

Fig. 12.1 Roof materials and constructions

The steel decking has a ribbed cross section, with ribs generally spaced at 150 mm (centre to centre) and 37.5 mm or 50 mm deep (see Fig. 12.2). The sloped-side ribs measure about 25 mm wide at the top for a narrow rib deck, 44 mm for an intermediate rib deck, and 62.5 mm for a wide rib deck. Wide rib decking is more popular, which can be used with 25-mm-thick insulation boards. Thinner insulation boards may require narrow deck rib opening. Wide rib decking also has higher section properties than other patterns, and hence can be used to span greater distances. These steel decks may be anchored to supporting flexural members by puddle welds by a welder, (power-activated and pneumatically driven fasteners, and self-drilling screws can also be used), as soon as the deck is placed properly on the rafters or top chord of the roof truss (Vinnakota 2006).

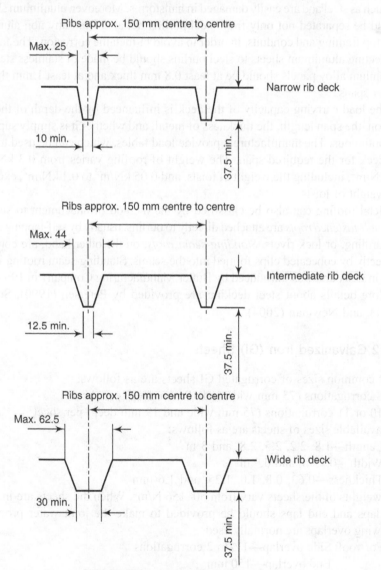

Fig. 12.2 Typical profiles of roof deck

Steel decks are available in different thicknesses, depths, rib spacing, widths, and lengths. They are available with or without stiffening elements, with or without acoustical material, as in cellular and non-cellular forms. The cellular decks can be used to provide electrical, telephone, and cable wiring and also serve as ducts for air distribution. They are also available with different coatings and in different colours. They are easy to maintain, durable, and aesthetically pleasing.

When properly anchored to supporting members, steel/aluminium decks provide lateral stability to the top flange of the structural member. They also resist the uplift forces due to wind during the construction stage. Steel decks may be considered as a simply supported or continuous depending on the purlin and joist spacing. Aluminium sheets also offer excellent corrosion resistance. But they expand approximately twice as much as steel and are easily damaged in hailstorms. Moreover, aluminium sheeting should be separated not only from steel purlins but also from any non-aluminium roof-top framing and conduits, in order to avoid bi-metallic corrosion. The fasteners connecting aluminium sheets to steel purlins should be made of stainless steel. The aluminium alloy panels should be at least 0.8 mm thick and at least 1 mm thick for longer spans.

The load carrying capacity of the deck is influenced by the depth of the cross section, the span length, the thickness of metal, and whether it is simply supported or continuous. The manufacturers provide load tables, which can be used to select the deck for the required span. The weight of roofing varies from 0.3 kN/m^2 to 1.0 kN/m^2, including the weight of joists, and 0.05 kN/m^2 to 0.1 kN/m^2, excluding the weight of joists.

Metal roofing can also be classified by the method of attachment to supports. *Through-fastened roofs* are attached directly to purlins, usually by self-tapping screws, self-drilling, or lock rivets. *Standing-seam roofs*, on the other hand, are connected indirectly by concealed clips formed into the seams. Standing-seam roofing is often used in USA and was introduced by Butler Manufacturing Company in 1969.

More details about steel decking are provided by Petersen (1990), Schittich (2001), and Newman (2004).

12.1.2 Galvanized Iron (GI) Sheets

Most common sizes of corrugated GI sheets are as follows:
(a) 8 corrugations (75 mm wide and 19 mm deep) per sheet
(b) 10 or 11 corrugations (75 mm wide and 19 mm deep) per sheet
The available sizes of sheets are as follows:
(a) Length—1.8, 2.2, 2.5, 2.8, and 3 m
(b) Width—0.75 m and 0.9 m
(c) Thickness—0.63, 0.8, 1.0, 1.25, and 1.6 mm
The weights of the sheets vary from 50–156 N/m^2. When the sheets are installed, side laps and end laps should be provided to make the joint water proof. The following overlaps are normally used:
(a) For roof: Side overlap—1½ to 2 corrugations
                 End overlap—150 mm

(b) For side cladding: Side overlap—1 corrugation
End overlap—100 mm

The sheets are fastened to purlins or side girts by 8-mm-diameter J- or L-type hook bolts with GI nuts along with GI and bituminous felt washers at a maximum pitch of 350 mm. Where laps do not occur over supports, 6-mm diameter bolts at a maximum pitch of 250 mm for roofs and 300–450 mm for sides are used.

Spacing of purlins and girts, which support the sheeting is governed by the length of the sheet, thickness of the sheet, and applied loading. The approximate section modulus of the corrugated GI sheeting may be taken as

$$Z = (4/15)bdt \qquad (12.1)$$

where b is the curvilinear width (equal to $1.13 \times$ covering width), d is the depth of the corrugation, and t is the thickness of the sheet.

Based on the above formula, the maximum purlin spacing is 1.8 m for a 2000-mm long and 750-mm wide sheet.

12.1.3 Asbestos Cement Sheets

Asbestos cement sheets may be used to cover the roof as an alternative to corrugated steel sheets. (These sheets are banned in many countries due to the risk of lung cancer caused by inhaling the fibres, while working with these sheets.) AC sheets are manufactured in two shapes, corrugated and Trafford, and are available in lengths of 1.75, 2.0, 2.5, and 3 m. They are manufactured in thicknesses of 6 mm or 7 mm. The maximum permissible spacing of purlins is as follows:
(a) for 6-mm sheet—1.4 m
(b) for 7-mm sheet—1.6 m
For side cladding, the spacing may be increased by 300 mm.

A side overlap of one corrugation is normally given. The end lap should not be less than 150 mm for slopes less than 18° and for flatter slopes this overlap may be increased. For side covering, an overlap of 100 mm is sufficient.

The weight of asbestos sheets varies from 160 N/m^2 to 170 N/m^2. The load per square metre of the sheet on the slope may be increased by 30% to get the load per square metre of the plan area, to account for the larger area on the slope and additional material in the side and end lapping. The sheets are fastened to purlins or girts by using 8-mm-diameter hook bolts at a maximum spacing of 350 mm.

In addition to steel, aluminium, GI and AC sheets, stainless steel and ferrocement roofing sheets can also be used. Ferrocement sheets can be produced in different shapes and sizes. Ferrocement sheets withstand heavy rainfall, cyclone, fire, and termite attack and are as durable as reinforced concrete. The fabrication does not involve any heavy machinery and the cost of ferrocement sheets is approximately 30% cheaper than conventional GI or corrugated AC roof sheeting (Mathews & Rao 1979).

Roof-top Equipment Roof-top mounted or suspended HVAC (Heating, Ventilation, and Air Conditioning) equipment may include anything from small fans and unit heaters to large air-conditioning units. They may be supported by a continuous curb or an elevated steel frame on legs. A properly designed and installed curb with sheet flashing may be less prone to leakage than discrete penetrations at frame legs (Newman 2004).

12.2 Selection of Bay Width

A bay is defined as the space between two adjacent bents (see Fig. 1.26). The roof truss along with the columns constitutes a bent. The space between two rows of columns of an industrial building is called an aisle or span. An industrial building may have a single span or multiple spans. Figure 12.3 shows industrial buildings with single, double, and multiple spans.

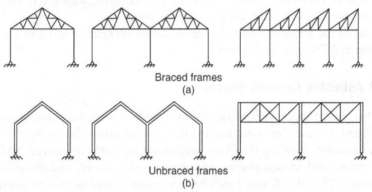

Braced frames
(a)

Unbraced frames
(b)

Fig. 12.3 Industrial buildings with single, double, and multiple spans

In most cases, the bay width may be dictated by owner requirements. Gravity loads generally control the bay size.

For crane buildings (for light and medium cranes), bays of approximately 4–8 m may be economical because of the cost of the crane gantry girders. Large bays may increase the cost of the tension flange bracing of the gantry girders. Though the bay widths in the range of 4–8 m provide economy, truss spans may range from 10–25 m or more.

12.3 Structural Framing

For the purpose of structural analysis and design, industrial buildings are classified as (see Fig. 12.3):
- Braced frames
- Unbraced frames

In braced buildings, the trusses rest on columns with hinge type of connections and the stability is provided by bracings in the three mutually perpendicular planes. These bracings are identified as follows:
(a) Bracings in the vertical plane in the end bays in the longitudinal direction [see Fig. 12.4(a)]
(b) Bracings in the horizontal plane at bottom chord level of the roof truss [see Fig. 12.4(c)]

(c) Bracings in the plane of upper chords of the roof truss [see Figs 12.4(a) and (b)]
(d) Bracings in the vertical plane in the end cross sections usually at the gable ends [see Figs 12.4(a) and (c)]

The function of a bracing is to transfer horizontal loads from the frames (such as due to wind or earthquake or horizontal surge due to acceleration and breaking of travelling cranes) to the foundation. The longitudinal bracing on each longitudinal end provides stability in the longitudinal direction. The gable bracings provide stability in the lateral direction. The tie bracings at the bottom chord level transfer lateral loads (due to wind or earthquake) of trusses to the end gable bracings. Similarly stability in the horizontal plane is provided by
- a rafter bracing in the end bays, which provide stability to trusses in their planes or
- a bracing system [see Fig. 12.4(c)] at the level of bottom chords of trusses, which provide stability to the bottom chords of the trusses.

Purlins act as lateral bracings to the compression chords of the roof trusses, which increase the design strength of the compression chords. The lateral ties provide similar functions to the bottom chord members when they are subjected to compression due to reversal of loading (see Section 12.5.8 also). X bracings (as shown in Fig. 12.4) are the commonly used bracing systems. K-type bracing systems may also be used. If the building is lengthy, bracings in the end bays alone may not be sufficient. In these cases, every fourth or fifth bay is braced and the roof upper chord bracings are also provided in these bays.

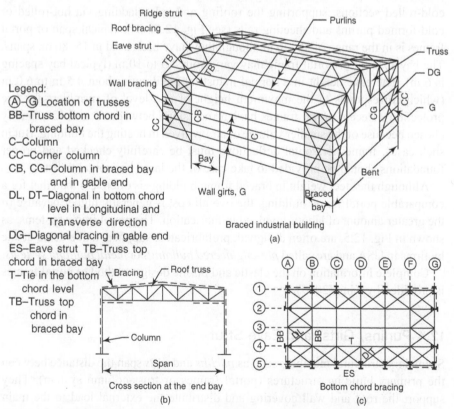

Legend:
Ⓐ–Ⓖ Location of trusses
BB–Truss bottom chord in braced bay
C–Column
CC–Corner column
CB, CG–Column in braced bay and in gable end
DL, DT–Diagonal in bottom chord level in Longitudinal and Transverse direction
DG–Diagonal bracing in gable end
ES–Eave strut TB–Truss top chord in braced bay
T–Tie in the bottom chord level
TB–Truss top chord in braced bay

Fig. 12.4 Structural framing for an industrial building

Braced frames are efficient in resisting the loads and do not sway. However, the braces introduce obstructions in some bays and may cause higher forces or uplift forces in some places. Wide flange columns are often used for exterior columns of braced frames. (For interior columns of braced frames with height less than 7 m, Square Hollow Section (SHS) columns may yield most economical solution because of their high radius of gyration about both axes.)

12.3.1 Unbraced Frames

Unbraced frames in the form of *portal frames* is the most common form of construction for industrial buildings, distinguished by its simplicity, clean lines, and economy. The frames can provide large column free areas, offering maximum adaptability of the space inside the building. Such large span buildings require less foundation, and eliminate internal columns, valley gutters, and internal drainage. Portal frame buildings offer many advantages such as more effective use of steel than in simple beams, easy extension at any time in the future, and ability to support heavy concentrated loads. The disadvantages include relatively high material unit cost and susceptibility to differential settlement and temperature stresses. In addition, these frames produce horizontal reaction on the foundation, which may be resisted by providing a long tie beam or by designing the foundation for this horizontal reaction.

Basically, a portal frame is a rigid jointed plane frame made from hot-rolled or cold-rolled sections, supporting the roofing and side cladding via hot-rolled or cold-formed purlins and sheeting rails (see Fig. 12.5). The typical span of portal frames is in the range of 30–40 m, though they have been used in 15–80 m, spans. The bay spacing of portal frame may vary from 4.5 to 10 m (typical bay spacing is 6 m). The eave height in a normal industrial building is about 4.5 m to 6.0 m (which corresponds to the maximum height of one level of sprinklers for fire protection). Recent portal frames have a roof slope between 6° and 12°, mainly chosen because of the smaller volume of air involved in heating the building. But in such cases, frame horizontal deflections must be carefully checked and proper foundations should be provided to take care of the large horizontal thrust.

Although the steel weight in braced frame buildings is often less than that for a comparable portal frame building, the overall cost is generally higher because of the greater amount of labour involved in fabrication. The portal frame systems, as shown in Fig. 12.5, are often designed, prefabricated, supplied, and erected at site by firms in USA and are called *pre-engineered buildings* or *metal building systems.*

Complete information on the elastic and plastic design method of portal frames is provided by King (2005).

12.4 Purlins, Girts, and Eave Strut

Secondary structural members such as *purlins* and *girts* span the distance between the primary building structures (portal frames or truss-column system). They support the roof and wall covering and distribute the external load to the main

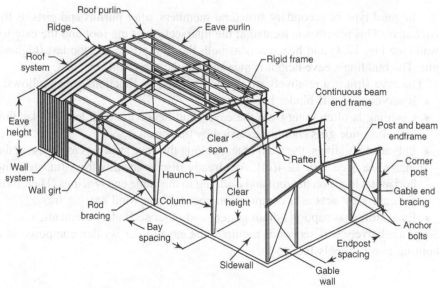

Fig. 12.5 Typical portal frame construction

frames or trusses (see Fig. 12.4). They also serve as the flange bracing for the rafters or columns and may function as a part of the building's lateral load resisting system. Purlin is a part of the roof bracing system and girts form a part of the wall bracing system of the building. The behaviour and design of purlins has been discussed in Section 6.11 and the design of sag rods in Section 3.9. Examples 6.10 and 3.10 illustrate the various steps involved in the design of purlins and sag rods, respectively. When sag rods are used for bracing the purlins top flange, it is advantageous to locate the sag rods 50–75 mm below the top of the compression flange. The weight of purlins in the total weight of the steel structure could vary from 10% to 25%. The weight of purlins may be equal to or even greater than the weight of the trusses. Hence they have to be designed properly. Usual members adopted for purlins include channels, angles, tubes, and cold-formed C- and Z-sections. When cold-formed sections are used, they should be properly protected with anti-corrosive treatment, since their thickness ranges from 1.6 mm to 4 mm only. Cold-formed C-, Z-, or sigma-purlins may be economical than hot-rolled purlins for spans of 5 m to 8 m. Angles and channel purlins without sag bars may be economical up to 5 m and tubes up to 6 m.

The main function of *girts* is to transfer wind loads from wall materials to the primary frame. Girts are positioned horizontally (see Fig. 12.4) to span between the columns. When the space between primary columns is more than 9 m, *wind columns* may be provided to reduce the girt span. Wind columns are essentially intermediate vertical girts spanning from the foundation to the eave. Since typical eave strut may not be capable of resisting the lateral reaction imposed by the wind column, a system of diagonal braces should be provided to transfer the lateral reaction to the adjacent primary framing columns. Similar to purlin spacing, girt spacing is governed by the load—resisting properties of wall panels.

The third type of secondary structural members, after purlins and girts, is the *eave strut*. This member is located at the intersection of the roof and the exterior wall (see Fig. 12.4) and hence acts as both the first purlin and the last (highest) girt. The building's eave height is measured to the top of this member.

The eave strut is a relatively strong member and its functions are as follows:

- It serves as a stiff binder beam.
- Cladding is often hung from the eave strut; hence the total load of cladding including side girts should be carried by this beam.
- In braced buildings, the wind bracing along the eave strut acts as a truss in the plan view [see Fig. 12.4(c)]. As already discussed, this truss transfers the horizontal loads on the roof and cladding to the gable end bracings. Therefore, the eave strut acts as a compression chord of the wind bracing truss.
- Eave strut also supports drain gutters and other secondary elements.

Since a relatively stiff section is required, the eave girder is often composed of a built-up two channels face-to-face.

12.5 Plane Trusses

A structure that is composed of a number of line members pin-connected at the ends to form a triangulated framework is called a *truss*. If all the members lie in a plane, the structure is a *planar truss*. In a truss, the members are so arranged that all the loads and reactions occur only at the joints (intersection points of the members). The centroidal axis of each member is straight, coincides with the line connecting the joint centres at each end of the member, and lies in a plane that also contains the lines of action of all the loads and reactions. The primary principle underlying the use of the truss as a load-carrying structure is that arranging elements into a triangular configuration results in a stable shape. Any deformations that occur in this stable structure are relatively minor and are associated with small changes in member length caused by the forces in the members by the external loads. Similarly, the angle formed between any two members remains relatively unchanged under load.

Trusses were found to have been constructed as early as 500 B.C. when Romans built a bridge using a form of timber truss across the Danube river. Though the potential of trusses were known and used in a few large public buildings in USA and Italy, the bridge builders of the early nineteenth century were responsible for the systematic use of the truss systems.

In simple roof systems the three-dimensional framework can be subdivided into planar components for analysis as planar trusses, purlins, etc., without seriously compromising the accuracy of the results. The external loads (which are applied at the joints) produce only tensile or compressive forces in the individual members of the truss. For common trusses with vertically acting loads, compressive forces are usually developed in the top chord members and tensile forces in the bottom chord members. Though the forces in the web members of a truss may be either tension or compression, there is often an alternating pattern of tensile and compressive forces present. Note that when the external loads reverse in direction (e.g., as in the case of wind loads) the top chords will be in tension and bottom chords will be in compression. Hence, it is often necessary to design the various members of a truss both for tension and compression and select the member size based on the critical force.

It is extremely important to note that when the loads are applied directly onto truss members themselves (as in the case of intermediate purlins), bending stresses will also develop in those members in addition to the basic tensile or compressive stresses. This results in complicated design procedures (they should be designed as per the provisions of beam-columns discussed in Chapter 9) and the overall efficiency of the truss is reduced (see Section 12.5.6 also).

12.5.1 Analysis of Trusses

The first step in the analysis of a truss is always to determine whether the truss under consideration is a stable configuration of members. There exists a relation between m, the number of members, j, the number of joints, and r, the reaction components. Thus, the expression

$$m = 2j - r \tag{12.2}$$

must be satisfied, if the truss is a internal statically determinate structure. The least number of reaction components required for external stability is r (equal to 3 for plane trusses). If m exceeds $(2j - r)$, then the excess members are called redundant members and the truss is said to be statically indeterminate. If fewer members, than those given by the expressions in Eqn (12.2), are present, then the truss will be unstable.

For a determinate truss, when purlins are located at the nodal points, the member forces can be found by employing the laws of statics to assure internal equilibrium of the truss. The process requires the repeated use of free-body diagrams, from which individual member forces are determined. *The method of joints* is a technique of truss analysis in which the member forces are determined by the sequential isolation of joints—the unknown member forces at one end of the truss are solved, which are then used to determine the member forces at subsequent joints. The other method is known as the *method of sections* in which the equilibrium of a part of the truss is considered, and the member forces are determined by using the three equations of equilibrium. $\Sigma F_x = 0$, $\Sigma F_y = 0$, and $\Sigma M = 0$. The forces in triangulated trusses may also be found by graphical means using force diagrams. The details of the method of joints and method of sections are provided by Thandavamoorthy (2005).

In statically indeterminate trusses (which have more number of members than a determinate truss), though the principles of statics are still valid, they can not be analysed using the method of joints or the method of sections. It is because we may have more number of unknowns than the equations of equilibrium. For analysing these indeterminate trusses, matrix methods of structural analysis are used (Livesley 1975). In this method of structural analysis, a set of simultaneous equations that describe the load-deformation characteristics of the structure under consideration is formed. These equations are solved using matrix algebra to obtain deformations or forces. From the deformations of joints, the member forces may be determined. Matrix algebra is ideally suited for setting up and solving equations in the computer. Two methods of matrix structural analysis are available. The flexibility or force method assumes the forces as unknowns and the displacement or stiffness method assumes the deformations as unknowns. Several commercial computer programs are available for the analysis of indeterminate trusses, which are based mainly on the stiffness method of structural analysis (e.g., SAP 2000 by Computers & Structures, Inc., California, STAAD Pro 2004 by Research Engineers International).

The analysis to find the forces in a multi-member truss by hand calculation can be tedious and time-consuming. Hence in practice, computer programs are often used to analyse determinate or indeterminate trusses. Most of these programs require the user to specify member sizes, so that the analysis can be performed. While forces in determinate structures do not depend upon the 'initial' member sizes given by the user, forces in indeterminate structures do depend upon initial member sizes. Hence it may be necessary to repeat the analysis two or three times, such that the 'initial' member sizes and the designed member sizes are the same. The member sizes given in Tables 12.1 and 12.2 may be used for giving the 'initial' member size input for the trusses shown in Fig. 12.6. These sizes are applicable to trusses with purlins placed at truss joints, having a minimum number of longitudinal ties, as shown in Fig. 12.6. These sizes are derived based on (a) the rise to span ratio being greater than 1/6 and (b) wind permeability of the building being $\leq 20\%$.

Table 12.1 Initial member size for roof trusses (angle sections) with $f_y = 250$ MPa

Span (m)	a	b	c, d	e	f
		Member Notations (see Fig. 12.6)			
		Truss spacing < 5 m			
10.0	1-65 × 6*	1-50 × 6	1-50 × 6	1-45 × 6	1-50 × 6
16.0	2-50 × 6	2-50 × 6	2-50 × 6	1-45 × 6	1-60 × 6
21.0	2-60 × 6**	2-50 × 6	2-50 × 6	1-50 × 6	1-65 × 6
26.0	2-65 × 6	2-60 × 6	2-60 × 6	1-60 × 6	1-65 × 6
		(Longitudinal ties: ISA 75 × 6)			
		5 m < Truss spacing < 6.5 m			
10.0	1-75 × 6	1-60 × 6	1-60 × 6	1-45 × 6	1-60 × 6
16.0	2-50 × 6	2-50 × 6	2-50 × 6	1-50 × 6	1-65 × 6
21.0	2-65 × 6	2-60 × 6	2-60 × 6	1-50 × 6	1-75 × 6
26.0	2-75 × 6	2-65 × 6	2-65 × 6	1-65 × 6	1-75 × 6
		(Longitudinal ties: ISA 90 × 6)			

* 1-65 × 6 means single angle ISA 65 × 65 × 6
** 2-60 × 6 means double angle ISA 60 × 60 × 6

12.5.2 Types of Trusses and Truss Configurations

Prior to analysing and selecting members for a roof truss, three engineering decisions are to be made: (a) the form of the chords must be determined, that is, whether they should be flat or sloping and whether they should be straight or curved; (b) the pattern of internal triangulation; (c) whether the trusses are simply supported or continuous. Important dimensional variables include the spans and depths of trusses, lengths of specific truss members (especially compression members), spacing of trusses, and transverse purlin spacing (this, in turn, dictates the way loads are applied on to the trusses and frequently, the placement of nodes within a truss).

Table 12.2 Initial member sizes for roof trusses with $f_y = 220$ MPa (Tubular members)

Span (m)	a	b	c, d	e	f
		Member Notations (see Fig. 12.6)			
		Truss spacing-5 m			
10.5	60.3 × 3.25*	48.3 × 3.25	48.3 × 3.25	42.4 × 3.25	42.5 × 3.25
16.0	76.1 × 3.25	48.3 × 3.25	60.3 × 3.65	42.4 × 3.25	42.5 × 3.25
21.0	88.9 × 4.05	60.3 × 3.25	76.1 × 3.25	42.4 × 3.25	48.3 × 3.25
26.0	101.6 × 4.85	76.1 × 3.25	88.9 × 4.05	48.3 × 3.25	60.3 × 3.25
		(Longitudinal tie = ISO 60.3 × 3.65)			
		5 m < truss spacing < 6.5 m			
10.5	76.1 × 3.25	48.3 × 3.25	60.3 × 3.65	42.4 × 3.25	42.4 × 3.25
16.0	88.9 × 4.05	60.3 × 3.25	76.1 × 3.65	42.4 × 3.25	48.3 × 3.25
21.0	101.6 × 4.85	76.1 × 3.25	88.9 × 4.05	48.3 × 3.25	48.3 × 3.25
26.0	114.3 × 4.50	76.1 × 3.65	88.9 × 4.85	48.3 × 3.25	60.3 × 3.25
		(Longitudinal ties = ISO 76.1 × 3.25)			

* 60.3 × 3.25 means tube having 60.3 mm outer diameter and 3.25 mm thickness.

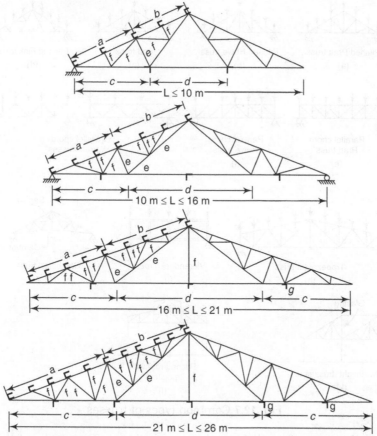

Fig. 12.6 Member notation: guidelines to member design (Refer Tables 12.1 and 12.2)

A variety of truss types have been used successfully and some common truss types are shown in Fig. 12.7. The designations *Pratt*, *Howe*, and *Warren* were originally used with parallel chord trusses [see Figs 12.7(e), (f), (g)], but now they are used more to distinguish between web systems in either flat or sloped chorded trusses. Pratt, Howe, and Warren were nineteenth century bridge designers who developed and popularized these forms. In Pratt truss the diagonals, which are longer and more heavily loaded than the adjacent verticals, are in tension under gravity loading; whereas in the comparable Howe truss they are in compression [for the loading shown in Fig. 12.7(f)]. However, the wind uplift may cause reversal of stresses in the members and nullify this benefit. Hence Pratt type trusses are more desirable than the Howe type trusses. Note that the diagonals having a slope of 40°–50° with the horizontal have been found to be the most efficient. Also the reversal of the direction of diagonals at mid-span in such trusses is characteristic of design for symmetrical loading. In a Warren truss, approximately half the diagonals are in compression. All except the end verticals are secondary members, and hence may be eliminated, without affecting the overall stability.

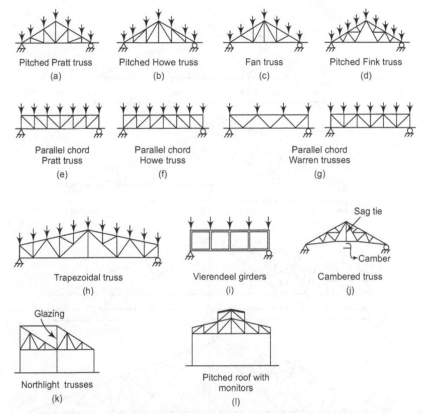

Pitched Pratt truss
(a)

Pitched Howe truss
(b)

Fan truss
(c)

Pitched Fink truss
(d)

Parallel chord
Pratt truss
(e)

Parallel chord
Howe truss
(f)

Parallel chord
Warren trusses
(g)

Trapezoidal truss
(h)

Vierendeel girders
(i)

Sag tie

Camber

Cambered truss
(j)

Glazing

Northlight trusses
(k)

Pitched roof with
monitors
(l)

Fig. 12.7 Common types of trusses

The advantage of parallel chord trusses is that they use webs of the same lengths and thus reduce fabrication costs for very long spans. The economical span to depth ratio of parallel chord trusses is in the range of 12 m to 24 m.

For very long span pitched roof, some depth of truss is provided at the ends, resulting in trapezoidal configuration for the trusses [see Fig. 12.7(h)]. Using this configuration results in the reduction of axial forces in the chord members adjacent to the supports. The secondary bending effects in those members are also reduced. The vertical members at supports with length about 1/10th of the truss height at mid-span are found to reduce the forces in the members adjacent to supports considerably (Sree Ramachandra Murthy et al. 2004). For very long spans (greater than 30 m), it may be economical to have trapezoidal trusses with sloping bottom and top chords. Such a configuration will reduce the length of web members and will result in uniform force in the chord members over the entire span. The slope of the bottom chord equal to about half the slope of the top chord is found to be more efficient.

The pitched Fink truss [Fig. 12.7(d)] usually proves to be economical for small spans (< 9 m), since the web members in such trusses are arranged in a fashion to obtain shorter members. As already mentioned, Pratt, Howe, and Warren trusses need not have the top chord parallel to the bottom one. Such an arrangement is used to provide a slope for drainage [Fig. 12.7(a) and 12.7(b)]. Pratt trusses with four or six panels are used for spans varying between 6 m to 15 m. The compound Fink truss shown in Fig. 12.7(d) may be used for a longer span. The simple fan truss [Fig. 12.7(c)] may be used to span 12 m and a compound fan truss can span up to 24 m. Fan trusses are often used when the rafter members of the roof trusses are to be subdivided into odd number of panels. A combination of fink and fan can also be used.

When the chords are parallel and diagonals are removed, as shown in Fig. 12.7(i), they are called as *Vierendeel girders*. In a Vierendeel girder, the loading is carried by a combination of pure flexure and flexure due to shear induced by the relative deformation between the ends of the top and bottom chord members, similar to that found in castellated beams. Though vierendeel girders are usually fabricated using I-sections, their load carrying capacity may be enhanced by using rolled hollow sections with butt or fillet welded connections. They may be analysed using elastic analysis and the moment capacities of the members checked using effective length factors for frames not braced against side sway (Martin & Purkiss 1992).

The lower chord of the trusses may be left straight as shown in Figs 12.7(a) and (d) or may be cambered, that is, fabricated with slight upward curve in the bottom chord member as shown in Fig. 12.7(j). The camber may be approximately in the range of 0.5 m to 1.0 m. Cambering is done for the sake of appearance so that in a lengthy room, a series of trusses, one behind the other, may not appear to sag. Cambering results in additional fabrication cost and cambered trusses involve careful assembly at site. A sag tie, as shown in Fig. 12.7(j), may be used to reduce the moment due to self weight in the long middle tie member and to reduce the resulting deflection of this member. The sag tie may also be used to carry the load due to the weight of the ceiling hung from the bottom chord, if any.

In roof trusses, the drainage, lighting, and ventilation requirements are the most important elements in establishing the upper chord slope, but occasionally, structural or aesthetic reasons may also control.

In single storey industrial buildings of the type shown in Figs 12.7(k) and (l), drainage is provided toward the eaves and valleys where longitudinal gutters and downspouts are used to carry off water. Artificial lighting is supplemented by windows in the sides of the monitors or the steeper slopes of north light or saw tooth roofs. Portions of the windows can be opened for natural ventilation.

For large column free areas, *lattice girders* are often used (see Fig. 12.8). Because of their greater depth, they usually provide greater stiffness against deflection. When lattice girders are employed as shown in Fig. 12.8, they normally span the width of the building. The saw-tooth, umbrella, and butterfly roofs span the length of the building and are supported by these lattice girders at frequent intervals. The lattice girders are parallel chord trusses as shown in Figs 12.7(e) to (g). The external loads applied through the roofing sheets and purlins are transferred to the lattice girders through the saw-tooth, umbrella, or butterfly trusses, which transfer the load to the end columns. Individual panel lengths are selected as per the spacing of the saw-tooth, umbrella, or butterfly trusses.

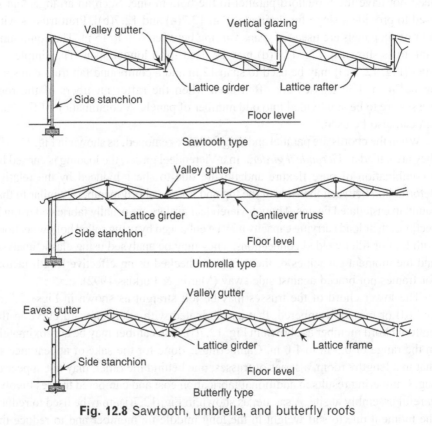

Fig. 12.8 Sawtooth, umbrella, and butterfly roofs

12.5.3 Pitches of Trusses

As seen in Fig. 12.7, most of the trusses are pitched. This is mainly done to drain off rain water on the sheeted slopes. In addition to providing the slope, the joints in the sheetings should be effectively sealed with mastic or washers. The pitch of a truss is defined as the ratio of the height of the truss to its span. The pitches usually provided for various types of roof coverings are given in Table 12.3.

Table 12.3 Pitch for roof trusses

Roof covering	Pitch
Corrugated GI sheets	1/3 to 1/6
Corrugated AC sheets	1/6 to 1/12
Lapped shingles (e.g. wood, asphalt, clay, and tile)	1/24 to 1/12
Flat roof and trapezoidal trusses	1/48 to 1/12

A pitch of 1/4 is found economical in cases where the roof has to carry snow loads in addition to wind loads. Where snow loads do not occur, lower pitches up to 1/6 are suitable. Lower pitches are advantageous since the wind pressure on the roof is reduced.

12.5.4 Spacing of Trusses

The spacing of trusses is mostly determined by the spacing of supporting columns, which in turn is determined by the functional requirements. Where there are no functional requirements, the spacing should be such that the cost of the roof is minimized. The larger the spacing, the smaller the cost of trusses, but larger is the cost of purlins and vice-versa. Roof coverings also cost more, if the spacing of the trusses is large.

Let us derive an approximate formula for arriving at the minimum cost, by considering the following variables.

S is the spacing of the trusses, C_t is the cost of trusses/unit area, C_p is the cost of purlins/unit area, C_r is the cost of roof coverings/unit area, and C is the overall cost of the roof system/unit area.

Since the cost of the truss is inversely proportional to the spacing of truss,

$$C_t = k_1/S$$

where k_1 is a constant. Similarly, the cost of purlins is directly proportional to the square of spacing of trusses. Thus,

$$C_p = k_2 S^2$$

The cost of roof coverings is directly proportional to the spacing of trusses. Thus, we have

$$C_r = k_3 S$$

Total cost $C = C_t + C_p + C_r$

$$= (k_1/S) + k_2 S^2 + k_3 S$$

For the overall cost is to be minimum, dC/dS should be zero. Thus,

$$-(k_1/S^2) + 2k_2 S + k_3 = 0$$

or $\quad (-k_1/S) + 2k_2 S^2 + k_3 S = 0$

or $\quad -C_t + 2C_p + C_r = 0$

Thus, we get $C_t = C_r + 2C_p$ $\hfill (12.3)$

Equation (12.3) shows that an economic system is obtained when the cost of trusses is equal to the cost of roof covering plus twice the cost of purlins. It has been found that the economic range of spacing is 1/5 to 1/3 of span. For lighter load, say, carrying no snow or superimposed load except wind, the larger spacing may be more economical. Spacing of 3–4.5 m for spans up to 15 m and 4.5–6 m for spans of 15–30 m may result in economy.

12.5.5 Spacing of Purlins

The spacing of purlins depends largely on the maximum safe span of the roof covering and glazing sheets. Hence, they should be less than or equal to their safe spans when they are directly placed on purlins. Thus for corrugated GI sheets, the purlin spacing may vary from 1.5 to 1.75 m, and for corrugated AC sheets, it is limited to 1.4 m, for 6-mm thick sheets, and 1.6 m, for 7-mm thick sheets. For larger spans, if the configuration of the truss is such that it is not feasible to place purlins at the nodes of upper chords, the purlins are placed between the nodes, thus introducing bending moments in the upper chords, in addition to the compressive force due to truss action (see Fig. 12.9). Hence in this case, the weight of the truss may be increased by about 10–15%. Therefore, it is preferable to place purlins at the nodal point of the truss, so that the upper chord members are subjected to only direct compression.

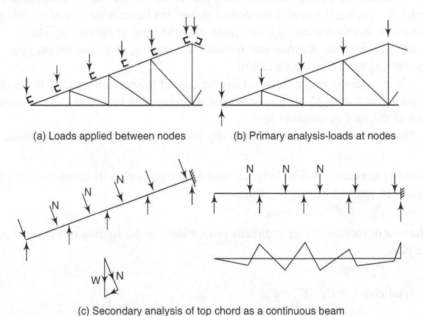

(a) Loads applied between nodes   (b) Primary analysis-loads at nodes

(c) Secondary analysis of top chord as a continuous beam

Fig. 12.9 Loads applied between nodes of truss

As discussed in Chapter 2, wind loading is not uniform over the roof; for example, the loading is much higher along the roof's perimeter and sometimes along the ridge. Instead of using structural roofing panels of heavier gauges, in the areas of higher localized loads, it is better to space the purlins closer.

12.5.6 Loads on Trusses

The main loads on trusses are dead, imposed, and wind loads. The dead load is due to sheeting or decking and their fixtures, insulation, felt, false ceiling (if provided), weight of purlins, and self weight. This load may range from 0.3 to 1.0 kN/m^2. Also the truss may be used for supporting some pipe line, fan, lighting fixtures, etc. Hence to take into account this probability, it may be worthwhile considering an occasional load of about 5 to 10 kN distributed at the lower panel points of the truss.

The weights of the purlins are known in advance as they are designed prior to the trusses. Since the weight of the truss is small compared to the total dead and imposed loads, considerable error in the assumed weight of the truss will not have a great impact on the stresses in the various members. For live load up to 2 kN/m^2, the following formula may be used to get an approximate estimate of the weight of the trusses:

$$w = 20 + 6.6L \qquad (12.4)$$

where w is the weight of the truss in N/m^2 and L is the span of the truss in m. For welded trusses, the self weight of the truss is given by

$$w = 53.7 + 0.53A \qquad (12.5)$$

where A is the area of one bay.

For live loads greater than 2kN/m^2, the value of w may be multiplied by the ratio of actual live load in kN/m^2/2. The dead weight of the truss, inclusive of lateral bracing, may also be assumed to be equal to about 10% of the load it supports. Long span trusses are likely to be heavier. The weight of bracings may be assumed to be 12–15 N/m^2 of the plan area. The weight of the truss should be computed after it has been designed to make sure that it is within 10% of the assumed weight.

The imposed load on roofs will be as per IS 875 (Part 2). The snow loads may be computed as per IS 875 (Part 4). The wind loads should be calculated as per IS 875 (Part 3). Wind loads are important in the design of light roofs where the suction can cause reversal of load in truss members. For example, a light angle member is satisfactory when used as a tie but may buckle when the reversal of load makes it to act as a strut.

Since earthquake load on a building depends on the mass of the building, earthquake loads calculated as per IS 1893 (Part 1), 2002, usually do not govern the design of light industrial buildings. Thus wind loads usually govern the design of normal trussed roofs.

12.5.7 Load Combination for Design

As mentioned earlier, the earthquake loads are not critical in the design of industrial building, since the weight of the roof is not considerable. Hence, the following combinations of loads are considered when there is no crane load:

1. Dead load + imposed load (live load)
2. Dead load + snow load
3. Dead load + wind load (wind direction being normal to the ridge or parallel to ridge whichever is severe)
4. Dead load + imposed load + wind load (which may not be critical in most of the cases)

The third combination should be considered with internal positive air pressure and internal suction air pressure separately to determine the worst combination of wind load. When crane load is present, the load combinations as given in IS 875 (Part 2) should be considered. The load combinations mentioned in this section should be considered along with appropriate partial load factors.

All the loads are assumed to act as concentrated loads at points where purlins are located on the upper chord. The weight of the truss is included in the purlin point loads.

12.5.8 Design of Truss Members

The members of the trusses are made of either rolled steel sections or built-up sections depending upon the span length and intensity of loading. Rolled steel single or double angles, T-sections, hollow circular, square, or rectangular sections are used in the roof trusses of industrial buildings [see Fig. 12.10(a)]. In long-span roof trusses and short span bridges, heavier rolled steel sections, such as channels and I-sections are used [see Fig. 12.10(b)]. Built-up I-sections, channels, angles, and plates are used in the case of long-span bridge trusses [Fig. 12.10(c)]. Access to the surface of the members for inspection, cleaning, and repainting during service are important considerations while using built-up sections. Hence in highly corrosive environments, fully closed welded box-sections or hollow sections are used, with their ends fully sealed to reduce the maintenance cost and improve the durability of the trusses.

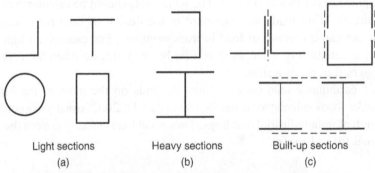

| Light sections | Heavy sections | Built-up sections |
| (a) | (b) | (c) |

Fig. 12.10 Cross section of truss members

The various steps involved in the design of truss members are as follows:
1. Depending upon the span, required lighting, and available roofing material, the type of truss is selected and a line diagram of the truss is sketched.

2. Various loads acting over the truss are calculated using IS 875 (Parts 1-5).

3. The purlins are designed and the loads acting on the truss at the purlin points are computed.

4. The roof truss is analysed for the various load combinations using the graphical method or the method of sections or joints or by a computer program and the forces acting on the members for various combinations are tabulated.

5. Each member may experience a maximum compressive or tensile force (called the design force) under a particular combination of loads. Note that a member which is under tension in one loading combination may be subjected to reversal of stresses under some other loading combination. Hence, the members have to be designed for both maximum compression and maximum tension and the size for the critical force has to be adopted. The design for compression is done as per Sections 5.8.4, 5.9.1, 5.10, and 5.12. The principal rafter is designed as a continuous strut and the other compression members are designed as discontinuous struts. The limiting slenderness ratios are discussed in Section 5.10.1. The effective length of compression members is taken as per Section 5.8.4.1. Similarly, the design for tension is done as per Section 3.7.2.

6. When purlins are placed at intermediate points, i.e., between the nodes of the top chord, the top chord will be subjected to bending moment in addition to axial compression. Since the rafter is a continuous member, the bending moments may be computed by any suitable method (say, moment distribution method or computer program). Then the member is designed for combined bending and axial compression as per Section 9.5.1.

7. The members meeting at a joint are so proportioned that their centroidal axes intersect at the same point, in order to avoid eccentricity. Then the joints of the trusses are designed either as bolted (see Section 10.6.2) or as welded joints (see Section 11.11.1). If the joint is constructed with eccentricity, then the members and fasteners must be designed to resist the moment that arises. The moment at the joint is divided between the members in proportion of their stiffness.

8. The maximum deflection of the truss may be computed by using either strain energy method or matrix stiffness analysis program. A computer analysis gives the value of deflection as part of the output. This deflection should be less than that specified in Table 6 of the code.

9. The detailed drawings and fabrication drawings are prepared and the material-take-off is worked out.

10. The lateral bracing members are then designed. When a cross braced wind girder is used, as shown in Fig. 12.11(a), it is necessary to use a computer analysis program, since the truss will be redundant. However, it is usual to neglect the compression diagonal and assume that the panel's shear is taken by the tension diagonals, as shown in Fig. 12.11(b). This idealization is useful to make the wind girder determinate and obtain the forces in various members by using method of sections or method of joints.

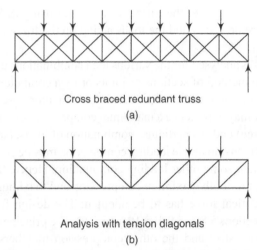

Cross braced redundant truss

(a)

Analysis with tension diagonals

(b)

Fig. 12.11   Cross-braced lattice wind girder

The design as per the procedure described here may result in very small angles that are sufficient to resist the forces in the various members of the truss. However, the members should be fairly stiff to avoid damage during loading, transport, off-loading, and erection. Since rafter is the primary compression member, a double angle (equal or unequal) is often preferred. Similarly, the main ties may be subjected to compression during handling or due to wind suction. Moreover, these ties often have a long unsupported length and hence double angle sections are used for these main ties also. All other web members can be designed as single angle members.

From practice, the following minimum sections are recommended for use in compound fink roof trusses.

 Rafters—2 ISA 75 × 50 × 6

 Main ties—2 ISA 75 × 50 × 6

 Centre tie—2 ISA 65 × 45 × 6

 Main sling—2 ISA 65 × 45 × 6

 Main strut—ISA 65 × 45 × 6

 All other members—ISA 50 × 50 × 6

The width of the members should be kept as minimum as possible, since wide members have greater secondary stresses.

While trusses are stiff in their plane, they are very weak out-of-plane. Consider the planar truss as shown in Fig. 12.12(a), in which, the top chord is braced at each panel point. The top chord when subjected to in-plane loading may buckle in the horizontal xz plane or the vertical xy plane. Unless we provide members having equal moment of inertia about both axes (e.g., square or round members including hollow sections), we have to calculate the buckling strength in the horizontal and vertical plane and adopt the least strength. When transverse members are provided, as shown in Fig. 12.12(b), it is still possible for the top chords to buckle in the horizontal plane. The effective lengths of the top chord members, as far as their resistance to buckling in the horizontal plane is concerned, is $2L$ and not just L, where L is the distance between the nodal points. Note that the members in the vertical plane do nothing to prevent this type of buckling in *xz* plane. Hence in order to make the buckling load in both the horizontal and vertical plane equal, we may have to provide members which are stiffer in the horizontal plane (such as rectangular sections, double angles, or H-sections) as shown in Fig. 12.12(b). When such lateral bracings in the form of purlins are not provided, the entire top chord of the truss may buckle laterally.

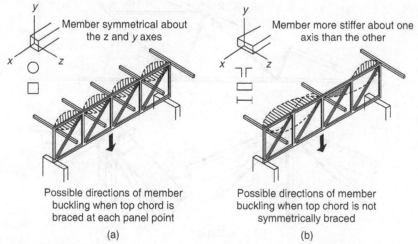

Fig. 12.12 Lateral buckling of truss members: use of transverse members for bracing

The above discussions are valid for the bottom chord member of the truss also, when it is subjected to compression due to reversal of stresses owing to wind suction. Hence lateral (longitudinal) ties are often provided at regular intervals in the bottom chords also (Fisher 1983). Depending upon the *L/r* ratio of the top and bottom chord members in the horizontal and vertical planes, it may be advantageous to adopt unequal angles. The longitudinal ties are under tension in most load situations but may be subjected to compression under wind loading condition depending upon the bracing orientation. There must be at least two longitudinal ties to form a truss action under wind load condition. The longitudinal ties may also be used to support false ceiling. It is desirable to restrict the slenderness ratio of such ties to 250, to avoid sagging.

12.5.9 Connections

Members of trusses can be jointed by riveting, bolting, or welding. As explained in Chapter 10, rivets have become obsolete and for important structures high-strength friction grip (HSFG) bolts and welds are often preferred. Trusses having short span are usually fabricated in shops using welding and transported to site as one unit. Longer span trusses are prefabricated in segments by welding in shop and assembled at site by bolting or welding. For example, fink or compound fink trusses are fabricated as two halves in the workshop. The two halves are transported and assembled at site, where the centre tie is also fitted up (see detail 1 and 3 of

Fig. 12.13). In such situations, the main slings will be subjected to severe handling stresses and hence are made of double angle sections.

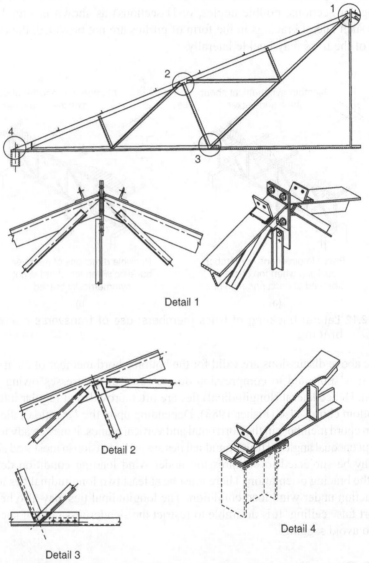

Detail 1

Detail 2

Detail 3

Detail 4

Fig. 12.13 Connection details of welded fink roof truss

If the rafter and tie members are made of T-sections, angle diagonals can be directly welded or bolted to the web of the T-sections. Often, it may not be possible to confine the connection within the width of the member due to inadequate space to accommodate the joint length. In such cases, gusset plates are used. The size, shape, and thickness of gusset plates depend upon the size of the members being joined, number and size of bolt or length of the required weld, and the force to be transmitted (see Section 10.6.2 for more discussion on gusset plate design). The connections should be so arranged that the centroidal axes of members meeting at the connection meet at a point. Example 10.12 shows the calculations required for the design of a bolted truss joint. Figure 12.14 shows typical bolted joints in trusses and lattice girders. Examples 11.8 and 11.9 showed the calculations required for the design of welded joints in trusses. The choice between welded and bolted connections depends on the equipment available with the fabricator and availability of electricity at site (if site welding is preferred). However, when large number of trusses are made, welded joints are economical. Morever, welded joints give better appearance and are easy to maintain. Standard joints should be used with as much repetitions of member shapes and sizes, end preparation, and fabrication operations as possible. This can be easily achieved with parallel chord lattice girders.

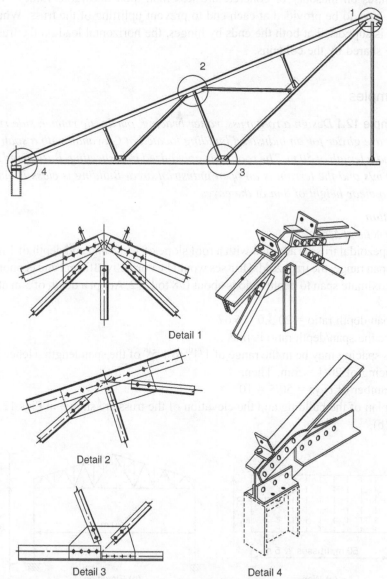

Detail 1

Detail 2

Detail 3

Detail 4

Fig. 12.14 Connections of bolted fink roof trusses

12.6 End Bearings

When the roof truss is supported on steel columns, suitable connections should be provided to transfer the reaction from the truss to the column. One end of the truss may be fixed to the column and the other end should be allowed to slide to account for the expansion of the truss. Slotted holes may be provided in the base angles (angles connecting the truss to the bearing plates at the top of the columns) so as to permit expansion of the truss due to differences in temperature. If concrete or masonry columns are used to support the roof truss, suitable bearing plates have to be used to distribute the load on these supporting members so that the pressures on masonry or concrete are less than their allowable values. Anchor bolts have to be provided at each end to prevent uplifting of the truss. When the truss is supported at both the ends by hinges, the horizontal load on the truss has to be shared by the columns.

Examples

Example 12.1 *Design a roof truss, rafter bracing, purlin, tie runner, side runner, and eave girder for an industrial building located at Guwahati with a span of 20 m and a length of 50 m. The roofing is galvanized iron sheeting. Basic wind speed is 50 m/s and the terrain is an open industrial area. Building is class B building with a clear height of 8 m at the eaves.*

Solution
1. *Structural Model*
A trapezoidal truss is adopted with a roof slope of 1 to 5 and end depth of 1 m. For this span range, the trapezoidal trusses would be normally efficient and economical. Approximate span to depth ratio is about L/8 to L/12. Adopt a depth of 3 m at mid-span.

Span/depth ratio = 20/3.0 = 6.67
Hence the span/depth ratio is fine.
Truss spacing may be in the range of $1/4^{th}$ to $1/5^{th}$ of the span length. Hence adopt a spacing of 20/4 = 5 m. Then,

Number of bays = 50/5 = 10
The plan of the building and the elevation of the truss are shown in Figs 12.15(a) and (b).

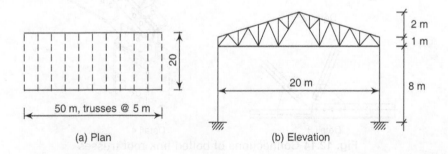

(a) Plan   (b) Elevation

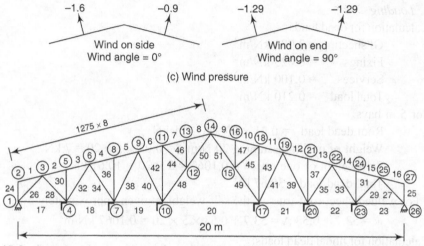

−1.6   −0.9   −1.29   −1.29

Wind on side
Wind angle = 0°

Wind on end
Wind angle = 90°

(c) Wind pressure

(d) Configuration of truss with member numbers and joint numbers adopted for the analysis

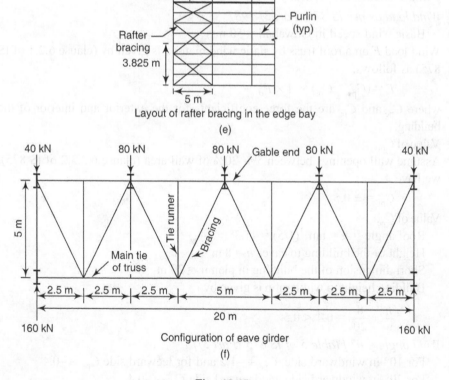

Ridge line   Truss (typ)

Purlin
(typ)

Rafter
bracing
3.825 m

5 m

Layout of rafter bracing in the edge bay

(e)

40 kN   80 kN   80 kN   Gable end 80 kN   40 kN

5 m

Tie runner

Bracing

Main tie
of truss

2.5 m   2.5 m   2.5 m   5 m   2.5 m   2.5 m   2.5 m

20 m

160 kN   160 kN

Configuration of eave girder

(f)

Fig. 12.15

2. *Loading*

Calculation for dead load:

$$\text{GI sheeting} = 0.085 \text{ kN/m}^2$$
$$\text{Fixings} \quad = 0.025 \text{ kN/m}^2$$
$$\text{Services} \quad = 0.100 \text{ kN/m}^2$$
$$\text{Total load} \quad = 0.210 \text{ kN/m}^2$$

For 5 m bays,

$$\text{Roof dead load} \quad = 0.21 \times 20 \times 5 = 21 \text{ kN}$$
$$\text{Weight of purlin (assuming 70 N/m}^2\text{)} = 0.07 \times 5 \times 20 = 7 \text{ kN}$$
$$\text{Self-weight of one truss*} = 0.1067 \times 5 \times 20 = 10.67 \text{ kN}$$
$$\text{Total dead load} = 38.67 \text{ kN}$$

* For welded sheet roof trusses, the self-weight is given approximately by

$$w = 53.7 + 0.53 \text{ A} = 53.7 + 0.53 \times 5 \times 20 = 0.1067 \text{ kN/m}^2$$

Calculation for nodal dead loads:

Since the truss has 16 internal nodes at the top chord [see Fig.12.15(b)],

$$\text{Intermediate nodal dead load } (W_1) = 38.67/16 = 2.42 \text{ kN}$$
$$\text{Dead load at end nodes } (W_1/2) = 2.42/2 = 1.21 \text{ kN}$$

(All these loads act vertically downwards at the nodes.)

Wind load as per IS 875 (Part 3)-1987

Basic wind speed in Guwahati = 50 m/s

Wind load F on a roof truss by static wind method is given by (clause 6.2.1 of IS 875) as follows:

$$F = (C_{pe} - C_{pi}) \times A \times P_d$$

where C_{pe} and C_{pi} are the force coefficients for the exterior and interior of the building.

Value of C_{pi}:

Assume wall openings between 5%–20% of wall area (clause 6.2.3.2 of IS 875), we have

$$C_{pi} = \pm 0.5$$

Value of C_{pe}:

Roof angle = $\alpha = \tan^{-1}(1/5) = 11.3°$

Height of the building to eaves $h = 8$ m

Short dimension of the building in plan $w = 20$ m

Building height to width ratio is given by

$$\frac{h}{w} = \frac{8}{20} = 0.4 < 0.5$$

Wind angle – 0° [Table 5 of IS 875 (Part 3)]

For 10° in windward side, $C_{pe} = -1.2$ and for leeward side $C_{pe} = -0.4$

For 20° in windward side and leeward side $C_{pe} = -0.4$

Roof angle $\alpha = 11.3°$

Then by interpolation we get

$$C_{pe} = -1.1 \text{ for windward and } -0.4 \text{ for leeward}$$

Wind angle – 90° [Table 5 of IS 875 (Part 3)]
    For 10° in windward and leeward, $C_{pe} = -0.8$
    For 20° in windward and leeward, $C_{pe} = -0.7$
    For 11.3°, $C_{pe} = -0.79$ for windward and leeward
Risk coefficient, $k_1 = 1.0$, assuming that the industrial building is under general category and its probable life is 50 years.
Terrain, height and structure size factor, k_2:
    Roof elevation: 8–11 m
Considering category 1 (exposed open terrain) and class B structure (length between 20–50 m) from Table 2 of IS 875 (Part 3)-1987, for 11 m, $k_2 = 1.038$
    Assume topography factor $k_3 = 1.0$ (because of flat land)
Wind pressure calculation
    Total height of the building = 11 m
    Basic wind speed V_b = 50 m/s
Design wind speed
$$V_z = k_1 \times k_2 \times k_3 \times V_b$$
$$k_1 = 1.0; \ k_2 = 1.038; \ k_3 = 1.0;$$
$$V_z = 1.038 \times 1 \times 1 \times 50 = 51.9 \text{ m/s}$$
    Design wind pressure $p_d = 0.6v_z^2 = 0.6 \times (51.9)^2$
$$= 1616.17 \text{ N/m}^2$$
$$= 1.616 \text{ kN/m}^2$$

Wind load on roof truss

Wind angle	Pressure coefficient		C_{pi}	$(C_{pe} \pm C_{pi})$		$A \times P_d$ (kN)	Wind load, F (kN)	
	C_{pe}			Wind-ward	Lee-ward		Wind-ward	Lee-ward
	Wind-ward	Lee-ward						
0°	−1.10	−0.4	−0.5	−1.6	−0.9	10.3	−16.48	−9.27
			0.5	−0.6	0.1	10.3	−6.18	1.03
90°	−0.79	−0.79	−0.5	−1.29	−1.29	10.3	−13.29	−13.29
			0.5	−0.29	−0.29	10.3	−2.987	−2.987

The critical wind pressure is shown in Fig.12.15(c).
3. *Design of Purlin*
        Span of purlin = 5 m
        Spacing of purlin = 1.275 m
        $\theta = 11.3°$
Load calculations:
    Live load = $0.75 - (11.3 - 10)0.02 = 0.724$ kN/m^2 > 0.4 kN/m^2
    Dead load = 0.21 kN/m^2
    Wind pressure = $1.616 \times 1.6 = 2.586$ kN/m^2
Load combinations:
1. DL + LL = $0.21 + 0.724 = 0.934$ kN/m^2
2. DL + WL
    Normal to slope = $-2.586 + 0.21\cos 11.3 = -2.38$ kN/m^2
    Parallel to slope = $0.21 \sin 11.3 = 0.041$ kN/m^2

(a) Load combination 1: DL + LL

$$w_z = (0.934 \times \cos 11.3) \times 1.275 = 1.168 \text{ kN /m}$$
$$w_y = (0.934 \times \sin 11.3) \times 1.275 = 0.233 \text{ kN/m}$$

where w_z is the load normal to z-axis, w_y is the load normal to y-axis, and 1.275 is the spacing of the purlin. Due to continuity of purlins, factored bending moments and shear force are as follows:

$$M_z = 1.5 \times 1.168 \times 5^2/10 = 4.38 \text{ kN m}$$
$$M_y = 1.5 \times 0.233 \times 5^2/10 = 0.874 \text{ kN m}$$
$$SF_z = 1.5 \times 1.168 \times 5/2 = 4.38 \text{ kN}$$

Try MC100 for which the properties are as follows:

$$D = 100 \text{ mm}; \ b_f = 50 \text{ mm}; \ t_w = 5 \text{ mm}; \ t_f = 7.7 \text{ mm}$$
$$I_{zz} = 192 \times 10^4 \text{ mm}^4$$
$$Z_{ez} = 37.3 \times 10^3 \text{ mm}^3, \ Z_{ey} = 7.71 \times 10^3 \text{ mm}^3$$
$$Z_{pz} = 43.83 \times 10^3 \text{ mm}^3, \ Z_{py} = 16.238 \times 10^3 \text{ mm}^3$$

Section classification:

$$b/t_f = 50/7.7 = 6.49 < 9.4$$
$$d/t_w = (100 - 2 \times 7.7)/5.0 = 16.92 < 42$$

Hence the section is plastic.

Check for shear capacity

As per clause 8.4 of IS 800,

$$A_v = (100 \times 5.0) = 500 \text{ mm}^2$$

$$\frac{A_v f_{yw}}{\sqrt{3}\gamma_{m0}} = \frac{500 \times 250}{\sqrt{3} \times 1.10 \times 10^3} = 65.6 \text{ kN} > 4.38 \text{ kN}$$

Hence shear capacity is very large compared to the shear force.

Check for moment capacity

$$M_{dz} = \frac{\beta_b Z_{pz} f_y}{\gamma_{m0}} = \frac{1 \times 43.83 \times 250 \times 10^3}{1.10 \times 10^6} = 9.96 \text{ kNm}$$

The above value should be less than

$$\frac{1.2 \times 37.3 \times 250 \times 10^3}{1.10 \times 10^6} = 10.17 \text{ kN m}$$

Hence $M_{dz} = 9.96 \text{ kN m} > M_z = 4.38 \text{ kN m}$
Hence the assumed section is safe.

$$M_{dy} = \frac{1 \times 16.238 \times 250 \times 10^3}{1.10 \times 10^6} = 3.69 \text{ kN m}$$

The above value should be less than

$$\frac{1.2 \times 7.71 \times 250 \times 10^3}{1.1} = 2.10 \text{ kN m}$$

Hence $M_{dy} = 3.69$ kN m $< M_y = 0.874$ kN m
Hence the section is satisfactory.
Check for biaxial bending

$$\frac{M_z}{M_{dz}} + \frac{M_y}{M_{dy}} \le 1$$

Thus, $\dfrac{4.38}{9.96} + \dfrac{0.874}{3.69} = 0.68 < 1.0$

Check for deflection
Calculation for deflection is based on the serviceability condition, i.e., with unfactored imposed loads.

$$W = 1.168 \times 5 = 5.84 \text{ kN}$$

$$\delta = \frac{5WL^3}{384EI_z}$$

$$= \frac{5 \times 5.84 \times 1000 \times 5000^3}{384 \times 2 \times 10^5 \times 192 \times 10^4}$$

$$= 24.75 \text{ mm}$$

As per IS 800, Table 6, deflection limit is $\dfrac{L}{150} = 33.33$ mm > 24.75 mm

Hence the deflection is within allowable limits.
(b) Load combination 2: DL + WL

$$w_z = 2.38 \times 1.275 = 3.035 \text{ kN/m}$$
$$w_y = 0.041 \times 1.275 = 0.052 \text{ kN/m}$$

Factored bending moments in this case are

$$M_z = 1.5 \times 3.035 \times 5^2/10 = 11.38 \text{ kN m} > M_{dz} = 9.96 \text{ kN m}$$
$$M_y = 1.5 \times 0.052 \times 5^2/10 = 0.195 \text{ kN m} < M_{dy} = 3.69 \text{ kN m}$$

Hence, the section is not safe. Let us adopt MC125, which has an

$$I_{zz} = 425 \times 10^4 \text{ mm}^2,$$
$$Z_{pz} = 77.88 \times 10^3 \text{ mm}^3 \text{ and } Z_{py} = 29.46 \times 10^3 \text{ mm}^3$$
$$M_{dz} = 1 \times 77.88 \times 250 \times 10^{-3}/1.1 = 17.7 \text{ kNm}$$
$$M_{dy} = 1 \times 29.46 \times 250 \times 10^{-3}/1.1 = 6.69 \text{ kNm}$$

Thus, the check for biaxial bending is

$$\frac{11.38}{17.7} + \frac{0.195}{6.69} = 0.67 < 1.0$$

Hence the section is safe.
Check for deflection

$$\delta = \frac{5 \times (3.035 \times 5) \times 1000 \times 5000^3}{384 \times 2.0 \times 10^5 \times 425 \times 10^4} = 29.06 < 33.33 \text{ mm}$$

Note that the purlins at the edges and ridge of the building will be subjected to a local pressure of $1.4p \pm 0.5p$ [as per Table 5 of IS 875 (Part 3)-1987] instead of $1.6p$ taken in the preceding calculations. Hence, the purlins at the edges or at the ridge of the building have to be checked for this local pressure or closer spacing of purlins may be adopted at these locations.

4. *Truss Analysis and Design*

Tributary area for each node of the truss:

Length of each panel along sloping roof

$$= \frac{1.25}{\cos 11.3°} = 1.275 \text{ m} < 1.4 \text{ m}$$

Spacing of trusses = 5 m

Tributary area for each node of the truss = $5 \times 1.275 = 6.375 \text{ m}^2$

Imposed load calculations:

From IS 875 (Part 2)-1987,

Live load = 0.75 kN/m^2

Reduction due to slope (see Table 2.3 and footnote 3)

$= (0.75 - 0.02 \times 1.3)2/3 = 0.483 \text{ kN/m}^2$

Load at intermediate nodes $W_2 = 0.483 \times 5 \times 1.25$

$= 3.02 \text{ kN}$

Load at end nodes $W_2/2 = 1.51 \text{ kN}$

(All these loads act vertically downwards.)

Maximum $C_{pe} \pm C_{pi}$ (critical wind loads to be considered for analysis):

Wind Angle	Windward side (W_3)		Leeward (W_4)	
	Intermediate nodes W_3	End and apex nodes $W_3/2$	Intermediate nodes W_4	End and apex nodes $W_4/2$
0°	−16.48	−8.24	−9.27	−4.64
90°	−13.29	−6.645	−13.29	−6.645

*Loads in kN

All these loads act perpendicular to the top chord member of the truss.

Forces in the members The truss has been modelled as a pin jointed plane truss as shown in Fig. 12.15(d) and analysed using the software PLTRUSS developed by the author. The analysis results are tabulated as follows [see truss configuration shown in Fig. 12.15(d) for member numbers]

Load factors and combinations (Table 4 of IS 800):

For dead + imposed load

$1.5 \times DL + 1.5 \times LL$

For dead + wind load

$1.5 \times DL + 1.5 \times WL$

Dead + imposed + wind loading case will not be critical as wind loads act in opposite direction to dead and imposed loads.

Table 12.4 Member forces under factored loads in kN

Member number	Dead load + Live load	Dead load + Wind load (0°)	Dead load + Wind load (90°)
1.	0	2.472	1.985
2.	–97.086	212.066	193.914
3.	–97.086	217.01	197.898
4.	–124.83	269.25	253.304
5.	–124.83	274.20	257.29
6.	–124.83	263.50	258.09
7.	–128.99	279.45	270.58
8.	–128.99	284.39	274.57
9.	–128.99	244.35	274.57
10.	–128.99	241.57	270.57
11.	–124.83	233.41	258.08
12.	–124.83	221.17	257.29
13.	–124.83	218.39	253.30
14.	–97.086	163.98	197.9
15.	–97.086	161.20	193.91
16.	0	1.39	1.99
17.	61.20	–141.17	–118.91
18.	113.66	–251.86	–219.05
19.	124.67	–261.51	–237.61
20.	108.801	–202.49	–201.68
21.	124.67	–212.69	–237.61
22.	113.66	–186.67	–219.06
23.	61.20	–97.66	–118.91
24.	–4.08	10.79	8.34
25.	–4.08	5.27	8.34
26.	–86.55	185.54	168.17
27.	–86.55	138.114	168.17
28.	48.084	–98.26	–92.45
29.	48.084	–79.65	–92.45
30.	–8.16	21.58	16.69
31.	–8.16	10.55	16.69
32.	–31.76	58.86	59.82
33.	–31.76	56.25	59.82
34.	15.04	–20.94	–26.92
35.	15.04	–30.85	–26.92
36.	–8.16	21.58	16.69
37.	–8.16	10.55	16.69
38.	–4.67	5.20	5.97
39.	–4.67	16.65	5.97
40.	–4.67	26.38	12.41

(contd)

(contd)

41.	−4.67	−2.46	12.41
42.	−12.24	32.38	25.03
43.	−12.24	15.82	25.03
44.	5.225	−13.82	−10.69
45.	5.225	−6.75	−10.69
46.	8.16	21.58	16.69
47.	8.16	10.55	16.69
48.	21.245	−72.17	−46.70
49.	21.245	−17.793	−46.70
50.	27.62	−89.024	−59.74
51.	27.62	−26.03	−59.74

Truss Reactions (kN)

Joint number	Case 1(DL + LL)		Case 2(DL + WL(0))		Case 3(DL + WL(90))	
	X	Y	X	Y	X	Y
1	0	43.52	11.31	−94.62	0	−84.84
26	0	43.52	0	−68.61	0	−84.84

5. *Design of Top Chord Member (Member No. 8)*

Factored compressive force = 128.99 kN

Factored tensile force = 284.39 kN

Trying two ISA 75 × 75 × 6 mm @ 0.136 kN/m
Sectional properties:

Area of cross section $A = 2 \times 866 = 1732$ mm^2

Radius of gyration $r_{zz} = 23$ mm, $I_y = 45.7 \times 10^4$ mm^4, $C_y = 20.6$ mm

Assuming 8-mm thick gusset plate,

$$I_y = 2[45.7 \times 10^4 + 866 (4 + 20.6)^2] = 196.21 \times 10^4 \text{ mm}^4$$

$$r_y = \sqrt{(196.21 \times 10^4 / 1732)} = 33.66 \text{ mm}$$

Section classification:

$$\varepsilon = (250/f_y)^{0.5} = (250/250)^{1/2} = 1.0$$

$$b/t = 75/6 = 12.5 < 15.7\varepsilon$$

∴ the section is semi-compact.

As no member in the section is slender, the full section is effective and there is no need to adopt reduction factor.

Maximum unrestrained length = L = 1275 mm

$KL = 0.85 \times L = 0.85 \times 12.75 = 1083.75$ mm

Note The effective length of top chord member may be taken as 0.7–1.0 times the distance between centres of connections as per clause 7.2.4 of IS 800. We have assumed the effective length factor as 0.85.

$$\lambda_y = 1083.75/23 = 47.12 < 180$$

Hence λ_y is within the allowable limits. From Table 9c of the code for $KL/r = 47.12$ and $f_y = 250$ MPa,

$$f_{cd} = 187.32 \text{ N/mm}^2$$

Axial capacity = $187.32 \times 1732/1000 = 324.4$ kN > 128.99 kN

Hence, section is safe against axial compression.

Axial tension capacity of the section = $1732 \times 250/1.10$
$$= 393.64 \text{ kN} > 284.39 \text{ kN}$$

Note: Design strength governed by tearing at net section should also be checked. Hence, section is safe in tension.

Note Though a smaller section may be chosen, this section is adopted to take care of handling stresses.

6. *Design of Bottom Chord Member (Member No. 20)*

Factored compressive force = 202.49 kN

Factored tensile force = 108.801 kN

Try two L $100 \times 100 \times 8$ @ 0.242 kN/m.

Sectional properties:

Area of cross section A = $2 \times 1540 = 3080$ mm^2

Radius of gyration $r_z = 30.7$ mm, $I_y = 145 \times 10^4$ mm^4, $C_y = 27.6$ mm

Assuming a 10-mm thick gusset plate,

$$I_y = 2[145 \times 10^4 + 1540 (5 + 27.6)^2] = 6.173 \times 10^6 \text{ mm}^4$$

$$r_y = \sqrt{(6.173 \times 10^6/ 3080)} = 44.77 \text{ mm}$$

Section classification:

$$b/t = 90/6 = 15 < 15.7\varepsilon$$

∴ the section is semi-compact.

Axial tension capacity of the selected section = $3080 \times (250/1.10) \times 10^{-3}$
$$= 700 \text{ kN} \gg 108.801 \text{ kN}$$

Hence, section is safe in tension. Providing longitudinal tie runner at every bottom node of the truss,

Maximum unrestrained length = $L = 5000$ mm

$r_y = 44.71$ mm

$\lambda_y = 5000 \times 0.85/44.71 = 95$

From Table 9c of IS 800 for $\lambda_y = 95$ and $f_y = 250$ MPa,

$$f_{cd} = 114 \text{ N/mm}^2$$

Axial capacity = $114 \times 3080/1000 = 351.12$ kN > 202.49 kN

Hence, section is safe against axial compression also.

Note It may be economical to adopt unequal angles for the top and bottom chord members. However, many unequal angle sections are not readily available in the market. Hence, equal angle sections have been used in the truss in this example.

7. *Design of Web Member (Member No. 28)*

Maximum compressive force = 98.26 kN

Maximum tensile force = 185.54 kN

Try ISA $90 \times 90 \times 6.0$ with

$$A = 1050 \text{ mm}^2; r_{zz} = 27.7 \text{ mm}; r_{vv} = 17.5 \text{ mm}$$

Section classification:

$$b/t = 90/6 = 15 < 15.7\varepsilon$$

Hence, the section is not slender.

Length of member = 1768 mm

The angle will be connected through one leg to the gusset. A minimum of two bolts will be provided at the ends to connect the angle. Assuming fixed condition (Table 12 of the code) and using the Table given in Appendix D for the capacity of eccentrically connected angles, we get

for 1.5 m length, capacity = 101 kN

for 2 m length, capacity = 89 kN

Hence   for 1.768 m, capacity = 94.56 kN ≈ 98.26 kN

Tensile capacity of the section = $(250/1.1) \times 1050/1000$

= 238.64 kN > 185.54 kN

Hence ISA $90 \times 90 \times 6.0$ is adequate for the web member.

Note Web members away from the support have less axial force. However, their length will be more. If desired, these may be designed and a smaller section be adopted.

8. *Check for Deflection*

Maximum deflection from computer output = 20.07 mm

Allowable deflection as per Table 6 of IS 800 : 2007

= span/240 = 20000/240 = 83.33 mm > 20.07 mm

Hence the deflection is within allowable limits.

9. *Design of Rafter Bracing Members*

Considering the layout of the rafter bracing as shown in Fig. 12.15(e),

Design wind pressure = 1.616 kN/m^2

Maximum force coefficient = -1.6

Factored wind load on rafter bracing

= $1.5 \times 1.616 \times 1.6 \times 3.825 \times 5/2 \times \sec 11.3°$

= 37.8 kN

Length of bracing = $\sqrt{(3825^2 + 5000^2)}$ = 6295.29 mm

Try $90 \times 90 \times 6$, $A = 1050 \text{ mm}^2$, $r_{min} = 17.5$ mm, and

$L/r = 6295.29 /17.5 = 359.7 < 400$ (Table 3 of IS 800 : 2007)

In the X bracing system, as shown in Fig. 12.15(e), the compression bracing will buckle and only the tension bracing will be effective. Also, the bracing members are usually bolted to the trusses at site.

Axial tensile capacity:

Design strength of member due to yielding of gross section

$$T_{dg} = A_g f_y / \gamma_{m0}$$

$$= 1050 \times (250/1.1)/1000$$

$$= 238.64 \text{ kN} > 37.8 \text{ kN}$$

Design strength due to rupture of critical section $T_{dn} = \alpha A_n f_y / \gamma_{m1}$

$\alpha = 0.6$ (Assuming two bolts of 16 mm diameter at the ends)

$A_n = 1050 - 18 \times 6 = 942$ mm^2

$T_{dn} = 0.6 \times 942 \times (410/1.25) \times 1000 = 185$ kN > 37.8 kN

Hence L $90 \times 90 \times 6$ is safe. The member has been found to be safe for block shear failure.

Note The forces in the bracing members are often small and rarely govern the design; but their slenderness limitations decide the size because of their long length.

10. *Design of Tie Runner*

Portion of wind load from gable end along the ridge will be transferred as axial load to tie runners provided along the length of building at tie level. Assume three intermediate gable end columns at a spacing of 5 m.

Wind load on cladding:

$L/w = 50/20 = 2.5 < 4$ and $h/w = 11/20 = 0.55 > 1/2$

From Table 4 of IS 875 (Part 3)-1987, external wind pressure $= +0.7$, with internal pressure of ± 0.5, maximum pressure $= 1.2p$

Factored wind load on intermediate runner $= 1.5 \times 1.616 \times 1.2 \times 5 \times (8/2 + 3/2)$

$= 80$ kN (Tension)

Min r required $= 5000 \times 0.85/250 = 17$ (maximum allowable $KL/r = 250$)

Try L $90 \times 90 \times 6$, with $A = 1050$ mm^2, $r_{min} = 17.5$ mm,

Design strength due to yielding of gross section

$T_{dg} = A_g f_y / \gamma_{m0}$

$= 1050 \times (250/1.1) / 1000 = 238.64$ kN > 80 kN

The design strength due to rupture of critical section (see the design of rafter bracing)

$= 185$ kN > 80 kN

Hence L $90 \times 90 \times 6$ is adequate.

11. *Design of Side Runner*

Assuming that the side sheetings are provided at a spacing of 1.25 m,

Span of side runner $= 5$ m

Calculation of loads

(a) Vertical loads

Self weight of side runner (MC100) $= 9.56$ kg/m

Weight of side sheeting $= 5$ kg/m^2

Thus weight of GI sheeting $= 5 \times 1.25 = 6.25$ kg/m

Total $w_y = 15.81$ kg/m

$= 0.16$ kN/m

(b) Wind loads

Maximum wind force co-efficient $= 1.2$

Wind load UDL $= 1.2 \times 1.616 \times 1.25$

$= 2.424$ kN/m

Factored loading

$$w_y = 1.5 \times 0.16 = 0.24 \text{ kN/m}$$
$$w_z = 1.5 \times 2.424 = 3.64 \text{ kN/m}$$

Try MC100, which has the following properties:

$D = 100$ mm; $b_f = 50$ mm; $t_w = 5.0$ mm; $t_f = 7.7$ mm; $I_{zz} = 192 \times 10^4$ mm^4;
$Z_{ez} = 37.3 \times 10^3$ mm^3; $Z_{pz} = 43.83 \times 10^3$ mm^3; $Z_{ey} = 7.71 \times 10^3$ mm^3
$Z_{py} = 16.238 \times 10^3$ mm^3

Assuming continuity of side runner,

$$BM_z = wl^2/8 = 3.64 \times 5^2/10 = 9.1 \text{ kN m}$$
$$BM_y = 0.24 \times 5^2/10 = 0.6 \text{ kN m}$$

Shear capacity:

From the design of purlin, shear capacity of the section = 65.6 kN

$$V = 3.64 \times 5/2 = 9.1 \text{ kN} < 65.6 \text{ kN}$$

Hence safe against shear.

Check moment capacity

From the calculations of purlin design

$$M_{dz} = 9.96 \text{ kN m} > 9.1 \text{ kN m}$$
$$M_{dy} = 3.69 \text{ kN m} > 0.6 \text{ kN m}$$

Check for biaxial bending

$$\frac{9.1}{9.96} + \frac{0.6}{3.69} = 1.08 > 1.0$$

Hence, the section may be revised to ISMC 125. The deflection has been calculated (similar to the purlin) and found to be within the codal limits. Note that for both the purlin and side runner, biaxial bending has been considered as per the code. If the sheeting is assumed to resist the in-plane loads, a smaller size of purlin and side runner may be sufficient to resist the bending moment in the z-direction.

12. Design of Eave Girder

Eave girders are provided at tie level and along both longitudinal and transverse directions to carry the wind forces at the tie level (see Fig. 12.4). Here the design of an eave girder along transverse direction is only considered. The design of an eave girder along the longitudinal direction is similar to the one in the transverse direction. The eave girder is in the form of a truss by connecting the bottom chord members of the truss to the gable columns at one end and main truss bottom points at the other end. The configuration of the eave girder is given in Fig. 12.15(f). In this example we are assuming a truss at both the ends of the building.

Wind loads on eave girders:

Design wind pressure = 1.616 kN/m^2

Maximum force coefficient = 1.2

Eaves girder bracings are connected between gable columns which are assumed to be spaced at 5 m centre to centre and truss bottom joints.

Wind load on internal joint $= [1.616 \times 1.2 \times 5 \times (8/2 + 3/2)] \times 1.5$
$$= 80 \text{ kN}$$

Wind load on end joints = 40 kN

Reaction = $(3 \times 80 + 2 \times 40) / 2 = 160$ kN

Using the method of joints, the maximum force in the bracing at the end of the eave girder $F_{br} = (160 - 40)/\cos 26.565 = \pm 134.2$ kN

Members of the girder are subjected to reversal of stresses and hence they have been checked for compression as well as tension.

$$\text{Length of the bracing} = \sqrt{2.5^2 + 5^2} = 5.59 \text{ m}$$

Required r_{min} = 5590/350 = 15.97 mm

Try section L $130 \times 130 \times 10$ with the following properties:

$$A = 2510 \text{ mm}^2; \; r_{min} = 25.7 \text{ mm}$$

Section classification:

$$\varepsilon = (250/f_y)^{0.5} = 1.0$$

Flange:

$$B/T = 130/10 = 13 < 15.7\varepsilon$$

∴ It is semi compact.

From the table of capacity of eccentrically connected angle with fixed end conditions (see Appendix D),

Capacity for a length of 5590 mm = 148.1 kN > 134.2 kN

Hence the section is safe.

Axial tensile capacity (due to yielding of gross section)

$$= 2510 \times 250/(1.1 \times 1000)$$

$$= 570 \text{ kN} > 134.2 \text{ kN}$$

Design strength due to rupture of critical section.

$$T_{dn} = \alpha A_n f_u/\gamma_{m1}$$
$$= 0.6 \times (2510 - 18 \times 10) \times 410 / (1.25 \times 1000)$$
$$= 458.5 \text{ kN} > 134.2 \text{ kN}$$

The section has to be checked for block shear failure also. Note that the tension capacity is very high compared to the capacity in compression. It is due to the long length of the member. In order to reduce the member size, X bracing could be adopted and the member may be designed to be effective only in tension.

Note that the top and bottom chords of the eave girder are the bottom chord members of the main truss and carry a force of ±160 kN. The bottom chord members have been designed for a compressive force of 202.49 kN and a tensile force of 108.8 kN (tensile capacity 700 kN). Hence, the sections are safe under gable wind loads. Similarly, the vertical members are tie runners (with a maximum force of 80 kN), which have been designed already.

Slotted holes may be provided in the base angles to permit expansion of the truss. For a change in temperature of 34°C (see IS 875–Part 5), the maximum change in length will be $0.000012 \times 34 \times 20,000 = 8.16$ mm. The slot for a 20-mm bolt may be 22 mm wide and 30 mm long.

Summary

Structural steel is often the material of choice for the construction of single storey industrial buildings, which constitute the major percentage of the total number of steel structures built around the world. The planning and design of these buildings require the knowledge of several items such as site condition, plant layout and work flow, availability of new materials and waste disposal facilities, HVAC equipment, crane types and capacity, future expansion plans, and budget and project schedule. In particular, the structural engineer should select items such as roofing and walling material, bay width, structural framing system, and type and shape of trusses.

A number of factors have to be considered while selecting the roofing (decking) or wall (cladding) material. Steel, aluminium, galvanized iron, asbestos, stainless steel, and ferrocement sheets can be used as cladding or decking material. Metal roofing can be classified by the method of attachment to supports. Through-fastened roofs are directly attached to purlins and hence provide lateral stability to the purlins. However, standing- seam roofing (which is used extensively in USA) is connected indirectly by concealed clips formed in to the seam and requires a separate system of purlin bracings. Some details about the various types of sheeting are provided. Some guidelines for fixing the bay width of industrial buildings are also given.

Depending on the structural framing system adopted, industrial buildings may be classified as braced frames and unbraced frames. In braced frames, trusses rest on columns with hinge type connections and stability is provided by bracings in three mutually perpendicular directions.

Since the weight of purlins may be equal to the weight of trusses, they should be properly designed. Channels, angles, tubes, cold-formed C-, Z- or sigma sections are employed as purlins. The functions of girts are similar to purlins except that they are used in the walls. The eave strut is located at the intersection of the roof and exterior wall and has to be designed carefully.

A triangulated framework of pin-ended members is called a truss. In most situations, the loads are applied at the nodal points of trusses by purlins. When the purlins are placed in between the nodal points, the top chord members have to be designed for the secondary bending moments. With the availability of digital computers and software packages, the trusses are often analysed using these software packages. However, hand methods will be quite useful to check the results of the computer output (especially the errors made in the input data to the programs). The software packages require the member sizes of the truss to be given as input. Hence, some guidelines to assume the initial member sizes are given.

The various types of trusses and their configurations are described. It is to be noted that these configurations may change from project to project and is often selected based on aesthetics, economy, and performance. Some guidelines are also provided to select the pitch, and spacing of trusses. The loads and load combination to be considered are briefly discussed and the various steps involved in the design of truss members are given. Sometimes it may be necessary to adopt a bigger section than those indicated by the actual design to take care of transportation and handling stresses. A brief discussion on the connections has been included. A few sketches showing the connection details of welded and bolted trusses are provided. The design of various members of trusses is illustrated through examples.

It has to be noted that with the knowledge of the behaviour and design of plates, beams, columns, tension members, compression members, and beam-columns, different types of structures (e.g., towers, multi-storey buildings, water tanks, bridges, chimneys, etc.) can be designed, using the appropriate code of practice.

Exercises

1. Design a roof truss for a railway platform of size 30×12 m situated in Chennai and as shown in Fig. 12.16. Assume asbestos cement sheetings.

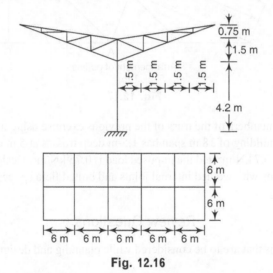

Fig. 12.16

2. An industrial building is shown in Fig. 12.17. The frames are at 5 m centres and the length of the building is 40 m. The purlin spacing of the roof is as shown in Fig. 12.17(b). The building is situated in Delhi. Assume live and wind loads as per IS 875 (part 2 and part 3) and the roof is covered with GI sheeting. Design the roof truss using angle members and gusseted joints. The truss is to be fabricated using welded joints in two parts for transport and assembled at site using bolted joints at A, B, and C as shown in Fig. 12.17.

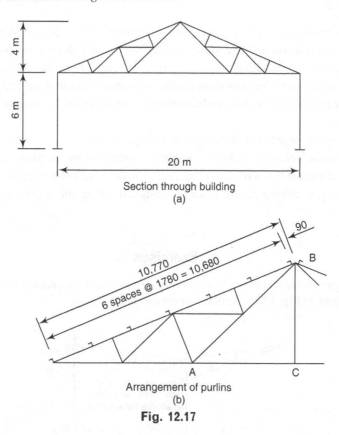

Section through building
(a)

Arrangement of purlins
(b)

Fig. 12.17

3. Design the members of the truss of the previous exercise using tubular members.
4. A flat roof building of 18 m span has 1.5-m deep trusses at 5 m centres. The total dead load is 0.7 kN/m² and the imposed load is 0.75 kN/m². Design the truss using angle sections with welded internal joints and bolted field splices.

Review Questions

1. List the items that are to be considered while planning and designing an industrial building.
2. List the items to be considered while selecting a cladding/decking system.
3. Name some of the cladding/decking materials that are used in practice.
4. What are the purposes of structural decking?
5. Under what condition will the decking provide lateral stability to the top flange of purlins?
6. What are the advantages and drawbacks of the following:
    (a) Aluminium decking
    (b) GI sheeting
    (c) Asbestos sheeting
    (d) Ferrocement sheeting
7. Why is it necessary to design cladding and fixtures for higher pressure coefficient than that used for the design of structural frameworks?

8. What are the different types of bracings used in a braced building?
9. What is the function of a bracing?
10. State the difference between a purlin and a girt.
11. What are the sections that are normally used as purlins or girts?
12. What are wind columns?
13. What are the functions of an eave strut?
14. How can one determine whether a given truss forms a stable configuration?
15. Why is it necessary to design truss members for both compression and tension forces?
16. Distinguish between determinate and indeterminate trusses?
17. When are bending moments to be considered in the design of the top chord of trusses?
18. Sketch the different truss configurations that are often used in practice.
19. Why are Pratt trusses more advantageous compared to Howe trusses?
20. What are the advantages of parallel chord trusses?
21. What are the requirements that are considered while fixing the upper chord slope of trusses?
22. State the advantage of north light roof trusses over other forms of trusses.
23. What is the economic range of spacing of a truss?
24. How is the spacing of purlins fixed?
25. What are the load combinations that are usually considered for truss analysis?
26. List the various steps involved in the design of truss members.
27. Why are the minimum sections recommended and adopted for truss members, even though a lighter section may be indicated by the design?
28. Describe the behaviour of top and bottom chord members of a truss when lateral purlins/ties are not provided at each node.
29. When are gusset plates used in a truss having T-section for rafters and bottom tie members?
30. Why is it necessary to provide connections that will allow movement in the supports of trusses?

Properties of Structural Steel Sections

The structural designer has choice of a variety of sections, which are available in the market. This appendix provides properties of structural steel sections often used in practice. For more complete details of I-sections, channels, equal and unequal angles, and T-sections refer to IS: 808-1989. Note that IS: 808 does not give values of the plastic section modulus. Hence these values for I-sections and channels have been provided based on IS: 800. Note that there are some small differences in the values given by IS: 808 and IS: 800. Only the values given by IS: 808 have been used in this book. However, these differences in values will not affect the design much. Also included in this appendix are the wide flange sections, which have been introduced recently (more information on these sections may be obtained from M/s. Jindal Vijayanagar Steel Limited). For the properties of castellated beams, circular tubes, square and rectangular hollow sections, and cold formed lipped channel and zed sections, refer Appendix A of Subramanian 2008.

Indian Standard Rolled Steel Plates

Steel plates are available in the following widths and thicknesses.

Widths: 160, 180, 200, 220, 250, 280, 320, 355, 400, 450, 500, 560, 630, 710, 800, 900, 1000, 1100, 1250, 1400, 1600, 1800, 2000, 2200, and 2500

Thickness: 5.0, 5.5, 6.0, 7.0, 8, 9, 10, 11, 12, 14, 16, 18, 20, 22, 25, 28, 32, 36, 40, 45, 50, 56, 63, 71, and 80

Table A.1 Sectional properties for beams

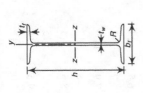

Designation	Mass N/m	Sectional dimensions						Sectional properties							
		Area (mm²)	h (mm)	R (mm)	b_f (mm)	t_w (mm)	t_f (mm)	I_z (cm⁴)	I_y (cm⁴)	r_z (mm)	r_y (mm)	Z_z (cm³)	Z_y (cm³)	Plastic modulus Z_{pz} (cm³)	Shape factor
MB 100	89	1140	100	9	50	4.7	7	183	12.9	40	10.5	36.6	5.16	41.24	1.1268
MB 125	133	1700	125	9	70	5	8	445	38.5	51.6	15.1	71.2	11	81.85	1.1399
MB 150	150	1910	150	9	75	8	8	718	46.8	61.3	15.7	95.7	12.5	110.48	1.1401
MB 175	196	2500	175	10	85	5.8	9	1260	76.7	71.3	17.6	144	18	166.08	1.1422
MB 200	242	3080	200	11	100	5.7	10	2120	137	82.9	21.1	212	27.4	253.86	1.1358
MB 225	311	3970	225	12	110	6.5	11.8	3440	218	93.1	23.4	306	39.7	348.27	1.1385
MB 250	373	4750	250	13	125	6.9	12.5	5130	335	104	26.5	410	53.5	465.71	1.1345
MB 300	460	5860	300	14	140	7.7	13.1	8990	486	124	28.6	599	69.5	651.74	1.1362
MB 350	524	6670	350	14	140	8.1	14.2	13600	538	143	28.4	779	76.8	889.57	1.1421
MB 400	615	7840	400	14	140	8.9	16	20500	622	162	28.2	1020	88.9	1176.18	1.1498
MB 450	724	9220	450	15	150	9.4	17.4	30400	834	182	30.1	1350	111	1533.36	1.15
MB 500	869	11100	500	17	180	10.2	17.2	45200	1370	202	35.2	1810	152	2074.67	1.1471
MB 550	1040	13200	550	18	190	11.2	19.3	64900	1830	222	37.3	2360	193	2711.98	1.1492
MB 600	1230	15600	600	20	210	12	20.3	91800	2650	242	41.2	3060	252	3510.63	1.1471

Table A.2 Sectional properties of columns and heavy weight beams

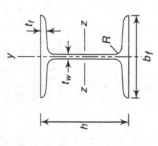

Designation	Mass (N/m)	Sectional dimensions						Sectional properties							
		Area (mm^2)	h (mm)	b_f (mm)	R (mm)	t_w (mm)	t_f (mm)	I_z (cm^4)	I_y (cm^4)	r_z (mm)	r_y (mm)	Z_z (cm^3)	Z_y (cm^3)	Plastic modulus, Z_{pz} (cm^3)	Shape factor
Column sections															
SC 100	200	2550	100	100	12	6	10	436	136	41.3	23.1	87.2	27.2	99.60	1.1422
SC 120	262	3340	120	120	12	6.5	11	842	255	50.2	27.6	140	42.6	159.49	1.1392
SC 140	333	4240	140	140	12	7	12	1470	438	58.9	32.1	211	62.5	238.59	1.1307
SC 150	371	4740	152	152	11.7	7.9	11.9	1970	700	64.5	38.4	259	91.9	285.87	1.1038
SC 160	419	5340	160	160	15	8	13	2420	695	67.4	36.1	303	86.8	341.67	1.1276
SC 180	505	6440	180	180	15	8.5	14	3740	1060	76.2	40.5	415	117	467.42	1.1263
SC 200	603	7680	200	200	18	9	15	5530	1530	84.8	44.6	553	153	620.03	1.1212
SC 220	704	8980	220	220	18	9.5	16	7880	2160	93.5	49	716	196	802.02	1.1201
SC 250	856	10900	250	250	23	10	17	12500	3260	107	54.6	997	260	1106.89	1.1102

(contd)

Table A.2 (*contd*)

Heavy weight beams/columns

HB 150	271	3450	150	8	150	5.4	9	1460	432	65	35.4	194	57.6	213.87	1.1024
HB 200	373	4750	200	9	200	6.1	9	3600	967	87.1	45.1	361	96.7	394.31	1.0923
HB 225	431	5490	225	10	225	6.5	9.1	5300	1350	98	49.6	469	120	511.55	1.0907
HB 250	510	6500	250	10	250	6.9	9.7	7740	1960	109	54.9	619	156	674.46	1.0896
HB 300	588	7480	300	11	250	7.6	10.6	12600	2200	130	54.1	836	175	914.60	1.0940
HB 350	674	8590	350	12	250	8.3	11.6	19200	2450	149	53.4	1090	196	1202.97	1.1036
HB 400	774	9870	400	14	250	9.1	12.7	28100	2730	169	52.6	1400	218	1548.92	1.1064
HB 450	872	11100	450	15	250	9.8	13.7	39200	3000	188	51.8	1740	239	1931.87	1.1103

Table A.3 Sectional properties of channel sections

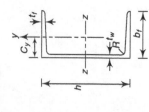

Designation	Mass (N/m)	Area (mm²)	h (mm)	R (mm)	b_f (mm)	t_w (mm)	t_f (mm)	C_y (mm)	I_z (cm⁴)	I_y (cm⁴)	r_z (mm)	r_y (mm)	Z_z (cm³)	Z_y (cm³)	Plastic modulus, Z_pz (cm³)	Shape factor
					Sectional dimensions							Sectional properties				
MC 75	71.4	910	75	8.5	40	4.8	7.5	13.2	78.5	12.9	29.4	11.9	20.9	4.81	24.57	1.1756
MC 100	95.6	1220	100	9	50	5	7.7	15.4	192	26.7	39.7	14.8	37.3	7.71	44.48	1.1584
MC 125	131	1670	125	9.5	65	5.3	8.2	19.5	425	61.1	50.5	19.1	68.1	13.4	77.88	1.1436
MC 150	168	2130	150	10	75	5.7	9	22	788	103	60.8	22	105	19.5	120.00	1.1429
MC 175	196	2490	175	10.5	75	6	10.2	21.9	1240	122	70.4	22.1	141	23	161.92	1.1484
MC 200	223	2850	200	11	75	6.2	11.4	22	1830	141	80.2	22.2	181	26.4	209.92	1.1598
MC 225	261	3330	225	12	80	6.5	12.4	23.1	2710	188	90.2	23.7	241	33	276.03	1.1453
MC 250	306	3900	250	12	80	7.2	14.1	23	3880	211	99.2	23.7	307	38.5	354.65	1.1552
MC 300	363	4630	300	13	90	7.8	13.6	23.5	6420	313	118	26	428	47.1	495.67	1.1581
MC 350	427	5440	350	14	100	8.3	13.5	24.4	10000	434	136	28.2	576	57.3	670.76	1.1645
MC 400	501	6380	400	15	100	8.8	15.3	24.2	15200	508	154	28.2	760	67	888.79	1.1695

Table A.4 Sectional properties of equal leg angles

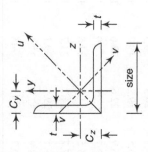

Designation	Mass (N/m)	Area (mm²)	Sectional dimensions				Sectional properties						
			C_z (mm)	C_y (mm)	I_z (cm⁴)	I_y (cm⁴)	r_z (mm)	r_y (mm)	$r_{u(max)}$ (mm)	$r_{v(min)}$ (mm)	Z_z (cm³)	Z_y (cm³)	Z_{pz} (cm³)
L20 20×3	9	112	5.9	5.9	0.4	0.4	5.8	5.8	7.3	3.7	0.3	0.3	0.52
×4	11	145	6.3	6.3	0.5	0.5	5.8	5.8	7.2	3.7	0.4	0.4	0.67
L25 25×3	11	141	7.1	7.1	0.8	0.8	7.3	7.3	9.3	4.7	0.4	0.4	0.84
×4	14	184	7.5	7.5	1	1	7.3	7.3	9.1	4.7	0.6	0.6	1.08
×5	18	225	7.9	7.9	1.2	1.2	7.2	7.2	9.1	4.7	0.7	0.7	1.31
L30 30×3	14	173	8.3	8.3	1.4	1.4	8.9	8.9	11.3	5.7	0.6	0.6	1.23
×4	18	226	8.7	8.7	1.8	1.8	8.9	8.9	11.2	5.7	0.8	0.8	1.59
×5	22	277	9.2	9.2	2.1	2.1	8.8	8.8	11.1	5.7	1.0	1.0	1.93
L35 35×3	16	203	9.5	9.5	2.3	2.3	10.5	10.5	13.3	6.7	0.9	0.9	1.69
×4	21	266	10.0	10.0	2.9	2.9	10.5	10.5	13.2	6.7	1.2	1.2	2.20
×5	26	327	10.4	10.4	3.5	3.5	10.4	10.4	13.1	6.7	1.4	1.4	2.68
×6	30	386	10.8	10.8	4.1	4.1	10.3	10.3	12.9	6.7	1.7	1.7	3.14

(contd)

Table A.4 (contd)

L40 40×3	18	234	10.8	10.8	3.4	3.4	12.1	12.1	15.4	7.7	1.2	1.2	2.23
×4	24	307	11.2	11.2	4.5	4.5	12.1	12.1	15.3	7.7	1.6	1.6	2.91
×5	30	378	11.6	11.6	5.4	5.4	12.0	12.0	15.1	7.7	1.9	1.9	3.56
×6	35	447	12.0	12.0	6.3	6.3	11.9	11.9	15.0	7.7	2.3	2.3	4.18
L45 45×3	21	264	12.0	12.0	5	5	13.8	13.8	17.4	8.7	1.5	1.5	2.85
×4	27	347	12.5	12.5	6.5	6.5	13.7	13.7	17.3	8.7	2	2	3.72
×5	34	428	12.9	12.9	7.9	7.9	13.6	13.6	17.2	8.7	2.5	2.5	4.56
×6	40	507	13.3	13.3	9.2	9.2	13.5	13.5	17.0	8.7	2.9	2.9	5.37
L50 50×3	23	295	13.2	13.2	6.9	6.9	15.3	15.3	19.4	9.7	1.9	1.9	3.54
×4	30	388	13.7	13.7	9.1	9.1	15.3	15.3	19.3	9.7	2.5	2.5	4.63
×5	38	479	14.1	14.1	11	11	15.2	15.2	19.2	9.7	3.1	3.1	5.68
×6	45	568	14.5	14.5	12.9	12.9	15.1	15.1	19.0	9.6	3.6	3.6	6.70
L55 55×5	41	527	15.3	15.3	14.7	14.7	16.7	16.7	21.1	10.6	3.7	3.7	6.93
×6	49	626	15.7	15.7	17.3	17.3	16.6	16.6	21.0	10.6	4.4	4.4	8.19
×8	64	818	16.5	16.5	22	22	16.4	16.4	20.7	10.6	5.7	5.7	10.58
×10	79	1000	17.2	17.2	26.3	26.3	16.2	16.2	20.3	10.6	7	7	12.83
L60 60×5	45	575	16.5	16.5	19.2	19.2	18.2	18.2	23.1	11.6	4.4	4.4	8.31
×6	54	684	16.9	16.9	22.6	22.6	18.2	18.2	22.9	11.5	5.2	5.2	9.82
×8	70	896	17.7	17.7	29	29	18.0	18.0	22.7	11.5	6.8	6.8	12.72
×10	86	1100	18.5	18.5	34.8	34.8	17.8	17.8	22.3	11.5	8.4	8.4	15.46
L65 65×5	49	625	17.7	17.7	24.7	24.7	19.9	19.9	25.1	12.6	5.2	5.2	9.81
×6	58	744	18.1	18.1	29.1	29.1	19.8	19.8	25.0	12.6	6.2	6.2	11.61
×8	77	976	18.9	18.9	37.4	37.4	19.6	19.6	24.7	12.5	8.1	8.1	15.06
×10	94	1200	19.7	19.7	45	45	19.4	19.4	24.4	12.5	9.9	9.9	18.34

(contd)

Table A.4 (*contd*)

L70 70×5	53	677	18.9	18.9	31.1	31.1	21.5	21.5	27.1	13.6	6.1	6.1	11.44
×6	63	806	19.4	19.4	36.8	36.8	21.4	21.4	27.0	13.6	7.3	7.3	13.54
×8	83	1060	20.2	20.2	47.4	47.4	21.2	21.2	26.7	13.5	9.5	9.5	17.60
×10	102	1300	21.0	21.0	57.2	57.2	21.0	21.0	26.4	13.5	11.7	11.7	21.46
L75 75×5	57	727	20.2	20.2	38.7	38.7	23.1	23.1	29.2	14.6	7.1	7.1	13.19
×6	68	866	20.6	20.6	45.7	45.7	23.0	23.0	29.1	14.6	8.4	8.4	15.63
×8	89	1140	21.4	21.4	59	59	22.8	22.8	28.8	14.5	11	11	20.34
×10	110	1400	22.2	22.2	71.4	71.4	22.6	22.6	28.4	14.5	13.5	13.5	24.84
L80 80×6	73	929	21.8	21.8	56	56	24.6	24.6	31.1	15.6	9.6	9.6	17.86
×8	96	1220	22.7	22.7	72.5	72.5	24.4	24.4	30.8	15.5	12.6	12.6	23.28
×10	118	1500	23.4	23.4	87.7	87.7	24.1	24.1	30.4	15.5	15.5	15.5	28.47
×12	140	1780	24.2	24.2	102	102	23.9	23.9	30.1	15.4	18.3	18.3	33.44
L90 90×6	82	1050	24.2	24.2	80.1	80.1	27.7	27.7	35.0	17.5	12.2	12.2	22.78
×8	108	1380	25.1	25.1	104	104	27.5	27.5	34.7	17.5	16	16	29.76
×10	134	1700	25.9	25.9	127	127	27.3	27.3	34.4	17.4	19.8	19.8	36.47
×12	158	2020	26.6	26.6	148	148	27.1	27.1	34.1	17.4	23.3	23.3	42.93
L100 100×6	92	1170	26.7	26.7	111	111	30.9	30.9	39.1	19.5	15.2	15.2	28.30
×8	121	1540	27.6	27.6	145	145	30.7	30.7	38.8	19.5	20	20	37.05
×10	149	1900	28.4	28.4	177	177	30.5	30.5	38.5	19.4	24.7	24.7	45.48
×12	177	2260	29.2	29.2	207	207	30.3	30.3	38.2	19.4	29.2	29.2	53.61
L110 110×8	134	1710	30.0	30.0	197	197	34.0	34.0	42.8	21.8	24.6	24.6	45.13
×10	166	2110	30.9	30.9	240	240	33.7	33.7	42.5	21.6	30.4	30.4	55.48
×12	197	2510	31.7	31.7	281	281	33.5	33.5	42.2	21.5	35.9	35.9	65.50
×16	257	3280	33.2	33.2	357	357	33.0	33.0	41.5	21.4	46.5	46.5	84.62

(*contd*)

(contd)

L130 130×8	159	2030	35.0	35.0	331	331	40.4	40.4	51.0	25.9	34.9	34.9	63.69
×10	197	2510	35.9	35.9	405	405	40.2	40.2	50.7	25.7	43.1	43.1	78.48
×12	235	2990	36.7	36.7	476	476	39.9	39.9	50.3	25.6	51	51	92.86
×16	307	3920	38.2	38.2	609	609	39.4	39.4	49.7	25.4	66.3	66.3	120.48
L150 150×10	229	2920	40.8	40.8	634	634	46.6	46.6	58.7	29.8	58	58	105.48
×12	273	3480	41.6	41.6	746	746	46.3	46.3	58.4	29.7	68.8	68.8	125.03
×16	358	4560	43.1	43.1	959	959	45.8	45.8	57.7	29.4	89.7	89.7	162.74
×20	441	5620	44.6	44.6	1160	1160	45.3	45.3	57.1	29.3	110	110	198.73
L200 200×12	369	4690	53.9	53.9	1830	1830	62.4	62.4	78.7	39.9	125	125	226.44
×16	485	6180	55.6	55.6	2370	2370	61.9	61.9	78.0	39.6	164	164	296.37
×20	600	7640	57.1	57.1	2880	2880	61.4	61.4	77.3	39.3	201	201	363.80
×25	739	9410	59.0	59.0	3470	3470	60.7	60.7	76.1	39.1	246	246	444.82

Table A.5 Sectional properties of unequal leg angles

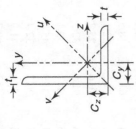

Designation	Mass (N/m)	Area (mm²)	Sectional dimensions						Sectional properties				
			C_z (mm)	C_y (mm)	I_z (cm⁴)	I_y (cm⁴)	r_z (mm)	r_y (mm)	$r_{u(max)}$ (mm)	$r_{v(min)}$ (mm)	Z_z (cm³)	Z_y (cm³)	Z_{pz} (cm³)
L30 20 × 3	11	141	9.8	4.9	1.2	0.4	9.2	5.4	9.9	4.1	0.6	0.3	1.15
× 4	14	184	10.2	5.3	1.5	0.5	9.2	5.4	9.8	4.1	0.8	0.4	1.48
× 5	18	225	10.6	5.7	1.9	0.6	9.1	5.3	9.7	4.1	1	0.4	1.78
L40 25 × 3	15	188	13	5.7	3	0.9	12.5	6.8	13.3	5.2	1.1	0.5	2.06
× 4	19	246	13.5	6.2	3.8	1.1	12.5	6.8	13.2	5.2	1.4	0.6	2.67
× 5	24	302	13.9	6.6	4.6	1.4	12.4	6.7	13.1	5.2	1.8	0.7	3.25
× 6	28	356	14.3	6.9	5.4	1.6	12.3	6.6	12.9	5.2	2.1	0.9	3.80
L45 30 × 3	17	218	14.2	6.9	4.4	1.5	14.2	8.4	15.2	6.3	1.4	0.7	2.67
× 4	22	286	14.7	7.3	5.7	2	14.1	8.4	15.1	6.3	1.9	0.9	3.48
× 5	28	352	15.1	7.7	6.9	2.4	14	8.3	15	6.3	2.3	1.1	4.25
× 6	33	416	15.5	8.1	8	2.8	13.9	8.2	14.9	6.3	2.7	1.3	4.98
L50 30 × 3	18	234	16.3	6.6	5.9	1.6	15.9	8.3	16.7	6.5	1.7	0.7	3.23
× 4	24	307	16.8	7	7.7	2.1	15.8	8.2	16.6	6.3	2.3	0.9	4.22

(contd)

(contd)

Table A.5 (contd)

× 5	30	378	17.2	7.4	9.3	2.5	15.7	8.1	16.5	6.3	2.8	1.1	5.16
× 6	35	447	17.6	7.8	10.9	2.9	15.6	8	16.4	6.3	3.4	1.3	6.05
L60 40 × 5	37	476	19.5	9.6	16.9	6	18.9	11.2	20.2	8.5	4.2	2	7.78
× 6	44	565	19.9	10	19.9	7	18.8	11.1	20.1	8.5	5	2.3	9.17
× 8	58	737	20.7	10.8	25.4	8.8	18.6	11	19.8	8.4	6.5	3	11.81
L65 45 × 5	41	526	20.7	10.8	22.1	8.6	20.5	12.8	22.2	9.6	5	2.5	9.28
× 6	49	625	21.1	11.2	26	10.1	20.4	12.7	22.1	9.5	5.9	3	10.96
× 8	64	817	21.9	12	33.2	12.8	20.2	12.5	21.8	9.5	7.7	3.9	14.15
L70 45 × 5	43	552	22.7	10.4	27.2	8.8	22.2	12.6	23.6	9.6	5.7	2.5	10.63
× 6	52	656	23.2	10.9	32	10.3	22.1	12.5	23.5	9.6	6.8	3	12.56
× 8	67	858	24	11.6	41	13.1	21.9	12.4	23.2	9.5	8.9	3.9	16.24
× 10	83	1050	24.8	12.4	49.3	15.6	21.6	12.2	22.9	9.5	10.9	4.8	19.69
L75 50 × 5	47	602	23.9	11.6	34.1	12.2	23.8	14.2	25.6	10.7	6.7	3.2	12.38
× 6	56	716	24.4	12	40.3	14.3	23.7	14.1	25.5	10.7	8	3.8	14.64
× 8	74	938	25.2	12.8	51.8	18.3	28.5	14	25.2	10.6	10.4	4.9	18.98
× 10	90	1150	26	13.6	62.2	21.8	23.3	13.8	24.9	10.6	12.7	6	23.06
L80 50 × 5	49	627	26	11.2	40.6	12.3	25.5	14	27	10.7	7.5	3.2	13.91
× 6	59	746	26.4	11.6	48	14.4	25.4	13.9	26.9	10.7	9	3.8	16.46
× 8	77	978	27.3	12.4	61.9	18.5	25.2	13.7	26.6	10.6	11.7	4.9	21.37
× 10	94	1200	28.1	13.2	74.7	22.1	24.9	13.6	26.3	10.6	14.4	6	26.00
L90 60 × 6	68	865	28.7	13.9	70.6	25.2	28.6	17.1	30.7	12.8	11.5	5.5	21.38
× 8	89	1140	29.6	14.8	91.5	32.4	28.4	16.9	30.4	12.8	15.1	7.2	27.85
× 10	110	1400	30.4	15.5	111	39.1	28.1	16.7	30.1	12.7	18.6	8.8	34.00
× 12	130	1660	31.2	16.3	129	45.2	27.9	16.5	29.8	12.7	22	10.3	39.85
L100 65 × 6	75	955	31.9	14.7	96.7	32.4	31.8	18.4	34	13.9	14.2	6.4	26.42
× 8	99	1260	32.8	15.5	126	41.9	31.6	18.3	33.8	13.9	18.7	8.5	34.48
× 10	122	1550	33.7	16.3	153	50.7	31.4	18.1	33.5	13.8	23.1	10.4	42.19

(contd)

Table A.5 (*contd*)

Section													
L100 75×6	80	1010	30.1	17.8	101	48.7	31.5	21.9	35	15.9	14.4	8.5	27.32
×8	105	1340	31	18.7	132	63.3	31.4	21.8	34.8	15.9	19.1	11.2	35.68
×10	130	1650	31.9	19.5	160	76.9	31.2	21.6	34.5	15.8	23.6	13	43.69
×12	154	1950	32.7	20.3	188	89.5	31	21.4	34.2	15.8	27.9	16.3	51.36
L125 75×6	92	1170	40.5	15.9	188	51.6	40.1	21	42.3	16.2	22.2	8.7	40.93
×8	121	1540	41.5	16.8	246	67.2	40	20.9	42.1	16.1	29.4	11.5	53.63
×10	149	1900	42.4	17.6	300	81.6	39.7	20.7	41.8	16.1	36.5	14.2	65.88
L125 95×6	101	1290	37.2	22.4	205	103	39.9	28.3	44.3	20.7	23.4	14.3	43.33
×8	134	1700	38	23.2	268	135	39.7	28.1	44.1	20.5	30.9	18.8	56.83
×10	165	2110	38.9	24	328	164	39.5	27.9	43.8	20.4	38.1	23.1	69.88
×12	197	2500	39.7	24.8	385	192	39.2	27.7	43.5	20.3	45.1	27.3	82.48
L150 75×8	137	1750	52.4	15.4	410	71.1	48.8	20.2	49.9	16.2	42	11.9	74.08
×10	170	2160	53.3	16.2	502	86.3	48.2	20	49.6	16.1	51.9	14.7	91.19
×12	202	2570	54.2	17	590	100	47.9	19.8	49.3	16	61.6	17.3	107.76
L150 115×8	163	2070	44.8	27.6	474	244	47.8	34.3	53.3	25	45.1	28	82.88
×10	201	2570	45.7	28.4	582	299	47.6	34.1	53.1	24.8	55.8	34.5	102.19
×12	240	3050	46.5	29.2	685	351	47.4	33.9	52.8	24.7	66.2	40.8	120.96
×16	314	4000	48.1	30.7	878	447	46.9	33.4	52.1	24.4	86.2	53	177.24
L200 100×10	229	2920	69.8	20.3	1230	215	64.8	27.1	66.8	21.7	94.3	26.9	165.25
×12	273	3480	70.7	21.1	1450	251	64.6	26.9	66.5	21.6	112	31.9	196.03
×16	358	4570	72.3	22.7	1870	320	64	26.6	65.9	21.3	147	41.3	255.42
L200 150×10	269	3430	60.2	35.5	1410	689	64.1	44.8	71	32.8	101	60.2	184.00

Table A.6 Sectional properties of rolled steel tee sections

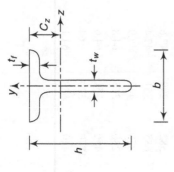

Designation	Weight (N/m)	Sectional area (mm²)	Depth of section h (mm)	Width of flange b (mm)	Thickness of flange t_f (mm)	Thickness of web t_w (mm)	Centre of gravity C_z (mm)	Moment of inertia I_z (cm⁴)	I_y (cm⁴)	Radius of gyration r_z (mm)	r_y (mm)	Moduli of section Z_z (cm³)	Z_y (cm³)
ISNT 20	9	113	20	20	3	3	6	0.4	0.2	5.9	3.9	0.3	0.2
ISNT 30	14	175	30	30	3	3	8.3	1.4	0.6	8.9	5.7	0.6	0.4
ISNT 40	35	448	40	40	6	6	12	6.3	3	11.8	8.2	2.2	1.5
ISNT 50	45	570	50	50	6	6	14.4	12.7	5.9	15	10.2	3.6	2.4
ISNT 60	54	690	60	60	6	6	16.7	22.5	10.1	18.1	12.1	5.2	3.4
ISNT 80	96	1225	80	80	8	8	22.3	71.2	32.3	24.1	16.2	12.3	8.1
ISNT 100	150	1910	100	100	10	10	27.9	173.8	79.9	30.2	20.5	24.1	16
ISNT 150	228	2908	150	150	10	10	39.5	608.8	257.5	45.6	30.3	54.6	35.7
ISHT 75	153	1949	75	150	9	8.4	16.2	96.2	230.2	22.2	34.4	16.4	30.1
ISHT 100	200	2547	100	200	9	7.8	19.1	193.8	497.3	27.6	44.2	24	49.3

(contd)

Table A.6 (contd)

ISHT 125	274	3485	125	250	9.7	8.8	23.7	415.4	1005.8	34.5	53.7	41	79.9
ISHT 150	294	3742	150	250	10.6	7.6	26.6	573.7	1096.8	39.2	54.1	46.5	87.7
ISST 100	81	1037	100	50	10	5.8	30.3	99	9.6	30.9	9.6	14.2	3.8
ISST 150	157	1996	150	75	11.6	8	37.5	450.2	37	47.5	13.6	43.9	9.9
ISST 200	284	3622	200	165	12.5	8	47.8	1267.8	358.2	59.2	31.5	83.3	43.4
ISST 250	375	4775	250	180	12.1	9.2	64	2774.4	532	76.2	33.4	149.2	59.1
ISLT 50	40	511	50	50	6.4	4	11.9	9.9	6.4	13.9	11.2	2.6	2.5
ISLT 75	71	904	75	80	6.8	4.8	17.2	41.9	27.6	21.5	17.5	7.2	6.9
ISLT 100	127	1616	100	100	10.8	5.7	21.3	116.6	75	26.9	21.5	14.8	15
ISJT 75	35	450	75	50	4.6	3	20	24.8	4.6	23.5	10.1	4.5	1.8
ISJT 87.5	40	514	87.5	50	4.8	3.2	25	39	4.8	27.5	9.7	6.2	1.9
ISJT 100	50	632	100	60	5	3.4	28.1	63.5	8.6	31.7	11.7	8.8	2.9
ISJT 112.5	64	814	112.5	80	5	3.7	30.1	101.6	20.2	35.3	15.8	12.3	5.1

Table A.7 Sectional properties of parallel flange beams and columns

IPE - European I-beams   H – Depth

HE - European wide flange beams   B - Flange width

W - American wide flange beams t_w - Web thickness

UC - British universal columns t_f - Flange thickness

HD - Wide flange columns   R - Fillet radius

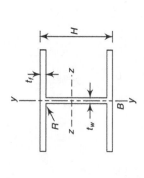

Designation	Mass (N/m)	Sectional area (mm²)	Main dimensions (mm)					Moment of inertia (cm⁴)		Radius of gyration (mm)		Modulus of section (cm³)	
			H	B	t_w	t_f	R	I_z	I_y	r_z	r_y	Z_z	Z_y
(1) Nominal size 200 mm													
IPE 200	224	2848	200	100	5.6	8.5	12	1943	142.4	82.6	22.4	194.3	28.47
HE 200 A	423	5383	190	200	6.5	10	18	3692	1336	82.8	49.8	388.6	133.6
HE 200 B	613	7808	200	200	9	15	18	5696	2003	85.4	50.7	569.6	200.3
HE 200 M	1030	13130	220	206	15	25	18	10640	3651	90.0	52.7	967.4	354.5
W 200 × 135 × 26.6	266	3400	207	133	5.8	8.4	10	2587	329.8	87.2	31.1	250	49.6
W 200 × 135 × 31.3	313	3992	210	134	6.4	10.2	10	3139	409.6	88.7	32	298.9	61.13
W 200 × 165 × 35.9	359	4575	201	165	6.2	10.2	10	3438	764.3	86.7	40.9	342.1	92.64
W 200 × 165 × 41.7	417	5317	205	166	7.2	11.8	10	4088	900.5	87.7	41.2	398.8	108.5
IPE 220	262	3337	220	110	5.9	9.2	12	2772	204.9	91.1	24.8	252	37.25
HE 220 A	505	6434	210	220	7	11	18	5410	1955	91.7	55.1	515.2	177.7
HE 220 B	715	9104	220	220	9.5	16	18	8091	2843	94.3	55.9	735.5	258.5
HE 220 M	1170	14940	240	226	15.5	26	18	14600	5012	98.9	57.9	1217	443.5

(contd)

Table A.7 (*contd*)

Column Sections

UC 200 × 203 × 46	461	5873	203.2	203.6	7.2	11	10.2	4568	1548	88.2	51.3	449.6	152.1
UC 200 × 203 × 52	520	6628	206.2	204.3	7.9	12.5	10.2	5259	1778	89.1	51.8	510.1	174
UC 200 × 203 × 60	600	7637	209.6	205.8	9.4	14.2	10.2	6125	2065	89.6	52	584.4	200.6
UC 200 × 203 × 71	710	9043	215.8	206.4	10	17.3	10.2	7618	2537	91.8	53	706	245.9
UC 200 × 203 × 86	860	10960	222.2	209.1	12.7	20.5	10.2	9449	3127	92.8	53.4	850.5	299.1
UC 200 × 200 × 100	1000	12670	229	210	14.5	27.3	10	11000	3660	93.2	53.7	961	349
(2) Nominal size 250 mm													
IPE 240	307	3912	240	120	6.2	9.8	15	3892	283.6	99.7	26.9	324.3	47.27
HE 240 A	603	7684	230	240	7.5	12	21	7763	2769	100.5	60	675.1	230.7
HE 240 B	832	10600	240	240	10	17	21	11260	3923	103.1	60.8	938.3	326.9
HE 240 M	1570	19960	270	248	18	32	21	24290	8153	110.3	63.9	1799	657.5
W 250 × 145 × 32.7	327	4175	258	146	6.1	9.1	13	4895	472.6	108.3	33.6	379.4	64.74
W 250 × 145 × 38.5	385	4929	262	147	6.6	11.2	13	6014	593.7	110.5	34.7	459.1	80.77
W 250 × 145 × 44.8	448	5732	266	148	7.6	13	13	7118	703.5	111.4	35	535.2	95.06
W250 × 200 × 49.1	491	6254	247	202	7.4	11	13	7070	1510	106	49	572	150
W250 × 200 × 58	580	7426	252	203	8	13.5	13	8740	1880	108	50.3	694	185
W250 × 200 × 67	670	8559	257	204	8.9	15.7	13	10400	2220	110	50.9	809	218
HD 260 × 68.2	682	8682	250	260	7.5	12.5	24	10450	3668	109.7	65	836.4	282.1
HD 260 × 93	930	11840	260	260	10	17.5	24	14920	5135	112.2	65.8	1148	395
HD 260 × 114	1140	14570	268	262	12.5	21.5	24	18910	6456	113.9	66.6	1141	492.8
HD 260 × 142	1420	18030	278	265	15.5	26.5	24	24330	8236	116.2	67.6	1750	621.6
HD 260 × 172	1720	21960	290	268	18	32.5	24	31310	10450	119.4	69	2159	779.7
HE 280 A	764	9726	270	280	8	13	24	13670	4763	118.6	70	1013	340.2

(*contd*)

Table A.7 (contd)

HE 280 B	1030	13140	280	280	10.5	18	24	19270	6595	121.1	70.9	1376	471
HE 280 M	1890	24020	310	288	18.5	33	24	39550	13160	128.3	74	2551	914.1
Column Sections													
W 250×250×73	730	9299	253	254	8.6	14.2	13	11290	3880	110.2	64.6	892.1	305.5
W 250×250×80	800	10210	256	255	9.4	15.6	13	12570	4314	111	65	982.4	338.3
W 250×250×89	890	11410	260	256	10.7	17.3	13	14260	4841	111.8	65.1	1097	378.2
W 250×250×101	1010	12900	264	257	11.9	19.6	13	16380	5549	112.7	65.6	1241	431.9
W 250×250×115	1150	14620	269	259	13.5	22.1	13	18940	6405	113.8	66.2	1408	494.6
W 250×250×131	1310	16700	275	261	15.4	25.1	13	22150	7446	115.2	66.8	1611	570.6
W 250×250×149	1490	18970	282	263	17.3	28.4	13	25940	8622	116.9	67.4	1840	655.7
W 250×250×167	1670	21320	289	265	19.2	31.8	13	30020	9879	118.7	68.1	2078	745.6
(3) Nominal size 300 mm													
IPE 300	422	5381	300	150	7.1	10.7	15	8356	603.8	124.6	33.5	557.1	80.5
HE 300 A	880	11250	290	300	8.5	14	27	18260	6310	127.4	74.9	1260	420.6
HE 300 B	1170	14910	300	300	11	19	27	25170	8563	129.9	75.8	1678	570.9
HE 300 M	2380	30310	340	310	21	39	27	59200	19400	139.8	80	3482	1252
W 310×100×23.8	238	3038	305	101	5.6	6.7	8	4280	115.6	118.7	19.5	280.7	22.89
W 310×100×28.3	283	3609	309	102	6	8.9	8	5431	158.1	122.7	20.9	351.5	30.99
W 310×100×32.7	327	4181	313	102	6.6	10.8	8	6507	191.9	124.7	21.4	415.8	37.62
W 310×165×38.7	387	4953	310	165	5.8	9.7	8	8527	726.8	131.2	38.3	550.1	88.1
W 310×165×44.5	445	5691	313	166	6.6	11.2	8	9934	854.7	132.1	38.8	634.8	103
W 310×165×52	520	6678	317	167	7.6	13.2	8	11851	1026	133.2	39.2	747.7	122.9
W 310×200×60	600	7588	303	203	7.5	13.1	15	12900	1830	130	49.1	851	180
W 310×200×67	670	8503	306	204	8.5	14.6	15	14500	2070	131	49.3	948	203

(contd)

Table A.7 (*contd*)

W 310 × 200 × 74	740	9484	310	205	9.4	16.3	15	16500	2340	132	49.7	1060	228
W 310 × 250 × 79	790	10046	306	254	8.8	14.6	15	17700	3990	133	63	1160	314
W 310 × 250 × 86	860	10998	310	254	9.1	16.3	15	19800	4450	134	63.6	1280	350
Column Sections													
W 310 × 310 × 97	970	12330	308	305	9.9	15.4	15	22240	7286	134.3	76.9	1444	477.8
W 310 × 310 × 107	1070	13620	311	306	10.9	17	15	24790	8123	134.9	77.2	1594	530.9
W 310 × 310 × 117	1170	14970	314	307	11.9	18.7	15	27510	9024	135.6	77.6	1753	587.9
W 310 × 310 × 129	1290	16510	318	308	13.1	20.6	15	30770	10040	136.5	78	1935	651.9
W 310 × 310 × 143	1430	18230	323	309	14	22.9	15	34760	11270	138.1	78.6	2153	729.4
W 310 × 310 × 158	1580	20050	327	310	15.5	25.1	15	38630	12470	138.8	78.9	2363	804.8
W 310 × 310 × 179	1790	22770	333	313	18	28.1	15	44530	14380	139.9	79.5	2675	918.7
W 310 × 310 × 202	2020	25800	341	315	20.1	31.8	15	51982	16588	141.9	80.2	3049	1053
W 310 × 310 × 226	2260	28880	348	317	22.1	35.6	15	59560	18930	143.6	81	3423	1194
HE 320 A	976	12440	310	300	9	15.5	27	22930	6985	135.8	74.9	1479	465.7
HE 320 B	1270	16130	320	300	11.5	20.5	27	30820	9239	138.2	75.7	1926	615.9
HE 320 M	2450	31200	359	309	21	40	27	68130	19710	147.8	79.5	3796	1276
(4) Nominal size 350 mm													
HE 340 A	1050	13350	330	300	9.5	16.5	27	27690	7436	144	74.6	1678	495.7
HE 340 B	1340	17090	340	300	12	21.5	27	36660	9690	146.5	75.3	2156	646
HE 340 M	2480	31580	377	309	21	40	27	76370	19710	155.5	79	4052	1276
IPE 360	571	7273	360	170	8	12.7	18	16270	1043	149.5	37.9	903.6	122.8
HE 360 A	1120	14280	350	300	10	17.5	27	33090	7887	152.2	74.3	1891	525.8
HE 360 B	1420	18060	360	300	12.5	22.5	27	43190	10140	154.6	74.9	2400	676.1
HE 360 M	2500	31880	395	308	21	40	27	84870	19520	163.2	78.3	4297	1268

(*contd*)

Table A.7 *(contd)*

Column Sections

W 360×370×134	1340	17060	356	369	11.2	18	20	41510	15080	156	94	2332	817.3
W 360×370×147	1470	18790	360	370	12.3	19.8	20	46290	16720	157	94.3	2572	903.9
W 360×370×162	1620	20630	364	371	13.3	21.8	20	51540	18560	158.1	94.9	2832	1001
W 360×370×179	1790	22830	368	373	15	23.9	20	57440	20680	158.6	95.2	3122	1109
W 360×370×196	1960	25030	372	374	16.4	26.2	20	63630	22860	159.4	95.6	3421	1222
W 360×410×216	2160	27550	375	394	17.3	27.7	20	71140	28250	160.7	101.3	3794	1434
W 360×410×237	2370	30090	380	395	18.9	30.2	20	78780	31040	161.8	101.6	4146	1572
W 360×410×262	2620	33460	387	398	21.1	33.3	20	89410	35020	163.5	102.3	4620	1760
W 360×410×287	2870	36630	393	399	22.6	36.6	20	99710	38780	165	102.9	5074	1944
W 360×410×314	3140	39920	399	401	24.9	39.6	20	110200	42600	166.2	103.3	5525	2125
W 360×410×347	3470	44200	407	404	27.2	43.7	20	124900	48090	168.1	104.3	6140	2380
(5) Nominal size 400 mm													
IPE 400	663	8446	400	180	8.6	13.5	21	23130	1318	165.5	39.5	1156	146.4
HE 400 A	1250	15900	390	300	11	19	27	45070	8564	168.4	73.4	2311	570.9
HE 400 B	1550	19780	400	300	13.5	24	27	57680	10820	170.8	74	2884	721.3
HE 400 M	2560	32580	432	307	21	40	27	104100	19340	178.8	77	4820	1260
W 410×140×38.8	388	4970	399	140	6.4	8.8	11	12620	403.5	159.3	28.5	632.6	57.65
W 410×140×46.1	461	5880	403	140	7	11.2	11	15550	513.6	162.6	29.5	771.9	73.37
W 410×180×53	530	6800	403	177	7.5	10.9	11	18600	1009	165.4	38.5	922.9	114
W 410×180×60	600	7580	407	178	7.7	12.8	11	21570	1205	168.7	39.9	1060	135.4
W 410×180×67	670	8580	410	179	8.8	14.4	11	24530	1379	169.1	40.1	1196	154.1
W 410×180×75	750	9520	413	180	9.7	16	11	27460	1559	169.8	40.5	1330	173.2
W 410×180×85	850	10830	417	181	10.9	18.2	11	31530	1803	170.6	40.8	1512	199.3

(contd)

Table A.7 (*contd*)

Section													
W 410 × 260 × 100	1000	12700	415	260	10	16.9	11	39800	4950	177	62.4	1920	381
W 410 × 260 × 114	1140	14600	420	261	11.6	19.3	11	46200	5720	178	62.6	2200	438
W 410 × 260 × 132	1320	16840	425	263	13.3	22.2	11	53900	6740	179	63.3	2540	513
W 410 × 260 × 149	1490	19030	431	265	14.9	25	11	61900	7770	180	63.9	2870	586
(6) Nominal size 450 mm													
IPE 450	776	9882	450	190	9.4	14.6	21	33740	1676	184.8	41.2	1500	176.4
HE 450 A	1400	17800	440	300	11.5	21	27	63720	9465	189.2	72.9	2896	631
HE 450 B	1710	21800	450	300	14	26	27	79890	11720	191.4	73.3	3551	781.4
HE 450 M	2630	33540	478	307	21	40	27	131500	19340	198	75.9	5501	1260
W 460 × 150 × 52	520	6620	450	152	7.6	10.8	11	21200	634	178.9	30.9	942	83.43
W 460 × 150 × 60	600	7580	455	153	8	13.3	10	25480	796.1	183.3	32.4	1120	104.1
W 460 × 150 × 68	680	8730	459	154	9.1	15.4	10	29680	940.5	184.4	32.8	1293	122.1
W 460 × 190 × 74	740	9460	457	190	9	14.5	10	33260	1661	187.5	41.9	1456	174.8
W 460 × 190 × 82	820	10440	460	191	9.9	16	10	37000	1862	188.3	42.2	1608	195
W 460 × 190 × 89	890	11390	463	192	10.5	17.7	10	40960	2093	189.6	42.9	1769	218
W 460 × 190 × 97	970	12350	466	193	11.4	19	10	44680	2282	190.2	43.1	1917	237.8
W 460 × 190 × 106	1060	13460	469	194	12.6	20.6	10	48790	2515	190.4	43.2	2081	259.2
Column Sections													
W 460 × 280 × 113	1130	14400	463	280	10.8	17.3	18	55600	6335	196.5	66.3	2402	452.5
W 460 × 280 × 128	1280	16360	467	282	12.2	19.6	18	63690	7333	197.3	67	2728	520.1
W 460 × 280 × 144	1440	18410	472	283	13.6	22.1	18	72600	8358	198.6	67.4	3076	590.7
W 460 × 280 × 158	1580	20080	476	284	15	23.9	18	79620	9137	199.1	67.5	3346	643.5
W 460 × 280 × 177	1770	22600	482	286	16.6	26.9	18	91040	10510	200.7	68.2	3777	734.7
W 460 × 280 × 193	1930	24820	489	283	17	30.5	18	103000	11500	204	68.1	4210	813

(*contd*)

Table A.7 *(contd)*

W 460 × 280 × 213	2130	27290	495	285	18.5	33.5	18	115000	13000	205	69	4650	912
W 460 × 280 × 235	2350	30100	501	287	20.6	36.6	18	128000	14500	206	69.4	5110	1010
(7) Nominal size 500 mm													
IPE 500	907	11550	500	200	10.2	16	21	48200	2142	204.3	43.1	1928	214.2
HE 500 A	1550	19750	490	300	12	23	27	86970	10370	209.8	72.4	3550	691.1
HE 500 B	1870	23860	500	300	14.5	28	27	107200	12620	211.9	72.7	4287	841.6
HE 500 M	2700	34430	524	306	21	40	27	161900	19150	216.9	74.6	6180	1252
W 530 × 210 × 92	920	11760	533	209	10.2	15.6	14	55240	2379	216.7	45	2073	227.7
W 530 × 210 × 101	1010	12940	537	210	10.9	17.4	14	61760	2692	218.5	45.6	2300	256.4
W 530 × 210 × 109	1090	13870	539	211	11.6	18.8	14	66730	2951	219.3	46.1	2476	279.7
W 530 × 210 × 123	1230	15690	544	212	13.1	21.2	14	76100	3377	220.2	46.4	2798	318.6
W 530 × 210 × 138	1380	17640	549	214	14.7	23.6	14	86160	3870	221	46.8	3139	361.7
W 530 × 310 × 150	1500	19220	543	312	12.7	20.3	14	101000	10300	229	73.2	3720	660
W 530 × 310 × 165	1650	21090	546	313	14	22.2	14	111000	11400	229	73.5	4070	728
W 530 × 310 × 182	1820	23170	551	315	15.2	24.4	14	124000	12700	231	74	4500	806
W 530 × 310 × 196	1960	25060	554	316	16.5	26.3	14	134000	13900	231	74.5	4840	880
W 530 × 310 × 213	2130	27920	560	318	18.3	29.2	14	151000	15700	233	75	5390	987
W 530 × 310 × 248	2480	31440	571	315	19	34.5	14	178000	18000	238	75.7	6230	1140
(8) Nominal size 550 mm													
IPE 550	1060	13440	550	210	11.1	17.2	24	67120	2668	223.5	44.5	2441	254.1
HE 550 A	1660	21180	540	300	12.5	24	27	111900	10820	229.9	71.5	4146	721.3
HE 550 B	1990	25410	550	300	15	29	27	136700	13080	232	71.7	4971	871.8
HE 550 M	2780	35440	572	306	21	40	27	198000	19160	236.4	73.5	6923	1252

(contd)

Table A.7 (contd)

(9) Nominal size 600 mm												
IPE 600	1220	15600	600	220	12	19	24	92080	3387	243	46.6	3069	307.9
HE 600 A	1780	22650	590	300	13	25	27	141200	11270	249.7	70.5	4787	751.4
HE 600 B	2120	27000	600	300	15.5	30	27	171000	13530	251.7	70.8	5701	902
HE 600 M	2850	36370	620	305	21	40	27	237400	18980	255.5	72.2	7660	1244
W 610 × 230 × 101	1010	12980	603	228	10.5	14.9	14	76470	2950	242.7	47.7	2536	258.8
W 610 × 230 × 113	1130	14440	608	228	11.2	17.3	14	87570	3425	246.2	48.7	2881	300.5
W 610 × 230 × 125	1250	15960	612	229	11.9	19.6	14	98650	3932	248.6	49.6	3224	343.4
W 610 × 230 × 140	1400	17850	617	230	13.1	22.2	14	111990	4514	250.5	50.3	3630	392.5
W 610 × 325 × 155	1550	19730	611	324	12.7	19	14	129000	10780	255.7	73.9	4222	666
W 610 × 325 × 174	1740	22200	616	325	14	21.6	14	147200	12370	257.4	74.6	4778	761
W 610 × 325 × 195	1950	24930	622	327	15.4	24.4	14	167900	14240	259.5	75.6	5398	871
W 610 × 325 × 217	2170	27760	628	328	16.5	27.7	14	190800	16310	262.1	76.7	6076	995
W 610 × 325 × 241	2410	30340	635	329	17.1	31	14	214200	18430	265.7	77.9	6746	1120
W 610 × 325 × 262	2620	33270	641	327	19	34	14	235990	19850	266.3	77.2	7363	1214
W 610 × 325 × 285	2850	36360	647	329	20.6	37.1	14	260700	22060	267.8	77.9	8059	1341
W 610 × 325 × 341	3410	43370	661	333	24.4	43.9	14	318300	27090	270.9	79	9630	1627
W 610 × 320 × 372	3720	47630	669	335	26.4	48	20	355000	30200	273	79.6	10600	1800
(10) Nominal size 650 mm												
HE 650 A	1900	24160	640	300	13.5	26	27	175200	11720	269.3	69.7	5474	781.6
HE 650 B	2250	28630	650	300	16	31	27	210600	13980	271.2	69.9	6480	932.3
HE 650 M	2930	37370	668	305	21	40	27	281700	18980	274.5	71.3	8433	1245
(11) Nominal size 700mm												
HE 700 A	2040	26050	690	300	14.5	27	27	215300	12180	287.5	68.4	6241	811.9
HE 700 B	2410	30640	700	300	17	32	27	256900	14440	289.6	68.7	7340	962.7
HE 700 M	3010	38300	716	304	21	40	27	329300	18800	293.2	70.1	9198	1237

B

Properties of Soils

This appendix presents some information and tables containing properties of soils which will be of interest to the structural designer.

B.1 Soil Tests

For low-rise buildings, depth of borings may be specified to be about 6 m below the anticipated foundation level, with at least one boring continuing deeper, to a lesser of 30 m, the least building dimension, or refusal. At least one soil boring should be specified for every 230 square metres of the building area for buildings over 12 m height, or having more than three storeys. For large buildings founded on poor soils, borings should be spaced at less than 15 m intervals. A minimum of five borings, one at the centre and the rest at the corners of the building, is recommended.

B.2 Order of Soil Suitability for Foundation Support

Best	:	Bed rock
Very good	:	Sand and gravel
Good	:	Medium to hard clay (that is kept dry)
Poor	:	Silts and soft clay
Undesirable	:	Organic silts and organic clay
Unsuitable	:	Peat

B.3 The Plasticity Index (PI)

The plasticity index (PI) of the soil provides an indication of how much clay will shrink or swell. The higher the PI, the greater is the shrink-swell potential.

PI of 0–15%	:	Low expansion potential
PI of 15–25%	:	Medium expansion potential
PI of 25% and above	:	High expansion potential

Table B.1 Typical mass densities of basic soil types

Type of Soil	Mass density ρ (Mg/m³)*			
	Poorly graded soil		Well-graded soil	
	Range	Typical value	Range	Typical value
Loose sand	1.70–1.90	1.75	1.75–2.00	1.85
Dense sand	1.90–2.10	2.00	2.00–2.20	2.10
Soft clay	1.60–1.90	1.75	1.60–1.90	1.75
Stiff clay	1.90–2.25	2.07	1.90–2.25	2.07
Silty soils	1.60–2.00	1.75	1.60–2.00	1.75
Gravelly soils	1.90–2.25	2.07	2.00–2.30	2.15

*Values are representative of moist sand, gravel, saturated silt, and clay.

Table B.2 Typical values of modulus of elasticity (E_s) for different types of soils

Type of Soil	E_s (N/mm²)
Clay	
Very soft	2–15
Soft	5–25
Medium	15–50
Hard	50–100
Sandy	25–250
Glacial till	
Loose	10–153
Dense	144–720
Very dense	478–1,440
Loess	14–57
Sand	
Silty	7–21
Loose	10–24
Dense	48–81
Sand and gravel	
Loose	48–148
Dense	96–192
Shale	144–14,400
Silt	2–20

Table B.3 Typical values of modulus of subgrade reaction (k_s) for different types of soils

Type of Soil	k_s (kN/m³)
Loose sand	4,800–16,000
Medium dense sand	9,600–80,000
Dense sand	64,000–1,28,000
Clayey medium dense sand	32,000–80,000
Silty medium dense sand	24,000–48,000

(contd)

(contd)

Clayey soil:	
$q_u \leq 200$ kN/m^2	12,000–24,000
$200 < q_u \leq 400$ kN/m^2	24,000–48,000
$q_u > 800$ kN/m^2	> 48,000
q_u – Safe bearing capacity	

Table B.4 Typical values of Poisson's ratio (μ) for soils

Type of soil	μ
Clay (saturated)	0.4 – 0.5
Clay (unsaturated)	0.1 – 0.3
Sandy clay	0.2 – 0.3
Silt	0.3 – 0.35
Sand (dense)	0.2 – 0.4
Course (void ratio = 0.4 – 0.7)	0.15
Fine grained (void ratio = 0.4 – 0.7)	0.25
Rock	0.1–0.4 (depends on type of rock)
Loess	0.1 – 0.3
Ice	0.36
Concrete	0.15

Table B.5 Allowable bearing pressures on soils (for preliminary design)

Type of rock/soil		Allowable bearing pressure (kN/m^2)	Standard penetration blow count (N)	Apparent cohesion c_u (kPa)
Hard rock without lamination and defects (e.g., granite, trap, and diorite)		3,200	>30	—
Laminated rocks (e.g., sandstone and lime-stone in sound condition)		1,600	>30	—
Soft or broken rock, hard shale, cemented material		900	30	—
Soft rock		450		
Gravel	Dense	450	>30	—
	Medium	96–285	>30	—
		Compact and dry	Loose and dry	
Sand*	Coarse	450	250	30–50
	Medium	250	48–120	15–30
	Fine or silt	150	100	<15
Clay+	Very stiff	190–450	15–30	100–200
	Medium stiff	200–250	4–15	25–100
	Soft	50–100	0–4	0–25
Peat, silts, made-up ground	To be determined after investigation			

Notes: * Reduce bearing pressures by half below the water table.

    + Alternatively, allow 1.2 times c_u for round and square footings, and 1.0 times c_u for length/width ratios of more than 4.0. Interpolate for intermediate values.

Table B.6 Typical interface friction angles (NAVFAC 1982)

	Interface materials	Interface friction angle δ
Mass concrete against	Clean sound rock	25
	Clean gravel, gravel-sand mixtures, coarse sand	29 – 31
	Clean fine to medium sand, silty medium to coarse sand, silty or clayey gravel	24 – 29
	Clean fine sand, silty or clayey fine to medium sand	19 – 24
	Fine sandy silt, nonplastic silt	17 – 19
	Medium-stiff, stiff and silty clay	17 – 19
Formed concrete against	Clean gravel, gravel-sand mixture, well-graded rock fill with spalls	22 – 26
	Clean gravel, silty sand-gravel mixture, single-size hard rock fill	17 – 22
	Silty sand, gravel, or sand mixed with silt, or clay	17
	Fine sandy silt, non-plastic silt	14
Steel sheet piles against	Clean gravel, gravel-sand mixture, well-graded rock fill with spalls	22
	Clean sand, silty sand-gravel mixture, single-size hard rock fill	17
	Silty sand, gravel, or sand mixed with silt or clay	14
	Fine sandy silt, nonplastic silt	11

Table B.7 Typical values of fundamental period for soil deposits (for rock motions with $a_{max} = 0.4g$) (SEAOC 1980)

Soil depth (m)	Dense sand (s)	5 m of fill over normally consolidated clay* (s)
10	0.3–0.5	0.5–1.0
30	0.6–1.2	1.5–2.3
60	1.0–1.8	1.8–2.8
90	1.5–2.3	2.0–3.0
150	2.0–3.5	—

*Representative of San Francisco bay area.

Table B.8 Mean shear wave velocities (m/s) for the top 30 m of ground (Borcherdt 1994)

General description	Mean shear–wave velocity		
	Minimum	*Average*	*Maximum*
Firm and hard rocks			
Hard rocks			
(e.g., metamorphic rocks with very widely spaced fractures)	1400	1620	—

(contd)

(contd)

Firm to hard rocks			
(e.g., granites, igneous rocks, conglomerates, sandstones, and shales with close to widely spaced fractures)	700	1050	1400
Gravelly soils and soft to firm rocks			
(e.g., soft igneous sedimentary rocks, sandstones, shales, gravels, and soils with > 20% gravel)	375	540	700
Stiff clays and sandy soils			
(e.g., loose to very dense sands, silt loams, sandy clays, and medium stiff to hard clays and silty clays (N > 5 blows/300mm)	200	290	375
Soft soils			
(e.g., loose submerged fills and very soft (N < 5 blows/300 mm) clays and silty clays < 37 m thick)	100	150	200
Very soft soils			
(e.g., loose saturated sand, marshland, recent reclamation)	50	75	100

Note: The fundamental time period T of soil layer of thickness H, having average shear wave velocity V_s is approximately

$$T = 4H/V_s$$

If we assume the weighted average shear wave velocity for 30 – 50 m soil layer as 290 m/s, then the fundamental period of soil layer will range from 0.41 to 0.69 second. The fundamental time period of 4 – 6 storey buildings, including the soil-structure interaction, should fall in the above range of time period of soil layers, i.e., 0.41 – 0.69 sec. That is, the seismic waves in this range of time period will be allowed only to pass and filter-out the other frequencies. Therefore, there will be quasi resonance of building and the soil layer. At this point the damaging energy from the seismic waves get into the buildings having similar time period of vibration as the soil layer. If the seismic damaging energy getting into the building is more than the capacity of the structure, then the building will show distress and may collapse.

Similarly, if we assume that the weighted average shear wave velocity for 150 – 300 m soil layer is around 500 m/s, then the fundamental time period will range from 1.2 – 2.4 s. The fundamental time period of 10 – 15 storey building, including soil structure interaction, will fall in the above range of time period of vibrations. Therefore, there will be quasi resonance of the buildings and the soil layer and the seismic waves will affect this group of buildings which will result in damage/collapse of buildings. Hence, it is important to know the depth of soil layers above the bedrock and its properties such as the shear wave velocities, which are related in the microzonation of a region.

C

Computer Programs

Introduction

With the advent of computers and software packages, it is now possible to analyse any structure for any geometry and given loading conditions quickly. Though several standard software packages are available for the linear, non-linear, buckling or dynamic analysis of structures, very few design packages are available. Nonetheless, some analysis software packages do have some design routines, which are based on specific codes. Two computer programs written in Visual Basic are provided in this appendix, which can be run on any standard personal computer under the Windows environment. Visual Basic has been chosen because it uses an interactive approach to the development of computer programs. Visual Basic interprets the code as we enter it, catching and highlighting syntax or spelling errors. It also partly compiles the code as it is entered. While compiling the whole code, if the compiler finds an error, it is highlighted in the code. We can fix the error and continue compiling without having to start all over. Visual Basic also provides a variety of user interfaces in the form of text boxes, labels, radio button, pictures, etc. Visual Basic version 6.0 is used in this example, because it is simple and popular though VB.NET is the most recent version. More details about Visual Basic may be found in the Microsoft web site http://msdn.microsoft.com/library/default.asp?url=/library/en-us/vbcon98/html/vbconpart1visualbasicbasics.asp.

C.1 Design Programs

When writing design programs, it is much easier if the code of practice follows a logical sequence of requirements and checks. Also, the design stresses and factors should be given in closed-form mathematical expressions. Fortunately, the current version of IS: 800, satisfies the above requirements.

C.1.1 Beam Design Program

The simply supported steel beam design program, given in this section, designs beams of compact, semi-compact, and plastic sections. The beam arrangement is shown in Fig. C.1 and the flow chart is shown in Fig. C.2.

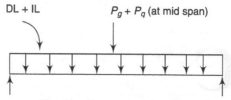

DL + IL $P_g + P_q$ (at mid span)

Fig. C.1 Beam and loading

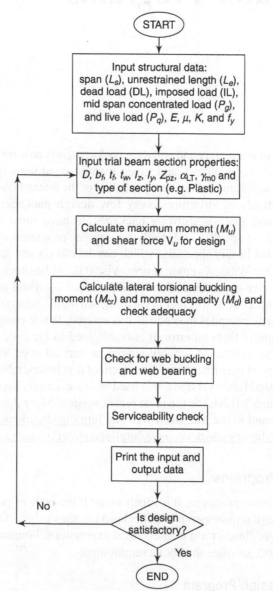

START

Input structural data:
span (L_s), unrestrained length (L_e),
dead load (DL), imposed load (IL),
mid span concentrated load (P_g),
and live load (P_q), E, μ, K, and f_y

Input trial beam section properties:
D, b_f, t_f, t_w, I_z, I_y, Z_{pz}, α_{LT}, γ_{m0} and
type of section (e.g. Plastic)

Calculate maximum moment (M_u)
and shear force V_u for design

Calculate lateral torsional buckling
moment (M_{cr}) and moment capacity (M_d) and
check adequacy

Check for web buckling
and web bearing

Serviceability check

Print the input and
output data

No

Is design
satisfactory?

Yes

END

Fig. C.2 Flowchart of beam design program

Only simple loading has been considered to reduce the length of the program (the user may modify it to suit his/her requirements). The notations used are the same as those given in IS: 800 (Section 8), and hence the program is easy to understand. The typical input screens of this program are given in Figs C.3 and C.4.

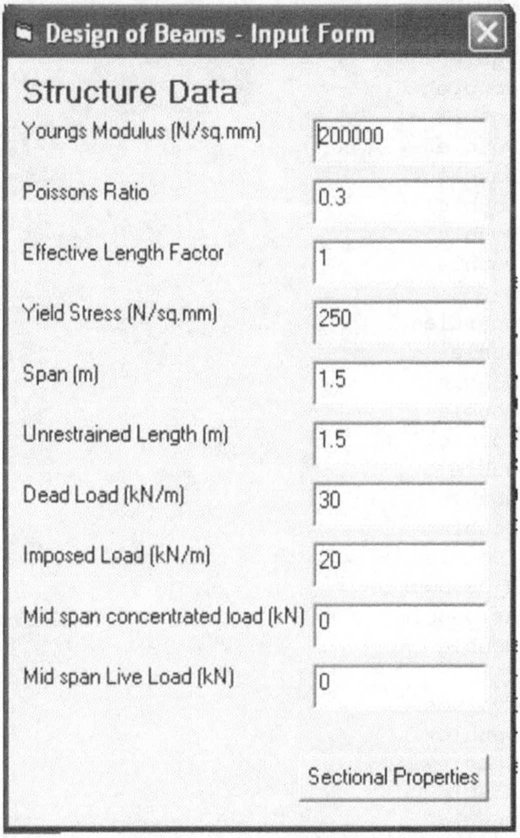

Fig. C.3 Input screen for structure data of beam design program

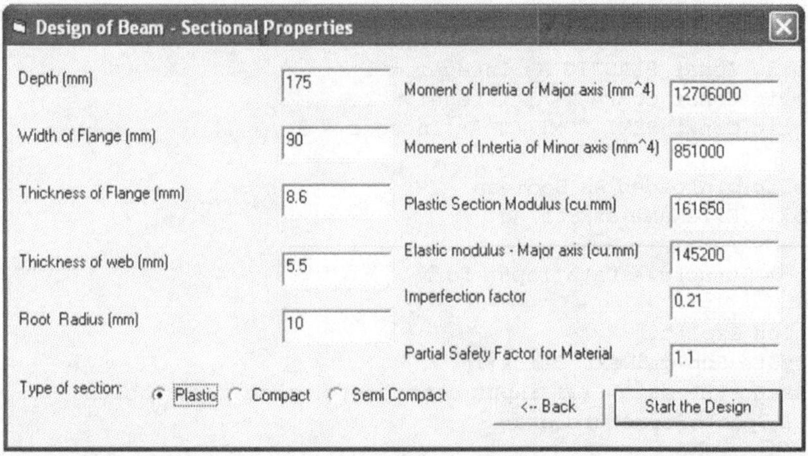

Fig. C.4 Input screen for sectional properties of beam design program

The following program is written in Visual Basic for beam design.

```
'Global Module

'Structure properties
Public E As Double
Public EMu As Double
Public K As Double
Public Fy As Double
Public LS As Double
Public Le As Double
Public DL As Double
Public IL As Double
Public PG As Double
Public PQ As Double

'Sectional Properties
Public D As Double
Public Bf As Double
Public tf As Double
Public tw As Double
Public R As Double
Public Iz As Double
Public Iy As Double
Public Zpz As Double
Public alphaLT As Double
Public GamaM As Double
Public Zz As Double

'intermediate values
Public Wu As Double
Public Vu As Double
Public Pu As Double
Public Mu As Double

Public Const PI As Double = 3.14159
Public sectionType As Integer

Public Const PLASTIC As Integer = 1
Public Const COMPACT As Integer = 2
Public Const SEMI_COMPACT As Integer = 3

Public blnLoaded As Boolean
Public FileName As String
```

```
'From Structure Data input form

Option Explicit
Private Sub cmdNext_Click()
'assign the data from input controls to global variables
Call UpdateStructureData
```

```
Unload Me
'Show the next input screen
frmSectionalProperties.Show
End Sub

Private Sub UpdateStructureData()
' transfer the data to global variables
E = Val(txtE)
EMu = Val(txtMu)
K = Val(txtK)
Fy = Val(txtFy)
LS = Val(txtSpan)
Le = Val(txtLE)
DL = Val(txtDeadLoad)
IL = Val(txtIL)
PG = Val(txtPG)
PQ = Val(txtPQ)

End Sub
— — — — — — — — — — — — — —

'From sectional properties input form, which also has design code
Option Explicit
Dim G As Double
Dim It As Double
Dim hf As Double
Dim Iw As Double
Dim Betaf As Double
Dim Mcr As Double
Dim LamdaLT As Double
Dim LamdaT1 As Double
Dim alphaLT As Double
Dim PhiLT As Double
Dim ChiLT As Double
Dim fbd As Double

Dim Md As Double
Dim Vd As Double
Dim Deltab As Double
Dim DeltaAll As Double

Dim b1 As Double
Dim n1 As Double
Dim Ab As Double
Dim I As Double
Dim A As Double
Dim rmin As Double
Dim Lamda As Double
Dim LamdaEff As Double
Dim Phi1 As Double
Dim Phi2 As Double
Dim fcd As Double
Dim SWB As Double
```

```
Dim n2 As Double
Dim Fw As Double

Dim blnSafe As Boolean

Private Sub cmdBack_Click()
    'Update the input controls with user entered Structure data
    Call LoadStructureData
    'Transfer user input sectional properties data to global variables
    Call UpdateSectionalProperties
    Unload Me

    frmBeamDesign1.Show

End Sub
Private Sub LoadStructureData()
    frmBeamDesign1.txtE = E
    frmBeamDesign1.txtMu = EMu
    frmBeamDesign1.txtK = Str(K)
    frmBeamDesign1.txtFy = Fy
    frmBeamDesign1.txtSpan = LS
    frmBeamDesign1.txtLE = Le
    frmBeamDesign1.txtDeadLoad = DL
    frmBeamDesign1.txtIL = IL
    frmBeamDesign1.txtPG = PG
    frmBeamDesign1.txtPQ = PQ
End Sub

Private Sub cmdDesign_Click()
'start the design process
'open the output file to print the results
    FileName = App.Path + "\BeamDesign.out"
    Open FileName For Output As 1
'pressume the design is safe
    blnSafe = True
    Call UpdateSectionalProperties
    Call Design
End Sub

Private Sub LoadSectionalProperties()
'if the data has been already transfered to the global variables
'then load them to input controls
'this check is necessary to sweap the default values with null
values
If blnLoaded Then
    frmSectionalProperties.txtDepth = Str(D)
    frmSectionalProperties.txtWidthOfFlange = Str(Bf)
    frmSectionalProperties.txtThickOfFlange = Str(tf)
    frmSectionalProperties.txtThickWeb = Str(tw)
    frmSectionalProperties.txtR = Str(R)
    frmSectionalProperties.txtMIMajor = Str(Iz)
    frmSectionalProperties.txtMIMinor = Str(Iy)
```

```
    frmSectionalProperties.txtPlastic = Str(Zpz)
    frmSectionalProperties.txtImpFactor = Str(alphaLT)
    frmSectionalProperties.txtSafetyFactor = Str(GamaM)
    frmSectionalProperties.txtZZ = Str(Zz)
    If sectionType = PLASTIC Then
       optPlastic = True
    ElseIf sectionType = COMPACT Then
       optCompact = True
    ElseIf sectionType = SEMI_COMPACT Then
       optSemiCompact = True
    End If
End If
End Sub

Private Sub UpdateSectionalProperties()
   D = Val(txtDepth)
   Bf = Val(txtWidthOfFlange)
   tf = Val(txtThickOfFlange)
   tw = Val(txtThickWeb)
   R = Val(txtR)
   Iz = Val(txtMIMajor)
   Iy = Val(txtMIMinor)
   Zpz = Val(txtPlastic)
   alphaLT = Val(txtImpFactor)
   GamaM = Val(txtSafetyFactor)
   Zz = Val(txtZZ)
   If optPlastic Then
      sectionType = PLASTIC
   ElseIf optCompact Then
      sectionType = COMPACT
   ElseIf optSemiCompact Then
      sectionType = SEMI_COMPACT
   End If
End Sub

Private Sub Form_Load()
   Call LoadSectionalProperties
   blnLoaded = True
End Sub

Private Sub Design()
   ' Print the input data into the output file
   Call PrintInput
   Call CalculateStdValues
   Call CalculateLTBucklingMomentCapacity
   ' continue the process only if so far the design is safe
   If blnSafe Then Call CalculateShearCapacity
   If blnSafe Then Call CalculateDeflection
   If blnSafe Then Call CheckForWebBuckling
   If blnSafe Then Call CheckForWebBearing
   Close #1
   If blnSafe Then End
End Sub
```

```
Private Sub CalculateStdValues()
   ' calculate maximum Mu and Vu for design
   Dim strMsg As String

   Wu = 1.5 * (DL + IL)
   Pu = 1.5 * (PG + PQ)
   Mu = Wu * LS * LS / 8# + Pu * LS / 4#
   Vu = (Wu * LS + Pu) / 2

   Print #1, Tab(5); "Output"
   Print #1, Tab(5); "----"

   Print #1, Tab(5); "Maximum bending moment for design (kNm)";
   Tab(60); Format(Mu, "#######0.0##")
   Print #1, Tab(5); "Maximum shear force for design (kN)"; Tab(60);
   Format(Vu, "######0.0##")
End Sub

Private Sub CalculateLTBucklingMomentCapacity()
Dim Betab As Double
Dim strMsg As String

   If sectionType = SEMI_COMPACT Then
      Betab = Zz / Zpz
   Else
      Betab = 1
   End If

   Betaf = 0.5
   G = E / (2 * (1 + EMu))
   hf = D - tf
   It = (2 * Bf * tf ^ 3 + hf * tw ^ 3) / 3
   Le = K * Le
   Iw = (1 - Betaf) * Betaf * Iy * hf ^ 2
   Mcr = (((PI ^ 2 * E * Iy) / (Le * 1000) ^ 2) * (G * It + (PI ^ 2
   * E * Iw) / (Le * 1000) ^ 2)) ^ 0.5
   LamdaLT = (Betab * Zpz * Fy / Mcr) ^ 0.5
   LamdaLT1 = (1.2 * Zz * Fy / Mcr) ^ 0.5
   If LamdaLT > LamdaLT1 then
        LamdaLT = LamdaLT1
   End If
   alphaLT = 0.21
   PhiLT = 0.5 * (1 + alphaLT * (LamdaLT - 0.2) + LamdaLT ^ 2)
   ChiLT = 1# / (PhiLT + (PhiLT ^ 2 - LamdaLT ^ 2) ^ 0.5)
   If (ChiLT > 1#) Then
      ChiLT = 1#
   End If
   fbd = ChiLT * Fy / GamaM


   Md = Betab * Zpz * fbd / 1000000#
   strMsg = "Torsional constant " + Str(It) + " mm^4" + vbCrLf
   strMsg = strMsg + "Warping constant " + Str(Iw) + " mm^6" + vbCrLf
   strMsg = strMsg + "Moment capacity of the section " + Str(Md) +
   " kNm" + vbCrLf
```

```
If Md < Mu Then
  MsgBox strMsg + "Moment capacity is less than " + Str(Mu) + "
  Section is unsafe. Revise the section", , "Unsafe section"
  Unload Me
  blnSafe = False
  frmBeamDesign1.Show
  Exit Sub
End If
Print #1, Tab(5); "Torsional constant (mm^4) "; Tab(60); Format(It,
"#######0.0##")
Print #1, Tab(5); "Warping constant (mm^6)"; Tab(60); Format(Iw,
"#######0.0##")
Print #1, Tab(5); "Moment capacity of the section (kNm)"; Tab(60);
Format(Md, "#######0.0##")
Print #1, Tab(5); "Moment capacity of Trial section is adequate"
Print #1, ""

End Sub

Private Sub CalculateShearCapacity()
  Dim strMsg As String

  Vd = ((Fy / (GamaM * Sqr(3))) * D * tw) / 1000
  strMsg = "Shear capacity = " + Str(0.6 * Vd) + " kN" + vbCrLf +
  "Shear force = " + Str(Vu) + " kN" + vbCrLf
  If 0.6 * Vd < Vu Then
    MsgBox strMsg + "Shear capacity is unsafe. Revise the section"
    Unload Me
    blnSafe = False
    frmBeamDesign1.Show
    Exit Sub
  Else
    Print #1, Tab(5); "Shear capacity (kN) "; Tab(60); Format(0.6
    * Vd, "#######0.0##")
    Print #1, Tab(5); "Shear force (kN) "; Tab(60); Format(Vu,
    "#######0.0##")
    Print #1, Tab(5); "Shear capacity of Trial section is adequate"
  End If

End Sub
Private Sub CalculateDeflection()

  Dim strMsg As String

  Deltab = (5 * (DL + IL) * (LS * 1000) ^ 4 / 384 + (PG + PQ) * LS
  ^ 3 / 48) / (E * Iz)
  DeltaAll = LS * 1000 / 300

  strMsg = "Delta allowable = " + Str(DeltaAll) + " mm" + vbCrLf +
  "Delta actual = " + Str(Deltab) + " mm" + vbCrLf
  If Deltab < DeltaAll Then
    'MsgBox strMsg + "Trial section is adequate for deflection
    check"
```

```
            Print #1, Tab(5); "Delta allowable (mm) "; Tab(60);
            Format(DeltaAll, "#######0.0##")
            Print #1, Tab(5); "Delta actual (mm) "; Tab(60); Format(Deltab,
            "#######0.0##")
            Print #1, Tab(5); "Trial section is adequate for deflection
            check"
            Print #1, ""
        Else
            MsgBox strMsg + "Revise the section for deflection consideration"
            Unload Me
            blnSafe = False
            frmBeamDesign1.Show
            Exit Sub
        End If
End Sub
Private Sub CheckForWebBuckling()
    Dim strMsg As String

    b1 = (Bf - tw) / 2
    n1 = D / 2
    Ab = (b1 + n1) * tw
    I = b1 * tw ^ 3 / 12
    A = b1 * tw
    rmin = Sqr(I / A)
    Lamda = 0.7 * (D - 2 * (tf + R)) / rmin
    LamdaEff = (Fy * Lamda ^ 2 / (PI ^ 2 * E)) ^ 0.5
    Phi1 = 0.5 * (1 + 0.49 * (LamdaEff - 0.2) + LamdaEff ^ 2)
    Phi2 = Phi1 + (Phi1 ^ 2 - LamdaEff ^ 2) ^ 0.5
    fcd = Fy / (Phi2 * GamaM)
    SWB = fcd * Ab / 1000

    strMsg = "Strength against web buckling = " + Str(SWB) + " kN" +
    vbCrLf + "Shear force = " + Str(Vu) + " kN"
    If SWB > Vu Then
        'MsgBox strMsg + "Safe against web buckling"
        Print #1, Tab(5); "Strength against web buckling (kN) ";
        Tab(60); Format(SWB, "#######0.0##")
        Print #1, Tab(5); "Shear force (kN) "; Tab(60); Format(Vu,
        "#######0.0##")
        Print #1, Tab(5); "Safe against web buckling"
    Else
        MsgBox strMsg + "Revise the section for web buckling"
        Unload Me
        blnSafe = False
        frmBeamDesign1.Show
        Exit Sub
    End If

End Sub
Private Sub CheckForWebBearing()
    Dim strMsg As String

    n2 = 2.5 * (tf + R)
    Fw = (b1 + n2) * tw * Fy / (GamaM * 1000)
```

```
   strMsg = "Strength against web bearing = " + Str(Fw) + " kN" +
   vbCrLf + "Shear force = " + Str(Vu) + " kN" + vbCrLf
   If Fw > Vu Then
      'MsgBox strMsg + "Safe against web bearing"
      MsgBox "Design is safe and " + vbCrLf + "The output is saved as
      BeamDesign.out in the working folder"
      Print #1, Tab(5); "Strength against web bearing (kN) "; Tab(60);
      Format(Fw, "#######0.0##")
      Print #1, Tab(5); "Shear force (kN) "; Tab(60); Format(Vu,
      "#######0.0##")
      Print #1, Tab(5); "Safe against web bearing"
      Print #1, ""
   Else
      MsgBox strMsg + "Revise the section for web bearing"
      Unload Me
      frmBeamDesign1.Show
      Exit Sub
   End If
End Sub

Private Sub optCompact_Click()
   If optCompact Then
   sectionType = COMPACT
   End If

End Sub

Private Sub optPlastic_Click()
   If optPlastic Then
      sectionType = PLASTIC
   End If
End Sub

Private Sub optSemiCompact_Click()
   If optSemiCompact Then
      sectionType = SEMI_COMPACT
   End If
End Sub
Private Sub PrintInput()

   Print #1, Tab(10); "Beam Design"
   Print #1, Tab(10); "------"
   Print #1, ""
   Print #1, Tab(5); "Structure Data"
   Print #1, Tab(5); "----------"

   Print #1, Tab(5); "Youngs Modulus (N/sq.mm) "; Tab(60); Format(E,
   "#######0.0##")
   Print #1, Tab(5); "Poisson's Ratio "; Tab(60); Format(EMu,
   "#######0.0##")
   Print #1, Tab(5); "Eff.Length factor "; Tab(60); Format(K,
   "#######0.0##")
```

```
      Print #1, Tab(5); "Yield Stress (N/sq.mm) "; Tab(60); Format(Fy,
      "#######0.0##")
      Print #1, Tab(5); "Span (m) "; Tab(60); Format(LS, "#######0.0##")
      Print #1, Tab(5); "Unrestrained length (m) "; Tab(60); Format(Le,
      "#######0.0##")
      Print #1, Tab(5); "Dead load (kN/m) "; Tab(60); Format(DL,
      "#######0.0##")
      Print #1, Tab(5); "Imposed load (kN/m) "; Tab(60); Format(IL,
      "#######0.0##")
      Print #1, Tab(5); "Mid span concentrated load (kN) "; Tab(60);
      Format(PG, "#######0.0##")
      Print #1, Tab(5); "Mid span live load (kN) "; Tab(60); Format(PQ,
      "#######0.0##")

      Print #1, Tab(5); ""
      Print #1, Tab(5); "Sectional Properties"
      Print #1, Tab(5); "------------"
      Print #1, Tab(5); "Depth (mm)"; Tab(60); Format(D, "#######0.0##")
      Print #1, Tab(5); "Width of flange (mm)"; Tab(60); Format(Bf,
      "#######0.0##")
      Print #1, Tab(5); "Thickness of flange (mm)"; Tab(60); Format(tf,
      "#######0.0##")
      Print #1, Tab(5); "Thickness of web (mm)"; Tab(60); Format(tw,
      "#######0.0##")
      Print #1, Tab(5); "Root radius (mm)"; Tab(60); Format(R,
      "#######0.0##")
      Print #1, Tab(5); "Moment of inertia of Major axis (mm^4)";
      Tab(60); Format(Iz, "#######0.0##")
      Print #1, Tab(5); "Moment of inertia of Minor axis (mm^4)";
      Tab(60); Format(Iy, "#######0.0##")
      Print #1, Tab(5); "Palstic section modulus (cu.mm)"; Tab(60);
      Format(Zpz, "#######0.0##")
      Print #1, Tab(5); "Elastic modulus about major axis (cu.mm)";
      Tab(60); Format(Zz, "#######0.0##")
      Print #1, Tab(5); "Imperfection factor "; Tab(60); Format(alphaLT,
      "#######0.0##")
      Print #1, Tab(5); "Partial safety factor for material "; Tab(60);
      Format(GamaM, "#######0.0##")
      Print #1, ""
End Sub
```

In order to show the use of this program, the following data was used and the resulting output is as follows:

```
      Beam Design
      -----------


Structure Data
--------------
Youngs Modulus <N/sq. mm>                          200000.0
Poisson's Ratio                                    0.3
Eff.Length factor                                  1.0
```

```
Yield Stress <N/sq.mm>                                      250.0
Span <m>                                                      1.5
Unrestrained length <m>                                       1.5
Dead load <kN/m>                                            30.0
Imposed load <kN/m>                                         20.0
Mid span concentrated load <kN>                              0.0
Mid span live load <kN>                                      0.0

Sectional Properties
--------------------
Depth <mm>                                                  175.0
Width of flange <mm>                                         90.0
Thickness of flange <mm>                                      8.6
Thickness of web <mm>                                         5.5
Root radius <mm>                                            10.0
Moment of inertia of Major axis <mm^4>                  12706000.0
Moment of inertia of Minor axis <mm^4>                    851000.0
Palstic section modulus <cu.mm>                           161650.0
Elastic modulus about major axis <cu.mm>                 145200.0
Imperfection factor                                          0.21
Partial safety factor for material                           1.1


Output
-----
Maximum bending moment for design <kNm>                      21.094
Maximum shear force for design <kN>                          56.25
Torsional constant <mm^4>                                 47391.627
Warping constant <mm^6>                                5890826240.0
Moment capacity of the section <kNm>                         31.044
Moment capacity of Trial section is adequate

Shear capacity <kN>                                          75.777
Shear force <kN>                                             56.25
Shear capacity of Trial section is adequate
Delta allowable <mm>                                          5.0
Delta actual <mm>                                             1.297
Trial section is adequate for deflection check

Strength against web buckling<kN>                          119.165
Shear force <kN>                                             56.25
Safe against web buckling
Strength against web bearing <kN>                          110.938
Shear force <kN>                                             56.25
Safe against web bearing
```

C.1.2 Beam-Column Design Program

The design of a beam-column, using a hot-rolled I-section, is considered in the program given in this section. In all cases, axial compression is taken as positive. The sign convention for the moments at the top and bottom ends of the column is positive for moments applied clockwise and negative for moments applied anti-clockwise.

Hot rolled I- or H- plastic, compact, or semi-compact sections can be designed using this program. There are two sets of input data in this program, as in the case of the beam program. The first is the structural data consisting of the following: factored axial force (N), factored bending moment at the top and bottom about the major axis (M_{Z1}, M_{Z2}), factored bending moment at the top and bottom about the minor axis (M_{Y1}, M_{Y2}), length in major and minor axis (L_z, L_y), effective length factors (K_z, K_y), Young's modulus (E), and Poisson's ratio (μ). The second set of data is the member data consisting of the following: depth of the section (D), area (A), breadth of flange (b_f), thickness of flange and web (t_f and t_w), root radius (R), radius of gyration in the major and minor axis (r_z, r_y), moment of inertia about minor axis (I_y), elastic modulus about major and minor axis (Z_z and Z_y), yield strength of material (f_y) and partial factor of safety of material (γ_{m0}) and type of section (plastic, compact, or semi-compact).

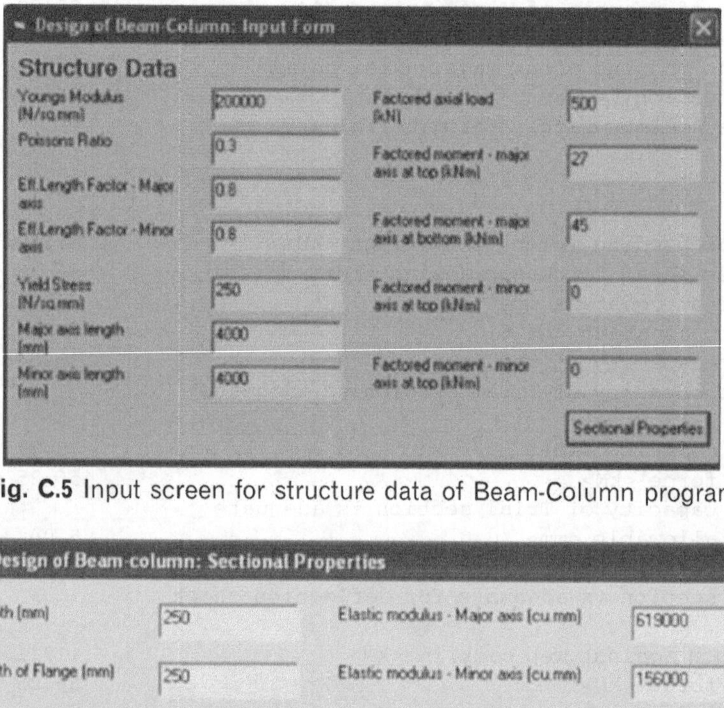

Fig. C.5 Input screen for structure data of Beam-Column program

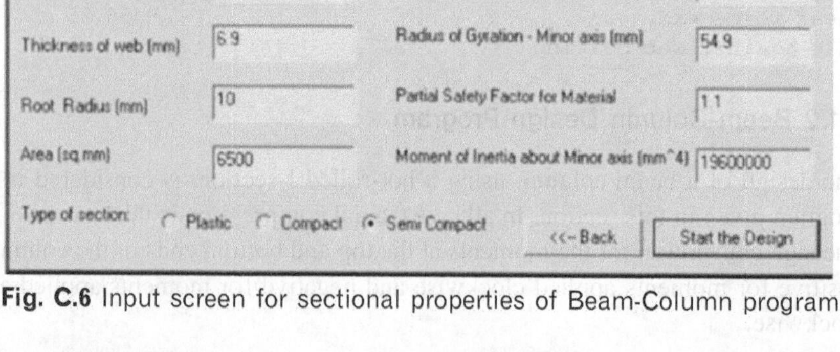

Fig. C.6 Input screen for sectional properties of Beam-Column program

Based on the preceding data, the plastic modulus about the major and minor axis (Z_{pz}, Z_{py}) and the warping and torsional rigidity (I_w, I_t) are calculated. The member buckling resistance in compression about the major axis and minor axis (P_{dz} and P_{dy}) are calculated based on clause 7.1.2 of the code. The member buckling resistance in bending is computed based on clause 8.2.2, about both the axes (M_{dz} and M_{dy}).

Using these values, the strength of the cross section is checked based on the interaction equations given in clause 9.3.1 of the code. Finally, the overall member strength is checked as per the interaction equation given in clause 9.3.2 of the code. If any one of the checks is not satisfied, the user has to change the trial section properties and run the program again. The program written in Visual Basic and typical input screens are given below:

```
Option Explicit
'Structure properties
Public E As Double
Public EMu As Double
Public Fy As Double
Public Lz As Double
Public Ly As Double
Public Kz As Double
Public Ky As Double
Public N As Double
Public Mz1 As Double
Public Mz2 As Double
Public My1 As Double
Public My2 As Double

'Sectional Properties
Public D As Double
Public Bf As Double
Public tf As Double
Public tw As Double
Public R As Double
Public rz As Double
Public ry As Double
Public Zz As Double
Public Zy As Double
Public Zpz As Double
Public Zpy As Double
Public GamaM As Double
Public A As Double
Public Iy As Double
Public Pd As Double

Public sectionType As Integer

Public Const PLASTIC As Integer = 1
Public Const COMPACT As Integer = 2
```

```
Public Const SEMI_COMPACT As Integer = 3
Public Const PI As Double = 3.14159

Public blnLoaded As Boolean
Public FileName As String
```

```
Option Explicit
Private Sub cmdNext_Click()
  Call UpdateStructureData
  Unload Me
  frmSectionalProperties.Show
End Sub

Private Sub UpdateStructureData()
  E = Val(txtE)
  EMu = Val(txtMu)
  Kz = Val(txtKz)
  Ky = Val(txtKy)
  Fy = Val(txtFy)
  Ly = Val(txtLY)
  Lz = Val(txtLZ)
  My1 = Val(txtMY1)
  My2 = Val(txtMY2)
  Mz1 = Val(txtMZ1)
  Mz2 = Val(txtMZ2)
  N = Val(txtN)
End Sub
```

```
Option Explicit
Dim G As Double
Dim It As Double
Dim hf As Double
Dim Iw As Double
Dim Betaf As Double

Dim Mcr As Double
Dim LambdaLT As Double
Dim alphaLT As Double
Dim PhiLT As Double
Dim ChiLT As Double
Dim fbd As Double

Dim Md As Double
Dim Vd As Double
Dim Deltab As Double
Dim DeltaAll As Double

Dim b1 As Double
Dim n1 As Double
Dim Ab As Double
Dim I As Double
Dim A As Double
```

```
Dim rmin As Double
Dim Lamda As Double
Dim LamdaEff As Double
Dim Phi1 As Double
Dim Phi2 As Double
Dim fcd As Double
Dim SWB As Double
Dim n2 As Double
Dim Fw As Double
Dim Mz As Double
Dim Mzm As Double
Dim My As Double
Dim Mym As Double
Dim Nd As Double
Dim Mdz, Mdz1 As Double
Dim Mdy As Double
Dim StressRatio As Double
Dim StressRatio1 As Double
Dim Lrz As Double
Dim Lry As Double
Dim AlphaZ As Double
Dim AlphaY As Double
Dim Betab As Double
Dim LambdaZ As Double
Dim LambdaY As Double
Dim Pdz As Double
Dim Pdy As Double
Dim blnSafe As Boolean

Private Sub cmdBack_Click()
   Call UpdateSectionalProperties
   Call LoadStructureData
   Unload Me
   frmBeamColumnDesign1.Show
End Sub
Private Sub LoadStructureData()
   frmBeamColumnDesign1.txtE = E
   frmBeamColumnDesign1.txtMu = EMu
   frmBeamColumnDesign1.txtKz = Str(Kz)
   frmBeamColumnDesign1.txtKy = Str(Ky)
   frmBeamColumnDesign1.txtFy = Fy
   frmBeamColumnDesign1.txtLY = Ly
   frmBeamColumnDesign1.txtLZ = Lz
   frmBeamColumnDesign1.txtMY1 = My1
   frmBeamColumnDesign1.txtMY2 = My2
   frmBeamColumnDesign1.txtMZ1 = Mz1
   frmBeamColumnDesign1.txtMZ2 = Mz2
   frmBeamColumnDesign1.txtN = N
End Sub

Private Sub cmdDesign_Click()
   FileName = App.Path + "\BeamColumn.out"
   Open FileName For Output As 1
```

```
      blnSafe = True
      Call UpdateSectionalProperties
      Call Design
  End Sub

  Private Sub LoadSectionalProperties()
      If blnLoaded Then
        frmSectionalProperties.txtDepth = Str(D)
        frmSectionalProperties.txtWidthOfFlange = Str(Bf)
        frmSectionalProperties.txtThickOfFlange = Str(tf)
        frmSectionalProperties.txtThickWeb = Str(tw)
        frmSectionalProperties.txtR = Str(R)
        frmSectionalProperties.txtRZ = Str(rz)
        frmSectionalProperties.txtRY = Str(ry)
        frmSectionalProperties.txtA = Str(A)
        frmSectionalProperties.txtZZ = Str(Zz)
        frmSectionalProperties.txtZY = Str(Zy)
        frmSectionalProperties.txtSafetyFactor = Str(GamaM)
        frmSectionalProperties.txtIy = Str(Iy)
        If sectionType = PLASTIC Then
           optPlastic = True
        ElseIf sectionType = COMPACT Then
           optCompact = True
        ElseIf sectionType = SEMI_COMPACT Then
           optSemiCompact = True
        End If
      End If
  End Sub

  Private Sub UpdateSectionalProperties()
      D = Val(txtDepth)
      Bf = Val(txtWidthOfFlange)
      tf = Val(txtThickOfFlange)
      tw = Val(txtThickWeb)
      R = Val(txtR)
      rz = Val(txtRZ)
      ry = Val(txtRY)
      A = Val(txtA)
      Zz = Val(txtZZ)
      Zy = Val(txtZY)
      GamaM = Val(txtSafetyFactor)
      Iy = Val(txtIy)
      If optPlastic Then
        sectionType = PLASTIC
      ElseIf optCompact Then
        sectionType = COMPACT
      ElseIf optSemiCompact Then
        sectionType = SEMI_COMPACT
      End If
  End Sub

  Private Sub Form_Load()
      Call LoadSectionalProperties
      blnLoaded = True
  End Sub
```

```
Private Sub Design()
   blnSafe = True
   Call PrintInput
   Call ComputeInteractionEqnLocal
   If blnSafe Then Call DesignCompressionResisMajor
   If blnSafe Then Call DesignCompressionResisMinor
   If blnSafe Then Call DesignBucklingResisMajor
   If blnSafe Then Call DesignResOfCrossSection
   If blnSafe Then Call DesignShearResistance
   If blnSafe Then Call DesignComBendingAxialForce
   Close #1
   If blnSafe Then End
End Sub
Private Sub PrintInput()

   Print #1, Tab(10); "Beam-Column Design"
   Print #1, Tab(10); "----------"
   Print #1, ""
   Print #1, Tab(5); "Structure Data"
   Print #1, Tab(5); "-------"
   Print #1, Tab(5); "Youngs Modulus (N/sq.mm)  "; Tab(60); Format(E,
   "#######0.0##")
   Print #1, Tab(5); "Poisson's Ratio "; Tab(60); Format(EMu,
   "#######0.0##")
   Print #1, Tab(5); "Yield Stress (N/sq.mm) "; Tab(60); Format(Fy,
   "#######0.0##")
   Print #1, Tab(5); "Major axis Length (mm) "; Tab(60); Format(Lz,
   "#######0.0##")
   Print #1, Tab(5); "Minor axis Length (mm) "; Tab(60); Format(Ly,
   "#######0.0##")
   Print #1, Tab(5); "Eff.Length factor - Major axis "; Tab(60);
   Format(Kz, "#######0.0##")
   Print #1, Tab(5); "Eff.Length factor - Minor axis "; Tab(60);
   Format(Ky, "#######0.0##")
   Print #1, Tab(5); "Factored axial load (kN) "; Tab(60); Format(N,
   "#######0.0##")
   Print #1, Tab(5); "Major axis factored moment at top (kNm)";
   Tab(60); Format(Mz1, "#######0.0##")
   Print #1, Tab(5); "Major axis factored moment at bottom (kNm)";
   Tab(60); Format(Mz2, "#######0.0##")
   Print #1, Tab(5); "Minor axis factored moment at top (kNm)";
   Tab(60); Format(My1, "#######0.0##")
   Print #1, Tab(5); "Minor axis factored moment at bottom (kNm)";
   Tab(60); Format(My2, "#######0.0##")

   Print #1, Tab(5); ""
   Print #1, Tab(5); "Sectional Properties"
   Print #1, Tab(5); "----------"
   Print #1, Tab(5); "Depth (mm)"; Tab(60); Format(D, "#######0.0##")
   Print #1, Tab(5); "Width of flange (mm)"; Tab(60); Format(Bf,
   "#######0.0##")
```

```
    Print #1, Tab(5); "Thickness of flange (mm)"; Tab(60); Format(tf,
    "#######0.0##")
    Print #1, Tab(5); "Thickness of web (mm)"; Tab(60); Format(tw,
    "#######0.0##")
    Print #1, Tab(5); "Root radius (mm)"; Tab(60); Format(R,
    "#######0.0##")
    Print #1, Tab(5); "Radius of Gyration major axis (mm)"; Tab(60);
    Format(rz, "#######0.0##")
    Print #1, Tab(5); "Radius of Gyration minor axis (mm)"; Tab(60);
    Format(ry, "#######0.0##")
    Print #1, Tab(5); "Elastic modulus about major axis (cu.mm)";
    Tab(60); Format(Zz, "#######0.0##")
    Print #1, Tab(5); "Elastic modulus about minor axis (cu.mm)";
    Tab(60); Format(Zy, "#######0.0##")
    Print #1, Tab(5); "Partial safety factor for material "; Tab(60);
    Format(GamaM, "#######0.0##")
    Print #1, Tab(5); "Area (sq.mm)"; Tab(60); Format(A, "#######0.0##")
    Print #1, Tab(5); "Moment of intertia about minor axis (mm^4)";
    Tab(60); Format(Iy, "#######0.0##")
    Print #1, ""
End Sub
Private Sub ComputeInteractionEqnLocal()
    G = E / (2 * (1 + EMu))
    If (Abs(Mz1) > Abs(Mz2)) Then
        Mz = Mz1
        Mzm = Mz2
    Else
        Mz = Mz2
        Mzm = Mz1
    End If

    If (Abs(My1) > Abs(My2)) Then
        My = My1
        Mym = My2
    Else
        My = My2
        Mym = My1
    End If
    'Properties of the Cross-section
    Zpz = 2 * Bf * tf * (D - tf) / 2 + tw * (D - 2 * tf) ^ 2 / 4
    Zpy = 2 * tf * Bf * Bf / 4 + (D - 2 * tf) * tw * tw / 4
    It = (2 * tf ^ 3 * Bf + (D - tf) * tw ^ 3) / 3
    Iw = (D - tf) ^ 2 * Iy / 4
    Nd = A * Fy / (GamaM * 1000#)
    ' Minor Axis buckling resistance in bending
    If sectionType = SEMI_COMPACT Then
        Mdy = Zy * Fy / (GamaM * 1000000#)
    Else
        Mdy = Zpy * Fy / (GamaM * 1000000#)
    End If
```

```
Dim strMsg As String
strMsg = "Plastic modulus about Major axis " + Str(Zpz) + " cu.mm"
+ vbCrLf
strMsg = strMsg + "Plastic modulus about Minor axis " + Str(Zpy)
+ " cu.mm" + vbCrLf
strMsg = strMsg + "Torsional constant" + Str(It) + " mm^4" +
vbCrLf
strMsg = strMsg + "Warping constant " + Str(Iw) + " mm^6"
'MsgBox strMsg
'MsgBox "Design Resistance in Bending about Minor Axis" + Str(Mdy)
+ " kNm" + vbCrLf + "Applied Moment about Minor Axis " + Str(My)
+ " kNm"
If sectionType = SEMI_COMPACT Then
   Betab = Zz / Zpz
Else
   Betab = 1
End If
Mdz = Betab * Zpz * Fy / (GamaM * 1000000#)
Print #1, Tab(5); "Output"
Print #1, Tab(5); "-----"

Print #1, Tab(5); "Plastic modulus about Major axis (cu.mm) ";
Tab(60); Format(Zpz, "#######0.0##")
Print #1, Tab(5); "Plastic modulus about Minor axis (cu.mm)";
Tab(60); Format(Zpy, "#######0.0##")
Print #1, Tab(5); "Torsional constant (mm^4)"; Tab(60); Format(It,
"#######0.0##")
Print #1, Tab(5); "Warping constant (mm^6)"; Tab(60); Format(Iw,
"#######0.0##")
Print #1, Tab(5); "Design Resistance in Bending about Minor Axis
(kNm)"; Tab(60); Format(Mdy, "#######0.0##")
Print #1, Tab(5); "Applied Moment about Minor Axis (kNm) ";
Tab(60); Format(My, "#######0.0##")
Print #1, ""
End Sub
Private Sub DesignResOfCrossSection()
Dim Alpha1, Alpha2, Nr, Mndz, Mndy As Double

'Local Capacity Check
   If sectionType = SEMI_COMPACT Then
      StressRatio = (N / Nd) + (Mz / Mdz) + (My / Mdy)
   Else
      Nr = N / Nd
      Mndz = 1.11 * Mdz * (1 - Nr)
      If Mndz > Mdz Then
         Mndz = Mdz
      End If
      If Nr <= 0.2 Then
         Mndy = Mdy
      Else
         Mndy = 1.56 * Mdy * (1 - Nr) * (Nr + 0.6)
      End If
```

```
     Alpha1 = 5 * Nr
     If Alpha1 < 1 Then
        Alpha1 = 1
     End If
     Alpha2 = 2
     ' section strength interaction equation
     StressRatio = (My / Mndy) ^ Alpha1 + (Mz / Mndz) ^ Alpha2

  End If
  If StressRatio < 1# Then
    'MsgBox "Interaction equation value = " + Str(StressRatio) +
    vbCrLf + " Section is safe"
    Print #1, Tab(5); "Interaction equation (section strength value
    "; Tab(60); Format(StressRatio, "#######0.0##"); " Section is
    safe"
    Print #1, ""
  Else
    MsgBox "Interaction equation value = " + Str(StressRatio) +
    vbCrLf + "Section unsafe due to member capacity check " +
    vbCrLf + "Revise the section"
    blnSafe = False
    Unload Me
    frmBeamColumnDesign1.Show
  End If
End Sub
Private Sub DesignCompressionResisMajor()
Dim Fcrz, Fcdz, Phiz, Phiz1 As Double
'Compression resistance - Major Axis
  Lz = Kz * Lz
  Ly = Ky * Ly
  Lrz = Lz / rz
  Lry = Ly / ry
  If (D / Bf) > 1.2 And tf <= 40 Then
     AlphaZ = 0.21
     AlphaY = 0.34
  ElseIf (D / Bf) > 1.2 And tf < 100 Then
     AlphaZ = 0.34
     AlphaY = 0.49
  ElseIf (D / Bf) <= 1.2 And tf <= 100 Then
     AlphaZ = 0.34
     AlphaY = 0.49
  Else
     AlphaZ = 0.76
     AlphaY = 0.76
  End If

  Fcrz = PI * PI * E / (Lrz * Lrz)
  LambdaZ = (Fy / Fcrz) ^ 0.5
  Phiz = 0.5 * (1 + AlphaZ * (LambdaZ - 0.2) + LambdaZ ^ 2)
  Phiz1 = Phiz + (Phiz ^ 2 - LambdaZ ^ 2) ^ 0.5
  Fcdz = Fy / (GamaM * Phiz1)
  If Fcdz > (Fy / GamaM) Then
```

```
      Fcdz = Fy / GamaM
   End If
   Pdz = Fcdz * A / 1000#
   Dim strMsg As String
   strMsg = "Compression resistance about Major Axis " + Str(Pdz) +
   " kN" + vbCrLf
   If Pdz > N Then
      'MsgBox strMsg + "Section is safe"
      Print #1, Tab(5); "Compression resistance about Major Axis
      (kN)"; Tab(60); Format(Pdz, "#######0.0##"); " Section is safe"
      Print #1, ""
   Else
      MsgBox strMsg + "Revise the Section"
      blnSafe = False
      Unload Me
      frmBeamColumnDesign1.Show
   End If

End Sub
Private Sub DesignCompressionResisMinor()
Dim Fcry, Fcdy, Phiy, Phiy1  As Double
Dim strMsg As String

   'Compression Resistance- Minor Axis
   Fcry = PI * PI * E / (Lry * Lry)
   LambdaY = (Fy / Fcry) ^ 0.5
   Phiy = 0.5 * (1 + AlphaY * (LambdaY - 0.2) + LambdaY ^ 2)
   Phiy1 = Phiy + (Phiy ^ 2 - LambdaY ^ 2) ^ 0.5
   Fcdy = Fy / (GamaM * Phiy1)
   If Fcdy > (Fy / GamaM) Then
      Fcdy = Fy / GamaM
   End If
   Pdy = Fcdy * A / 1000#

   strMsg = "Compression resistance about Minor Axis " + Str(Pdy) +
   " kN" + vbCrLf
   If Pdy > N Then
      Print #1, Tab(5); "Compression resistance about Minor Axis
      (kNm)"; Tab(60); Format(Pdy, "#######0.0##"); " Section is
      safe"
      Print #1, ""
   Else
      MsgBox strMsg + "Revise the section"
      blnSafe = False
      Unload Me
      frmBeamColumnDesign1.Show
   End If
   If Pdz > Pdy Then
      Pd = Pdy
   Else
      Pd = Pdz
   End If
```

```
      Print #1, Tab(5); "Compression Resistance (kN)"; Tab(60); Format(Pd,
      "#######0.0##")
      Print #1, Tab(5); "Compressive Force (kN)"; Tab(60); Format(N,
      "#######0.0##")
      Print #1, ""
End Sub

Private Sub DesignBucklingResisMajor()
Dim PhiLT, PhiLT1, chi, C1, Chi1, Mcr1, fcrb As Double
Dim strMsg As String

'Member Buckling resistance
' Major Axis
  If Abs(Mz1) < Abs(Mz2) Then
     Chi1 = Mz1 / Mz2
  Else
     Chi1 = Mz2 / Mz1
  End If

  'Eqn suggested by Gardner and Nethercot
   C1 = 1.88 - 1.4 * Chi1 + 0.52 * Chi1 ^ 2
   If C1 > 2.7 Then C1 = 2.7

  fcrb = C1 * (1473.5 / Lry) ^ 2 * ((1 + (1 / 20) * (Lry / (D / tf))
   ^ 2) ^ 0.5)
   Mcr = fcrb * Zz
   LambdaLT = (Betab * Zpz * Fy / Mcr) ^ 0.5
   LambdaLT1 = (1.2 * Zz * Fy / Mcr) ^ 0.5
   If LambdaLT > LambdaLT1 Then
      LambdaLT = LambdaLT1
   End If
   LambdaLT1 = (1.2 * Zz * Fy / Mcr) ^ 0.5
   If LambdaLT > LambdaLT1 Then
      LambdaLT = LambdaLT1
   End If
  alphaLT = 0.21
  PhiLT = 0.5 * (1 + alphaLT * (LambdaLT - 0.2) + LambdaLT ^ 2)
  PhiLT1 = PhiLT + (PhiLT ^ 2 - LambdaLT ^ 2) ^ 0.5
  chi = 1# / PhiLT1
  If chi > 1# Then
     chi = 1#
  End If
  fbd = chi * Fy / GamaM

  Mdz1 = fbd * Betab * Zpz / 1000000#

  strMsg = "Buckling Resistance in Bending about Major Axis" +
  Str(Mdz) + "kNm" + vbCrLf + "Applied Moment about Major axis " +
  Str(Mz) + "kNm" + vbCrLf
  If Mdz1 > Mz Then
     Print #1, Tab(5); "Buckling Resistance in Bending about Major
     Axis (kNm)"; Tab(60); Format(Mdz1, "#######0.0##")
```

```
      Print #1, Tab(5); "Applied Moment about Major axis (kNm)";
      Tab(60); Format(Mz, "#######0.0##")
      Print #1, Tab(5); "Buckling Resistance in Bending about Major
      Axis is safe"
      Print #1, ""
   Else
      MsgBox strMsg + "Revise the Section"
      blnSafe = False
      Unload Me
      frmBeamColumnDesign1.Show
   End If

End Sub


Private Sub DesignShearResistance()

Dim MSF, Av, Vp As Double
Dim strMsg As String

   'Shear Resistance of the cross-section
   MSF = Abs(Mz1 - Mz2) / (Lz / 1000)
   Av = D * tw
   Vp = Av * Fy / (Sqr(3) * GamaM * 1000#)

   strMsg = "Shear Resistance Parallel to Web " + Str(Vp) + " kN" +
   vbCrLf + " Max.S.F. " + Str(MSF) + " kN" + vbCrLf
   If Vp > MSF Then
      Print #1, Tab(5); "Shear Resistance Parallel to Web (kN)";
      Tab(60); Format(Vp, "#######0.0##")
      Print #1, Tab(5); "Maximum Shear Force (kN)"; Tab(60); Format(MSF,
      "#######0.0##")
      Print #1, Tab(5); "Shear Resistance Parallel to Web is safe"
      Print #1, ""
   Else
      MsgBox strMsg + "Revise the Section"
      blnSafe = False
      Unload Me
      frmBeamColumnDesign1.Show
   End If

End Sub

Private Sub DesignComBendingAxialForce()
Dim Psiz, BetaMz, MulT, Mufz, Muz, Muy, KKz, Kky, psiy, BetaMy, nz,
Cmz, Cmy, CmLT, ny, klt, klt1 As Double
Dim strMsg As String
'Member Bending Resistance in Combined Bending and axial Compression
'Major Axis
   Psiz = Mzm / Mz
   nz = N / Pdz
   Kz = 1 + (LambdaZ - 0.2) * nz
   Cmz = 0.6 + 0.4 * Psiz
```

```
If Cmz < 0.4 Then
   Cmz = 0.4
End If
If Kz > 1 + 0.8 * nz Then
   Kz = 1 + 0.8 * nz
End If
klt = 0
Cmy = 0
'Minor axis
If Not (Mym = 0 And My = 0) Then
    psiy = Mym / My
    ny = N / Pdy
    Ky = 1 + (LambdaY - 0.2) * ny

    If Ky > 1 + 0.8 * ny Then
       Ky = 1 + 0.8 * ny
    End If
    Cmy = 0.6 + 0.4 * psiy
    If Cmy < 0.4 Then
       Cmy = 0.4
    End If
    CmLT = 0.6 + 0.4 * psiy
    If CmLT < 0.4 Then
       CmLT = 0.4
    End If
    klt = 1 - 0.1 * LambdaLT * ny / (CmLT - 0.25)
    klt1 = 1 - 0.1 * ny / (CmLT - 0.25)
    If klt > klt1 Then
        klt = klt1
    End If
End If
'Check with Interaction Formula for overall buckling

StressRatio1 = (N / Pdy) + (Ky * Cmy * Abs(My) / Mdy) + (klt *
Abs(Mz) / Mdz1)
StressRatio2 = (N / Pdz) + (0.6 * Ky * Cmy * Abs(My) / Mdy) + (Kz
* Cmz * Abs(Mz) / Mdz1)
strMsg = "Interaction equation value  = " + Str(StressRatio1) +
vbCrLf
If StressRatio1 < 1# And StressRatio2 < 1 Then
   Print #1, Tab(5); "Interaction equation value 1 "; Tab(60);
   Format(StressRatio1, "#######0.0##")
   Print #1, Tab(5); "Interaction equation value 2 "; Tab(60);
   Format(StressRatio2, "#######0.0##")
 Print #1, Tab(5); "Member is safe under combined Axial Force
 and B.M."
 Print #1, ""
 MsgBox "Member is safe" + vbCrLf + "The output is stored as
 BeamColumn.out under the working folder", , "Desgin Over"
Else
 MsgBox strMsg + "Revise the section"
 blnSafe = False
```

```
     Unload Me
   frmBeamColumnDesign1.Show

 End If

End Sub

Private Sub optCompact_Click()
   If optCompact Then
   sectionType = COMPACT
   End If

End Sub

Private Sub optPlastic_Click()
   If optPlastic Then
     sectionType = PLASTIC
   End If
End Sub

Private Sub optSemiCompact_Click()
   If optSemiCompact Then
     sectionType = SEMI_COMPACT
   End If
End Sub
```

To illustrate the use of the program, a trial run of the program was conducted for example 9.1 and the resulting output is as follows:

```
Beam-Column Design
------------------


Structure Data
-------------
Youngs Modulus <N/sq.mm>                              200000.0
Poisson's Ratio                                       0.3
Yield Stress <N/sq.mm>                                250.0
Major axis Length <mm>                                4000.0
Minor axis Length <mm>                                4000.0
Eff.Length factor - Major axis                        0.8
Eff.Length factor - Minor axis                        0.8
Factored axial load <kN>                              500.0
Major axis factored moment at top <kNm>               27.0
Major axis factored moment at bottom <kNm>            45.0
Minor axis factored moment at top <kNm>               0.0
Minor axis factored moment at bottom <kNm>            0.0


Sectional Properties
-------------------
Depth <mm>                                            250.0
Width of flange <mm>                                  250.0
Thickness of flange <mm>                              9.7
Thickness of web <mm>                                 6.9
```

```
Root radius <mm>                                     10.0
Radius of Gyration major axis <mm>                   109.0
Radius of Gyration minor axis <mm>                   54.9
Elastic modulus about major axis <cu.mm>             619000.0
Elastic modulus about minor axis <cu.mm>             156000.0
Partial safety factor for material                   1.1
Area <sq.mm>                                         6500.0
Moment of intertia about minor axis <mm^4>           19600000.0

Output
-----
Plastic modulus about Major axis <cu.mm>             674456.721
Plastic modulus about Minor axis <cu.mm>             305869.717
Torsional constant <mm^4>                            178425.738
Warping constant <mm^6>                              282946041000.0
Design Resistance in Bending about
Minor Axis <kNm>                                     35.455
Applied Moment about Minor Axis <kNm>                0.0
Compression resistance about Major Axis <kN>         1407.624 Section
                                                     is safe
Compression resistance about Minor Axis <kN>         1110.522 Section
                                                     is safe
Compression Resistance <kN>                          1110.522
Compressive Force <kN>                               500.0

Buckling Resistance in Bending about
Major Axis <kNm>                                     128.525
Applied Moment about Major axis <kNm>                45.0
Buckling Resistance in Bending about
Major Axis is safe

Interaction equation value                           0.658 Section is
                                                     safe

Shear Resistance Parallel to Web <kN>                226.348
Maximum Shear Force <kN>                              5.625
Shear Resistance Parallel to Web is safe

Interaction equation value                           0.45
Interaction equation value2                          0.663
Member is safe under combined Axial Force and B.M.
```

The inclusion of these computer programs is only to illustrate the usefulness of computers in repetitive and complex calculations, which are time consuming if done manually. However, the publisher and the author do not provide any warranty for their use to design problems. The user is advised to test the programs thoroughly, before using them.

D

Design Aids

This appendix contains design aids in the form of tables which will be useful to select bolts, welds, compression members, beams, and tension members for the required capacity.

Table D.1 Strength of bolts in clearance holes

	Bolt grade Gr-4.6 (N/mm²)	Bolts grade Gr-8.8 (N/mm²)	Other grades of bolts (N/mm²)
Shear strength, v_{nsb}	185	370	$f_u/(\sqrt{3} \times 1.25)$
Bearing strength, v_{npb}	400	800	$2.5k_b^* f_u/1.25$
Tension strength, t_{nb}	211	563	$0.9 f_u/1.25$ but not greater than $f_{yb}(1.25/1.1)$

* Assuming $k_b = 0.5$

Table D.2 Bearing strength of connected parts for ordinary bolts in clearance holes, v_{npb}, in N/mm²

Grade 410	Grade 540	Grade 570	Other grades
410	540	570	$2.5k_b^* f_u/1.25$

Table D.3 Design capacity of ordinary bolts based on net tensile area (grade 4.6)

Bolt size, d(mm)	Tensile stress Area of bolt, (mm²)	Tension capacity, T_b(kN)t_{nb} = 272 MPa	Single shear capacity, V_{sb}(kN) v_{nsb} =185 MPa	Minimum thickness of ply for bolt bearing v_{npb}= 400 MPa t_{bb}= t_e, mm
(12)	84.3	24.6	15.6	3.2
16	157	43.8	29.0	4.5
20	245	68.5	45.3	5.6
(22)	303	82.9	56.0	6.3
24	353	98.7	65.3	6.8
(27)	459	124.9	84.9	7.8
30	561	154.2	103.8	8.6
36	817	222.0	151.1	10.5

$v_{sb}= A_{nb}v_{nsb}$; $T_b= A_{nb}t_{nb}$; $t_{bb}=V_{npb}/(dv_{npb})$; Sizes in brackets are not preferred

Table D.4 Capacities of high strength friction grip bolts in clearance holes (k_h=1, μ=0.48)

Bolt diameter, (mm)	Stress area of bolt (mm²)		Proof load of bolt (kN) Property class		Tensile capacity, (kN) Property class		Slip resistance in single shear (n_e=1,γ_{mf}=1.25) (kN) Property class	
	Thread	Shank	8.8	10.9	8.8	10.9	8.8	10.9
(12)	84.3	113	47.2	61.3	48.5	63	18.1	23.5
16	157	201	87	114	90.4	117	33.7	43.8
20	245	314	137	178	141	183	52.6	68.4
(22)	303	380	169	220	174	226	65.1	84.7
24	353	452	197	256	203	264	75.9	98.6
(27)	459	572	257	334	264	343	98.7	128
30	561	706	314	408	323	420	120	156
36	817	1017	457	594	470	611	175	228

Tension capacity = $0.9 f_u A_{nb}/1.25$

Slip resistance = $\mu_f n_e K_h F_o/\gamma_{mf}$

γ_{mf}=1.1(at service load) = 1.25(at ultimate load)

F_o = minimum bolt tension (proof load) = $A_{nb} f_o$, $f_o = 0.7 f_{ub}$

n_e = Number of effective interfaces offering frictional resistance to slip

Table D.5 Gauge distance specified in special publication 1 (SP-1) of BIS

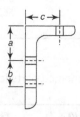

Leg size (mm)	Double of row of bolts		Single row of bolts	Maximum Bolt size for double row of bolts (mm)
	a (mm)	b (mm)	c (mm)	
200	75	85	115	27
150	55	65	90	22
130	50	55	80	20
125	45	55	75	20
115	45	50	70	12
110	45	45	65	12
100	40	40	60	12
95	-	-	55	-
90	-	-	50	-
80	-	-	45	-

(contd)

Table D.5 (*contd*)

75	-	-	40	-
70	-	-	40	-
65	-	-	35	-
60	-	-	35	-
55	-	-	30	-
50	-	-	28	-
45	-	-	25	-
40	-	-	21	-
35	-	-	19	-
30	-	-	17	-
25	-	-	15	-
20	-	-	12	-

Table D.6 Design strength f_{wd} in weld material for f_u = 410 MPa

Type of stress	Design stress, N/mm^2
Direct tension or compression (shop welding)	189
Direct tension or compression (site welding)	158

For other electrodes $f_{wd} = f_u /(\sqrt{3} \times \gamma_{mw})$

γ_{mw} = 1.25 (shop welding) = 1.5 (site welding)

Capacity in kN for fillet weld(shop welded) = 0.7 × size × 189 × effective length/1000

Table D.7 Design capacity of fillet welds

Leg length, s (mm)	Design Capacity per unit length, R_{nw} (kN/mm)			
	Fe 410 with E43 and E51 electrodes		Fe 540 , shop welded	
	Shop welde f_{wd} = 189 MPa	Site welded, f_{wd} = 158 MPa	With E43 f_{wd} =189 MPa	With E51 f_{wd} = 235 MPa
4	0.529	0.442	0.529	0.658
5	0.661	0.553	0.661	0.822
6	0.793	0.663	0.793	0.987
8	1.058	0.884	1.058	1.316
10	1.323	1.106	1.323	1.645
12	1.587	1.327	1.587	1.974
15	1.984	1.659	1.984	2.467
18	2.381	1.990	2.381	2.961
20	2.646	2.212	2.646	3.290
22	2.910	2.433	2.910	3.619
25	3.307	2.765	3.307	4.112

$R_{nw} = 0.7 \times s \times f_{wd} / 1000$

Table D.8 Capacity of angle tension members with single row of bolts (f_y = 250 MPa) — Clause 6.2 to 6.4 of Code

A (mm)	B (mm)	t (mm)	A_g (mm²)	No. of bolts	Gauge distance (mm)	Dia. of bolt (mm)	Pitch (mm)	Edge Distance (mm)	T_{dg} (kN)	T_{dn} (kN)	T_{db1} (kN)	T_{db2} (kN)	T_d (kN)
35	35	5	327	2	19	12	30	25	74.3	60.9	50.1	48.4	48.4
35	35	5	327	3	19	12	30	25	74.3	70.7	69.8	62.9	62.9
35	35	5	327	4	19	12	30	25	74.3	74.0	89.5	77.4	74.0
35	35	6	386	2	19	12	30	25	87.7	75.9	60.1	58.1	58.1
35	35	6	386	3	19	12	30	25	87.7	85.3	83.7	75.5	75.5
35	35	6	386	4	19	12	30	25	87.7	88.5	107.4	92.9	87.7
40	40	5	378	2	21	12	30	25	85.9	66.3	54.5	51.8	51.8
40	40	5	378	3	21	12	30	25	85.9	81.1	74.2	66.3	66.3
40	40	5	378	4	21	12	30	25	85.9	86.0	93.9	80.8	80.8
40	40	6	447	2	21	12	30	25	101.6	84.6	65.4	62.2	62.2
40	40	6	447	3	21	12	30	25	101.6	98.9	89.1	79.6	79.6
40	40	6	447	4	21	12	30	25	101.6	103.6	112.7	97.0	97.0
45	45	5	428	2	25	12	30	25	97.3	77.3	56.0	53.0	53.0
45	45	5	428	3	25	12	30	25	97.3	89.3	75.7	67.5	67.5
45	45	5	428	4	25	12	30	25	97.3	96.6	95.4	82.0	82.0
45	45	6	507	2	25	12	30	25	115.2	91.5	67.2	63.6	63.6
45	45	6	507	3	25	12	30	25	115.2	110.3	90.8	81.0	81.0
45	45	6	507	4	25	12	30	25	115.2	117.4	114.5	98.3	98.3
50	50	5	479	2	28	16	40	35	108.9	81.3	68.4	65.9	65.9
50	50	5	479	3	28	16	40	35	108.9	96.3	94.6	84.7	84.7
50	50	5	479	4	28	16	40	35	108.9	103.9	120.9	103.4	103.4
50	50	6	568	2	28	16	40	35	129.1	96.5	82.1	79.1	79.1

(contd)

Table D.8 (*contd*)

50	50	6	568	3	28	16	40	35	129.1	118.8	113.6	101.6	101.6
50	50	6	568	4	28	16	40	35	129.1	126.2	145.1	124.1	124.1
55	55	5	527	2	30	16	40	35	119.8	92.7	72.8	69.3	69.3
55	55	5	527	3	30	16	40	35	119.8	104.0	99.1	88.1	88.1
55	55	5	527	4	30	16	40	35	119.8	114.2	125.3	106.8	106.8
55	55	6	626	2	30	16	40	35	142.3	109.9	87.4	83.2	83.2
55	55	6	626	3	30	16	40	35	142.3	129.7	118.9	105.7	105.7
55	55	6	626	4	30	16	40	35	142.3	139.7	150.4	128.2	128.2
55	55	8	818	2	30	16	40	35	185.9	150.9	116.5	110.9	110.9
55	55	8	818	3	30	16	40	35	185.9	179.3	158.5	140.9	140.9
55	55	8	818	4	30	16	40	35	185.9	188.8	200.5	170.9	170.9
55	55	10	1000	2	30	16	40	35	227.3	199.2	145.6	138.6	138.6
55	55	10	1000	3	30	16	40	35	227.3	226.4	198.1	176.1	176.1
55	55	10	1000	4	30	16	40	35	227.3	235.5	250.6	213.6	213.6
60	60	5	575	2	35	16	40	35	130.7	104.0	72.8	69.3	69.3
60	60	5	575	3	35	16	40	35	130.7	108.9	99.1	88.1	88.1
60	60	5	575	4	35	16	40	35	130.7	122.5	125.3	106.8	106.8
60	60	6	684	2	35	16	40	35	155.5	123.5	87.4	83.2	83.2
60	60	6	684	3	35	16	40	35	155.5	137.8	118.9	105.7	105.7
60	60	6	684	4	35	16	40	35	155.5	151.2	150.4	128.2	128.2
60	60	8	896	2	35	16	40	35	203.6	161.0	116.5	110.9	110.9
60	60	8	896	3	35	16	40	35	203.6	193.8	158.5	140.9	140.9
60	60	8	896	4	35	16	40	35	203.6	206.6	200.5	170.9	170.9
60	60	10	1100	2	35	16	40	35	250.0	210.4	145.6	138.6	138.6
60	60	10	1100	3	35	16	40	35	250.0	247.3	198.1	176.1	176.1

(*contd*)

Table D.8 (*contd*)

60	60	10	1100	4	35	16	40	35	250.0	259.6	250.6	213.6	213.6
65	65	5	625	2	35	16	40	35	142.0	115.4	80.2	75.0	75.0
65	65	5	625	3	35	16	40	35	142.0	115.4	106.4	93.7	93.7
65	65	5	625	4	35	16	40	35	142.0	131.2	132.7	112.5	112.5
65	65	6	744	2	35	16	40	35	169.1	137.1	96.2	90.0	90.0
65	65	6	744	3	35	16	40	35	169.1	146.4	127.7	112.5	112.5
65	65	6	744	4	35	16	40	35	169.1	163.0	159.2	135.0	135.0
65	65	8	976	2	35	16	40	35	221.8	179.2	128.3	120.0	120.0
65	65	8	976	3	35	16	40	35	221.8	208.8	170.3	150.0	150.0
65	65	8	976	4	35	16	40	35	221.8	224.8	212.3	180.0	180.0
65	65	10	1200	2	35	16	40	35	272.7	222.5	160.4	150.0	150.0
65	65	10	1200	3	35	16	40	35	272.7	268.7	212.9	187.5	187.5
65	65	10	1200	4	35	16	40	35	272.7	284.1	265.4	225.0	225.0
70	70	5	677	2	40	16	40	35	153.9	126.8	80.2	75.0	75.0
70	70	5	677	3	40	16	40	35	153.9	126.8	106.4	93.7	93.7
70	70	5	677	4	40	16	40	35	153.9	136.9	132.7	112.5	112.5
70	70	6	806	2	40	16	40	35	183.2	150.7	96.2	90.0	90.0
70	70	6	806	3	40	16	40	35	183.2	150.7	127.7	112.5	112.5
70	70	6	806	4	40	16	40	35	183.2	171.9	159.2	135.0	135.0
70	70	8	1060	2	40	16	40	35	240.9	197.4	128.3	120.0	120.0
70	70	8	1060	3	40	16	40	35	240.9	219.3	170.3	150.0	150.0
70	70	8	1060	4	40	16	40	35	240.9	240.0	212.3	180.0	180.0
70	70	10	1300	2	40	16	40	35	295.5	242.2	160.4	150.0	150.0
70	70	10	1300	3	40	16	40	35	295.5	285.7	212.9	187.5	187.5
70	70	10	1300	4	40	16	40	35	295.5	305.6	265.4	225.0	225.0

(*contd*)

Table D.8 *(contd)*

75	5	727	2	40	16	40	35	165.2	138.1	87.6	80.7	80.7
75	5	727	3	40	16	40	35	165.2	138.1	113.8	99.4	99.4
75	5	727	4	40	16	40	35	165.2	143.3	140.1	118.2	118.2
75	6	866	2	40	16	40	35	196.8	164.4	105.1	96.8	96.8
75	6	866	3	40	16	40	35	196.8	164.4	136.6	119.3	119.3
75	6	866	4	40	16	40	35	196.8	181.4	168.1	141.8	141.8
75	8	1140	2	40	16	40	35	259.1	215.5	140.1	129.1	129.1
75	8	1140	3	40	16	40	35	259.1	230.9	182.1	159.1	159.1
75	8	1140	4	40	16	40	35	259.1	255.9	224.1	189.1	189.1
75	10	1400	2	40	16	40	35	318.2	264.9	175.2	161.4	161.4
75	10	1400	3	40	16	40	35	318.2	303.7	227.7	198.9	198.9
75	10	1400	4	40	16	40	35	318.2	327.8	280.1	236.3	236.3
80	6	929	2	45	20	50	40	211.1	170.9	113.4	106.0	106.0
80	6	929	3	45	20	50	40	211.1	170.9	152.7	134.7	134.7
80	6	929	4	45	20	50	40	211.1	192.9	192.1	163.3	163.3
80	8	1220	2	45	20	50	40	277.3	224.3	151.2	141.4	141.4
80	8	1220	3	45	20	50	40	277.3	246.1	203.6	179.5	179.5
80	8	1220	4	45	20	50	40	277.3	271.0	256.1	217.7	217.7
80	10	1500	2	45	20	50	40	340.9	275.8	188.9	176.7	176.7
80	10	1500	3	45	20	50	40	340.9	322.4	254.6	224.4	224.4
80	10	1500	4	45	20	50	40	340.9	346.6	320.2	272.1	272.1
80	12	1780	2	45	20	50	40	404.5	325.8	226.7	212.0	212.0
80	12	1780	3	45	20	50	40	404.5	396.3	305.5	269.3	269.3
80	12	1780	4	45	20	50	40	404.5	419.8	384.2	326.6	326.6
90	6	1050	2	50	20	50	40	238.6	198.2	122.2	112.8	112.8

(contd)

724 *Steel Structures: Design and Practice*

Table D.8 (*contd*)

90	90	6	1050	3	50	20	50	40	238.6	198.2	161.6	141.5	141.5
90	90	6	1050	4	50	20	50	40	238.6	207.5	201.0	170.1	170.1
90	90	8	1380	2	50	20	50	40	313.6	260.6	163.0	150.4	150.4
90	90	8	1380	3	50	20	50	40	313.6	262.4	215.5	188.6	188.6
90	90	8	1380	4	50	20	50	40	313.6	298.3	267.9	226.8	226.8
90	90	10	1700	2	50	20	50	40	386.4	321.2	203.7	188.1	188.1
90	90	10	1700	3	50	20	50	40	386.4	351.7	269.3	235.8	235.8
90	90	10	1700	4	50	20	50	40	386.4	386.6	334.9	283.5	283.5
90	90	12	2020	2	50	20	50	40	459.1	380.0	244.4	225.7	225.7
90	90	12	2020	3	50	20	50	40	459.1	438.4	323.2	282.9	282.9
90	90	12	2020	4	50	20	50	40	459.1	472.4	401.9	340.2	340.2
100	100	6	1170	2	60	20	50	40	265.9	225.4	122.2	112.8	112.8
100	100	6	1170	3	60	20	50	40	265.9	225.4	161.6	141.5	141.5
100	100	6	1170	4	60	20	50	40	265.9	225.4	201.0	170.1	170.1
100	100	8	1540	2	60	20	50	40	350.0	296.9	163.0	150.4	150.4
100	100	8	1540	3	60	20	50	40	350.0	296.9	215.5	188.6	188.6
100	100	8	1540	4	60	20	50	40	350.0	316.7	267.9	226.8	226.8
100	100	10	1900	2	60	20	50	40	431.8	366.6	203.7	188.1	188.1
100	100	10	1900	3	60	20	50	40	431.8	367.7	269.3	235.8	235.8
100	100	10	1900	4	60	20	50	40	431.8	417.7	334.9	283.5	283.5
100	100	12	2260	2	60	20	50	40	513.6	434.5	244.4	225.7	225.7
100	100	12	2260	3	60	20	50	40	513.6	467.4	323.2	282.9	282.9
100	100	12	2260	4	60	20	50	40	513.6	516.3	401.9	340.2	340.2
110	110	8	1710	2	65	20	50	40	388.6	333.3	174.8	159.5	159.5
110	110	8	1710	3	65	20	50	40	388.6	333.3	227.3	197.7	197.7

(*contd*)

Table D.8 (*contd*)

110	110	8	1710	4	65	20	50	40	388.6	333.3	279.7	235.9	235.9
110	110	10	2110	2	65	20	50	40	479.5	412.1	218.5	199.4	199.4
110	110	10	2110	3	65	20	50	40	479.5	412.1	284.1	247.1	247.1
110	110	10	2110	4	65	20	50	40	479.5	445.3	349.7	294.9	294.9
110	110	12	2510	2	65	20	50	40	570.5	489.0	262.2	239.3	239.3
110	110	12	2510	3	65	20	50	40	570.5	491.2	340.9	296.6	296.6
110	110	12	2510	4	65	20	50	40	570.5	556.6	419.6	353.8	353.8
110	110	16	3280	2	65	20	50	40	570.5	637.5	349.5	319.1	319.1
110	110	16	3280	3	65	20	50	40	745.5	709.2	454.5	395.4	395.4
110	110	16	3280	4	65	20	50	40	745.5	771.9	559.5	471.8	471.8
130	130	8	2030	2	80	20	50	40	745.5	406.0	186.6	168.6	168.6
130	130	8	2030	3	80	20	50	40	461.4	406.0	239.1	206.8	206.8
130	130	8	2030	4	80	20	50	40	461.4	406.0	291.6	245.0	245.0
130	130	10	2510	2	80	20	50	40	570.5	502.9	233.2	210.8	210.8
130	130	10	2510	3	80	20	50	40	570.5	502.9	298.8	258.5	258.5
130	130	10	2510	4	80	20	50	40	570.5	502.9	364.4	306.2	306.2
130	130	12	2990	2	80	20	50	40	679.5	598.1	279.9	252.9	252.9
130	130	12	2990	3	80	20	50	40	679.5	598.1	358.6	310.2	310.2
130	130	12	2990	4	80	20	50	40	679.5	610.7	437.3	367.5	367.5
130	130	16	3920	2	80	20	50	40	890.9	782.9	373.2	337.3	337.3
130	130	16	3920	3	80	20	50	40	890.9	782.9	478.1	413.6	413.6
130	130	16	3920	4	80	20	50	40	890.9	877.4	583.1	490.0	490.0
150	150	10	2920	2	90	20	50	40	663.6	593.8	262.7	233.5	233.5
150	150	10	2920	3	90	20	50	40	663.6	593.8	328.4	281.2	281.2
150	150	10	2920	4	90	20	50	40	663.6	593.8	394.0	329.0	329.0

(*contd*)

Table D.8 (*contd*)

150	150	12	3480	2	90	20	50	40	790.9	707.1	315.3	280.2	280.2
150	150	12	3480	3	90	20	50	40	790.9	707.1	394.0	337.5	337.5
150	150	12	3480	4	90	20	50	40	790.9	707.1	472.8	394.8	394.8
150	150	16	4560	2	90	20	50	40	1036.4	928.2	420.4	373.6	373.6
150	150	16	4560	3	90	20	50	40	1036.4	928.2	525.4	450.0	450.0
150	150	16	4560	4	90	20	50	40	1036.4	954.7	630.3	526.3	526.3
150	150	20	5620	2	90	20	50	40	1277.3	1142.1	525.5	467.0	467.0
150	150	20	5620	3	90	20	50	40	1277.3	1142.1	656.7	562.5	562.5
150	150	20	5620	4	90	20	50	40	1277.3	1263.2	787.9	657.9	657.9
200	200	12	4690	2	115	20	50	40	1065.9	979.7	403.9	348.4	348.4
200	200	12	4690	3	115	20	50	40	1065.9	979.7	482.6	405.7	405.7
200	200	12	4690	4	115	20	50	40	1065.9	979.7	561.3	462.9	462.9
200	200	16	6180	2	115	20	50	40	1404.5	1291.7	538.5	464.5	464.5
200	200	16	6180	3	115	20	50	40	1404.5	1291.7	643.4	540.9	540.9
200	200	16	6180	4	115	20	50	40	1404.5	1291.7	748.4	617.2	617.2
200	200	20	7640	2	115	20	50	40	1736.4	1596.4	673.1	580.7	580.7
200	200	20	7640	3	115	20	50	40	1736.4	1596.4	804.3	676.1	676.1
200	200	20	7640	4	115	20	50	40	1736.4	1596.4	935.5	771.6	771.6
200	200	25	9410	2	115	20	50	40	2138.6	1967.1	841.4	725.8	725.8
200	200	25	9410	3	115	20	50	40	2138.6	1967.1	1005.4	845.1	845.1
200	200	25	9410	4	115	20	50	40	2138.6	1967.1	1169.4	964.4	964.4

Table D.9 Capacity of concentrically loaded angles in compression (f_y = 250 mpa) – Clause 7.1.2 of the code

Designation	Mass (N/m)	Sectional Dimensions		Capacity in KN for effective length KL (m)													
		Area (mm²)	r_v(Min) (mm)	0.5	1	1.5	2	2.5	3	3.5	4	4.5	5	5.5	6	6.5	7
L35 35×5	26	327	6.7	47	20	10	6	4	3	2	2	1	1	1	1	1	1
×6	30	386	6.7	56	23	11	7	4	3	2	2	1	1	1	1	1	1
L40 40×5	30	378	7.7	61	28	14	9	6	4	3	2	2	2	1	1	1	1
×6	35	447	7.7	72	33	17	10	7	5	4	3	2	2	1	1	1	1
L45 45×5	34	428	8.7	74	38	20	12	8	6	4	3	3	2	2	2	1	1
×6	40	507	8.7	87	45	24	14	10	7	5	4	3	3	2	2	2	1
L50 50×5	38	479	9.7	87	49	27	16	11	8	6	5	4	3	2	2	2	2
×6	45	568	9.6	102	58	31	19	13	9	7	5	4	3	3	2	2	2
L55 55×5	41	527	10.6	99	60	34	21	14	10	8	6	5	4	3	3	2	2
×6	49	626	10.6	117	72	41	25	17	12	9	7	6	5	4	3	3	2
×8	64	818	10.6	153	94	53	33	22	16	12	9	7	6	5	4	4	3
×10	79	1000	10.6	188	115	65	40	27	19	15	11	9	7	6	5	4	4
L60 60×5	45	575	11.6	111	73	43	27	18	13	10	8	6	5	4	4	3	3
×6	54	684	11.5	132	86	51	32	21	15	12	9	7	6	5	4	4	3
×8	70	896	11.5	173	112	66	41	28	20	15	12	9	8	6	5	5	4
×10	86	1100	11.5	212	138	81	51	34	25	19	14	12	9	8	7	6	5
*L65 65×5	49	625(600)	12.6	119	82	51	32	22	16	12	9	8	6	5	4	4	3
×6	58	744	12.6	148	102	63	40	27	20	15	12	9	8	6	5	4	4
×8	77	976	12.5	193	133	82	52	35	26	19	15	12	10	8	7	6	5
×10	94	1200	12.5	237	164	100	64	44	31	24	18	15	12	10	9	7	6

(contd)

Table D.9 (*contd*)

*L70 70×5	53	677(600)	13.6	121	88	57	37	25	18	14	11	9	7	6	5	4	4
×6	63	806	13.6	163	118	76	49	34	25	19	15	12	10	8	7	6	5
×8	83	1060	13.5	214	154	99	64	44	32	24	19	15	12	10	9	8	7
×10	102	1300	13.5	262	189	121	79	54	39	30	23	19	15	13	11	9	8
*L75 75×5	57	727(600)	14.6	123	93	62	41	29	21	16	12	10	8	7	6	5	4
×6	68	866	14.6	178	134	90	59	41	30	23	18	14	12	10	8	7	6
×8	89	1140	14.5	234	175	117	77	54	39	30	23	19	15	13	11	9	8
×10	110	1400	14.5	288	215	144	95	66	48	36	28	23	19	16	13	11	10
*L80 80×6	73	929(864)	15.6	180	140	97	66	46	34	26	20	16	13	11	9	8	7
×8	96	1220	15.5	254	196	136	92	64	47	36	28	23	19	15	13	11	10
×10	118	1500	15.5	313	242	167	113	79	58	44	34	28	23	19	16	14	12
×12	140	1780	15.4	370	285	197	132	93	68	51	40	32	27	22	19	16	14
*L90 90×6	82	1050(864)	17.5	184	149	110	78	55	41	31	25	20	16	14	12	10	9
×8	108	1380	17.5	294	238	176	124	88	65	50	39	32	26	22	19	16	14
×10	134	1700	17.4	362	293	215	151	108	80	61	48	39	32	27	23	20	17
×12	158	2020	17.4	430	348	256	180	128	95	73	57	46	38	32	27	23	20
*L100 100×6	92	1170(864)	19.5	188	157	122	90	66	49	38	30	24	20	17	14	12	11
×8	121	1540	19.5	334	280	217	160	117	88	68	53	43	36	30	26	22	19
×10	149	1900	19.4	412	344	267	196	143	107	83	65	53	44	37	31	27	23
×12	177	2260	19.4	490	409	318	233	170	127	98	78	63	52	44	37	32	28
*L110 110×8	134	1710(1536)	21.8	339	291	237	182	137	104	81	65	53	44	37	31	27	24
×10	166	2110	21.6	465	399	323	247	186	141	110	88	71	59	50	42	36	32
×12	197	2510	21.5	553	473	383	293	220	167	130	103	84	70	59	50	43	37
×16	257	3280	21.4	722	618	499	381	285	217	168	134	109	90	76	65	56	48
*L130 130×8	159	2030(1536)	25.9	346	307	263	216	172	135	107	87	71	59	50	43	37	32

(*contd*)

Table D.9 *(contd)*

*L130 130×10	197	2510(2400)	25.7	540	478	410	335	266	209	166	134	110	91	77	66	57	50
×12	235	2990	25.6	673	595	509	416	329	259	205	166	136	113	96	82	71	62
×16	307	3920	25.4	881	779	665	542	428	336	266	214	176	146	124	106	91	80
L150 150×10	229	2920(2400)	29.8	549	496	439	376	313	255	207	170	141	118	100	86	75	65
×12	273	3480	29.7	795	718	636	545	452	368	299	245	203	170	145	124	108	94
×16	358	4560	29.4	1041	939	830	709	586	476	386	316	261	219	186	160	138	121
×20	441	5620	29.3	1234	1116	989	848	705	575	468	384	318	267	227	195	169	148
*L200 200×12	369	4690(3456)	39.9	809	753	695	633	567	497	430	369	316	271	234	204	179	158
×16	485	6180	39.6	1446	1344	1241	1129	1009	884	763	653	559	480	414	360	315	278
×20	600	7640	39.3	1718	1599	1477	1347	1207	1060	918	788	675	580	502	437	382	337
×25	739	9410	39.1	2116	1968	1817	1656	1481	1300	1124	964	826	709	613	533	467	412

*For these sections the limiting width to thickness ratio as given in Table 2 of IS 800 are exceeded; hence only the effective width is taken in the calculation of capacity.

Note: For thicknesses ≥ 20, f_y has been reduced to 240 MPa as per Table 1 of IS 800:2007

Table D. 10 Capacity of eccentrically connected angles in compression (f_y = 250 MPa) with fixed ends (Clause 7.5.1.2, k_1 = 0.20, k_2 = 0.35 and k_3 = 20)

Designation	Mass (N/m)	Sectional dimensions		Capacity in KN for effective length KL (m)													
		Area (mm^2)	r_v (Min) (mm)	0.5	1	1.5	2	2.5	3	3.5	4	4.5	5	5.5	6	6.5	7
L35 35×5	26	327	6.7	51	34	21	14	10	7	5	4	3	3	2	2	2	1
×6	30	386	6.7	62	41	25	16	11	8	6	5	4	3	3	2	2	2
L40 40×5	30	378	7.7	60	44	29	20	14	10	8	6	5	4	3	3	2	2
×6	35	447	7.7	73	53	35	23	16	12	9	7	6	5	4	3	3	3
L45 45×5	34	428	8.7	68	53	37	26	19	14	11	9	7	6	5	4	3	3
×6	40	507	8.7	84	65	45	32	23	17	13	10	8	7	6	5	4	4
L50 50×5	38	479	9.7	75	61	46	33	25	19	14	12	9	8	7	6	5	4
×6	45	568	9.6	93	75	56	40	29	22	17	14	11	9	8	7	6	5
L55 55×5	41	527	10.6	81	68	53	40	30	23	18	15	12	10	8	7	6	5
×6	49	626	10.6	102	85	66	49	37	28	22	18	14	12	10	9	7	6
×8	64	818	10.6	141	118	90	66	49	38	29	23	19	16	13	11	10	9
×10	79	1000	10.6	177	148	113	83	61	46	36	29	23	19	16	14	12	10
L60 60×5	45	575	11.6	86	75	60	47	36	29	23	18	15	13	11	9	8	7
×6	54	684	11.5	109	94	75	58	44	34	27	22	18	15	13	11	9	8
×8	70	896	11.5	153	132	104	79	60	46	36	29	24	20	17	14	12	11
×10	86	1100	11.5	195	167	132	100	75	58	45	36	30	25	21	18	15	13
*L65 65×5	49	625(600)	12.6	87	77	65	52	41	33	26	22	18	15	13	11	10	8
×6	58	744	12.6	117	103	85	68	53	42	34	27	23	19	16	14	12	11
×8	77	976	12.5	166	146	119	94	72	57	45	36	30	25	21	18	16	14
×10	94	1200	12.5	212	186	152	118	91	71	56	45	37	31	26	22	19	17

(contd)

Table D.10 (contd)

(contd)

Designation																	
*L70 70×5	53	677(600)	13.6	84	76	65	54	44	36	29	24	20	17	15	13	11	10
×6	63	806	13.6	124	111	95	77	62	50	40	33	28	23	20	17	15	13
×8	83	1060	13.5	178	160	134	108	86	68	55	45	37	31	26	23	20	17
×10	102	1300	13.5	228	204	171	137	108	85	68	55	46	38	32	28	24	21
*L75 75×5	57	727(600)	14.6	81	75	65	55	46	38	31	26	22	19	16	14	12	11
×6	68	866	14.6	130	118	103	86	70	58	47	39	33	28	24	21	18	16
×8	89	1140	14.5	189	172	148	122	99	80	65	53	44	37	32	27	24	21
×10	110	1400	14.5	244	222	190	156	125	100	81	66	55	46	39	34	30	26
*L80 80×6	73	929(864)	15.6	126	116	103	88	73	61	51	42	36	31	26	23	20	18
×8	96	1220	15.5	200	184	161	136	112	91	75	62	52	44	38	33	29	25
×10	118	1500	15.5	260	239	208	175	143	116	95	78	65	55	47	41	35	31
×12	140	1780	15.4	318	291	254	212	172	139	113	93	78	66	56	48	42	37
*L90 90×6	82	1050(864)	17.5	118	111	101	89	77	66	56	48	41	36	31	27	24	21
×8	108	1380	17.5	218	205	184	161	137	115	97	82	70	60	52	45	39	35
×10	134	1700	17.4	288	270	242	210	177	148	124	104	88	75	64	56	49	43
×12	158	2020	17.4	356	333	298	258	217	181	150	126	106	90	77	67	58	51
*L100 100×6	92	1170(864)	19.5	110	105	97	88	78	69	60	52	46	40	35	31	28	25
×8	121	1540	19.5	234	222	204	183	160	139	119	103	89	77	67	59	52	46
×10	149	1900	19.4	314	298	273	243	211	181	155	132	113	97	84	74	65	57
×12	177	2260	19.4	392	371	340	302	261	223	189	161	137	118	102	89	78	69
*L110 110×8	134	1710(1536)	21.8	223	214	200	183	165	146	129	113	99	87	77	68	60	54
×10	166	2110	21.6	339	325	303	276	246	216	188	163	142	124	109	96	85	75
×12	197	2510	21.5	427	409	380	345	306	267	232	200	173	150	131	115	102	90
×16	257	3280	21.4	591	565	525	475	419	364	314	270	232	201	174	153	134	119

Table D.10 (contd)

Section																	
*L130 130×8	159	2030(1536)	25.9	201	195	187	176	163	150	136	123	111	100	91	82	74	67
*L130 130×10	197	2510(2400)	25.7	360	350	333	312	287	261	235	211	189	168	150	135	121	109
×12	235	2990	25.6	485	471	447	418	383	347	311	277	246	218	194	173	155	139
×16	307	3920	25.4	689	667	633	589	538	485	432	382	337	297	263	233	207	185
*L150 150×10	229	2920(2400)	29.8	332	325	314	299	282	263	243	223	204	186	170	155	141	129
×12	273	3480	29.7	534	522	503	478	449	416	383	349	318	288	261	236	214	195
×16	358	4560	29.4	777	758	729	691	646	596	545	494	445	400	360	324	292	264
×20	441	5620	29.3	966	943	906	857	800	737	671	606	545	488	437	392	352	318
*L200 200×12	369	4690(3456)	39.9	445	440	432	421	407	391	374	355	336	317	298	280	262	246
×16	485	6180	39.6	951	939	919	893	861	824	783	739	695	650	606	564	524	487
×20	600	7640	39.3	1230	1214	1188	1152	1109	1058	1003	944	884	823	764	708	655	605
×25	739	9410	39.1	1603	1581	1546	1499	1440	1373	1299	1220	1139	1059	980	905	834	768

*For these sections the limiting width to thickness ratio as given in Table 2 of IS 800 are exceeded; hence only the effective width is taken in the calculation of capacity

Note: For thicknesses ≥ 20, f_y has been reduced to 240 MPa as per Table 1 of IS 800:2007

Table D. 11 Capacity of eccentrically connected angles in compression (f_y = 250 MPa) with hinged ends (Clause 7.5.1.2, k_1 = 1.25, k_2 = 0.5 and k_3 = 60)

Designation	Sectional dimensions			Capacity in KN for effective length KL (m)													
	Mass (N/m)	Area (mm²)	r_v (Min) (mm)	0.5	1	1.5	2	2.5	3	3.5	4	4.5	5	5.5	6	6.5	7
L35 35 ×5	26	327	6.7	26	18	12	9	6	5	4	3	2	2	2	1	1	1
×6	30	386	6.7	32	22	15	10	7	5	4	3	3	2	2	2	1	1
L40 40 ×5	30	378	7.7	30	23	16	12	9	7	5	4	3	3	2	2	2	1
×6	35	447	7.7	37	28	20	14	10	8	6	5	4	3	3	2	2	2
L45 45 ×5	34	428	8.7	33	26	20	15	11	9	7	6	5	4	3	3	2	2
×6	40	507	8.7	41	33	25	19	14	11	8	7	6	5	4	3	3	3
L50 50 ×5	38	479	9.7	35	30	24	19	14	11	9	7	6	5	4	3	3	3
×6	45	568	9.6	45	38	30	23	18	14	11	9	7	6	5	4	4	3
L55 55 ×5	41	527	10.6	37	32	27	22	17	14	11	9	8	7	6	5	4	4
×6	49	626	10.6	48	42	34	27	21	17	14	11	9	8	6	6	5	4
×8	64	818	10.6	71	60	48	38	29	23	19	15	13	11	9	8	7	6
×10	79	1000	10.6	91	77	61	47	37	29	23	19	15	13	11	9	8	7
L60 60 ×5	45	575	11.6	38	34	29	24	20	17	14	11	10	8	7	6	5	5
×6	54	684	11.5	51	45	38	31	25	20	17	14	12	10	8	7	6	6
×8	70	896	11.5	76	66	55	44	35	28	23	19	16	13	11	10	8	7
×10	86	1100	11.5	99	86	70	56	44	35	28	23	19	16	14	12	10	9
*L65 65 ×5	49	625(600)	12.6	38	35	30	26	22	18	15	13	11	10	8	7	6	6
×6	58	744	12.6	53	48	42	35	29	24	20	17	14	12	11	9	8	7
×8	77	976	12.5	81	72	61	50	41	33	28	23	19	16	14	12	11	9
×10	94	1200	12.5	107	95	79	65	52	42	35	29	24	20	17	15	13	12
*L70 70 ×5	53	677(600)	13.6	36	33	30	26	23	19	16	14	12	11	9	8	7	6

(contd)

Table D.11 (contd)

×6	63	806	13.6	55	51	45	39	33	28	23	20	17	15	13	11	10	9
×8	83	1060	13.5	85	78	67	57	47	39	33	27	23	20	17	15	13	11
×10	102	1300	13.5	114	103	88	74	61	50	41	35	29	25	21	18	16	14
*L75 75×5	57	727(600)	14.6	34	32	29	26	23	20	17	15	13	11	10	9	8	7
×6	68	866	14.6	57	53	48	42	36	31	27	23	20	17	15	13	12	10
×8	89	1140	14.5	89	82	73	63	53	45	38	32	27	24	20	18	16	14
×10	110	1400	14.5	121	110	97	82	69	58	48	41	35	30	26	22	20	17
*L80 80×6	73	929(864)	15.6	54	51	47	42	37	32	28	24	21	18	16	14	13	11
×8	96	1220	15.5	93	87	78	68	59	50	43	37	32	27	24	21	19	16
×10	118	1500	15.5	127	117	104	90	77	65	55	47	40	35	30	26	23	21
×12	140	1780	15.4	160	147	130	112	94	80	67	57	48	42	36	31	28	24
*L90 90×6	82	1050(864)	17.5	49	47	44	40	37	33	29	26	23	20	18	16	15	13
×8	108	1380	17.5	98	93	86	78	69	61	53	46	41	36	32	28	25	22
×10	134	1700	17.4	137	129	118	105	92	80	69	60	52	46	40	35	31	28
×12	158	2020	17.4	176	165	150	132	115	99	86	74	64	56	49	43	38	34
*L100 100×6	92	1170(864)	19.5	44	43	41	38	35	33	30	27	24	22	20	18	16	15
×8	121	1540	19.5	102	98	92	85	77	70	62	56	49	44	39	35	32	29
×10	149	1900	19.4	146	139	130	118	106	94	83	74	65	57	51	45	41	36
×12	177	2260	19.4	189	180	167	151	135	119	104	92	80	71	63	55	49	44
*L110 110×8	134	1710(1536)	21.8	95	92	88	82	76	70	64	58	53	48	43	39	36	32
×10	166	2110	21.6	153	148	140	130	119	108	98	88	79	70	63	57	51	46
×12	197	2510	21.5	202	194	183	169	153	138	123	110	98	87	78	70	63	57

(contd)

Table D.11 *(contd)*

×16	257	3280	21.4	296	283	264	242	218	194	172	152	134	119	106	94	84	76
*L130 130×8	159	2030(1536)	25.9	81	80	77	74	71	67	63	59	55	51	47	44	40	37
*L130 130×10	197	2510(2400)	25.7	155	152	146	139	132	123	114	105	97	89	82	75	69	63
×12	235	2990	25.6	220	215	206	195	183	169	156	143	131	119	109	99	90	83
×16	307	3920	25.4	334	325	310	291	270	247	226	205	185	167	151	137	124	113
*L150 150×10	229	2920(2400)	29.8	137	135	132	128	123	117	111	105	98	92	86	81	75	70
×12	273	3480	29.7	232	228	222	214	204	193	182	170	159	148	137	127	118	109
×16	358	4560	29.4	365	358	346	330	312	293	273	253	233	215	198	182	167	154
×20	441	5620	29.3	476	465	448	426	401	374	346	319	293	268	245	225	206	188
*L200 200×12	369	4690(3456)	39.9	178	177	175	172	168	164	159	154	149	143	137	132	126	120
×16	485	6180	39.6	414	410	403	394	384	371	358	343	328	313	297	282	267	253
×20	600	7640	39.3	569	563	552	538	521	502	481	459	436	413	390	368	346	326
×25	739	9410	39.1	780	770	754	733	708	679	647	614	581	547	515	483	453	424

*For these sections the limiting width to thickness ratio as given in Table 2 of IS 800 are exceeded; hence only the effective width is taken in the calculation of capacity.

Note: For thicknesses ≥ 20, f_y has been reduced to 240 MPa as per Table 1 of IS 800:2007

Table D.12 Capacity of I-section compression members ((Clause 7.1.2.1)

Compression Resistance P_{dz} and P_{dy} (kN) for effective length L_e (m)

Section Designation		2	2.5	3	3.5	4	4.5	5	5.5	6	6.5	7
MB100	P_{dz}	234	219	199	175	150	127	107	91	79	68	59
	P_{dy}	48	32	23	17	13	10	8	7	6	5	4
MB125	P_{dz}	364	351	335	315	290	262	233	205	180	158	140
	P_{dy}	134	92	66	50	39	31	25	21	18	15	13
MB150	P_{dz}	417	407	394	379	361	339	313	286	258	232	208
	P_{dy}	160	110	80	60	47	38	31	26	22	19	16
MB175	P_{dz}	553	542	530	515	499	479	456	429	399	368	337
	P_{dy}	248	175	128	97	76	61	50	42	35	30	26
MB200	P_{dz}	689	678	666	652	637	620	600	577	551	522	490
	P_{dy}	389	288	216	166	131	105	87	73	62	53	46
MB225	P_{dz}	894	881	868	854	838	821	801	779	754	725	693
	P_{dy}	561	430	329	255	202	164	135	114	97	83	72
MB250	P_{dz}	1076	1063	1049	1035	1019	1002	983	963	940	913	884
	P_{dy}	749	603	475	375	301	245	203	171	146	125	109
MB300	P_{dz}	1337	1324	1310	1296	1282	1266	1249	1231	1212	1190	1167
	P_{dy}	975	809	651	521	421	345	287	242	207	178	155
MB350	P_{dz}	1530	1517	1504	1490	1476	1462	1447	1431	1414	1396	1376
	P_{dy}	1105	914	734	587	474	388	323	272	232	200	175
MB400	P_{dz}	1805	1792	1779	1765	1751	1736	1722	1706	1690	1673	1654
	P_{dy}	1293	1066	855	683	551	451	375	316	269	232	203
MB450	P_{dz}	2130	2116	2102	2088	2073	2059	2044	2028	2012	1995	1978
	P_{dy}	1584	1339	1095	886	721	593	494	418	357	308	269

(contd)

Table D.12 (*contd*)

Section												
MB500	P_{dz}	2571	2556	2541	2525	2510	2494	2478	2462	2445	2428	2410
	P_{dy}	2060	1829	1572	1322	1103	923	778	662	569	493	432
MB550	P_{dz}	3064	3048	3031	3015	2998	2981	2964	2947	2930	2911	2893
	P_{dy}	2506	2257	1974	1686	1423	1200	1017	869	749	651	570
MB600	P_{dz}	3484	3467	3450	3434	3417	3400	3383	3366	3348	3330	3312
	P_{dy}	2955	2726	2460	2170	1882	1619	1392	1201	1042	910	800
SC100	P_{dz}	501	460	414	363	314	269	231	199	172	150	132
	P_{dy}	322	246	190	149	119	97	80	68	58	50	43
SC120	P_{dz}	688	650	606	557	503	449	397	350	308	272	241
	P_{dy}	495	401	321	258	209	172	144	122	105	90	79
SC140	P_{dz}	899	861	819	772	719	662	603	545	490	439	394
	P_{dy}	698	592	492	406	336	280	236	201	173	151	132
SC150	P_{dz}	1019	982	941	896	846	790	731	671	611	555	503
	P_{dy}	855	759	660	566	482	410	351	303	263	230	202
SC160	P_{dz}	1155	1115	1073	1025	973	915	853	788	723	660	601
	P_{dy}	936	819	702	593	500	423	360	308	267	233	205
SC180	P_{dz}	1414	1373	1330	1283	1232	1176	1115	1049	981	911	843
	P_{dy}	1187	1065	938	814	700	601	517	447	389	341	301
SC200	P_{dz}	1705	1662	1618	1570	1519	1463	1403	1338	1268	1196	1121
	P_{dy}	1467	1339	1203	1065	933	813	708	617	541	477	422
SC220	P_{dz}	2011	1967	1921	1873	1821	1766	1706	1642	1574	1501	1424
	P_{dy}	1766	1634	1493	1346	1201	1063	937	826	730	647	576
SC250	P_{dz}	2468	2422	2374	2325	2274	2219	2162	2100	2034	1964	1890
	P_{dy}	2207	2067	1919	1762	1602	1444	1294	1156	1032	923	827

(*contd*)

Table D.12 (contd)

Section												
HB150	P_{dz}	742	716	686	654	618	578	536	492	449	408	370
	P_{dy}	599	522	445	374	314	265	225	193	167	145	128
HB200	P_{dz}	1057	1032	1005	977	946	914	878	840	798	755	711
	P_{dy}	911	833	750	665	584	510	444	388	341	300	266
HB225	P_{dz}	1235	1209	1182	1155	1125	1094	1060	1024	986	944	901
	P_{dy}	1084	1004	919	831	743	659	582	513	454	403	359
HB250	P_{dz}	1474	1447	1419	1390	1360	1329	1295	1260	1222	1181	1139
	P_{dy}	1318	1235	1147	1055	960	866	776	694	620	555	497
HB300	P_{dz}	1716	1690	1664	1637	1610	1581	1552	1521	1488	1454	1418
	P_{dy}	1511	1414	1311	1202	1091	982	879	784	699	625	560
HB350	P_{dz}	1986	1960	1934	1907	1881	1853	1825	1796	1766	1734	1701
	P_{dy}	1729	1616	1496	1369	1240	1114	994	886	789	704	630
HB400	P_{dz}	2296	2270	2243	2217	2190	2163	2135	2107	2078	2048	2016
	P_{dy}	1979	1847	1705	1557	1407	1260	1123	998	888	791	707
HB450	P_{dz}	2595	2568	2541	2514	2487	2460	2433	2405	2376	2347	2317
	P_{dy}	2217	2065	1903	1733	1561	1395	1240	1100	977	870	777
UC 152 × 152 × 23	P_{dz}	630	608	583	556	525	492	456	420	383	349	316
	P_{dy}	519	457	394	335	283	240	205	176	152	133	117
UC 152 × 152 × 30	P_{dz}	828	800	769	735	698	657	612	566	520	475	432
	P_{dy}	689	611	531	455	388	330	282	243	211	185	163
UC 152 × 152 × 37	P_{dz}	1021	987	950	909	864	815	761	705	649	594	542
	P_{dy}	853	758	660	567	484	412	353	304	264	231	204
UC 203 × 203 × 46	P_{dz}	1309	1277	1245	1211	1174	1134	1091	1045	995	943	888
	P_{dy}	1170	1089	1002	911	819	731	649	575	510	454	405

(contd)

Table D.12 (*contd*)

Section												
UC 203×203×52	P_{dz}	1478	1444	1407	1369	1328	1284	1237	1185	1130	1072	1011
	P_{dy}	1324	1233	1136	1035	932	833	741	657	583	519	464
UC 203×203×60	P_{dz}	1702	1662	1621	1577	1531	1481	1426	1367	1305	1238	1169
	P_{dy}	1525	1421	1310	1194	1077	962	856	760	675	601	537
UC 203×203×71	P_{dz}	2022	1977	1929	1879	1826	1769	1707	1640	1569	1493	1414
	P_{dy}	1817	1697	1569	1434	1297	1164	1038	924	822	733	656
UC 203×203×86	P_{dz}	2453	2399	2342	2282	2219	2151	2077	1998	1913	1822	1728
	P_{dy}	2207	2062	1908	1747	1582	1421	1269	1130	1007	898	804
UC 254×254×73	P_{dz}	2113	2075	2036	1996	1954	1910	1863	1813	1760	1704	1644
	P_{dy}	1957	1860	1759	1652	1540	1426	1311	1198	1092	993	902
UC 254×254×89	P_{dz}	2574	2528	2482	2434	2383	2331	2275	2216	2153	2086	2015
	P_{dy}	2386	2270	2148	2020	1886	1748	1610	1474	1345	1224	1114
UC 254×254×107	P_{dz}	3101	3047	2991	2934	2874	2812	2746	2676	2601	2522	2438
	P_{dy}	2876	2737	2592	2439	2279	2114	1948	1785	1630	1485	1351
UC 254×254×132	P_{dz}	3145	3091	3036	2979	2921	2859	2795	2726	2654	2576	2494
	P_{dy}	2920	2782	2638	2486	2327	2163	1997	1834	1678	1531	1395
UC 254×254×167	P_{dz}	4856	4775	4692	4608	4520	4429	4333	4232	4124	4010	3889
	P_{dy}	4517	4308	4090	3862	3622	3375	3124	2876	2636	2410	2201
UC 305×305×97	P_{dz}	2837	2795	2753	2711	2667	2622	2576	2527	2476	2422	2366
	P_{dy}	2672	2567	2458	2345	2227	2104	1977	1848	1720	1595	1475
UC 305×305×118	P_{dz}	3455	3405	3354	3303	3251	3197	3141	3083	3022	2958	2890
	P_{dy}	3258	3131	3000	2864	2723	2575	2422	2268	2113	1961	1816
UC 305×305×137	P_{dz}	4014	3956	3899	3840	3780	3718	3655	3588	3518	3445	3368
	P_{dy}	3787	3641	3491	3335	3171	3002	2826	2648	2470	2295	2126

(*contd*)

Table D.12 (contd)

UC 305 × 305 × 158	P_{dz}	4684	4618	4551	4484	4415	4344	4270	4194	4114	4030	3942
	P_{dy}	4423	4254	4081	3901	3713	3517	3315	3109	2902	2699	2504
UC 305 × 305 × 198	P_{dz}	5821	5740	5659	5578	5494	5409	5321	5229	5133	5033	4928
	P_{dy}	5503	5297	5087	4868	4640	4403	4157	3907	3655	3406	3165
UC 305 × 305 × 240	P_{dz}	7060	6964	6868	6772	6673	6572	6468	6360	6248	6130	6007
	P_{dy}	6681	6435	6184	5923	5652	5370	5078	4779	4478	4179	3889
UC 305 × 305 × 283	P_{dz}	8329	8219	8109	7997	7884	7768	7648	7525	7396	7262	7122
	P_{dy}	7890	7605	7314	7013	6699	6373	6035	5689	5339	4991	4652
UC 356 × 368 × 129	P_{dz}	3808	3761	3713	3665	3617	3567	3517	3465	3411	3355	3296
	P_{dy}	3661	3547	3433	3316	3195	3070	2940	2806	2669	2529	2390
UC 356 × 368 × 153	P_{dz}	4517	4461	4406	4349	4292	4234	4175	4114	4051	3985	3917
	P_{dy}	4343	4210	4075	3937	3795	3648	3496	3338	3177	3013	2848
UC 356 × 368 × 177	P_{dz}	5231	5167	5103	5038	4973	4906	4838	4768	4696	4621	4543
	P_{dy}	5031	4878	4722	4564	4400	4231	4056	3875	3689	3500	3311
UC 356 × 368 × 202	P_{dz}	5969	5896	5824	5751	5677	5602	5525	5446	5365	5280	5192
	P_{dy}	5743	5569	5393	5213	5028	4837	4639	4434	4224	4010	3795

Note: For thicknesses ≥ 20, f_y has been reduced to 240 MPa as per Table 1 of IS 800:2007

Table D. 13 Capacity of square hollow section compression members (Clause 7.1.2.1)

Square Hollow Sections D × B (mm×mm)	Thickness t (mm)	Unit weight, w (N/m)	Sec. Area A (mm²)	Radius of gyration r_z (mm)	Length (m)													
					0.5	1	1.5	2	2.5	3	3.5	4	4.5	5	5.5	6	6.5	7
25 × 25	1.6	11.2	143	9.4	28	17	9	5	3	2	2	1	1	1	1	1	1	0
	2	13.6	174	9.2	34	20	10	6	4	3	2	2	1	1	1	1	1	1
	2.6	16.9	216	8.9	42	24	12	7	5	3	2	2	1	1	1	1	1	1
	3.2	19.8	253	8.6	48	27	13	8	5	4	3	2	2	1	1	1	1	1
32 × 32	2	18	230	12.1	47	36	22	13	9	6	5	4	3	2	2	2	1	1
	2.6	22.6	288	11.8	59	45	26	16	10	7	6	4	3	3	2	2	2	2
	3.2	26.9	342	11.5	69	52	30	18	12	8	6	5	4	3	3	2	2	2
38 × 38	2	21.8	278	14.6	58	50	35	22	15	11	8	6	5	4	3	3	2	2
	2.6	27.5	351	14.3	73	62	43	27	18	13	10	8	6	5	4	3	3	3
	3.2	32.9	419	14	87	73	50	31	21	15	11	9	7	6	5	4	3	3
	4	39.5	503	13.6	104	87	57	36	24	17	13	10	8	6	5	4	3	3
40 × 40	2.6	29.2	372	15.1	78	68	49	32	21	15	11	9	7	6	5	4	3	3
	2.9	32.1	409	14.9	86	74	53	34	23	16	12	9	8	6	5	4	4	3
	3.2	34.9	445	14.8	93	80	57	37	25	18	13	10	8	7	5	5	4	3
	4	42	535	14.4	112	95	66	42	28	20	15	12	9	8	6	5	5	4
49.5 × 49. 5	2.6	36.9	470	19	100	92	77	57	40	29	22	17	14	11	9	8	7	6
	2.9	40.7	519	18.8	111	101	85	62	44	32	24	19	15	12	10	9	7	6
	3.6	49.3	628	18.5	134	122	101	73	52	37	28	22	18	14	12	10	9	7
	4.5	59.5	758	18	161	146	119	85	59	43	32	25	20	16	14	12	10	9
60 × 60	2.6	45.5	580	23.3	126	118	107	89	68	52	40	31	25	21	17	15	12	11
	2.9	50.3	641	23.1	139	130	117	97	75	56	43	34	27	22	19	16	14	12

(contd)

Table D.13 (contd)

Size																		
	3.2	55	701	23	152	142	128	106	81	61	47	37	30	24	20	17	15	13
	4	67.1	855	22.6	185	173	155	127	97	72	55	43	35	29	24	20	17	15
	4.8	78.5	1001	22.2	216	202	180	146	110	82	63	49	40	32	27	23	20	17
72 × 72	3.2	67.1	854	27.9	186	178	166	149	126	100	79	63	51	42	36	30	26	23
	4	82.2	1047	27.5	228	218	203	182	152	121	95	76	61	51	42	36	31	27
	4.8	96.6	1231	27.1	268	255	238	212	176	139	109	87	70	58	49	41	35	31
91.5 × 91.5	3.6	96.7	1232	35.6	271	262	251	237	218	192	163	136	113	95	80	69	59	52
	4.5	118.8	1514	35.2	333	322	308	291	266	234	198	164	136	114	97	83	71	62
	5.4	140.1	1785	34.8	393	379	363	341	312	273	230	190	158	132	112	95	82	72
113.5 × 113.5	4.8	159.2	2028	44	450	438	424	409	389	364	331	294	255	220	190	164	143	125
	5.4	177.4	2260	43.8	501	487	473	455	433	404	368	326	283	244	210	181	158	138
132 × 132	4.8	187.1	2383	51.6	530	518	506	492	475	454	428	397	360	321	284	250	220	195
	5.4	208.8	2659	51.3	592	578	564	548	529	506	477	441	399	356	314	276	243	215
180 × 180	5	269.7	3436	71.1	770	757	744	731	717	701	683	661	636	607	573	535	495	454
	6	320.5	4083	70.6	914	899	884	868	851	832	810	784	754	718	677	632	583	535
	7	370.3	4718	70.1	1056	1039	1021	1003	983	960	934	904	869	827	779	725	669	613
220 × 220	6	395.9	5043	87	1133	1118	1103	1087	1071	1054	1035	1015	991	965	934	899	860	817
	7	458.3	5838	86.5	1311	1294	1276	1258	1239	1219	1198	1173	1146	1115	1079	1038	992	941
250 × 250	6	452.4	5763	99.2	1297	1282	1266	1251	1236	1219	1202	1183	1162	1139	1114	1085	1052	1016
	7	524.2	6678	98.7	1503	1485	1467	1449	1431	1412	1392	1370	1345	1319	1289	1255	1217	1174

Table D. 14 Capacity of circular section compression members (Clause 7.1.2.1)

Nominal bore (mm)	Outside diameter (mm)	Thickness (mm)	Weight (N/m)	Area of cross section (mm²)	Radius of gyration (mm)	Length (m)													
						0.5	1	1.5	2	2.5	3	3.5	4	4.5	5	5.5	6	6.5	7
15	21.3	2.65	12.2	155	6.7	26	11	5	3	2	1	1	1	1	0	0	0	0	0
20	26.9	2.65	15.8	202	8.6	39	21	11	6	4	3	2	2	1	1	1	1	1	1
25	33.7	3.25	24.6	311	10.8	62	44	24	15	10	7	5	4	3	3	2	2	1	1
32	42.4	3.25	31.5	400	13.9	83	70	47	30	20	14	10	8	6	5	4	4	3	3
40	48.3	3.25	36.1	460	16	97	86	64	43	29	21	16	12	10	8	7	6	5	4
50	60.3	3.65	51	650	20.1	139	129	111	85	61	45	34	27	21	17	15	12	11	9
65	76.1	3.65	65.3	831	25.6	181	171	158	138	111	86	67	53	43	35	29	25	21	19
80	88.9	4.05	84.8	1080	30	236	226	214	196	170	140	113	91	74	61	51	44	38	33
90	101.6	4.05	97.5	1240	34.5	273	263	252	237	215	188	158	131	108	90	76	65	56	49
100	114.3	4.5	121	1550	38.9	342	332	320	305	285	259	227	194	164	139	118	101	88	76
110	127	4.85	146	1860	43.2	412	401	388	374	355	331	300	264	229	196	169	146	127	111
125	139.7	4.85	162	2050	47.7	455	444	432	418	402	381	354	321	285	250	218	190	166	146
135	152.4	4.85	177	2250	52.2	501	490	478	465	449	430	407	378	344	307	272	240	212	187
150	165.1	4.85	192	2440	56.7	544	533	521	509	494	477	456	430	399	365	328	293	261	233
150	168.3	4.85	196	2490	57.8	556	544	533	520	506	489	469	443	413	379	343	307	275	245
175	193.7	5.4	250	3190	66.6	714	701	689	675	661	644	625	602	575	542	506	466	426	387
200	219.1	5.6	294	3760	75.5	843	830	817	803	789	773	755	735	711	683	651	615	575	533

Table D. 15 Capacity of I-section beams (Clause 8.2.2)

Section	Moment capacity; M_p (kNm)	Shear capacity; V_p (kN)	Buckling resistance moment, M_d (kNm) for effective length L_e (m)								
			2	3	4	5	6	7	8	9	10
MB100	9.473	61.672	5.792	4.246	3.310	2.706	2.288	1.981	1.748	1.563	1.414
MB125	18.602	82.010	13.196	10.019	7.884	6.467	5.475	4.747	4.189	3.749	3.394
MB150	25.109	98.412	17.072	12.377	9.510	7.703	6.475	5.587	4.915	4.389	3.966
MB175	37.745	133.184	27.431	20.324	15.661	12.681	10.651	9.185	8.078	7.212	6.515
MB200	57.695	149.586	44.646	33.884	26.103	21.048	17.617	15.154	13.303	11.861	10.705
MB225	79.152	191.903	65.387	52.983	42.205	34.508	29.075	25.103	22.089	19.726	17.825
MB250	105.843	226.348	90.590	75.709	61.160	50.110	42.169	36.347	31.934	28.484	25.714
MB300	148.123	303.109	129.896	110.593	89.711	73.022	60.959	52.188	45.610	40.518	36.465
MB350	202.175	371.997	175.290	145.636	114.790	91.656	75.616	64.241	55.850	49.426	44.356
MB400	267.314	467.129	231.557	191.771	150.487	119.756	98.572	83.612	72.607	64.203	57.579
MB450	348.491	555.044	307.514	260.805	207.922	165.844	136.178	115.145	99.703	87.952	78.727
MB500	471.516	669.201	429.080	379.051	313.549	252.260	205.892	172.451	147.967	129.504	115.165
MB550	616.359	808.290	566.823	508.859	430.880	352.256	289.488	243.063	208.692	182.649	162.381
MB600	797.870	944.755	717.170	659.588	580.164	489.117	407.324	343.003	294.137	256.756	227.604
UB 203 × 133 × 25	57.537	151.980	50.915	43.531	35.130	28.294	23.387	19.866	17.258	15.261	13.686
UB 203 × 133 × 30	70.409	173.667	62.895	55.140	46.237	38.433	32.451	27.964	24.540	21.860	19.712
UB 254 × 146 × 31	88.032	197.926	79.139	68.737	55.827	44.617	36.451	30.630	26.374	23.157	20.649
UB 254 × 146 × 37	108.525	211.625	98.376	87.311	73.679	60.998	51.099	43.698	38.104	33.771	30.330
UB 254 × 146 × 43	127.411	245.258	116.092	104.461	90.408	76.785	65.556	56.805	49.997	44.614	40.273
UB 305 × 165 × 40	139.452	238.866	128.422	115.440	97.830	79.901	65.530	54.904	47.051	41.114	36.503

(contd)

Table D.15 (*contd*)

Section											
UB 305 × 165 × 46	161.491	269.546	149.153	135.148	116.614	97.349	81.296	69.042	59.774	52.643	47.028
UB 305 × 165 × 54	190.134	321.763	176.148	160.967	141.533	120.981	103.080	88.859	77.781	69.069	62.094
HEA 320	351.038	366.093	345.540	334.814	323.010	309.635	294.429	277.535	259.567	241.435	223.996
HEB 320	469.481	482.875	462.578	449.140	435.108	420.107	403.942	386.629	368.413	349.728	331.092
HEM 320	988.978	989.237	979.867	958.635	939.444	921.400	903.889	886.513	869.018	851.249	833.127
UB 356 × 171 × 45	172.728	322.765	158.812	141.837	118.196	94.453	75.997	62.702	53.078	45.924	40.448
UB 356 × 171 × 51	200.314	344.704	184.916	166.693	141.723	115.844	94.881	79.334	67.850	59.182	52.466
UB 356 × 171 × 57	226.255	380.500	209.384	189.960	163.876	136.382	113.387	95.884	82.710	72.624	64.720
UB 356 × 171 × 67	271.873	433.923	252.686	231.623	204.398	175.051	149.116	128.391	112.229	99.531	89.381
UB 406 × 178 × 54	235.872	406.772	217.831	195.782	164.637	132.225	106.418	87.654	74.036	63.916	56.183
UB 406 × 178 × 60	268.788	421.277	249.204	225.851	193.381	158.591	129.759	108.199	92.256	80.243	70.960
UB 406 × 178 × 67	302.097	472.734	280.619	255.595	221.492	184.557	153.114	129.054	110.958	97.144	86.358
UB 406 × 178 × 74	337.274	514.577	314.016	287.578	252.342	213.757	179.899	153.282	132.864	117.048	104.558
IPEA 360	195.762	309.691	180.750	162.879	138.252	112.686	92.027	76.757	65.512	57.047	50.504
IPE 360	285.562	377.902	259.050	231.872	199.383	168.580	143.612	124.323	109.379	97.586	88.086
IPEO 360	259.244	439.416	240.185	218.924	191.098	161.610	136.307	116.560	101.403	89.621	80.275
IPEA 400 / NPB 400 × 180 × 57.4	244.335	364.649	227.241	206.660	177.666	145.802	118.964	98.802	83.907	72.719	64.107
IPE 400 / NPB 400 × 180 × 66.3	281.437	451.383	261.627	238.390	206.502	171.742	142.109	119.474	102.496	89.572	79.505
IPEO 400 / NPB 400 × 180 × 75.7	325.761	514.209	303.850	278.903	245.617	208.747	175.991	150.051	130.082	114.592	102.350
UB 457 × 191 × 67	330.026	505.693	307.960	281.157	242.965	199.805	162.695	134.612	113.855	98.300	86.368
UB 457 × 191 × 74	371.322	539.691	347.368	318.792	278.891	233.235	192.750	161.319	137.670	119.714	105.794

(*contd*)

Table D.15 (contd)

UB 457 × 191 × 82	411.904	597.558	385.908	355.432	313.726	265.853	222.471	188.058	161.744	141.515	125.679
UB 457 × 191 × 89	453.349	638.458	425.536	393.501	350.601	301.191	255.311	218.002	188.932	166.275	148.349
UB 457 × 191 × 98	503.068	698.867	473.090	439.265	395.104	344.259	295.885	255.431	223.207	197.677	177.223
IPEA 450/NPB	321.913	445.767	301.335	276.388	240.947	199.970	163.754	135.887	115.115	99.478	87.450
450 × 190 × 67.2											
IPE 450/NPB	369.073	555.044	345.163	316.810	277.468	232.635	192.849	161.865	138.476	120.663	106.814
450 × 190 × 77.6											
IPEO 450/NPB	447.362	658.179	419.854	388.338	346.349	298.129	253.299	216.729	188.147	165.809	148.095
450 × 190 × 92.4											
IPEA 500/NPB	422.546	547.800	397.865	367.775	325.160	273.787	225.965	188.000	159.284	137.531	120.761
500 × 200 × 79.4											
IPE 500/NPB	478.934	669.201	450.553	416.600	369.455	313.517	261.455	219.752	187.872	163.474	144.486
500 × 200 × 90.7											
IPEO 500/NPB	574.131	796.743	541.570	503.666	453.039	393.120	335.313	287.044	248.922	219.025	195.310
500 × 200 × 107.3											
UB 533 × 210 × 82	460.061	665.485	433.418	400.480	353.096	295.295	241.533	199.169	167.393	143.515	125.243
UB 533 × 210 × 92	528.576	706.507	499.445	463.865	413.947	352.583	293.459	245.322	208.402	180.223	158.403
UB 533 × 210 × 101	585.832	760.575	554.440	516.515	464.388	400.417	337.540	285.157	244.269	212.657	187.930
UB 533 × 210 × 109	635.017	821.176	601.600	561.622	507.588	441.569	375.854	320.148	276.043	241.571	214.379
UB 533 × 210 × 122	718.464	907.378	681.946	638.948	582.506	514.149	444.684	383.951	334.583	295.223	263.700
UB 610 × 229 × 101	645.876	830.243	614.443	574.726	518.677	446.130	371.585	308.560	259.583	222.178	193.360
UB 610 × 229 × 113	736.712	884.968	702.478	659.412	600.104	523.660	443.151	372.939	317.098	273.760	239.973
UB 610 × 229 × 125.1	826.501	955.932	789.365	742.933	680.421	600.601	515.262	438.936	376.914	328.011	289.421
UB 610 × 229 × 139.9	932.455	1060.923	891.812	841.432	775.270	691.963	601.952	519.437	450.783	395.642	351.507

(contd)

Table D.15 (*contd*)

Section											
IPEA 600/NPB 600 × 220 × 107.6	682.930	767.692	649.805	608.476	551.200	478.021	402.573	337.980	287.146	247.900	217.377
IPE 600/NPB 600 × 220 × 122.4	767.298	944.755	729.114	682.307	618.297	538.013	456.145	386.045	330.562	287.412	253.599
IPEO 600/NPB 600 × 220 × 154.5	985.171	1200.626	939.500	884.801	814.008	727.185	635.624	552.540	483.304	427.301	382.082
HEAA 600/WPB 600 × 300 × 128.8	785.881	899.092	768.552	739.592	704.276	659.270	603.120	539.327	475.488	417.853	368.762
HEA 600/WPB 600 × 300 × 177.8	1178.455	1006.426	1156.580	1117.432	1072.511	1019.154	956.079	884.886	810.314	737.982	671.792
HEB 600/WPB 600 × 300 × 211.9	1422.716	1220.309	1396.995	1351.225	1300.204	1241.674	1174.676	1100.424	1022.437	945.224	872.474
HEM 600/WPB 600 × 300 × 285	1956.114	1708.432	1924.662	1867.074	1806.600	1741.597	1671.276	1595.877	1516.727	1435.976	1356.027
HEAA 700/WPB 700 × 300 × 149.9	1055.662	1142.891	1030.819	990.553	940.736	876.359	795.669	705.015	616.212	537.701	471.887
HEA 700/WPB 700 × 300 × 204.5	1553.772	1312.816	1522.742	1469.153	1406.540	1330.683	1239.696	1136.856	1030.580	929.744	839.547
HEB 700/WPB 700 × 300 × 240.5	1848.161	1561.470	1812.108	1750.114	1679.484	1596.501	1499.698	1391.743	1279.440	1170.573	1070.545
HEM 700/WPB 700 × 300 × 300.7	2350.855	1972.963	2308.919	2235.044	2154.469	2064.323	1963.454	1853.196	1737.469	1621.562	1510.281
HEAA 800/WPB 800 × 300 × 172	1351.465	1414.508	1317.266	1263.609	1196.131	1107.653	996.560	873.837	756.697	655.474	571.995

(*contd*)

Table D.15 (*contd*)

HEA 800/WPB 800×300×224	1913.894	1554.909	1873.312	1804.952	1723.528	1622.715	1499.858	1360.893	1219.502	1088.456	973.910
HEB 800/WPB 800×300×262	2261.445	1837.024	2214.411	2135.511	2043.559	1932.751	1800.918	1653.217	1501.565	1358.050	1229.614
HEM 800/WPB 800×300×317	2774.852	2243.006	2720.809	2628.695	2524.975	2404.825	2266.487	2113.358	1953.921	1798.146	1653.409
HEAA 900/WPB 900×300×198	1746.222	1712.368	1700.188	1629.254	1539.287	1420.540	1271.687	1109.159	956.308	825.769	718.951
HEA 900/WPB 900×300×252	2385.364	1868.515	2332.190	2244.678	2139.174	2006.900	1844.607	1661.876	1478.586	1311.536	1167.614
HEB 900/WPB 900×300×291	2788.332	2184.746	2727.263	2627.103	2508.681	2363.762	2189.599	1994.761	1797.282	1613.739	1452.291
HEM 900/WPB 900×300×333	3445.381	3223.976	3365.226	3240.920	3097.075	2926.227	2727.255	2509.314	2289.156	2082.207	1896.813

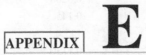

E

Conversion Factors

E.1 Conversion Factors

Some conversion factors useful in structural design, especially when the designer is using the American standard codes (which are still in FPS units), are provided in this appendix.

Quantity	To convert	To	Multiply by
Length	inch (in.)	mm	25.4
	foot (ft)	m	0.3048
	metre (m)	ft	3.2808
	mile	km	1.609
Area	in.^{2}	mm^2	645.16
	ft^2	m^2	0.0929
	m^2	ft^2	10.764
Volume	in.^{3}	mm^3	16,387
	ft^3	m^3	0.02832
	m^3	ft^3	35.315
	gallon	litre	3.7854
	litre	gallon	0.2642
Mass per unit volume	lb/ft^3	kg/m^3	16.0185
	kg/m^3	lb/ft^3	0.062428
Force	kilo pound (kip)	kN	4.448
	lb	N	4.448
	ton (2000 lb)	kN	8.896
	N	lb	0.2248
	kN	Kip	0.2248
Pressure, stress	psi	MPa	0.006895
	ksi	MPa	6.895
	kN/m^2	kip/ft^2	0.02089

(contd)

(contd)

	psf	N/m²	47.88
	N/m²	psf	0.02089
	MPa	ksi	0.145
	MPa	psi	145.0
Moments	in.-lb	N m	0.1130
	kip-in.	kN m	0.1130
	kip-ft	kN m	1.3558
	kN m	ft-kip	0.7376
Uniform loading	kip/ft	kN/m	14.59
	kip/in.	kN/m	175.2
	kN/m	kip/ft	0.06852
Speed	mile/h	m/s	0.447
Acceleration	ft/s²	m/s²	0.3048
Density	lb/in.³	kg/m³	27,680
	lb/ft³	kg/m³	16.02
Temperature	degree Fahrenheit (°F)	Degree celsius (°C)	$(t°-32)/1.8$
Inertia	in.⁴	mm⁴	416,231
Energy	ft-lb	Joule (N m/9.81)	1.356

1 Pa = 1N/m², 1 MPa = 10⁶ Pa = 1 N/mm², g = 32.17 ft/s² = 9.807m/s², 1 erg = 10⁻⁷ J, 1 Hz (Hertz) = 1 cycle/s, °K (Kelvin) = °C + 273

E.2 Basic SI units Relating to Structural Steel Design

Quantity	Unit	Symbol
Length	metre	m
Mass	kilogram	kg
Time	second	s

E.3 Derived SI Units Relating to Structural Design

The SI unit of force is Newton (N). 1 Newton is the force which causes a mass of 1 kg to have an acceleration of 1m/s². The acceleration due to gravity is 9.807 m/s² approximately, and hence the weight of a mass of 1 kg is 9.807 N.

Quantity	Unit	Symbol	Formula
Force	Newton	N	kgm/s²
Pressure, Stress	Pascal	Pa	N/m²
Energy or work	Joule	J	N m

Bibliography

Adluri, S.M.R. and M.K.S. Madugula, 'Development of column curves for single angles', *Journal of Structural Engineering*, vol. 122, no. 3, March 1996, pp. 318–325.

Agerskov, H., 'High strength bolted connection subject to prying', *Journal of Structural Div.*, ASCE, vol. 102, no. ST1, January 1979.

Arya, A.S. and J.L. Ajmani, *Design of Steel Structures*, 4th edn, Nem Chand & Bros., Roorkee, 1989.

Astanesh, A., 'Procedure for design and analysis of hanger type connections', *Engineering Journal*, AISC, vol. 22, 2nd quarter, 1985, pp. 63–66.

AISC 360-05, *Specification for Structural Steel Buildings*, American Institute of Steel Construction, Chicago, IL, 2005 (www.aisc.org).

AISC, *Specification for Load and Resistance Factor Design of Single – Angle Members*, American Institute of Steel Construction, Chicago, USA, December 1993.

Allen, D.E., *Structural Safety*, Division of Building Research, National Research Council of Canada (CBD 147) Ottawa, 1972.

Allen, D.E. and T.M. Murray, 'Design Criteria for Vibration due to Walking', *Engineering Journal*, AISC, vol. 30, no. 4, 4th quarter 1993, pp. 117–129.

Allen, H.G. and P.S. Bulson, *Background to Buckling*, McGraw-Hill Book Company (UK) Ltd, London, 1980, 582 pp.

Alpsten, G.A., *Variation in Mechanical and Cross-sectional Properties of Steel*, Swedish Institute of Steel Construction, Publication No. 42, 1973.

ANSI/ASCE, 10-90, *Design of Latticed Steel Transmission Structures*, ASCE, New York, NY, 1992.

ASCE/ANSI, 8–90, *Specification for the Design of Cold-formed Stainless Steel Structural Members*, ASCE, 1991.

Ballio, G. and F.M. Mazzolani, *Theory and Design of Steel Structures*, Chapman and Hall, London, 1983, 632 pp.

Basler, K., 'Strength of plate girders in shear', *Journal of the Structural Div.*, Proc. ASCE, 87, no. ST 7, October 1961, pp. 151–180.

Basler, K., 'Strength of plate girders under combined bending and shear, Part II,' Transactions, ASCE, vol. 128, 1963, pp. 720–735.

Bathon, L., W.H. Mueller III, and L. Kempner, Jr., 'Ultimate load capacity of single steel angles', *Journal of Structural Engineering*, ASCE, vol. 119, no. 1, January 1993, pp. 279–300.

Becker, R. and M. Ishler, *Seismic Design Practice for Eccentrically Braced Frames*, Structural Steel Education Council, Morga, CA, 1996.

Becker, R., *Seismic Design of Special Concentrically Braced Frames*, Structural Steel education Council, Morga, CA, 1995.

Beedle, L.S., *Plastic Design of Steel Frames*, John Wiley & Sons, Inc., 1958.

Billington, D.P., *The Tower and the Bridge, The New Art of Structural Engineering*, Princeton University Press, Princeton, NJ, 1983.

Bjorhovde, R., *Deterministic and Probabilistic Approaches to the Strength of Steel Columns*, Ph.D Dissertation, Lehigh University, Bethlehem, PA, May 1972.

Bjorhovde, R. and S.K. Chakrabarti, 'Tests of full-size gusset plates', *Journal of Structural Engineering*, ASCE, vol. 111, no. 3, March 1985, pp. 667–684.

Bjorhovde, R., T.V. Galambos, and M.K. Ravindra, 'LRFD Criteria for steel beam-columns', *Journal of the Structural Div.*, ASCE, vol. 104, no. ST9, September 1978, pp. 1371–1387.

Bleich, H., *Buckling Strength of Metal Structures*, McGraw-Hill, New York, 1952.

Blodgett, O.W., *Design of Welded Structures*, The James P. Lincoln Arc Welding Foundation, Cleveland, Ohio, 1966.

Borcherdt, R., 'Estimates of site dependent response spectra for design (Methodology and justification)', *Earthquake Spectra*, vol.10, no.4, 1994, pp. 617–53.

Bowles, J.E., *Structural Steel Design*, McGraw-Hill International Book Co., New York, 1980, 536 pp.

Bozorgnia, Y. and V.V. Bertero, *Earthquake Engineering: From Engineering Seismology to Performance-Based Engineering*, CRC Press, Boca Raton, Florida, 2004, 1152 pp.

British Steel, *The Prevention of Corrosion on Structural Steelwork*, British Steel Corporation, 1996.

Brockenbrough, R.L. and F.S. Merrit, *Structural Steel Designers' Handbook*, 3rd edn, McGraw-Hill, New York, 1999, 1208 pp.

Brown, C.B. and X. Yin, 'Errors in Structural Engineering', *Journal of Structural Engineering*, ASCE, vol. 114, no. 11, November 1998, pp. 2575–2593.

Bruneau, M., C.M. Uang, and A. Whittaker, *Ductile Design of Steel Structures*, McGraw-Hill, New York, 1997, 485 pp.

BS 5950-1:2000 *Structural use of Steelwork in Building – Part 1: Code of Practice for Design – Rolled and Welded Sections*, British Standard Institution, London, August 2001.

Buchanan, A.H., *Structural Design for Fire Safety*, John Wiley & Sons Ltd., Chichester, England, 2001, 421 pp.

Caccese, V., M. Elgaaly, and R. Chen, 'Experimental study of thin steel plate shear walls under cyclic load', *Journal of Structural Engineering*, ASCE, vol. 119, no. 2, February 1993, pp. 573–587 (also post buckling of steel plate shear walls under cyclic loads, pp. 588–605).

Callele, L.J., R.G. Driver, and G.Y. Grondin, 'Design and behavior of Multi-orientation fillet weld connections', *Engineering Journal*, AISC, 4th quarter, vol. 46, no. 4, 2009, pp. 257–272.

Carter, C.J. and K.A. Grubb, 'Prequalified moment connections (revisited)', *Modern Steel Construction*, vol. 50, no.1, January 2010, pp. 54–57.

Chen, W.F. and T. Atsuta, *Theory of Beam-Columns, Vol. 1: In-plane Behavior and Design*, McGraw-Hill, New York, 1976.

Chen, W.F. and T. Atsuta, *Theory of Beam-Columns, Vol. 2: Space Behavior and Design*, McGraw-Hill, New York, 1977.

Chen, W.F., Y. Goto, and J.Y.R. Liew, *Stability Design of Semi-Rigid Frames*, Wiley Inter Science, 1995, 488 pp.

Chen, W.F. and S.E. Kim, *LRFD Steel Design Using Advanced Analysis*, CRS Press, Boca Raton and New York, 1997, 441 pp.

Chen, W.F. and E.M. Lui, *Stability Design of Steel Frames*, CRC Press, Boca Raton, Florida, 1991, 380 pp.

Chen, W.F. and S. Toma (ed.), *Advanced Analysis of Steel Frames: Theory, Software and Applications*, CRC Press, Boca Raton and New York, 1994, 400 pp.

Chen, W.F. and S. Zhou, 'C$_m$ factor in load and resistance factor design', *Journal of Structural Engineering*, ASCE, vol. 113, no.8, August 1987, pp. 1738–1754.

Cheng, J.J.R., G.Y. Grondin, and M.C.H. Yam, 'Design and behaviour of gusset plate connections', *Fourth International Workshop on Connections in Steel Structures*, October 22–25, 2000, Roanoke, VA, USA, pp. 307–317.

Chesson, E. Jr., N.L. Faustino, and W.H. Munse, 'High-strength bolts subject to tension and shear', *Journal of the Structural Div.*, ASCE, vol. 91, no. ST5, October 1965, pp. 155–180.

Chopra, A.K., *Dynamics of Structures, Theory and Applications to Earthquake Engineering*, 2nd edn, Prentice Hall Inc., Englewood Cliffs, NJ, 2000, 844 pp.

Clough, R.W. and J. Penzien, *Dynamics of Structures*, McGraw-Hill, New York, 1993.

Cochrane, V.H., 'Rules for rivet hole deductions in tension members', *Engineering News Record*, vol. 89, November 16, 1922, pp. 847–848.

Cook, R.A., *Strength Design of Anchorage to Concrete*, Portland Cement Association, Skokie, IL, 1999, 86 pp.

Corr, D.J., D.M. McCann, and B.M. Mcdonald, 'Lessons learned from Marcy Bridge collapse', *Forensic Engineering*, 2009, pp. 395–403.

Coulbourne, W.L., E.S. Tezak, and T.P. McAllister, 'Design guidelines for community shelters for extreme wind events', *Journal of Architectural Engineering*, ASCE, vol. 8, no.2, June 2002, pp. 69–77.

CAN/CSA-5.16.1–09, *Limit States Design of Steel Structures*, Canadian Standards Association, Rexdale, Ontario, Canada, 2009.

Dafedar, J.B., Y.M. Desai, and M.R. Shiyekar, 'Review of code provisions for effective length of framed columns', *The Indian Concrete Journal*, vol. 75, no. 6, June 2001, pp. 402–407.

Davies, J.M. and B.A. Brown, *Plastic Design to BS 5950*, Blackwell Science, 1996.

Dawe, J.L. and G.L. Kulak, 'Welded connections under combined shear and moment', *Journal of the Structural Div.*, ASCE, vol.100, no. ST4, April 1974, pp. 727–741.

Dexter, R.J. and S.A. Altstadt, 'Strength and ductility of tension flanges in girders', *Recent Developments in Bridge Engineering*, Mahmoud, K.M. (ed.), A.A. Balkema/ Sweto & Zeitlinger, Lisse, The Netherlands, 2003, pp. 67–81.

DeWolf, J.T. and D.T. Ricker, *Column Base Plates, Steel Design Guide Series No. 1*, 2nd edn, AISC, Chicago, IL, 2006.

DIN 18800, Teil 2, *Stahlbauten: Stabilitätsfälle, Knicken von Stäben und Stabwerben,*, Deutsches Institut fur Normung e.v., Berlin, December 1980.

Douty, R.T. and W. McGuire, High-strength bolted moment connections, *Journal of the Structural Div.*, ASCE, vol. 91, no. ST2, February 1965.

Dowling, P.J., P. Knowles, and G.W. Owens, *Structural Steel Design*, The Steel Construction Institute and Butterworths, London, 1988, 399 pp.

Drake, R.M. and S.J. Elkin, 'Beam-column base-plate design, LRFD method', *Engineering Journal*, AISC, vol. 36, no. 1, 1999, pp. 29–38, Discussion by Doyle, J.M. and J.M. Fisher, vol. 42, no. 4, 2005, pp. 273–274.

Driver, R.G., G.L. Kulak, D.J.L. Kennedy, and A.E. Elwi, 'Cyclic test of a four storey steel plate shear wall', *Journal of Structural Engineering,* ASCE, vol. 124, no. 2, February 1998, pp. 111–120.

Driver, R.G., G.Y. Grondin, and G.L. Kulak, 'Unified block shear equation for achieving consistent reliability', *Journal of Constructional Steel Research,* vol. 62, no.3, March 2006, pp. 210–222.

Duan, L. and W.F. Chen, 'Design interaction equation for steel beam-columns', *Journal of Structural Engineering,* ASCE, vol. 115, no. 5, May 1989, pp. 1225–1243.

Duggal, S.K., *Design of Steel Structures,* 3rd edn, Tata McGraw-Hill, 2010, 890 pp.

Elgaaly, M. and H. Dagher, 'Beams and girders with corrugated webs', *Proceedings of the SSRC Annual Technical Session, Lehigh University,* Bethlehem, PA, 1990, pp. 37–53.

Epstein, H.I. and B. Thacker, 'Effect of bolt stagger for block shear tension failures in angles', *Computers and Structures,* vol. 39, no.5, 1991, pp. 571–76.

Epstein, H.I. and L.J. Aleksiewicz, 'Block shear equations revisited.... Again', *Engineering Journal,* AISC, vol. 45, no.1, 1st quarter, 2008, pp. 5–12.

ENV 1993-1– 4, *Euro code 3 Design of Steel Structures,* Part 1.4, General Rules and supplementary rules for Stainless Steels, CEN, 1992.

ESDEP (the European Steel Design Education Programme) Lectures, http:// www.esdep.org/members/master/wg08/l0410.htm, *Katholieke Universiteit, Department of Building Materials and Building Technology, Nederland,* April 15, 2006.

Euler, L., Sur le Force de Colonnes, *Memoires de l'Academie Royale des Sciences et Belles, Lettres,* vol. 13, Berlin, 1959. (English translation by J.A.Van den Broek, *American Journal of Physics,* vol. 15, 1947, pp. 309).

Euro Inox, European Stainless Steel Development Group, *Design Manual for Structural Stainless Steel,* Nickel Development Institute, Toronto, Canada, 1994.

Fan, Y.L, Y.L. Mo. and R.S. Herman, 'Hybrid bridge girders–A preliminary design example', *Concrete International, ACI,* vol. 28, no.1, January 2006, pp. 65–69.

FEMA 55, *Coastal Construction Manual,* Federal Emergency Management Agency, 2000.

FEMA 450, NEHRP *Recommended Provisions for New Buildings and Other Structures Part 1–Provisions, Part 2–Commentary,* Building Seismic Safety Council, National Institute of Building Sciences, Washington DC, 2003 (www.bssconline.org/ NEHRP2003/comments/provisions/).

Feld, J. and K. Carper, *Construction Failures,* 2nd ed., Wiley, New York, 1997.

Fisher, J.M, 'The importance of tension chord bracing', *Engineering Journal,* AISC, vol. 20, 3rd quarter, 1983, pp.103–106, Discussions, 2nd quarter, 1984, pp.122–123.

Fisher, J.M., 'Industrial buildings', *Constructional Steel Design, An International Guide,* Dowling, P.J., J.E. Harding, and R. Bjorhovde (ed.), *Elsevier Applied Science,* London/New York, 1991, pp. 627–644.

Fuchs, W., R. Eligehausen, and J.E. Breen, 'Concrete capacity design (CCD) approach for fastening to concrete', *ACI Structural Journal,* Jan–Feb. 1995, vol. 92, no. 1, pp. 73–94.

Gardner, L. and D.A. Nethercot, *Designers' Guide to EN 1993-1-1, Euro code 3: Design of Steel Structures–General Rules and Rules for Buildings,* Thomas Telford, London and Steel Construction Institute, 2005.

Galambos, T.V., 'Load and resistance factor design', *Engineering Journal,* AISC, 3rd quarter, vol.18, no.3, pp.74–82, 1981.

Galambos, T.V. (ed.), *Guide to Stability Design Criteria for Metal Structures*, 5th edn, John Wiley & Sons, 1998.

Galvery, Jr. W.L. and F.M. Marlow, *Welding Essentials; Questions and Answers*, Expanded edn, Industrial Press, 2001, 480 pp.

Gaylord, E.H., C.N. Gaylord, and J.E. Stallmeyer, *Design of Steel Structures*, 3rd edn, McGraw-Hill, New York, 1992, 792 pp.

Gaylord, E.H., C.N. Gaylord, and J.E. Stallmeyer, *Structural Engineering Handbook*, 4th edn, McGraw-Hill, New York, 1996, 1248 pp.

Gordon, J.E., *Structures (or why Things Don't Fall Down)*, Penguin Books, New York, 1978.

Greiner, R. and J. Lindner, 'Interaction formulae for members subjected to bending and axial compression in Eurocode 3 – The method 2 approach', *Journal of Constructional Steel Research*, vol. 62, 2006.

Gupta, A.K., 'Buckling of coped steel beams', *Journal of the Structural Engineering*, ASCE, vol. 110, no. 9, September 1984, pp. 1977–1987, also see discussion by Cheng, J.J. and Y.A. Yura, vol. 112, no. 1, January 1986, pp. 201–204.

Hamburger, R.O., H. Krawinkler, J.O. Malley, and S.M. Adan, *Seismic design of steel special moment frames: A guide for practicing Engineers*, NEHRP Seismic design technical brief no.2, National Institute of Standards and Technology, Gaithersburg, USA (http://www.nehrp.gov/pdf/nistgcr9-917-3.pdf).

Hardash, S. and R. Bjorhovde, 'New design criteria for gusset plates in tension', *Engineering Journal*, AISC, vol. 22, no. 2, 2nd quarter, 1985, pp. 77–94.

Hatfield, F.J., 'Design chart for vibration of office and residential floors', *Engineering Journal*, AISC, vol. 29, no.4, 4th quarter 1992, pp. 141–144.

Hill, H.N., 'Lateral buckling of channels and Z-beams', *Transactions*, ASCE, vol. 119, 1954, pp. 829–841.

Holmes, M. and L.H. Martin, *Analysis and Design of Structural Connections*, Ellies Horwood Ltd., 1983.

Horne, M.R., *Plastic Theory of Structures*, Pergamon Press Ltd., Oxford, 1979.

Horne, M.R. and L.J. Morris, *Plastic Design of Low-rise Frames*, Granada, London, 1981.

IS : 800–2007, *Indian Standard Code for General Construction in Steel*, 3rd revision, Indian Standards Institution, New Delhi, 2008.

IS : 801–1975, *Code of Practice for Cold-Formed Light Gauge Steel Structural Members in General Building Construction*, 1st Revision, Bureau of Indian Standards, New Delhi.

IS : 802 (Part 1, Sec. 1) – 1995 *Use of Structural Steel in Overhead Transmission Line Towers - Code of Practice, Part 1 Material, Loads and Permissible Stresses, Sec.1.Materials and loads and Sec.2 – Permissible Stresses*, (1992), Bureau of Indian Standards, New Delhi.

IS : 807–1976, *Code of Practice for Design, Manufacture, Erection and Testing (Structural portion) of Cranes and Hoists*, Bureau of Indian Standards, New Delhi.

IS : 814–1991, *Covered electrodes for manual metal arc welding of carbon and carbon manganese steel* (5th revision), Bureau of Indian Standards, New Delhi.

IS : 816–1969, *Code of practice for use of metal arc welding for general construction in mild steel* (5th revision), Bureau of Indian Standards, New Delhi.

IS : 875–1987, *Code of Practice for Design Loads (other than earthquake) for Buildings and Structures* (Part 1–Part 5), Bureau of Indian Standards, New Delhi.

IS : 1024–1999, *Code of Practice for use of Welding in Bridges and Structures Subjected to Dynamic Loading (2nd revision)*, Bureau of Indian Standards, New Delhi.

IS : 1149–1982, *High tensile steel rivet bars for structural purposes* (3rd revision), BIS, New Delhi.

IS : 1161–1979, *Specification for Steel Tube for Structural Purposes*, 3rd edn, Bureau of Indian Standards, New Delhi.

IS : 1363–2002, *Hexagon head bolts, screws and nuts of product grade C*:
Part 1 Hexagon head bolts (size range M5 to M64) (3rd revision)
Part 2 Hexagon head screws (size range M5 to M64) (3rd revision)
Part 3 Hexagon nuts (size range M5 to M64) (3rd revision)

IS : 1364–2002, *Hexagon head bolts screws and nuts of product grades A and B*:
Part 1 Hexagon head bolts (size range M1.6 to M64) (3rd revision)
Part 2 Hexagon head screws (size range M1.6 to M64) (3rd revision)
Part 3 Hexagon nuts (size range M1.6 to M64) (3rd revision)
Part 4 Hexagon thin nuts (chamfered) (size range M1.6 to M64) (3rd revision)
Part 5 Hexagon thin nuts (unchamfered) (size range M1.6 to M10) (3rd revision)

IS : 1367–1992 (Parts 1 to 18), *Technical supply conditions for threaded steel fasteners*, Bureau of Indian Standards, New Delhi.

IS : 1852–1985, *Rolling and Cutting Tolerances for Hot-Rolled Steel Products*, 4th revision, Bureau of Indian Standards, New Delhi.

IS : 1893–2002, *Criteria for earthquake resistant Design of Structures, Part 1: General Provisions and buildings* (5th revision), Bureau of Indian Standards, New Delhi.

IS : 2062–1995, *Steel for General Structural Purposes – Specification*, 5th revision, Bureau of Indian Standards, New Delhi.

IS : 3757–1985, *Specification for high strength structural bolts* (2nd revision), Bureau of Indian Standards, New Delhi.

IS : 4000–1992, *Code of practice for high strength bolts in steel structures* (1st revision), Bureau of Indian Standards, New Delhi.

IS : 5624–1993, *Foundation Bolts – Specification*, Bureau of Indian Standards, New Delhi.

IS : 6533–1989, *Code of Practice for Design and Construction of Steel Chimneys Part 1, Mechanical Aspects, Part 2 – Structural Aspects*, Bureau of Indian Standards, New Delhi.

IS : 8500–1991, *Structural Steel–Micro Alloyed (medium and high strength qualities) – Specifications*, 1st revision, Bureau of Indian Standards, New Delhi.

IS : 9172–1979, *Recommended Design Practice for Corrosion Protection of Steel Structures*, Bureau of Indian Standards, New Delhi.

IS : 9178–1979, *Criteria for Design of Steel Bins for Storage of Bulk Materials Part 1 – General Requirements and Assessment of Loads, Part 2 – Design Storage Criteria*, Bureau of Indian Standards, New Delhi.

Jain, S.K., 'A proposed draft of IS: 1893 Provisions on Seismic design of buildings – Part II. Commentary and examples', *Journal of Structural Engineering*, SERC, vol. 22, no.2, July 1995, pp. 73–90.

Jamshidi, M., S. Fallaha, A. Azizinamini, K. Price, and M. Cress, 'Application of high performance steel in steel bridge construction' *Building to Last Structures Congress – Proceedings*, vol. 1, 1997, ASCE, New York, NY.

Jeffus, L., *Welding: Principles and Applications*, 5th edn, Thomson Delmar learning, 2002, 944 pp.

Johnson, R.P., *Composite Structures of Steel and Concrete, Vol. 1, Beams, Columns, Frames and Applications in Buildings*, Blackwell Scientific Publications, London, 1994.

Kalayanaraman, V., B.N. Sridhara, and K. Mahadevan, Sleeved Column System, *Proc. of the SSRC Task Group Meetings and Task Force Sessions*, Lehigh Univ., Bethlehem. Pa., 1994.

Kaminetzky, D., *Design and Construction Failures: Lessons from Forensic Investigations*, McGraw-Hill, New York, 1991.

Kavanagh, T., 'Some Aesthetic Considerations in Steel Design', *Journal of the Structural Division*, ASCE, vol. 101, no. 11, 1975, pp. 2257–2275.

Kavanagh, T.C., 'Effective length of framed columns', *Transactions, ASCE*, vol. 127, 1962, Part II, pp. 81–101.

Kerensky, O.A., A.R. Flint, W.C. Brown, 'The basis for design of beams and plate girders in the revised British standard 153', *Proceedings of the Institution of Civil Engineers (London)*, vol. 5, no. PtIII, August 1956, pp. 396–461.

Kevin, L.R., M. Clark, and A.W. Knott, 'Failure Analysis Case Information Disseminator', *Journal of Performance of Constructed Facilities*, ASCE, vol. 16, no. 3, August 2000, pp. 127–131.

Khalil, S.H. and C.H. Ho, 'Black bolts under combined tension and shear', *The Structural Engineer*, vol. 57 B, December 1979, pp. 69–76.

King, C.M., 'Design of Steel Portal frames for Europe', *Steel Construction Institute*, Ascot, 2005.

Kirby, P.A. and D.A. Nethercot, *Design for Structural Stability*, Granada Publishing, London, 1979.

Kitipornchai, S. and H.W. Lee, 'Inelastic buckling of single angle, tee and double angle struts', *Journal of Constructional Steel Research*, London, vol. 6, no. 1, 1986, pp. 3–20, also vol. 6, no. 3, 1986, pp. 219–236.

Kulak, G.L., J.W. Fisher, and J.H.A. Struik, *Guide to Design Criteria for Bolted and Riveted Joints*, 2nd edition, John Wiley and Sons, New York, 1987, 309 pp.

Kulak, G.L. and G.Y. Grondin, 'Block shear failure in steel members- a review of design practice', *Fourth International Workshop on Connection in Steel Structures*, Roanoke, VA, October 22–25, 2000, pp. 329–339.

Kulak, G.L., and G.Y. Grondin, *Limit States Design in Structural Steel*, 7th edn, Canadian Institute of Steel Construction, 2002.

Kulak, G.L. L. Kennedy, R.G. Driver, and M. Medhekar, 'Steel plate shear walls – An overview', *Engineering Journal*, AISC, vol. 39, no.1, 2001, pp.50–62.

Kulak, G.L. and E.Y. Wu, 'Shear lag in bolted angle tension members', ASCE, *Journal of Structural Engineering*, vol. 123, no. 9, September1997, pp. 1144–1152.

Lawson, R.M., D.L. Mullett, and J.W. Rackham, *Design of Asymmetric Slimflor Beams using Deep Composite Decking*, The Steel Construction Institute, UK, SCI P 175 1997.

Lay, M.G. and T.V. Galambos, 'Bracing requirements for inelastic steel beams', *Journal of the Structural Div.*, ASCE, vol. 92, no. ST2, April 1966, pp. 207–228.

Levy, M. and M. Selvadori, *Why Buildings Fall Down: How Structures Fail*, W.W. Norton, New York, 1992.

Lindner, J., 'Design of steel beams and beam columns', *Engineering Structures*, vol. 19, no. 5, 1997, pp. 378–384.

Lindner, J., and N. Subramanian, 'Allowable stress provisions of IS: 800–1962', *Journal of the Institution of Engineers (India)*, vol. 62, no. Pt.Cl4, January 1982, pp. 246–248.

Lindner, J., J. Scheer, and H. Schmidt, Stahlbauten, *Erläuterungen zu DIN 18800, Teil 1 bis Teil 4*, 2nd edn, Beuth und Ernst & Sohn, Berlin, 1994.

Lwin, M.M., *High Performance Steel Designers Guide*, 2nd edn, U.S. Department of Transportation, San Francisco, USA, 2002 (http://www.steel.org/infrastructure/bridges/high performance/hps designersguide/intro.htm), 30 November 2004.

MacGinley, T.J., *Steel Structures: Practical Design Studies*, 2nd edn, E&FN Span, London, 1997.

MacGinley, T.J. and P.T.C. Ang, *Structural Steelwork–Design to Limit State Theory*, Butterworths/Heinemann, Oxford, 1992.

Machacek, J. and M. Tuma, 'Fatigue life of girders with undulating web', *Journal of Constructional Steel Research*, vol. 62, 2006, pp. 168–177.

Madugula, M.K.S. and J.B. Kennedy, *Single and Compound Angle Members – Structural Analysis and Design*, Elsevier Applied Science, London, England, 1985.

Maquoi, R., Plate Girders, In *Constructional Steel Design – An International Guide*, Dowling, P.J., J.E. Harding, and R. Bjorhovde (ed.), Elsevier Applied Science Publishers, London, 1992, pp. 133–173.

Martin, L.H. and J.A. Purkiss, *Structural Design for Steelwork to BS 5950*, Edward Arnold, London, 1992, 468 pp.

Mazzolani, F.M. and V. Piluso, *Theory and Design of Seismic Resistant Steel Frames*, E & FN Spon, 1996, 544 pp.

McGuire, W., *Steel Structures*, Prentice Hall, Englewood Cliffs, New Jersey, 1968.

Minor, J.E., 'Tornado Technology and Professional Practice', *Journal of Structural Div.* ASCE, vol. 108, no.11, November 1982, pp. 2411–2422.

Morris, L.J., and D.R. Plum, *Structural Steelwork Design to BS5950*, 2nd edn, Addison Wesley Longman, Essex, 1996.

Munse and Chesson, 'Riveted and bolted joints: Net section design', *Journal of Structural Engineering*, ASCE, vol. 89, no. 1, 1963, pp. 107–126.

NAVFAC, *Foundations and Earth Structures, Design Manuel 7.2*, Naval Facilities Engineering Command, Dept. of Navy, Alexandria, Virginia, USA, 1982.

Narayanan, R. (ed.), *Plated Structures: Stability and Strength*, Elsevier Applied Science Publishers, London, 1983.

Narayanan, R. (ed.), *Axially Compressed Structures: Stability and Strength*, Elsevier Applied Science Publishers, Essex, 1982.

Narayanan, R., (ed.), *Beams and Beam Columns: Stability and Strength*, Applied Science Publishers, Barking, Essex, England, 1983.

Narayanan, R. et al., *Teaching Material On Structural Steel Design* (45 Chapters based on Limit State design of Steel Structures), INSDAG, New Delhi, http://www.steel-insdag.org/new/contents.asp.

Narayanan, R and T.M. Roberts (ed.), *Structures Subjected to Repeated Loading: Stability and Strength*, Spon Press, UK, 1990, 291 pp.

Nethercot, D.A., 'Elastic lateral buckling of beams', *in Beams and Beam Columns: Stability and Strength*, R. Narayanan (ed.), Applied Science Publishers, Essex, England, 1983.

Nethercot, D.A., *Limit States Design of Structural Steelwork*, 3rd edn, Spon Press, London, 2001, 270 pp.

Newman, A., *Metal Building Systems: Design and Specifications*, 2nd edn, McGraw-Hill, New York, 2004, 576 pp.

Owens, G.W. and B.D. Cheal, *Structural Steelwork Connections*, Butterworths, London, 1989, 330 pp.

Owens, G.W., P.J. Driver, and G.J. Kriege, 'Punched holes in structural steel- work', *Journal of Constructional Steel Research*, vol. 1, no. 3, 1981.

Petersen, C., *Statik und Stabilität der Baukonstruktionen*, Friedr.Vieweg & Sohn, Braunschweig/Wiesbaden, Germany, 1980, 960 pp.

Petersen, C., *Stahlbau*, Friedr. Vieweg & Sohn Verlag, Braunschweig, 1990, (Chapter 17, Trapezprofil-Bauweise, pp. 719–745).

Pillai, U., 'An assessment of CSA Standard equation for beam-column design', *Canadian Journal of Civil Engineering*, vol. 8, 1981.

Porter, D.M., K.C. Rockey, and H.R. Evans, 'The collapse of plate girders loaded in shear', *The Structural Engineer*, vol. 53, no. 8, 1975, pp. 313–325.

Rangwala, S.C., K.S. Rangwala, and P.S. Rangwala, *Engineering Materials*, 22nd edn, Charotar Publishing House, Anand, 1997.

Ricker, D.T., 'Tips for avoiding crane runway problems', *Engineering Journal*, AISC, vol. 19, no. 4, 1982, pp. 181–205.

Ricker, D.T., 'Some practical aspects of column base selection', *Engineering Journal*, AISC, 3rd quarter, vol. 26, no. 3, 1989, pp. 81–89.

Rockey, K.C., G. Valtinat, and K.H. Tang, 'The design of transverse stiffeners on webs loaded in shear – An ultimate load approach', *Proceedings Institution of Civil Engineers*, Part 2, vol. 71, no. 2, December 1981, pp. 1069–1099.

Roeder and Lehman, 'Seismic design and behavior of concentrically braced steel frames', *Structures Magazine*, February 2008, pp. 37–39.

Roik, K., J. Carl, and J. Lindner, *Biege Torsions Probleme Gerader Dünnwandiger Stäbe*, Verlag von Wilhelm Ernst & Sohn, Berlin, 1972.

Sabelli, R., 'Recommended provisions for Buckling-Restrained Braced Frames', AISC *Engineering Journal*, 4th quarter, 2004, pp 155–175.

Salmon, C.G., J.E. Johnson, and F.A. Malhas, *Steel Structures, Design and Behavior*, 5th edn, Harper Collins College Publishers, New York, 2008, 888 pp.

Salmon, C.G. and J.E. Johnson, *Steel Structures – Design and Behaviour*, 4th edn, Harper Collins College Publishers, New York, 1996, 1024 pp.

Salvadori, M.G., 'Lateral buckling of eccentrically loaded I-columns', *Transactions*, ASCE, vol. 121, 1956, pp. 1163.

Schilling, C.G., 'Design of hybrid steel beams, Report of the subcommittee 1 on beams and girders, Joint ASCE-AASHO committee on flexural members', *Journal of the Structural Div.*, ASCE, vol. 94, no. ST6, June 1968, pp. 1397–1426.

Schittich, C., *In detail: Building Skins, Concepts, Layers, Materials*, Birkhouser, Switzerland, 2001, 120 pp.

Schodek, D.L., *Structures*, 4th edn, Prentice Hall, Upper Saddle River, New Jersey, 2001, 581 pp.

SEAOC, *Recommended lateral force requirements and commentary*, Structural Engineers Association of California, 1980.

Seilie, I.F. and J.D. Hooper, 'Steel plate shear walls! Practical design and construction', *Modern Steel Construction*, AISC, vol. 45, no. 4, April 2005, pp. 37–42.

Shanmugam, N.E., J.Y.R. Liew, and S.L. Lee, 'Ultimate strength design of biaxially loaded steel box beam-columns', *Journal of Constructional Steel Research*, vol. 26, no. 2–3, 1993, pp. 99–123.

Shedd, T.C., *Structural Design in Steel*, John Wiley & Sons, New York, 1934.

Smith, E. and H. Epstein. H., 'Hartford Coliseum roof collapse: Structural collapse and lessons learned', *Civil Engineering*, ASCE, April 1980, pp. 59–62.

Sree Ramachandra Murthy, D., D. Bhanu Prasad, K.V. Ramesh, and D.L. Narasimha Rao, 'Economical aspects in the design of steel trusses', *Civil Engineering and Construction Review*, November 2004, pp. 30–38.

Sreevidya, S. and N. Subramanian, 'Aesthetic appraisal of antenna towers', *Journal of Architectural Engineering*, ASCE, vol. 9, no.3, September 2003, pp.103–108.

Structural Engineering Institute, *Effective Length and Notional Load Approaches for Assessing Frame Stability: Implications for American Steel Design*, American Society of Civil Engineers, 1997, 442 pp.

Struik, J.H.A., A.O. Oyeledun, and J.W. Fisher, 'Bolt tension control with a direct tension indicator', *Engineering Journal*, AISC, vol. 10, no. 1, 1st quarter, pp.1–5.

Subramanian, N., *Versuche an geschraubten Flanchverbindungen in L-Form*, Unpublished report, Lehrstuhl fuer Stahlbau, Der Universitaet Bundeswehr, Munich, Germany, 1984.

Subramanian, N., 'Why buildings collapse?' *Science Reporter*, vol. 24, no. 5-6, May–June 1984, pp. 232–238.

Subramanian, N., 'Aesthetics of non-habitat structures', *The Bridge and Structural Engineer*, Journal of ING/IABSE, vol.17, no.4, 1987, pp.75–100.

Subramanian, N., 'Diagnosis of the causes of failures', *The Bridge and Structural Engineer*, vol.19, no.1, March 1989, pp. 24–42.

Subramanian, N., 'Computer analysis and design of tall structures', *Civil Engineering and Construction Review*, vol. 8, no. 4, April 1995, pp. 42–46.

Subramanian, N. and S. Mangalam, 'Restoration of a factory building', *Journal of the Institution of Engineers (India)*, vol. 79, May 1998, pp. 14–17.

Subramanian, N., and S. Mangalam, 'Distress to green houses due to improper execution', *Civil Engineering and Construction Review*, vol. 10, no. 8, August 1997, pp. 45–48.

Subramanian, N., 'Improper detailing results in delayed commissioning', *Journal of Performance of Constructed Facilities*, ASCE, vol. 13, no.1, February 1999, pp. 29–33.

Subramanian, N., 'Restoration of cold storage factory building', *Journal of Performance of Constructed Facilities*, ASCE, vol. 14, no. 4, November 2000, pp.155–159.

Subramanian, N., 'Corrosion protection of steel structures', *The Master Builder*, vol. 2, no.1, January–February 2000, pp. 63–66 and no. 2, April–May 2000, pp. 14–17.

Subramanian, N., 'Recent developments in the design of anchor bolts', *The Indian Concrete Journal*, vol. 74, no. 7, July 2000, pp. 407–414.

Subramanian, N., 'Collapse of WTC – Its impact on skyscraper construction', *The Indian Concrete Journal*, vol. 76, no. 3, March 2002, pp. 165–169.

Subramanian, N., *Space Structures: Principles and Practice*, Multi-Science Publishing Co. Ltd., Essex, UK, 2006, vol. 1 & 2, 844 pp.

Subramanian, N., *Design of Steel Structures*, Oxford University Press, New Delhi, 2008, 1211 pp.

Subramanian, N., 'India's first stainless steel space frame roof', *Fifth International Conference on Space Structures*, University of Surrey, UK, August 2002.

Subramanian, N. and R.A. Cook, 'Behaviour of grouted anchors', *The Indian Concrete Journal*, vol. 78, no. 4, April 2004, pp. 14–21.

Subramanian, N. and R.A. Cook, 'Installation, behaviour and design of bonded anchors', *The Indian Concrete Journal*, vol. 76, no.1, January 2002, pp. 47–56.

Subramanian, N. and V. Vasanthi, 'Design of anchor bolts in concrete', *The Bridge and Structural Engineer*, vol. 21, no. 3, September 1991, pp. 47–73.

Sfintesco, D., 'Experimental basis of the European column curves', *Construction Metallique*, no. 3, September 1970, pp. 5.

Swiatek, D. and E. Whitbeck, 'Anchor rods', *Modern Steel Construction*, vol. 44, no. 12, December 2004, pp. 31–33.

Tall, L., *Structural Steel Design*, Ronald Press, New York, NY, 1974.

Taranath, B.S., *Steel, Concrete and Composite Design of Tall Buildings*, 2nd edn, McGraw Hill, New York, 1998, 998 pp.

Tebedge, N. and W.F. Chen, 'Design criteria for H-columns under biaxial loading', *Journal of the Structural Div.*, ASCE, vol. 100, no. ST3, March 1974, pp. 579–598.

Thornton, W.A., 'Prying action–A general treatment', *Engineering Journal*, AISC, vol. 22, 2nd quarter, 1985, pp. 67–75.

Thornton, W.A., 'Connections–Art, science and information into the quest for economy and safety', *Engineering Journal*, AISC, vol. 32, no. 4, 4th quarter, 1995, pp. 132–144.

Thornton, W.A., L.S. Muir, 'Design of vertical bracing connections for high-seismic drift', *Modern Steel Construction*, vol. 49, no.3, March 2009, pp.61–65.

Timoshenko, S.P., *History of the Strength of Materials*, McGraw-Hill, New York, 1953.

Timoshenko, S.P. and J.M. Gere, *Theory of Elastic Stability*, 2nd edn, McGraw-Hill, New York, 1961.

Trahair, N.S., M.A. Bradford, and D.A. Nethercot, *The Behaviour and Design of Steel Structures* to BS 5950, 3rd edn, Spon Press, London, 2001, 448 pp.

Usha and Kalyanaraman, V., 'Strength of steel tension members', *Proc. All India Seminar on Modern Trends in Steel Structures*, Nagpur, 8–9 February 2002, pp. 21–29.

Usha, *Analytical Study of Nonlinear Behaviour of Steel Angle Tension Members*, Thesis submitted for M.S. (by Research), IIT, Madras, September 2003.

Valitnat, G., Tragsicherheitsnachweise für spezielle trägerformen-vollwandträger mit schlanken stegen und vertikalsteifen, *Stahlbau Handbuch für Studium und Praxis*, Band 1, Stahlbau-Verlags *GMBH*, Köln, 1982, pp. 531–542.

Veljkovic, M. and B. Johansson, 'Design of hybrid steel girders', *Journal of Constructional Steel Research*, vol. 60, 2004, pp. 535–547.

Vinnakota, S., *Steel Structures: Behavior & LRFD*, McGraw-Hill Higher Education, New York, 2006.

Warner, J., *Practical Handbook of Grouting: Soil, Rock and Structures*, John Wiley & Sons Inc., New York, 2004, 720 pp.

Weaver, W.M., *Crane Handbook: Design, Data and Engineering Information used in the Manufacture and Application of Overhead and Gantry Cranes*, 4th edn, Whiting Corporation, Harvey, Ill., 1979.

Whalen, T.M., S. Gopal, and D.M. Abraham, 'Cost–benefit model for the construction of tornado shelters', *Journal of Construction Engineering and Management*, ASCE, vol. 130, no. 6, November—December 2004, pp. 772–779.

Whitmore, R.E., *Experimental Investigation of Stresses in Gusset Plates*, Bulletin no. 16, Engineering Experiment Station, University of Tennessee, USA, 1952.

Williams, A., *Seismic Design of Buildings and Bridges*, 3rd edn, Engineering Press, California, 2004.

Winter, G., 'Lateral bracing of columns and beams', *Transactions*, ASCE, vol. 125, part 1, 1960, pp. 809–825.

Wood, R.H., 'Effective lengths of columns in multistory frames', *The Structural Engineer*, vol. 52, no. 7, July 1974, pp. 235–244, no. 8, August 1974, pp. 295–302, no. 9, September 1974, pp. 341–346.

Wong, E. and R.G. Driver, 'Critical evaluation of equivalent moment factor procedures for laterally unsupported beams', *Engineering Journal*, AISC, 1st quarter, vol. 47, no.1, 2010, pp. 1–20.

Woolcock, S.T. and S. Kitipornchai, 'Design of single angle web struts in trusses', *Journal of Structural Engineering*, ASCE, vol. 113, no. 9, September 1987, pp. 2102–07.

Yoo, C.H. and S.C. Lee, 'Mechanics of web panel postbuckling behavior in shear', *Journal of Structural Engineering*, ASCE, vol. 132, no. 10, October 2006, pp.1580–89.

Yu, W.W., *Cold Formed Steel Design*, 3rd edn, Inter Science, 2000, 848 pp.

Yura, J.A., *Elements for Teaching Load and Resistance Factor Design*, University of Texas at Austin, TX, 1998, pp. 55–71.

Yura, J., 'Fundamental of beam bracing', *Engineering Journal*, AISC, vol. 38, no.1, 2001, pp. 11–26.

Yura, J.A., P.C. Birkemoe, and J.M. Ricles, 'Beam web shear connections: An experimental study', *Journal of the Structural Div.*, ASCE, vol. 108, no. ST2, February 1982, pp. 311–325.

Ziemian, R.D. (ed.), *Guide to Stability Design Criteria for Metal Structures*, 6th edn, John Wiley & Sons, 2010, 1078 pp.

http://ethics.tamu.edu/ethics/hyatt/hyatt2.htm (12 March 2010)

http://failurebydesign.info (15 March 2010)

http://www.starseismic.net [Buckling Restrained Braces, (2 February 2010)]

www.steelcastconnections.com [Source for the Kaiser Bolted Bracket (KBB), 30 December 2009]

http://www.sacsteel.org [The SAC steel project funded by FEMA to solve the brittle behavior of welded steel frame structures that occurred on 17 January 1994, Northridge, California Earthquake, (19 March 2010)]

www.corusconstruction.com (25 May 2010)

http://nptel.iitm.ac.in/courses/IIT-MADRAS/Design_Steel_Structures_I/index.php [Web-based course developed by the Indian Institute of Technology, (10 January 2010)]

http://nptel.iitm.ac.in/courses/IIT-MADRAS/Design_Steel_Structures_II/index.php [Web-based course developed by the Indian Institute of Technology, (15 January 2010)]

www.nptel.iitm.ac.in/video.php?courseId=1118 (20 June 2010)

www.steeltools.org [(Contains many useful tools for the steel designer), 19 June 2010]

www.aisc.org [(American Institute of Steel Construction's website), 20 June 2010]

www.steel-sci.org [(The Steel Construction Institute, UK website), 22 June 2010]

www.thestructuralengineer.info/onlinelibrary (Contains many freely downloadable articles on steel design)

Index